CALCULUS

LYNN H. LOOMIS
Harvard University

THIRD EDITION

 ADDISON-WESLEY PUBLISHING COMPANY
Reading, Massachusetts ▪ Menlo Park, California
London ▪ Amsterdam ▪ Don Mills, Ontario ▪ Sydney

SPONSORING EDITOR: *Stephen H. Quigley*
PRODUCTION EDITOR: *Marion E. Howe*
DESIGNER: *Vanessa Piñeiro*
ILLUSTRATOR: *B. J. and F. W. Taylor*
COVER DESIGN: *Richard W. Hannus*

Library of Congress Cataloging in Publication Data
Loomis, Lynn H., 1915–
  Calculus.

  Includes index.
  1. Calculus. I. Title.
QA303.L872 1982   515   81-14937
ISBN 0-201-05045-5   AACR2

PREFACE

This book is intended as an intuitive, but mathematically sound, treatment of the standard calculus sequence. Its approach to the basic concepts—limit, continuity, derivative, integral—is mainly geometrical. Variables and "y is a function of x" are used extensively but carefully; Leibniz notation and function notation receive about equal time.

There is some emphasis on approximation and computation (though this material can be considered to be optional). For the most part this appears as estimation, directed to the question "How good an answer do I have?" However, Appendix 4 goes on to the natural follow-up, "What must I do to get the accuracy I want?," and in this computational context there is an introduction to the ϵ, δ theory of limits.

Most of the traditional topics from analytic geometry are covered in Chapter 1 and Appendix 3. Chapter 1 treats lines, circles, translation of axes, and completing the square to simplify quadratic equations, while Appendix 3 develops the conic sections from their standard locus definitions. Some related material involving polar coordinates and parametric equations will be found in Chapter 10.

The changes for this edition are extensive.

The development has been reorganized, for greater smoothness and flexibility. It will fit courses given in three terms and also courses given in four quarters. See below.

There is much new writing. The discussions of limits, continuity, and the definite integral are wholly new, as are sections on l'Hôpital's rule, improper integrals, certain integral applications, and the natural logarithm function. Many other sections have been almost completely rewritten.

The exposition has been simplified. Duplicated discussions have been eliminated. Some of the more complicated topics have been postponed (e.g., centers of mass, curvature) and several difficult arguments have been dropped. The integral is discussed entirely in terms of sequential convergence.

The problem collection has been reorganized and strengthened. The symbol ▦ has been introduced to suggest the use of a hand calculator.

The organization now provides convenient break points for courses given either in three terms or in four quarters. At six chapters per term, the book divides as follows:

> Term I (Chapters 1–6).† Differential calculus; introduction to the integral.
>   II (7–12). Integral calculus; infinite series.
>   III (13–17; 18 and/or 19). Vectors; functions of several variables.

For a course in four quarters (at four or five chapters per quarter) the breaks occur naturally at subject changes:

> Quarter I (Chapters 1–5). Differential calculus.
>   II (6–10). Integral calculus.
>   III (10–14). Infinite series; vectors and vector functions of one variable.
>   IV (15–17; 18 and/or 19). Functions of several variables; (differential equations).

Chapter 10 is mentioned twice: it can go into either quarter or it can be split between them.

---

† Strong classes would presumably cover Chapter 1 only lightly. Classes that are less well prepared could omit some or all of the sections on estimation and theory (5.7–5.10, 6.9).

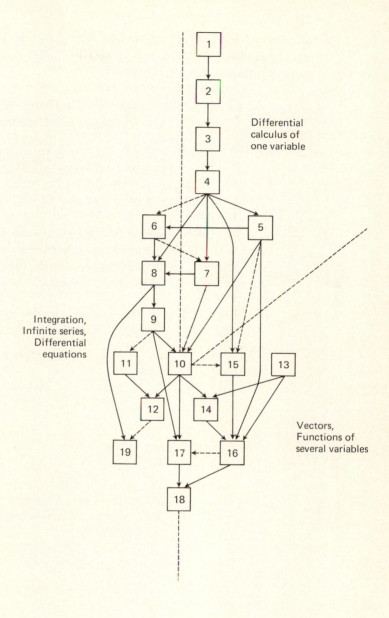

Differential
calculus of
one variable

Integration,
Infinite series,
Differential
equations

Vectors,
Functions of
several variables

Integration can be reached in the first quarter by putting off the trigonometric functions and some applications of the derivative, as follows:

Quarter I (Chapters 1–3, 4.1, 4.2, 5.1–5.5, 6). Differential calculus of algebraic functions; introduction to the integral.
 II (4.3–4.5, 5.6 (5.7–5.10 optional), 7–10). Transcendental functions; further applications of the derivative; integral calculus.
 III and IV as before.

For other possible re-alignments see the flow chart on page v. It shows the major dependencies between chapters, and any reordering of material that is consistent with the chart should involve only minor problems of accommodation. A dotted arrow indicates a single point of contact.

I would like to express my sincere appreciation and gratitude to my colleagues at Addison-Wesley for their constant help and encouragement.

Concord, Massachusetts   L. H. L.
January 1982

contents

CHAPTER 3 THE DERIVATIVE

CHAPTER 4 TECHNIQUE OF DIFFERENTIATION

CHAPTER 5 APPLICATIONS OF THE DERIVATIVE

CHAPTER 6 THE INTEGRAL

CHAPTER 7 THE NATURAL LOGARITHM AND EXPONENTIAL FUNCTIONS; INVERSE FUNCTIONS

CHAPTER 8 TECHNIQUE OF INTEGRATION

CHAPTER 9 THE DEFINITE INTEGRAL: FURTHER DEVELOPMENT AND APPLICATIONS

CHAPTER 14 VECTOR-VALUED FUNCTIONS

CHAPTER 15 FUNCTIONS OF SEVERAL VARIABLES

CHAPTER 16 VECTORS IN CALCULUS OF SEVERAL VARIABLES; HIGHER PARTIAL DERIVATIVES

INTRODUCTION

NEWTON, LEIBNIZ, AND THE CALCULUS

The invention of calculus is attributed to two geniuses of the 17th century, Isaac Newton in England, and, independently, Gottfried Leibniz in Germany. Earlier mathematicians had uncovered bits and pieces of the subject. Newton and Leibniz discovered its *pattern*. They thereby created an algorithmic discipline of enormous power, applicable to all sorts of fundamental questions about the nature of the world.

Although the new calculus obviously worked, Newton and Leibniz did not have a clear idea of *why* it worked. They tried to explain its successes by geometric reasoning, since at that time all mathematical phenomena were viewed in terms of geometry, but their explanations were unsatisfactory. In fact, the logical foundations of calculus remained a mystery for another century and a half. Some fragmentary progress occurred, and a new point of view gradually emerged, based on numbers, variables, and functional relationships between variables. Then, around 1820, the French mathematician, Augustin Cauchy, settled the matter by showing that calculus rests on the properties of the limit operation. This was still not what we today call rigor, and it took another fifty years of deeper probing to reach the bedrock of ϵ, δ reasoning and the completeness of the real-number system.

The chronological development of calculus was thus marked at several points by leaps in precision and sophistication. Now, three hundred years after Newton and Leibniz, we can start our study of the subject at practically any level we wish. Since there seems to be little point in repeating the confusions of the first one hundred fifty years, we shall approach calculus at about the level of Cauchy, which is still very intuitive. We can then increase our precision in a natural manner as the subject unfolds.

CHAPTER 1

GRAPHS AND FUNCTIONS (PRECALCULUS REVIEW)

Calculus is about functions, and it is important before starting calculus to have a reasonably good understanding of what a function is and what its graph can be like. This preliminary chapter reviews the necessary background material about graphs and functions.

1
COORDINATES AND GRAPHS

Figure 1 illustrates a coordinate system on a line l.

First we have chosen on l an origin point O and a unit point E distinct from O. The origin O divides l into two half-lines, or *rays*. The half-line containing E is called *positive*, and the other one *negative*. The segment OE is taken as the unit of length. Then each point P on l is assigned a number x, called the coordinate of P, as follows:

If P is on the positive side of O, then x is the length of the segment OP (in terms of the unit OE).

Figure 1

If P is on the negative side of O, then x is *minus* the length of OP.

Conversely, if we are given any real number x, then there is a unique point P having x as its coordinate. For example, if x is negative, then we obtain P by measuring off the distance $-x$ from the origin O on the negative side of O. Thus, each point P determines a unique number x, and each number x determines a unique point P. This is what we mean when we call the coordinate assignment a *one-to-one correspondence* between the points of l and the real numbers.

Because of the coordinate correspondence, we can think of the real number system $\mathbb{R}$ in its entirety as though it were a geometric line, and we often refer to numbers like 3, π, and $-\sqrt{2}$ as being "points" on the "number line."

The coordinate plane involves two number lines. We choose a pair of perpendicular lines intersecting in a point O, and on each line we set up a coordinate system with O as zero point. We call O the *origin of coordinates*, or simply the *origin;* the intersecting number lines are the *coordinate axes*. We choose one of the axes as the *horizontal* axis, and picture it horizontal, with the *positive direction to the right*. The positive direction of the other axis, called the *vertical* axis, is pictured *upward*.

The axes divide the plane into four parts called *quadrants*. These are conventionally numbered in the counterclockwise direction, starting from the quadrant that is above the positive half of the horizontal axis (Fig. 2).

Figure 2

Given a coordinate system as described above, then each point in the plane determines, and is determined by, an ordered pair of coordinate numbers, as indicated in Fig. 3. We shall feel free to call $(3, 2)$ a point in the coordinate plane, just as we call 3 a point on the number line.

Notice that the point $(0, 2)$ is different from the point $(2, 0)$. It is crucial that the coordinates form an *ordered* pair. The *first* coordinate

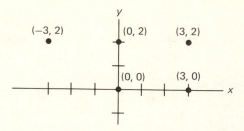

Figure 3

is obtained by dropping a perpendicular to the *horizontal* axis; it is the coordinate on the horizontal axis of the foot of this perpendicular. The second coordinate is found similarly on the vertical axis.

We normally label the axes x and y, with the horizontal axis being taken as the x-axis. Then the point $(3, 2)$ has the x-coordinate 3 and the y-coordinate 2.

A *solution* of an equation in x and y is a pair of values (x, y) that "satisfies" the equation, i.e., that makes it true. For example, $(x, y) = (3, 2)$ is a solution of the equation

$$x^2 + 4y^2 = 25,$$

because it is true that $3^2 + 4(2^2) = 25$. Another solution is $(5, 0)$. The pair $(x, y) = (2, 3)$ is not a solution because

$$2^2 + 4(3^2) \neq 25.$$

The collection of all solution pairs is called the *solution set* of the equation.

The fact that the ordered pair $(3, 2)$ *is* a solution while $(2, 3)$ is *not* a solution depends on our understanding that the first number in an ordered pair is a value of x and the second number a value of y. This goes back to the basic conventions that the first coordinate of a point in the plane is the one measured horizontally, and that the horizontal axis is the x-axis.

The *graph* of an equation is just the geometric representation of its solution set. That is, the graph is the curve or other configuration in the plane consisting of the points whose coordinates satisfy the equation. The graph of the above equation is shown in Fig. 4.

Two equations having the same solution set (i.e., the same graph) are said to be *equivalent*. Here are some other equations equivalent to the equation $x^2 + 4y^2 = 25$:

$$\frac{x^2}{25} + \frac{y^2}{(25/4)} = 1, \qquad \sqrt{x^2 + 4y^2} = 5, \qquad x^2 + 4y^2 - 25 = 0.$$

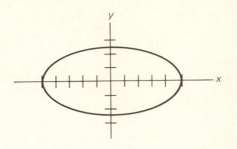

Figure 4

On the other hand, the two equations

$$x = y \qquad \text{and} \qquad x^2 = y^2$$

are *not* equivalent. The point $(1, -1)$ is on one graph but not on the other.

If the curve C is the graph of the equation E, then we call E *an* equation of C, because, as we saw above, *other* equations can have the same graph. However, if E has been singled out from various equivalent equations because it has a *standard form*, then we often call it *the* equation of C. For example (see Section 6),

$$\frac{x^2}{a^2} + \frac{y^2}{b^2} = 1$$

is called *the* equation of an ellipse with center at the origin, meaning that this is the standard form for the equation.

The *intercepts* of a graph are the points where the graph intersects the axes. An intercept on the x-axis is a point of the form $(a, 0)$, but since the second coordinate of an x-intercept is automatically zero, it is customary to say that the x-intercept is a. Any intercept $(a, 0)$ of

$$x^2 + 4y^2 = 25$$

must satisfy the equation

$$a^2 + 4(0^2) = 25,$$

giving $a^2 = 25$ and $a = \pm 5$. Thus *we obtain the x-intercepts of an equation by setting $y = 0$ and solving for x.* Similarly, we get the y-intercepts by setting $x = 0$ and solving for y. The above equation has the y-intercepts $b = \pm 5/2$.

Intercepts can be important "positioning" aids. For example, the equation $3x - 4y - 5 = 0$ has the x-intercept $a = 5/3$ and the y-

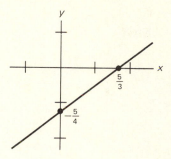

Figure 5

intercept $b = -5/4$, and once we know that the graph of this equation is a straight line, we can just draw it through these two points (Fig. 5).

The points whose coordinates satisfy two equations simultaneously, say

$$y = x^2 \quad \text{and} \quad y = 2x - 1,$$

are the points lying on both graphs, i.e., the points of intersection of the graphs. We find these points by the usual techniques for solving simultaneous equations. Here we can eliminate y by subtracting the second equation from the first, getting

$$0 = x^2 - 2x + 1 = (x - 1)^2.$$

This gives the solution $x = 1$. The corresponding y value is then obtained from either of the original equations, say $y = x^2 = (1)^2 = 1$, so $(1, 1)$ is the only point common to the two curves (Fig. 6).

But now suppose that we want to consider how the two graphs are related to each other overall. Suppose, for instance, that we want to compare the points on the two graphs that have the same x-

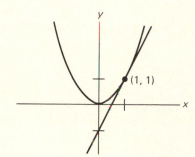

Figure 6

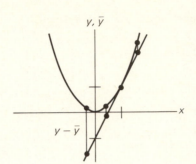

Figure 7

coordinates. To do this it is convenient to use two different y variables, such as in

$$y = x^2 \quad \text{and} \quad \bar{y} = 2x - 1,$$

since the two equations will generally determine different y values for a given value of x. Then we can conclude from

$$y - \bar{y} = x^2 - (2x - 1) = (x - 1)^2 \geq 0$$

that the first graph lies everywhere above the second graph, with only the point $(1, 1)$ in common, as shown in Fig. 7.

PROBLEMS FOR SECTION 1

1. Draw a plane coordinate system and plot the points $(-2, 1)$, $(2, -1)$, $(-1, 0)$, $(4, 1)$, $(4, -1)$.

2. Draw the collection of all points whose first coordinate is 2 (in a plane coordinate system). Describe this collection geometrically.

3. Draw and describe the collection whose second coordinate is -3.

4. Draw the collection of all points (x, y) for which $y/x = 2$. Is this the same as the collection for which $y = 2x$?

5. Draw the collection of all points (x, y) for which $y = -x$.

6. a) Given the point $P = (3, 4)$, find a point $Q_1 = (x, y)$ such that the perpendicular bisector of the segment PQ_1 is the y-axis.

b) Now find Q_2 such that the perpendicular bisector of PQ_2 is the x-axis.

c) Finally, find Q_3 such that the midpoint of the segment PQ_3 is the origin.

d) What kind of geometric figure is the quadrilateral $PQ_1Q_3Q_2$?

7. Answer the same four questions for the point $P = (-2, 1)$.

8. The same for $P = (-1, -3)$

The graph of an equation of the form

$$ax + by + c = 0$$

is always a straight line, provided that a and b are not both zero. The graph of

$$x^2 + y^2 + Ax + By + C = 0$$

is always a circle (or nothing). Assuming these facts, sketch the graphs of the equations in Problems 9 through 17 from their intercepts.

9. $x + y = 1$

10. $x - y = 2$

11. $2x + y = 2$

12. $3p + 2q = 6$ (with the p-axis horizontal)

13. $3p + 2q = 6$ (q-axis horizontal)

14. $x^2 + y^2 = 2$

15. $x^2 + y^2 + 2x - 2y = 0$

16. $x^2 + y^2 - 4x = 0$ (assume the center is on the x-axis)

17. $x^2 + y^2 + 2x - 3 = 0$

18. a) Draw the graphs of Problems 9 and 10 above on the same coordinate system, and estimate their point of intersection.

   b) Compute the point of intersection by solving the equations simultaneously.

19. Same for (9) and (11)

20. Same for (10) and (11)

21. Same for (10) and (14)

22. Same for (9) and (14)

23. Same for (16) and (17)

24. Same for (15) and (16)

25. Graph the circle $x^2 + y^2 = 4$ and the line $x + y = 3$. Do these graphs appear to intersect? Show by algebra that they do not.

26. Graph the equation $y = x^2$. Here we don't know the shape of the graph, and the only way we can start is to plot a few points on the graph. Make up a little table of values of $y = x^2$, when x has the values $-2, -\frac{3}{2}, -1, -\frac{1}{2}, 0, \frac{1}{2}, 1, \frac{3}{2}, 2$. Then plot these points (x, y) and draw a smooth curve through them.

27. Graph the equation $y = x^2 - 2$, proceeding as in the above problem.

28. Graph the equation $y = x^2 - 2x - 2$, using the x-values of Problem 26.

29. Graph the equation $y = 2 - x^2$.

30. Graph the equation $y^2 = x + 2$. (The easiest way to do this is to give y some simple values and determine the corresponding x values from the equation.)

31. Graph the equation $y^2 = -x$.

32. a) Graph the equations $y = x^2$ and $y = x + 2$ on the same axes. Estimate their points of intersection.

   b) Compute the points of intersection algebraically by solving the equations simultaneously.

33. The graph of the equation $x^2 + 4y^2 = 4$ is a smooth oval-shaped curve called an *ellipse*. It will be longest in the direction of the x-axis and shortest in the direction of the y-axis. Use this information to sketch the graph, starting with only its intercepts.

34. The graph of $9x^2 + 4y^2 = 36$ is also an ellipse, but this time it is longest in the y direction. Plot its intercepts and sketch and graph.

35. Graph $y = x^2/2$, using integer values of x from $x = -3$ to $x = 3$.

36. Graph $y = x^2/4$, using integer values of x from $x = -5$ to $x = 5$.

2
THE
STRAIGHT
LINE

Horizontal and vertical lines have the simplest equations. Consider the horizontal line 3 units above the x-axis (Fig. 1). A point (x, y) lies on this line if and only if its y-coordinate is 3, so this line is the graph of the equation $y = 3$.

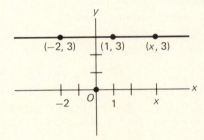

Figure 1

In general, the graph of

$$y = b$$

is the horizontal line with y-intercept b. If b is negative, say $b = -4$, then the line lies 4 units *below* the x-axis.

Similarly,

$$x = a$$

is the equation of the vertical line with x-intercept a (Fig. 2).

Note that although x is missing from an equation like $y = 3$, we can think of it as being there with a zero coefficient:

$$0x + y = 3.$$

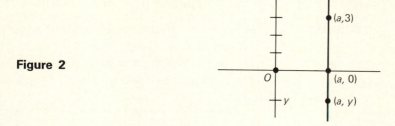

Figure 2

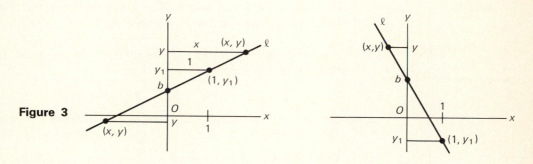

Figure 3

Now let l be any nonvertical line, with y-intercept b. Let $(1, y_1)$ be the point on l having x-coordinate 1. The similar triangles shown in Fig. 3 have corresponding sides proportional. That is,

$$\frac{y - b}{x} = \frac{y_1 - b}{1}$$

for any other point (x, y) on l. Some of the coordinate differences in this formula may be the negatives of the triangle side lengths, so signs have to be checked, but the equation above is always correct. Moreover, if the point (x, y) is not on l, then it does not determine similar triangles and the above equation does not hold. Now set

$$m = y_1 - b$$

in the above equation and solve for y. The end result is

THEOREM 1 *A point (x, y) is on l if and only if its coordinates satisfy the equation*

$$\boxed{y = mx + b.}$$

The coefficient m is called the *slope* of the line, and the equation $y = mx + b$ is called the *slope-intercept* form of the equation for l.

Example   Identify the graph of the equation $2x + 3y - 6 = 0$.

Solution   Solving for y we obtain the equivalent equation

$$y = -\frac{2}{3}x + 2.$$

This is of the form

$$y = mx + b,$$

with $m = -2/3$ and $b = 2$. The graph is therefore a straight line, with slope $-2/3$ and y-intercept 2.   □

The point of intersection of two nonvertical lines can be found by simultaneously solving their equations:

$$y = mx + b, \qquad y = m'x + b'.$$

We eliminate y by subtracting one equation from the other and get $x = (b' - b)/(m - m')$. But this works only if $m \neq m'$. Thus,

Two nonvertical lines have a unique point of intersection if and only if their slopes are different.

Consequently,

Two nonvertical lines are parallel (or identical) if and only if they have the same slope.

The slope of a line tells how steep the line is. Consider, for example, the line through the origin with slope m. Its equation is $y = mx$ and it contains the point $(1, m)$. Three such lines are shown in Fig. 4.

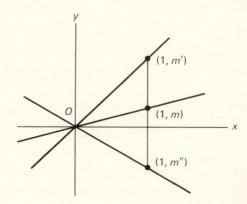

Figure 4

Notice that $m > 0$ for a line *rising* to the right and that $m < 0$ for a line *falling* to the right. A line with large positive slope rises steeply and a line with small positive slope rises gently.

The slope m can be calculated as the change in y divided by the change in x between any two points on the line.

That is,

THEOREM 2 *If (x_1, y_1) and (x_2, y_2) are any two distinct points on a line with slope m, then*

$$m = \frac{y_2 - y_1}{x_2 - x_1}.$$

Proof Since each pair of numbers satisfies the equation $y = mx + b$, we have

$$y_2 = mx_2 + b, \qquad y_1 = mx_1 + b;$$

and so, subtracting,

$$y_2 - y_1 = m(x_2 - x_1).$$

Since the line is not vertical, the two x-coordinates must be different and we can divide by the nonzero number $x_2 - x_1$, getting

$$m = \frac{y_2 - y_1}{x_2 - x_1}. \qquad \blacksquare$$

This slope formula can be interpreted as saying that the slope m is the constant rate of change of y with respect to x along the line. If $m = 3$, then y increases 3 units per unit increase in x, and if $m = -\frac{1}{2}$, then y *decreases* one-half unit per unit *increase* in x.

Two distinct points (x_1, y_1) and (x_2, y_2) determine a line, and if the line is not vertical then the above formula gives its slope m. Thus the line through $(1, 2)$ and $(3, -1)$ has slope

$$m = \frac{y_2 - y_1}{x_2 - x_1} = \frac{(-1) - 2}{(3 - 1)} = -\frac{3}{2}.$$

A line is determined if we know its slope m and one point (x_1, y_1) on it, and we can calculate its equation from this data. For example, a line with slope 2 has the equation

$$y = 2x + b,$$

with b still to be determined. If we know that $(2, 1)$ is on the line then

$$1 = 2 \cdot 2 + b,$$

so $b = -3$, and the equation of the line is

$$y = 2x - 3.$$

When this calculation is repeated in general terms, it yields a new standard form for the equation of a line. Thus, if (x_1, y_1) is on the line

$$y = mx + b,$$

then it is true that

$$y_1 = mx_1 + b,$$

and if we subtract the second equation from the first then the b's cancel and we get the *point–slope* form of the equation:

$$\boxed{y - y_1 = m(x - x_1).}$$

We can write this form down directly whenever we know the slope of a line and the coordinates of a point on it. For example, the line through $(2, 1)$ with slope 2 has the point–slope equation

$$y - 1 = 2(x - 2).$$

Solving this for y, we get back to the slope–intercept form

$$y = 2x - 4 + 1 = 2x - 3.$$

For the purposes of geometry it is useful to have a single form for the equation of a line that covers all lines, vertical or nonvertical. The equation

$$\boxed{Ax + By + C = 0}$$

does this (it being assumed that at least one of the coefficients A and B is not zero).

In order to see that the graph of the above equation is always a line, we consider two cases. If $B \neq 0$ we can solve the equation for y and get an equivalent equation of the form $y = mx + b$. In this case the graph is a nonvertical line. For example, the equation

$$x + 2y - 6 = 0$$

can be solved for y, turning into the equivalent equation

$$y = \left(-\frac{1}{2}\right)x + 3.$$

Its graph is thus the nonvertical line with slope $-1/2$ and y-intercept 3.

If B is 0, then A cannot be 0 and we can solve for x, getting an equation of the form $x = a$. In this case the graph is the vertical line with x-intercept a. Thus $3x + 4 = 0$ becomes $x = -\frac{4}{3}$, the equation of the vertical line $\frac{4}{3}$ units to the left of the y-axis.

Conversely, the standard forms

$$y = mx + b \qquad \text{and} \qquad x = a$$

can be rewritten

$$mx - y + b = 0 \qquad \text{and} \qquad x - a = 0,$$

respectively, both of which are of the form $Ax + By + C = 0$.

PROBLEMS FOR SECTION 2

1. Write the equation of the line with y-intercept -6 and slope 3. Find its x-intercept. Find the point where it intersects the vertical line $x = 1$. Same for the horizontal line $y = 1$.

Find the slope m and the y-intercept b of each of the following lines.

2. $x + y = 0$ **3.** $x + 2y = 4$ **4.** $3x - 4y = 4$

5. $2x + 3y - 6 = 0$ **6.** $2y - 3x - 6 = 0$ **7.** $x - y + 1 = 0$

8. Show that the lines $4x + 2y = 1$ and $y = -2x + 3$ are parallel

a) by drawing both graphs;

b) by proving algebraically that there is no point of intersection.

Draw the graphs of the following equations.

9. $x + y + 1 = 0$ **10.** $y + 2x = 0$ **11.** $3y = x$ **12.** $y = x - 1$

13. $2x - 3y = 6$ **14.** $x + y = 1$ **15.** $x + 2y = 3$

16. Determine the slope and y-intercept for the line whose equation is $Ax + By + C = 0$ where A, B, and C are arbitrary constants, and $B \neq 0$.

Determine the slope of the line through the given pair of points, and then write the equation of the line.

17. $(0, 0)$, $(2, 1)$ **18.** $(2, -1)$, $(-4, 0)$ **19.** $(-2, 1)$, $(-2, 4)$

20. $(3, 1)$, $(-2, 6)$ **21.** $(2, 1)$, $(-3, 1)$ **22.** $(1, 2)$, $(-2, -1)$

Find the (equation of the) line.

23. Through $(1, 2)$, with slope $-(1/2)$ **24.** Through the points $(1, 2)$ and $(2, -1)$

25. Through the points $(-2, -1)$ and $(1, 1)$ **26.** Through the points $(1, 2)$ and $(1, 4)$

27. Through $(2, 3)$, with x-intercept -1

Find the point of intersection of the two lines.

28. $x + y = 1$, and $y = x$ **29.** $x + y = 1$, and $x + 2y = 4$

30. $3x + y = 2$, and $x = -2$ **31.** Find the line through the point $(-1, 2)$ parallel to the line through $(1, 1)$ and $(0, 2)$.

32. Which of the following pairs of lines are parallel?

  a) $2x + y - 2 = 0$, $4x + 2y + 18 = 0$

  b) $x + 2y - 3 = 0$, $2x - y + 3 = 0$

  c) $x + y - 1 = 0$, $y - x + 3 = 0$

  d) $x + 3y + 2 = 0$, $x + 3y - 2 = 0$

  e) $y + x = 1$, $2x + 2y + 5 = 0$

33. Determine the equation of the line passing through the point $(2, 1)$ and parallel to the line through points $(4, 2)$ and $(-2, 3)$.

34. Determine the equation of the line passing through the point $(-2, 5)$ and parallel to the line $2x + y + 6 = 0$.

In Problems 35–38, determine algebraically whether or not the three given points lie on a straight line.

35. $(0, 0)$, $(5, 4)$, $(-10, -8)$ **36.** $(0, -1)$, $(1, 0)$, $(3, 2)$

37. $(6, 5)$, $(3, 3)$, $(1, 2)$ **38.** $(7, 5)$, $(3, 3)$, $(-1, 1)$

39. Find the fourth vertex of the parallelogram with vertices at $(-1, 0)$, $(0, -1)$, $(3, 0)$ in that order.

40. A parallelogram has two sides lying on the lines $y = 2x$ and $3y = x$, and a vertex at $(3, 3)$. Find the equations of the lines containing the other two sides, and find the remaining vertices.

41. If a line has nonzero intercepts a and b on the x and y axes, respectively, show that its equation can be written in the form

$$\frac{x}{a} + \frac{y}{b} = 1.$$

42. Show that the points $(1, 1)$ and $(2, 4)$ are on opposite sides of the line $x - y + 1 = 0$, by reasoning as follows: First, check the sign of $x_0 - y_0 + 1$ for each of the above two points (x_0, y_0). What, then, must happen to the sign of $x - y - 1$ as (x, y) moves along the line segment joining $(1, 1)$ to $(2, 4)$?

43. Show, by reasoning in a manner suggested by the above problem, that the points $(-1, 1)$ and $(2, 4)$ are on the same side of the line $x - y + 1 = 0$.

44. Suppose that the line $Ax + By + C$ does not go through the origin. Show that a point (x_0, y_0) and the origin are on the same side (opposite sides) of this line if $Ax_0 + By_0 + C$ and C have the same sign (opposite signs).

45. Determine the graph of $y^2 = x^2$.

46. Determine the graph of $y^2 - 4x^2 = 0$.

47. Determine the graph of $x^2 + 2xy + y^2 = 1$.

3
DISTANCE
IN THE
COORDINATE
PLANE

We regard the numbers x and $-x$ as having the same *magnitude*. They are at the same distance from 0 on the number line. This common distance is x or $-x$, whichever one is positive. It is called the *absolute value* of x and is designated $|x|$. Thus,

$$|x| = |-x|,$$

$$|x| = \begin{cases} x & \text{if } x \text{ is positive or zero} \\ -x & \text{if } -x \text{ is positive, i.e., if } x \text{ is negative.} \end{cases}$$

Also,

$$|x| = \sqrt{x^2},$$

since by definition $\sqrt{a}$ is the positive square root of a.

Example Graph the equation $y = |x|$.

Solution By the definition of absolute value, the equation $y = |x|$ says that

$$y = x \qquad \text{if } x \geq 0,$$
$$y = -x \qquad \text{if } x \leq 0.$$

Thus when $x \geq 0$ the graph runs along the straight line $y = x$, and when $x \leq 0$ it runs along the line $y = -x$, as shown in Fig. 1.

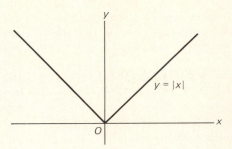

Figure 1

Any distance on the number line can be expressed as an absolute value.

THEOREM 3   *The distance d between any two numbers x and t, considered as points on the number line, is*

$$d = |x - t|.$$

This is a consequence of the additivity of geometric length:

If a segment is subdivided, then the length of the whole is the sum of the length of its parts.

We assume this. Then, to verify the theorem, we suppose that neither x nor t is zero and consider three cases, depending on whether x and t are both negative, both positive, or of opposite signs. Figure 2 illustrates how the additivity law is used in the third case. Further details are omitted.

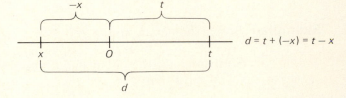

Figure 2

The difference $t - x$ is called the *signed* distance *from* the point x *to* the point t. It is ± the actual distance, depending on whether t is on the positive or negative side of x. It follows that

The coordinate of a point P is the signed distance from the origin to P.

Note also that for *any* three numbers x, y, and z, the identity

$$z - x = (z - y) + (y - x)$$

can be read as saying that the signed distance from x to y plus the signed distance from y to z equals the signed distance from x to z.

The laws relating absolute value to addition and multiplication are

$$|x + y| \le |x| + |y|,$$
$$|xy| = |x| \cdot |y|.$$

These properties of $|x|$ are important for numerical computations and will be used off and on throughout the book. See Appendix 1 for a brief review of inequalities.

We now turn to the coordinate plane. Up until now we have not needed the same units of distance on the two axes, and it is frequently convenient to use different units, especially when we are graphing the relationship between a pair of unlike quantities. For example, if we wish to show graphically how the average daily temperature varies through the months of the year, we would probably use scales somewhat as in Fig. 3.

Figure 3

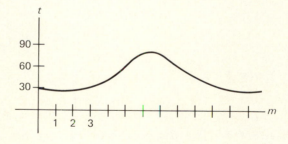

However, from now on a common unit will ordinarily be assumed. Such a coordinate system is said to be *Cartesian*.

The Pythagorean theorem says that a triangle is a right triangle if and only if the square (of the length) of one side is equal to the sum of the squares (of the lengths) of the other two sides. Thus the

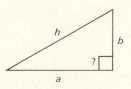

Figure 4

triangle in Fig. 4 is a right triangle with hypotenuse length h if and only if

$$h^2 = a^2 + b^2.$$

This numerical statement of the theorem depends on a unit of length common to the whole plane and is the basic reason for using a Cartesian coordinate system. When we combine the Pythagorean theorem with Theorem 3, as shown in Fig. 5, we obtain a formula for the distance d between any two points (x_1, y_1) and (x_2, y_2) in the plane:

$$a = |x_2 - x_1|,$$
$$b = |y_2 - y_1|,$$
$$d^2 = a^2 + b^2$$
$$= (x_2 - x_1)^2 + (y_2 - y_1)^2.$$

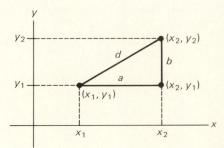

Figure 5

THEOREM 4   *The distance between any two points (x_1, y_1) and (x_2, y_2) in the Cartesian plane is given by the formula*

$$\boxed{d = \sqrt{(x_1 - x_2)^2 + (y_1 - y_2)^2}.}$$

Strictly speaking, the Pythagorean theorem applies only when there is a genuine right triangle. However, the distance formula is correct in all cases. For example, if the two points lie on a horizontal line, then $y_1 - y_2 = 0$ and the formula reduces to

$$d = \sqrt{(x_1 - x_2)^2} = |x_1 - x_2|,$$

which is the correct one-dimensional distance formula.

The slope condition for perpendicular lines also follows from the Pythagorean theorem.

THEOREM 5   *Two nonvertical lines are perpendicular if and only if their slopes m and n satisfy*

$$\boxed{mn = -1.}$$

Proof   Consider first the special case of two nonvertical perpendicular lines through the origin, with equations $y = mx$ and $y = nx$. These lines intersect the vertical line $x = 1$ in the points $(1, m)$ and $(1, n)$, respectively, and will be perpendicular if and only if the triangle thus formed is a right triangle (Fig. 6). By the Pythagorean theorem, this will be the case if and only if

$$c^2 = a^2 + b^2,$$

or

$$(m - n)^2 = (1 + m^2) + (1 + n^2).$$

If we expand the square on the left, cancel the terms m^2 and n^2, and finally divide by -2, we end up with

$$mn = -1.$$

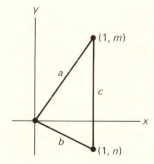

Figure 6

Conversely, if this equation holds, then we can work backward through the above steps and conclude that $c^2 = a^2 + b^2$ so that the a, b, c triangle is a right triangle. This proves the theorem for two lines through the origin. In the general case we replace two given lines by the lines parallel to them through the origin. Since this replacement changes neither the slopes nor the angle between the lines, the general result follows from the special case.   ∎

PROBLEMS FOR SECTION 3

Determine the distance between each of the following pairs of points.

1. $(2, 1)$, $(3, 3)$ **2.** $(-1, 2)$, $(2, -3)$ **3.** $(0, 0)$, $(0, 4)$ **4.** $(2, -5)$, $(2, 1)$

5. $(8, 3)$, $(-2, 1)$ **6.** $(3, 4)$, $(-6, -1)$ **7.** $(0, 1)$, $(-2, 0)$ **8.** $(5, 2)$, $(3, -1)$

9. $(1, 1)$, $(-1, -1)$ **10.** $(2, -1)$, $(-1, 3)$

Find the lengths of the sides of the triangles with the given points as vertices.

11. $A(4, 1)$, $B(2, -1)$, $C(-1, 5)$

12. $A(1, 2)$, $B(3, 1)$, $C(4, 2)$

13. $A(3, -4)$, $B(2, 1)$, $C(6, -2)$

14. $A(0, 0)$, $B(2, 1)$, $C(1, 2)$

15. Show that the triangle with vertices at $P_1(1, -2)$, $P_2(-4, 2)$, and $P_3(1, 6)$ is isosceles.

16. Prove that the triangle with vertices at $(2, 1)$, $(1, 3)$, and $(8, 4)$ is a right triangle, by checking the Pythagorean theorem.

17. Prove that the above triangle is a right triangle by computing slopes.

18. Find (the equation of) the line through the point $(1, 1)$ perpendicular to the line $x + 2y = 0$.

19. Find the line through $(2, 0)$ perpendicular to the line through $(2, 0)$ and $(-1, 1)$.

20. Find the equation of the line passing through point $(-2, 3)$ and perpendicular to the line $2x - 3y + 6 = 0$.

21. Find the line through $(1/2, 1)$ and perpendicular to $4x - 7y = 8$.

22. Suppose that $a < b$ and consider the segment ab on the number line. Show that its midpoint is $(a + b)/2$. [*Hint*: If m is the midpoint, then the signed distance from a to m should equal the signed distance from m to b.]

23. Find the point (number) that is two-thirds of the way from a to b. We say that this point divides the segment ab in the ratio two to one.

24. Find the point (number) that divides the segment ab in the ratio r to s, where r and s are any positive numbers.

25. The midpoint of the line segment joining the points (a_1, b_1) and (a_2, b_2) is the point

$$\left(\frac{a_1 + a_2}{2}, \frac{b_1 + b_2}{2} \right).$$

Verify this analytically in two steps:

a) Show that the three points are collinear.

b) Show that the segment endpoints are equidistant from the claimed midpoint.

Given the vertices of a triangle $P_1(7, 9)$, $P_2(-5, -7)$, and $P_3(12, -3)$, find

27. The equation of side P_1P_2.

29. The equation of the altitude through P_2.

31. The points $P_1(3, -2)$, $P_2(4, 1)$, and $P_3(-3, 5)$ are the vertices of a triangle. Show that the line through the midpoints of the sides P_1P_2 and P_1P_3 is parallel to the base P_2P_3 of the triangle.

33. Show that the points equidistant from two given points form a straight line (by equating two distances and simplifying algebraically).

35. Graph the equation $y = x + |x|$.

37. Graph the equation $|x| + |y| = 1$.

39. Prove that the three perpendicular bisectors of the sides of a triangle are concurrent.

41. Find the distance from $(1, -1)$ to the line $2x - y + 1 = 0$. (As above, first find the foot of the perpendicular dropped from the point to the line.)

26. Using the formula in Problem 25, write the equation of the perpendicular bisector of the segment between $(7, 4)$ and $(-1, -2)$.

28. The equation of the median through P_1.

30. The equation of the perpendicular bisector of the side P_1P_2.

32. Prove the same result for the triangle with vertices (x_1, y_1), (x_2, y_2), (x_3, y_3).

34. Show that if two medians of a triangle are equal, then the triangle is isosceles. (Let the vertices be $(a, 0)$, $(b, 0)$, and $(0, c)$, and equate the lengths of the medians through the vertices $(a, 0)$ and $(b, 0)$.)

36. Graph the equation $y = x - |x|$.

38. Prove algebraically that the three medians of a triangle are concurrent. (We can simplify the algebra by taking the y-axis through one vertex and the x-axis along the opposite side. The vertices are then $(a, 0)$, $(b, 0)$, $(0, c)$.)

40. Find the distance from the point $(2, 3)$ to the line $x + 2y - 1 = 0$. (Find the point of intersection of the given line and the line perpendicular to it through the given point.)

42. Prove that the distance d from the point (x_0, y_0) to the line $Ax + By + C = 0$ is given by

$$d = \frac{|Ax_0 + By_0 + C|}{\sqrt{A^2 + B^2}}.$$

4
THE CIRCLE

A point (x, y) lies on the circle of radius r and center (a, b) if and only if the distance between (x, y) and (a, b) is r (Fig. 1); that is, if and only if

$$\sqrt{(x - a)^2 + (y - b)^2} = r.$$

So this is an equation for the circle. We usually eliminate the radical by squaring, to get the equivalent equation

$$\boxed{(x - a)^2 + (y - b)^2 = r^2.}$$ (*)

This is the standard form. In particular, the equation for the circle of radius r about the origin is

$$x^2 + y^2 = r^2.$$

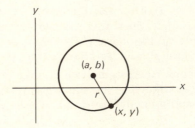

Figure 1

Example 1   The circle about the origin with radius $r = 3$ has the equation

$$x^2 + y^2 = 9.$$ □

Example 2   The equation

$$x^2 + y^2 = 3$$

has the form $x^2 + y^2 = r^2$ with $r = \sqrt{3}$. Therefore its graph is the circle about the origin with radius $r = \sqrt{3}$.   □

Example 3   The circle with center $(1, -2)$ and radius 5 has the equation

$$(x - 1)^2 + (y + 2)^2 = 25.$$ □

The equation $(x - a)^2 + (y - b)^2 = r^2$ can be directly interpreted as saying that a certain distance is r. But this form is artificial from the point of view of a general polynomial equation. We would more likely find the squares expanded and the constants collected. These steps transform the equation in Example 3 into

$$x^2 + y^2 - 2x + 4y - 20 = 0.$$

This is of the general form

$$\boxed{x^2 + y^2 + Ax + By + C = 0,}$$ (**)

and leads us to wonder if every equation of this form has a circle as its graph. Simple examples show that this is too much to expect. Thus

$$x^2 + y^2 = 0$$

is satisfied only for $(x, y) = (0, 0)$ and hence has this point for its graph, while

$$x^2 + y^2 + 1 = 0$$

is *never* satisfied and has an empty graph.

So there is a problem about the graph of an equation having the general form (**). Is it a circle or not? And what circle, if so? To answer these questions we try to work back from the general form (**) to the standard form (*). We gather together the x terms and make them into a perfect square by adding a suitable constant; then we do the same with the y terms. The only remaining question is whether the resulting constant on the right side of the equation is positive, zero, or negative.

Example 4 If we try to recapture Example 3 from its expanded form by following this prescription, we start with

$$x^2 + y^2 - 2x + 4y - 20 = 0,$$

and first rewrite it as

$$(x^2 - 2x + K) + (y^2 + 4y + L) = 20 + K + L,$$

where we have to discover what values for the constants K and L will make the parentheses into perfect squares. Notice that we have to add these constants to *both sides* of the equation in order for the new equation to be equivalent to the old. In order to find K, we note that $(x^2 - 2x + K)$ must turn out to be $(x - a)^2 = x^2 - 2ax + a^2$. Thus the two expressions

$$x^2 - 2x + K \qquad \text{and} \qquad x^2 - 2ax + a^2$$

must be the same, so that $2 = 2a$ and $K = a^2$. This gives $a = 1$ and $K = 1^2 = 1$.

Similarly, the two expressions

$$y^2 + 4y + L \qquad \text{and} \qquad (y - b)^2 = y^2 - 2by + b^2$$

must be the same, so that $4 = -2b$ and $L = b^2$, giving $b = -2$, and $L = (-2)^2 = 4$.

This process is known as *completing the square*. For the x terms, the end result is given by the rule

To get the constant that must be added, divide the coefficient of x by 2 and then square.

At this point we have

$$(x - 1)^2 + (y + 2)^2 = 20 + 1 + 4 = 25,$$

which is of the form (*) with $(a, b) = (1, -2)$ and $r = 5$. Now we can identify the graph as the circle with center at $(1, -2)$ and radius 5.

□

Example 5   Find the graph of $x^2 + y^2 - 6x + 2y + 10 = 0$.

Solution   Following the above procedure, we write the equation in the form

$$(x^2 - 6x + K) + (y^2 + 2y + L) = -10 + K + L$$

and see that $K = 9$ and $L = 1$, so that the equation becomes

$$(x - 3)^2 + (y + 1)^2 = 0.$$

Since a sum of squares is zero only if each square is zero, the only solution of this equation is the point $(x, y) = (3, -1)$.

If 10 were replaced by 15 in this example, we would end up with -5 on the right, and since a sum of squares can never be negative there is no graph.

□

PROBLEMS FOR SECTION 4

Write the equation of each of the following circles in *standard* form. Then expand and collect coefficients to get the general form. Sketch each circle.

1. $C(2, 1)$ and $r = 3$

2. $C(-1, 0)$ and $r = \sqrt{2}$

3. $C(2, -3)$ and $r = 1$

5. $C(-2, 4)$ and $r = 5$

7. $C(0, 0)$ and $r = |-4|$

4. $C(-2, -3)$ and $r = 4$

6. $C(2, -1)$ and $r = 0$

8. $C(2, 5)$ and $r = \sqrt{3}$

By completing the square, convert each of the following equations into the standard form for the equation of a circle. Specify the coordinates of the *center* and the *radius* of the circle.

9. $x^2 + y^2 + 6x - 8y = 0$

11. $x^2 + y^2 + 3x - 5y - \frac{1}{2} = 0$

10. $x^2 + y^2 - 4x + 2y + 5 = 0$

12. $2x^2 + 2y^2 + 3x + 5y + 2 = 0$

Identify the graph of each of the following equations. That is, determine whether the graph is a circle, and if so, what circle.

13. $x^2 + y^2 + 2x = 0$

15. $x^2 + y^2 + y = 1$

17. $x^2 + y^2 + 2x + 2y + 4 = 0$

14. $x^2 + y^2 + 4x - 2y = 4$

16. $x^2 + y^2 + 2y + 1 = 0$

18. Determine the value of k such that $x^2 + y^2 - 8x + 10y + k = 0$ is the equation of a circle with radius 5.

19. Derive the equation of the circle whose center is at $(5, -2)$ and that passes through the point $(-1, 5)$.

20. Find the equation of the circle having as a diameter the line segment joining $P_1(-4, -3)$ and $P_2(2, 5)$.

21. Find the equation of the circle passing through the origin and the point $(4, 2)$ and having its center on the x-axis.

22. Find the equation of and identify the circle passing through the three points $(1, 0)$, $(-1, 0)$, and $(0, 2)$.

23. Same question for the points $(0, 0)$, $(1, 0)$, and $(2, 1)$.

24. Identify the locus of a point $P = (x, y)$ moving in such a way that the sum of the squares of its distances from $(-2, 0)$ and $(2, 0)$ is a constant k greater than 8.

25. Find the locus of a point $P = (x, y)$ whose distance from the origin is twice its distance from $(3, 0)$.

26. Let Q_1 and Q_2 be fixed points and let k be a positive constant. Prove that the locus of a point P, which is moving in such a way that its distance from Q_1 is k times its distance from Q_2, is either a circle or a straight line.

27. Prove algebraically that the line through the origin and the point (a, b) intersects the unit circle $x^2 + y^2 = 1$ in the point $(a/r, b/r)$, where $r = \sqrt{a^2 + b^2}$.

28. Show that the point $(1, -3)$ lies inside the circle $x^2 + y^2 + 6x + 8y = 0$.

29. Show that $x_0^2 + y_0^2 + Ax_0 + By_0 + C < 0$ if the point (x_0, y_0) lies inside the circle

$$x^2 + y^2 + Ax + By + C = 0.$$

30. Show that if $x_0^2 + y_0^2 + Ax_0 + By_0 + C < 0$, then the point (x_0, y_0) lies inside the circle

$$x^2 + y^2 + Ax + By + C = 0.$$

(Reason as follows: If x_1 and y_1 are very large, then

$$x_1^2 + y_1^2 + Ax_1 + By_1 + C$$

is positive. What, therefore, must happen to the sign of

$$x^2 + y^2 + Ax + By + C$$

as the point (x, y) moves along the line segment from (x_0, y_0) to (x_1, y_1)?)

5
SYMMETRY

Frequently we face the problem of drawing a good graph. The crudest approach is to calculate a few solution points, draw as smooth a curve as we can through them, and hope that we have a reasonable approximation to the graph. Unfortunately we can make gross errors this way, because the few points we choose may fail to indicate some essential feature of the graph, which we therefore miss completely. We shall see later on how calculus helps to find the correct general shape. Here we shall look at symmetry properties of graphs that can be deduced from the algebraic form of the equation.

If a point P is not on a line l, then the *symmetric image* of P in l is the "mirror image" of P in l. It is the unique point P' such that l is the perpendicular bisector of the segment PP'. We obtain P' by dropping the perpendicular from P to l and then continuing an equal distance across l (Fig. 1).

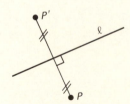

Figure 1

Similarly, a geometric figure F has a mirror image F' consisting of all the symmetric images in l of the points of F (Fig. 2). We sometimes call F' the *reflection* of F in the line l.

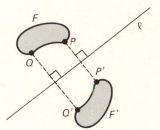

Figure 2

Also, F is the mirror image of F'', so F and F'' are mirror images of each other in l. A geometric figure F has l as a *line of symmetry* if F is its own mirror image in l, i.e., if F contains the symmetric image in l of each of its points P (Fig. 3). Then the 180° rotation of the plane over the axis l just interchanges each symmetric pair of points and carries the figure F exactly into itself.

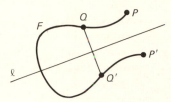

Figure 3

An equilateral triangle has three lines of symmetry; an isosceles triangle that is not equilateral has one line of symmetry; and a nonisosceles (scalene) triangle has no line of symmetry. A square has four lines of symmetry, and a rectangle that is not square has two lines of symmetry. A circle has an *infinite number* of lines of symmetry.

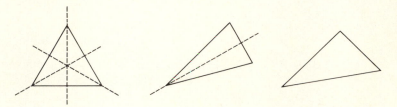

The symmetric (mirror) image of the point (a, b) in the y-axis is the point $(-a, b)$. The following example shows how this carries over to equations.

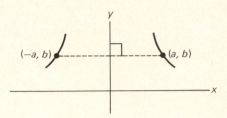

Example 1   Show that the graphs of the equations $xy = 1$ and $-xy = 1$ are mirror images in the y-axis.

Solution   The point (a, b) satisfies the equation $xy = 1$ if and only if the point $(-a, b)$ satisfies the equation $-xy = 1$. That is, the points on the graph of $-xy = 1$ are exactly the mirror images in the y-axis of the points on the graph of $xy = 1$. Figure 4 shows the top halves of the graphs.

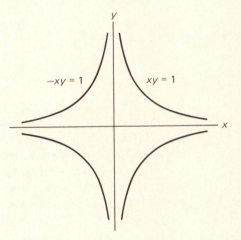

Figure 4

The same reasoning shows in general that

If the equation E' is obtained from the equation E by replacing x by $-x$, then the graphs of E and E' are mirror images in the y-axis.

In particular,

The graph of an equation is symmetric in the y-axis (i.e., is its own mirror image) if and only if replacing x by $-x$ in the equation results in an equivalent equation.

Similarly, *a graph is symmetric about the x-axis if replacing y by −y in its equation yields an equivalent equation.*

Example 2 Since $y = x^2$ is equivalent to $y = (-x)^2$, the graph of $y = x^2$ is symmetric about the y-axis (Fig. 5a). □

Example 3 The graph of

$$\frac{x^2}{a^2} + \frac{y^2}{b^2} = 1$$

is symmetric about each axis since replacing either x or y by its negative leaves the equation unchanged (Fig. 5b).

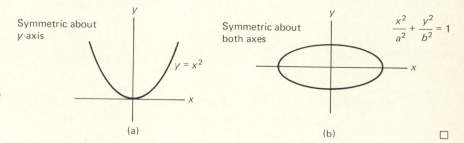

Figure 5

Reflections in the line $y = x$ are also easy to characterize, since the mirror image of the point (a, b) is the point (b, a) (Fig. 6). Thus,

If the equation E′ is obtained from the equation E by interchanging x and y, then the graphs of E and E′ are mirror images in the line y = x. It follows that if the equations E and E′ are equivalent, then their common graph is symmetric about the line y = x.

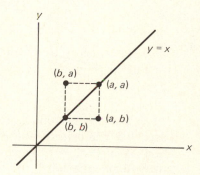

Figure 6

Example 4   The graphs of $y = x^2$ and $x = y^2$ are mirror images in the line $y = x$, as shown in Fig. 7(a).

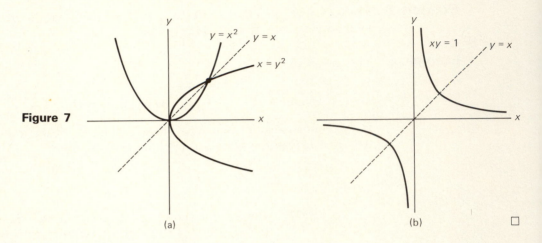

Figure 7

(a)

(b)

Example 5   Since $yx = 1$ is equivalent to $xy = 1$, the graph of $xy = 1$ is symmetric about the line $y = x$ (see Fig. 7b.)   □

Another type of symmetry centers about a point Q. Two points P and P' are symmetric *in* Q, or *with respect to* Q, if Q is the midpoint of the segment PP'. Then P' is the symmetric image of P in Q, and vice versa. A figure F has Q as a *point of symmetry* if F contains the symmetric image in Q of each of its points P.

A parallelogram has a point of symmetry. A triangle does not.

If a figure F has a pair of perpendicular lines of symmetry, then their intersection is necessarily a point of symmetry for F (Fig. 8).

Figure 8

When we introduce coordinates, we see that symmetry with respect to the origin can be stated as follows: If (x, y) is on the graph then so is $(-x, -y)$. That is, replacing (x, y) by $(-x, -y)$ yields an equivalent equation.

Example 6 Since $y = 1/x$ is equivalent to $-y = 1/(-x)$, its graph is symmetric in the origin (Fig. 9).

Figure 9

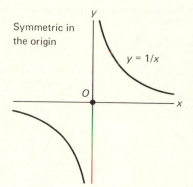

Symmetric in the origin

$y = 1/x$

As the above example shows, a curve symmetric in the *origin* need not be symmetric in *either axis*. However, a curve that is symmetric about *both* axes must of necessity also be symmetric in the origin.

PROBLEMS FOR SECTION 5

State what symmetry, if any, the graphs of the following equations have with respect to the axes, the line $y = x$, and the origin. In each case, if the graph is unsymmetrical, write down an equation for the mirror image graph.

1. $9x^2 + 16y^2 = 144$

2. $x^2 + 2xy + y^3 = 0$

3. $2x^3 + 3y + 2y^3 = 0$

4. $2y = x$

5. $2x^2 + y = 0$

6. $y^2 + 2x + 2x^2 + 5 = 0$

7. $x^2 - y^2 = 0$

8. $xy = 1$

9. $x^2 + xy + y^2 = 1$

10. $x^5 - xy + y^3 = 0$

11. $x(1 - y^2) = 1$

12. $\dfrac{x}{y} + \dfrac{y}{x} = 3$

13. $xy^2 - y = x^3$

14. $x - y + xy = 0$

15. Given an equation E in x and y, let E' be obtained from E by replacing x by $-y$ and y by $-x$. Show that the two graphs are mirror images of each other in the line $y = -x$.

16. Determine which of the equations in Problem 1 through 14 have graphs that are symmetric about the line $y = -x$. (Use Problem 15.)

17. Referring to Problem 15, write down the equation E' for the equations in Problems 1 through 6.

18. Determine when the graph of an equation is symmetric in the line $x = 1$.

19. Suppose that the equations E and E' have graphs that are symmetric about the point $(2, -1)$. How is E' determined from E?

**6
SECOND-
DEGREE
EQUATIONS**

Graphs of second-degree equations are treated in Appendix 3. The present section contains a few useful preliminary facts. The standard equations of degree two are

$$y = kx^2 \qquad \text{Parabola}$$

$$\frac{x^2}{a^2} + \frac{y^2}{b^2} = 1 \qquad (a > b > 0) \qquad \text{Ellipse}$$

$$\frac{x^2}{a^2} - \frac{y^2}{b^2} = 1 \qquad \text{Hyperbola}$$

The general shapes of the graphs are shown in Fig. 1. The ellipse is symmetric in both axes and has a center of symmetry at the origin. It has a diameter of maximum length $2a$ along the x-axis and a diameter of minimum length $2b$ along the y-axis. The maximum diameter and the axis along which it lies are each referred to as the *major axis* of the ellipse. The minimum diameter and its axis are each called the *minor axis*. The intercepts a and b are thus the lengths of the semimajor and semiminor axes, respectively.

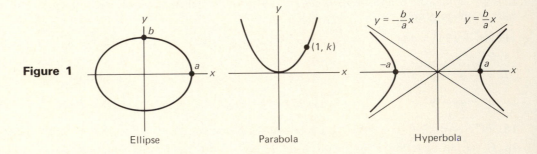

Figure 1

Ellipse   Parabola   Hyperbola

To graph an ellipse, we can calculate a few points in the first quadrant and then use symmetry to obtain the corresponding points in the other quadrants. But for a quick sketch we just plot the intercepts and draw a smooth, symmetric oval through them.

Example 1   Quickly sketch the graph of

$$9x^2 + 16y^2 = 144.$$

Solution   Dividing by 144 yields the standard form

$$\frac{x^2}{16} + \frac{y^2}{9} = 1,$$

so we have an ellipse with $a^2 = 16$, $b^2 = 9$. The intercepts are thus

$$\pm a = \pm 4 \qquad \text{and} \qquad \pm b = \pm 3,$$

and the graph looks as shown in Fig. 2.

Figure 2

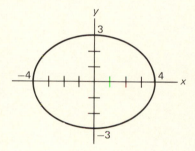

The parabola has one line of symmetry, here the y-axis, and one intersection with its line of symmetry, called its *vertex*, here the origin. Starting at the vertex it opens up along its line of symmetry in one direction, here the direction of the positive y-axis.

The hyperbola has two axes of symmetry, but it intersects only one of them and this is called the axis of the hyperbola. To sketch the hyperbola

$$\frac{x^2}{a^2} - \frac{y^2}{b^2} = 1$$

quickly, we use the x-intercepts $\pm a$ and the straight lines

$$y = \pm \frac{b}{a} x.$$

The hyperbola is *asymptotic* to these lines, in the sense that it approaches them more and more closely as $|x|$ gets large. (This is a limit statement that we will take up in the next chapter.)

Example 2   Sketch the hyperbola

$$\frac{x^2}{9} - \frac{y^2}{4} = 1.$$

Solution   Here $a^2 = 9$ and the x-intercepts are ± 3. The asymptotes are

$$y = \pm \frac{2}{3} x.$$

We plot the intercepts and asymptotes and then sketch the hyperbola (Fig. 3).

Figure 3

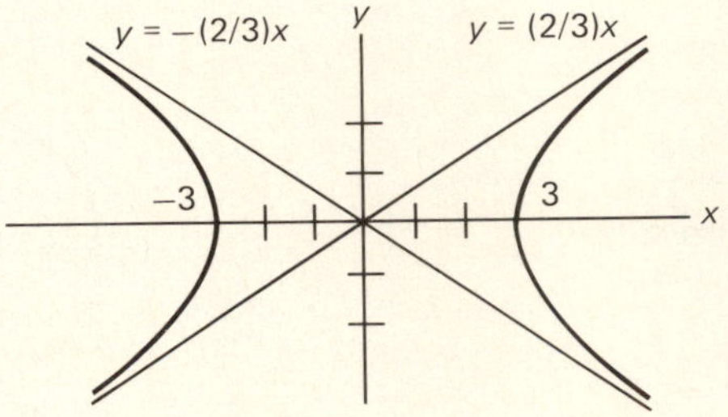

If we interchange x and y in the standard equations we just interchange the roles of the coordinate axes. For example, $x = y^2$ is a parabola opening up along the positive x-axis; $x^2 + y^2/4 = 1$ is an ellipse with major axis along the y-axis; and $y^2 - x^2 = 1$ is a hyperbola with its intercepts on the y-axis. See Fig. 4. These interchanged forms are equally simple equations and can also be considered to be standard equations.

Figure 4

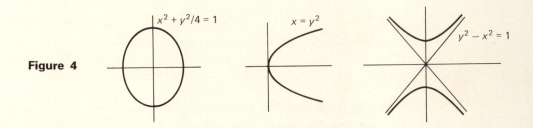

PROBLEMS FOR SECTION 6

Use intercepts, symmetry, and (possibly) asymptotes to make a quick sketch of the graph of each of the following equations.

1. $x^2 + 4y^2 = 1$ **2.** $x^2 - 4y^2 = 1$ **3.** $x^2 - 4y^2 = -1$ **4.** $x^2 - 4y = 0$

5. $4x^2 + 9y^2 = 36$ **6.** $2x^2 - y^2 = 4$ **7.** $x + 4y^2 = 0$ **8.** $x^2 + 4y^2 = 0$

9. $y + x^2 = 0$ **10.** $y^2 + x^2 = 4$ **11.** $x^2 + 9y^2 = 9$ **12.** $9x^2 + 4y^2 = 36$

13. $3x^2 + 2y^2 = 6$ **14.** $3x^2 - 2y^2 = 6$ **15.** $4x^2 + y^2 = 4$ **16.** $x^2 - 4y^2 = 4$

17. $4x^2 - y^2 = 4$ **18.** $x^2 + 4y = 0$ **19.** $y^2 + x^2 = 0$ **20.** $9x - y^2 = 0$

21. $4x^2 - y^2 = 0$

7
TRANSLATION OF AXES†

In Section 5 we saw that the graph of

$$x^2 + y^2 - 2x + 4y - 20 = 0$$

is the circle of radius 5 centered at the point $(1, -2)$. So if we choose a new XY-axis system, with origin O' at $(1, -2)$, as shown in Fig. 1, then the new equation having the same graph is $X^2 + Y^2 = 25$. This is a much simpler equation, and one that shows immediately what the curve is.

Figure 1

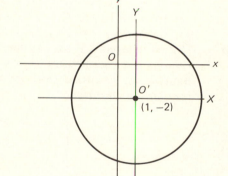

---
† This section can be postponed; it is useful, but not essential. The first reference to it occurs in Chapter 10.

It is useful to be able to recognize in the same way "off-center" ellipses, hyperbolas, and parabolas. Consider, therefore, what happens when we change to a new set of axes, having the same unit of distance and the same directions as the original axes, but a new origin O' at the point having original coordinates (h, k). These new XY-axes can be viewed as obtained by *translating* the old axes, i.e., by sliding them parallel to themselves to a new position.

For the point P shown in Fig. 2, the new and old coordinates are related by

$$x = h + X,$$
$$y = k + Y.$$

The same is true for any point P, as shown by the following general argument.

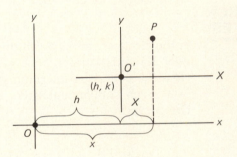

Figure 2

THEOREM 6   *If a translated XY-axis system is chosen with its origin O' at the point having old coordinates (h, k), then the new coordinates (X, Y) of a point P are obtained from its old coordinates (x, y) by the change-of-coordinate equations*

$$\boxed{\begin{array}{l} X = x - h, \\ Y = y - k. \end{array}}$$

Proof   We look at the x and X coordinates as one-dimensional coordinate systems on a single horizontal line, say the horizontal line through $O' = (h, k)$. We know that when the new origin is put at h, then the new coordinate of x is the signed distance from h to x, namely $X = x - h$ (by the remark after Theorem 3). Since the first coordinates x and X remain constant along every vertical line in the plane, the identity $X = x - h$ holds throughout the plane. The identity $Y = y - k$ is proved in the same way. We thus have the theorem. ■

Example 1 Consider the equation

$$y = 2x^2 - 4x.$$

We wonder whether it can be simplified by using a translated coordinate system. Our experience with circles suggests that completing the square may lead to a simplifying new origin, and following this lead we get the equivalent equation

$$y + 2 = 2(x^2 - 2x + 1)$$
$$= 2(x - 1)^2.$$

This equation becomes

$$Y = 2X^2$$

under the change of coordinates

$$X = x - 1,$$
$$Y = y + 2.$$

That is, a point has new coordinates (X, Y) which satisfy the equation

$$Y = 2X^2$$

if and only if its old coordinates (x, y) satisfy

$$y = 2x^2 - 4x.$$

The graphs of the two equations are the same curve in the plane, and the common graph is now seen to be a parabola (Fig. 3), located in the standard way with respect to the *new axes*, with new origin O' at $(1, -2)$.

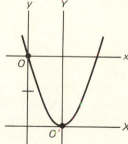

Figure 3

□

Example 2 Identify the graph of the equation

$$x^2 + 4y^2 + 4x = 0.$$

Solution We gather together the x terms and complete the square, just as we did for the parabola above and for circles in Section 5. Thus,

$$(x^2 + 4x + 4) + 4y^2 = 4,$$
$$(x + 2)^2 + 4y^2 = 4,$$
$$\frac{(x + 2)^2}{4} + y^2 = 1.$$

The equation is of the form

$$\frac{X^2}{4} + \frac{Y^2}{1} = 1,$$

and it is thus the standard equation of an ellipse in the (X, Y) coordinate system, with semimajor axis 2 and semiminor axis 1. Since

$$X = x + 2 = x - h,$$
$$Y = y = y - k,$$

we see that

$$(h, k) = (-2, 0).$$

These are the (x, y) coordinates of the new origin, and we now have enough information to draw the graph.

Figure 4

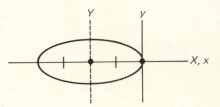

Proceeding just as we did in the above examples, completing squares and then setting $X = x - h$ and $Y = y - k$, we can identify the graph of any equation of the form

$$\boxed{Ax^2 + By^2 + Cx + Dy + E = 0.}$$

PROBLEMS FOR SECTION 7

1. a) Draw an xy-coordinate system and then draw new XY axes with a new origin at the point $(1, 4)$. From your sketch determine the new coordinates of the point P that has old coordinates $(3, 3)$.

  b) Write down the change-of-coordinate equations and compute the new coordinates of $(3, 3)$ from them.

Follow the above instructions in each of the following situations.

2. New origin at $(-2, 3)$; $P = (-3, 3)$

3. New origin at $(2, -4)$; $P = (5, 5)$

4. New origin at $(-5, -2)$; $P = (-6, -3)$

5. New origin at $(3, 2)$; $P = (1, 1)$

6. New origin at $(6, -1)$; $P = (6, 0)$

7. New origin at $(-3, -2)$; $P = (-1, -1)$

8. New origin at $(-3, 2)$; $P = (2, 1)$

Reduce each of the following equations to a standard form by a translation of axes. Identify and sketch the graph.

9. $x^2 + y^2 + 4x - 6y = 5$

10. $25x^2 - 16y^2 = 400$

11. $x^2 + 9y^2 = 9$

12. $y^2 + 2y - 8x - 3 = 0$

13. $4x^2 + 9y^2 + 16x - 18y - 11 = 0$

14. $x^2 + y^2 - 2x + 3y + 3 = 0$

15. $4x^2 - y^2 - 8x + 2y + 7 = 0$

16. $x^2 + 2x + 4y - 7 = 0$

17. $x^2 + 4y^2 - 2x - 16y + 13 = 0$

18. $9x^2 - 16y^2 + 36x + 96y - 144 = 0$

Find the new equation for the graph of each of the following equations when a new axis system is chosen with new origin O' at the given point.

19. $x^2 + y^2 = 4;$ $O' = (2, -1)$

20. $xy = 1;$ $O' = (1, 3)$

21. $xy - 3x + 2y = 6;$ $O' = (-2, 3)$

22. $x^2y - y^3 = 0;$ $O' = (1, -1)$

8
FUNCTIONS

Many different operations come up in mathematics. Some apply to pairs of numbers, like the operation of addition, or multiplication, or forming the sum of two squares. Others apply to single numbers, like the operation of squaring, or the operation of taking the positive square root, or the operation of multiplying by 3 and then adding 7. It is these single-number operations that we shall be concerned with here. The numbers to which such an operation can be applied make up its *domain*. For example, a number has a square root if and only if it is nonnegative, so the domain of the square root operation is the set of all nonnegative numbers. We can think of an operation of this sort as being applied to "input" numbers and determining corresponding "output" numbers. Its domain is then its pool of permissible input numbers.

These operations are all examples of mathematical functions. Moreover, any function can be thought of this way.

DEFINITION   A *function* is an operation that determines a unique "output" number y when applied to a given "input" number x. The collection of numbers to which the operation can be applied makes up its *domain*.

Example 1   The operation of taking the reciprocal of a number, i.e., the operation of going from x to $1/x$, is a function defined for all numbers except 0. Its domain is thus the collection of all nonzero numbers.   □

We use letters such as f and g to represent functions. If f is a function and x is a number in its domain, then we designate by $f(x)$ the result of applying the operation f to x. The symbol $f(x)$ is read "f of x."

Example 2   If f is the reciprocal function considered above, then $f(x) = 1/x$. Therefore

$$f(3) = \frac{1}{3}, \qquad f(-1) = -1, \qquad f(1 + t) = \frac{1}{(1 + t)}, \qquad \text{and} \quad f\left(\frac{1}{a}\right) = a.$$

Also, $1/(f(x)) = x$ and

$$f(x) + \frac{1}{f(x)} = \frac{1}{x} + x = \frac{1 + x^2}{x}.$$ □

Example 3   Let g be the square-root operation, $g(x) = \sqrt{x}$, where $\sqrt{x}$ is always the positive square root of x, as in

$$\sqrt{(-3)^2} = \sqrt{9} = +3.$$

Then

$$g(4) = 2, \qquad g(x^8) = x^4, \qquad g(y^2) = |y|.$$

The domain of g is the collection of all numbers having square roots, i.e., the collection of all nonnegative numbers.   □

From another point of view, a function f can be considered to be a *rule of assignment*. Instead of thinking of f as *operating* on the input number x and producing the output number $f(x)$, we view f as *assigning* the number $f(x)$ to the number x. For example, the square-root function is then viewed as the rule: assign the number $\sqrt{x}$ to the number x.

Example 4   An equation in two variables may determine a function, either directly or indirectly. Consider the following equations:

$$y = x^2,$$
$$y + yx^2 = 1,$$
$$y^3 = x,$$
$$x + y + x^5 + y^7 = 0.$$

Each of these equations has the property that *for each value of x there is a unique value of y making the equation true.* We then have the function (rule): Assign to each number x the unique number y that it determines. And we say that *the equation determines y as a function of x.*

The first equation is in the form

$$y = f(x) \qquad \text{with} \quad f(x) = x^2.$$

It expresses y *explicitly* as a function of x, the function being the squaring function.

The second equation does not present y explicitly as a function of x. However, we can solve it for y, obtaining the equivalent equation

$$y = \frac{1}{1 + x^2},$$

which *is* of the explicit form $y = f(x)$. Solving for y is the easiest way to show that an equation determines y as a function of x.

The third equation is like the second. It is not in the explicit form, but it can be solved for y, giving the equivalent explicit equation

$$y = x^{1/3}.$$

The last equation is different. It cannot readily be solved for y. Nevertheless, we can prove (see p. 88) that it does determine y as a function of x.

In each case but the first we say that the equation determines y *implicitly* as a function of x. In two of the cases it was easy to find an equivalent explicit expression for the function.   □

In general, whenever two variable quantities x and y are so related that the value of y depends on and is uniquely determined by the value of x, we say that y *is a function of* x. The function f in question is the rule: Assign to each x the unique y it determines. We call x the *independent variable*, because its value is chosen arbitrarily, and we call y the *dependent variable*, because its value depends on the value of x.

The "y is a function of x" language is very useful, since in practice functions frequently arise as relationships among variables.

Example 5   In a manufacturing process the cost C of producing x items is a function of x. In a simple situation there might be a fixed initial cost I to purchase a machine, and then an additional cost per item of amount k, for material, labor, etc. The total cost of producing x items is then

$$C = I + kx.$$

Thus, $C = f(x)$, where $f(x) = I + kx$. The average cost, or cost per item, is given by

$$c = \frac{C}{x},$$

and this also is a function of x:

$$c = g(x) = \frac{I}{x} + k.$$

Note that since x is the number of items manufactured, x is a positive integer. The domain of the cost function $f(x) = I + kx$ is thus a set of positive integers. However, if we were to graph the cost

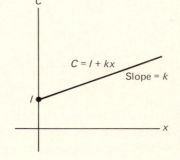

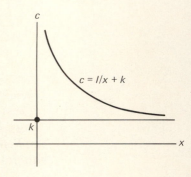

Figure 1

equation $C = I + kx$ in order to visualize the relationship between C and x, we would generally draw the smooth curve for all positive x, as in Fig. 1.   □

Example 6 If a flexible container of gas is kept at a constant temperature, then it is observed that the volume V occupied by the gas is inversely proportional to the pressure p applied to it. That is,

$$pV = \text{constant.}$$

This relationship determines V as a function of p, and also determines p as a function of V. Here, again, the function domains are the *positive* real numbers, because of the physical interpretations of the variables.   □

Function domains are generally described in terms of *intervals*. An interval is simply a segment on the number line. If its endpoints are the numbers a and b, then the interval consists of the numbers lying between a and b. Supposing that a is to the left of b, these are the points that are simultaneously to the right of a and to the left of b. However, we may or may not want to include the endpoints themselves in the interval. The *closed interval* from a to b, designated $[a, b]$, includes the endpoints and so contains exactly those numbers x such that $a \le x \le b$. Thus $[1, 2]$ consists of the real numbers from 1 to 2, *inclusive*.

We use a parenthesis to indicate an *excluded* endpoint. Therefore the interval $[a, b)$ contains a but not b and so contains exactly those numbers x such that $a \le x < b$. The interval (a, b), with both endpoints excluded, is called the *open* interval from a to b. There is ambiguity here since the same symbol (a, b) is used for a point in the coordinate plane, but the context will always make it clear which meaning is intended. Figure 2 shows how these intervals are indicated geometrically.

Figure 2

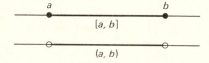

A half-line is considered to be an interval extending to infinity. We use the infinity symbol, ∞, to indicate the unterminated direction of such an interval. Thus $[a, \infty)$ consists of those points x such that $x \geq a$. And $(-\infty, \infty)$ is the whole real-number system. There is no question of having $[a, \infty]$, since ∞ is not a number.

The domains of the functions that we meet in calculus always consist of one or more intervals. For example:

the domain of $f(x) = \sqrt{x}$ is $[0, \infty)$;

the domain of $f(x) = \sqrt{1 - x^2}$ is $[-1, 1]$;

the domain of $f(x) = 1/x$ is the union of $(-\infty, 0)$ and $(0, \infty)$.

We frequently refer to intervals by letters such as I and J, as in Fig. 3. We do this when we want to talk about an interval but don't need to refer specifically to its endpoints; and we may not care particularly whether the interval is open or closed.

Figure 3

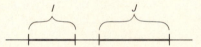

Any point x in an interval I that is not an endpoint of I is called an *interior point* of I (Fig. 4). We also say that x is *in the interior of I*. An interval extends *across* an interior point but lies entirely *on one side of* an endpoint. An open interval consists only of interior points.

Figure 4

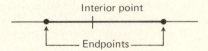

Sometimes $f(x)$ is called the *image* of x under f, or the f-image of x. This terminology is also applied to whole sets. Thus the image of the set A under f, designated $f(A)$, is the set of all images $f(x)$ of points x in A.

Example 7   If $f(x) = 1/x$, then

$$f([1, 2]) = [1/2, 1],$$
$$f((0, 1)) = (1, \infty).$$

If g is the squaring function, $g(x) = x^2$, then $g([1, 2]) = [1, 4]$.   □

If D is the domain of f, then its image

$$R = f(D)$$

is called the *range* of f. It is the set of all output numbers for f.

Combinations of Functions There is nothing new about adding, multiplying, and dividing functions. These are part of ordinary algebra.

Example 8 If $f(x) = 1 + x^3$ and $g(x) = x^2$, then

$$f(x) + 2g(x) = (1 + x^3) + 2x^2 = x^3 + 2x^2 + 1;$$
$$f(x)g(x) = (1 + x^3)x^2 = x^5 + x^2;$$
$$\frac{f(x)}{g(x)} = \frac{1 + x^3}{x^2} = \frac{1}{x^2} + x. \qquad \square$$

The only thing that may need attention is the *domain* of the resulting function.

Example 9 If $f(x) = \sqrt{x}$ and $g(x) = \sqrt{1 - x}$, then

$$f(x) + g(x) = \sqrt{x} + \sqrt{1 - x}$$

is defined only when $f(x)$ and $g(x)$ are *both* defined, i.e., when *simultaneously* $x \geq 0$ and $x \leq 1$. The domain is thus $[0, 1]$.

The domain of

$$\frac{g(x)}{f(x)} = \frac{\sqrt{1 - x}}{\sqrt{x}}$$

must further exclude any values of x for which $f(x) = 0$. There is only one such, namely, $x = 0$, so the domain is $(0, 1]$. $\qquad \square$

Algebraic simplification may enlarge a domain.

Example 10 Offhand, we might write

$$\sqrt{x - 1} \cdot \sqrt{x + 1} = \sqrt{x^2 - 1}$$

but this is true only when both sides are defined, i.e., when $x \geq 1$. However, the right side is defined also when $x \leq -1$.

Another example is given by

$$\frac{x^2}{x} = x. \qquad \square$$

A less familiar way of combining functions is *putting one function inside another.*

Example 11   If $f(x) = \sqrt{x}$ and $g(x) = 1 + x^2$, then

$$f(g(x)) = \sqrt{g(x)} = \sqrt{1 + x^2},$$
$$g(f(x)) = 1 + (f(x))^2 = 1 + (\sqrt{x})^2 = 1 + x.$$

Note that $g(f(x))$ makes sense for a particular number x only if f is defined at x and also g is defined at $f(x)$. Here $f(x) = \sqrt{x}$, so x cannot be negative, and the final equation

$$g(f(x)) = 1 + x$$

is valid only for $x \geq 0$, even though the right side is defined for all x.

$\square$

This method of combining functions is called *composition* of functions. It is very important in calculus, largely because many complicated functions can be *decomposed* into simpler constituents.

PROBLEMS FOR SECTION 8

For each of the following functions, determine the domain and express it as an interval or a union of intervals.

1. $f(x) = \sqrt{x}$ **2.** $f(x) = \sqrt[3]{x}$ **3.** $f(x) = \dfrac{1}{x - 2}$

4. $f(x) = x^2 + 2x + 1$ **5.** $f(x) = \dfrac{x + 2}{x^2 - 4}$ **6.** $f(x) = \sqrt{1 - x^2}$

7. $y = \dfrac{x + 4}{(x - 2)(x + 3)}$ **8.** $y = \sqrt{36 - x^2}$ **9.** $y = \sqrt{x^2 - 16}$

10. $y = \dfrac{1}{9 - x^2}$

For each of the following functions find $f(z)$ when z is the given value.

11. $f(t) = t^2 + 2t - 9$; $z = 2$ **12.** $f(x) = x^3 + 2x^2 - \sqrt{x} + 3$; $z = \pi$

13. $f(u) = u^3 + 3u^2 + 2u$; $z = a + 1$ **14.** $f(x) = \sqrt{2x + 11}$; $z = 2.5$

15. $f(x) = \dfrac{x^2 - 1}{x + 1}$; $z = u - 1$

If $H(t) = (1 - t)/(1 + t)$, find

16. $H(-x)$ **17.** $H(1/t)$ **18.** $H(1/(1 + x))$ **19.** $H(H(t))$

Show that $x - y$ is a factor of $g(x) - g(y)$ for each of the following functions.

20. $g(x) = x^2$ **21.** $g(x) = 1/x$ **22.** $g(x) = x^3$ **23.** $g(x) = 1/(x^2 + 1)$

24. Let f be the reciprocal function, $f(x) = 1/x$. Show that $f(f(x)) = x$ for all x different from 0.

25. Let g be the function $g(x) = \sqrt{1 - x^2}$. Determine the domain D of g and show that $g(g(x)) = x$ for every x in "one-half" of D.

26. If f is the square-root function and g is the squaring function, show that

$$g(f(x)) = x \qquad \text{for every } x \text{ in domain } f;$$

$$f(g(x)) = |x| \qquad \text{for every } x.$$

27. If f is the cube-root function, $f(x) = x^{1/3}$, what is the function g such that $g(f(x)) = x$ for all x?

28. Find a function f such that

$$\frac{f(x) + 1}{f(x) - 1} = x.$$

(Solve the equation for $f(x)$.) What is the domain of f?

In each of the following cases, decide whether or not the equation determines y as a function f of x. If not, show why not. If so, find a formula for f and give its domain.

29. $x^2 + 4y^2 = 1$ **30.** $x^2 + 4y = 1$ **31.** $x + 4y^2 = 1$

32. $xy + x - y = 2$ **33.** $\sqrt{x - y} = 1$ **34.** $x^2 + y^3 = 0$

35. $x^3 + y^2 = 0$ **36.** $x^3 + y^3 = 0$ **37.** $\dfrac{x + y}{x - y} = x$

38. $y^2 + 4x^2 + 4xy = 0$

39. The area A of an equilateral triangle is a function of the side length s, $A = f(s)$. Find this function f.

A function f is said to be *even* if

$$f(-x) = f(x)$$

for every number x in its domain. Note that if x is in the domain of f, then $-x$ must also be there if this condition is to make sense. A function f is *odd* if

$$f(-x) = -f(x)$$

for every x in its domain. Again it is understood that $-x$ must be in the domain for each x in the domain.

40. Suppose that

$$f = g + h$$

where g is even and h is odd. Prove that

$$g(x) = \frac{f(x)+f(-x)}{2},$$

and

$$h(x) = \frac{f(x)-f(-x)}{2}.$$

Now show that an *arbitrary* function f, defined on an interval of the form $[-a, a]$, can be written in a *unique* way as a sum of an even function g and an odd function h. We shall call g the *even component* of f, and h the *odd component* of f.

42. Find the even and odd components of the polynomial $p(x) = a_0 + a_1x + \cdots + a_nx^n$. When is $p(x)$ itself even? odd? Your answer here should suggest where the words *even* and *odd* come from.

41. Find the even and odd components of the polynomial $1 + x + x^2$.

43. Show that the sum of two odd functions is odd, and that the product of two odd functions is even. Write down the rest of the similar statements about even and odd functions.

Each of the following problems gives two of the three functions f, g, and h. The problem is to find the third function so that

$$f(x) = g(h(x)).$$

44. $g(x) = \dfrac{1}{x}$, $h(x) = 1 + x^2$

45. $g(x) = \dfrac{1}{x^2}$, $h(x) = 1 - x$

46. $g(x) = \sqrt{1 + x^2}$, $h(x) = \sqrt{1+x}$

47. $g(x) = \sqrt{1 + x^2}$, $f(x) = \sqrt{1 + x}$

48. $g(x) = \dfrac{1}{1 - x}$, $f(x) = \dfrac{-x}{1 - x}$

49. $f(x) = x + \dfrac{1}{x}$, $h(x) = \dfrac{1}{x}$

50. $f(x) = x^2 + 1$, $h(x) = \dfrac{1}{x + 1}$

In each of the following problems, find functions g and h that are simpler than the given function f and such that

$$f(x) = g(h(x)).$$

51. $f(x) = \sqrt{x^3 - 1}$

52. $f(x) = \dfrac{x^{1/3}}{1 + x^{2/3}}$

53. $f(x) = (x - 1)\sqrt{x^2 - 2x + 2}$

9
THE GRAPH
OF A
FUNCTION

As x runs across the domain of a function f, the point $(x, f(x))$ traces a curve in the xy-coordinate plane called the *graph* of f. The graph is thus made up of the points (x, y) satisfying the equation $y = f(x)$:

The graph of the function f in the xy-plane is the graph of the equation

$$y = f(x).$$

Example 1

If $f(x) = x^3$, then the graph of f is the set of points (x, x^3), i.e., the set of points (x, y) such that $y = x^3$, i.e., the graph of the equation $y = x^3$.

If we wanted to graph f using axes that are labeled differently, say a horizontal u-axis and a vertical v-axis (the uv-plane), then we would graph the equation $v = f(u)$. See Fig. 1.

Figure 1

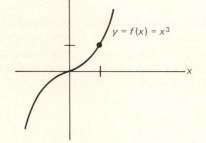
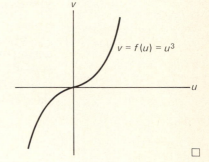

The convention is that

The independent variable is plotted horizontally.

Example 2

The equation $y^2 - 2y = x$ expresses x as a function of y,

$$x = f(y) = y^2 - 2y.$$

The graph of this equation is shown at the left in Fig. 2. It is not the graph of f because the independent variable y is plotted vertically. So we make x the independent variable by reversing the variables in the equation defining f.

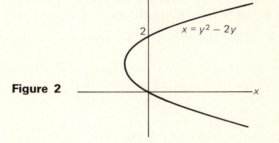

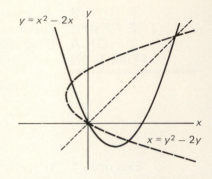

Figure 2

The graph of f, shown at the right, is thus the graph of the equation

$$y = f(x) = x^2 - 2x,$$

which is obtained from the given equation by interchanging the variables. The two graphs are mirror images of each other in the line $y = x$. (See Section 6.)   □

Our definition of a function as an operation leads us to view it as acting on one number at a time. A function graph, on the other hand, spreads the whole input–output relationship out before us as a geometric curve. The two points of view reinforce each other, and are part of our total concept of a function.

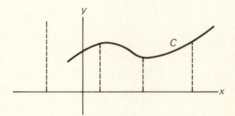

Figure 3

It is possible to take the graphical representation as the basic notion. We note that a curve in the coordinate plane determines y as a function of x if each vertical line cuts it in only one point (or not at all for x not in the domain) as in Fig. 3. For this just says geometrically that only one y value goes with a given x value. We can then define a function by this kind of configuration in the coordinate plane.

DEFINITION   A *function* is given by a configuration C in the coordinate plane (a collection C of ordered pairs of numbers) having the property that each first number x occurring in the points of C is paired with only one second number y.

This definition is more abstract than the definition of a function as a rule. Rules can be different but have the same graph. For example, if

$$f(x) = x^2 - 1 \quad \text{and} \quad g(x) = (x - 1)(x + 1),$$

then $f(x) = g(x)$ for all x, so f and g have the same graph. But they are different rules.

PROBLEMS FOR SECTION 9

1. Graph the squaring function $f(x) = x^2$ by plotting the points over $x = 0, \pm\frac{1}{2}, \pm1, \pm\frac{3}{2}, \pm2$. Name this graph from the list of standard equations in Section 6.

2. The graph of $f(x) = \sqrt{x}$ is half of a parabola. How do we know this? Sketch the graph.

3. Show that the graph of $f(x) = \sqrt{1 - x^2}$, is a semicircle about the origin. Then plot one point on the graph and sketch it.

4. Show that the graph of $f(x) = \frac{1}{2}\sqrt{x^2 + 4}$ is half of a hyperbola (by comparing with a nearly equivalent equation in Section 6). Plot the intercept and asymptotes and quickly sketch the graph.

5. The equation $y = \sqrt{x}$ determines x as a function f of y, $x = f(y)$. What are $f(0)$ and $f(1)$? Using these two values only, sketch the graph of f, being sure that it has the right general shape. (The equation $y = \sqrt{x}$ is equivalent to $x = y^2$ and $y \geq 0$.)

6. The equation

$$2x - y + 1 = 0$$

determines y as a function f of x, and also determines x as a function g of y. Draw the graphs of these two functions f and g.

7. For positive x and y the equation

$$yx^2 = 1$$

determines each of the variables x and y as a function of the other. Draw the graphs of these two functions.

8. Show that f is even if and only if its graph is symmetric about the y-axis. Show that f is odd if and only if its graph is symmetric with respect to the origin.

9. Graph $f(x) = x/|x|$. What symmetry does the graph have?

10. Graph $f(x) = |x| + x$. Find and graph the even and odd components of f.

11. Sketch the graphs of $f(x) = x^2$ and $g(x) = x^4$ on the same axis system, using the function values at $x = 0, 1/2, 1, 3/2$, and symmetry. What configuration in the coordinate plane do the graphs of the even powers x^{2n} approach as n gets very large?

12. Answer the same question for $f(x) = x^3$, $g(x) = x^5$, and the family of all odd powers x^{2n+1}.

13. A polynomial of degree three, $p(x) = x^3 + a_2x^2 + a_1x + a_0$, behaves like x^3 when $|x|$ is very large. Moreover, for *any* polynomial p, $x - a$ is a factor of $p(x)$ if and only if $p(a) = 0$. (This is a theorem of polynomial algebra.) Classify the possible graphs of cubic polynomials p by the ways in which they can cross or touch the x-axis, and draw a sketch of each possibility. (Show that there are four possibilities, as follows:

a) cross 3 times;

b) cross once and touch once from above;

c) cross once and touch once from below;

d) cross once.)

The following polynomials exhibit the above four possibilities. Determine which is which and sketch very roughly. [*Hint*: Factor.]

14. $x^3 + x$ **15.** $x^3 - x$ **16.** $x^3 + 2x^2 + x$ **17.** $x^3 - 2x^2 + x$

18. Make the same analysis of the possible graphs of a fourth-degree polynomial

$$p(x) = x^4 + a_3x^3 + a_2x^2 + a_1x + a_0.$$

Sketch the following polynomials.

19. $y = 3 - x + x^2$ **20.** $y = 3x - x^2 + x^3$ **21.** $y = (x + 1)^3$

22. $y = x^3 + 2x^2 - x - 2$ **23.** $y = x^4 - 1$ **24.** $y = x^4 - 2x^2 + 1$

25. A rubber strand occupying the segment $[a, b]$ is stretched to the new position $[c, d]$. If y is the new position in $[c, d]$ of the point originally at the position x in $[a, b]$, then y is a function of x. Derive a formula for this function. Draw its graph.

26. Suppose a rubber line segment is held under tension between endpoints a and b. We attach clips to the segment at points $x_1, x_2, \ldots, x_n$, and then move these clips to the new positions (along the same line) $y_1, y_2, \ldots, y_n$. If y is the new position of the point originally at x, then y is a function of x. Describe the graph of this function.

27. The graph of g is obtained by translating the graph of f four units to the right. Show that $g(x) = f(x - 4)$.

28. Suppose that the graph of f is translated 3 units to the left. If g is the function having this new graph, find the relationship between f and g.

29. Suppose that f assumes its largest value at $x = 3$. Where does $g(x) = f(x + 7)$ have its largest value?

30. Suppose that $f(x) = 0$ at $x = \pm 2$ and nowhere else. If $g(x) = f(x - a)$, where is $g(x) = 0$?

31. A function f satisfies the equation $f(2 - x) = f(x)$ for all x. Show that its graph is symmetric about the line $x = 1$.

32. Find the equation that says the graph of f is symmetric about the line $x = -2$.

CHaPter 2
LIMITS AND CONTINUITY

Determining a limit is a new kind of number process that underlies calculus and much of advanced mathematics. This chapter contains a brief intuitive introduction to limits and to the closely related notion of continuity. The *theory* of limits (ϵ and δ) is discussed in Appendix 4.

1 LIMITS

As a first example of a limit consider a number and its infinite decimal expansion. What does it mean to say that π is given exactly by its infinite decimal expansion $3.14159\ldots$? What we have in mind is that the finite decimals

$$3, \quad 3.1, \quad 3.14, \quad 3.141, \quad 3.1415, \quad 3.14159,\ldots$$

come closer and closer to π, and in their totality they determine π exactly. The above list continues indefinitely, forming an *infinite sequence*, and π is the *limit* of this infinite sequence.

A limit process is a method of determining a number by coming closer and closer to it. Or, it is a way of calculating a number by finding better and better approximations to it. The number that is specified this way is called the *limit* of the process.

The Slope of a Tangent Line

Example 1 illustrates the type of limit we need for the first part of calculus. In geometric terms, the problem is to find the tangent line to a curve at a given point on the curve (Fig. 1). Geometric constructions can be used for a few types of curves, including circles, but they do not work in general. So we have to look for some new idea. We introduce coordinates and note that all we then need is the slope of the tangent line, since we already have a point on the line. Surprisingly, we can find it for practically any function we can write down.

Figure 1

Example 1

We shall compute the slope of the tangent line to the graph of $f(x) = x^3$ at the point $(1, 1)$. We start by choosing a second nearby point (x, x^3) on the graph, and we write down the slope of the *secant* line through these two points on the graph:

$$m_{\text{sec}} = \frac{x^3 - 1}{x - 1}.$$

This is just the slope formula $(y_1 - y_0)/(x_1 - x_0)$ for the two points $(x_0, y_0) = (1, 1)$ and $(x_1, y_1) = (x, x^3)$.

Now let x approach 1 and imagine the point (x, x^3) sliding along the graph toward the fixed point $(1, 1)$. As this happens the secant line rotates about $(1, 1)$, and we visualize it approaching the tangent line as its limiting position (see Fig. 2). Then the slope m_{sec} of the secant line approaches the slope m of the tangent line as its limiting value.

The arrow $\rightarrow$ is the symbol for the word *approaches*, so in symbols the above statement is

$$m_{\text{sec}} \rightarrow m \qquad \text{as } x \rightarrow 1.$$

Alternatively, we write

$$m = \lim_{x \to 1} m_{\text{sec}},$$

which is read "m is the limit of m_{sec} as x approaches 1." When we put

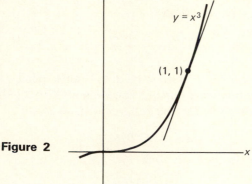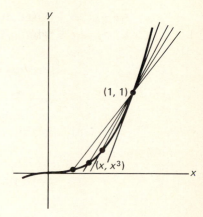

Figure 2

in the formula for m_{sec}, we have

$$m = \lim_{x \to 1} \frac{x^3 - 1}{x - 1}$$

as our conclusion about the slope of the tangent line. The question is, Can we *compute m* from this description of it as a limit?

Note that we can't compute the limit by simply setting $x = 1$ in the formula $m_{\text{sec}} = (x^3 - 1)/(x - 1)$ because this would just give the meaningless expression $0/0$. Nor can we see what happens as $x \to 1$ by just looking at the formula. We have the quotient of two quantities each of which becomes arbitrarily small, and we can't see how the quotient behaves.

Before settling the question, let us try to guess the limit by computing the secant slopes for a few values of x near 1. This experiment is very easy to carry out on a hand calculator, and it can give a concrete feeling for what is happening as x moves closer and closer to 1. Here are the results (to four decimal places).

x	0.9	0.99	0.999	0.9999	1.0001	1.001	1.01	1.1
m_{sec}	2.7100	2.9701	2.9970	2.9997	3.0003	3.0030	3.0301	3.3100

It seems clear from the table that m_{sec} is getting close to 3 as x gets close to 1. We shall now verify this conjecture by an algebraic calculation.

The reason that the numerator in m_{sec} becomes small is that it contains the denominator $x - 1$ as a factor:

$$x^3 - 1 = (x - 1)(x^2 + x + 1).$$

After this common factor is canceled from top and bottom we *can* see what goes on. We have

$$m_{\text{sec}} = \frac{x^3 - 1}{x - 1} = \frac{(x - 1)(x^2 + x + 1)}{(x - 1)} = x^2 + x + 1 \qquad \text{for } x \neq 1,$$

and when x approaches 1, then $x^2 + x + 1$ approaches $1^2 + 1 + 1 = 3$ as its limiting value. The slope of the tangent line at the point $(1, 1)$ is thus 3. The equation for the tangent line is then

$$y - 1 = m(x - 1),$$

with $m = 3$, and we end up with

$$y = 3x - 2.$$

The tangent-line slope appears to be the slope of the graph itself at the point of tangency. So the graph of $y = x^3$ also has the slope 3 at the point $(1, 1)$.

In limit notation, the above computation of m is written

$$m = \lim_{x \to 1} m_{\text{sec}} = \lim_{x \to 1} \frac{x^3 - 1}{x - 1} = \lim_{x \to 1} (x^2 + x + 1) = 1^2 + 1 + 1 = 3. \quad \square$$

Example 2   We calculate the limit of

$$f(x) = x^3 - 3x^2 + 5$$

as $x \to 2$.

Here we can see the answer by inspection, as in the final evaluation in Example 1. As $x \to 2$, we have, simultaneously, that

$$x^3 \to 2^3 = 8$$

and

$$3x^2 \to 3 \cdot 2^2 = 12.$$

So

$$x^3 - 3x^2 + 5 \to 2^3 - 3 \cdot 2^2 + 5 = 8 - 12 + 5 = 1. \qquad \square$$

Two points about the above examples need emphasizing.

(I)   When we say in Example 1 that m_{sec} *approaches* m, we mean that m_{sec} *comes right up to* m. It may never get there, but it does come *arbitrarily close:* however small we would like the difference $m_{\text{sec}} - m$ to be, it eventually becomes and remains at least that small.

"Eventually" means *after x has been restricted to a suitably small interval about* 1.

(II)   However, we cannot allow x to be equal to 1. The equation

$$m_{\text{sec}} = \frac{x^3 - 1}{x - 1} = x^2 + x + 1$$

is not valid when $x = 1$. Geometrically, this is the fact that we need two *distinct* points to determine a secant line. So when we look for the limiting value of m_{sec} as $x \to 1$, we must keep x distinct from 1.

When we go over to the right side, $x^2 + x + 1$, it might seem that now we can let $x = 1$. Nevertheless, it is only when $x \neq 1$ that the right side equals the secant slope, and it is the limiting value of the secant slope that we want to evaluate. Thus the exclusion $x \neq 1$ remains in force.

In Example 2 we are asking a similar question: What is happening to the value of $f(x)$ as x moves *toward* 2? We are not *asking* about the value of $f(x)$ at $x = 2$.

This exclusion is an important part of the notion of a limit. When we are defining what we mean by

$$\lim_{x \to a} f(x),$$

it is immaterial whether or not f is defined at $x = a$. In either case, we consider only the values of $f(x)$ for $x \neq a$.

Informal definition   We say that $f(x)$ *approaches the limit A as x approaches a* if we can cause $f(x)$ to be as close to A as we wish merely by restricting x to a suitably small interval about a, but excluding the value $x = a$.

As already indicated, the standard notations for this are

$$f(x) \to A \qquad \text{as } x \to a;$$
$$\lim_{x \to a} f(x) = A.$$

One final comment about the limit notion: The closeness of x to a is measured by how small (in magnitude) the difference $x - a$ is. And the magnitude of a number is its absolute value. So the following re-phrasal is a somewhat more basic way of stating the definition.

We say that $f(x)$ approaches the limit A as x approaches a if we can cause $|f(x) - A|$ to be as small as we wish merely by taking $|x - a|$ sufficiently small, but always different from zero.

The Limit Laws

The calculations in Examples 1 and 2 depend formally on certain laws of algebra for limits. For example, the final step of Example 2, which added up the limits of the individual terms, was an application of the following law:

$$\left. \begin{array}{lll} \text{If} & f(x) \to A, \\ \text{and} & g(x) \to B, \\ \text{then} & f(x) + g(x) \to A + B \end{array} \right\} \text{ as } x \to a.$$

This should seem plausible. Surely, if we take x so close to a that we have simultaneously $f(x)$ very close to A and $g(x)$ very close to B, then $f(x) + g(x)$ must be close to $A + B$.

In any event, we shall simply assume the limit properties listed below. Their proofs are discussed in Appendix 4. To start things off, we state in L0 a couple of limit facts that may seem pretty trivial.

L0. $\lim\limits_{x \to a} x = a$; $\lim\limits_{x \to a} c = c$, where c is a constant function.

L1. Suppose that

$$\lim_{x \to a} f(x) = A \qquad \text{and} \qquad \lim_{x \to a} g(x) = B.$$

Then

a) $\lim\limits_{x \to a} (f(x) + g(x)) = A + B$,

b) $\lim\limits_{x \to a} f(x)g(x) = AB$,

c) $\lim\limits_{x \to a} \dfrac{g(x)}{f(x)} = \dfrac{B}{A}$, provided $A \neq 0$.

d) $\lim\limits_{x \to a} \sqrt{f(x)} = \sqrt{A}$, provided $f \geq 0$.

L2. Suppose that $f(x) = g(x)$ for all $x \neq a$ (in some interval about a). Then

$$\lim_{x \to a} f(x) = \lim_{x \to a} g(x),$$

in the sense that if either limit exists then so does the other, with the equality holding.

REMARKS   **1.** By adding further terms and repeatedly applying L1a, we get an extended version of L1a involving a finite sum of any length. In the "approaching" notation it is,

If

$$f_1(x) \to A_1, \quad f_2(x) \to A_2, \quad \ldots, \quad f_n(x) \to A_n,$$

in each case as $x \to a$, then

$$f_1(x) + f_2(x) + \cdots + f_n(x) \to A_1 + A_2 + \cdots + A_n$$

as $x \to a$.

The product law L1b has a similar extended version. We regard these more general forms as parts of L1.

2. The laws of algebra in L1 are often written somewhat more loosely as follows:

$$\text{a) } \lim_{x \to a} (f(x) + g(x)) = \lim_{x \to a} f(x) + \lim_{x \to a} g(x)$$

and so on. This must be understood as saying, *If* both limits on the right exist, *then* the limit on the left exists and has the indicated value.

3. L1 can be paraphrased in words, as follows:

The limit of a sum is the sum of the limits;
The limit of a product is the product of the limits;
The limit of a quotient is the quotient of the limits, provided that the denominator limit is not zero.
The limit of a square root is the square root of the limit.

4. In these laws there are implicit assumptions about domains. Generally it is understood that all the functions in question are defined throughout some interval about the limit point a except possibly at a itself. Occasionally we want to consider a function defined only on *one side* of a. For example, the domain of f might be an interval $[a, b]$ or (a, b). Then the other functions in question will be defined on the *same* side of a, so that the combinations $f(x) + g(x)$, $f(x)g(x)$, etc., can still be formed.

Limit computations such as in Examples 1 and 2 become completely formal if we justify each step by one of the limit laws. We do this for Example 2 in the next example.

Example 3   Rework Example 2, using the limit laws explicitly.

Solution

$$\lim_{x \to 2} (x^3 - 3x^2 + 5) = \lim_{x \to 2} x^3 + \lim_{x \to 2} (-3x^2) + \lim_{x \to 2} 5 \qquad \text{by L1a}$$

$$= \left(\lim_{x \to 2} x\right)^3 + \lim_{x \to 2} (-3)\left(\lim_{x \to 2} x\right)^2$$

$$+ \lim_{x \to 2} 5 \qquad \text{by L1b}$$

$$= 2^3 - 3 \cdot 2^2 + 5 = 1 \qquad \text{by L0.} \quad \square$$

We would normally omit the references at the right in Example 3, since it is clear from the context which laws are being applied.

Limits of quotients require particular care. If the denominator limit is nonzero, then we apply L1c.

Example 4   Compute the limit of $x/(1 + \sqrt{x})$, as $x \to 2$.

Solution

$$\lim_{x \to 2} \frac{x}{1 + \sqrt{x}} = \frac{\lim_{x \to 2} x}{\lim_{x \to 2} (1 + \sqrt{x})} \qquad \text{by L1c}$$

$$= \frac{2}{1 + \sqrt{2}} \qquad \text{by L0, L1a, and L1d.} \quad \square$$

However, the quotient limit law L1c cannot be applied when the denominator limit turns out to be zero. In this case the quotient may not *have* a limit. That was our initial uncertainty in Example 1. Fortunately, an algebraic simplification allowed us to proceed in a different way.

Here is a similar example.

Example 5   Find the limit

$$\lim_{x \to 3} \frac{3 - x}{9 - x^2}.$$

Solution   We cannot compute this limit by the quotient limit law because the denominator limit is zero. However,

$$9 - x^2 = (3 - x)(3 + x),$$

so

$$\frac{3-x}{9-x^2} = \frac{(3-x)}{(3-x)(3+x)} = \frac{1}{3+x}$$

for all $x \neq 3$. Therefore,

$$\lim_{x \to 3}\frac{3-x}{9-x^2} = \lim_{x \to 3}\frac{1}{x+3} \qquad \text{by L2.}$$

Now we can apply L1c (and L0) to get

$$\frac{1}{\lim_{x \to 3}(x+3)} \qquad \text{by L1c}$$

$$= \frac{1}{3+3} = \frac{1}{6} \qquad \text{by L1a and L0.} \qquad \square$$

PROBLEMS FOR SECTION 1

Evaluate each of the following limits, citing the relevant reasons in each case.

1. $\lim_{x \to 2} \sqrt{x}$

2. $\lim_{x \to 2} \dfrac{x+1}{x^2-1}$

3. $\lim_{x \to 8} x^{1/3}$

4. $\lim_{x \to -1} 5x^3 + 4x^2 + 1$

5. $\lim_{x \to 0} \dfrac{1-\sqrt{x}}{1+\sqrt{x}}$

In each of the following problems use a calculator to make a table of approximations to the limit, as was done in Example 1 in the text. Then guess the limit.

▦ 6. $\lim_{x \to 1} \dfrac{x^2-1}{x^3-1}$

▦ 7. $\lim_{x \to 1} \dfrac{x^4-1}{x-1}$

▦ 8. $\lim_{x \to 4} \dfrac{x\sqrt{x}-8}{x-4}$

▦ 9. $\lim_{x \to 1} \dfrac{\sqrt{x}-1}{x^2-1}$

▦ 10. $\lim_{x \to 8} \dfrac{x-8}{x^{1/3}-2}$

▦ 11. $\lim_{x \to 3} \dfrac{\sqrt{x}-\sqrt{3}}{x^2-9}$

Use the trick of Example 5 to evaluate each of the following limits.

12. $\lim_{x \to 1} \dfrac{x-1}{x^2-1}$

13. $\lim_{x \to 2} \dfrac{x^3-8}{x-2}$

14. $\lim_{x \to 4} \dfrac{\sqrt{x}-2}{x-4}$

15. $\lim_{x \to 1} \dfrac{x^2-1}{x^3-1}$

16. $\lim_{x \to 1} \dfrac{x^4-1}{x-1}$

17. $\lim_{x \to 4} \dfrac{x-\sqrt{x}-2}{x-4}$

18. $\lim_{x \to 1} \dfrac{x^3-1}{x+\sqrt{x}-2}$

19. $\lim_{x \to 0} \dfrac{\sqrt{1+x}-1}{x}$

20. $\lim_{x \to 4} \dfrac{x\sqrt{x}-8}{x-4}$

Evaluate the following limits (if possible).

21. $\lim\limits_{x \to 1} \dfrac{\sqrt{x} - 1}{x^2 - 1}$

22. $\lim\limits_{x \to -1} \dfrac{2x^2 - x - 3}{x + 1}$

23. $\lim\limits_{x \to 0} \dfrac{2x^2 - x - 3}{x + 1}$

24. $\lim\limits_{x \to 8} \dfrac{x - 8}{x^{1/3} - 2}$

25. $\lim\limits_{x \to 3} \dfrac{\sqrt{x} - \sqrt{3}}{x^2 - 9}$

26. $\lim\limits_{x \to 4} \dfrac{x - 4}{3(\sqrt{x} - 2)}$

27. $\lim\limits_{x \to 4} \dfrac{x - 4}{x^2 - x - 12}$

28. $\lim\limits_{x \to -3} \dfrac{x - 4}{x^2 - x - 12}$

29. $\lim\limits_{x \to 4} \dfrac{\sqrt{x^2 + 9} - 5}{x^2 - 4x}$

30. $\lim\limits_{x \to 2} \dfrac{\sqrt{x^2 + 9} - 5}{x^2 - 4x}$

31. $\lim\limits_{x \to 0} \dfrac{\sqrt{x^2 + 9} - 5}{x^2 - 4x}$

32. $\lim\limits_{x \to 2} \left(\dfrac{x^4 - 16}{x^3 - 8} \right)^{1/2}$

33. $\lim\limits_{x \to -4} \dfrac{x^2 + x - 6}{x^2 + 7x + 12}$

34. $\lim\limits_{x \to -3} \dfrac{x^2 + x - 6}{x^2 + 7x + 12}$

35. $\lim\limits_{x \to -2} \dfrac{x^2 + x - 6}{x^2 + 7x + 12}$

36. $\lim\limits_{x \to 1} \dfrac{x^2 - \sqrt{x}}{x - 1}$

37. Compute the slope of the tangent line to the graph of $f(x) = x^2$ at the point $(1, 1)$. (Imitate the text discussion of $f(x) = x^3$.)

38. Compute the slope of the tangent line to the graph of $y = x^2$ at the point over $x_0 = 2$. Write down the equation of the tangent line.

In each of Problems 39 through 45, compute the slope of the tangent line to the given graph at the point having the given x-coordinate. Then write down the equation of the tangent line.

39. $y = x^2$, $x_0 = -1$

40. $y = x^2$, $x_0 = a$ (For what value of a will the tangent line have slope 1?)

41. $y = 1/x$, $x_0 = 1$ **42.** $y = 1/x$, $x_0 = 2$

43. $y = x - x^2$, $x_0 = 0$ **44.** $y = x - x^2$, $x_0 = 2$

45. $y = x - x^2$, $x_0 = a$ (For what value of a will the tangent be horizontal?

46. Presumably a straight line $y = mx + b$ is its own tangent line at any point. Verify this by the slope-determining procedure, applied to the function $f(x) = mx + b$ at $x_0 = a$.

**2
CONTINUITY**

Something else was going on in the last section. We were often computing a limit $\lim_{x \to a} f(x)$ in a situation where $f(a)$ was defined (unlike our beginning limit problem), and, checking back, we see that

$$\lim_{x \to a} f(x) = f(a).$$

A function having this property is said to be continuous at the point a.

DEFINITION   A function f is *continuous at* $x = a$ if f is defined on some interval containing a and

$$\boxed{f(x) \to f(a) \qquad \text{as } x \to a.}$$

A function f is *continuous* if f is continuous at every point of its domain.

Properties of Continuous Functions

Properties of continuous functions are of great importance in the development of calculus, as will be seen. For the moment, continuity provides the following shortcut: If we had known that we were dealing with a continuous function in Examples 3 and 4 in the last section, then we could have omitted all the details of the limit calculation, simply calculating $f(a)$ instead. The continuity of f then guarantees that this is the limit that we wanted. Limits that can be evaluated this way are often called *trivial* limits.

For the above and other reasons we want to be able to recognize on sight as many continuous functions as possible. Fortunately, most functions that come up are continuous (at all points of their domains). In order to see why, we look back at the limit laws and note that L0 and L1 can be restated for continuous functions as follows:

L0. The functions $f(x) = x$ and $g(x) = c$ are everywhere continuous.

L1′. If f and g are continuous at a, then so are $f + g$, fg, g/f (provided $f(a) \neq 0$), and $\sqrt{f}$ (provided $f \geq 0$).

By repeatedly applying these principles, we establish the continuity of any functions that can be built up from x and constants by using the two operations of addition and multiplication, i.e., any *polynomial*. Examples are

$$p(x) = x^2 + 2x + 5,$$
$$q(x) = 4x^{13} - x^9 + 2x^5 + 3x.$$

Knowing that every polynomial is continuous, we can then conclude, from L1′ again, that the quotient

$$r(x) = \frac{p(x)}{q(x)}$$

of any two polynomials is continuous, except where it is undefined (i.e., where $q(x) = 0$). A quotient of two polynomials is called a *rational function*.

We thus have the theorem,

THEOREM 1   *A polynomial function p(x) is everywhere continuous. A rational function*

$$r(x) = \frac{p(x)}{q(x)},$$

where p and q are polynomials, is continuous (where defined).

We can also mix in the square-root operation with the other operations to get more complicated continuous functions.

Many limit evaluations now become trivial.

Example 1   Any polynomial $p(x)$ is everywhere continuous, so $\lim_{x \to a} p(x) = p(a)$ for any a. For example,

$$\lim_{x \to 4} (x^3 - 8x) = 4^3 - 8 \cdot 4 = 32. \qquad \square$$

Example 2   A rational function

$$r(x) = \frac{p(x)}{q(x)}$$

is continuous where defined. That is, $\lim_{x \to a} r(x) = r(a)$, provided $q(a) \neq 0$. For example, if

$$r(x) = \frac{x - 1}{x + 1},$$

then

$$\lim_{x \to 3} r(x) = r(3) = \frac{3 - 1}{3 + 1} = \frac{1}{2}. \qquad \square$$

Example 3
$$\lim_{x \to 3} \sqrt{\frac{x + 1}{x - 1}} = \sqrt{2}. \qquad \square$$

Returning now to Example 2 in the last section, we recognize that we have a continuous function (a polynomial), so

$$\lim_{x \to 2} f(x) = f(2) = 8 - 12 + 5 = 1.$$

However, the starting limits of Examples 1 and 5 in the last section were not trivial in this sense. Preliminary manipulation was needed to reduce them to trivial limits.

Example 4   Here is another nontrivial limit of this sort. We wish to show that if

$$f(x) = \frac{1 - \sqrt{x}}{1 - x},$$

then

$$\lim_{x \to 1} f(x) = \frac{1}{2}.$$

Solution   If we try substituting $x = 1$, we get 0/0, which is meaningless; so f is not defined at $x = 1$ and we do not have a trivial limit. We can nevertheless calculate the limit as follows. Factoring $1 - x$ as a difference of squares, $1 - x = (1 - \sqrt{x})(1 + \sqrt{x})$, we see that

$$\frac{1 - \sqrt{x}}{1 - x} = \frac{1 - \sqrt{x}}{(1 - \sqrt{x})(1 + \sqrt{x})} = \frac{1}{1 + \sqrt{x}}$$

when $x \neq 1$. Now

$$g(x) = \frac{1}{1 + \sqrt{x}}$$

is continuous on $[0, \infty)$ by L1′, so its limit at 1 is its value there:

$$\lim_{x \to 1} g(x) = g(1) = \frac{1}{1 + \sqrt{1}} = \frac{1}{2}.$$

Since $f(x) = g(x)$ for $x \neq 1$, we have

$$\lim_{x \to 1} f(x) = \lim_{x \to 1} g(x) = g(1) = \frac{1}{2},$$

by L2. In brief,

$$\lim_{x \to 1} \frac{1 - \sqrt{x}}{1 - x} = \lim_{x \to 1} \frac{1}{1 + \sqrt{x}} = \frac{1}{1 + \sqrt{1}} = \frac{1}{2}.$$

Thus, although f is not defined at $x = 1$, it is equal *everywhere else* to a function g that is continuous at 1, and we use this fact to show that f has a limit at $x = 1$.   □

REMARK   The function $f(x)$ above is the secant slope for the graph of $y = \sqrt{x}$ at the point $(1, 1)$. The limit calculation shows that the tangent line to the graph at that point has slope $1/2$.

The continuity of simple functions such as polynomials seems firmly imbedded in our intuitions. When we graph such a function we draw it as smoothly as we can, and by just looking at the graph we can see that

$$f(x) \to f(a) \qquad \text{as } x \to a.$$

A calculator provides another way to watch $f(x)$ approach $f(a)$.

▦ Example 5   Use a hand calculator—the simplest four-function calculator will do—to construct a small table illustrating the limit

$$x^4 \to 16 \qquad \text{as } x \to 2.$$

Solution   Values of x^4 can be calculated in various ways. On some calculators we just enter a number x and then press the keys $\times$, $=$, $\times$, $=$. Rounding the calculated value to three decimal places gives the following values.

x	x^4
1.9	13.032
1.99	15.682
1.999	15.968
1.9999	15.997
2.0001	16.003
2.001	16.032
2.01	16.322
2.1	19.448

At one or two points we shall need the following simple fact. If f has the limit l at a without being continuous there, and if we define (or redefine) $f(a)$ to be l, then f becomes continuous at a. Strictly speaking, the modified function is a new function, so a more precise statement is,

If f has the limit l at a, and if we define a new function g by

$$g(x) = f(x), \qquad \text{for } x \neq a;$$
$$g(a) = l,$$

then g is continuous at a and equal to f elsewhere.

We call g the *continuous modification* of f at a. Sometimes we can find a formula for g. For example,

$$g(x) = \frac{1}{1 + \sqrt{x}}$$

is the continuous modification of

$$f(x) = \frac{1 - \sqrt{x}}{1 - x}$$

at $x = 1$. The general idea is that if $f(x)$ has a limit as $x \to a$, then the limit l is the number that f *ought* to have as its value at a, whether or not it does.

We now complete the list of limit laws.

L2′. (Squeeze limit law). If $0 \le f(x) \le g(x)$ on some interval about a (except, possibly, at $x = a$) and if $g(x) \to 0$ as $x \to a$, then $f(x) \to 0$ as $x \to a$.

L3. (Composition of continuous functions). If $f(x)$ is continuous at $x = a$ and if $g(y)$ is continuous at $y = f(a)$, then $g(f(x))$ is continuous at $x = a$.

L4. (Contagiousness of positivity). If $f(x) \to l$ as $x \to a$ and if $l > 0$, then $f(x) > 0$ on some interval about a (possibly excluding $x = a$).

Like the others, these laws require eventual proof. However, we assume them for now.

Example 6 We sketch a proof that $x^{1/n}$ is continuous, using the factorization

$$y^n - z^n = (y - z)(y^{n-1} + y^{n-2}z + \cdots + z^{n-1})$$

together with the squeeze limit law. (For a proof that is less complicated, but also less elementary, see Section 5.)

Proof Suppose, to begin with, that x and a are both larger than 1. If we write

$$x - a = (x^{1/n})^n - (a^{1/n})^n$$

and apply the above factorization with $y = x^{1/n}$ and $z = a^{1/n}$, we get

$$x - a = (x^{1/n} - a^{1/n})S,$$

where S is a sum of n terms, each larger than 1. Therefore, $S > n$, so

$$|x - a| = |x^{1/n} - a^{1/n}| \cdot |S| \geq n \, |x^{1/n} - a^{1/n}|,$$

or

$$|x^{1/n} - a^{1/n}| \leq \frac{1}{n} |x - a|.$$

The right side approaches 0 as $x \to a$, and the squeeze limit law then implies that the left side does too. That is,

$$x^{1/n} \to a^{1/n} \qquad \text{as } x \to a,$$

so $x^{1/n}$ is continuous at $x = a$. This method of proof extends to any positive a, but not to $a = 0$. The special case

$$x^{1/n} \to 0 \qquad \text{as } x \to 0$$

can be seen directly. If we want to make $x^{1/n}$ less than a small positive number d, we only have to take x less than d^n. Because, for all positive x,

$$\text{if } x < d^n \qquad \text{then } x^{1/n} < d. \qquad \blacksquare$$

Finally, here is a simple example in which L3 is needed.

Example 7   Justify the limit

$$\lim_{x \to 3} (x^2 - 1)^{1/3} = 2.$$

Solution   The function $x^{1/3}$ is continuous by Example 6, and $x^2 - 1$ is continuous by Theorem 1. Therefore, $f(x) = (x^2 - 1)^{1/3}$ is continuous, by L3, and

$$\lim_{x \to 3} f(x) = f(3) = (9 - 1)^{1/3} = 2. \qquad \square$$

PROBLEMS FOR SECTION 2

Verify the continuity of each of the following functions by applying the continuity principles stated in this section.

1. $3x^2 + 5x$

2. $x/(1 + x^2)$

3. $1/\sqrt{x}$

4. $\sqrt{x^2 + 4}$

5. $\sqrt{x} + \dfrac{1}{\sqrt{x}}$

6. $\dfrac{1}{x^{1/3} + 1}$

7. $\dfrac{\sqrt{x}}{1 + x^2}$

8. $x^{1/2} + x^{1/3} + x^{1/4}$

9. $\sqrt{x^2}$

10. $|x|$

11. $|x^{1/3}|$ (sketch this graph)

12. $x\sqrt{1 - \sqrt{x}}$

13. In the manner of Example 5, use a calculator to investigate the continuity of $f(x) = x^3$, at $x = 1.5$

14. The same for $f(x) = 3x^2 - x^3 = x^2(3 - x)$, at $x = 2$

15. The same for $f(x) = 1/x^2$, at $x = 3$

16. The same for $f(x) = x/(1 + x^2)$, at $x = 1$

17. a) Compute how close $f(x) = 2x - x^2$ is to $f(3) = -3$ when x is near 3. Do this first algebraically by writing $x - 3 = h$ and factoring h from $f(x) - f(3) = f(3 + h) - f(3)$. Then give $x - 3 = h$ the values .1, .01, .001, .0001, and compute the difference $f(x) - f(3)$.

b) Now compute how close $f(x)$ is to $f(2) = 0$, following the same program as above.

18. Show that when x is within a distance d of 2, then x^2 is within a distance $d(d + 4)$ of 4. [*Hint*: Write $x = 2 + u$, where $|u| < d$, and then expand x^2.]

19. Suppose that x is within a distance d of the point 2, where $d < 1$. Show that then x^2 is within a distance $5d$ of 4.

20. Suppose that x is within a distance d of 2, where $d < 1$. Show that then x^3 is within a distance $19d$ of 8.

21. Strictly speaking, the function $f(x) = (x^2 - x)/(x - 1)$ is not defined at $x = 1$. Draw its graph, indicating the missing domain point. Show that f becomes continuous if $f(1)$ is suitably defined.

22. Suppose that $f(x) = (x^2 + x - 6)/(x - 2)$ for $x \neq 2$. Define $f(2)$ so that f will be everywhere continuous.

23. The same problem for

$$f(x) = \frac{x^2 - 1}{x^3 - 1}$$

and the point $x = 1$

24. The same problem for

$$f(x) = \frac{\sqrt{x} - 2}{x - 4}$$

at the point $x = 4$

25. Use L2′ and earlier limit laws to prove

If

$$f(x) \leq g(x) \leq h(x)$$

for all x in some interval about a, except possibly at $x = a$, and if $f(x)$ and $h(x)$ have equal limits as $x \to a$, then the limit

$$\lim_{x \to a} g(x)$$

exists and has the same value.

27. Evaluate $\lim_{x \to 0} f(x)$ if $\sqrt{1 + x^2} \leq f(x) \leq 1 + |x|$.

29. It might seem that L3 ought to generalize as follows:

If $g(y) \to B$ as $y \to A$, and if $f(x) \to A$ as $x \to a$, then $g(f(x)) \to B$ as $x \to a$.

Show that a counterexample can be constructed from the function f defined as follows:

$$f(x) = 0, \quad \text{if } x \neq 0;$$
$$f(0) = 1.$$

26. Show that it is possible for a function f to satisfy the inequality $2x \leq f(x) \leq x^2 + 1$ on an interval containing $x = 1$. Supposing that this inequality does hold near $x = 1$, prove that $f(x) \to 2$ as $x \to 1$.

28. Use L4 to prove

If a nonnegative function has a limit at $x = a$, then the limit is nonnegative.

30. The following generalization of L3 is true:

If g is continuous at A and if $f(x) \to A$ as $x \to a$, then $g(f(x)) \to g(A)$ as $x \to a$.

Prove this from L3 by making use of the continuous modification of f.

---

3 DISCONTINUITY AND ONE-SIDED LIMITS

Some functions that we meet in everyday life and in mathematics are not continuous everywhere.

DEFINITION

We say that f is *discontinuous* at a if a is a number in the domain of f at which f is not continuous.

Example 1

When we mail a letter we have to use one stamp of a certain denomination for anything up to and including one ounce in weight, then two stamps up to and including two ounces, etc. Thus, if $p(x)$ is the number of stamps required to mail a letter weighing x ounces, then the graph of p looks like Fig. 1.

The empty dot ∘ shows a point *not* on the graph, whereas solid dots show points on the graph. The function $p(x)$ is called the postage-stamp function. It is defined for $x > 0$ and has a *jump*

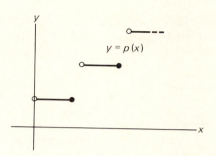

Figure 1

discontinuity at each of the points $x = 1, 2, 3, \ldots$. For example, the change in $p(x)$ *cannot* be made as small as we wish when x changes from slightly below 1 to slightly above 1, so $p(x)$ is *not* continuous as x moves through the value 1. □

Example 2　The greatest integer function $[x]$ is an analogue of the postage-stamp function in pure mathematics: $[x]$ is defined as the largest integer that is less than or equal to x. Thus $[2.5] = 2$, $[2] = 2$, and $[1.9] = 1$. We define $[x]$ for negative x in the same way. The integers less than -2.5 are all negative but among them there is a greatest one, namely -3. Thus $[-2.5] = -3$, $[-3] = -3$, etc. The graph of $[x]$ looks like Fig. 2.

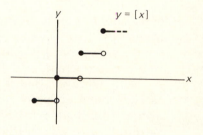

Figure 2

□

Example 3　The graph of $x - [x]$ looks like Fig. 3.

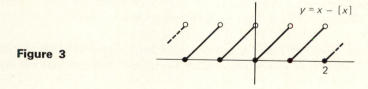

Figure 3

□

One-sided Limits   Although the function f in Example 3 does not have a limit at the origin (or at any other integer n), it does have *one-sided* limits there. Looking at the graph, we see that

$$f(x) \to 1 \qquad \text{as } x \to 0 \text{ from the } \textit{left} \text{ (or from } \textit{below}),$$

and

$$f(x) \to 0 \qquad \text{as } x \to 0 \text{ from the } \textit{right} \text{ (or from } \textit{above}).$$

We indicate that x is to approach c from the left by writing $x \to c^-$ or $x \uparrow c$. Thus for the left-limit statement above we can write

$$\lim_{x \to 0^-} f(x) = 1 \qquad \text{or} \qquad f(x) \to 1 \qquad \text{as } x \uparrow 0.$$

Right limits are indicated in a similar way, by writing $x \to c^+$ or $x \downarrow c$.

We now have *three* numbers that a function may associate with a point c:

$$\boxed{\lim_{x \to c^-} f(x), \qquad f(c), \qquad \lim_{x \to c^+} f(x).}$$

Using them we can dissect some earlier notions a little more finely. Thus:

The ordinary (two-sided) limit $\lim_{x \to c} f(x)$ *exists if and only if both one-sided limits exist and are equal.*

This should be part of our notion of what a limit is and we simply assume it.

It follows that *f is continuous at c if and only if all three numbers exist and are equal.*

Note, however, that if the domain of f is an interval having c as its right-hand endpoint, then there can be no right limit at c. The ordinary limit at c in this case *is* the left limit (if it exists), and continuity at c means that c is in the domain and

$$f(x) \to f(c) \qquad \text{as } x \uparrow c.$$

Jump Discontinuities   If f is discontinuous at c, then either the one-sided limits do not both exist or both exist but the three numbers are not all the same. The second case is the more important.

DEFINITION   We say that f has a *jump discontinuity* at c if all three of the above numbers exist (two if c is an endpoint) but are *not* all equal.

The functions in Examples 1 through 3 have jump discontinuities at every integer $x = n$.

This is the only type of discontinuity that comes up in practice. To see why, consider first any function f that is *increasing*, in the sense that the graph of f *rises* from left to right. We imagine x moving to the right toward a point c $(x{\uparrow}c)$. Then the values $f(x)$ steadily increase, but are blocked from above by $f(c)$. So it would seem that they have to "pile up" at some point $\leq f(c)$. But this just means that $\lim_{x\uparrow c} f(x)$ exists and is $\leq f(c)$.

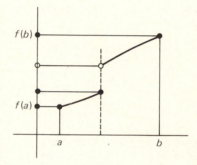

Figure 4

The same type of reasoning show that f has a right limit at c. So f either is continuous at c or has a jump discontinuity there (Fig. 4). Although this is not a real proof, it should seem persuasive enough to give us confidence in the following theorem:

THEOREM 2   *If f is increasing on an interval I and discontinuous at a point c in I, then the discontinuity at c is necessarily a jump discontinuity.*

Decreasing functions are defined similarly, and for them Theorem 2 remains true. Normally a function will be neither increasing nor decreasing over its whole domain but will alternate between increasing for a while and decreasing for a while. That is, normally a function will *oscillate*. But if it has only a *finite number* of oscillations on a closed interval $[a, b]$, then Theorem 2 and its decreasing counterpart imply that any discontinuity of f in $[a, b]$ must be a jump discontinuity. This is the usual situation.

As a corollary of Theorem 2 we have the following test for continuity. It will be useful when we consider inverse functions.

THEOREM 3   *If f is increasing on an interval I, and if the range of f is an interval J, then f is continuous on I.*

     Proof   The range of f has no gaps, so f makes no jumps, and hence f is everywhere continuous, by Theorem 2.   ■

PROBLEMS FOR SECTION 3

1. Graph the function $f(x) = |x|/x$. What are its one-sided limits at $x = 0$?

Determine each of the one-sided limits below.

2. $\lim\limits_{x \to 1^+} \dfrac{|x - 1|}{x - 1}$

3. $\lim\limits_{x \to 1^-} \dfrac{|x - 1|}{x - 1}$

4. $\lim\limits_{x \to 2^-} \dfrac{|x - 2|}{x - 2}$

5. $\lim\limits_{x \to 3^+} \dfrac{x - 3}{|x - 3|}$

6. $\lim\limits_{x \to 4^-} \dfrac{x^2 - 16}{|x - 4|}$

7. $\lim\limits_{x \to 1^+} \dfrac{|1 - x^2|}{1 - x}$

8. What would it mean to say that f is continuous from the left at $x = a$? Write down the definition for continuity at the right at $x = a$.

9. Examine the greatest integer function for one-sided continuity and state your conclusions. The same for the postage-stamp function $p(x)$.

10. What are the values of the following one-sided limits?

     a) $\lim\limits_{x \to 4^-} [x]$

     b) $\lim\limits_{x \to 4^+} [x]$

     c) $\lim\limits_{x \to 4^-} p(x)$

     d) $\lim\limits_{x \to 4^+} p(x)$

     e) $\lim\limits_{x \to 4^-} [x] + p(x)$

     f) $\lim\limits_{x \to 4^+} [x] + p(x)$

11. Show that $[x] + p(x)$ is not one-sidedly continuous at $x = n$, for every positive integer n.

12. Suppose that

$$\lim_{x \to a^-} f(x), \qquad f(a), \qquad \lim_{x \to a^+} f(x)$$

all exist. List the possibilities for equality and/or inequality among these three values, giving a name to each case.

13. Show that the following function is continuous everywhere. The crucial point is $t = -1$, where you have to show that the left- and right-hand limits are both equal to $h(-1)$. Graph the function h.

$$h(t) = \begin{cases} 3t + 5 & \text{if } t < -1, \\ t^2 + 1 & \text{if } t \geq -1. \end{cases}$$

14. Determine the constant c so as to make the function

$$f(x) = \begin{cases} x^2 & \text{if } x < 1/2, \\ c - x^2 & \text{if } x \geq 1/2 \end{cases}$$

continuous at $1/2$. Sketch its graph.

15. Define a function (by drawing its graph) that is constant over each of the intervals $(-\infty, -1)$, $(-1, 1)$, $(1, \infty)$, and has discontinuities at $x = -1$ and $x = 1$.

16. Draw the graph of $f(x) = x - [x/2]$. What are its points of discontinuity?

17. Let f be defined for $x \geq 0$ and set $g(x) = f(|x|)$ for all x. Show that f has a limit (is continuous) at 0 from the right if and only if g has a limit (is continuous) at 0.

18. Show that $y^2 - 4y$ is continuous at $x = 2$ if $y = f(x)$ and

$$\lim_{x \to 2} f(x) = 1, \qquad f(2) = 3.$$

19. Show that $y^3 - 4y + 1$ is continuous at $x = a$ if $y = f(x)$ and

$$\lim_{x \to a^-} f(x) = -2, \qquad f(a) = 0, \qquad \lim_{x \to a^+} f(x) = +2.$$

20. Prove that

$$\lim_{x \to 0} [1 - x^2] = 0$$

from the limit laws and the known behavior of $[x]$. Graph the function $f(x) = [1 - x^2]$.

4

INFINITE LIMITS AND LIMITS AT INFINITY

Although the symbol $+\infty$ does not represent a number, it is nevertheless used in limit statements with a special convention about what is being said.

Example 1   The equation

$$\lim_{x \to 0} \frac{1}{x^2} = +\infty,$$

read literally, says that $1/x^2$ approaches and comes arbitrarily close to $+\infty$ as x approaches 0. We interpret this as meaning that $1/x^2$ *grows without bound* as x approaches 0, in the sense that it becomes and remains larger than any given number when x is taken suitably small. This is true. For example, if we want

$$\frac{1}{x^2} > 10,000,$$

then we only have to take $|x| < 1/100$. Figure 1 shows the graph.
Similarly,

$$\lim_{x \to 0} -\frac{1}{x^2} = -\infty,$$

meaning that $-1/x^2$ becomes and remains less than any given negative number when $|x|$ is taken suitably small. For example,

$$-\frac{1}{x^2} < -10{,}000$$

when $|x| < 1/100$.

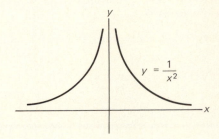

Figure 1

The "approaching" symbolism seems especially appropriate for infinite limits. Thus,

$$\frac{1}{x^2} \to +\infty \qquad \text{as } x \to 0.$$

Example 2   The function $f(x) = 1/x$ is more complicated because its behavior as x approaches 0 depends on whether x is positive or negative:

$$\lim_{x \to 0^-} \frac{1}{x} = -\infty, \qquad \lim_{x \to 0^+} \frac{1}{x} = +\infty.$$

See Fig. 2.

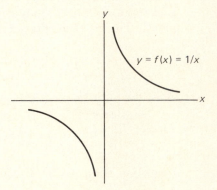

Figure 2

Example 3   Consider next the statement

$$\lim_{x \to +\infty} \frac{1}{x} = 0.$$

Here the limit is a number but the independent variable x is tending to ∞, i.e., growing without bound. The statement means that $1/x$ can be made as close to zero as desired by taking x sufficiently *large*. For example, to get

$$\frac{1}{x} < \frac{1}{1000},$$

we take $x > 1000$.   □

Example 4   Similarly

$$\lim_{x \to +\infty} \frac{1}{\sqrt{x}} = 0.$$

But $1/\sqrt{x}$ gets small much more slowly than $1/x$. In order to ensure that

$$\frac{1}{\sqrt{x}} < \frac{1}{1000},$$

we have to take $x > (1000)^2 = 1{,}000{,}000$.

It is customary to combine the two statements

$$\lim_{x \to +\infty} \frac{1}{x} = 0 \quad \text{and} \quad \lim_{x \to -\infty} \frac{1}{x} = 0$$

into the single statement

$$\lim_{x \to \infty} \frac{1}{x} = 0 \quad \text{or} \quad \frac{1}{x} \to 0 \quad \text{as } x \to \infty.$$ □

Limits at infinity obey the same limit laws as limits at a, but sometimes a preliminary revision is needed before a computation can be made.

Example 5   Verify that

$$\lim_{x \to \infty} \frac{3 + x^2}{3x^2 + 1} = \frac{1}{3}.$$

Solution First rewrite the expression by dividing top and bottom by x^2, getting

$$\frac{\dfrac{3}{x^2} + 1}{3 + \dfrac{1}{x^2}}$$

(for $x \neq 0$). Now $3/x^2$ and $1/x^2$ both approach 0 as x tends to ∞. Thus

$$\lim_{x \to \infty} \frac{3 + x^2}{3x^2 + 1} = \lim_{x \to \infty} \frac{\dfrac{3}{x^2} + 1}{3 + \dfrac{1}{x^2}}$$

$$= \frac{0 + 1}{3 + 0} = \frac{1}{3}. \qquad \square$$

For another way of handling problems like this see the discussion following Example 6.

Example 6 Show that the hyperbola

$$y = \frac{b}{a}\sqrt{x^2 - a^2}$$

is asymptotic to the line

$$y' = \frac{b}{a}x$$

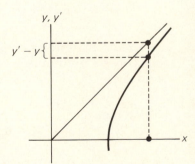

Figure 3

in the strong sense that

$$y' - y \to 0 \qquad \text{as } x \to +\infty.$$

Solution We rationalize a numerator:

$$y' - y = \frac{b}{a}[x - \sqrt{x^2 - a^2}]$$

$$= \frac{b}{a}[x - \sqrt{x^2 - a^2}]\frac{(x + \sqrt{x^2 - a^2})}{(x + \sqrt{x^2 - a^2})}$$

$$= \frac{ab}{x + \sqrt{x^2 - a^2}}.$$

In particular,

$$0 < y' - y < \frac{ab}{x},$$

and since $1/x \to 0$ as $x \to +\infty$, it follows that

$$y' - y \to 0$$

by the squeeze limit law. $\square$

If

$$\lim_{x \to \infty} \frac{f(x)}{g(x)} = 1,$$

then f and g must behave in essentially the same way "at ∞." For example, if g has a limit at infinity, then f must have the same limit:

If $g(x) \to l$, then

$$f(x) = \frac{f(x)}{g(x)} \cdot g(x) \to 1 \cdot l = l$$

as $x \to \infty$. This holds for the limits $+\infty$ and $-\infty$ as well.

A polynomial

$$p(x) = a_n x^n + a_{n-1} x^{n-1} + \cdots + a_0$$

behaves like its highest-degree term $a_n x^n$ at ∞. For (supposing

$a_n \neq 0$, of course)

$$\frac{p(x)}{a_n x^n} = 1 + \frac{a_{n-1}}{a_n} \cdot \frac{1}{x} + \cdots + \frac{a_0}{a_n x^n}$$

$$\rightarrow 1 + 0 + \cdots + 0$$

$$= 1$$

as $x \to \infty$.

In the same way we can show that any rational function

$$r(x) = \frac{p(x)}{q(x)}$$

$$= \frac{a_n x^n + a_{n-1} x^{n-1} + \cdots + a_0}{b_m x^m + b_{m-1} x^{m-1} + \cdots + b_0}$$

behaves like $(a_n/b_m) x^{n-m}$ at ∞.

Example 7 Redo Example 5 by this method, i.e., by computing the limits of $(a_n/b_m) x^{n-m}$ as $x \to \infty$.

Solution Here $n - m = 2 - 2 = 0$, and

$$\lim_{x \to \infty} \frac{3 + x^2}{3x^2 + 1} = \lim_{x \to \infty} \frac{1}{3} x^{2-2}$$

$$= \lim_{x \to \infty} \frac{1}{3}$$

$$= \frac{1}{3}. \qquad \square$$

Example 8

$$\lim_{x \to \infty} \frac{3 + x^3}{3x^2 + 1} = \lim_{x \to \infty} \frac{1}{3} x^{3-2}$$

$$= \lim_{x \to \infty} \frac{x}{3}$$

$$= \infty. \qquad \square$$

PROBLEMS FOR SECTION 4

Evaluate the following limits.

1. $\lim_{x \to +\infty} \dfrac{7x + 2}{3 - x}$

2. $\lim_{x \to +\infty} (\sqrt{x^2 + 4} - x)$

3. $\lim_{x \to -\infty} \left(1 + \dfrac{2}{x}\right)$

4. $\lim_{x \to +\infty} \sqrt{x^2 + x + 1} - x$

5. $\lim_{x \to \infty} \dfrac{2x^2 + 1}{6 + x - 3x^2}$

6. $\lim_{x \to \infty} \dfrac{x}{x^2 + 5}$

7. $\lim_{x \to +\infty} \dfrac{3^x - 3^{-x}}{3^x + 3^{-x}}$

8. $\lim_{x \to -\infty} \dfrac{3^x - 3^{-x}}{3^x + 3^{-x}}$

9. $\lim_{x \to \infty} \dfrac{\sqrt{x + 1}}{\sqrt{4x - 1}}$

10. $\lim_{x \to \infty} \left(\dfrac{1}{x} + 1\right)\left(\dfrac{5x^2 - 1}{x^2}\right)$

11. $\lim_{x \to \infty} \dfrac{1 + 99x + 2x^2}{3x^2 + 4}$

Evaluate the following limits.

12. $\lim_{x \to -1^-} \dfrac{3x^2 - 4x - 7}{x^2 + 2x + 1}$

13. $\lim_{x \to 5^+} \dfrac{5x^2 + x^3}{25 - x^2}$

14. $\lim_{x \to 1^+} \dfrac{x^2}{1 - x^2}$

15. $\lim_{x \to 2^+} \dfrac{\sqrt{x^2 - 4}}{x - 2}$

16. $\lim_{x \to 0} \dfrac{x + 1}{|x|}$

17. Evaluate

$$\lim_{x \to 3} \dfrac{2x^2 - 5x - 3}{x^2 - x - 6}$$

and

$$\lim_{x \to \infty} \dfrac{2x^2 - 5x - 3}{x^2 - x - 6}.$$

Justify both calculations.

18. Prove the assertion made above about $r(x)$ being asymptotically equal to

$$(a_n/b_m)x^{n-m} \qquad \text{at } \infty.$$

19. How large must a positive number x be taken in order to ensure that $1/x^3$ is less than .001?

20. How large must x be taken to ensure that $x^{-1/3}$ is less than .1? .01? .001?

21. Let d be a small positive number. Determine how large a positive number x must be taken in order to ensure that $1/x^3 < d$.

22. Determine how large x must be taken to ensure that $x^{-1/3} < d$.

23. Show that

$$\dfrac{1}{\sqrt{x}} > 1000$$

for all x smaller than a certain number d. (Find an explicit value of d that will work.)

24. If M is any positive number, show that

$$\dfrac{1}{\sqrt{x}} > M$$

for all x smaller than a certain number d. (Find d in terms of M.)

25. Given a positive number d, determine a positive number M such that

$$\frac{1}{x^2} < d$$

for all x larger than M.

26. Given a positive number d, determine a positive number M such that

$$\frac{1}{x^{1/3}} < d$$

for all x larger than M.

5
THE
INTERMEDIATE-
VALUE
PRINCIPLE

Much of calculus rests ultimately on a few simple properties of continuous functions. This section concerns one of them, the *intermediate-value principle*, together with its surrounding ideas. Our discussion will be based entirely on geometric plausibility. Proofs will be given later, mostly in the appendices.

When we draw a graph, a discontinuity requires a break in the drawing: the pencil must be lifted and put down at a different point. On the other hand, when we draw a continuous graph (over an interval on the x-axis), the pencil need not be lifted; the continuous graph is a single connected strand. In particular, the graph of a continuous function cannot "jump over" a horizontal line $y = k$.

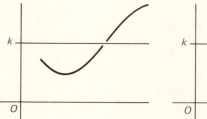

Figure 1

If the graph crosses from a value below the line to a value above it, then in the process the graph must intersect the line. (Otherwise, the part above $y = k$ would be disconnected from the part below.) This is the intermediate-value principle.

> **Intermediate-value principle** *If f is continuous on the closed interval $[a, b]$, and if k is a number between $f(a)$ and $f(b)$, then there is a number X in $[a, b]$ such that $f(X) = k$.*

Note that this is an *existence* principle, asserting that a number having a certain property exists, but not specifying where it is, or how to calculate it. We shall take up the proof in Appendix 5.

Starting from the intermediate-value principle and arguing a little further one can obtain the following stronger theorem. It is the best general result we have about the range of a continuous function.

THEOREM 4 **The Range Principle** *If f is continuous on an interval I, then the range of f over I is also an interval.*

Example 1 Find the range of $f(x) = x^4 + 1$.

Solution Note that $f(0) = 1$ and that $f(x) \geq 1$ for all x (because $x^4 \geq 0$). Also that $f(x) \to +\infty$ as $x \to \pm\infty$. The range therefore extends from 1 to $+\infty$. Since the domain of f is an interval (namely $(-\infty, \infty)$), and since polynomials are continuous, it follows that the range must be the interval $[1, +\infty)$, by the theorem. □

Example 2 Find the range of $f(x) = x^5 - x^2$.

Solution Again the domain is the interval $(-\infty, \infty)$, so the range must be an interval, by the theorem. Here $f(x) \to -\infty$ as $x \to -\infty$, and $f(x) \to +\infty$ as $x \to +\infty$, so the range interval is $(-\infty, \infty)$.

Note that this conclusion is equivalent to the assertion that the polynomial equation

$$x^5 - x^2 = c$$

must have at least one solution for each value of the constant c. □

Example 3 The domain of $f(x) = 1/x^3$ is not a single interval, but is the union of the two intervals $(-\infty, 0)$ and $(0, +\infty)$.

On $(0, +\infty)$, $f(x) \to +\infty$ as $x \to 0$ and $f(x) \to 0$ as $x \to +\infty$. So the range of f over this interval is $(0, +\infty)$. Similarly, the range over $(-\infty, 0)$ is $(-\infty, 0)$. So here the range equals the domain, a union of two intervals.

Here the theorem could be avoided by solving the equation $y = 1/x^3$ for x, getting $x = 1/y^{1/3}$. The two equations are equivalent, so the range of x^{-3} is the domain of $y^{-1/3}$, and hence is the above pair of intervals. □

Example 4   The function

$$f(x) = \frac{1+x}{\sqrt{x}} = \frac{1}{\sqrt{x}} + \sqrt{x}$$

is defined and continuous on the interval $(0, \infty)$, and $f(x) \to +\infty$ as x approaches either 0 or ∞. The graph must therefore be something like that shown in Fig. 2.

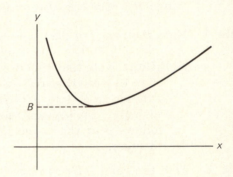

Figure 2

It appears that f must have a positive minimum value B, in which case the range interval is $[B, \infty)$. But finding B is a nontrivial problem that we shall solve in Chapter 5 as an application of calculus. (This particular problem can also be solved by algebra.) □

It is the intermediate-value principle (or the range principle) that guarantees that polynomial equations have roots.

Example 5   Does the equation

$$x^6 + x^2 - 2 = 0$$

have any solutions? Yes, because the equation is of the form

$$f(x) = 0,$$

where f is continuous,

$$f(0) = -2 \quad \text{and} \quad f(x) \to +\infty \quad \text{as } x \to \pm\infty.$$

It follows from Theorem 4 that the range of f over $[0, \infty)$ is the interval $[-2, \infty)$. Therefore, there is at least one point X in $[0, \infty)$ for which $f(X) = 0$. (Then $f(-X) = 0$, also, because f is an even function. See p. 47.) □

Example 6   Show that the equation

$$x^3 = 2$$

has a unique solution.

Solution   That a solution *exists* follows from the intermediate-value principle. In fact, there is a solution in the interval $[\frac{5}{4}, \frac{4}{3}]$, since

$$\left(\frac{5}{4}\right)^3 = \frac{125}{64} < 2 < \frac{64}{27} = \left(\frac{4}{3}\right)^3.$$

The solution is *unique* because the graph of $f(x) = x^3$ is rising and crosses the horizontal line $y = 2$ only once. (See Fig. 1 on p. 49. Also, see Example 7.)   □

The unique solution in Example 6 is designated $2^{1/3}$ and is called the *cube root* of 2.

In the same way, the equation $x^3 = k$ has a unique solution for each number k. And a larger k determines a larger solution. That is, the equation

$$y = x^3$$

determines x as an increasing function of y. This solution function is called the *cube-root function* and is designated

$$x = y^{1/3}.$$

Its domain and range are both the interval $(-\infty, \infty)$, and it is continuous, by Theorem 3 in Section 3 (or Example 6 in Section 2).

Unique Solutions and Increasing Functions   The question of *unique* solutions thus brings us back to increasing and decreasing functions (Fig. 3). We didn't state a formal definition before, so we give it now.

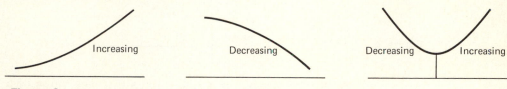

Increasing   Decreasing   Decreasing   Increasing

Figure 3

DEFINITION   The function f is *increasing* on an interval I in its domain if $f(x_1) < f(x_2)$ whenever $x_1 < x_2$ in I. Similarly, f is *decreasing* on I if

$f(x_1) > f(x_2)$ whenever $x_1 < x_2$ in I. Finally, f is *monotone* on I if *either f is increasing on I or f is decreasing on I.*

Example 7   If n is a positive odd integer, then $f(x) = x^n$ is increasing on $(-\infty, \infty)$. If n is even, then $f(x) = x^n$ is decreasing on $(-\infty, 0]$ and increasing on $[0, \infty)$. (These facts can be seen geometrically by sketching the graphs of $y = x^n$ for various integers n. Algebraic proofs are outlined in Problems 25 and 26. But the easiest proofs use calculus. See Section 2 in Chapter 5.)

It follows, just as in the discussion of $y = x^3$ and $x = y^{1/3}$ above, that the nth-root function

$$x = y^{1/n}$$

is defined and continuous, with domain $(-\infty, \infty)$ if n is odd, and domain $[0, \infty)$ if n is even.   □

For future use, we note that the same conclusions hold for *any* function f that is continuous and increasing on an interval I. We follow the same steps:

1. The range of f, $J = f(I)$, is also an interval, by the range principle (Theorem 4).

2. For each k in J, the equation

$$f(x) = k$$

has a *unique* solution in I (because f is increasing).

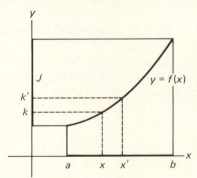

Figure 4

3. A larger value of k determines a larger value of x (see Fig. 4). The equation

$$y = f(x)$$

thus determines x as an *increasing* function of y, say $x = g(y)$.

4. The solution function g has domain J and range I, and is *continuous*, by Theorem 3 in Section 3.

Similar conclusions hold for decreasing functions. We shall continue this discussion in Chapter 7.

PROBLEMS FOR SECTION 5

Use Theorem 4 to determine the ranges of the following functions.

1. $f(x) = x^6$ **2.** $f(x) = x^7$ **3.** $f(x) = x^{1/6}$ **4.** $f(x) = x^{1/7}$

5. $f(x) = \dfrac{1}{9 - x^2}$ **6.** $f(x) = \dfrac{1}{x^2 + 2x}$ **7.** $f(x) = \sqrt{36 - x^2}$ **8.** $f(x) = \sqrt{x^2 - 16}$

9. $f(x) = (8 - x^2)^{1/3}$ **10.** $f(x) = (16 - x^2)^{1/4}$

11. $f(x) = x + 1/x$ (Solve $y = x + 1/x$ for x in terms of y to see what values of y are possible.)

Use the intermediate-value principle to show that each of the following equations has a solution on the given interval.

12. $x^3 - x^2 + x = 2, [1, 2]$ **13.** $x^5 - x^4 - x^2 = 0, [1, 2]$ **14.** $x^4 - x^3 - x^2 - x = -1, [0, 1]$

15. $x^{1/3} + x^{1/2} = 2, [0, 4]$ **16.** $x^{1/2} - x^{1/3} = 3, [32, 64]$ **17.** $x^3 = 3(1 + \sqrt{x}), [0, 2]$

Find the domain and range of each of the following functions.

18. $y = 1/(\sqrt{x} - 1)$ **19.** $y = \sqrt{4x^2 - 1}$

The following problems involve working with inequalities. See Appendix 1 for a review of the inequality laws. Later we will find it easier to prove that particular functions are increasing by showing in each case that the derivative is positive.

For Problems 20 through 27 let f and g be functions that are increasing on an interval I. Show that

20. $f + g$ is increasing   **21.** fg is increasing if f and g are both positive

22. $1/f$ is decreasing if f is positive   **23.** $-f$ is decreasing

24. $f(-x)$ is decreasing on $-I$ **25.** Use Problem 21 to show that the functions $x, x^2, x^3, \ldots$ are all increasing on $I = [0, \infty)$.

26. Use Problems 23 through 25 to show that x^n is decreasing on $I = (-\infty, 0]$ if n is even; x^n is increasing on $I = (-\infty, 0]$ if n is odd.   **27.** Show that $f(g(x))$ is increasing (if the domain of f includes the range of g).

28. Suppose that $f(g(x)) = x$ and that f is increasing. Show that g is increasing. [*Hint*: Use the "reverse-increasing" property of f.]

30. Show that f^n and $f^{1/n}$ are increasing, if f is positive and increasing.

32. Suppose that f is continuous on $[a, b]$ and increasing on (a, b). Prove that f is increasing on $[a, b]$.

What must be shown here is that

$$f(a) < f(x) < f(b)$$

for each x in the open interval (a, b). To start, choose some number y between x and b, $x < y < b$. Show that $f(y) \leq f(b)$ by applying some version of L4 to f on the interval $[y, b]$.

34. In the same way, show that the equation

$$y^3 + yx^2 + x^3 = 1$$

determines y as a function of x.

29. Show from Problems 25 and 28 that $x^{1/n}$ is increasing on $[0, \infty)$.

31. List the answers to Problems 20 through 24 if f and g are decreasing on I.

33. Show that the equation

$$x + y + x^5 + y^7 = 0$$

determines y as a function of x. (Fix the value of x. Show that there is then a unique number y satisfying the equation, in the manner of the examples and problems in this section.)

CHAPTER 3
THE DERIVATIVE

We shall rely heavily on geometric reasoning in our development of the calculus. This does not mean we have a particular interest in geometry, although it is one of the subjects to which calculus can be applied. (For example, we shall be able to compute areas and volumes that are otherwise inaccessible to us.) Rather, the reason for the special role of geometry is that geometric figures help us understand calculus itself, in much the way that the graph of a function f lets us see in a very vivid manner how $f(x)$ varies with x. Furthermore, geometric reasoning will help us to apply calculus to *other* disciplines. Here, for example, is a problem of a type that we shall take up in Chapter 5.

The efficiency E of a certain machine varies with the viscosity x of the lubricating oil that is used. The machine runs badly if the oil is either too thin or too thick, i.e., if x is either too small or too large. Our theoretical calculations (or perhaps our empirical observations) show us that the relationship between E and x is given by

$$E = x - x^3.$$

The problem is to determine the viscosity x that gives the *maximum* efficiency E. When we plot E against x we find that the graph in the first quadrant has the slope shown on the next page.

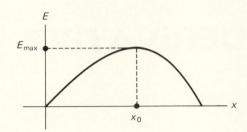

Now we see that our problem can be rephrased geometrically as that of finding the *highest point* on this graph.

This example shows how a nongeometric problem can be cast into geometric terms, but it doesn't suggest the role of calculus. We shall see, however, that calculus will let us compute the slope m of the tangent line as a function of the point of tangency x. At the highest point on the above graph, the tangent line appears to be horizontal, so we want the value of x for which $m = 0$. Therefore, in order to solve our maximum problem, we first compute the tangent slope function $m(x)$ by calculus, and then solve the equation $m(x) = 0$ to find the value $x = x_0$ giving maximum efficiency.

Thus, when we approach the derivative through the tangent-line problem, we will not be solving *only* this geometric problem. We will also be creating a geometric interpretation of the derivative that will help us in calculus itself and in the applications of calculus to other disciplines.

Later in this chapter we shall take up the other major interpretation of the derivative, as a rate of change. Otherwise, the chapter treats the computation of derivatives from scratch and the beginning rules of the systematic differentiation procedure.

1
THE
DERIVATIVE

We begin by reviewing, with some changes, the tangent-slope calculation for the curve $y = x^3$ that was carried out in Example 1 of Chapter 2. This will lead to the derivative.

Let $(x_0, y_0) = (x_0, x_0^3)$ be any fixed point on the graph of $f(x) = x^3$. (Before we used $(x_0, x_0^3) = (1, 1)$.) We propose to find the slope of the tangent line at this point. We start with a second, nearby point on the graph, (x, x^3), and write down the slope of the secant line joining these two points:

$$m_{\text{sec}} = \frac{x^3 - x_0^3}{x - x_0}.$$

But now we shall change notation. We set

$$h = x - x_0,$$

or $x = x_0 + h$, and we make h the new independent variable. This gives

$$m_{\text{sec}} = \frac{(x_0 + h)^3 - x_0^3}{h}.$$

We have seen that the tangent slope m is the limit of the secant slope m_{sec} as the running point approaches the fixed point along the curve. In the new notation this means that we want the limit as $h \to 0$. As things stand there is no possibility for a trivial limit evaluation, because setting $h = 0$ in the m_{sec} quotient gives the meaningless 0/0. So, as before, we have to manipulate m_{sec} into a more usable form. This time we will expand the cube $(x_0 + h)^3$, and the algebra will be a little different. But it turns out, as before, that the numerator has h as a factor:

$$(x_0 + h)^3 - x_0^3 = (x_0^3 + 3x_0^2 h + 3x_0 h^2 + h^3) - x_0^3$$
$$= h(3x_0^2 + 3x_0 h + h^2).$$

So

$$m = \lim_{h \to 0} \frac{(x_0 + h)^3 - x_0^3}{h} = \lim_{h \to 0} (3x_0^2 + 3x_0 h + h^2) = 3x_0^2.$$

We have thus obtained the formula

$$m = 3x_0^2$$

for the slope m of the graph and its tangent line at the point (x_0, x_0^3). This tells us, for example, that at $x_0 = 2$ the curve has slope

$$m = 3 \cdot 2^2 = 12,$$

and that at $x_0 = -1/2$ its slope is

$$m = 3\left(-\frac{1}{2}\right)^2 = \frac{3}{4}.$$

We can now write down the equation of any tangent line to the curve $y = x^3$.

Example 1   Find the equation of the tangent to the curve $y = x^3$ at the point $(x_0, y_0) = (-1, -1)$.

Solution The formula $m = 3x_0^2$ gives the value

$$m = 3(-1)^2 = 3$$

for the slope of the tangent line at this point. The point-slope equation

$$y - y_0 = m(x - x_0)$$

for the tangent line is thus

$$y - (-1) = 3(x - (-1)),$$

or

$$y + 1 = 3(x + 1),$$

or

$$y = 3x + 2. \qquad\qquad \square$$

A slight change in viewpoint brings us to the derivative. The formula $m = 3x_0^2$ expresses the tangent slope m as a function of the point of tangency. Dropping the subscript on x_0, we thus have the equation

$$m = 3x^2$$

expressing the varying slope m as a new function of x. This new function is derived from the original function

$$y = x^3$$

by the limit calculation, and is called the *derivative* of x^3.

The calculation that we have just made for the function $f(x) = x^3$ can be described for an arbitrary function f. We first compute the slope of the secant line through the two points on the graph of f over x and $x + h$,

$$m_{\text{sec}} = \frac{f(x + h) - f(x)}{h}.$$

Then we calculate the limit of m_{sec} as h approaches 0. This gives a number that we interpret geometrically as the slope of the graph and its tangent line at the point $(x, f(x))$. See Fig. 1. The limit depends on x and is therefore a new function of x, called the *derivative* of the function f, and designated f'. In this notation we calculated above that $f'(x) = 3x^2$ when $f(x) = x^3$.

Omitting the geometric motivation, we have the following formal definition.

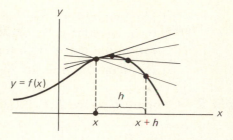

Figure 1

DEFINITION Given any function f, its *derivative* f' is the function whose value at x is defined by

$$f'(x) = \lim_{h \to 0} \frac{f(x+h) - f(x)}{h}.$$

The domain of f' then consists of those numbers x for which this limit exists.

The formal definition makes no reference to geometry. The fact that $(f(x+h) - f(x))/h$ and $f'(x)$ can be viewed as slopes gives a geometric *interpretation* of what we are doing. There are also other interpretations related to the notion of velocity (see Sections 3 and 4). It is therefore important to understand that the above definition is a purely *analytic* definition, independent of interpretations. Given a function f, we derive a new function f' from f by making a certain limit calculation. The instructions are given in the above formula. In this neutral frame of mind we call

$$\frac{f(x+h) - f(x)}{h}$$

the *difference quotient* of the function f.

Calculating Derivatives In applying this definition we always have the same problem: we can't directly evaluate the limit because of the $0/0$ difficulty that we experienced earlier. Thus we always have to manipulate the difference quotient into a new form that we can handle. (This may be simple or complicated. It is usually the heart of the problem; the subsequent limit evaluation is usually easy.) So here is our scheme:

a) Write down the difference quotient $(f(x+h) - f(x))/h$ for the particular function in question;

b) Manipulate the difference quotient into a form allowing the limit to be evaluated (in these first examples this will always mean cancelling out h from top and bottom);

c) Evaluate the limit as $h \to 0$.

Example 2   Compute $f'(x)$ by this procedure when $f(x) = 3x^2$.

Solution

a) $$\frac{f(x + h) - f(x)}{h} = \frac{3(x + h)^2 - 3x^2}{h}$$

b) $$= \frac{3x^2 + 6xh + 3h^2 - 3x^2}{h}$$

$$= \frac{6xh + 3h^2}{h} = 6x + 3h.$$

c) $$f'(x) = \lim_{h \to 0} \frac{f(x + h) - f(x)}{h} = \lim_{h \to 0} (6x + 3h) = 6x.$$

Thus $f(x) = 3x^2$ has the derivative $f'(x) = 6x.$ $\square$

Example 3   We compute $f'(x)$ when $f(x) = 1/x$.

a) $$\frac{f(x + h) - f(x)}{h} = \frac{\dfrac{1}{x + h} - \dfrac{1}{x}}{h}$$

b) $$= \frac{\dfrac{x - (x + h)}{(x + h)x}}{h} = \frac{-h}{(x + h)x} \cdot \frac{1}{h} = -\frac{1}{(x + h)x}.$$

c) $$f'(x) = \lim_{h \to 0} \frac{f(x + h) - f(x)}{h} = \lim_{h \to 0} \left(-\frac{1}{(x + h)x} \right) = -\frac{1}{xx} = -\frac{1}{x^2}.$$

Thus $f(x) = 1/x$ has the derivative $f'(x) = -1/x^2.$ $\square$

Example 4   Find $f'(x)$ for $f(x) = \sqrt{x}$.

Solution

a) $$\frac{f(x + h) - f(x)}{h} = \frac{\sqrt{x + h} - \sqrt{x}}{h}.$$

b) Here we shall "rationalize the numerator" by multiplying top and bottom by $\sqrt{x + h} + \sqrt{x}$, giving

$$\frac{\sqrt{x + h} - \sqrt{x}}{h} \cdot \frac{\sqrt{x + h} + \sqrt{x}}{\sqrt{x + h} + \sqrt{x}} = \frac{h}{h(\sqrt{x + h} + \sqrt{x})}$$

$$= \frac{1}{\sqrt{x + h} + \sqrt{x}}.$$

c) Since

$$\frac{1}{\sqrt{x + h} + \sqrt{x}} \rightarrow \frac{1}{\sqrt{x} + \sqrt{x}} = \frac{1}{2\sqrt{x}}$$

as $h \to 0$, we see that $f(x) = \sqrt{x}$ has the derivative $f'(x) = 1/(2\sqrt{x})$. $\qquad\square$

Example 5 Find $f'(x)$ when $f(x) = x + x^2$.

Solution

$$\frac{f(x + h) - f(x)}{h} = \frac{((x + h) + (x + h)^2) - (x + x^2)}{h}$$

$$= \frac{h + 2xh + h^2}{h}$$

$$= 1 + 2x + h,$$

which approaches $1 + 2x$ as its limit as h approaches 0. Thus $f'(x) = 1 + 2x$. $\qquad\square$

One might guess offhand that a continuous function is necessarily differentiable. Not so, as the following example shows.

Example 6 The function

$$f(x) = |x|$$

is continuous everywhere, but not differentiable at the origin. This is geometrically plausible, since the graph (Fig. 2) does not appear to have a tangent line at the origin.

Checking the algebra, we see that

$$\frac{f(x + h) - f(0)}{h} = \frac{|h|}{h} = \begin{cases} +1 & \text{if } h > 0, \\ -1 & \text{if } h < 0. \end{cases}$$

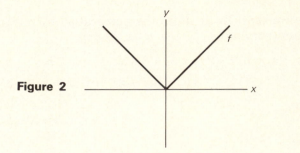

Figure 2

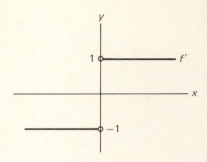

The difference quotient thus has the one-sided limits −1 on the left and +1 on the right. We could say that the graph has one-sided tangents at the origin, with slopes −1 on the left and +1 on the right. But the two-sided limit, and hence the derivative, does not exist. □

PROBLEMS FOR SECTION 1

Establish the following derivative formulas, using the three-step scheme in each case to evaluate the limit of the difference quotient.

1. If $f(x) = x^2$, then $f'(x) = 2x$.   **2.** If $f(x) = x^4$, then $f'(x) = 4x^3$.

3. If $f(x) = 1/x^2$, then $f'(x) = -2/x^3$.   **4.** If $f(x) = 1/x^{1/2}$, then $f'(x) = -1/2x^{3/2}$.

5. If $f(s) = -3s$, then $f'(s) = -3$.   **6.** If $f(x) = 2x - x^2$, then $f'(x) = 2(1 - x)$.

7. If $f(x) = mx + b$, then $f'(x) = m$.   **8.** If $f(x) = ax^2 + bx + c$, then $f'(x) = 2ax + b$.

9. Each of the first four functions above is of the form $f(x) = x^a$. Examine the form of $f'(x)$ in each of these cases, and then conjecture what the general rule is for the derivative of a power of x: If $f(x) = x^a$, then $f'(x) = $ ____.

Calculate the derivatives of the following functions:

10. $f(x) = x + 1$ **11.** $f(x) = x^2 + 4$ **12.** $f(x) = x^3/3$

13. $f(x) = \sqrt{2x}$ **14.** $g(t) = \sqrt{3t - 2}$ **15.** $f(y) = 1/(y - 1)$

16. $f(x) = x + (1/x)$ **17.** $g(x) = (x^2 + 1)/x$ **18.** $f(s) = 1/(s^2 - 1)$

19. $f(x) = \sqrt{x^2 + 1}$ **20.** $f(x) = (x - 4)/(x + 1)$ **21.** $h(x) = 1/\sqrt{x + 1}$

22. $f(x) = \sqrt{ax + b}$ **23.** $f(x) = x/(x^2 + 1)$ **24.** $f(x) = 1/(1 + \sqrt{x})$

**2
SOME
GENERAL
RESULTS
ABOUT
DERIVATIVES**

**Leibniz
Notation**

In situations where y is a function of x, Leibniz wrote the derivative

$$\frac{dy}{dx},$$

which is read "the derivative of y with respect to x." This notation allows us to write down the derivatives of particular functions of x without using the function symbols f and f'. For example,

$$\text{if } y = x^3, \qquad \text{then } \frac{dy}{dx} = 3x^2.$$

Even more simply, we can write

$$\frac{d}{dx}\,x^3 = 3x^2.$$

We must remember, however, that only a *function* can have a derivative. The symbol dy/dx implies that y is a certain function of x, and dy/dx is the derivative of that function. Thus,

$$\text{if } y = f(x), \qquad \text{then } \frac{dy}{dx} = f'(x).$$

Warning: We do not yet have separate meanings for the symbols dx and dy, and dy/dx is not (yet) to be considered as a quotient. It is just another symbol for the derivative. The letters x and y in dy/dx are not to be thought of as the variables x and y, but suggest that dy/dx is related to these variables.

Because dy/dx doesn't display the variable x in the way that $f'(x)$ does, substitutions are harder to indicate. We have to use some such notation as

$$\frac{dy}{dx}\bigg|_{x=5}$$

for the value of dy/dx when $x = 5$. Thus, if $dy/dx = 3x^2$, then $dy/dx|_{x=5} = 75$. In general,

$$\text{if } y = f(x), \qquad \text{then } \frac{dy}{dx}\bigg|_{x=a} = f'(a).$$

From now on we shall use the notations dy/dx and $f'(x)$ interchangeably.

A function f is *differentiable* at a point x_0 if its derivative exists at x_0. If f is differentiable at every point of its domain we say simply that f is differentiable.

Some Derivative Rules The several special cases that have been worked out in Section 1 in the examples and problems suggest that x^k is differentiable and

$$\frac{d}{dx} x^k = kx^{k-1}$$

for any constant k. This is true, but we won't be able to give a completely general proof, covering all cases, until Chapter 7. Meanwhile, we can point out ways of proving various special cases as we go along. For example, the device used for x^3 in the first section works just as well when k is any positive integer m. It requires knowing something about the binomial expansion of $(x + h)^m$. We need the first two terms explicitly, plus the fact that all the remaining terms contain h to powers higher than 1.

THEOREM 1 If $f(x) = x^m$, where m is a positive integer, then f is differentiable and $f'(x) = mx^{m-1}$.

Proof When we multiply out $(x + h)^m$, we get

$$(x + h)^m = x^m + mx^{m-1}h + \begin{pmatrix} \text{terms containing} \\ h^2 \text{ as a factor} \end{pmatrix}.$$

So

$$(x + h)^m - x^m = mx^{m-1}h + \begin{pmatrix} \text{terms containing} \\ h^2 \text{ as a factor} \end{pmatrix},$$

and

$$\frac{(x + h)^m - x^m}{h} = mx^{m-1} + \begin{pmatrix} \text{terms containing} \\ h \text{ as a factor} \end{pmatrix}.$$

We now let h approach 0. Then every term containing h approaches 0, and the sum of all these terms approaches 0. The right side above thus approaches $mx^{m-1} + 0 = mx^{m-1}$. That is,

$$\frac{d}{dx} x^m = \lim_{h \to 0} \frac{(x + h)^m - x^m}{h} = mx^{m-1} + 0 = mx^{m-1}. \qquad \blacksquare$$

Another thing suggested by the examples and problems is that the derivative of the *sum* of two functions is always the *sum* of the two derivatives. We now check this.

THEOREM 2 *If the functions f and g are differentiable at x, then so is their sum F = f + g, and*

$$F'(x) = f'(x) + g'(x).$$

Proof

$$\frac{F(x + h) - F(x)}{h} = \frac{[f(x + h) + g(x + h)] - [f(x) + g(x)]}{h}$$

$$= \frac{f(x + h) - f(x)}{h} + \frac{g(x + h) - g(x)}{h}.$$

As $h \to 0$ the two terms on the right approach $f'(x)$ and $g'(x)$, respectively, so their sum approaches $f'(x) + g'(x)$. Thus

$$F'(x) = \lim_{h \to 0} \frac{F(x + h) - F(x)}{h}$$

$$= f'(x) + g'(x). \qquad \blacksquare$$

In this limit computation, the difference quotient of F rearranges itself into the sum of the difference quotients of f and g, and this turns into $F'(x) = f'(x) + g'(x)$ upon taking limits. In a similar way, we can show that the derivative of a constant times f is that constant times the derivative of f:

$$(cf)' = cf'.$$

THEOREM 3 *If the function f is differentiable at x, then so is g = cf, and*

$$g'(x) = cf'(x).$$

For example,

$$\frac{d}{dx} 4x^5 = 4 \frac{d}{dx} x^5$$

$$= 4 \cdot 5x^4$$

$$= 20x^4.$$

We repeat the formulas of the above three theorems in Leibniz notation:

$$\frac{d}{dx}x^m = mx^{m-1}$$

$$\frac{d}{dx}(u + v) = \frac{du}{dx} + \frac{dv}{dx}$$

$$\frac{d}{dx}cu = c\frac{du}{dx}$$

From these theorems we can write down the derivative of any polynomial (showing, incidentally, that every polynomial is differentiable). For example,

$$\frac{d}{dx}(5x^3 + 3x^2) = \frac{d}{dx}(5x^3) + \frac{d}{dx}(3x^2)$$

$$= 5\frac{d}{dx}x^3 + 3\frac{d}{dx}x^2$$

$$= 5 \cdot 3x^2 + 3 \cdot 2x$$

$$= 15x^2 + 6x.$$

After a little practice one can omit the middle steps in this type of computation and just write the answer down.

Tangent Lines   As a first application, we can find the tangent line to a polynomial graph at any point on the graph.

Example   Find the equation of the line tangent to the graph of $y = x^3 - 7x$ at the point given by $x_0 = 2$.

Solution   When $x_0 = 2$,

$$y_0 = x_0^3 - 7x_0 = 8 - 14 = -6.$$

Thus the point of tangency is $(x_0, y_0) = (2, -6)$. The slope of the graph at that point is

$$m = \left.\frac{dy}{dx}\right|_{x=2} = \left.3x^2 - 7\right|_{x=2} = 12 - 7 = 5.$$

The tangent line $y - y_0 = m(x - x_0)$ is thus

$$y - (-6) = 5(x - 2), \quad \text{or} \quad y = 5x - 16. \qquad \square$$

Note that these calculations have all been made without going back to the three-step scheme. All necessary computations using the three-step scheme were gathered into the proofs of Theorems 1 through 3, after which we simply refer to the theorems.

This use of Theorems 1 through 3 to differentiate polynomials is an example of *systematic differentiation*. We work out once and for all the derivatives of various special functions, like x^n. Also, we develop formulas for the derivatives of *combinations* of functions, such as the formula $(f + g)' = f' + g'$. We can then write down the derivative of practically any function, using only these basic formulas and combining rules. This program is carried out in Chapter 4.

We saw at the end of Section 1 that a continuous function is not necessarily differentiable. However, a *differentiable function is necessarily continuous*. Geometrically, this is the fact that if a graph has a tangent line at a point, then it *cannot* be discontinuous there. The analytic proof is just a limit evaluation.

THEOREM 4   *If f is differentiable at x, then f is continuous at x.*

Proof

$$\lim_{h \to 0} [f(x + h) - f(x)] = \lim_{h \to 0} \left[\frac{f(x + h) - f(x)}{h} \right] h$$

$$= \lim_{h \to 0} \left[\frac{f(x + h) - f(x)}{h} \right] \cdot \lim_{h \to 0} h$$

$$= f'(x) \cdot 0 = 0.$$

Therefore,

$$\lim_{h \to 0} f(x + h) = f(x).$$

This is the same thing as saying that $f(t) \to f(x)$ as $t \to x$, so f is continuous at x. ∎

Higher Derivatives   The derivative f' may again be a differentiable function and we naturally use the notation f'' for *its* derivative $(f')'$. For example, if $f(x) = x^5$, then

$$f'(x) = 5x^4, \quad \text{and} \quad f''(x) = \frac{d}{dx} f'(x) = \frac{d}{dx} 5x^4 = 20x^3.$$

Then $f'''(x) = 60x^2$, $f''''(x) = 120x$, $f'''''(x) = 120$, and $f''''''(x) = 0$.

When the number of primes begins to be unwieldy, we use a numeral in parentheses instead. Thus,

$$f^{(4)}(x) = f''''(x) = 120x, \qquad f^{(5)}(x) = 120, \qquad \text{and } f^{(6)}(x) = 0.$$

in the above example.

The Leibniz notation for the second derivative is

$$\frac{d^2y}{dx^2} = \frac{d}{dx}\left(\frac{dy}{dx}\right).$$

Note that the superscript 2 is on d in the upper part of the term, and on dx in the lower, and that this is consistent with the right side of the equation. In Leibniz notation, the above example is

$$\frac{d}{dx}x^5 = 5x^4, \qquad \frac{d^2}{dx^2}x^5 = 20x^3, \qquad \frac{d^3}{dx^3}x^5 = 60x^2,$$

$$\frac{d^4}{dx^4}x^5 = 120x, \qquad \frac{d^5}{dx^5}x^5 = 120, \qquad \frac{d^6}{dx^6}x^5 = 0.$$

The higher-order derivatives of f play an important role in calculus and its applications. The second derivative d^2y/dx^2 is particularly important because it can be directly interpreted. We shall see in Chapter 5 that the sign of d^2y/dx^2 determines which way the graph of $y = f(x)$ is turning. Later on, in Chapter 15, this fact will be sharpened into a formula for the *curvature* of a graph. In the next section, the second derivative will be interpreted as acceleration.

PROBLEMS FOR SECTION 2

Find dy/dx and d^2y/dx^2 for each of the following functions:

1. $y = x^7$

2. $y = 5x^{120}$

3. $y = x^3 + 3x^2$

4. $y = x^3 + x - 1$

5. $y = mx + b$

6. $y = ax^2 + bx + c$

7. $y = 4x^4 - x^2 + 2$

8. $y = \dfrac{x^5}{5} + \dfrac{x^4}{4} + \dfrac{x^3}{3} + \dfrac{x^2}{2} + x + 1$

9. $y = x(x - 1)$

10. $y = (x - 2)(x + 3)$

11. What is d^4y/dx^4 if $y = x^3$?

12. The symbol $n!$ (n factorial) represents the product

$$n(n - 1)(n - 2) \cdots 2 \cdot 1.$$

Thus $4! = 4 \cdot 3 \cdot 2 \cdot 1 = 24$. Show that

$$\frac{d^n}{dx^n} x^n = n!$$

for $n = 1, 2, 3, 4, 5$.

13. Show that if the above formula holds for $n = m$, it holds for $n = m + 1$. It must therefore hold for every value of n. Why?

14. Prove that if m is a positive number for which the power formula holds then

$$\frac{d}{dx} x^{-m} = (-m)x^{(-m)-1}.$$

(After a first step simplifying the difference quotient, use the limit already assumed.)

Using this new formula, differentiate the following functions:

15. $f(x) = x + \dfrac{1}{x}$ **16.** $f(x) = x^3 + x^{-3}$

17. $f(x) = \dfrac{x^3 + 1}{x^2}$ **18.** $f(x) = \dfrac{x + 1}{\sqrt{x}}$

19. If $f(x) = x - (1/x)$, prove that

$$f(x^2) = xf(x)f'(x).$$

20. Find the formula for $d^n y/dx^n$ when $y = 1/x$.

21. If $f(x) = x^{1/2}$, show that $f''(x)f(x) = -1/4x$.

22. Find the points on the graph of $y = x^3 + 3x^2 - 9x$ at which the tangent line is horizontal.

23. Sketch the graph of $y = x + (1/x)$ in the first quadrant. Find its lowest point. (Your sketch should suggest that the tangent line must be horizontal at the lowest point.)

24. The parabola $y = ax^2 + bx - 2$ is tangent to the line $y = 4x + 7$ at the point $(-1, 3)$. Find a and b.

25. Find the coordinates of the vertex of the parabola $y = x^2 - 6x + 1$. (Make use of the fact that the slope of the tangent to the curve is zero at the vertex.)

26. Find the points on the curve

$$y = x^3 + x^2$$

where the tangent has slope 1.

27. Find the equation of the line *normal* to the curve $y = x^2 + 3x + 2$ at the point where $x = 3$. (The normal line is perpendicular to the tangent.)

28. Prove that the tangents to the parabola $y = x^2$ at the ends of any chord through $(0, \frac{1}{4})$ are perpendicular to each other.

29. Find the vertex of the parabola

$$y = ax^2 + bx + c.$$

(See Problem 25.)

31. Prove that the tangent line to the parabola $y = x^2$ at the point (x_0, y_0) has y-intercept $-y_0(= -x_0^2)$.

33. Find the lines tangent to the parabola $y = x^2$ through the external point $(2, 3)$. [*Hint*: Equate two expressions for the slope of the tangent line.]

35. State the limit law (from Chapter 2) needed to make the proof of Theorem 2 logically complete.

37. Suppose that $f(x) = xg(x)$ on an interval I about the origin, and that $g(x)$ is continuous at $x = 0$. Prove that $f'(0)$ exists and equals $g(0)$.

39. Suppose that $g(x) \leq f(x) \leq h(x)$ on an interval about x_0, and that

$$g(x_0) = h(x_0), \qquad g'(x_0) = h'(x_0).$$

Show that then $f'(x_0)$ exists and has the same value.

30. Show that the area of the triangle cut off in the first quadrant by a tangent line to the graph of $f(x) = 1/x$ is always 2, no matter what the point of tangency is. (Find the tangent line at the point $(a, 1/a)$ on the curve, then find the area of the triangle cut off, etc.)

32. Find the equation of the tangent line to the graph of $y = x^3$ at the point $(x_0, y_0) = (x_0, x_0^3)$. Then show that the tangent line intersects the graph again at the point where $x = -2x_0$.

34. Show that a line can be drawn through the point (a, b) tangent to the parabola $y = x^2$ if and only if $b \leq a^2$. Interpret this requirement on (a, b) geometrically.

36. Determine the derivative of $f(x) = |x|$, and show that it can be expressed

$$\frac{d}{dx}|x| = \frac{|x|}{x}, \qquad x \neq 0.$$

38. Suppose that $-x^2 \leq f(x) \leq x^2$ on an interval I about the origin. Show that $f'(0)$ exists and has the value 0.

3
INCREMENTS;
VELOCITY

When considering the difference quotient $(f(x + h) - f(x))/h$, it is often useful to think of $x + h$ as a second value of x and the difference h as a *change in* x. If we wish to emphasize this viewpoint we replace h by Δx, read "delta x." Here Δx is not the product of a number Δ times a number x, but a wholly new numerical variable, called the *increment* in x. Its value is always to be thought of as the *change* in the value of x, from a *first* value to a *second* value, or from an *old* value to a *new* value. The most consistent notation in this situation would be to use subscripts on x, say x_0 and x_1, for the old and new values, respectively. Then

$$\Delta x = x_1 - x_0$$

is the change in x in going from the old to the new value. We can also consider the new value as being obtained from the old value by adding the change:

$$x_1 = x_0 + \Delta x.$$

An increment Δx can be of either sign. Thus if $x_0 = 2$ and $\Delta x = -3$, then $x_1 = 2 + (-3) = -1$.

As often as not we discard one of the subscripts, say the zero subscript. In this case the old and new values are x and $x + \Delta x (= x_1)$. That is, we use x itself to represent the old value of x.

Now suppose that y is a function of x, $y = f(x)$. When x changes from x to $x + \Delta x$, then y changes from $f(x)$ to $f(x + \Delta x)$. The increment in y thus has the basic formula

$$\boxed{\Delta y = f(x + \Delta x) - f(x).}$$

Example 1 If $y = x^2$, then

$$\Delta y = (x + \Delta x)^2 - x^2 = 2x \, \Delta x + (\Delta x)^2 = \Delta x(2x + \Delta x). \qquad \square$$

Example 2 When $y = x^3$, we have

$$\Delta y = f(x + \Delta x) - f(x) = (x + \Delta x)^3 - x^3$$
$$= x^3 + 3x^2 \, \Delta x + 3x(\Delta x)^2 + (\Delta x)^3 - x^3$$
$$= \Delta x(3x^2 + 3x \, \Delta x + (\Delta x)^2). \qquad \square$$

In terms of increments, the difference quotient $(f(x + h) - f(x))/h$ becomes

$$\boxed{\frac{f(x + \Delta x) - f(x)}{\Delta x} = \frac{\Delta y}{\Delta x},}$$

which tells us, literally, that the difference quotient is the *change in y divided by the change in x*. The derivative can now be expressed

$$\boxed{\frac{dy}{dx} = \lim_{\Delta x \to 0} \frac{\Delta y}{\Delta x}.}$$

(Although of later origin, the Δ symbolism obviously fits well with Leibniz's symbolism.)

From the above expression for dy/dx we can obtain

$$\lim_{\Delta x \to 0} \Delta y = \lim_{\Delta x \to 0} \left(\frac{\Delta y}{\Delta x}\right)\Delta x$$

$$= \left(\lim_{\Delta x \to 0} \frac{\Delta y}{\Delta x}\right)\cdot\left(\lim_{\Delta x \to 0} \Delta x\right) = \frac{dy}{dx}\cdot 0 = 0.$$

This is just a reformulation in terms of increments of our earlier proof that a differentiable function is continuous.

Example 3 We recompute the derivative formula for $y = x^3$ in increment notation. From Example 2,

$$\frac{\Delta y}{\Delta x} = 3x^2 + 3x\,\Delta x + (\Delta x)^2.$$

Therefore

$$\frac{dy}{dx} = \lim_{\Delta x \to 0} \frac{\Delta y}{\Delta x} = 3x^2 + 3x\cdot 0 + 0^2 = 3x^2. \qquad \square$$

Thinking of the difference quotient as a ratio of changes leads to interpreting the derivative as a rate of change. Velocity is the most familiar such rate of change, and we shall take it up here. Other examples will be considered in the next section.

Velocity Suppose that a particle is moving along a coordinate line. At each instant of time t the particle has a unique position coordinate s, so its position s is a function of t,

$$s = f(t).$$

If the particle is you in your car, and if it takes you half an hour to drive ten miles through a city, then you say that your average velocity is 20 miles per hour through the city, because

$$\frac{\text{Distance traveled}}{\text{Elapsed time}} = \frac{10}{1/2} = 20.$$

In general,

$$\boxed{\text{Average velocity} = \frac{\text{Distance traveled}}{\text{Elapsed time}} = \frac{\Delta s}{\Delta t},}$$

a formula that ought to agree with your idea of average velocity.

Since

$$\frac{\Delta s}{\Delta t} = \frac{f(t + \Delta t) - f(t)}{\Delta t},$$

we see that the average velocity is the difference quotient of position as a function of time.

Example 4   An object released from rest falls

$$s = 16t^2$$

feet in t seconds. Here the position function is $f(t) = 16t^2$, measured downward from the point of release. The average velocity of the object during the first second of fall is

$$\frac{\Delta s}{\Delta t} = \frac{f(1) - f(0)}{1} = \frac{16 \cdot 1^2 - 16 \cdot 0^2}{1} = 16 \text{ feet per second.}$$

During the *second* second it is

$$\frac{\Delta s}{\Delta t} = \frac{f(2) - f(1)}{1} = \frac{16 \cdot 2^2 - 16 \cdot 1^2}{1} = \frac{64 - 16}{1} = 48 \text{ ft/sec.}$$

Its average velocity during the *first two seconds* is

$$\frac{\Delta s}{\Delta t} = \frac{16 \cdot 2^2 - 16 \cdot 0^2}{2} = \frac{64}{2} = 32 \text{ ft/sec.} \qquad \square$$

We turn next to the notion of *instantaneous velocity* v. It is the number that measures how fast something is moving at a given instant. It is the varying reading on a perfect speedometer. In order to obtain a formula for v, let v_0 be its value at time t_0 and let Δt be a very small change in t. Then over the time interval between t_0 and $t_0 + \Delta t$ the velocity v varies only slightly from v_0, so the average velocity over this time interval is very close to v_0. That is,

$$\frac{\Delta s}{\Delta t} \approx v_0,$$

where $\approx$ is read "is approximately equal to." For smaller Δt, the difference between v_0 and the average value is smaller still. Therefore,

$$v_0 = \lim_{\Delta t \to 0} \frac{\Delta s}{\Delta t}.$$

So v_0 is given by the derivative $ds/dt|_{t_0}$. Dropping subscripts, we have

$$\boxed{\text{Instantaneous velocity} = v = \frac{ds}{dt}.}$$

We thus have a second interpretation of the derivative. For Newton, who was struggling to understand the motions of the planets and other heavenly bodies, this was undoubtedly the major interpretation of the derivative.

Example 5 If $s = 3t - 10$, where s is measured in feet and t in seconds, then $v = ds/dt = 3$ feet per second. The particle is moving in *uniform motion* with constant velocity $v = 3$ ft/sec.   □

Example 6 If the motion of the particle is described by $s = t^3 - t$, in the same units as above, then the velocity v is

$$v = \frac{ds}{dt} = (3t^2 - 1) \text{ feet per second.}$$

Thus, at time $t = 2$, its velocity is $v = 3 \cdot 2^2 - 1 = 11$ ft/sec, and at time $t = 0$ it is $v = 3 \cdot 0^2 - 1 = -1$ ft/sec. Note that the velocity is negative at $t = 0$. Thus $\Delta s/\Delta t < 0$ for small Δt and the particle is moving *backward* at time $t = 0$. If the particle was moving and the clock running before the zero setting on the clock, we see that at time $t = -1$ the velocity was $v = 3(-1)^2 - 1 = 2$.   □

The *speed* of a particle is the magnitude of its velocity:

$$\boxed{\text{Speed} = |v|.}$$

Thus, in Example 3, the particle velocity at $t = 0$ was -1 ft/sec and its speed was 1 ft/sec.

Although the particle in the above example is moving backward and forward on the line, the *graph* of the motion is the graph of the equation $s = t^3 - t$ in the ts-plane.

The particle is *not* moving along this curve, but along the s-coordinate line, pictured in Fig. 1 as the vertical axis. Its position at time t is obtained from the graph by the "up-and-over" procedure indicated in the figure. (The graph is called the "world line" of the particle in space-time, space being one-dimensional here.)

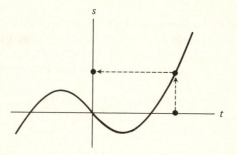

Figure 1

Acceleration We say that a particle is *accelerating* if its velocity is increasing. Its *average acceleration* over an interval of time is given by the formula

$$\text{Average acceleration} = \frac{\text{Change in velocity}}{\text{Elapsed time}} = \frac{\Delta v}{\Delta t}.$$

We also say that a particle is *decelerating* if its velocity is decreasing. In this case, the above formula gives a *negative* average acceleration. Following this clue, we equate *deceleration* to *negative acceleration* and forget about deceleration.

Instantaneous acceleration a is defined to be the limit of the average acceleration as $\Delta t \to 0$. The acceleration of the particle is thus the derivative of its velocity:

$$\text{Acceleration} = a = \lim_{\Delta t \to 0} \frac{\Delta v}{\Delta t} = \frac{dv}{dt}.$$

In the above example we had $v = 3t^2 - 1$ and so the acceleration is

$$a = \frac{dv}{dt} = 6t.$$

The particle is moving with increasing acceleration.

Since the velocity is so many feet per second, the acceleration, which is the *change in velocity per unit of time*, must be so many (feet per second) per second. We designate this ft/sec². Thus in the problem above the acceleration was $6t$ ft/sec².

Note that acceleration is the *second* derivative of the position s:

$$a = \frac{dv}{dt} = \frac{d}{dt}\left(\frac{ds}{dt}\right) = \frac{d^2s}{dt^2}.$$

Example 7   An object is thrown vertically upward, and its position above the ground (after t seconds have elapsed) is given by $s = 50t - 16t^2$ feet. Find its initial velocity, the time when it reaches its highest point, and its velocity when it strikes the ground, as well as its acceleration.

Solution   The velocity is

$$v = \frac{ds}{dt} = 50 - 32t \text{ ft/sec.}$$

The initial velocity v_0 is the velocity at time $t_0 = 0$. Thus

$$v_0 = 50 \text{ ft/sec.}$$

The highest point will be reached just when the upward velocity has fallen away to zero. Setting $v = 0$, we have

$$0 = 50 - 32t,$$

$$t = \frac{50}{32} = \frac{25}{16} \text{ sec.}$$

The particle returns to ground at the moment when s becomes 0 again, i.e., when

$$0 = s = 50t - 16t^2.$$

This gives $t = 50/16$ seconds, and then

$$v = 50 - 32t = 50 - 32 \cdot \frac{50}{16} = -50 \text{ ft/sec.}$$

The acceleration of the object is

$$a = \frac{dv}{dt} = \frac{d}{dt}(50 - 32t) = -32 \text{ ft/sec}^2.$$

The acceleration is constant and it is *negative*. The velocity is *decreasing* at a constant rate.   □

PROBLEMS FOR SECTION 3

Recompute the derivatives of the following functions in increment notation:

1. $f(x) = x^2$ **2.** $f(x) = 1/x$ **3.** $f(x) = \sqrt{x}$

4. $f(x) = x^4$ **5.** $f(x) = x + x^2$ **6.** $f(x) = 1/(1 + x)$

7. $f(x) = 1/(1 + x^2)$

8. If $y = x^2$ and x is positive, then y can be interpreted as the area of a square of side x. Then Δy is the *increase in area* when the side is increased by Δx. Draw a figure and interpret the two terms of Δy as areas.

9. If $y = x^3$ and x is positive, then y can be interpreted as the volume of a cube of side x. Then Δy is the *increase in volume* when the side is increased by Δx. Draw a figure and interpret the three terms of Δy as volumes.

10. Write out the proof of Theorem 2 in Section 2 in increment notation. (Note first that if $u = g(x)$, $v = h(x)$, and $y = u + v$, then $\Delta y = \Delta u + \Delta v$.)

11. A particle moves along the x-axis and its position as a function of time is given by $x = t^3 - 2t^2 + t + 1$. Find its velocity. At what moments is it standing still? When is it moving forward and when backward?

12. A particle is moving along a horizontal line on which the positive direction is to the right. The equation of motion of the particle is $s(t) = t^3 - 9t^2 + 24t + 1$.

a) What is the velocity of the particle at $t = 1$?

b) Is the particle moving to the right or left at $t = 1$?

c) What is the acceleration of the particle at $t = 1$?

d) Is the speed of the particle increasing or decreasing at $t = 1$? Explain your reasoning.

e) At what times, if any, is the velocity of the particle zero?

13. A ball thrown directly upward with a speed of 96 ft/sec moves according to the law

$$y = 96t - 16t^2,$$

where y is the height in feet above the ground, and t is the time in seconds after it is thrown. How high does the ball go?

14. The movement of an airplane from the time it releases its brakes until takeoff is governed by the equation $s = 2t^2$, where s is the distance from the starting point in feet and t is the time in seconds since brake release.

a) If takeoff velocity is 120 mph (176 fps), determine the time elapsed from brake release until takeoff.

b) What is the distance covered during the time?

15. A coin is thrown straight up from the top of a building which is 300 feet tall. After t seconds, its position is described by $s = -16t^2 + 24t + 300$, where s is its height above the ground in feet. When does the coin begin to descend? What is its velocity when it is 305 feet above the ground?

16. In the first t seconds after a space shot is launched, the rocket reaches a height of $40t^2$ feet above the earth.

a) What is the rocket's velocity after 5 seconds?

b) If the speed of sound is 1050 ft/sec, calculate the height at which the space craft attains supersonic speed.

17. Given that the motion of a particle is expressed by the equation

$$s(t) = t^3 - 6t^2 + 2,$$

a) determine the velocity of the particle at $t = 2.3$;

b) determine the acceleration of the particle at 2.3;

c) determine the *average* velocity of the particle in the time interval $t = 1.3$ to $t = 3.3$.

19. A ball is thrown down from the top of a 200 ft tower with an equation of motion

$$y = -16t^2 - 40t + 200.$$

What is the *average* velocity of the ball in its trip to the ground? At what time t does the velocity of the ball equal its average velocity?

21. The distance, rate, and time problems of elementary algebra refer to motions of constant velocity. The formula $d = rt$ would then be restated in our language as

$$\Delta s = r \, \Delta t,$$

where r is the constant velocity. Show that the most general description of such a motion is given by

$$s = rt + c,$$

where c is any constant.

18. The height above the ground of a bullet shot vertically upward with an initial velocity of 320 ft/sec is given by $s(t) \times 320t - 16t^2$, where $s(t)$ is the height in ft and t is time in sec.

a) Determine the velocity of the bullet 4 sec after it is fired.

b) Determine the time required for the bullet to reach its maximum height, and the maximum height attained.

20. A particle whose velocity is zero during an interval of time is standing still during that time interval. Restate this obvious fact as a theorem about differentiable functions. (The analytic proof will be given in the first section of Chapter 5.)

4
RATE OF CHANGE; MARGINAL QUANTITIES IN ECONOMICS

So far we have interpreted the derivative as the *slope* of the tangent line to a graph and as the *velocity* of a moving particle. Now we claim that we can *always* interpret $f'(x_0)$ as the *rate of change of y with respect to x as x passes through the value x_0*, or *the rate of change of the function f at the point x_0*. The discussion below is intended to make this interpretation seem plausible.

Consider first a use of the word *rate* that is probably familiar. If f is linear, say

$$y = f(x) = 3x - 5,$$

then

$$\Delta y = f(x_0 + \Delta x) - f(x_0)$$
$$= [3(x_0 + \Delta x) - 5] - [3x_0 - 5]$$
$$= 3\,\Delta x.$$

Thus the change in y is always exactly 3 times the change in x, no matter what values we use for x_0 and Δx. We say that y changes exactly three times *as fast as* x changes, even though we are not necessarily talking about *motion*. All we mean is that if x is changed, then y will change, and the change in y will always be three units per unit change in x. In this situation it is customary to say that 3 is the constant *rate* at which y changes as compared to x, or the *constant rate of change of y with respect to x*.

For the general linear function

$$y = ax + b$$

we have, similarly,

$$\Delta y = a\,\Delta x$$

and a is the constant rate of change of y with respect to x. Knowing the rate a we can obtain Δy from Δx by the above equation. Knowing Δx and Δy we obtain the rate a from

$$\frac{\Delta y}{\Delta x} = a,$$

(provided $\Delta x \neq 0$).

When $f(x)$ is not a linear function, then the increment ratio

$$\frac{\Delta y}{\Delta x} = \frac{f(x_0 + \Delta x) - f(x_0)}{\Delta x}$$

depends on x_0 and Δx. If its value is r (for given x_0 and Δx), then the equation

$$\Delta y = r\,\Delta x$$

says that the total change in y is exactly r times the total change in x. We therefore say that y has been changing *on the average r times as fast* as x over the interval from x_0 to $x_0 + \Delta x$, and we call r the average rate of change of y with respect to x.

DEFINITION   The *average rate of change of y with respect to x*, over the interval between x_0 and $x_1 = x_0 + \Delta x$, is

$$\boxed{\;Average\ rate\ of\ change\ = \frac{\Delta y}{\Delta x} = \frac{f(x_0 + \Delta x) - f(x_0)}{\Delta x}.\;}$$

The derivative

$$\frac{dy}{dx} = \lim_{\Delta x \to 0} \frac{\Delta y}{\Delta x}$$

is now viewed as the limit of the average rate of change as $\Delta x \to 0$. We therefore interpret

$$\left. \frac{dy}{dx} \right|_{x=x_0} = f'(x_0)$$

as the *true rate of change of y with respect to x at the point x_0, or the true rate of change as x passes through the value $x = x_0$.*

Average velocity and true (or instantaneous) velocity are the special case when y is distance and x is time. We saw in Section 3 that intuitions about velocity lead to the formula $v = ds/dt$ for the true velocity; and this supports the general interpretation of $dy/dx = f'(x)$ as the true rate of change.

The rate of change of any quantity with respect to time is called its *time rate of change.* A time rate of change answers the question of how *fast* something is changing.

Example 1   A stone is dropped into a pond and it is observed that the spreading circular ripple has the radius $r = 2t$ ft after t seconds. How fast is the area increasing at the end of 5 seconds?

Solution   We have the formula

$$A = \pi r^2 = \pi(2t)^2 = 4\pi t^2$$

for the area at the end of t seconds. Its rate of change at time t is therefore

$$\frac{dA}{dt} = 8\pi t,$$

and at $t = 5$ the area is increasing at the rate of

$$\frac{dA}{dt}\bigg|_{t=5} = 40\pi,$$

or about 126 sq. ft/sec. □

We now give some further examples of the general rate-of-change interpretation, starting with some words on rate of change as a *coefficient*.

If a differentiable function f has a constant rate of change over an interval I, then f must be linear over I,

$$f(x) = mx + b,$$

and its rate of change $f'(x)$ is the coefficient m. (The reason is that if f' has the constant value m, then $f(x) - mx$ has derivative zero. And any function with derivative everywhere zero must be a constant. (See Problem 20 in Section 3. The proof will be given later.)

Because of the above fact, when a rate of change is constant or nearly constant in the applications, it is often called a *coefficient*.

Example 2 We consider a metal rod that would be one unit long at zero degrees (centigrade). When it is heated it expands. The rate of change of its length l with respect to the temperature T is called the *coefficient of thermal expansion* for the given metal. This rate of change is the derivative dl/dT, and calling it a *coefficient* implies that it is very nearly a constant c, so that l is given approximately by

$$l = cT + k$$

(where $k = 1$ since $l = 1$ when $T = 0$). Actually, the coefficient of thermal expansion is variable to the extent that its variability is mentioned in tables and handbooks. The notion of a variable coefficient seems paradoxical, but if we keep in mind that what we really are talking about is the derivative dl/dT, the fact that it turns out to be a nonconstant function of T ceases to be troublesome. □

Example 3 We experiment with an electrical circuit and plot the voltage E required to produce a current I. For simple circuits we find that E is proportional to I, so that

$$E = RI,$$

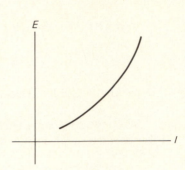

Figure 1

where the constant of proportionality R is called the *resistance* of the circuit. Of course the rate of change dE/dI then has the constant value R, and R is a true *coefficient* in the above sense.

However, if E and I are the plate voltage and plate current of a vacuum tube (for a given grid voltage), then the graph of E as a function of I looks like Fig. 1. It is called a *characteristic* of the tube.

Now we must define the resistance R as the rate of change dE/dI, and we note that it increases as the current I increases. □

Example 4 In economics the word *marginal* signifies a rate of change. For example, if it costs C dollars to produce x tons of coal, then the *marginal* cost is the increase in total cost C *per extra ton produced*, i.e., the rate of change of C with respect to x, dC/dx. If the relationship between C and x has the simple form

$$C = I + kx,$$

then the marginal cost is the coefficient k. Here I is a fixed initial cost, for machinery, etc., and the marginal cost k is the constant running cost per ton, for labor, expended materials like fuel, etc.

Normally the relationship between C and x is more complicated than this. We impose one more condition on our problem, namely, that C is the cost to produce x tons of coal *in a given fixed time interval*, say one week. There will generally be a fixed cost of I dollars per week. But the *extra* running cost per ton, instead of being constant, will probably decrease as production increases, because it is possible to achieve greater internal efficiency with greater volume. The trend will continue only up to a certain point, after which the cost per ton will begin to increase again, because the larger demand will make it necessary to utilize older machinery, pay overtime wages, etc. The marginal cost rises to very high values if the producer

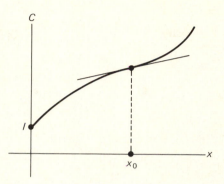

Figure 2

strains the capacity in an effort to put out a very large weekly tonnage. Thus a typical cost curve is shown in Fig. 2.

For example, $C = x^3 - 3x^2 + 4x + 1$ has a graph like this. (We shall not try to invent a precise cost situation that would lead to this formula.) The marginal cost is

$$c_M = \frac{dC}{dx} = 3x^2 - 6x + 4.$$

It is variable, and has a minimum value at the point labeled x_0. We shall see later that the minimum value of the marginal cost occurs where *its* derivative dc_M/dx is zero. Since

$$\frac{dc_M}{dx} = 6x - 6 = 6(x - 1),$$

it follows that c_M has its minimum value at $x_0 = 1$, and that its minimum value is

$$\text{Min } c_M = 3(1)^2 - 6 \cdot 1 + 4 = 1.$$

This is the point at which the producer is operating most efficiently, but, because of other factors, it may *not* be the point at which the producer will choose to operate. $\square$

Example 5   We continue the above example. Our producer must also consider how many tons of coal can be sold at a given price. Presumably, the lower the price the more coal will sell (per week), so the *demand curve* is the graph of a decreasing function (Fig. 3).

In order to compare this with the cost function, we plot this relationship with axes interchanged, and so consider p as a function of x.

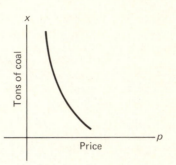
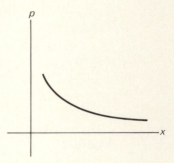

Figure 3

This function is the producer's *demand function*. Its value $p = d(x)$ is the price the producer must charge in order to sell exactly x tons (per week).

Then $R = px = d(x)x$ is the *total weekly revenue* from selling x tons. Figure 4 shows a typical revenue curve. The rate of change of R with respect to x, dR/dx, is called the *marginal revenue*.

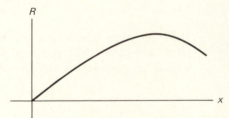

Figure 4

Our scales are all wrong. In order to plot a curve that will react to moderate changes in x, the unit of quantity would probably be a thousand tons, and the unit of money might be $10,000. The revenue R would then be the number of $10,000 units obtained from selling x thousands of tons. (Continued in Chapter 5.)   □

PROBLEMS FOR SECTION 4

1. What is the rate of change of the volume of a sphere with respect to its radius?

2. Find the rate of change of the volume of a sphere with respect to its radius when the radius is 5 inches. (The answer should be expressed in cubic inches per inch.) Find the rate of change when $r = 10$ inches.

3. a) What is the rate of change of the area of a circle with respect to its radius?

b) What is the value of this rate of change when the radius is 5 inches? (The answer should be expressed in square inches per inch.)

c) What is the rate when $r = 10$ cm?

5. A growing tree increases its diameter at the rate of $\frac{1}{4}$ inch per year, and its height at the rate of 1 foot per year. Assuming that the shape of the tree is approximately conical, at what rate is new wood being added when the tree is 10 years old? 50 years old?

7. If l_0 is the length of a piece of platinum wire at 0° centigrade, then its length l_t at t°C is given by

$$l_t = l_0(1 + \alpha t + \beta t^2)$$

where $\alpha = 0.0868 \times 10^{-4}$ and $\beta = 0.013 \times 10^{-7}$. Discuss the sense in which platinum has a coefficient of thermal expansion.

4. What is the rate of change of the volume of a cube with respect to its edge length, the unit of length being the centimeter? By how much must the volume be increased in order to double this rate of change?

6. A growing cubical crystal increases its edge length at the rate of one millimeter per day. How fast is its surface area increasing at the end of the first week? How fast is its volume increasing then?

8. A manufacturing firm finds that it costs

$$10^{-4}(x^3 - 1500x^2) + 150x + 5000$$

dollars per month to produce x items per month.

a) What is the marginal cost c_M if the firm is producing 250 items per month? (The answer should be in dollars per item.) What is the marginal cost if they produce 500 items per month? 1000 items per month?

b) Now compute the *average* cost per item when monthly production is 250 items; 500 items; 1000 items.

9. The marginal cost of production will generally have a graph like the one at the left below. That is, the marginal cost will normally decrease with increasing production until it reaches a minimum value, and then will increase. Show by an intuitive argument that the total cost function $C(x)$ must necessarily look like the graph at the right.

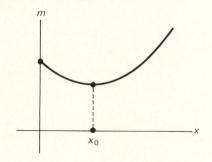

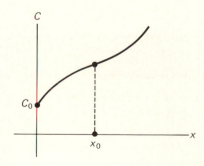

10. A marginal cost graph like the one shown in the preceding problem is approximately a quadratic graph, i.e., approximately the graph of

$$m(x) = ax^2 + bx + c$$

for suitable values of the coefficients a, b, and c.

a) Find this quadratic function if $m(0) = 10$ and if m has the minimum value $m(1000) = 8$ at $x_0 = 1000$. [*Hint*: Cast the general expression for $m(x)$ above into a new form by completing the square.]

b) Find the corresponding cost function $C(x)$, assuming that there is a fixed initial cost of 500 dollars even if no items are produced. (The problem is to find a cubic function $C(x)$ whose derivative is a given quadratic function $m(x)$, and such that $C(0) = 500$.)

11. A manufacturer's average cost per item is $f(x)$ dollars. Show that the marginal cost is

$$f(x) + xf'(x)$$

dollars per item. (This is really a problem for Section 2. You are asked to prove that if $C(x) = xf(x)$, then $C'(x) = f(x) + xf'(x)$.)

In the remaining problems, assume that the power formula holds for all exponents k.

12. Find the rate of change for the surface area of a sphere with respect to its volume.

13. Find the rate of change for the surface area of a cube with respect to its volume.

14. A balloon is being filled with air at the rate of 10 cubic feet per minute. How fast is its diameter expanding when its volume is 8 cubic feet? 125 cubic feet?

CHAPTER 4
THE TECHNIQUE OF DIFFERENTIATION

Newton and Leibniz showed that derivatives can be computed systematically. Starting with the derivative formulas for a few basic functions, and with the derivative rules for four or five ways of combining functions, we can compute in a systematic, and almost automatic, way the derivative of practically any function that is likely to come up.

So far we have established the formula

$$\frac{d}{dx} x^k = kx^{k-1},$$

for $k =$ a positive integer, and the rules

$$(f + g)' = f' + g', \qquad (cf)' = cf'.$$

This chapter will complete the machinery. The remaining rules are the product and quotient rules

$$(fg)' = f \cdot g' + g \cdot f', \qquad \left(\frac{f}{g}\right)' = \frac{g \cdot f' - f \cdot g'}{g^2},$$

and the chain rule and the inverse-function rule, which we shall state later.

The list of basic functions will be extended by adding $\sin x$ and $\cos x$, and their derivative formulas

$$\frac{d}{dx}\sin x = \cos x, \qquad \frac{d}{dx}\cos x = -\sin x.$$

Finally, there is the problem of differentials. Although modern mathematics develops differentials conceptually, it seems better in a first course to take a simpler approach. In Section 5 differentials are treated in a purely formal, or symbolic, manner.

1 THE PRODUCT AND QUOTIENT RULES

One might guess offhand that the derivative of the product of two functions ought to be the product of their derivatives. But trying this out on even the simplest functions shows that it doesn't work. For instance,

$$x^2 = x \cdot x, \qquad \frac{d(x^2)}{dx} = 2x, \qquad \text{and} \qquad \frac{dx}{dx} = 1,$$

but $2x$ is not $1 \cdot 1$. The quotient rule can't be guessed either.

These rules were stated above in function notation. Here are their Leibniz forms.

THEOREM 1   *If u and v are differentiable functions of x, then so are uv and u/v (where defined), and*

$$\frac{d}{dx}(uv) = u\frac{dv}{dx} + v\frac{du}{dx};$$

$$\frac{d}{dx}\left(\frac{u}{v}\right) = \frac{v\dfrac{du}{dx} - u\dfrac{dv}{dx}}{v^2}.$$

It may be easier to remember "word" versions, such as the following:

The derivative of a product is the first *times the derivative of the* second *plus the* second *times the derivative of the* first.

The derivative of a quotient is the denominator *times the derivative of the* numerator, *minus the* numerator *times the derivative of the* denominator, *all divided by the* denominator *squared.*

The proof of Theorem 1 will be given later in the section.

These rules enable us to compute derivatives of more complicated functions than we could handle before. But not right away, because each rule needs the other rules as sources for interesting examples. At the moment we have no functions at all that need the product rule! So we shall briefly illustrate the product rule and then use it in a bootstrap manner to get more interesting functions to practice the rule on.

Example 1 Find the derivative of $f(x) = (x - 1)(x + 3)$.

Solution By the product rule,

$$\frac{d}{dx}(x - 1)(x + 3) = (x - 1)\frac{d}{dx}(x + 3) + (x + 3)\frac{d}{dx}(x - 1)$$

$$= (x - 1) \cdot 1 + (x + 3) \cdot 1 = 2x + 2.$$

If we multiply the product out, we can use the polynomial rules instead:

$$\frac{d}{dx}(x - 1)(x + 3) = \frac{d}{dx}(x^2 + 2x - 3) = 2x + 2. \qquad \square$$

Example 2 Differentiate $(x^7 - 3x^2)(2x^5 + 1)$.

Solution

$$\frac{d}{dx}(x^7 - 3x^2)(2x^5 + 1)$$

$$= (x^7 - 3x^2)\frac{d}{dx}(2x^5 + 1) + (2x^5 + 1)\frac{d}{dx}(x^7 - 3x^2)$$

$$= (x^7 - 3x^2) \cdot 10x^4 + (2x^5 + 1)(7x^6 - 6x).$$

This is already a correct answer, and it might be good enough. But normally we would simplify as far as possible. Here we would multiply out and collect terms, to get

$$24x^{11} - 35x^6 - 6x.$$

Again the product rule is not absolutely necessary, because we could multiply to begin with and then differentiate a polynomial.

$\square$

The product rule can be rewritten

$$(fg)' = f'g + fg' = fg\left(\frac{f'}{f} + \frac{g'}{g}\right).$$

These forms easily extend to three or more factors. Thus

$$(fgh)' = f'(gh) + f(gh)' = f'gh + fg'h + fgh' = fgh\left(\frac{f'}{f} + \frac{g'}{g} + \frac{h'}{h}\right).$$

Note the symmetrical appearance, each term being the product fgh with one factor differentiated. In general,

$$(f_1 f_2 \cdots f_k)' = f_1' f_2 \cdots f_k + f_1 f_2' \cdots f_k + \cdots + f_1 f_2 \cdots f_{k-1} f_k'$$
$$= f_1 f_2 \cdots f_k\left[\frac{f_1'}{f_1} + \frac{f_2'}{f_2} + \cdots + \frac{f_k'}{f_k}\right].$$

The second of these two general forms is easier to use, but it excludes zero as a possible value of the functions.

The General Power Rule   If we take the k functions $f_1, f_2, \ldots, f_k$ all to be the same function f,

$$f_1 = f_2 = \cdots = f_k = f,$$

then we have, in Leibniz notation,

THEOREM 2   *If $u = f(x)$ is any differentiable function of x and k is any positive integer, then u^k is a differentiable function of x and*

$$\frac{d}{dx}\, u^k = ku^{k-1}\frac{du}{dx}.$$

This is called the *general power rule*. We shall see later on (in Chapter 7, Section 4) that it remains true when k is *any real number*.

Example 3   Find the derivative of $(1 + x^2)^{20}$.

Solution   By the general power rule,

$$\frac{d}{dx}\,(1 + x^2)^{20} = 20(1 + x^2)^{19}\frac{d}{dx}\,(1 + x^2)$$
$$= 20(1 + x^2)^{19} \cdot 2x = 40x(1 + x^2)^{19}.$$

Note that this derivative *could* be calculated by first expanding $(1 + x^2)^{20}$ by the binomial theorem, getting a polynomial of degree 40, and then using the polynomial rules. As a practical matter this is out of the question. □

Example 4 If $u = x$, then $du/dx = 1$, and Theorem 2 reduces to

$$\frac{d}{dx} x^k = kx^{k-1}.$$

This is a new proof of Theorem 1 in Chapter 3. □

Example 5 Find dy/dx when $y = (x - 3)^9(x^3 + 2)^{15}$.

Solution Here again we could expand the binomials and so express y as a polynomial of degree 54. But the product rule and general power rule let us proceed directly:

$$\frac{dy}{dx} = (x - 3)^9 \frac{d}{dx} (x^3 + 2)^{15} + (x^3 + 2)^{15} \frac{d}{dx} (x - 3)^9$$

$$= (x - 3)^9 \cdot 15(x^3 + 2)^{14} \cdot 3x^2 + (x^3 + 2)^{15} \cdot 9(x - 3)^8$$

$$= 9(x - 3)^8(x^3 + 2)^{14}[6x^3 - 15x^2 + 2].$$

If we know how to write down the derivatives of the two factors we would not generally bother with the first line above but would start by writing out the second line. □

It is common to write y' for dy/dx. This brief notation is less precise, since it does not display the independent variable, but it is frequently used when there is no confusion about the independent variable. So if $y = x^5$ then $y' = 5x^4$. And $y'' = 20x^3$ and $y''' = 60x^2$. The product rule becomes,

$$\text{If } y = uv, \text{ then } y' = uv' + vu' = y\left(\frac{u'}{u} + \frac{v'}{v}\right).$$

Example 6 Redo Example 2, using the product rule in this last form.

Solution Given

$$y = \underbrace{(x^7 - 3x^2)}_{u}\underbrace{(2x^5 + 1)}_{v},$$

then

$$y' = y\left(\frac{u'}{u} + \frac{v'}{v}\right)$$

$$= y\left(\frac{7x^6 - 6x}{x^7 - 3x^2} + \frac{10x^4}{2x^5 + 1}\right).$$

When the terms in the parentheses are combined, we get the same answer as before.   □

The next examples illustrate the quotient rule.

Example 7

$$\frac{d}{dx}\frac{(x^2 - 2x)}{x - 1} = \frac{(x - 1)\frac{d}{dx}(x^2 - 2x) - (x^2 - 2x)\frac{d}{dx}(x - 1)}{(x - 1)^2}$$

$$= \frac{(x - 1)(2x - 2) - (x^2 - 2x)\cdot 1}{(x - 1)^2}$$

$$= \frac{x^2 - 2x + 2}{(x - 1)^2}.$$ □

Example 8   Calculate

$$\frac{d}{dx}\left(\frac{x}{1 + x^2}\right)^{10}$$

in two ways—first as a power of a quotient, then as a quotient of powers.

Solution

a) $\dfrac{d}{dx}\left(\dfrac{x}{1 + x^2}\right)^{10} = 10\left(\dfrac{x}{1 + x^2}\right)^9 \dfrac{(1 + x^2)1 - x\cdot 2x}{(1 + x^2)^2}$

$$= \frac{10x^9(1 - x^2)}{(1 + x^2)^{11}}.$$

b) $\dfrac{d}{dx}\dfrac{x^{10}}{(1 + x^2)^{10}} = \dfrac{(1 + x^2)^{10}\cdot 10x^9 - x^{10}\cdot 10(1 + x^2)^9\cdot 2x}{(1 + x^2)^{20}}$

$$= \frac{10[(1 + x^2) - 2x^2]x^9}{(1 + x^2)^{11}} = \frac{10x^9(1 - x^2)}{(1 + x^2)^{11}}.$$ □

Example 9   When $u = 1$, the quotient rule can be written

$$\frac{d}{dx} v^{-1} = (-1)v^{-2} \frac{dv}{dx} .$$

Note that this is the general power rule for $k = -1$.   □

We turn now to the proofs of the product and quotient rules. These proofs could be given in the h-notation of Chapter 3, but we shall use increment notation. Recall from Chapter 3 that if $y = f(x)$, then Δy depends on Δx (and x) according to the formula

$$\Delta y = f(x + \Delta x) - f(x).$$

If we add the equation $y = f(x)$, we obtain the equivalent equation

$$y + \Delta y = f(x + \Delta x).$$

This just says that $y + \Delta y$ is the new value of y when $x + \Delta x$ is the new value of x.

Similarly, we can express the new value of any variable u as its old value plus its increment, i.e., as $u + \Delta u$.

Example 10   If the variables u and v are given increments Δu and Δv, what is the increment in the product $y = uv$?

Solution   The new values of u and v are $u + \Delta u$ and $v + \Delta v$, so the new value $y + \Delta y$ is given by

$$y + \Delta y = (u + \Delta u)(v + \Delta v).$$

Then

$$\Delta y = (y + \Delta y) - y$$
$$= (u + \Delta u)(v + \Delta v) - uv$$
$$= u \, \Delta v + v \, \Delta u + \Delta u \, \Delta v.$$

This expression for $\Delta(uv)$ can be pictured when everything is positive by interpreting uv as the area of a rectangle with sides u and v (Fig. 1). Note that the added area is shown in three pieces, corresponding to the three terms in the above sum.

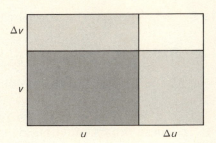

Figure 1

Proof of Theorem 1 We are assuming that the functions $u = g(x)$ and $v = h(x)$ are differentiable. If we give x an increment Δx, getting the new values

$$u + \Delta u = g(x + \Delta x) \qquad \text{and} \qquad v + \Delta v = h(x + \Delta x),$$

then our hypothesis is that the limits

$$\lim_{\Delta x \to 0} \frac{\Delta u}{\Delta x} = \frac{du}{dx}, \qquad \lim_{\Delta x \to 0} \frac{\Delta v}{\Delta x} = \frac{dv}{dx}$$

both exist. We computed above that the product $y = uv$ has the increment

$$\Delta y = u \, \Delta v + v \, \Delta u + \Delta u \, \Delta v.$$

Thus,

$$\frac{\Delta y}{\Delta x} = u \frac{\Delta v}{\Delta x} + v \frac{\Delta u}{\Delta x} + \Delta u \frac{\Delta v}{\Delta x}.$$

As $\Delta x \to 0$, the right side of this equation has the limit

$$u \frac{dv}{dx} + v \frac{du}{dx} + 0 \frac{dv}{dx}.$$

That is, the limit $dy/dx = \lim_{\Delta x \to 0} \Delta y/\Delta x$ exists, and

$$\frac{dy}{dx} = u \frac{dv}{dx} + v \frac{du}{dx}.$$

Similarly, for the quotient $y = u/v$ we have

$$\Delta y = (y + \Delta y) - y$$

$$= \frac{u + \Delta u}{v + \Delta v} - \frac{u}{v}$$

$$= \frac{v \, \Delta u - u \, \Delta v}{v(v + \Delta v)}.$$

Therefore,

$$\frac{\Delta y}{\Delta x} = \frac{v\dfrac{\Delta u}{\Delta x} - u\dfrac{\Delta v}{\Delta x}}{v(v + \Delta v)},$$

and taking the limit as $\Delta x \to 0$ we get the quotient rule

$$\frac{dy}{dx} = \frac{v\dfrac{du}{dx} - u\dfrac{dv}{dx}}{v^2}. \qquad \blacksquare$$

Note in each case how the formula for dy/dx is anticipated by the form of the expression for $\Delta y/\Delta x$. This is why we used increments here.

Also note that the quotient rule can be derived by combining the product rule and the power rule with exponent $a = -1$:

$$\frac{d}{dx}\left(\frac{u}{v}\right) = \frac{d}{dx}\, u \cdot v^{-1} = \cdots$$

(computation to be finished as an exercise). The quotient rule is thus not really a basic rule, but it is so useful that it should be learned anyway.

PROBLEMS FOR SECTION 1

Differentiate the following functions.

1. $y = (x - 1)^3(x + 2)^4$

2. $y = x^m \cdot x^n$
(Do it two ways and compare the answers.)

3. $s = 3t^5(t^2 + 2t)$
(Two ways and compare)

4. $y = x^{21}(1 + x)^{21}$
(Two ways, using $x(1 + x) = x + x^2$)

5. $y = (x^3 + 6x^2 - 2x + 1)(x^2 + 3x - 5)$

6. $h(x) = (x + 1)^2(x^2 + 1)^{-3}$

7. $u = v^{-4}(1 + 2v)^4$ (Two ways)

8. $s = (t + 1)^2(t + 1)^{-2}$ (Two ways)

9. $y = (x^2 - 1)(x + 1)^{-1}$ (Two ways)

10. $y = (2x - 3)(3x + 1)(x + 2)$

11. $y = x^3(2x - 1)^4(x + 2)^2$

12. $s = (2t^2 + t^{-2})(t^2 - 3)(4t + 1)$

13. $y = \dfrac{x}{1 + 4x}$

14. $s = \dfrac{t}{1 + t^2}$

15. $y = \dfrac{x^2 - 1}{x^2 + 2}$

16. $y = \dfrac{x^3}{1 + 3x^2}$

17. $s = \dfrac{2t}{(1 + t^2)^2}$

18. $s = \dfrac{3t - 4}{2t + 5}$

19. $u = \dfrac{2v + 1}{(2 - 3v)^2}$

20. $y = \dfrac{f(x)}{x}$

21. $y = \dfrac{2 + x^2}{1 + x + x^2}$

22. $y = \dfrac{4x + 1}{x^2 + 2x + 5}$

23. a) $y = \dfrac{1}{1 - x^2}$ b) $y = \dfrac{x^2}{1 - x^2}$ (Compare the answers and explain.)

24. $y = \dfrac{6x^2 - 2}{(x^2 + 1)^3}$

25. $y = \left(\dfrac{1 - x}{1 + x}\right)^2$

26. $y = \left(\dfrac{1 - x}{1 + x}\right)^n$

27. $y = \dfrac{1 - f(x)}{1 + f(x)}$

28. $y = \left(\dfrac{f(x)}{1 + f(x)}\right)^n$

29. $y = \dfrac{f(x) - g(x)}{f(x) + g(x)}$

30. Show that if $y = u/v$, then

$$\frac{y'}{y} = \frac{u'}{u} - \frac{v'}{v}$$

(where $y' = dy/dx$, etc.). Show, also, that if $y = u^k$, then

$$\frac{y'}{y} = k\frac{u'}{u}.$$

31. Use Problem 30 and the general product formula from the text to show that if

$$y = \frac{u_1 u_2 \cdots u_m}{v_1 v_2 \cdots v_n}$$

then

$$\frac{y'}{y} = \frac{u_1'}{u_1} + \frac{u_2'}{u_2} + \cdots + \frac{u_m'}{u_m} - \frac{v_1'}{v_1} - \frac{v_2'}{v_2} - \cdots - \frac{v_n'}{v_n}.$$

Problems can be worked using Problems 30 and 31 if desired.

32. $y = \dfrac{1 + x}{1 - x}(3x + 2)$

33. $y = \dfrac{1 - 2x}{(x + 3)(3x + 1)}$

34. $s = \dfrac{t^3 + 2}{(3t + 1)(2t^2 + 1)}$

35. $y = \dfrac{x^2(x^2 + 2)}{(x^2 - 1)(x^2 + 3)}$

36. $y = \dfrac{(x - 1)(x + 2)}{(x - 2)(x + 1)}$

37. $u = \dfrac{(v^2 + 1)\left(1 + \dfrac{1}{v}\right)^{10}}{1 - 3v}$

38. $y = \left(\dfrac{x^2 - x - 1}{x^2 + 1}\right)^3$

39. $y = \dfrac{x(1/x - 1)^4}{(1 - x)^5(3x - 4)^2}$

40. $y = \dfrac{(5x^3 - 4x^2 + 1)^9}{(4x^5 + x^4 - 3x^3)^8}$

41. $s = \dfrac{1 + t^3}{(1 + t^2)(1 + t^4)}$

42. $u = (v^3 + av^2 + bv + c)(v^3 - av^2 - bv - c)$

43. $y = \dfrac{(1 + x^2)^5(x^5 + x^4 + x^3 + x^2 + 1)^2}{(1 + x)^{20}}$

44. $y = \dfrac{x}{\left(x + \dfrac{1}{x}\right)^2\left(x^2 + \dfrac{1}{x^2}\right)^3}$

45. If $f(x) = g(x)h(x)$, show that

$$f'' = g''h + 2g'h' + gh'';$$
$$f''' = g'''h + 3g''h' + 3g'h'' + gh'''.$$

46. Find d^2y/dx^2 when $y = x/(1 - x)$.

47. Find d^2y/dx^2 when $y = 1/(1 + x^2)$.

48. Call $(m - n)$ the degree of the rational function

$$\frac{a_m x^m + a_{m-1}x^{m-1} + \cdots + a_0}{b_n x^n + b_{n-1}x^{n-1} + \cdots + b_0},$$

where a_m and b_n are both nonzero. Show that if $r(x)$ is a rational function with *nonzero* degree p, then $r'(x)$ is of degree $(p - 1)$.

49. Show by a simple example that if $r(x)$ is a rational function of degree zero, then the degree of $r'(x)$ can be any negative integer except -1.

50. If $y = uv$, what is the formula for $\Delta y/y$ (when u and v are given increments Δu and Δv, respectively)?

51. If $y = uvw$, find the formula for $\Delta y/y$, when u, v, and w are given the increments Δu, Δv, and Δw, respectively.

52. Formalize the proof of the product law given in the text by quoting the limit law from Section 1 of Chapter 2 that justifies each step in the limit computation. (You will also have to quote Theorem 4 of Chapter 3.)

53. Formalize the proof of the quotient law in the same way.

54. Compute $f'(a)$ if $f(x) = (x - a)g(x)$. Can this result be true if g is not differentiable at $x = a$? Discuss.

55. Show by using the product law that the derivative of $(x - a)^m (x - b)^n$ is zero at a point between a and b.

56. Suppose that f is a differentiable function such that

$$tf'(t) = nf(t).$$

Show that $f(t) = ct^n$. (Compute the derivative of $f(t)/t^n$.)

57. Write out the proof of the product rule using function notation.

2
ALGEBRAIC FUNCTIONS; IMPLICIT DIFFERENTIATION

In this section we shall calculate the derivatives of algebraic functions. The method used, called *implicit differentiation*, assumes that the function in question is differentiable. At the end, we shall prove the differentiability of certain of these implicitly defined functions, by proving the inverse function rule.

A function $y = f(x)$ is *algebraic* if it satisfies a polynomial equation in x and y. For example, the nth root function

$$y = f(x) = x^{1/n}$$

is algebraic because it satisfies the polynomial equation

$$y^n = x$$

or

$$x - y^n = 0.$$

Generally, it is difficult or impossible to find an explicit expression for an algebraic function by solving its polynomial equation for y in terms of x. For example, the equation

$$y^3 + yx^2 + x^3 = 1$$

determines y as a function of x (see page 88), but it would be hard to find an explicit formula for this algebraic function.

Implicit Differentiation   We can nevertheless find a formula for the derivative of an algebraic function *without* solving for y. We just differentiate the original polynomial equation *with respect to x, treating y as a differentiable function of x*. This involves computing dy^k/dx for various integers k, and hence involves the general power rule.

Example 1   Find a formula for dy/dx, supposing that y is a differentiable function of x that satisfies the equation

$$y^3 + x^3 = 1.$$

Solution   Differentiating the equation with respect to x, we have

$$3y^2 \frac{dy}{dx} + 3x^2 = 0,$$

$$\frac{dy}{dx} = -\frac{x^2}{y^2}. \qquad \square$$

Example 2   Find a formula for dy/dx, supposing that y is a differentiable function of x that satisfies the equation

$$y^3 + yx^2 + x^3 = 1.$$

Solution   We differentiate the equation with respect to x, noting that the term yx^2 is a product and hence requires the product rule. We get

$$3y^2 \frac{dy}{dx} + \left[y \cdot 2x + x^2 \frac{dy}{dx} \right] + 3x^2 = 0,$$

$$\frac{dy}{dx} = -\frac{2xy + 3x^2}{3y^2 + x^2}.$$

If $y = h(x)$ is the function we are talking about, then this equation becomes

$$h'(x) = -\frac{2xh(x) + 3x^2}{3(h(x))^2 + x^2}.$$ □

It is possible for a function $y = f(x)$ to satisfy an equation in x and y without being the only such function. For example, the function

$$y = f(x) = \sqrt{1 - x^2}$$

satisfies the equation

$$x^2 + y^2 = 1.$$

Figure 1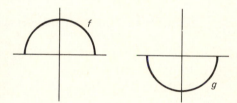

However, the function graph is only the upper half of the equation graph, and there is a second function satisfying the same equation whose graph is the lower half of the equation graph. This is the negative of the above function,

$$y = g(x) = -\sqrt{1 - x^2}.$$

Since both functions satisfy the same equation, we can't say that either one of them is uniquely determined by the equation. We can call them *function solutions* of the equation.

If we allow discontinuity, then we can find still more function solutions of $x^2 + y^2 = 1$. For example, the function h whose graph is drawn at the left in Fig. 2 is one. This function h is made up artificially by choosing part of the function f above and part of g. However, we shall consider only continuous solution functions.

The functions defined *implicitly* by an equation are simply its continuous solution functions. Thus, if an equation in x and y has the graph shown at the right in Fig. 2, then there are three continuous functions defined implicitly by the equation.

A derivative formula obtained by implicit differentiation is correct for *all* functions defined implicitly by the equation.

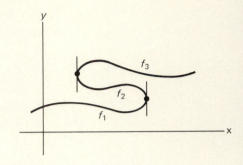

Figure 2

Example 3   Suppose that y is a differentiable function of x defined implicitly by the equation

$$x^2 + y^2 = 1.$$

Differentiating with respect to x, we get

$$2x + 2y\frac{dy}{dx} = 0,$$

and so

$$\frac{dy}{dx} = -\frac{x}{y}.$$

To check that this is *always* correct, we borrow from the next page the general power rule for $k = 1/2$. Thus, if $y = f(x) = \sqrt{1 - x^2}$, we have

$$\frac{dy}{dx} = \frac{1}{2}\frac{-2x}{\sqrt{1 - x^2}} = \frac{-x}{\sqrt{1 - x^2}} = -\frac{x}{y};$$

while if $y = g(x) = -\sqrt{1 - x^2}$, then

$$\frac{dy}{dx} = -\frac{1}{2}\frac{-2x}{\sqrt{1 - x^2}} = \frac{x}{\sqrt{1 - x^2}} = -\frac{x}{y}.$$ □

Higher derivatives of implicitly defined functions can be computed similarly. The following example illustrates the process.

Example 4   Compute d^2y/dx^2 when y is a differentiable function of x implicitly defined by the equation

$$\frac{x^2}{a^2} - \frac{y^2}{b^2} = 1.$$

Solution   We first compute dy/dx implicitly as in our earlier examples. We have

$$\frac{2x}{a^2} - \frac{2y\dfrac{dy}{dx}}{b^2} = 0,$$

$$\frac{dy}{dx} = \frac{b^2x}{a^2y}.$$

Differentiating this equation with respect to x gives

$$\frac{d^2y}{dx^2} = \left[\frac{b^2}{a^2}\,\frac{y - x\dfrac{dy}{dx}}{y^2}\right].$$

Now all we have to do is substitute the formula already obtained for dy/dx in the right side above. Thus

$$\frac{d^2y}{dx^2} = \frac{b^2}{a^2}\left[\frac{y - x\left(\dfrac{b^2x}{a^2y}\right)}{y^2}\right]$$

$$= \frac{b^2}{a^2}\left[\frac{a^2y^2 - b^2x^2}{a^2y^3}\right]$$

$$= \frac{b^2}{a^2}\left[-\frac{a^2b^2}{a^2y^3}\right]$$

$$= \frac{-b^4}{a^2y^3}.$$

In order to reach the last line above we used the equation

$$a^2y^2 - b^2x^2 = -a^2b^2,$$

which is another form of the equation of the hyperbola

$$\frac{x^2}{a^2} - \frac{y^2}{b^2} = 1. \qquad\qquad \square$$

The General Power Rule Again   Implicit differentiation is not restricted to algebraic functions. We shall now use it to extend the general power rule to any positive rational exponent k, *assuming the differentiability of u^k.*

THEOREM 2′ *Let $u = f(x)$ be a positive differentiable function of x, and suppose that u^k is also differentiable, where k is a positive rational number. Then*

$$\boxed{\frac{d}{dx}\, u^k = ku^{k-1}\frac{du}{dx}.}$$

Proof By hypothesis, $k = m/n$, where m and n are positive integers. So the function

$$y = u^k = u^{m/n} = [f(x)]^{m/n}$$

is defined implicitly by the equation

$$y^n = u^m = [f(x)]^m.$$

We now differentiate with respect to x, by the integer power rule (Theorem 2), and get

$$ny^{n-1}\frac{dy}{dx} = mu^{m-1}\frac{du}{dx}.$$

Therefore,

$$\frac{dy}{dx} = \frac{m}{n}\frac{u^{m-1}}{y^{n-1}}\frac{du}{dx}$$

$$= \frac{m}{n}\frac{y}{u}\frac{du}{dx} \quad \left(\text{since } \frac{u^m}{y^n} = 1\right)$$

$$= \frac{m}{n}\, u^{m/n-1}\frac{du}{dx} = ku^{k-1}\frac{du}{dx}. \qquad \blacksquare$$

Example 5 Find the derivative of $\sqrt{1 + x^3}$.

Solution We use the general power rule with $k = 1/2$.

$$\frac{d}{dx}\sqrt{1 + x^3} = \frac{d}{dx}(1 + x^3)^{1/2}$$

$$= \frac{1}{2}(1 + x^3)^{-1/2}\frac{d}{dx}(1 + x^3)$$

$$= \frac{3x^2}{2\sqrt{1 + x^3}}. \qquad \square$$

Example 6 Calculate the derivative of $x(ax + b)^{1/5}$.

Solution

$$\frac{d}{dx} x(ax + b)^{1/5} = x \cdot \frac{1}{5}(ax + b)^{-4/5} \cdot a + (ax + b)^{1/5}$$

$$= \frac{ax}{5(ax + b)^{4/5}} + \frac{ax + b}{(ax + b)^{4/5}}$$

$$= \frac{6ax + 5b}{5(ax + b)^{4/5}}. \qquad \square$$

Example 7 Find y' if $y = \sqrt{x - \sqrt{x}}$.

Solution

$$y' = \frac{1}{2}(x - \sqrt{x})^{-1/2}\left(1 - \frac{1}{2}x^{-1/2}\right).$$

This rearranges to

$$\frac{2\sqrt{x} - 1}{4\sqrt{x^2 - x^{3/2}}}. \qquad \square$$

Example 8 Find a formula for dy/dx if y is a differentiable function of x defined implicitly by the equation

$$y + y^{1/3} = x^2.$$

Solution We differentiate with respect to x, using the general power rule. This gives

$$\frac{dy}{dx} + \frac{1}{3}y^{-2/3}\frac{dy}{dx} = 2x.$$

Thus,

$$\frac{dy}{dx} = \frac{6x}{3 + y^{-2/3}}. \qquad \square$$

The above examples all involved algebraic functions. The next section has examples of the general power rule applied to non-algebraic functions.

The Inverse Function Rule

The calculations above assumed the differentiability of an implicitly defined function, and a proof of differentiability ought now to be given. However, the general proof is too advanced to be considered here. Instead, we give a proof for the special case when the defining equation has the form

$$g(y) = x.$$

We are thus considering a continuous function $y = f(x)$ such that

$$g(f(x)) = x.$$

In this situation we say that g inverts f, and the theorem in question is called the *inverse function rule*. (Inverse functions will be discussed in Chapter 7.)

The nth root function $f(x) = x^{1/n}$ is covered by the theorem, since it is continuous (from Chapter 2) and is inverted by the nth power function $g(x) = x^n$:

$$g(f(x)) = (x^{1/n})^n = x.$$

THEOREM 3

The Inverse Function Rule   *If g is differentiable and f is continuous, and*

$$g(f(x)) = x$$

for all x in an interval I, then f is differentiable and

$$\boxed{f'(x) = \frac{1}{g'(f(x))}}$$

at all points x for which the denominator $g'(f(x))$ is not zero.

Proof of Theorem   Let Δy be the increment in $y = f(x)$ produced by a nonzero increment Δx in x at x_0. That is,

$$y_0 + \Delta y = f(x_0 + \Delta x), \qquad \text{where } y_0 = f(x_0).$$

The identity $x = g(f(x))$ then gives us

$$\Delta x = (x_0 + \Delta x) - x_0 = g(f(x_0 + \Delta x)) - g(f(x_0))$$
$$= g(y_0 + \Delta y) - g(y_0).$$

In particular, Δy cannot be zero, since $\Delta x \neq 0$. If we divide this equation by Δy and then take reciprocals, the result can be written

$$\frac{\Delta y}{\Delta x} = \frac{1}{\dfrac{g(y_0 + \Delta y) - g(y_0)}{\Delta y}}.$$

Now let $\Delta x \to 0$. Then $\Delta y \to 0$ since the function $y = f(x)$ is continuous at x_0. But as $\Delta y \to 0$, the right side above approaches the limit $1/g'(y_0)$, provided $g'(y_0) \neq 0$. Thus the derivative

$$f'(x_0) = \lim_{\Delta x \to 0} \frac{\Delta y}{\Delta x}$$

exists and has the value $1/g'(y_0)$. ∎

As noted earlier, we have the corollary:

COROLLARY   *The nth root function $f(x) = x^{1/n}$ is differentiable (for $x > 0$).*

It follows that $x^{m/n} = (x^{1/n})^m$ is differentiable for any positive integer m (Theorem 2), and that the power formula continues to hold (Theorem 2′). Finally, $x^{-(m/n)}$ is covered by the quotient law. (Check this.) We can thus conclude

THEOREM 4   *For any rational number k, x^k is differentiable and*

$$\boxed{\frac{d}{dx}\, x^k = kx^{k-1}}$$

for $x > 0$.

PROBLEMS FOR SECTION 2

For Problems 1 through 9 find dy/dx by

a) Solving explicitly for y and then differentiating,

b) Using implicit differentiation.

Show that your two answers are equivalent.

1. $2xy + 5 = 0$ **2.** $y^3 = 4x^2 + 2x + 1$ **3.** $x^2 + y^2 = 16$

4. $y^2 = 2x + 1$

5. $3x^2 + 2x + y^2 = 10$

6. $y^5 = x^3$

7. $y^2 = \dfrac{x^2 - 1}{x^2 + 1}$

8. $x^2 + 2xy = 3y^2$

9. $x^{3/2} + y^{3/2} = 2$

In Problems 10 through 17 find dy/dx by implicit differentiation.

10. $2x^3 + 3y^3 + 6 = 0$

11. $x^2y - xy^2 + x^2 + y^2 = 0$

12. $x^4y^3 - 5xy = 100$

13. $y^5 + y^2x^3 + x^5 = 0$

14. $xy + x - 2y - 1 = 0$

15. $y^{1/3} + y^{1/2} = x^6$

16. $x + \sqrt{xy} + y = 1$

17. $(xy)^4 + (xy)^{1/3} = x + y$

18. Write the equation of the line tangent to the curve $xy + y^2 - 2x = 0$ at the point $(1, 1)$.

19. Find y'' in terms of x and y when $x^2 + 4y^2 = 25$.

20. Find the equation of the tangent line to the curve $x^3 - 3xy^3 + xy^2 = xy + 6$ at the point $(1, -1)$.

21. Find the equation of the normal line to $2x^2 - y^2 = 1$ at the point $(1, 1)$.

22. Find y' and y'' when $2xy - x^2 + 3y^2 + 6 = 0$.

23. Prove by implicit differentiation that the tangent line to the ellipse

$$\frac{x^2}{a^2} + \frac{y^2}{b^2} = 1$$

at the point (x_0, y_0) has the equation

$$\frac{xx_0}{a^2} + \frac{yy_0}{b^2} = 1.$$

24. Find the equation for the tangent line to the hyperbola

$$\frac{x^2}{a^2} - \frac{y^2}{b^2} = 1$$

at the point (x_0, y_0) in a form analogous to the one given above for the ellipse.

25. Find the horizontal tangents to the curve

$$y^3 + x^2y + 2x^3 = 28.$$

26. Find the tangents to the curve

$$y^4 + x^2y^2 - 2x^4 = 2$$

that pass through the origin.

Differentiate the following functions.

27. $\sqrt{1 - x}$

28. $(ax + b)^{7/6}$

29. $x\sqrt{1 + x^2}$

30. $(1 + 2x)\sqrt{1 - 3x}$

31. $1/(5x + 1)^{2/3}$

32. $\sqrt{1 - 4x^3}$

33. $(2x + x^2)^{2/3}$

34. $(1 + 4x)^{1/4}$

35. $x^2\sqrt{x - (1/x)}$

36. $(ax^2 + bx + c)^k$

37. $y = (1 + x)^{1/3}/(1 - x)^{1/2}$

38. $\dfrac{1}{3x + \sqrt{7x^2 - 5}}$

39. $\dfrac{u}{\sqrt{1 + 5u^4}}$

40. $(ax^2 + bx + c)\sqrt{2ax + b}$

41. $y = \dfrac{x(1 + 2x)^{-1/2}}{2 + 3x^2}$

42. $y = (1 - 3x)^{2/3}(1 + 3x)^{1/3}$

43. $y = \left(x^2 - 8x + \dfrac{1}{x}\right)^{1/2}(7x^{3/2} - 4x^3 + 1)^{3/4}$

44. $y = (4x + 9x^2 - 32)^{5/2}(x - 8x^4)^{1/2}$

45. $y = \dfrac{1 + (1 + x^{1/2})^{1/3}}{1 + (1 + x^{1/3})^{1/2}}$

46. Prove that if the power formula holds for x^a, then it holds for x^{-a}. (This completes the proof of Theorem 4.)

3
THE
TRIGONOMETRIC
FUNCTIONS

Any function that is not algebraic is said to be *transcendental*. It can be shown that the trigonometric functions are transcendental, and they are the most familiar examples of this class of functions. (Further transcendental functions will be introduced in Chapter 7.)

We assume familiarity with the trigonometric functions and review here only what is immediately needed. Appendix 2 contains further material, including graphs, identities, and proofs.

Radian Measure

First recall how the radian measure of an angle is defined. We choose a circle with center at the vertex—any radius will do (Fig. 1). Then the radian measure θ is given by

$$\boxed{\theta = \frac{l}{r} = \frac{\text{length of intercepted arc}}{\text{radius of circle}}.}$$

Although this appears to vary with r, in fact it does not. For example, a right angle cuts off one fourth of the circle circumference, so $l = (2\pi r)/4 = \pi r/2$, and $\theta = \pi/2$, as shown in Fig. 1.

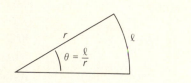 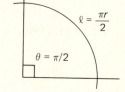

Figure 1

We will also need the sector area formula. In a given circle, the area A of a sector is proportional to the length l of its intercepted arc.

Using the full circle for comparison gives the proportionality

$$\frac{A}{\pi r^2} = \frac{l}{2\pi r} = \frac{\theta}{2\pi}.$$

Thus

$$A = \frac{1}{2} r^2\theta.$$

Sin θ and Cos θ   The trigonometric functions $\sin \theta$ and $\cos \theta$ must be defined for all real numbers θ, but it is convenient to start with the familiar right triangle definitions for an acute angle. Figure 2 shows the standard way of setting this up in terms of coordinates.

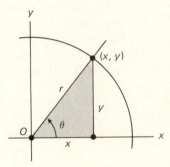

Figure 2

The vertex of the angle is placed at the origin, and the angle opens out into the first quadrant, with one side along the positive x-axis. If (x, y) is a point on the other side, at the positive distance

$$r = \sqrt{x^2 + y^2}$$

from the origin, and if θ is the radian measure of the angle, then

DEFINITION

$$\cos \theta = \frac{x}{r} \quad \left(= \frac{\text{adjacent}}{\text{hypotenuse}}\right),$$

$$\sin \theta = \frac{y}{r} \quad \left(= \frac{\text{opposite}}{\text{hypotenuse}}\right),$$

$$\tan \theta = \frac{y}{x} \quad \left(= \frac{\text{opposite}}{\text{adjacent}}\right).$$

Here is a table of values for a few popular angles. (See Appendix 2.)

θ	0	$\pi/6$	$\pi/4$	$\pi/3$	$\pi/2$
$\sin\theta$	0	1/2	$\sqrt{2}/2$	$\sqrt{3}/2$	1
$\cos\theta$	1	$\sqrt{3}/2$	$\sqrt{2}/2$	1/2	0
$\tan\theta$	0	$\sqrt{3}/3$	1	$\sqrt{3}$	—

We emphasize: Radian measure is used in calculus; $\cos\theta$ is the cosine of the angle having *radian* measure θ.

The next step is to extend the above definitions to all real numbers θ. To do this we use *directed* angles. We view an angle as extending *from* an *initial* ray *to* a *terminal* ray. In the standard setup shown in Fig. 2, the positive x-axis is taken as the initial ray. We consider the angle to extend *counterclockwise from* the positive x-axis *to* the terminal ray containing the point (x, y). We can think of θ as the amount of angle that is swept out by a ray that rotates counterclockwise from the initial position to the terminal position (Fig. 3). Angles in the counterclockwise direction are assigned positive radian measure, as we did above. Angles that are directed *clockwise* are assigned *negative* radian measure. Thus a right angle measured clockwise is $-\pi/2$ radians.

Figure 3

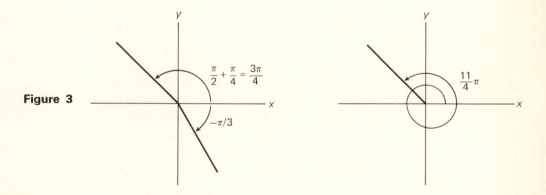

Very large angles can be thought of as being swept out by a rotating ray that makes more than one complete revolution. For example, we might consider the total angle swept out by a spoke on a revolving wheel during a certain interval of time. If the wheel makes exactly $3\frac{1}{2}$ counterclockwise revolutions, then the spoke has revolved through

$$\frac{7}{2}2\pi = 7\pi \text{ radians.}$$

No matter what the angle, however, there is always a terminal ray, and the definition

$$\cos\theta = \frac{x}{r}$$

is correct no matter how large (or small) a number θ is. The same goes for $\sin\theta$. (But $\tan\theta$ is only defined when $x \neq 0$.) The graphs of $\sin\theta$ and $\cos\theta$ are shown in Fig. 4.

Figure 4

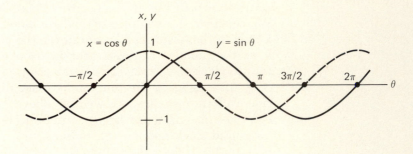

Finally, we need the identities

$$\boxed{\begin{aligned} \sin(-t) &= -\sin t, \\ \cos(-t) &= \cos t \end{aligned}}$$

($\sin t$ is an *odd* function; $\cos t$ is *even*), and

$$\boxed{\begin{aligned} \sin(s+t) &= \sin s \cos t + \cos s \sin t, \\ \cos(s+t) &= \cos s \cos t - \sin s \sin t \end{aligned}}$$

(the addition laws). Other identities, such as

$$\sin^2 t + \cos^2 t = 1,$$

can be derived from these. Thus,

$$1 = \cos 0 = \cos(t - t)$$
$$= \cos t \cos(-t) - \sin t \sin(-t)$$
$$= \cos^2 t + \sin^2 t.$$

This ends our brief review. See Appendix 2 for a more complete list of standard identities and their proofs.

Derivatives We now show that the functions $\sin x$ and $\cos x$ are differentiable.
of Sin θ We first show that they are differentiable at the origin. Differentia-
and Cos θ bility elsewhere, and the derivative formulas, then follow from the addition laws. The starting limits are

$$\sin'(0) = \lim_{h \to 0} \frac{\sin(0 + h) - \sin(0)}{h} = \lim_{h \to 0} \frac{\sin h}{h} = 1,$$

$$\cos'(0) = \lim_{h \to 0} \frac{\cos(0 + h) - \cos(0)}{h} = \lim_{h \to 0} \frac{\cos h - 1}{h} = 0.$$

The $(\sin h)/h$ limit is the basic one. We prove it geometrically, using inequalities between areas in the unit circle.

LEMMA

$$\boxed{\frac{\sin h}{h} \to 1 \qquad \text{as } h \to 0.}$$

Proof In the unit circle the sector area formula becomes $A = \theta/2$. So $A = h/2$ in Fig. 5. Also, A lies between the two triangular areas shown in the figure, so

$$\frac{\sin h}{2} < \frac{h}{2} < \frac{\tan h}{2}.$$

Dividing throughout by $(\sin h)/2$, we obtain the inequality

$$1 < \frac{h}{\sin h} < \frac{1}{\cos h}.$$

Figure 5

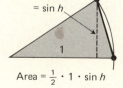

Altitude = $\sin h$

Area = $\frac{1}{2} \cdot 1 \cdot \sin h$

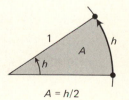

1 h A h

$A = h/2$

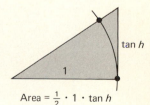

$\tan h$ 1

Area = $\frac{1}{2} \cdot 1 \cdot \tan h$

Assuming that $\cos t$ is continuous at $t = 0$, so that $\lim_{t\to 0} \cos t = \cos 0 = 1$, we conclude that

$$\frac{h}{\sin h} \to 1 \qquad \text{as } h \to 0$$

by the squeeze limit law. (For a proof that $\cos t$ is continuous at $t = 0$, see Problem 40.)

We have been implicitly assuming h to be positive. However,

$$\frac{\sin(-h)}{-h} = \frac{\sin h}{h},$$

so we do not need a separate argument for negative h. ∎

The other limit is 0:

$$\cos' 0 = \lim_{h\to 0} \frac{\cos h - 1}{h} = 0.$$

This limit is left as an exercise. It says that the graph of the cosine function has a horizontal tangent at $t = 0$, which should seem true.

Now the addition laws for $\sin x$ and $\cos x$ give the following theorem.

THEOREM 5   *The functions $\sin x$ and $\cos x$ are everywhere differentiable, and*

$$\boxed{\begin{aligned} \frac{d}{dx} \sin x &= \cos x, \\ \frac{d}{dx} \cos x &= -\sin x. \end{aligned}}$$

Proof   We have

$$\frac{\sin(x + h) - \sin x}{h} = \frac{\sin x \cos h + \cos x \sin h - \sin x}{h}$$

$$= \sin x \left(\frac{\cos h - 1}{h}\right) + \cos x \left(\frac{\sin h}{h}\right).$$

We have just seen that the quantities in parentheses have limits 0 and 1 respectively. Thus

$$\frac{d}{dx}\sin x = \lim_{h \to 0} \frac{\sin(x+h) - \sin x}{h}$$

$$= (\sin x) \cdot 0 + (\cos x) \cdot 1$$

$$= \cos x.$$

The proof for $\cos x$ goes in exactly the same way, starting from the addition law for $\cos(x+h)$. ∎

The remaining trigonometric functions are

$$\tan x = \frac{\sin x}{\cos x}, \qquad \cot x = \frac{\cos x}{\sin x},$$

$$\sec x = \frac{1}{\cos x}, \qquad \csc x = \frac{1}{\sin x}.$$

These are the tangent, secant, cotangent, and cosecant functions. The derivative formulas for these four functions are calculated from the sine and cosine formulas by the quotient rule. They are

$$\frac{d}{dx}\tan x = \sec^2 x,$$

$$\frac{d}{dx}\sec x = \tan x \sec x,$$

$$\frac{d}{dx}\cot x = -\csc^2 x,$$

$$\frac{d}{dx}\csc x = -\cot x \csc x.$$

The tangent and secant derivative formulas come up often enough to be worth memorizing, but they can always be worked out if forgotten.

Example 1 Prove that

$$\frac{d}{dx}\tan x = \sec^2 x.$$

Solution By definition

$$\tan x = \frac{\sin x}{\cos x}.$$

We can therefore compute its derivative by the quotient rule:

$$\frac{d}{dx}\tan x = \frac{\cos x \cdot \cos x - \sin x(-\sin x)}{\cos^2 x}$$

$$= \frac{\cos^2 x + \sin^2 x}{\cos^2 x} = \frac{1}{\cos^2 x} = \sec^2 x$$

(since $\sec x = 1/\cos x$).   □

Example 2 $$\frac{d}{dx}x\tan x = x\sec^2 x + \tan x$$ □

Example 3 $$\frac{d}{dx}(\sin x)^3 = 3(\sin x)^2 \cos x$$

This would normally be written

$$\frac{d}{dx}\sin^3 x = 3\sin^2 x \cos x.$$ □

Example 4 $$\frac{d}{dt}\frac{1 - \cos t}{\sin t} = \frac{\sin t \cdot \sin t - (1 - \cos t)\cos t}{\sin^2 t}$$

$$= \frac{1 - \cos t}{\sin^2 t} = \frac{1}{1 + \cos t},$$

(since $\sin^2 t = 1 - \cos^2 t$).   □

Example 5 Find $$\frac{d}{dx}\frac{1}{\sqrt{\cos x}}.$$

Solution By the quotient rule,

$$\frac{d}{dx}\frac{1}{\sqrt{\cos x}} = \frac{-\dfrac{1}{2}(\cos x)^{-1/2}(-\sin x)}{\cos x}$$

$$= \frac{1}{2}\frac{\sin x}{(\cos x)^{3/2}}.$$

Or, by the general power rule,

$$\frac{d}{dx}(\cos x)^{-1/2} = \left(-\frac{1}{2}\right)(\cos x)^{-3/2}(-\sin x),$$

which equals the first answer.   □

Dividing the identity $\sin^2 x + \cos^2 x = 1$ first by $\cos^2 x$ and then by $\sin^2 x$ gives the useful identities

$$\tan^2 x + 1 = \sec^2 x,$$
$$1 + \cot^2 x = \csc^2 x.$$

Frequently derivatives involving trigonometric functions can be simplified by using these and other trigonometric identities.

Example 6 Find the derivative of

$$y = \frac{\tan x}{1 + \sec x}.$$

Solution

$$y' = \frac{(1 + \sec x)\sec^2 x - \tan x(\sec x \tan x)}{(1 + \sec x)^2}$$

$$= \frac{\sec^2 x + \sec x}{(1 + \sec x)^2} \qquad (\text{because } \sec^2 x - \tan^2 x = 1)$$

$$= \frac{\sec x}{1 + \sec x}.$$ □

Note that

$$\frac{d^2}{dx^2}\sin x = -\sin x \qquad \text{and} \qquad \frac{d^2}{dx^2}\cos x = -\cos x.$$

Thus the functions $\sin x$ and $\cos x$ are both solutions of the *second-order differential equation*

$$\frac{d^2 y}{dx^2} = -y.$$

Furthermore, the two functions together are essentially the only solutions of this equation, in the following sense: Every other solution f is of the form

$$f(x) = A \sin x + B \cos x$$

for some constants A and B. A proof of this important fact is outlined below, but the details of the individual steps are left as exercises.

a) If g and h are any two solutions of the differential equation $y'' = -y$, then so is every linear combination of g and h, i.e., every function f of the form

$$f(x) = Ag(x) + Bh(x),$$

where A and B are constants.

b) If f is any solution then

$$f^2 + (f')^2 = \text{constant}.$$

Thus every solution f satisfies an identity of the type

$$\sin^2 x + \cos^2 x = 1.$$

c) It follows from (b) that if ϕ is a solution such that $\phi(0) = \phi'(0) = 0$, then ϕ is the zero function: $\phi(x) = 0$ for all x.

d) If f is any solution, then f must be of the form

$$f(x) = f(0)\cos x + f'(0)\sin x$$

(set $\phi = f - f(0)\cos - f'(0)\sin$ and apply (c)). Thus the general solution of the differential equation $d^2y/dx^2 + y = 0$ is of the form $A \sin x + B \cos x$.

The chain rule (Section 4) will let us show that essentially the same conclusions hold for the differential equation

$$\frac{d^2y}{dx^2} + b^2y = 0,$$

the general solution now being

$$A \sin bx + B \cos bx.$$

PROBLEMS FOR SECTION 3

Prove the following formulas for the derivatives of the remaining trigonometric functions:

1. $\dfrac{d}{dx} \sec x = \sec x \tan x$ 2. $\dfrac{d}{dx} \cot x = -\csc^2 x$ 3. $\dfrac{d}{dx} \csc x = -\csc x \cot x$

Differentiate the following functions:

4. $x \sin x$

5. $\sqrt{x} \cos x$

6. $\sin x / x$

7. $\sin^2 x$

8. $\sin x \cos x$

9. $\cos^2 x + \sin^2 x$

10. $\dfrac{\sin x}{1 + \cos x}$

11. $\dfrac{1 - \cos x}{\sin x}$

12. $\cos^2 x - \sin^2 x$

13. $\tan^2 x$

14. $\sec^2 x$

15. $\sqrt{1 + \cos x}$

16. $\sqrt{1 + \sin^2 x}$

17. $\sin^3 x \cos^3 x$

18. $\cos^4 x - \sin^4 x$

19. $\sec^2 x - \tan^2 x$

20. $x \sin x - \cos x$

21. $\sin^a x \cos^b x$, (where a and b are rational)

22. $x^2 \cos x - 2x \sin x - 2 \cos x$

23. $(1/2)(x + \sin x \cos x)$

24. Show that $y = x \sin x$ satisfies the fourth-order differential equation

$$\frac{d^4 y}{dx^4} + 2 \frac{d^2 y}{dx^2} + y = 0.$$

Some more derivatives:

25. $\dfrac{\sin x + \cos x}{\sin x - \cos x}$

26. $\sqrt{\dfrac{1 - \cos x}{1 + \cos x}}$

27. $\dfrac{\sec x}{1 + x^2}$

28. $\sqrt{1 - \cos^2 x}$

29. $\dfrac{\tan x}{1 + \tan^2 x}$

30. $\dfrac{1 - \sin 2x}{1 + \sin 2x}$

31. $(1/5)\sin^5 x - (2/3)\sin^3 x + \sin x$
(Simplify the answer.)

32. $(1/3)\tan^3 x - \tan x + x$
(Simplify the answer.)

33. $(1/5)\tan^5 x + (2/3)\tan^3 x + \tan x$
(Simplify.)

34. $y = (x \cos x)^4$

35. $y = \sqrt{1 + (\sin x / x)}$

36. $s = \sqrt{1 - \sqrt{\sec t}}$

37. $y = \dfrac{x^{1/3}}{\sin x - x}$

38. $r = \dfrac{\tan \theta}{\theta^2 + \cos^2 \theta}$

39. $y = (k - 1)\cos^{k+1} x - (k + 1)\cos^{k-1} x$

40. $s = (1 + t^2)t^n \cos^m(t)$

41. $y = \dfrac{x \sin x + \cos x}{x \cos x - \sin x}$

42. $y = (1 + \sec^2 x)^{2/3}$

43. $u = \dfrac{v}{(\tan v + \sec v)^k}$

44. We saw in the text that $0 < \sin t < t$ when t is positive. Use this inequality to prove that $\cos t \to 1$ as $t \to 0$. (There are a couple of ways to start this. For one thing, $\sin t = \sqrt{1 - \cos^2 t}$.)

46. Use the identity for $\sin A - \sin B$ (see Appendix 2) to give another proof that the derivative of $\sin x$ is $\cos x$.

48. Show that if $g(x)$ and $h(x)$ are any two solutions of the differential equation

$$\frac{d^2 y}{dx^2} = -y, \qquad \text{(I)}$$

then so is any linear combination

$$f(x) = Ag(x) + Bh(x),$$

where A and B are constants.

50. Show that every solution f of (I) is of the form

$$f(x) = A \cos x + B \sin x.$$

[*Hint*: Set $\phi(x) = f(x) - f(0)\cos x - f'(0)\sin x$ and apply the above problems.]

45. Show that the unit circle about the origin lies entirely below the graph of $\cos x$ except for the point of tangency of the two graphs at $x = 0$.

47. Complete the proof of Theorem 5. That is, prove that $d \cos x/dx = -\sin x$ from scratch, using the addition law for $\cos(x + h)$.

49. Show that $f^2 + (f')^2$ is constant for any solution f of (I). Conclude that if ϕ is a solution such that $\phi(0) = \phi'(0) = 0$, then ϕ is identically zero.

51. Use the addition law for

$$\cos h = \cos\left(\frac{h}{2} + \frac{h}{2}\right)$$

and the $(\sin h)/h$ limit to prove that

$$\lim_{h \to 0} \frac{\cos h - 1}{h} = 0.$$

**4
THE CHAIN
RULE**

The chain rule tells us how to calculate derivatives when we put functions *inside* of other functions. We have already met one special case: The general power rule

$$\frac{d}{dx}[f(x)]^k = kf(x)^{k-1} f'(x)$$

tells us how to differentiate when we put a function $f(x)$ inside the kth power function. But there are very simple combinations of this sort that we can't yet differentiate, such as $\sin(3 + x^2)$. In order to handle this general method of combination we need one final rule, called the *chain rule*.

If y is a function of u and if u is a function of x, then y is a function of x. For example, if $y = \sin u$ and $u = 3 + x^2$, then $y = \sin(3 + x^2)$. The question is, How does the derivative dy/dx of such a final composite function relate to the derivatives dy/du and du/dx of its two constituent functions?

It is easy to see from the rate-of-change interpretation of the derivative what the answer ought to be. Suppose that when $x = a$, the derivative dy/du has the value r and du/dx has the value s. Thus, y is increasing r times as fast as u and u is increasing s times as fast as x (as x passes through a). Therefore, altogether y is increasing rs times as fast as x. That is,

$$\frac{dy}{dx} = rs = \left(\frac{dy}{du}\right)\left(\frac{du}{dx}\right).$$

The following theorem is thus plausible.

THEOREM 6   **Chain Rule** *If y is a differentiable function of u and if u is a differentiable function of x, then y is a differentiable function of x and*

$$\boxed{\frac{dy}{dx} = \frac{dy}{du} \cdot \frac{du}{dx}.}$$

This is the Leibniz version of the chain rule. We shall prove it later in the section.

Example 1   Compute dy/dx when

$$y = \sin u \qquad \text{and} \qquad u = 3 + x^2.$$

Solution   We have

$$\frac{dy}{du} = \cos u, \qquad \frac{du}{dx} = 2x,$$

and the chain rule gives

$$\frac{dy}{dx} = \frac{dy}{du} \cdot \frac{du}{dx} = (\cos u)2x = 2x \cos(3 + x^2). \qquad \square$$

The intermediate variable u may not be part of the original problem. For instance, the above example would typically arise as the problem of computing dy/dx when $y = \sin(3 + x^2)$. Here we have a function that is too complicated for our earlier rules to handle. But setting $u = 3 + x^2$ breaks the complicated function down into two simpler stages that we do know how to differentiate,

$$y = \sin u \qquad \text{and} \qquad u = 3 + x^2,$$

and we can now compute dy/dx by the chain rule, as above.

Here is the same example without commentary.

Example 2   Compute dy/dx when $y = \sin(3 + x^2)$.

Solution   Setting $u = 3 + x^2$, we have $y = \sin u$. Then by the chain rule,

$$\frac{dy}{dx} = \frac{dy}{du} \cdot \frac{du}{dx} = (\cos u) \cdot 2x = 2x \cos(3 + x^2). \qquad \square$$

Here the derivatives dy/du and du/dx are so simple that there is no need to compute them separately before multiplying them in the chain rule. But if one of them has to be worked out, then computing it separately may help to keep track of what is going on.

Example 3   Find dy/dx when $y = \tan(x/(1 + x))$.

Solution   Setting $u = x/(1 + x)$ breaks y up suitably, into

$$y = \tan u \qquad \text{and} \qquad u = \frac{x}{1 + x}.$$

Then

$$\frac{dy}{du} = \sec^2 u, \qquad \frac{du}{dx} = \frac{(1 + x) \cdot 1 - x \cdot 1}{(1 + x)^2} = \frac{1}{(1 + x)^2},$$

and so

$$\frac{dy}{dx} = \frac{dy}{du} \cdot \frac{du}{dx} = \sec^2 u \cdot \frac{1}{(1 + x)^2}$$

$$= \frac{\sec^2 \left(\dfrac{x}{1 + x} \right)}{(1 + x)^2}. \qquad \square$$

Note that the above answers are functions of x alone. The intermediate variable u could be left in the first answer *only* if the answer also contained the statement $u = x^2 + 3$.

With more practice, and in situations where the two derivatives dy/du and du/dx are known, the chain-rule derivative can be written down in one or two steps.

Example 4   As we look at

$$y = (\underbrace{1 + x^2}_{u})^{3/2},$$

we mentally take $1 + x^2$ as u and see that we then have $y = u^{3/2}$. Therefore, as we put down the righthand side of the Leibniz formula, we first think dy/du, see that it is $(3/2)u^{1/2}$ and write this down, and then think du/dx, see that it is $2x$ and write that down. Thus, we would probably first write

$$\frac{dy}{dx} = \frac{3}{2} u^{1/2} \cdot 2x,$$

or even immediately

$$\frac{dy}{dx} = \frac{3}{2} (1 + x^2)^{1/2} \cdot 2x. \qquad \square$$

If we use function notation for the outside function in the chain rule, say $y = g(u)$, then Theorem 6 can be restated in terms of g and g'.

THEOREM 6'   *If $y = g(u)$ and u is a function of x, and if both functions are differentiable, then $y = g(u)$ is a differentiable function of x, and*

$$\boxed{\frac{d}{dx} g(u) = g'(u) \frac{du}{dx}.}$$

Theorem 6' is a very useful reformulation of the chain rule. For example, if we take $g(u)$ to be one of the functions u^k, $\sin u$, and $\cos u$, we obtain the *general* differentiation formulas

$$\boxed{\begin{aligned} \frac{d}{dx} u^k &= ku^{k-1} \frac{du}{dx}, \\[1em] \frac{d}{dx} \sin u &= \cos u \frac{du}{dx}, \\[1em] \frac{d}{dx} \cos u &= -\sin u \frac{du}{dx}. \end{aligned}}$$

The first of these formulas is the general power rule, now *completely* proved for *any* rational exponent k (by Theorem 4 in Section 2). These general derivative formulas are sometimes referred to as having "built-in" chain rule.

Example 5
$$\frac{d}{dx}(x^4 + 5x)^{8/7} = \frac{8}{7}(x^4 + 5x)^{1/7}(4x^3 + 5).$$ □

Example 6
$$\frac{d}{dx}\cos\left(\frac{1}{x}\right) = -\sin\left(\frac{1}{x}\right) \cdot \left(-\frac{1}{x^2}\right) = \frac{\sin\left(\frac{1}{x}\right)}{x^2}.$$ □

In the theorem above, u is some function of x that may still be quite complicated, and one or more further applications of the chain rule may be needed to find *its* derivative. The next two examples illustrate this.

Example 7
$$\frac{d}{dx}\sqrt{1 + \sqrt{1 + x^2}} = \frac{d}{dx}[1 + (1 + x^2)^{1/2}]^{1/2}$$

$$= \frac{1}{2}[1 + (1 + x^2)^{1/2}]^{-1/2}\frac{d}{dx}[1 + (1 + x^2)^{1/2}]$$

$$= \frac{1}{2}[1 + (1 + x^2)^{1/2}]^{-1/2} \cdot \frac{1}{2}(1 + x^2)^{-1/2} \cdot 2x$$

$$= \frac{x}{2\sqrt{1 + x^2}\,\sqrt{1 + \sqrt{1 + x^2}}}.$$ □

Example 8   Find the derivative of $\sqrt{1 + \sin^2(at)}$.

Solution
$$\frac{d}{dt}(1 + \sin^2(at))^{1/2} = \frac{1}{2}(1 + \sin^2(at))^{-1/2} \cdot 2\sin(at) \cdot \cos(at) \cdot a$$

$$= \frac{a\sin(2at)}{2\sqrt{1 + \sin^2(at)}}$$

(using the $\sin 2x$ formula).   □

In Example 8 we applied Theorem 6′ three times: first with $g(u) = \sqrt{u}$, then with $g(u) = 1 + u^2$, and finally with $g(u) = \sin u$.

Composition of   In Theorem 6′, u is still some unspecified function of x. If $u = h(x)$, so
Functions   that $du/dx = h'(x)$, then Theorem 6′ can be rewritten

$$\boxed{\frac{d}{dx}g(h(x)) = g'(h(x))h'(x).}$$

This allows the most precise statement of the chain rule:

THEOREM 6″ *If the function h is differentiable at x_0 and if g is differentiable at $u_0 = h(x_0)$, then the function $f(x) = g(h(x))$ is differentiable at x_0 and*

$$f'(x_0) = g'(u_0) \cdot h'(x_0)$$
$$= g'(h(x_0))h'(x_0).$$

The function

$$f(x) = g(h(x))$$

is called the *composition product* of g and h. As a function of x it is formed by applying the outside function g to the inside function of x, $h(x)$.

Example 9 The function $f(x) = (3 + x^2)^{3/2}$ is the composition product of the outside function $g(u) = u^{3/2}$ and the inside function $h(x) = 3 + x^2$.

$\square$

The composition product of g and h is sometimes designated by the special notation $g \circ h$. This means that $g \circ h$ is the function defined by

$$(g \circ h)(x) = g(h(x)).$$

This notation allows us to write the function form of the chain rule without any variables at all:

$$\boxed{(g \circ h)' = (g' \circ h)h'.}$$

Before taking up the proof of the chain rule, we recall an earlier remark. Suppose that a function $f(x)$ has a limit l as $x \to 0$ but that $f(0)$ is not defined. Then f can be extended to be continuous at $x = 0$. Its extension F is defined by

$$F(x) = f(x) \qquad \text{when } x \neq 0,$$
$$F(0) = l.$$

F is called the continuous modification of f at $x = 0$. This came up in Section 2 of Chapter 2. Here it will help in the proof of the chain rule. We will have a function $g(x)$ that is differentiable at a point $x = a$. That is, the difference quotient of g has the limit $g'(a)$ as $h \to 0$. So the difference quotient has a continuous modification $D(h)$ at $h = 0$, defined by

$$D(h) = \frac{g(a + h) - g(a)}{h} \qquad \text{when } h \neq 0,$$

$$D(0) = g'(a).$$

Note that the cross-multiplied equation

$$g(a + h) - g(a) = hD(h)$$

is then true for *all* h, including $h = 0$.

We now apply this remark to the proof of the chain rule.

Proof of Theorem 6′ We are considering

$$y = g(u),$$

where u is a differentiable function of x. An increment Δx in x produces an increment Δu in u, and this in turn produces the increment

$$\Delta y = g(u + \Delta u) - g(u)$$

in y. We rewrite the right side of this equation by the above remark and have

$$\Delta y = D(\Delta u)\,\Delta u,$$

where $D(\Delta u)$ is the continuous modification of the difference quotient function for g. Now divide by Δx,

$$\frac{\Delta y}{\Delta x} = D(\Delta u)\frac{\Delta u}{\Delta x},$$

and let $\Delta x \to 0$. Then $\Delta u/\Delta x \to du/dx$ by hypothesis. Also, $\Delta u \to 0$ because a differentiable function is continuous, and $D(\Delta u) \to g'(u)$ by the definition of D. The right side above thus approaches $g'(u)(du/dx)$ as Δx approaches 0. Thus the limit

$$\frac{dy}{dx} = \lim_{\Delta x \to 0} \frac{\Delta y}{\Delta x}$$

exists, and

$$\frac{dy}{dx} = g'(u)\frac{du}{dx}.$$

PROBLEMS FOR SECTION 4

Differentiate the following functions:

1. $\sin 3x$

2. $\cos(1 - x)$

3. $\tan(1 + 2x^2)$

4. $x\sqrt{1 + 4x^3}$

5. $(2x + x^5)^{2/9}$

6. $(1 + 4x)^{1/4}(1 - 4x)$

7. $\sin(ax + b)$

8. $\cos(ax^2 + bx + c)$

9. $a \sin b\theta + b \sin a\theta$

10. $\sqrt{\cos x}$

11. $\sin \sqrt{x}$

12. $(1 + 2\cos x)^{1/3}$

13. $\sec \dfrac{1}{2x}$

14. $\sqrt{1 + \cos at}$

15. $x\sqrt{1 + x \sin 4x}$

16. $\dfrac{1}{x^2 - \sqrt{x^2 - 1}}$

17. $\tan \dfrac{1}{x^2}$

18. $\sin(x + \sqrt{x})$

19. $\cos(\tan \theta)$

20. $(1 - 3x)^a(1 + 3x)^b$

21. $\cos(\sqrt{1 - x^2})$

22. $\sqrt{\sin x} - \sqrt{\cos x}$

23. $(ax^2 + bx + c)^k(2ax + b)$

24. $x^2 \sin \dfrac{1}{x}$

25. $y = \sqrt{\dfrac{1 + x^m}{1 - x^n}}$

26. $\dfrac{\sin x^2}{1 + \sqrt{\cos x}}$

27. $\sin(\sin x)$

28. $y = \dfrac{x + 1}{(x^2 + 2x + 2)^{3/2}}$

29. $y = \dfrac{x\sqrt{1 + 2x}}{1 + x^2}$

30. $\dfrac{1 + \sin(x^2)}{1 + \cos(x^2)}$

31. $x \sec x^2$

32. $(x^a + 1)^{1/a}$

33. $(x^{2/3} + a^{2/3})^{3/2}$

34. $w = \dfrac{u}{\sqrt{1 + 5u^4}}$

35. $y = \dfrac{(\sin x)^{1/3}}{1 + (\cos x)^{1/3}}$

36. $y = x\sqrt{(ax + b)/(cx + d)}$

37. $y = \dfrac{\sqrt{x}\sin \sqrt{x}}{\cos x}$

38. $\dfrac{\sin(\cos t)}{\cos(\sin t)}$

39. $u = (\sin v)(\cos v^2)(\sin v^3)$

40. $f(x + g(x))$

41. $f(x^n g(x))$

42. $f([g(x)]^n)$

43. $\sqrt{1 - (g(x))^2}$

44. $f(\cos[g(x)])$

45. $\sin^3(x^2)$

46. $\sin \sqrt{\cos x}$

47. $\cos^2(x^3)$

48. $\sqrt{x + \sqrt{1 - x^4}}$

49. $x^2 \sec \dfrac{2}{x}$

50. $(x \tan x)^{1/4}$

51. $\cos x \sin(\tan x)$

52. $\sin(\cos(\sin 2x))$

53. $\sqrt{1 + \sin^2(\sqrt{t})}$

54. $(\tan x^3)^{1/3}$

55. $y = \sqrt{\cos\left(\dfrac{1}{1 + x^2}\right)}$

56. $\dfrac{t}{t^4 + \cos^2(t^4)}$

57. $[\sin^3(ax) + \cos^3(ax)]^{2/3}$

58. $\sqrt{1 + x^2} \tan(1 + x^2)$

59. $y = \left[\cos\left(\sin \dfrac{4}{x}\right)\right]^{1/8}$

60. $\sin^2(\tan^2(x^2))$

61. $\sqrt{1 + \sqrt{1 + \sqrt{1 + x^2}}}$

In Problems 62 through 67, find dy/dx by implicit differentiation.

62. $\sin xy = y^2$

63. $\cos y = x$

64. $\tan y - \tan x = a$

65. $\sin y = xy$

66. $y \cos \sqrt{y} = x$

67. $y^3 + y = \sin x$

68. Suppose that there is a differentiable function f such that $\tan f(x) = x$ for all x. Show that $f'(x) = 1/(1 + x^2)$.

69. Suppose that g is a differentiable function such that $\sin(g(x)) = x$ for all x in $(-1, 1)$. Show that $g'(x) = 1/\sqrt{1 - x^2}$.

70. Show that

$$\frac{d}{dx} \sin(f(x)) = \cos(f(x))$$

only if $f(x)$ is of the form

$$f(x) = x + b.$$

71. From the sketches of graphs of even and odd functions decide how such functions behave under differentiation, and then prove your conjecture from the chain rule.

72. Prove that a function $y = f(x)$ satisfies the differential equation

$$\frac{d^2y}{dx^2} + b^2 y = 0,$$

where b is a constant, if and only if $f(x)$ is of the form

$$y = A \sin bx + B \cos bx$$

for some constants A and B.

**5
THE
DIFFERENTIAL
FORMALISM**

Although we have insisted that the Leibniz symbol dy/dx is not a quotient, nevertheless the chain rule

$$\frac{dy}{dt} = \frac{dy}{dx} \cdot \frac{dx}{dt}$$

looks as though it could be obtained by canceling the two occurrences of dx on the right, and this "cancellation" is a good way of remembering the rule.

A similar situation will be discovered in Chapter 8, when an integration is performed by changing variables.

Thus the chain rule permits us to treat dy/dx as though it were a quotient in several important situations. We could let it go at that, but terms such as dx and dy are used frequently in mathematical writing, and we should say something about them. They will be treated here as new variables that are related to each other by certain equations.

Suppose that up to this point we had been using only function notation. In particular, we would then have the chain rule only in the following form:

If $y = f(x)$ and $x = g(s)$, and if we set $h(s) = f(g(s))$, then

$$h'(s) = f'(g(s)) \cdot g'(s)$$
$$= f'(x) \cdot g'(s).$$

We now reintroduce the Leibniz symbols, but in a different way. Suppose there is some underlying independent variable, say t. We associate with t a new independent variable dt, called the differential of t. Like Δt, this is not a product of something d times something t, but a wholly new variable, independent of t, but to be used along with t. We can give dt any value we wish, and in practice we often set $dt = \Delta t$.

Now suppose that x is a differentiable function of t, say

$$x = g(t).$$

Then the differential dx of the dependent variable x is defined by

$$\boxed{dx = g'(t)\, dt.}$$

Note that dx is a function of the two variables t and dt. The differential of any other variable depending on t is defined in the same way. Thus if

$$z = h(t),$$

then

$$dz = h'(t)\, dt.$$

Finally, suppose that

$$y = f(x),$$

where $x = g(t)$, as above. Then y is a function of t,

$$y = h(t),$$

where $h(t) = f(g(t))$. Since

$$h'(t) = f'(x)g'(t)$$

by the chain rule, and since $g'(t)\, dt = dx$, we have

$$dy = h'(t)\, dt = f'(x)g'(t)\, dt = f'(x)\, dx.$$

We thus see that the equation

$$dy = f'(x)\,dx$$

holds when $y = f(x)$, *regardless of whether x is independent or both x and y depend on some other variable t. So long as* $y = f(x)$, *then also* $dy = f'(x)\,dx$, *no matter what variable ultimately turns out to be the underlying independent variable.*

Moreover, if $dx \neq 0$, then we can divide by it and have

$$\frac{dy}{dx} = f'(x),$$

where now the Leibniz symbol dy/dx is a quotient, a quotient of two related differentials.

The Differential Rules

The derivative rules now acquire very useful differential formulations. For example, if we multiply the product rule

$$\frac{d}{dx}(uv) = u\frac{dv}{dx} + v\frac{du}{dx}$$

by dx, it becomes the differential product rule

$$d(uv) = u\,dv + v\,du.$$

This form has the advantage of being "independent of the independent variable." Moreover, it turns out to be just as valid for functions of several variables as it is for functions of one variable.

Here is our basic list of derivatives in differential form:

$$d(u^k) = ku^{k-1}\,du \qquad \text{(for any rational exponent } k\text{),}$$
$$d(\sin u) = \cos u\,du,$$
$$d(\cos u) = -\sin u\,du,$$
$$d(u + v) = du + dv,$$
$$d(cu) = c\,du,$$
$$d(uv) = u\,dv + v\,du,$$
$$d\left(\frac{u}{v}\right) = \frac{v\,du - u\,dv}{v^2}.$$

We recover the derivative forms of these rules by dividing through by the differential of the independent variable.

Example 1 Find the differential, and hence the derivative, of

$$y = \frac{x^2}{1 + x^3}.$$

Solution

$$dy = \frac{(1 + x^3)\, d(x^2) - x^2\, d(1 + x^3)}{(1 + x^3)^2}$$

$$= \frac{(1 + x^3)2x\, dx - x^2 3x^2\, dx}{(1 + x^3)^2}$$

$$= \frac{2x - x^4}{(1 + x^3)^2}\, dx.$$

$$\frac{dy}{dx} = \frac{2x - x^4}{(1 + x^3)^2}.$$

□

Example 2 Find the differential and derivative of

$$y = \sin(x^{1/5}).$$

Solution

$$dy = \cos(x^{1/5})\, d(x^{1/5})$$

$$= \cos(x^{1/5})\frac{1}{5} x^{-4/5}\, dx$$

$$= \frac{\cos(x^{1/5})}{5x^{4/5}}\, dx$$

$$\frac{dy}{dx} = \frac{\cos(x^{1/5})}{5x^{4/5}}.$$

□

PROBLEMS FOR SECTION 5

Find the following differentials.

1. $d(x^3 + 2x + 3)$ **2.** $d(3x + 2)$ **3.** $d(x^2 + 2)(2x + 1)^3$ **4.** $d\!\left(\dfrac{x - 3}{x + 2}\right)$

5. $d \cos \sqrt{1 + x^2}$ **6.** $d\sqrt{1 + \sqrt{1 + x}}$ **7.** $d\!\left(\dfrac{x^3 + 2x + 1}{x^2 + 3}\right)$ **8.** $d(\cos^2 2x + \sin 3x)$

9. $d(x^{-1} + 2x + 5)$ **10.** $d\left(\dfrac{\sin x}{\cos x}\right)$

Using the method of differentials, find the derivatives of the following functions.

11. $y = 2 + 2x$ **12.** $y = \dfrac{1}{(2 + 3x)^3}$ **13.** $y = (x + 2)^{2/3}(x - 1)^{1/3}$

14. $y = \dfrac{x^{2/3}}{(x + 1)^{2/3}}$

Use differentials to find the derivative dy/dx in each of the following situations, assuming in each case that the equation defines y as a differentiable function of x. Also find the derivative dx/dy, assuming that the equation defines x as a differentiable function of y.

15. $\tan y = x^2$ **16.** $y^3 + x = \sin xy$ **17.** $y^3 + xy - x^5 = 5$

18. $2x^2 + xy - y^2 + 2x - 3y + 5 = 0$

CHAPTER 5
APPLICATIONS OF THE DERIVATIVE

Throughout this chapter we shall apply two new facts, called the *mean-value principle* and the *extreme-value principle*. Both appear to be true when we draw figures, and we assume them on this basis. The last section then discusses why "obvious" things do need to be proved, after all, and proves the mean-value principle from the extreme-value principle.

1
THE MEAN-VALUE PRINCIPLE

Graphs such as those in Fig. 1 show that:

If we draw a secant line through two points on the graph of a differentiable function, then there is at least one point in between where the tangent line is parallel to the secant.

We visualize moving the secant line parallel to itself, and we see that it turns into the tangent line at its last point of contact with the graph.

When we state this analytically as an equality of two slopes, it becomes what is called the mean-value principle.

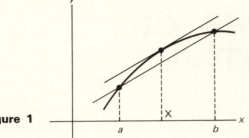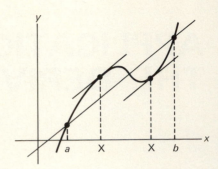

Figure 1

Mean-Value Principle *If f is differentiable on the closed interval [a, b], then there is at least one point X strictly between a and b at which*

$$f'(X) = \frac{f(b) - f(a)}{b - a}.$$

The mean-value principle also has a velocity interpretation. Consider again a particle moving along a coordinate line with position coordinate given by $s = f(t)$. We saw earlier that its instantaneous velocity is given by $ds/dt = f'(t)$, and that the average velocity over the time interval $[t_0, t_1]$ is

$$\frac{\Delta s}{\Delta t} = \frac{f(t_1) - f(t_0)}{t_1 - t_0}.$$

We assume that the velocity varies continuously with t; so as it varies from above its average value to below (or vice versa), it will cross the average value at some time $t = T$ by the intermediate-value principle. That is, there must be some number T between t_0 and t_1 such that

$$f'(T) = \frac{f(t_1) - f(t_0)}{t_1 - t_0},$$

which is again the mean-value principle. (This interpretation explains the phrase *mean-value*, because average values are also called mean values.)

The mean-value principle thus appears to be true in the contexts of the two principal interpretations of the derivative. It will be proved at the end of the chapter.

Replacing a and b by x and $x + \Delta x$, we have the increment form

$$f'(X) = \frac{f(x + \Delta x) - f(x)}{\Delta x} = \frac{\Delta y}{\Delta x},$$

where X lies strictly between x and $x + \Delta x$. Also useful are the cross-multiplied forms:

$$f(b) - f(a) = f'(X)(b - a)$$
$$\Delta y = f'(X)\,\Delta x;$$
$$f(x + \Delta x) = f(x) + f'(X)\,\Delta x.$$

REMARK The cross-multiplied version is trivially true when $b = a$, becoming then $0 = f'(X) \cdot 0$, except for the fine point that X cannot lie *strictly* between a and b if $b = a$. When we want to include this case, as we do in some applications, we interpret *between* as allowing $a = X = b$.

Example 1 Apply the mean-value principle to $f(x) = 4x^2 - 5x$ over $[1, 2]$ and then solve for X.

Solution By the mean-value principle,

$$f(2) - f(1) = f'(X)(2 - 1)$$

for some number X between 1 and 2. Here $f(2) = 16 - 10 = 6$, $f(1) = -1$, and $f'(x) = 8x - 5$. So

$$6 - (-1) = (8X - 5)(1).$$

Solving this equation for X, we find that

$$X = \frac{7 + 5}{8} = \frac{3}{2}. \qquad \square$$

Example 2 When the mean-value principle is applied to the function $f(x) = x^3$, it says that for any two numbers a and b there is a number X lying between a and b, such that

$$b^3 - a^3 = 3X^2(b - a).$$

Again we can solve for X. Since

$$b^3 - a^3 = (b - a)(b^2 + ab + a^2),$$

it follows that

$$X = \sqrt{\frac{b^2 + ab + a^2}{3}}. \qquad \square$$

Example 3 The mean-value principle for $f(x) = \sqrt{x}$ guarantees that there is a number X between a and b such that

$$\sqrt{b} - \sqrt{a} = \frac{1}{2\sqrt{X}}(b - a).$$

Here, also, we can find X by solving this mean-value equation:

$$2\sqrt{X} = \frac{b - a}{\sqrt{b} - \sqrt{a}} = \sqrt{b} + \sqrt{a},$$

$$X = \left(\frac{\sqrt{b} + \sqrt{a}}{2}\right)^2.$$

Note that if a is replaced by b, then the value of this final expression is increased (since $a < b$) to $(\sqrt{b})^2 = b$. This shows directly that $X < b$. Similarly, it can be directly checked that $a < X$. $\qquad \square$

It is clear from the above figure that there may be more than one such point X. Moreover, the mean-value principle says nothing about where X is, except that it lies somewhere between a and b. One might feel that anything so vague as this couldn't be very helpful, but we shall see that the mean-value principle is extremely useful. In the usual context, $f'(x)$ is known to have some property throughout the interval (a, b), and it therefore has this property at the point X, no matter where it is. So the mere fact that there must *be* such a point X is often enough to get us off the ground. We shall give an example below to show how this works, but most of the applications of the mean-value principle will come later.

THEOREM 1 *If f' exists and is everywhere zero on an interval I, then f is a constant function on I. That is, there is a constant c such that $f(x) = c$ for every x in I.*

Geometrically, we can reason that if every tangent line to a graph is horizontal, then the graph can never rise or fall and so must be a horizontal straight line.

Proof Analytically, the theorem is an immediate corollary of the mean-value principle. Choose any point x_0 in I and set

$c = f(x_0)$. For any other x the mean-value principle says that

$$f(x) - f(x_0) = f'(X)(x - x_0),$$

where X is some number between x_0 and x. But since f' is everywhere 0 this shows that $f(x) - f(x_0) = 0$, and so

$$f(x) = f(x_0) = c$$

for all x. ∎

This fact is crucially important when it comes to reversing the differentiation procedure. We saw a simple example of its use in Section 4 of Chapter 3.

PROBLEMS FOR SECTION 1

For each of the following functions and the given value of a and b, find a value for X such that $f'(X) = (f(b) - f(a))/(b - a)$.

1. $f(x) = x^2 - 6x + 5, a = 1, b = 4$

2. $f(x) = 3x^2 + 4x - 3, a = 1, b = 3$

3. $f(x) = x^3, a = 0, b = 1$

4. $f(x) = -\dfrac{2}{x}, a = 1, b = 3$

5. $f(x) = x^{-2}, a = 1, b = 2$

6. Show that if you average 40 miles per hour on a trip from Chicago to St. Louis, then for at least one instant during your trip your speedometer reads 40 miles per hour.

7. Given that $f(x) = (x + 2)/(x + 1)$ and $a = 1$, $b = 2$, find all values X in the interval $(1, 2)$ such that

$$f'(X) = \frac{f(b) - f(a)}{b - a}.$$

8. Prove that a particle whose velocity is zero during an interval of time is standing still during that time interval.

9. Prove that if $f'(x) = 3x^2$, then $f(x) = x^3 + c$ for some constant c. [*Hint*: What is $(f - g)'$ if $g(x) = x^3$?]

10. Prove that if f and g have the same derivative on an interval I, then $f(x) = g(x) + c$ for some constant c.

11. Prove that if f' is constant, say $f' = m$, then $f(x) = mx + b$.

12. What is the most general function f such that $f''(x) = x$? [*Hint*: Use Problem 10 twice, first finding what f' must be, and then f.]

13. A body moving vertically under the influence of gravity has constant acceleration $d^2s/dt^2 = -32$ ft/sec^2 (if the positive s direction is upward).

a) Show that its motion must be of the form

$$s = -16t^2 + bt + c$$

for some constants b and c.

b) Determine the motion explicitly if $s = 0$ and

$$v = \frac{ds}{dt} = 50 \text{ ft/sec}$$

when $t = 0$.

14. If $f(x) = x^{1/3}$, find the unique number X (depending on x) for which

$$f(x) - f(0) = f'(X)(x - 0).$$

15. If $f(x) = x^{\alpha}$ ($\alpha \neq 1$ and >0), find the unique number X (depending on x) for which

$$f(x) - f(0) = f'(X)(x - 0).$$

16. Show that there is at most one function f defined on the interval $(0, \infty)$ such that

$$f(1) = 0 \qquad \text{and} \qquad f'(x) = 1/x$$

for all x. (Show that if f and g both have these properties, then $f(x) = g(x)$ for all x.)

17. Show that a function f is uniquely determined on an interval I if we know its derivative f' and its value at one point. That is, if $f' = g'$ on I and if $f(x_0) = g(x_0)$, then $f = g$ on I.

18. Complete Example 1 in the text by proving algebraically that $a < X < b$.

19. A motorist enters a toll road on which the speed limit is 55 mph and receives a ticket stamped 12:00 noon. He drives 140 miles on the road and exits at 2:00 P.M. Upon paying his toll he is arrested for speeding. Why?

---

2 ELEMENTARY GRAPH SKETCHING

In this section and the next we shall see how calculus helps us to sketch graphs.

If we look at some random graphs of differentiable functions and a few tangent lines to them (Fig. 1), it appears that the behavior of f is related to the *sign* of f', as follows.

Figure 1

First, a graph with positive slope is rising, and a graph with negative slope is falling. Thus we conjecture:

THEOREM 2 *If $f' > 0$ on an interval I, then f is increasing on I. If $f' < 0$ on an interval I, then f is decreasing on I.*

Second, it appears that if a graph changes from sloping up to sloping down, then in between it must have gone over a summit point where the slope is zero. Similarly, if it changes from sloping down to sloping up, then it must have gone through a bottom point where the slope is zero. That is,

If f is differentiable on an interval I, and f' changes sign on I, then $f' = 0$ somewhere in I.

We conclude that if f' is never zero, then the sign of f' never changes, and hence (by Theorem 2) the graph is either everywhere rising or everywhere falling. That is, f is monotone on I. Thus,

THEOREM 3 *If f' exists and is never zero on an interval I, then f is monotone on I.*

Critical Points The values of x for which $f'(x) = 0$ are called the *critical points* of f. They are critical for determining the shape of the graph, since they divide the axis into a succession of intervals on each of which f is monotone. This idea lets us find the overall shape of a graph with a minimum of calculation.

The Shape Principle We call Theorem 3 the shape principle. Its proof will be discussed after some examples.

Example 1 Apply Theorem 2 to $f(x) = x^n$, where n is a positive integer.

Solution If n is even, then $n - 1$ is odd, so the derivative $f'(x) = nx^{n-1}$ is negative on $(-\infty, 0)$ and positive on $(0, \infty)$. Therefore, by Theorem 2, f is decreasing on $(-\infty, 0)$ and increasing on $(0, \infty)$.

Similarly, if n is odd then we find that f is increasing on both $(-\infty, 0)$ and $(0, \infty)$, and hence everywhere.

These familiar facts about x^n can also be proved by algebra. (See the problems at the very end of Chapter 2.) □

Example 2 Sketch the graph of $f(x) = (x^3 - 3x + 2)/3$ using Theorem 3.

Solution   The critical points of f are the values of x for which $f'(x) = 0$. Since

$$f'(x) = \frac{3x^2 - 3}{3} = x^2 - 1,$$

the critical points of f are the roots of $x^2 - 1 = 0$, i.e., the two points $x = -1, +1$.

We make a little table showing the values of f at the critical points and we add the behavior of f at $\pm\infty$. We have

$$f(-1) = \frac{-1 + 3 + 2}{3} = \frac{4}{3} \quad \text{and} \quad f(1) = \frac{1 - 3 + 2}{3} = 0.$$

Also, since $f(x)$ behaves like its highest power $x^3/3$ for large x, we see that $f(x) \to -\infty$ as $x \to -\infty$, and $f(x) \to +\infty$ as $x \to +\infty$. Our table is thus

x	$-\infty$	-1	1	$+\infty$
$f(x)$	$-\infty$	$4/3$	0	$+\infty$

The shape principle says that f is monotone on each of the intervals

$$(-\infty, -1), \quad (-1, 1), \quad (1, \infty).$$

So if we plot f at the critical points, then we can just "connect the dots" to see how f behaves. Connecting $(1, 0)$ to $(+\infty, +\infty)$ means drawing to the right from $(1, 0)$ with positive slope. We thus get the scheme shown in Fig. 2.

Figure 2

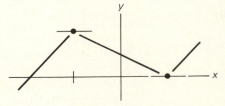

Note that the table shows all of this. It shows, for example, that $f(x)$ goes from $4/3$ to 0 as x increases from -1 to 1, so f is decreasing over $(-1, 1)$.

Now consider how we would actually sketch the graph from this information. There aren't going to be any sharp

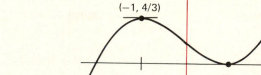

Figure 3

corners; the graph is going to be as smooth and simple as possible subject to the above conditions. It changes from rising to falling as it passes over $x = -1$ and so is smoothly turning downward there. (In Section 3 we shall consider what this really means.) Similarly, it is smoothly turning upward as it passes over $x = +1$. Its overall shape must therefore be something like Fig. 3.

Example 3 Sketch the graph of $f(x) = x^4 - 2x^2$ in the above way.

Solution We have $f'(x) = 4x^3 - 4x = 4x(x^2 - 1)$, giving the critical points $x = -1, 0, +1$. When $|x|$ is very large, $f(x)$ behaves like x^4 and so approaches $+\infty$ as x approaches $-\infty$ or $+\infty$. We thus get the following determining table, marked for monotonicity.

x	$-\infty$	-1	0	$+1$	∞
$f(x)$	∞	-1	0	-1	∞

decr. incr. decr. incr.

In this case we also can find the x-intercepts (i.e., the solutions of $x^4 - 2x^2 = 0$), which are $-\sqrt{2}, 0, \sqrt{2}$. Plotting the critical points and intercepts and using the shape principle, we get the graph in Fig. 4.

Figure 4

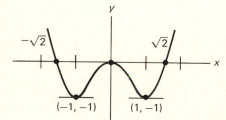

Another method for discovering whether f is increasing or decreasing on one of its intervals of monotonicity is to check the sign of $f'(x)$ *at one point* in the interval. We know that f' has constant sign over the whole interval, and this check will tell us what that constant sign is, and hence whether f is increasing or decreasing there.

Consider the above example from this point of view. Instead of finding how f behaves at $-\infty$, we note that

$$f'(-2) = 4x(x^2 - 1)|_{x=-2} < 0.$$

Since this is negative, f must be decreasing on $(-\infty, -1)$. Similarly, the evaluation $f'(2) > 0$ shows that f must be increasing over $(1, \infty)$.

The shape principle may seem just as obvious geometrically as the mean-value principle. Logically, it is a consequence of the mean-value principle and properties of continuous functions. The proof is immediate (except for one fine point).

Proof of Theorem 2 By the mean-value principle and the hypothesis that f' is everywhere positive,

$$\frac{f(x_2) - f(x_1)}{x_2 - x_1} = f'(X) > 0$$

for any two distinct points x_1 and x_2 in I. Thus $f(x_2) - f(x_1)$ is positive whenever $x_2 - x_1$ is positive, so f is increasing on I.

It follows in the same way that f is decreasing on I if f' is everywhere negative on I.   ■

Partial Proof of Theorem 3 In practice it will always be the case that f' is continuous (on its domain). Then f' cannot change sign on an interval without going through the value 0, by the intermediate-value principle, and the shape principle follows as noted earlier.   ■

In theory, f' can have discontinuities. However, it is still true that f' cannot change sign without going through 0 (see Section 9), so Theorem 3 holds in any event.

REMARK (I) Sometimes our information is that f' is positive at a certain point, $f'(x_0) > 0$, and we know nothing else. What then can we conclude? Only the following:

THEOREM 4 *If $f'(x_0) > 0$, then $f(x) - f(x_0)$ changes sign from $-$ to $+$ as x crosses the point x_0. That is, there is a small interval about x_0 on which we have*

$$f(x) < f(x_0) \qquad if \ x < x_0,$$
$$f(x) > f(x_0) \qquad if \ x > x_0.$$

(If $f'(x_0) < 0$, the f inequalities are interchanged.)

Proof Since $f'(x_0)$ is the limit of the difference quotient $(f(x) - f(x_0))/(x - x_0)$ as $x \to x_0$, it follows that if $f'(x_0)$ is positive then so is the difference quotient for all x sufficiently close to x_0. (See the limit law L4 in Chapter 2, Section 2.) And since a quotient is positive exactly when its numerator and denominator have the same sign, we conclude that

$$f(x) - f(x_0) \qquad and \qquad x - x_0$$

have the same sign when x is close to x_0. ∎

REMARK (II) In Example 1 we concluded from the shape principle that f was decreasing on the open interval $(-1, 1)$. In fact, f was decreasing on the closed interval $[-1, 1]$, even though $f' = 0$ at the two endpoints. In general,

If f is continuous on $[a, b]$ and monotone on (a, b), then f is monotone on $[a, b]$.

This should seem reasonable. It can be proved from the limit laws. (See Problem 32 in Section 5 of Chapter 2.)

PROBLEMS FOR SECTION 2

Use calculus to sketch the graphs of the following equations:

1. $y = x^2 - x$
2. $y = 2 - x - x^2$
3. $y = x^2 - 4x + 4$
4. $y = x^2 - 2x + 2$
5. $s = t^3 - t$
6. $y = x^4 - 3x^2 + 2$
7. $u = 3v^4 + 4v^3$
8. $2y = x^3 - 3x$
9. $6y = x^3 + 3x$
10. $y = x^3 - 4x^2 - 3x$
11. $y = x^3 - x^2 - x$
12. $2y = x^3 - 3x^2$
13. $y = x^3 - 3x^2 + 3x$
14. $y = x^3 - 4x^2 + 3x$
15. $y = x + \dfrac{1}{x}$ (Be careful around the origin.)
16. $y = x^2 - 3x^{2/3}$ (Same warning)
17. $3y = 3x^5 - 10x^3 + 15x + 3$
18. $y = 3x^4 - 8x^3 + 6x^2 + 1$

19. $4y = x^4 - 4x$

21. Discuss the graph of

$$y = ax^2 + bx + c$$

in the light of the shape principle.

23. a) Prove that $\sin x \leq x$ for all nonnegative x (use Theorem 2).

  b) Use part (a) and Theorem 2 again to conclude that

$$1 - \cos x \leq \frac{x^2}{2}.$$

25. Prove that if $f'(x) \geq 0$ for every x in an interval I, then $f(x) \leq f(y)$ whenever $x < y$ in I.

27. Carefully sketch the graphs of $y = x$ and $y = x + x^2$ over the interval $[-1, 1]$. Then draw part of the graph of a function f that will lie between these two curves but will not be monotone on any interval containing $x = 0$. You won't be able to draw the whole graph of f, but describe what feature it will have that will prevent it from being monotone on any interval about 0.

29. Rewrite the inequality proved in Problem 23(b) as

$$1 - \frac{x^2}{2} \leq \cos x,$$

and continue applying Theorem 2, showing first that

$$x - \frac{x^3}{3!} \leq \sin x$$

for all nonnegative x, and then that

$$\frac{x^2}{2} - \frac{x^4}{4!} \leq 1 - \cos x.$$

20. $4y = x^4 - 10x^2 + 9$

22. Sketch the graph of a function f defined for positive x and having the properties

$$f(1) = 0,$$
$$f'(x) = 1/x, \quad \text{all } x > 0.$$

24. Suppose that $g(x)$ is defined on an interval containing the origin, with

$$g(0) = 0 \quad \text{and} \quad g'(0) = 1.$$

Show that there is an interval I about 0 on which

$$g(x) > \frac{x}{2} \text{ when } x > 0; \quad g(x) < \frac{x}{2} \text{ when } x < 0.$$

[*Hint*: Apply Theorem 4.]

26. Let f be any function at all satisfying the inequality

$$x \leq f(x) \leq x + x^2.$$

Prove that $f'(0)$ exists and has the value 1.

28. What conclusion can you draw from the results in the above two problems?

30. Carry this process on for one more step, proving that

$$\sin x \leq x - \frac{x^3}{3!} + \frac{x^5}{5!}$$

for all nonnegative x.

31. a) Use inequalities from the above two problems to show that

$$\sin x \approx x - \frac{x^3}{3!}$$

with a positive error that is at most $x^5/5!$.

b) Show that $\sin 0.3 \approx 0.2955$ to the nearest four decimal places.

33. Suppose that we know the following things about a function f:

1. $f'' > 0$ on $[0, \infty)$;

2. $f(0) > 0$; $\lim_{x \to \infty} f(x) = 0$.

Prove that $f > 0$ on $[0, \infty)$. (Show, first, that if it were ever the case that $f'(x_0) > 0$, then $f(x) \to +\infty$ as $x \to +\infty$. So $f' \le 0$ on $[0, \infty)$.)

32. A function may have a derivative from the right at a point x_0 and also a derivative from the left at x_0, and yet $f'(x_0)$ may fail to exist because the one-sided derivatives are not equal. The function $|x|$ is an example. Show, however, that any such function is continuous at x_0. (See the proof of Theorem 4 in Chapter 3.)

34. Show that $f(\theta) = (a - b\cos^2\theta)/\sin^2\theta$ is monotone on $(0, \pi/2)$, provided that $a \ne b$.

3
CONCAVITY; SINGULAR POINTS

One of the most distinctive features of a graph is the direction in which it is turning or curving. The first two sections of graphs shown in Fig. 1 are *concave up* (turning upward) and the second two are *concave down* (turning downward). Note that the graph of a decreasing function can be turning in *either* direction (first and third patches), as can the graph of an increasing function (second and fourth patches).

Figure 1

Finding the concavity of a graph contributes as much to getting a correct overall picture as locating the critical points. Knowing how the graph is curving is really what tells us its shape. For example, if

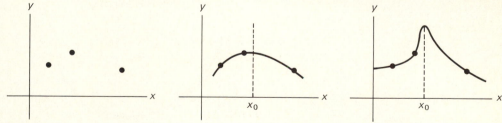

Figure 2

we know that the three points shown in Fig. 2 lie on a graph that is concave down, then we know how the graph must look. But if we don't know that the graph is concave down then it may look entirely different, as at the right in Fig. 2. Roughly speaking, *constant* concavity rules out oscillatory, or wiggly, behavior (Fig. 3).

Figure 3

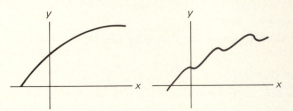

Most graphs will be concave up over some intervals and concave down over other intervals. Points across which the direction of concavity changes are called *inflection points*. In Fig. 4 the graph of Fig. 3, Section 2 is relabeled for concavity. The only point of inflection is at $x = 0$, the curve being concave down over $(-\infty, 0]$ and concave up over $[0, \infty)$. We shall see below how we know this.

Figure 4

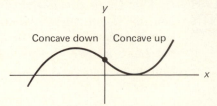

Geometrically, it appears that a graph is *concave up* if its tangent line rotates *counterclockwise* and its slope *increases* as the point of tangency runs from left to right along the curve. You should be able to "see" the slope increasing. If you visualize a succession of tangent lines as in Fig. 5, then you see that each slope is greater than the one before. Or you can visualize holding a straightedge against

Figure 5

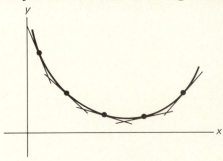

the bottom of the curve and "rolling" it counterclockwise along the curve. We make this property our definition.

DEFINITION A function f is *concave up* over the interval I if f' exists and is an increasing function on I. Similarly, f is *concave down* on I if f' is decreasing on I.

In view of the above definition, the results of Section 1 relate the *concavity* of f to its *second* derivative f'', as follows.

1. If $f''(x) > 0$ on an interval I, then f concave up on I (since then f' is increasing on I), and if $f''(x) < 0$, then f is concave down on I.

2. If f'' exists and is never zero on an interval I, then f is of constant concavity on I (since then f' is monotone on I). The intervals of constant concavity of f (the intervals of monotonicity of f') are thus marked off by the values of x where $f''(x) = 0$, that is, by the critical points of f'.

Example 1 Sketch the graph of $f(x) = x^2 - 2x - 3$.

Before using calculus, we recall from Chapter 1 that the graph is a parabola opening upward, and that the vertex can be found algebraically by completing the square.

Solution We have

$$f'(x) = 2x - 2 = 2(x - 1),$$
$$f''(x) = 2.$$

There is a single critical point, at $x = 1$ (since $f'(x) = 0$ at

Figure 6

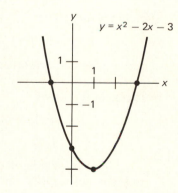

just that point). And the graph is everywhere concave up (since f'' is everywhere positive), thus confirming a feature of a parabolic graph that we have always taken for granted. Since $f(x) = (x - 3)(x + 1)$, the x-intercepts are at 3 and -1. Also, $f(0) = -3$ and $f(1) = -4$. Figure 6 summarizes all this information.   □

Example 2   Show that Fig. 4 has the correct concavity.

Solution   Figure 4 is the graph of $f(x) = (x^3 - 3x + 2)/3$, so
$$f''(x) = 2x.$$

Since $f''(x) = 2x$ is negative on $(-\infty, 0)$, the graph is concave down there. And since f'' is positive on $(0, \infty)$, the graph is concave up there. Figure 4 shows all this. (In fact, Fig. 4 was originally drawn with these concavity facts in mind.)   □

Example 3   Since $d^2 \sin x/dx^2 = -\sin x$, it follows that the graph of $\sin x$ is concave down whenever it is above the x-axis and concave up whenever it is below (Fig. 7). We have always drawn the graph this way, but until now we couldn't be absolutely certain that there wasn't some extra wiggle we were overlooking.

Figure 7

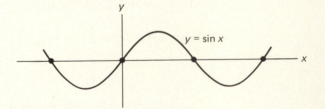

□

Example 4   Sketch the graph of
$$f(x) = x^4 - 2x^3 = x^3(x - 2),$$

using the concavity shape principle.

Solution   We have
$$f'(x) = 4x^3 - 6x^2 = 2x^2(2x - 3),$$
$$f''(x) = 12x^2 - 12x = 12x(x - 1).$$

The intervals of constant concavity for f are marked off by the values of x for which $f''(x) = 0$ as we noted above, and we see that these are $x = 0$ and $x = 1$. The intervals of constant concavity are thus $(-\infty, 0]$, $[0, 1]$, and $[1, \infty)$.

Next, we check the sign of f'' at a point in each interval. We see, for example, that $f''(2)$ is positive, so f is concave up on $[1, \infty)$. And since $f''(1/2)$ is negative, f is concave down on $[0, 1]$. In this way we get the concavity scheme of Fig. 8.

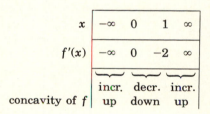

Figure 8

Another method is to compute the value of f' at each of its critical points and its limits at $\pm\infty$, to see whether f' increases or decreases from one critical point to the next. This would give the table

x	$-\infty$	0	1	∞
$f'(x)$	$-\infty$	0	-2	∞
	incr.	decr.	incr.	
concavity of f	up	down	up	

However, we would generally use the first method, and hence not compute values of f' (except to find where $f' = 0$).

Having found the concavity intervals the only things further we need are the critical points of f. We see from the formula for $f'(x)$ that $f'(x) = 0$ when $x = 0$ and $x = 3/2$.

We do *not* need to determine where f is increasing or decreasing because that information is already present. For example, a curve that is concave up must be increasing just after a horizontal tangent (Fig. 9).

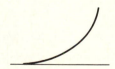

Figure 9

We compute the values of f at all the special points and end up with the following combined table.

x	$-\infty$	0	1	3/2	$+\infty$
$f(x)$	$+\infty$	0	-1	$-27/16$	$+\infty$
$f'(x)$	$-\infty$	0	-2	0	$+\infty$
$f''(x)$		0	0		
Concavity		Up	Down	Up	

These data determine the features of the graph (Fig. 10).

Figure 10

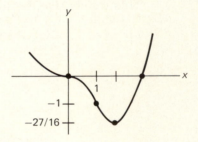

Example 5   We sketch the graph of

$$f(x) = \frac{1}{1 + x^2}.$$

The first two derivatives are

$$f'(x) = \frac{-2x}{(1 + x^2)^2},$$

$$f''(x) = \frac{(1 + x^2)^2(-2) - (-2x)2(1 + x^2)(2x)}{(1 + x^2)^4}$$

$$= \frac{-2(1 + x^2) + 8x^2}{(1 + x^2)^3}$$

$$= \frac{2(3x^2 - 1)}{(1 + x^2)^3}.$$

This time we go directly to the combined table, listing the zeros of both f' and f'', the values of f at these critical points, and its limits at infinity.

x	$-\infty$	$-1/\sqrt{3}$	0	$1/\sqrt{3}$	$+\infty$
$f(x)$	0	3/4	1	3/4	0
$f'(x)$			0		
$f''(x)$		0		0	
Concavity		Up	Down	Up	

The direction of concavity depends on whether f' is increasing or decreasing, and this was determined by checking the sign of $f''(x)$ at a single interior point of each concavity interval.

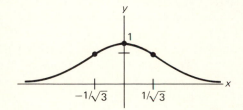

Figure 11

In plotting the above values in Fig. 11 we use the approximation $1/\sqrt{3} \approx 3/5$. As x tends to infinity, the graph approaches the x-axis. Two curves related this way appear to be "tangent at infinity." We say they are *asymptotic*.   □

Singular Points   If a function f is defined by an expression involving fractional exponents or a nontrivial denominator, then there are apt to be one or more isolated points x_0 at which f' (and perhaps f) is undefined. Such a point is called a *singularity* of f.

DEFINITION   A function f has a *singularity* at x_0 if $f'(x_0)$ fails to exist but f is defined and differentiable at all other points of an interval about x_0.

Each of the functions $g(x) = x^{2/3}$ and $h(x) = 1/x^2$ has a singularity at the origin, as shown in the middle and at the right in Fig. 12. Note that f itself may or may not be defined at a singularity.

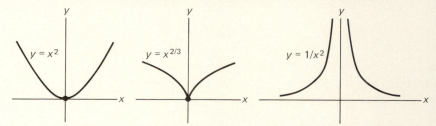

Figure 12

When we compare these two functions with $f(x) = x^2$, shown at the left, we see that a singularity can be just as critical for the behavior of a graph as a point where $f'(x) = 0$. We therefore must list singularities along with critical points as points across which the behavior of f may change. In fact, it is customary to extend the notion of critical point to include the first type of singularity shown above, as follows:

DEFINITION   A *critical point* of a continuous function f is an interior point x in its domain at which either $f'(x) = 0$ or $f'(x)$ is not defined.

A singularity at which f itself is undefined is not a critical point; we want a critical point to correspond to a point on the graph of f. Thus $f(x) = 1/x^2$ has a singularity at the origin, but this is not called a critical point of f.

In view of this definition and the remark at the end of Section 2, the shape principle can be reformulated as follows:

> **Shape Principle**   *If f is continuous on an interval I and without critical points interior to I, then f is monotone on I.*

Example 6   Graph the function

$$f(x) = \frac{1}{x^2 - 1}.$$

Solution   This function has the added complication of infinite singularities at ± 1. Now we have to determine whether $f(x)$ approaches $+\infty$ or $-\infty$ as x approaches one of these points from the left and from the right. But again concavity considerations do it for us, as we shall see in a minute. We have

$$f'(x) = \frac{-2x}{(x^2 - 1)^2},$$

$$f''(x) = \frac{(x^2 - 1)^2(-2) - (-2x)2(x^2 - 1)2x}{(x^2 - 1)^4}$$

$$= \frac{-2(x^2 - 1) + 8x^2}{(x^2 - 1)^3}$$

$$= \frac{2(3x^2 + 1)}{(x^2 - 1)^3}.$$

Notice that f' has *no* critical points (since f'' is never 0), but has singularities at $x = \pm 1$. The intervals of constant concavity are thus $(-\infty, -1)$, $(-1, 1)$, $(1, \infty)$. As usual, we determine the direction of concavity from the sign of f'' at one point in each of these intervals ($x = 0, \pm 2$). Then we have the following table, including these check points as well as the points where the function is not defined (nd):

x	$-\infty$	-2	-1	0	$+1$	$+2$	$+\infty$
$f(x)$	0		nd	-1	nd		0
$f'(x)$			nd	0	nd		
$f''(x)$		$+$	nd	$-$	nd	$+$	
Concavity		Up		Down		Up	

We now know that the graph is concave down over $(-1, 1)$ and that it blows up at each endpoint. It therefore must approach $-\infty$ at both endpoints. Similarly it is concave up over $(-\infty, -1)$ and blows up at -1 and so must approach $+\infty$ there. Reasoning this way, we get the graph shown in Fig. 13. The graph is asymptotic to the x-axis and to the two vertical lines at $x = 1$ *and* $x = -1$. Note how these three asymptotes combine with the known concavity to govern the overall shape of the graph.

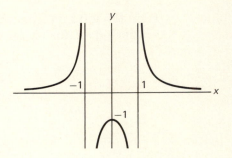

Figure 13

Example 7 Graph

$$f(x) = (x^2 - 1)^{2/3}.$$

Solution We have

$$f'(x) = \frac{4}{3} x(x^2 - 1)^{-1/3} = \frac{4}{3} \frac{x}{(x^2 - 1)^{1/3}},$$

$$f''(x) = \frac{4}{3} \frac{(x^2 - 1)^{1/3} - x \frac{1}{3} (x^2 - 1)^{-2/3} \cdot 2x}{(x^2 - 1)^{2/3}}$$

$$= \frac{4}{9} \frac{x^2 - 3}{(x^2 - 1)^{4/3}}.$$

Thus, f is everywhere defined and continuous;

f has critical points at $-1, 0, 1$;

f' has critical points at $-\sqrt{3}, \sqrt{3}$ and singularities at $-1, 1$.

Since f is even, we can restrict the table to $[0, \infty)$.

x	0	1	$\sqrt{3}$	∞
$f(x)$	1	0	$2^{2/3}$	∞
$f'(x)$	0	$\mp\infty$	$2 \cdot 2^{2/3}/\sqrt{3}$	∞
$f''(x)$	$-$	∞	0	
Concavity	Down	Down	Up	

Using a pocket calculator, we find

$$f(\sqrt{3}) = 2^{2/3} \approx (1.26)^2 \approx 1.59,$$
$$f'(\sqrt{3}) \approx 1.83.$$

The graph is shown in Fig. 14.

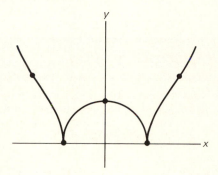

Figure 14

The Notion of Concavity Probably you would agree that a concave upward graph has all four of the properties suggested by Fig. 15.

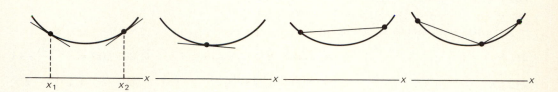

Figure 15

These properties are

1. *The slope $f'(x)$ of the tangent line is an increasing function of x ($f'(x_1) < f'(x_2)$ if $x_1 < x_2$).*

2. *The graph of f lies strictly above each tangent (except for the point of tangency).*

3. *Each arc of the graph of f lies strictly below its chord (except for the two endpoints).*

4. *The second of two nonoverlapping chords always has the larger slope.*

However, we do not have to check all four of these properties for a function since it can be shown that they are all logically equivalent to each other. So if the graph of a differentiable function has any one of these properties, then of necessity it has them all. It is by virtue of this theorem that the definition of concavity given earlier is adequate.

PROBLEMS FOR SECTION 3

Determine the intervals of constant concavity, critical points, and singularities (if any) for the graph of each of the following equations. Then sketch the graph.

1. $x^2 - 4 = y$ **2.** $y = x^2 + x$ **3.** $3y = 8x^3 - 6x + 1$

4. $y = 8x^3 - 6x^2 + 1$ **5.** $y = x^3 + x$ **6.** $y(x - 2) = 1$

7. $x^2y - 2x^2 - 16y = 0$ **8.** $xy - y - x - 2 = 0$ **9.** $y = \sin 3x$

10. $x^2y - x^2 + 4y = 0$ **11.** $y = x^3 - 3x + 2$ **12.** $x^2y - x - 4y = 0$

13. $y = x^2 - 2x + 3$ **14.** $x^2y + y - 4x = 0$ **15.** $y = x^3 - (21/4)x^2 + 9x - 4$

16. $y = x^4 - 2x^2$ **17.** $9y = (x + 2)(x - 2)^3$ **18.** $y = (x - 1)(x + 1)^3$

19. $y = (1/3)x^2 - 4x^2 + 12x - 8$

20. $y = x - x^3$, over the interval $[0, 1]$ (Be sure to have the right slopes at the endpoints.)

21. $f(x) = x^{1/3}$ **22.** $f(x) = x^{2/3}$ ▦ **23.** $f(x) = x^{2/3}\left(\dfrac{5}{2} - x\right)$

24. $f(x) = x^{1/3} + x^{2/3}$ **25.** $f(x) = 1/(x^2 - x)$ **26.** $f(x) = 1/(x^3 - x^2)$

27. $f(x) = \sin |x|$ **28.** $f(x) = (x - 1)^{1/3} + (x + 1)^{1/3}$ ▦ **29.** $f(x) = (x - 1)^{2/3} - (x + 1)^{2/3}$

▦ **30.** $y = x/2 - \sin x$, over the interval $[0, 2\pi]$.   ▦ **31.** $y = 2\sin x + \sin 2x$, over $[0, 2\pi]$.

32. $y = x/\sqrt{1 + x^2}$. (This function has finite limits as $x \to \pm\infty$, which determine horizontal asymptotes toward which the graph tends as $x \to +\infty$ and $x \to -\infty$, respectively. Be sure to get these values right.)

33. $y = x^3/(1 + x^2)$

34. $y = \tan x$, over $(-\pi/2, \pi/2)$. Your results should justify the general features shown in Section 3 of Chapter 4.

35. $y = \sec x$, over $(-\pi/2, 3\pi/2)$.

36. $y = \dfrac{1}{x} + x$. Note that this graph is asymptotic to the line $y = x$, since the difference between the two y-coordinates $(1/x)$ approaches 0 as $x \to \infty$.

37. $y = \dfrac{1}{x} - x$

38. $y = \dfrac{1}{x^2} + 2x$

39. $y = \dfrac{1}{x} + \sqrt{x}$. (This graph is asymptotic to the half parabola $y = \sqrt{x}$.)

40. A function graph turns in the same direction whether it is traced forward or backward, so $f(x)$ and $f(-x)$ should show the same concavity behavior. State this fact more carefully and verify it analytically.

41. Show that the cubic graph

$$y = x^3 + ax^2 + bx + c$$

has three possible shapes, depending on whether $a^2 < 3b$, $a^2 = 3b$, or $a^2 > 3b$. Sketch the three possible shapes.

In the following four problems we consider a differentiable function f over an interval I and show that the various possible definitions for f being *concave up* are all equivalent.

42. Show that if $f'(x)$ is increasing, then the graph of f lies strictly above each tangent line (except for the point for tangency). [*Hint*: Try to use the fact that if $g'(x) > 0$, then g is increasing, for a suitably chosen function g.]

43. Show that if the graph of f lies strictly above each of its tangent lines, then the second of any pair of nonoverlapping chords always has the larger slope.

44. Show that if the graph of f has the property that the second of any pair of nonoverlapping chords always has the larger slope, then each arc of the graph of f lies strictly below its chord (except for the endpoints).

45. Show, finally, that if each arc of the graph of f lies strictly below its chord, then $f'(x)$ is increasing.

46. Show that a rational function of degree at most 1 has an asymptote as x approaches infinity.

4

FINDING A MAXIMUM OR MINIMUM VALUE

We turn now to the problem of finding the maximum (or minimum) value that a varying quantity can have, and "where" it occurs. Historically, this was one of the very first successes of calculus.

Suppose, for example, that a rancher has only one mile of fence with which to enclose a rectangular grazing area along the bank of a straight river. There is no need to fence the bank of the river itself, so the fence will run along only three sides of the rectangle. The problem is to choose the dimensions of the rectangle so as to maximize its area. Note that if a *square* were chosen, then each of the three fenced sides would be 1/3 of a mile long and the area would be 1/9 square mile. However, there is a better solution.

If x is the fence length *perpendicular* to the river (Fig. 1), then the length available to run parallel to the river is $1 - 2x$ and the

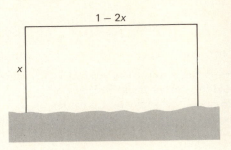

Figure 1

area enclosed is

$$A = x(1 - 2x) = x - 2x^2.$$

The physical limitation on x is that $0 \leq x \leq 1/2$. The graph of A as a function of x over this interval must look something like Fig. 2, because $A = 0$ when $x = 0$ or when $x = 1/2$, and $A > 0$ in between. It seems clear that A will have its maximum value at a point where the tangent line is horizontal, i.e., where $dA/dx = 0$. Assuming that this is correct, we have

$$\frac{dA}{dx} = 1 - 4x,$$

and $dA/dx = 0$ when $x = 1/4$. The maximum area is thus

$$A = x - 2x^2\big|_{x=1/4} = \frac{1}{4} - \frac{1}{8} = \frac{1}{8} \text{ sq. mi.}$$

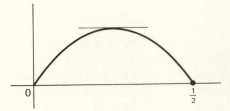

Figure 2

Before considering more examples, we shall show that the simple procedure used above is correct.

Suppose that a quantity to be maximized is called y and that we have found an expression for y as a continuous function of some other variable, say $y = f(x)$, where the conditions of the problem restrict x to an interval I.

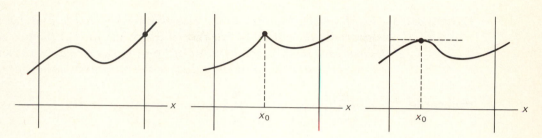

Figure 3

Figure 3 illustrates the possible ways in which $f(x)$ can achieve a maximum value over an interval I:

1. At an endpoint of the interval; or

2. At an interior point where the derivative is not defined; or

3. At an interior point x_0 where the derivative exists.

In the third case it probably seems geometrically clear that $f'(x_0)$ must be 0, and this is easily verified analytically.

Theorem 5 *If a function f assumes a maximum (or minimum) value at an interior point x_0 in its domain, and if $f'(x_0)$ exists, then $f'(x_0) = 0$.*

Proof Theorem 3 of Section 1 shows that if $f'(x_0)$ is *not* zero, then $f(x_0)$ is *not* a maximum value. So if $f(x_0)$ is the maximum value, then $f'(x_0)$ must be zero. ∎

Thus, if f has a maximum value on I, then it must occur at an endpoint or at a critical point. To finish up, we need to know under what circumstances we can be sure that f does have a maximum value. Here is the guarantee.

> **Extreme-Value Principle** If f is continuous on a finite closed interval $[a, b]$, then f assumes maximum and minimum values there.

The proof is in Appendix 5. Like the intermediate-value principle, the extreme-value principle underlies a lot of calculus and will come up frequently from now on. Here it combines with the preceding analysis of possibilities to give the theorem we are after.

THEOREM 6   *Let f be continuous on the closed interval* $[a, b]$, *and let M be the largest of the values of f at its critical points and at the two endpoints. Then M is the maximum value of f on* $[a, b]$. *Similarly, the smallest of these critical and endpoint values is the minimum value of f on* $[a, b]$.

Example 1   Find the maximum and minimum values of $f(x) = x^4 - 4x$ on the interval $[0, 2]$.

Solution   The derivative $f'(x) = 4x^3 - 4 = 4(x^3 - 1)$ is zero when $x^3 - 1 = 0$, i.e., when

$$x^3 = 1, \quad \text{or} \quad x = 1.$$

Thus $x = 1$ is the only critical point of f. Theorem 6 says that the maximum value of f on $[0, 2]$ is either this critical point value, $f(1) = -3$, or one of the two endpoint values, $f(0) = 0$ and $f(2) = 16 - 8 = 8$. So the maximum value is 8. Similarly, the minimum value is -3.   □

Example 2   Find the extreme values (the maximum and minimum values) of $f(x) = x^3 - x^2$ on the interval $[1, 4]$.

Solution   Since

$$f'(x) = 3x^2 - 2x = x(3x - 2),$$

the critical points of f are $x = 0, \frac{2}{3}$. Neither of these critical points belongs to the interval $[1, 4]$, so the maximum and minimum values of f on this interval are the two endpoint values,

$$f(4) = 48 \quad \text{and} \quad f(1) = 0,$$

by Theorem 6.   □

Example 3   Find the extreme values of

$$f(x) = x^3 - 2x^2 + x + 1$$

on $[-1, 1]$.

Solution   We first find the critical points:

$$f'(x) = 3x^2 - 4x + 1 = (3x - 1)(x - 1) = 0 \quad \text{at } x = 1/3, 1.$$

Next we compute the values of f at the critical point(s) and endpoints:

$$f(-1) = -3,$$

$$f(1/3) = \frac{1}{27} - \frac{2}{9} + \frac{1}{3} + 1 = \frac{31}{27},$$

$$f(1) = 1.$$

Then $31/27$ and -3 are the maximum and minimum values, respectively, by Theorem 6. □

If I is missing an endpoint, or if I extends to infinity, then the extreme-value principle is not applicable and in fact f may not *have* a maximum (minimum) on I. Three such situations are shown in Fig. 4. In the middle and left figures, f has a minimum but not a maximum value. In the right figure it has neither.

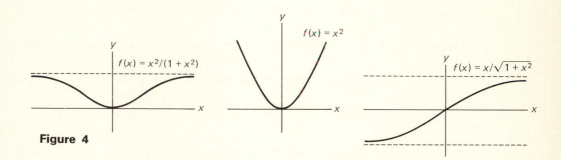

Figure 4

The easiest case to treat is when there is only one critical point x_0 in I. Then we have the following tests.

THEOREM 7 *Suppose that a continuous function f has exactly one critical point x_0 on an interval I. Then $f(x_0)$ is the maximum value of f on I if either one of the following is true:*

1. (**First-derivative test**) *$f'(x)$ is positive at some trial point before x_0 and negative at some trial point after x_0; or*

2. (**Second derivative test**) *$f''(x_0)$ exists and is negative.*

The same tests with "signs reversed" guarantee that $f(x_0)$ is the minimum value of f on I.

Proof   The critical point x_0 divides I into two subintervals, I_1 and I_2, on each of which f' has constant sign (since neither contains an interior critical point). So if we find that $f'(x) > 0$ at some trial point x in I_1, then $f' > 0$ everywhere and f is increasing on I_1. Similarly, if there is a trial point x in I_2 at which $f'(x) < 0$, then f is decreasing on I_2. Therefore, f is increasing up to x_0 and decreasing after x_0, so $f(x_0)$ is the maximum value of f on I.

Finally, if $f''(x_0) < 0$, then f' changes sign from $+$ to $-$ as x crosses x_0 by Theorem 3 and we are back to the first test.   ∎

Example 4   Apply Theorem 7 to $f(x) = x + (1/x)$ on the interval $(0, \infty)$.

Solution   Since $f'(x) = 1 - (1/x^2)$, the only critical point of f in the interval $(0, \infty)$ is at $x = 1$. At this point $f''(x) = 2/x^3$ is positive. Therefore $f(1) = 2$ is the minimum value of f, by the second-derivative test. There is no maximum value.   □

Example 5   Find the maximum and minimum values of $f(x) = x/(1 + x^2)^2$.

Solution   Using the quotient rule, we find that

$$f'(x) = \frac{1 - 3x^2}{(1 + x^2)^3},$$

so there are two critical points, at $x = \pm\sqrt{3}/3$. We now seem to be stymied, since neither Theorem 6 nor Theorem 7 can be applied. However, on $[0, \infty)$ there is a single critical point, at $x = \sqrt{3}/3$, and we can apply the first test in Theorem 7. Since $f'(0) = 1$ and $f'(1) = -1/4$, we can conclude that $f(\sqrt{3}/3) = 3\sqrt{3}/16$ is the maximum value of f on $[0, \infty)$. But f is negative on $(-\infty, 0]$, so the above value is the absolute maximum of f.

Finally, since f is an odd function, $f(-\sqrt{3}/3) = -3\sqrt{3}/16$ is the absolute minimum value of f.   □

Relative Extrema   We say that f has a *local maximum*, or a *relative maximum*, at x_0 if $f(x_0)$ is the maximum value of f on *some* interval $[u, v]$ centered at x_0 (Fig. 5). This may not be the absolute maximum value of f because f may very well have larger values outside of $[u, v]$.

Similarly, x_0 may be a *relative minimum* point for f.

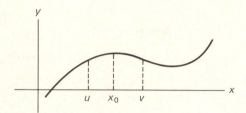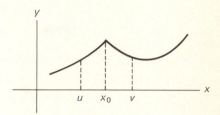

Figure 5

To test for a relative maximum we may be able to apply Theorem 7. For example:

If x_0 is a critical point at which $f''(x_0) < 0$, then $f(x_0)$ is the maximum value of f on any interval $[u, v]$ about x_0 that contains no other critical point.

If f has neither a local maximum nor a local minimum at a critical point x_0, then f is monotone in any interval $[u, v]$ about x_0 that contains no other critical point. That is, f is then monotone "in the neighborhood of x_0." This can happen whether $f'(x_0)$ is zero or is undefined, as in Fig. 6.

Determining which of these several possibilities occur at each of the critical points of f is called *classifying the critical points* of f.

Figure 6

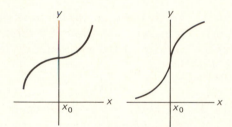

PROBLEMS FOR SECTION 4

Use Theorem 6 to find the maximum and minimum values of each of the following functions over the specified interval.

1. $f(x) = 3 + x - x^2$; $[0, 2]$
2. $f(x) = x^3 - x^2 - x + 2$; $[0, 2]$
3. $F(x) = x^3 - 5x^2 - 8x + 20$; $[-1, 5]$
4. $g(x) = x\sqrt{x + 3}$; $[-3, 3]$
5. $h(x) = x^{2/3}(x - 5)$; $[-1, 1]$
6. $G(x) = \sqrt[3]{x^2 - 2x}$; $[0, 2]$
7. $y = x\sqrt{2x - x^2}$; $[0, 2]$

8. $y = x^{3/2}(x - 8)^{-1/2}$; $[10, 16]$ **9.** $f(x) = 2x^4 + x$; $[-1, 1]$ **10.** $f(x) = \sin x + \cos x$; $[0, \pi]$

11. $f(x) = x^5 - 10x$; $[-2, 2]$ **12.** $f(x) = (x - 1)/(x^2 + 1)$; $[0, 3]$ **13.** $f(x) = \sin x + \cos^2 x$; $[0, 2\pi]$

Use Theorem 7 to find the maximum or minimum value of $f(x)$ in each of the following situations.

14. $f(x) = 2x^2 + x$ **15.** $f(x) = 3 + x - x^2$ **16.** $f(x) = x^3 - 3x$, over $[-1, 2]$.

17. $f(x) = x^3 - x^2 - x + 2$, over $[-1, 0]$.   **18.** $f(x) = x\sqrt{x + 3}$, over $[-3, 3]$.

19. $f(x) = 2x^4 + x$ **20.** $f(x) = \sin x + \cos x$, over $[0, \pi]$.

21. $f(x) = x^2/(1 + x^3)$ over $[0, \infty)$.   **22.** $f(x) = 2 \sec x + \tan x$, over $(-\pi/2, \pi/2)$.

23. Show that if $f'(x_0) = f''(x_0) = 0$ but $f'''(x_0) > 0$, then f is increasing on an interval about x_0. (Apply Theorem 7 to $f'(x)$.)

24. Show that if $f'(x_0) = f''(x_0) = f'''(x_0) = 0$ but $f''''(x_0) > 0$, then f has a relative minimum at x_0. (Apply Theorem 7 to $f'''(x)$ and work backward, or apply Problem 23 to $f'(x)$.)

25. Reread Problems 23 and 24. Now state and prove the next in this chain of results.

26. State the analogs of Problems 23 and 24 having reversed inequalities.

Find and classify all critical points of the following functions, using the higher derivative tests from Problems 23 through 26.

27. $y = x^3 - 3x^2$ **28.** $y = x^3 - x^2 - x + 2$ **29.** $y = (x - 1)^2(x + 1)$ **30.** $y = x^4 - 2x^3 + 3$

31. $y = 2x^3 - 24x$ **32.** $y = 3x^4 - 4x^3 - 6x^2 + 4$ **33.** $y = x^3(x + 3)^2$

34. $y = x^4 - 4x^3 + 16x$ **35.** $y = x^6 - 2x^3 + 1$ **36.** $y = (x^3 - 1)^2$

5
MAXIMUM–MINIMUM WORD PROBLEMS

In this section we begin with several examples, and then give a few general suggestions for solving word problems. The first requirement is to read the problem very carefully.

Example 1

We wish to make an open box out of a square piece of cardboard measuring 12 inches on a side. To do this we cut out equal squares from the four corners and bend up the four resulting flaps. What size squares should be cut out if we wish the volume of the resulting box to be as large as possible?

Solution   Let the edge of the cut-out squares be x inches. Then the box will have a square base measuring $12 - 2x$ inches on a side, and its height will be x inches. Now the volume V of a rectangular box is the product of its three dimensions, so

$$V = x(12 - 2x)^2 = 144x - 48x^2 + 4x^3.$$

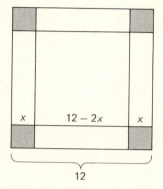

Figure 1

We want to maximize V, so we compute

$$\frac{dV}{dx} = 144 - 96x + 12x^2 = 12(12 - 8x + x^2)$$

$$= 12(6 - x)(2 - x).$$

The conditions of the problem restrict x to the interval $[0, 6]$. We note that V is zero at each endpoint and positive in between, and that V has a single interior critical point, at $x = 2$. This therefore gives the maximum value of V, by Theorem 6 or Theorem 7. $\square$

Example 2 We continue with our coal producer from Examples 4 and 5 in Section 4 of Chapter 3. Suppose that the revenue function for the operation is

$$R = 9x - 2x^2,$$

and that the cost function is

$$C = x^3 - 3x^2 + 4x + 1.$$

Then

$$P = R - C = -x^3 + x^2 + 5x - 1$$

is the profit, and the producer proposes to operate so as to maximize profit. Since

$$\frac{dP}{dx} = -3x^2 + 2x + 5 = -(3x - 5)(x + 1),$$

and since the production x is always positive, we see that the only critical point for profit occurs at $x = 5/3$. The corresponding value of the profit works out to be $P = 148/27 \approx 5\frac{1}{2}$. It seems clear that this must be the maximum profit, and this is easily checked. For one thing, $d^2P/dx^2 = -6x + 2$ is negative at $x = 5/3$, so Theorem 7 guarantees that this is the maximum point. In the units we chose before, the maximum profit will be $5500 per week (we are probably dealing with a corporation) and this will be achieved by producing 1667 tons of coal per week. □

Example 3   Any entrepreneur will experience losses when production is too low, because of fixed costs, and also when production is too high, because of very high marginal costs. Unless the operation can be profitable at some in-between production, there won't be a business at all. We can therefore suppose the profit curve looks like this:

Figure 2

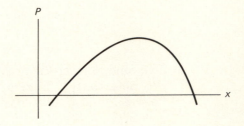

Thus the maximum profit will always occur at a critical point. Now $P = R - C$, so if all functions are differentiable, then

$$0 = \frac{dP}{dx} = \frac{dR}{dx} - \frac{dC}{dx}$$

if and only if $dR/dx = dC/dx$. Therefore,

Profit will be maximum when the marginal revenue equals the marginal cost. □

It will often happen, in setting up a maximum problem, that the variable to be maximized (or minimized) is most naturally expressed in terms of more than one independent variable. But then the conditions of the problem will allow all but one of these variables to be eliminated, as in the next example.

Example 4   Consider the problem of finding the most efficient shape for a one-cubic-foot cylindrical container (Fig. 3), the most efficient cylinder being the one using the least material, i.e., having the *smallest total*

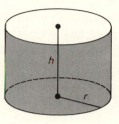

Figure 3

surface area. Now the volume of a cylinder is $V = \pi r^2 h$, and its total surface area is $A = 2\pi r^2 + 2\pi rh$. The problem, therefore, is to determine r and h so that A has the smallest possible value subject to the requirement that $V = 1$. The restriction

$$V = \pi r^2 h = 1$$

allows us to solve for either h or r as a function of the other, say,

$$h = \frac{1}{\pi r^2};$$

and when this is substituted in the area formula, the area becomes a function of r alone,

$$A = 2\pi r^2 + \frac{2}{r}.$$

The physical limitation on r is that it be positive. To find the critical points we compute dA/dr and set it equal to 0,

$$\frac{dA}{dr} = 4\pi r - \frac{2}{r^2} = 0,$$

which gives $4\pi r^3 = 2$. Thus $r = (1/2\pi)^{1/3}$ is the only critical point. Since

$$\frac{d^2A}{dr^2} = 4\pi + \frac{4}{r^3} > 0,$$

we know from Theorem 7 that A has its minimum value at the critical point.

Since we asked for the *shape* of the most efficient container, we want h also, and we compute

$$h = \frac{1}{\pi r^2} = \frac{(2\pi)^{2/3}}{\pi} = \frac{2}{(2\pi)^{1/3}} = 2r.$$

The most efficient shape is therefore that for which the altitude h equals the diameter $2r$ of the base. The can has a square cross section along its vertical axis.   □

Here are some general suggestions for tackling max–min problems.

1. Assign variables and constants to all relevant quantities—let $x =$ this, let $y =$ that, let $A =$ the other, etc. If possible, do this by drawing a figure and labeling it.

2. Express the quantity to be maximized (or minimized) as a function f of *one* other variable.

If you can't do this right away, then try to find two equations involving three variables, and then eliminate one variable between them.

These equations may come from formulas of geometry or from the statement of the problem.

3. Determine the interval I over which the *independent* variable is allowed to range.

4. Find the maximum or minimum value of f on I by Theorem 6 or Theorem 7.

Sometimes an angle can be taken as the independent variable, as in the following example.

Example 5   A corridor of width a meets a corridor of width b at right angles. Workers wish to push a heavy beam of length c on dollies around the corner, but they want to be sure it will be able to make the turn before starting. How long a beam will go around the corner (neglecting the width of the beam)?

Solution   This is essentially the problem of minimizing the length l of the segment cut off by the corridor on a varying line through the corner point P. The beam will go around the corner if its length c is less than the minimum of l.

In terms of the angle θ shown in Fig. 4, the length l is given by

$$l = \frac{a}{\sin \theta} + \frac{b}{\cos \theta}.$$

The domain of θ is $(0, \pi/2)$, and l approaches $+\infty$ as θ approaches 0 or $\pi/2$. The minimum of l will therefore occur

Figure 4

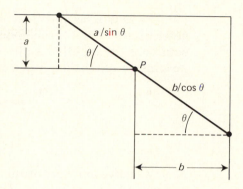

at a point where $dl/d\theta = 0$. We see that

$$\frac{dl}{d\theta} = -\frac{a \cos \theta}{\sin^2\theta} + \frac{b \sin \theta}{\cos^2\theta}$$

$$= \frac{b \sin^3\theta - a \cos^3\theta}{\sin^2\theta \cos^2\theta},$$

so $dl/d\theta = 0$ if and only if

$$b \sin^3\theta - a \cos^3\theta = 0 \quad \text{or} \quad \tan \theta = \left(\frac{a}{b}\right)^{1/3}.$$

The corresponding value of l then works out to be

$$l = (a^{2/3} + b^{2/3})^{3/2}.$$

The beam will go around the corner if its length is not greater than this value of l. □

The Method of Auxiliary Variables

This is a different procedure that can be used when we initally have two equations in three variables.

Example 6 In Example 4 we wanted to minimize

$$A = 2\pi r^2 + 2\pi rh = 2\pi[r^2 + rh]$$

subject to the condition that

$$V = \pi r^2 h = 1.$$

This restriction determines either h or r as a function of the other, and what we did before was to solve for h, substitute in the first equation, and go on.

The new procedure will be to treat h as a function of r that is determined by the second equation, but leave both h and r in both equations. We differentiate the two equations with respect to r, keeping in mind that h is a function of r:

$$\frac{dA}{dr} = 2\pi\left[2r + r\frac{dh}{dr} + h\right],$$

$$\frac{dV}{dr} = \pi\left[r^2\frac{dh}{dr} + 2rh\right] = 0.$$

Now we solve and substitute. We solve the second equation for dh/dr and substitute in the first. This gives $dh/dr = -2h/r$ and

$$\frac{dA}{dr} = 2\pi\left[2r + r\left(\frac{-2h}{r}\right) + h\right] = 2\pi[2r - h].$$

The critical-point equation $dA/dr = 0$ now appears as $2r - h = 0$, or

$$h = 2r. \qquad\qquad \square$$

Note two things.

First, the answer appears as a relationship among the supporting variables. For some problems this is really all we want. The above problem, for instance, asked for the most efficient shape and this is exactly what the answer $h = 2r$ gives us. If we want to know what the minimum area is, and for what value of r it occurs, then we have to substitute $h = 2r$ in the equation $\pi r^2 h = 1$, and solve for r.

The second point about this method is that it really gives us only the critical point configuration. Further work is needed to prove that it gives the maximum (or minimum) value. Thus, to complete the problem above, we calculate

$$\frac{d^2A}{dr^2} = 2\pi\left[2 - \frac{dh}{dr}\right] = 2\pi\left[2 + \frac{2h}{r}\right].$$

Since this is everywhere positive, A is concave up as a function of r, and its only critical point gives the minimum value of A.

This final step may not be feasible in some problems.

Example 7 Find the rectangle of largest area (with sides parallel to the axes) that can be inscribed in the ellipse

$$\frac{x^2}{a^2} + \frac{y^2}{b^2} = 1.$$

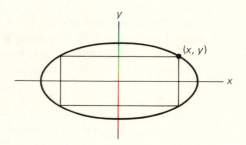

Figure 5

Solution If (x, y) is the rectangle vertex lying in the first quadrant (Fig. 5), then the area of the rectangle is

$$A = 4xy,$$

where

$$\frac{x^2}{a^2} + \frac{y^2}{b^2} = 1.$$

The second equation determines y as a function of x, and we can therefore differentiate both equations with respect to x:

$$\frac{dA}{dx} = 4x\frac{dy}{dx} + 4y,$$

$$\frac{2x}{a^2} + \frac{2y\,dy/dx}{b^2} = 0.$$

We can then solve for dy/dx from the second equation and substitute in the first, obtaining

$$\frac{dy}{dx} = -\frac{b^2x}{a^2y},$$

$$\frac{dA}{dx} = 4\left(-\frac{b^2x^2}{a^2y} + y\right).$$

The critical-point equation $dA/dx = 0$ is thus

$$0 = \frac{-b^2x^2 + a^2y^2}{a^2y},$$

from which we conclude that

$$\frac{y}{x} = \frac{b}{a}.$$

at the critical point. That is, the critical rectangle has sides proportional to the axes of the ellipse. In order to find the rectangular area, we have to substitute $y = bx/a$ in the equation of the ellipse and solve for x and y. We find that

$$x = \frac{a}{\sqrt{2}}, \qquad y = \frac{b}{\sqrt{2}},$$

so the critical (maximum) area is

$$A = 4xy = 2ab. \qquad \square$$

PROBLEMS FOR SECTION 5

1. In a certain physical situation, it is found that a variable quantity Q can be expressed in the form $Q = x + y$, where x and y are related by

$$x^2 + y^2 = 1.$$

Find the maximum value of Q.

3. Find the maximum value of $Q = xy$, if x and y are related by the condition

$$2x^2 + y^2 = 1.$$

5. Find the point on the graph $y = x^3 - 6x^2 - 3x$ at which the tangent line has minimum slope.

7. Find the minimum value of $3x + y^3$ on the circle $x^2 + y^2 = 2$.

9. Find the circular cylinder of maximum volume that can be inscribed in the cone of altitude H and base radius R.

11. A fenced rectangular garden is to be laid out with one side adjoining a neighbor's lot and is to contain 48 square yards. If the neighbor pays for half the dividing fence, what dimensions of the garden will minimize your cost for the fence?

2. In a certain situation it is found that a variable quantity Q that is to be maximized is expressed in terms of variable quantities x and y by $Q = xy^2$, where x and y are related by

$$x^2 + y^2 = 1.$$

Find the maximum value of Q.

4. A quantity Q is given by $Q = x^3 + 2y^3$, where x and y are positive variables related by the equation

$$x + y = 1.$$

Show that the minimum value of Q is $6 - 4\sqrt{2}$, and that this is greater than $1/3$.

6. Find the maximum value of $x - 2y$ on the unit circle $x^2 + y^2 = 1$.

8. Find the rectangle of maximum area that can be inscribed in the circle

$$x^2 + y^2 = 1.$$

10. Find the circular cylinder of maximum volume that can be inscribed in a sphere of radius r.

12. The post office places a limit of 120 in. on the combined length and girth of a package. What are the dimensions of a rectangular box with square cross section, that will contain the largest mailable volume?

13. Find the proportions for a rectangle of given area A that will minimize the distance from one corner to the midpoint of a nonadjacent side.

14. You are the owner of an 80-unit motel. When the daily charge for a unit is $40, all units are occupied. If the daily charge is increased by $2d$ dollars then $3d$ of the units become vacant. Each occupied unit requires $6 daily for service and repairs. What should be your daily charge to realize the most profit?

15. A wire 24 in. long is cut in two, and then one part is bent into the shape of a circle and the other into the shape of a square. How should it be cut if the sum of the areas of the circle and the square is to be a *minimum*?

16. Find the positive number for which the sum of its reciprocal and four times its square is the smallest possible.

17. Find the shortest distance from the point $(0, 2)$ to the hyperbola $x^2 - y^2 = 1$.

18. Find the rectangle of maximum perimeter that can be inscribed in the ellipse $4x^2 + 9y^2 = 36$. The sides of the rectangle are parallel to the axes of the ellipse.

19. The strength of a wooden beam of rectangular cross section is proportional to the width of the beam and the square of its depth. Determine the proportions of the strongest beam that can be cut from a circular log.

20. A small loan company is limited by law to an 18% interest charge on any loan. The amount of money available for loans is proportional to the interest rate the company will pay investors. If the company can loan out all the money that is invested with it, what interest rate should it pay its investors in order to maximize profits?

21. Find the point(s) on the parabola $4 = x^2$ closest to the point $(0, 2)$.

22. A rectangular box with a square base and a cover is to be built to contain 640 cu. ft. If the cost per square foot for the bottom is 30¢ and for the top and sides 20¢, what are the dimensions for a minimum cost?

23. Find the minimum distance from the point $(1, 4)$ to the parabola $x^2 = 3y$.

24. In connecting a water line to a building at A (see Fig. 6), the contractor finds that he must connect to a certain point C on the water main which lies under the paved parking lot of a shopping center. It will cost him $40 a foot to dig, lay pipe, fill, and resurface the parking lot, but only $24 a foot to lay pipe along the edge. Find the distance from the store water inlet to the point B that will minimize his costs, where B is the point at which he should turn the water line and go directly to point C.

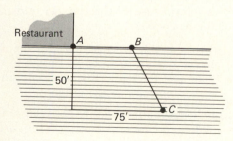

Figure 6

25. A frame for a cylindrically shaped lamp shade is made from a piece of wire 16 ft long. The frame consists of two equal circles, two diametral wires in the upper circle, and four equal wires from the upper to the lower circle. For what radius will the volume of the cylinder be a maximum?

26. A rectangular box is to be made from a sheet of tin 20 in. by 20 in. by cutting a square from each corner and turning up the sides. Find the edge of this square which makes the volume a maximum.

27. Two towers 40 ft apart are 30 and 20 ft high respectively. A wire fastened to the top of each tower is guyed to the ground at a point between the towers, and is tightened so that there is no sag. How far from the taller tower will the wire touch the ground if the length of the wire is a minimum?

28. A steel mill is capable of producing x tons per day of a low-grade steel and y tons per day of a high-grade steel, where

$$y = \frac{40 - 5x}{10 - x}.$$

If the fixed-market price of low-grade steel is half that of the high-grade steel, find the amount of low-grade steel to be produced each day that will yield maximum receipts.

29. A window in the shape of a rectangle surmounted by a semicircle has a perimeter of 24 ft. If the amount of light entering through the window is to be as large as possible, what should the dimensions of the window be?

30. Two roads intersect at right angles, and a spring is located in an adjoining field 10 yd from one road and 5 yd from the other. How should a straight path just passing the spring be laid out from one road to the other so as to cut off the least amount of land? How much land is cut off?

31. Find the shortest path in the above situation.

32. Which is the more efficient container, a cube or a circular cylinder? (Find the most efficient cylinder and then compare with the cube. See page 198.)

33. Solve the Norman window problem (Problem 29) if colored glass is used for the semicircle that admits only half as much light per unit area as does the clear glass in the rectangle.

34. A silo consists of a circular cylinder with a hemispherical top. Find its most efficient shape. That is, find the relative dimensions that maximize the volume for a given total area (base area plus lateral area of the cylinder, plus the area of the hemisphere).

35. Find the most economical shape for a silo like that in the above problem if the construction of the hemisphere costs twice as much per unit area as the construction of the cylinder.

36. A smooth graph not passing through the origin always has a point (x_0, y_0) closest to the origin. Show that the segment from the origin to (x_0, y_0) is perpendicular to the graph.

37. A tank has hemispherical ends and a cylindrical center. Find its most efficient shape. That is, find the proportions of the cylinder that will maximize the volume for a given surface area.

38. If the hemispherical ends to the tank in the above problem cost twice as much (per unit area) as the cylindrical center, find its most economical shape.

39. It costs a manufacturer $x^3 - 3x^2 + 4x + 1$ hundreds of dollars to produce x thousands of an item per week; the items sell for $10x$ hundreds of dollars. How many should be sold per week in order to maximize profit, and what is this maximum profit?

40. A manufacturer's cost function is $4\sqrt{x} + 1$, and the revenue function is $9x - x^2$, both in thousands of dollars per thousand items. Find the maximum profit.

41. The illumination from a light source is inversely proportional to the square of the distance from the light and directly proportional to the sine of the angle of incidence. How high should a light be placed on a pole in order to maximize the illumination on the ground along the circumference of a circle of radius 25 feet?

42. Find the maximum value of $f(\theta) = (a - b\cos\theta)/\sin\theta$ on $(0, \pi/2)$, where $0 < b < a$.

This problem comes up repeatedly in discussions of why certain angles seem to occur in nature. See E. Batschelet, *Introduction to Mathematics for Life Scientists* (New York–Heidelberg–Berlin: Springer-Verlag, 1974), Chapter 9.

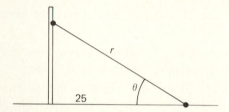

6
RELATED
RATES

Suppose that y is a function of x, say $y = f(x)$, and that x varies with time t. Because y depends on x, y will also vary with time. That is, if y is a function of x and x is a function of t, then y is a function of t. The chain rule says that

$$\frac{dy}{dt} = \frac{dy}{dx}\frac{dx}{dt} = f'(x)\frac{dx}{dt}.$$

Thus the rate of change of y is related to the rate of change of x because of the relationship between y and x.

Example 1 Suppose that a circular ripple is spreading out over a pond, and that the radius r is increasing at the rate of 2 feet per second at the moment when $r = 5$ feet. How fast is the disturbed area increasing at that moment?

Solution The disturbed area A and the radius r are varying with time, but at all times they are related by the equation

$$A = \pi r^2.$$

We take the time t to be the underlying independent variable, and differentiate this identity with respect to t:

$$\frac{dA}{dt} = 2\pi r \frac{dr}{dt}.$$

At the moment in question we are given $dr/dt = 2$ and $r = 5$. Therefore, at that moment

$$\frac{dA}{dt} = 2\pi 5 \cdot 2 = 20\pi,$$

or approximately 63 square feet per second.   □

There is no sure-fire rule for solving rate problems, but keeping the following points in mind will help.

1. Assign variables and constants to all relevant quantities. To the extent possible, do this by drawing and labeling a figure.

2. Find an equation in *two* variables, the variable whose rate is wanted and the variable whose rate is given. The equation must be true for all values of the variables.

3. Differentiate with respect to t, regarding both variables as functions of t. The final result will be an equation connecting the wanted rate and the given rate.

4. Substitute known values and solve for the wanted rate.

2′. If a single equation in the two key variables is hard to find, try to find two equations in three variables, and then eliminate one variable, either before or after the next step. (See Examples 2 and 3.)

2″. If more than one rate is known then find an equation connecting the variable whose rate is wanted with the two or more variables whose rates are given. Then proceed as in steps 3 and 4. (See Example 4.)

WARNING   We do not substitute specific values (Step 4) until after we have differentiated (Step 3).

Example 2   A tank in the shape of an inverted cone with equilateral cross section is being filled with water at the constant rate of 10 cubic feet per minute. How fast is the water rising when its depth is 5 feet?

Solution   The amount of water in the cone is related to its depth h and its surface radius r by the formula for the volume of a cone:

$$V = \frac{1}{3}\,\pi r^2 h.$$

Here, V, r, and h are all increasing with time as water pours in, but are always related by the above equation. We are given $dV/dt = 10$, and we want to know the value of dh/dt when $h = 5$. Our general principle requires an equation relating V to h. The above volume equation isn't quite right because it contains the extra variable r. However, we can eliminate r by using the fact that the tank cross section is equilateral (Fig. 1): We have

$$h^2 + r^2 = 4r^2$$

by the Pythagorean theorem, so

$$h^2 = 3r^2.$$

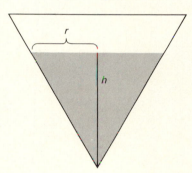

Figure 1

Therefore

$$V = \frac{1}{3}\,\pi\frac{h^2}{3}\cdot h = \frac{\pi}{9}h^3,$$

which is the required equation. Then

$$\frac{dV}{dt} = \frac{\pi}{9}\cdot 3h^2\cdot\frac{dh}{dt}.$$

We are given $dV/dt = 10$, and we want dh/dt when $h = 5$. At that moment

$$10 = \frac{\pi 25}{3}\frac{dh}{dt},$$

and

$$\frac{dh}{dt} = \frac{6}{5\pi}.$$

So the water is rising at $6/5\pi$ ft/min when the depth is 5 ft. $\square$

Example 3 A spherical rubber balloon is being inflated at the rate of 4 cubic inches per second. How fast is its surface area increasing when its radius is 5 inches?

Solution We use the volume and surface area formulas for a sphere:

$$V = \frac{4}{3}\pi r^3, \qquad A = 4\pi r^2.$$

The simplest procedure here is to differentiate these equations as they stand, since we can see that dr/dt will disappear upon dividing the resulting equations. Thus, differentiating both equations with respect to t, we get

$$\frac{dV}{dt} = 4\pi r^2 \frac{dr}{dt}, \qquad \frac{dA}{dt} = 8\pi r \frac{dr}{dt},$$

and therefore,

$$\frac{dA}{dt} = \frac{2}{r}\frac{dV}{dt} = \frac{8}{r}\,\text{in}^2/\text{sec}.$$

Thus

$$\frac{dA}{dt} = \frac{8}{5}\,\text{in}^2/\text{sec}$$

when $r = 5$. $\square$

Example 4 At noon, ship A is moving due north and ship B due east, and A is 30 nautical miles northwest of B. At that moment A is moving at 20 knots (nautical miles per hour) and B at 15 knots. How fast are the ships separating at noon?

Solution We plot the ships' paths as coordinate axes, as shown in Fig. 2.

The distance D between the ships is given by

$$D = \sqrt{x^2 + y^2},$$

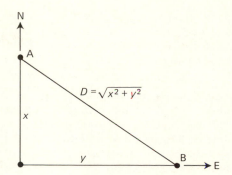

Figure 2

where x and y are the distances from the ships to the point where their paths crossed. As time goes on,

$$\frac{dD}{dt} = \frac{1}{2}\frac{2x\dfrac{dx}{dt} + 2y\dfrac{dy}{dt}}{\sqrt{x^2 + y^2}}$$

$$= \frac{x\dfrac{dx}{dt} + y\dfrac{dy}{dt}}{D}.$$

At the moment in question, $D = 30$, $x = y = 30/\sqrt{2}$, $dx/dt = 20$, and $dy/dt = 15$. Substituting these values in the above equation gives

$$\frac{dD}{dt} = \frac{(30/\sqrt{2})20 + (30/\sqrt{2})15}{30} = \frac{35}{\sqrt{2}} \approx 25 \text{ knots.} \qquad \square$$

PROBLEMS FOR SECTION 6

1. Two variable quantities Q and R are found to be related by the equation

$$Q^3 + R^3 = 9.$$

What is the rate of change dQ/dt at the moment when $Q = 2$, if $dR/dt = 3$ at that moment?

2. Two variable quantities x and y are found to be related by the equation

$$x^2 + y^2 = 25.$$

If x increases at the constant rate of 3 units per unit of time, what is the rate at which y is increasing at the moment when $x = 0$? When $x = 3$? When $x = 5$?

3. A ladder 26 ft long leans against a vertical wall. If the lower end is being moved away from the wall at the rate of 5 ft/sec, how fast is the top descending when the lower end is 10 ft from the wall?

5. A rope 32 ft long is attached to a weight and passed over a pulley 16 feet above the ground. The other end of the rope is pulled away along the ground at the rate of 3 ft/sec. At what rate does the weight rise at the instant when the end being pulled is 12 ft from its initial point?

7. Two cars start from the same point at the same time. One travels north at 25 mi/hr, and the other travels east at 60 mi/hr. How fast is the distance between them increasing as they start out? At the end of an hour?

9. A drawbridge with two 10 ft spans is being raised at a rate of 2 radians per minute. How fast is the distance increasing between the ends of the spans when they are at an elevation of $\pi/4$ radians?

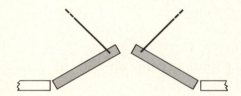

11. If, in Problem 10, the man reaches the center of the bridge 5 sec before the boat passes under it, find the rate at which the distance between them is changing 4 sec later.

13. A hemispherical bowl of diameter 18 inches is being filled with water. If the depth of the water is increasing at the rate of 1/8 in/sec when it is 8 in. deep, how fast is the water flowing in? (The volume of a segment of a sphere of radius r is

$$\pi h^2\left(r - \frac{h}{3}\right),$$

where h is the height of the segment.)

4. A conical funnel is 8 in. across the top and 12 in. deep. A liquid is flowing in at a rate of 60 cu. in/sec and flowing out at a rate of 40 cu. in/sec. Find how fast the surface is rising when the liquid is 6 in. deep.

6. A man 6 ft tall walks away from a street light 15 ft high at the rate of 3 mi/hr.

a) How fast is the far end of his shadow moving when he is 30 ft from the light pole?

b) How fast is his shadow lengthening?

8. A revolving beacon located 3 miles from a straight shore line makes 2 revolutions/min. Find the speed of the spot of light along the shore when it is two miles away from the point on the shore nearest the light.

10. A bridge is 30 ft above a canal. A motor boat traveling at 10 ft/sec passes under the center of the bridge at the same instant that a man walking 5 ft/sec reaches that point. How rapidly are they separating 3 sec later?

12. A conical reservoir 8 ft across and 4 ft deep is filled at a rate of 2 cu. ft/sec. A leak allows fluid to run out at a rate depending on the depth of the water, the loss being at the rate of $h^4/8$ cubic feet per minute when its depth is h feet. At what depth will the surface level stabilize?

14. A balloon is rising vertically over a point A on the ground at the rate of 15 ft/sec. A point B on the ground is level with and 30 ft from A. The angle of elevation is being measured at B. At what rate is the elevation changing when the balloon is $30\sqrt{3}$ ft above A?

15. Water is being pumped into a trough 8 ft long with a triangular cross section 2 ft × 2 ft × 2 ft at the rate of 16 cu. ft/min. How fast is the surface level rising when the water is 1 ft deep?

16. Gas is escaping from a spherical balloon at the rate of 1/2 cu. in/sec. At what rate is the surface area decreasing when the radius is 6 in.?

17. If $uv = 1$ and if v is increasing at the constant rate of 2 units per unit time, what is the rate of change of u when $v = 4$?

18. A particle moves along the parabola $y = 3x^2$ in such a way that its x-coordinate is increasing at the constant rate of 5 inches per second (the inch being the unit of distance in the coordinate plane). What is the rate at which its y-coordinate is changing when $x = -1$? When $x = 0$? When $x = 1/3$?

19. A particle moves along the parabola $y = x^2$, and dy/dt is found to have the value 3 when $x = 2$. How fast is x changing at that moment?

20. A circular wheel of radius 5 feet lies in the xy-plane, with its center at the origin, and revolves steadily in the counterclockwise direction at the rate of 25 revolutions per minute. How fast is the y-coordinate of a particle on the wheel increasing as the particle goes through the point $(3, 4)$? The same question for its x-coordinate.

21. Referring to the above problem, suppose that the wheel is rotating at the constant rate of one radian per minute. Show that then $dy/dt = x$, for all positions of the particle.

22. Suppose a particle traces a circle about the origin in such a way that $dy/dt = x$. Show that the particle is moving along the circle at the constant rate of 1 radian per unit of time.

23. A rolling snowball picks up new snow at a rate proportional to its surface area. Assuming that the snowball always remains spherical, show that its radius is increasing at a constant rate.

24. A ladder 20 feet long leans against a wall 10 feet high. If the lower end of the ladder is pulled away from the wall at the rate of 3 feet per second, how fast is the angle between the ladder and the top of the wall changing at the moment when the angle is $45° = \pi/4$ radians?

25. A light is at the top of a tower 80 ft high. A ball is dropped from the same height from a point 20 ft from the light. Assuming that the ball falls according to the law $s = 16t^2$, how fast is the shadow of the ball moving along the ground two seconds after release?

7

SOME ELEMENTARY ESTIMATION

We shall now apply the mean-value principle directly to some elementary approximations.

In practical computations we almost always replace each number by a finite decimal that approximates it. For example, we know that

$$3.14 < \pi < 3.15,$$

and on the basis of this inequality we know that each of 3.14 and 3.15 approximates π with an error less than 0.01, that is, with an error less than 1 in the second decimal place. The question is; How well does $f(3.14)$ approximate $f(\pi)$ for some given function f?

Approximation and Error

We consider first some notions and notations about approximation. When we regard a number x as an *approximation of a number a*, we call the difference $(a - x)$ the *error in the approximation;* it is *how much we miss the mark by.* The choice of sign is arbitrary, and some people would say that $(x - a)$ is the error; but since we are usually concerned with the *magnitude* of the error, $|a - x|$, this ambiguity in sign is seldom important.

Now if we subtract 3.14 from all terms of the inequality

$$3.14 < \pi < 3.15,$$

we obtain the inequality

$$0 < \pi - 3.14 < 0.01,$$

and this says *exactly* that the error in the approximation of π by 3.14 is positive and less than .01. If we subtract 3.15 instead, we get

$$-.01 < \pi - 3.15 < 0.$$

So this time the error is *negative* but less than .01 in magnitude. These conclusions depend on the basic inequality law:

If $x < y$, then $x + c < y + c$, for any number c.

Other inequality laws also will be needed in these examples. They are mostly in accord with common sense, but can be looked up in Appendix 1 in case of doubt.

We also know that

$$3.141 < \pi < 3.142.$$

(This is what we mean when we write $\pi = 3.141 \cdots$.) This inequality shows that

$$3.14 < \pi < 3.142,$$

so 3.14 approximates π with an error less than 0.002. It follows, in particular, that 3.14 is the two-place decimal *closest* to π, i.e., that

$$|\pi - 3.14| < 0.005.$$

In general, we shall say that

The number x approximates the number a accurately to two decimal places if $|x - a| \leq 0.005$. More generally, the approximation is accurate to n decimal places if

$$|x - a| \leq \frac{10^{-n}}{2} \; (= 5 \cdot 10^{-(n+1)}).$$

Decimals are discussed further in Chapter 11, Section 1.

Applying the Mean-Value Principle

We come now to the use of the mean-value principle. The general question is how well an approximation survives the application of a differentiable function. What connection can we establish between how well x approximates a and how well $f(x)$ approximates $f(a)$?

Example 1 Show that $(3.14)^3$ approximates π^3 with an error less than 1 in the first decimal place, i.e., with an error E less than 0.1. Use the inequality $\pi - 3.14 < 0.002$ that we noted above.

Solution We use the mean-value principle with $f(x) = x^3$ and the x-values 3.14 and π. It guarantees a number X lying between these two x values such that

$$E = \pi^3 - (3.14)^3 = 3X^2(\pi - 3.14).$$

Since $X < 4$ and $\pi - 3.14 < 0.002$, we can conclude that $\pi^3 - (3.14)^3 < 48(0.002) = 0.096 < .1$, as we claimed.

We could have picked a smaller bound for X, such as $X < 3.2$, but $X < 4$ was easier to use in the calculations. ☐

Example 2 A table gives $a = 1.414$ as the value of $\sqrt{2}$ to the nearest three decimal places. Justify this, using the calculator result $(1.414)^2 = 1.999396$.

Solution We use $g(x) = \sqrt{x}$ and the x-values 2 and 1.999396. Then there is a number X lying between these two values such that

$$\sqrt{2} - 1.414 = g'(X)[2 - 1.999396]$$

$$= \frac{1}{2\sqrt{X}} (0.000604) < \frac{0.000604}{2.8} < 0.0003,$$

where we have used the fact that $\sqrt{X} > 1.4$. This meets our criterion for three-decimal-place accuracy. ☐

Example 3   Show that $1/3.14$ approximates $1/\pi$ accurately to three decimal places. Use the fact that $\pi - 3.14 < 0.002$.

Solution   We use the mean-value principle for $f(x) = 1/x$ and $f'(x) = -1/x^2$. Thus

$$\frac{1}{3.14} - \frac{1}{\pi} = \frac{1}{X^2}(\pi - 3.14) < \frac{0.002}{X^2} < \frac{0.002}{9} < 0.0003.$$

Since this is less than 0.0005, the approximation is accurate to three decimal places.   □

Example 4   How closely should we approximate π by a number r in order to ensure that r^3 approximates π^3 accurately to two decimal places?

Solution   We want to know how small to take $|\pi - r|$ in order to ensure that

$$|\pi^3 - r^3| < 0.005.$$

But

$$\pi^3 - r^3 = 3X^2(\pi - r)$$

for some X between π and r, by the mean-value principle. If we restrict r to be less than 4, then $3X^2 < 3(4)^2 = 48$, so

$$|\pi^3 - r^3| < 48\,|\pi - r|.$$

Therefore, in order to ensure that $|\pi^3 - r^3| < 0.005$, it is sufficient to require that

$$48\,|\pi - r| < 0.005 \qquad \text{or} \qquad |\pi - r| < \frac{0.005}{48}.$$

Since

$$\frac{0.005}{48} > \frac{0.005}{50} = 0.0001,$$

it is sufficient to take r so that

$$|\pi - r| < 0.0001,$$

that is, to take r as an approximation to π that is accurate to within 1 in the fourth decimal place. You may know that $\pi = 3.1415\cdots$, which means that

$$3.1415 < \pi < 3.1416.$$

We can thus take r as either 3.1415 or 3.1416, and be certain that r^3 approximates π^3 accurately to two decimal places. ☐

Example 4 is harder than the preceding examples, where we were asked to verify only that $f(x) - f(r)$ is as small as we claimed for two given numbers x and r. Here r is *not* given, and we have to *find* it: we have to find how close r should be taken to x in order to ensure that $f(x) - f(r)$ is as small as required.

The above examples can also be worked out by algebra, but this will not be possible for our later applications of the mean-value principle to computations.

PROBLEMS FOR SECTION 7

1. We know that 3.1416 approximates π with an error less than 10^{-5} in magnitude. Show, then, that $(3.1416)^5$ approximates π^5 to the nearest two decimal places. (Use the fact that $3.15 < \sqrt{10}$.)

2. Assume again that 3.1416 approximates π with an error less than 10^{-5} in magnitude. We find by long division that $1/3.1416 = 0.31831$ with an error less than 10^{-6}. Show, therefore, that 0.31831 approximates $1/\pi$ to the nearest five decimal places.

3. Show that 10 approximates $(100,001)^{1/5}$ to the nearest four decimal places.

4. Show that, if r is the closest eight-place decimal to π, then $r^{1/3}$ approximates $\pi^{1/3}$ with an error less than 1 in the ninth decimal place.

5. Show that, if r is the closest n-place decimal to π, then $r^{1/3}$ approximates $\pi^{1/3}$ with an error less than 1 in the $(n + 1)$st decimal place.

6. a) How closely should we approximate a positive number a by a finite decimal r $(r < a)$ in order to ensure that r^3 approximates a^3 with an error less than 0.01? The answer will be in terms of a.

   b) Same question if the tolerable error in the approximation of a^3 by r^3 is d.

7. How closely should the finite decimal r approximate π in order to ensure that $1/r$ approximates $1/\pi$ with an error at most d?

8. Prove from the mean-value principle that
$$x^n - 1 > n(x - 1)$$
if $x > 1$ and $n > 1$.

9. Prove that if $b > 1$ then $b^n \to \infty$ as $n \to \infty$. [*Hint*: Use the preceding problem, with x fixed and n variable, to show that b^n is larger than any preassigned number M when n is chosen large enough. This is what we mean when we say that $b^n \to \infty$ as $n \to \infty$.

10. Prove that if $0 < a < 1$, then $a^n \to 0$ as $n \to \infty$. (Use the preceding problem.)

11. Prove that if $x > 0$, then

$$\sqrt{1 + x} < 1 + \frac{x}{2}.$$

12. Prove that if $x > 1$, then

$$x^{1/5} < \frac{4}{5} + \frac{x}{5}.$$

8
THE
TANGENT-LINE
APPROXIMATION

The line tangent to the graph of

$$y = f(x)$$

at the point $(a, f(a))$ has the equation

$$y_t = f(a) + f'(a)(x - a).$$

This is the point-slope equation of the tangent line, slightly modified by shifting the y_0 term to the right side. We have used the different dependent variable y_t so that we can compare the two graphs for all values of x. See Fig. 1.

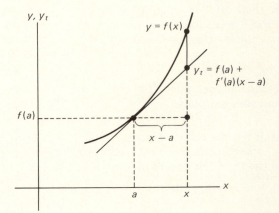

Figure 1

The tangent line has the direction of the graph *at* the point of tangency, and therefore the tangent line is a good approximation to the graph *near* the point of tangency. In analytic terms, the linear function $f(a) + f'(a)(x - a)$ is a good approximation to $f(x)$ near $x = a$:

$$\boxed{f(x) \approx f(a) + f'(a)(x - a).}$$

This tangent-line approximation is related to the calculations in the last section, but here the approximation occurs at the next stage.

There we approximated $f(x)$ by $f(a)$. The error was $f(x) - f(a)$, and we used the mean-value principle to estimate it. Here we are using the better approximation $f(a) + f'(a)(x - a)$ for $f(x)$. This is equivalent to approximating $f(x) - f(a)$ by $f'(a)(x - a)$, so *here we are, in effect, approximating the former error term*. The error in *this* approximation is

$$[f(x) - f(a)] - f'(a)(x - a),$$

and when we use the mean-value principle to estimate this "second generation" error, it will be the *second* derivative f'' that will be involved.

For the moment, let us put aside questions about the error and simply use the above approximate equality as a rough estimate of $f(x)$ when x is near a. Since the new estimating formula contains $(x - a)$, it will be most easily used in situations where we know $x - a$ exactly, and the following examples are of this sort. This temporarily steers us away from the problems in the last section, where we only had a bound on $x - a$. More on this later.

Example 1 When $f(x) = \sqrt{x}$ and $a = 25$, the approximation

$$f(x) \approx f(a) + f'(a)(x - a)$$

becomes

$$\sqrt{x} \approx \sqrt{25} + \frac{1}{2\sqrt{25}}(x - 25).$$

If $x = 27$, this gives us the rough estimate

$$\sqrt{27} \approx 5 + \frac{2}{10} = 5.2.$$

The actual value is 5.196 to three places. ☐

This example shows the circumstances under which the above approximation is most easily used. *If we know the values of f and f' at a, then the formula estimates the value of f at a nearby point x.*

Example 2 Estimate $63^{1/3}$.

Solution Since $64^{1/3} = 4$, we set $f(x) = x^{1/3}$ and evaluate the formula at $a = 64$ and $x = 63$. Here $f'(x) = \frac{1}{3}x^{-2/3}$, and the approximation becomes

$$63^{1/3} \approx 64^{1/3} + \frac{1}{3 \cdot 64^{2/3}} \cdot (-1) = 4 - \frac{1}{48} \approx 4 - \frac{1}{50} = 3.98.$$

(When we replace 1/48 by 1/50 we increase the error by

$$\frac{1}{48} - \frac{1}{50} = \frac{2}{48 \cdot 50} = \frac{1}{1200} < 0.001.)$$

The estimate above is correct to two decimal places, although we have no way of showing this at the moment.   □

If we set $\Delta x = x - a$, then $f(x) - f(a) = \Delta y$, and the approximate equality

$$f(x) \approx f(a) + f'(a)(x - a)$$

turns into

$$\boxed{\Delta y \approx f'(a)\, \Delta x}$$

upon subtracting $f(a)$ from both sides. This form also can be used to make rough estimates.

Example 3   By approximately how much does the volume of a sphere increase when its radius is increased from 10 to 11 feet?

Solution   Here $V = (4/3)\pi r^3$ and $dV/dr = 4\pi r^2$, so the formula is

$$\Delta V \approx 4\pi a^2 \, \Delta r.$$

When $a = 10$ and $\Delta r = 11 - 10 = 1$, we have the estimate

$$\Delta V \approx 4\pi \cdot 100 \cdot 1 = 400\pi \text{ cu. ft.}$$

This is just an order-of-magnitude estimate, but such rough approximations can be very useful in giving a feel for what is happening. The actual value of ΔV here is

$$\Delta V = \frac{4}{3}\pi(11^3 - 10^3)$$

$$= \frac{4}{3}\pi[331] = 441\frac{1}{3}\pi \text{ cu. ft.}$$ □

The Error Now let us consider the error E in these approximations. It is the difference $E = y - y_t$, for which we have the two expressions

$$E = f(x) - [f(a) + f'(a)(x - a)]$$
$$= \Delta y - f'(a)\, \Delta x.$$

Note that the second form of E is obtained from the first by rebracketing. The error E has a very simple geometric interpretation as shown in Fig. 2. Of course, $E \to 0$ as $\Delta x \to 0$, but it appears from the figure that $E = \Delta y - f'(a)\,\Delta x$ becomes small *even in comparison to* Δx, and thus approaches 0 *faster than* Δx.

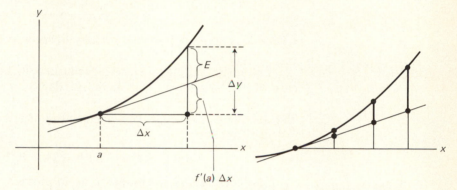

Figure 2

Let us calculate E explicitly for $y = f(x) = x^3$. Then

$$\Delta y = (a + \Delta x)^3 - a^3 = 3a^2\,\Delta x + 3a(\Delta x)^2 + (\Delta x)^3,$$
$$f'(a)\,\Delta x = 3a^2\,\Delta x,$$
$$E = \Delta y - f'(a)\,\Delta x = (\Delta x)^2[3a + \Delta x].$$

Such calculations make it appear that $E = \Delta y - f'(a)\,\Delta x$ approaches 0 faster than Δx by virture of having $(\Delta x)^2$ as a factor. But

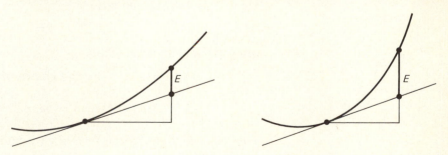

Figure 3

something else is involved too. It is apparent that E will be comparatively small for a graph that bends away from its tangent line slowly (Fig. 3), and our experience with concavity suggests that this will occur if the *second* derivative $f''(x)$ is small near $x = a$.

The following theorem verifies these reasonable conjectures.

THEOREM 8   *If f'' exists on an interval I, then for any two points a and $x = a + \Delta x$ in I, the error E in the tangent line approximation can be written in the form*

$$E = \frac{f''(X)}{2}(\Delta x)^2,$$

where X is some point lying strictly between a and x.

Later (in Chapter 12), this formula will be seen to be a special case of Taylor's formula with remainder.

Trick Proof   Let b be any fixed value of x ($\neq a$) and consider the function

$$h(t) = f(t) + f'(t)(b - t) + K(b - t)^2.$$

The constant K is chosen so as to make

$$h(a) = f(b).$$

We don't care about the actual value of K, so long as it is clear that K exists. Inspection shows that we also have

$$h(b) = f(b).$$

Since $h(b) = h(a)$, it follows from the mean-value principle that $h'(X) = 0$ for some point X strictly between a and b. This turns into the equation

$$f''(X) = 2K$$

when h' is calculated. Thus

$$f(b) = h(a) \qquad \text{(by the choice of } K\text{)}$$

$$= f(a) + f'(a)(b - a) + \frac{f''(X)}{2}(b - a)^2.$$

This is equivalent to the formula we had to prove.   ∎

Example 4   We reconsider Example 1. Taking $f(x) = \sqrt{x}$, $a = 25$, and $\Delta x = 2$, we had

$$\sqrt{27} \approx 5 + \frac{1}{2}\frac{1}{\sqrt{25}} \cdot 2 = 5.2.$$

But now we can go on. Since $f''(x) = -1/(4x^{3/2})$, the theorem says that the error E in this approximation can be written

$$E = -\frac{1}{8X^{3/2}}(2)^2 = -\frac{1}{2X^{3/2}},$$

where X lies between 25 and 27. In particular,

$$|E| \le \frac{1}{2(25)^{3/2}} = \frac{1}{250} = 0.004,$$

so

$$\sqrt{27} \approx 5.20,$$

correct to the nearest two decimal places. $\square$

Finally, we show that the tangent-line approximation would be of little use in the problems of the last section. There we neither knew nor needed to know $x - a$ exactly; only an upper bound was required in estimating the error. Here $x - a$ is part of the approximation formula. We have to know it exactly or else face the problem of approximating the approximation.

Consider, say, the earlier problem of estimating π^3 starting from the estimate 3.14 for π. We now would write

$$\pi^3 \approx (3.14)^3 + 3(3.14)^2(\pi - 3.14),$$

with an error

$$E = \frac{6X}{2}(\pi - 3.14)^2 < 10(.002)^2 = 4 \cdot 10^{-5}.$$

The trouble is that in order to use the above estimate we have to approximate the factor $\pi - 3.14$ on the right side. If we use the approximation $\pi \approx 3.1416$, then $\pi - 3.14 \approx 0.0016$ with an error $< 10^{-5}$. However, by the time the outside factor is multiplied in, this error bound increases to about $30 \cdot 10^{-5} = 3 \cdot 10^{-4}$ and becomes the major error. On the other hand, by the simpler method of the last section, $\pi^3 \approx (3.1416)^3$, with an error

$$3X^2(\pi - 3.1416) < 30 \cdot 10^{-5} = 3 \cdot 10^{-4}$$

of the same order of magnitude, and with about the same amount of arithmetic needed to multiply everything out. So the tangent-line approximation offers no great advantage in this problem.

PROBLEMS FOR SECTION 8

Use the tangent-line approximation to estimate each of the following values.

1. $\sqrt{65}$ **2.** $\sqrt[3]{124}$ **3.** $\sqrt[4]{80}$ **4.** $\sqrt[5]{31}$

5. $(.98)^{-1}$ **6.** $(17)^{1/4}$ **7.** $26^{1/3}$ **8.** $\sqrt{24}$

9. $1/\sqrt{15}$ **10.** $1/80$

11. Find the approximate change in the volume of a sphere of radius r caused by increasing the radius by 2%.

12. A metal cylinder is found by measurement to be 2 ft in diameter and 5 ft long. What, approximately, is the error in the computed volume of this cylinder if the error in measuring the diameter was 1/2 in.?

13. Let V and A be the volume and surface area of a sphere of radius r. If the radius is changed by a small amount Δr, show that

$$\Delta V \approx A \cdot \Delta r.$$

14. Let V and A be the volume and surface area of a cube of edge length x. If x is given a small increment Δx, show that

$$\Delta V \approx A \frac{\Delta x}{2}.$$

▦ **15.** Solve Problem 1 again, along the lines of Example 4 in the text. That is, obtain an expression for the error E by applying Theorem 8, and then state the tangent-line estimate in decimal form, with some statement about E.

▦ **16.** Same for Problem 2.   ▦ **17.** Same for Problem 3.   ▦ **18.** Same for Problem 4.

▦ **19.** Same for Problem 5.   ▦ **20.** Same for Problem 6.   ▦ **21.** Same for Problem 7.

▦ **22.** Using the fact that $(9/2)^2 = 20\frac{1}{4}$, show that

$$\sqrt{20} = 4.4722\ldots$$

with an error less than 1 in the fourth decimal place.

▦ **23.** Using the fact that $(3/2)^4 = 5\frac{1}{16}$, and assuming that $5^{3/4} > 3$, show that

$$5^{1/4} \approx \frac{3}{2} - \frac{1}{216}$$

with an error less than $1/10 \cdot 2^{12}$, and hence less than $3(10)^{-5}$.

24. Show that

$$\sin t \approx t$$

with an error less than $t^3/2$. (Use the fact that $|\sin t| \leq |t|$.)

25. Show that

$$1 - \cos x \leq \frac{x^2}{2}$$

for all x.

26. Prove for any positive number x that

$$\sqrt{1 + x} \approx 1 + \frac{x}{2}$$

with an error less than $x^2/8$.

28. Draw the figure for Theorem 8 showing E as the length of a vertical line segment for the case when Δx is negative (and f'' is positive.)

30. By comparing $\sqrt{10}$ to $\sqrt{9} = 3$, and referring to the above problem, show that $\sqrt{10} \approx 19/6$, with an error that is negative and less than $1/200$ in magnitude. Conclude that

$$3.16 < \sqrt{10} < 3.17.$$

32. In Chapter 4 we defined the differential dy in terms of a new independent variable dx (supposing that $y = f(x)$). Look up that definition, and show that

If we set $dx = \Delta x$, then $\Delta y - dy$ is the error E in the tangent-line approximation.

Draw the figure for the error E and relabel it so as to exhibit the identity $E = \Delta y - dy$.

27. Show that if $h > 0$ and $a < 2$, then

$$(1 + h)^a \approx 1 + ah$$

with an error at most $|a(a - 1)h^2/2|$ in magnitude.

29. Theorem 8 shows that the error E has the same sign as f'' (supposing that f'' has constant sign). The figures in the text and Problem 28 show this geometrically for the concave up case. Draw the corresponding figures for the case that E and f'' are both negative.

31. By comparing $\sqrt{3}$ to $\sqrt{25/9} = 5/3$, show that $\sqrt{3} \approx 26/15$, with an error that is negative and less than $1/750$ in magnitude. Conclude that

$$1.732 < \sqrt{3} < 1.734.$$

33. Find the approximate change in the surface area of a sphere if the radius is increased from 10 in. to 10.1 in.; if the volume is decreased from 27π in^3 to 26π in^3.

The *relative change* in the value of $y = f(x)$, or the *relative error* in the measurement of y, is

$$\frac{\Delta y}{y} \approx \frac{f'(x)}{f(x)} \Delta x.$$

For the volume of a cube of edge-length x, it is

$$\frac{\Delta V}{V} \approx \frac{3x^2}{x^3} \Delta x = 3\frac{\Delta x}{x}.$$

Thus the relative change in the volume is approximately 3 times the relative change in the edge length.

This is a dimensionless quantity (e.g., cubic inches/cubic inches), and it is frequently converted to a percent by multiplying by 100. Thus,

The *percentage change* in the value of $y = f(x)$, or the *percentage error* in the measurement of y, is

$$100\frac{\Delta y}{y} \approx 100\frac{f'(x)}{f(x)} \Delta x.$$

For example, a one percent increase in the edge length of a cube results (approximately) in a 3 percent increase in its volume.

34. The intensity I of radiation given off by a body is proportional to T^4, where T is the absolute temperature (in degrees Kelvin) of the body. What is the percent change in the radiation intensity I if there is a 0.2% increase in its temperature T?

35. What is the error in the computation of the volume of a sphere if there is a one percent error in the measurement of its diameter?

36. The resistance R to the flow of blood through a blood vessel of a given length is proportional to $1/d^4$, where d is its diameter. What is the percent change in R resulting from a one percent decrease in the diameter d?

37. Find the relative change in the area of a sphere if the relative change in its volume is 0.08.

38. The gravitational attraction between two astronomical bodies is proportional to $1/r^2$, where r is the distance between their centers (Newton's inverse square law). What is the relative change in this force if the relative change in r is ϵ?

9
NEWTON'S METHOD FOR ROOTS OF EQUATIONS

The approximate equality

$$f(x) \approx f(a) + f'(a)(x - a)$$

was used in the last section to estimate the value of $f(x)$ for a given x, in terms of known values at a nearby point a.

It can also be used in the opposite direction to estimate the point x where $f(x)$ has a given value k, that is, to estimate a root of the equation $f(x) = k$. We just replace $f(x)$ by k in the approximate equality and solve for x. Again a must be reasonably close to the point x in question, so here a is obtained by a rough initial estimate of the root.

The solution takes a slightly simpler form if we want a root of the equation $f(x) = 0$, and problems can always be set up this way. In this case the approximate equation becomes

$$0 \approx f(a) + f'(a)(x - a),$$

and solving for x gives

$$x \approx a - \frac{f(a)}{f'(a)}.$$

Thus

$$x_1 = a - \frac{f(a)}{f'(a)}$$

is our new estimate of the root x of the equation $f(x) = 0$ in terms of an initial rough estimate $x_0 = a$.

Example 1 Estimate the positive solution of the equation $x^2 = 27$.

> **Solution** We want a root of $x^2 - 27 = 0$, so here $f(x) = x^2 - 27$. Since 27 lies between $5^2 = 25$ and $6^2 = 36$, it is reasonable to take $a = 5$ as the initial rough estimate. So
>
> $$x_1 = a - \frac{f(a)}{f'(a)} = 5 - \frac{-2}{10} = 5.2$$
>
> is our new estimate of the positive root of the equation $x^2 = 27$.   □

Note that we have obtained the same estimate of $\sqrt{27}$ as in the last section, but by a different method. This time we are approximating the root r of the equation

$$f(x) = 0$$

by the solution $x = x_1$ of the equation

$$f(a) + f'(a)(x - a) = 0.$$

Again, there is a very simple geometric interpretation. The root r and its approximation x_1 are the intersection points on the x-axis of the graph of f and its tangent line at $(a, f(a))$, as pictured in Fig. 1.

Figure 1

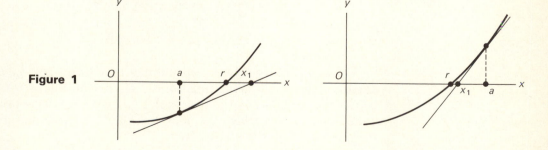

In practice, a will be a preliminary, very rough estimate of the root r, and x_1 will be an improved estimate. The figures show that *if the graph of f is increasing and concave up, then the estimate x_1 will always be larger than the root r, regardless of whether the initial estimate a is larger or smaller than r.*

Referring again to Fig. 1, we can see that there is an important safeguard present in the right configuration that is absent in the left:

If $f(a) > 0$, as at the right, then x_1 lies between r and a and hence is necessarily a better approximation to r than a is.

But if $f(a) < 0$, as at the left, then there is no control on the location of x_1, and it might be so far away as to be useless.

In the remaining discussion we shall assume the essential features shown in Fig. 1, namely, that f is monotone and of constant concavity on a given interval I, and that f has a root in I. We shall also suppose that the first calculated approximation x_1 lies in I, so that the possible difficulty mentioned above does not occur.

We can now *iterate* the approximation, replacing the initial estimate $x_0 = a$ by x_1, and thus obtaining a "second-generation" estimate

$$x_2 = x_1 - \frac{f(x_1)}{f'(x_1)}.$$

It appears from Fig. 2 that x_2 is an exceedingly close estimate of r.

Figure 2

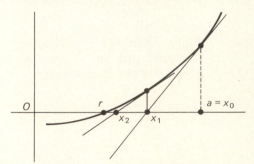

If we go on in this way, we obtain a sequence $x_0, x_1, x_2, x_3, \ldots$, of rapidly improving approximations, i.e., an infinite sequence that converges very fast on the true root $x = r$. This iterative procedure is called *Newton's method.* We shall see that it can be extraordinarily accurate at even the second step.

Example 2 Compute a "second-generation" approximation to the solution of $x^2 = 27$ by starting from the estimate x_1 obtained in Example 1.

Solution

$$x_2 = x_1 - \frac{f(x_1)}{f'(x_1)}$$

$$= 5.2 - \frac{(5.2)^2 - 27}{2(5.2)}$$

$$= 5.2 - \frac{0.04}{10.4}$$

$$= 5.2 - 0.0038461 \cdots.$$

Thus

$$x_2 = 5.196154$$

is our improved estimate. We carried the long division .04/10.4 out to six decimal places because it can be shown that *this second estimate is correct to within two units in the sixth decimal place.* That is, the error $x_2 - r$ is less than $2(10)^{-6}$. □

Estimating the Error We shall now find a bound for the error $E = r - x_1$, where r and x_1 are defined by

$$f(r) = 0,$$
$$f(a) + f'(a)(x_1 - a) = 0.$$

To start with, we subtract from $f(x_1)$ the above two expressions for 0; to get

$$f(x_1) - f(r) = f(x_1) - f(a) - f'(a)(x_1 - a).$$

We now apply the mean-value principle on the left and Theorem 8 on the right, and conclude that

$$f'(X)(x_1 - r) = \frac{f''(Y)}{2}(x_1 - a)^2,$$

where Y is some number between a and x_1, and X is some number between x_1 and r. Now suppose we know that $|f'(x)| \geq L > 0$ and $|f''(x)| \leq B$ for all the x-values in question. Then we have

$$\boxed{|x_1 - r| \leq \frac{B}{2L}(x_1 - a)^2}$$

as a bound for the error $E = r - x_1$ in the estimate of r by x_1. In applying this formula, remember that

a) a is a preliminary, rough estimate of the root r of the equation $f(x) = 0$;

b) x_1 is the new estimate obtained from a by one application of Newton's method;

c) B is an upper bound for $|f''(x)|$ over the values of x in question;

d) L is a positive *lower* bound for $|f'(x)|$ over these values of x.

Example 3   Use this formula to estimate the error in Example 2.

Solution   Here $f(x) = x^2 - 27$, and we are operating with values of x larger than 5. Therefore $f'(x) = 2x \geq 10$, while $f''(x) \equiv 2$. Using these constants in the error formula above, we get

$$|E| \leq \frac{2}{2 \cdot 10} (x_2 - x_1)^2 = \frac{(x_2 - x_1)^2}{10}.$$

(We have replaced a and x_1 by x_1 and x_2.) Since $x_1 - x_2 = 0.0038 \cdots < 0.004$, we see that

$$|E| < \frac{(0.004)^2}{10} = 0.0000016 = 1.6(10)^{-6}.$$

Therefore

$$\sqrt{27} \approx 5.196154,$$

with an error less than 2 in the sixth decimal place.   □

PROBLEMS FOR SECTION 9

In each of the following problems, locate the nearest integer a to the solution of $f(x) = 0$ by sketching the graph of f, and then estimate the root by one application of Newton's formula.

1. $f(x) = x^2 - 10, x > 0$ **2.** $f(x) = x^2 - 5, x > 0$ **3.** $f(x) = x^3 + x - 1$ **4.** $f(x) = x^3 + 2x - 5$

▦ **5.** a) In Problem 4 above the answers are $a = 1$, $x_0 = 1\frac{2}{5} = 1.4$. Now use this estimate for x_0 as a new value of a, and obtain the second-generation estimate

$$x_1 = 1.331.$$

b) Use the error formula, with $f''(x) \leq 6(1.4) = 8.4$ and $f'(x) \geq 5$, to conclude that $x = 1.33$ is the solution of the equation $x^3 + 2x = 5$ to the nearest two decimal places.

▦ **7.** Show that the positive solution of the equation

$$\sin x = \frac{2}{3}x$$

is given by $x = 3/2$, with an error at most 0.01 in magnitude. (Take $a = \pi/2$ and use the approximation once.)

▦ **9.** Starting from the nearest integer solution of the equation $x^2 - 10 = 0$, and applying the reverse tangent line approximation twice (Newton's method) show that

$$\sqrt{10} = 3.16228,$$

correct to five decimal places. (The error is less than 5 in the sixth decimal place.)

▦ **11.** Compute $\sqrt{101}$ to three decimal places by a single application of Newton's method. (Start with $a = 10$.)

▦ **6.** The answers in Problem 3 above are $a = 1$, $x_0 = 3/4$. Now shown that

$$x = 0.686$$

is the solution of the equation $x^3 + x = 1$ with an error at most 5 in the third decimal place. (Use x_0 as a new starting estimate a, to obtain a second-generation estimate x_1, and then apply the error formula.)

8. Show by two figures that if the graph of f is increasing and concave down, then Newton's estimate of the solution $f(x) = 0$ is always too small, regardless of whether the initial estimate is too small or too large.

▦ **10.** Since $(2.2)^2 = 4.84$ and $(2.3)^2 = 5.29$, we know that $2.2 < \sqrt{5} < 2.3$. Using 2.2 as the initial estimate, apply Newton's method once. Show that your answer is accurate to three decimal places.

In the following problems apply Newton's method twice, starting from the given rough approximation. Estimate the error in the final approximation.

12. $\sqrt{3} \approx 2$ ▦ **13.** $\sqrt{2} \approx \dfrac{3}{2}$ ▦ **14.** $\sqrt{5} \approx \dfrac{5}{2}$ ▦ **15.** $\sqrt{3} \approx \dfrac{7}{4}$

16. Let $x_0, x_1, x_2, \ldots$ be the sequence obtained by applying Newton's method to the computation of $\sqrt{c}$, starting from an initial rough estimate x_0. Show that

$$x_{n+1} = \frac{1}{2}\left(x_n + \frac{c}{x_n}\right)$$

for $n = 0, 1, 2, \ldots$.

▦ **17.** Using a pocket calculator, compute $\sqrt{2}$ by repeatedly applying the formula in the above problem, starting from $x_0 = 2$. Continue until two successive values are equal. The result will be $\sqrt{2}$ accurate to as many places as the calculator shows (except for a possible error of 1 in the last place).

18. Justify the claim made at the end of Problem 17 by applying the error formula.

20. Calculate $\sqrt{5}$ in the same manner, starting from $x_0 = 3$.

22. Show that Newton's method for computing $c^{1/3}$ proceeds according to the iterative step

$$x_{n+1} = \frac{1}{3}\left(2x_n + \frac{c}{x_n^2}\right).$$

24. Compute $7^{1/3}$, starting from $x_0 = 2$.

19. Calculate $\sqrt{11}$ following the scheme of Problems 16 and 17 starting from $x_0 = 4$.

21. Calculate $\sqrt{3}$, starting from $x_0 = 2$.

23. Compute $2^{1/3}$ by Newton's method (the formula above), starting from $x_0 = 2$.

25. Find the iterative formula for solving $x^k = c$ by Newton's method.

**10
PROOF OF THE
MEAN-VALUE
THEOREM**

In this section we shall prove the mean-value principle and related results from the extreme-value and intermediate-value principles. These two results, in turn, are proved in Appendix 5. The reader may wonder why it is necessary to prove facts that appear self-evident on the basis of geometric figures. There seems to be little point in proving the obvious. The reason lies in the uncertain and tenuous nature of the obvious. Something that appeared obvious yesterday may seem less so today, and proofs become relevant once one has seen how geometric intuition can be misleading. For example, consider again our (correct) geometric observation that a positive slope must imply an increasing graph. Similar considerations might in the beginning have led us to conjecture conversely that if f is strictly increasing over an interval I, then f' is positive everywhere in I. But the example $f(x) = x^3$ shows that this isn't so. Here f is a strictly increasing function, and yet $f'(x) = 3x^2$ is zero at the origin.

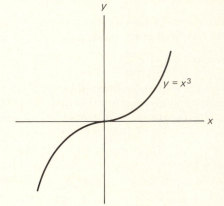

Figure 1

This conjecture and the "counterexample" showing the conjecture to be false together illustrate why geometric intuition cannot be trusted absolutely. Something may seem to be true principally because we have not yet noticed, and cannot imagine, a situation in which it is false. If we can't imagine how something might fail to happen, we are tempted to conclude that it must always happen. Of course, according to this principle of reasoning, the poorer our imagination the more facts we could establish!

A more serious example concerns the nature of a derivative. Every derivative we have met so far is continuous on its domain. That is, every function f that we have met so far has the property that if f is differentiable on an interval I, then f' is continuous on I. This is true whether f is given by a formula or by a graph that we draw: we find it impossible to draw an everywhere differentiable graph (over I) for which the tangent line varies discontinuously. So it would be reasonable to conjecture, on the basis of intuition bolstered by experience, that a derivative is *necessarily* continuous on its domain.

Yet in Problem 4 we shall concoct a function f with the property that f' exists everywhere on $(-1, 1)$ and is discontinuous at the origin! So our conjecture is false. It can be proved that the graph of any such function must contain an infinite number of wiggles, and this is why such a graph cannot be drawn.

This conjecture and counterexample affect geometric intuition in the following way. Since any differentiable graph we draw is of necessity smooth, it follows that a geometric property, like the mean-value principle, that we have inferred from looking at graphs may actually depend on the *continuity* of f', and not just the *existence* of f'.

So we are led to investigate what (if any) versions of the geometric principles we really can prove.

There is a slight refinement of the mean-value principle that is standard and useful. First, we note that the mean-value property can be spoiled by a single interior singularity. For example, consider $f(x) = x^{2/3}$ over the interval $[a, b] = [-1, 1]$. The secant slope $[f(1) - f(-1)]/(1 - (-1))$ is then 0, but there is no point where the derivative $f'(x) = 2/3x^{-1/3}$ is 0. See Fig. 2(a). This failure of the mean-value property occurs because f has a singularity at the interior point where we would expect f' to be zero, namely, at the origin. On the other hand, Fig. 2(b) shows that a singularity at an endpoint of $[a, b]$ does no damage, provided f is continuous there.

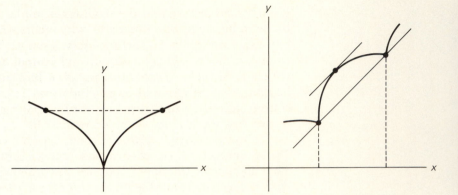

Figure 2

THEOREM 9 **The Mean-Value Theorem** *If f is continuous on the closed interval [a, b] and differentiable everywhere except possibly at one or both endpoints, then there is an interior point X at which*

$$f'(X) = \frac{f(b) - f(a)}{b - a}.$$

To prove the mean-value theorem we have to find a point x interior to the interval $[a, b]$ at which $f'(x) = m$, where m is the secant slope. But $f'(x) = m$ just when the derivative of $f(x) - mx$ is zero. So what we have to find is a critical point of the function $h(x) = f(x) - mx$. See Fig. 3. This is a special case of the mean-value theorem, called Rolle's Theorem, and we prove it first.

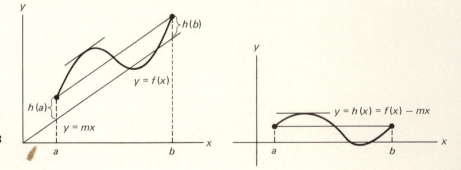

Figure 3

THEOREM 10 **Rolle's Theorem** *Suppose that f is continuous on [a, b], and differentiable on (a, b), and that f(a) = f(b). Then there is an interior point X at which f'(X) = 0.*

Proof By the extreme-value principle, f assumes maximum and minimum values on $[a, b]$. If the maximum value of f on $[a, b]$ is not the common endpoint value $f(a) = f(b)$, then a point X where the maximum value is assumed must necessarily be an interior point, and so $f'(X) = 0$, by Theorem 5 in Section 4. In the same way, if the minimum value of f is not the common endpoint value, then we get an interior point W where $f'(W) = 0$. The only remaining possibility is that the common endpoint value $f(a) = f(b)$ is both the maximum and minimum value of f on $[a, b]$. In this case, f is the constant function $f(x) = f(a)$, and $f'(x) = 0$ at *all* interior points. ∎

Proof of the Mean-Value Theorem Determine the constant m so that the function

$$h(x) = f(x) - mx$$

has equal values at a and b (so that Rolle's Theorem can be applied):

$$f(a) - ma = f(b) - mb.$$

Solving for m gives

$$m = \frac{f(b) - f(a)}{b - a},$$

which is the secant slope, as of course it should be. Since now $h(a) = h(b)$, it follows from Rolle's Theorem that $h'(X) = 0$ for some point X interior to $[a, b]$. That is, $f'(X) - m = 0$, or

$$f'(X) = m = \frac{f(b) - f(a)}{b - a}.$$
∎

A stronger form of the increasing-function theorem is an immediate corollary:

THEOREM 2′ *If f is continuous on the interval I and if f' exists and is positive throughout the interior of I, then f is increasing on I.*

Proof If x_1 and x_2 are any two distinct points of I then, by the mean-value theorem, there exists a point X strictly between

x_1 and x_2 (and hence in the interior of I) such that

$$\frac{f(x_2) - f(x_1)}{x_2 - x_1} = f'(X) > 0.$$

Thus $f(x_2) - f(x_1)$ always has the same sign as $x_2 - x_1$, so f is increasing.   ■

Consider, finally, the zero-crossing principle for f'. It seems evident geometrically that a graph can change from sloping upward to sloping downward only by passing through a point where its slope is zero. Now, try to examine your visualization of the situation to see *why* this must be so. There seem to be two ways to think about it. First, one can visualize a varying tangent line and watch its inclination changing as the point of tangency moves along the curve. If it starts with a positive slope and rotates, as it slides along, to a position with negative slope, then it had to go through a position of zero slope. When we think this way, we are visualizing the tangent line changing position *continuously*, and our conclusion is that if $f'(x)$ changes continuously from a positive value to a negative value, then it has to go through a zero value in the process. In this case the underlying principle is the intermediate-value principle, applied to f'.

There is another way to think about what is happening. When we start with a graph sloping upward and end with the graph sloping downward it seems clear that somewhere in between the graph had to go over a highest point, and that at this peak point the slope must be zero.

Such reasoning can be developed into a formal proof of the zero-crossing property for f' from the extreme-value property for f. (See Problem 5.) This is a more general result, since it doesn't require the continuity of f'. In practice, this is an empty generality, because in practice any differentiable function we work with will always have a continuous derivative. But in theory, the stronger proof is needed; there *do exist* functions having discontinuous derivatives. (See Problem 4.)

We are now down to the intermediate-value and extreme-value principles for a continuous function. We might feel that these properties are inherently a part of our notion of continuity, and are less in need of analytic justification than the other properties we have discussed. Indeed, until about a hundred years ago they were so taken for granted that they weren't even separately formulated. To Newton and other early workers in calculus, a continuous graph was the path traced by a moving particle, and that such a graph had these properties went without saying.

Nevertheless, this was another case of equating intuitive plausibility with proof. Something was accepted as being true because it was unimaginable that it be false. And if we have agreed that a weak imagination is not a suitable basis for mathematical truth, then the intermediate- and extreme-value principles need analytic proof, just as everything else does.

These proofs were discovered only relatively recently (during the last hundred years or so). They turn out to involve more sophisticated techniques than are needed elsewhere in beginning calculus, and they really belong to the foundations of a larger discipline, called analysis, in which calculus is imbedded. The proofs are given in Appendix 5.

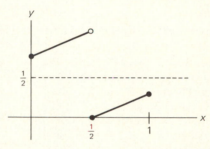

Figure 4

The critical nature of these two properties becomes clearer if we note that even one discontinuity may cause them both to fail. The function graphed in Fig. 4 is continuous everywhere on $[0, 1]$ except at $x = 1/2$, but it has no maximum value, nor does it assume the intermediate value of $1/2$. Thus we have to know that f is continuous

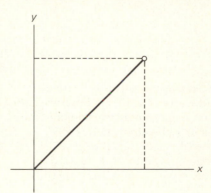

Figure 5

at every single point of the interval $[a, b]$ in order to apply these properties.

Moreover, we can't drop an endpoint from $[a, b]$ without losing, possibly, an extreme value. The function $f(x) = x$ loses its maximum value on $[0, 1]$ if we drop off $x = 1$, and it loses its minimum if we drop $x = 0$. It has neither a maximum value nor a minimum value on the *open* interval $(0, 1)$. See Fig. 5.

PROBLEMS FOR SECTION 10

Along with Theorem 2′, there is of course its "change-of-sign" counterpart:

Theorem 2″ *If f is continuous on I and f′ is negative at every interior point of I, then f is decreasing on I.*

1. Prove Theorem 2″ from the mean-value theorem.

3. Prove the mean-value theorem from the zero-crossing property for $f′$ and the two monotone properties Theorems 2′ and 2″. [*Hint:* Suppose there is no point X, as claimed in the mean-value theorem. Then concoct a function g whose derivative is never zero in $I = (a, b)$, and end up with a contradiction.]

5. Prove the zero-crossing property of $f′$ from the extreme-value property of f. [*Hint:* Suppose $f′$ does change sign on I, and choose two points c, d, in I, where $f′$ has opposite signs, with $c < d$. Apply Theorem 4 to show that if $f′(c) > 0 > f′(d)$, then the maximum value of f on $[c, d]$ cannot be assumed at either c or d. Now apply Theorem 5, etc.]

2. Prove Theorem 2″ directly from Theorem 2′.

4. Let f be the function defined as follows:

$$f(x) = x^2 \sin(1/x), \quad \text{if } x \neq 0;$$

$$f(0) = 0.$$

a) Find the formula for $f′(x)$ when $x \neq 0$, by the differentiation rules

b) Show that $f′(x)$ does not have a limit as x approaches 0.

c) Show that the difference quotient at 0,

$$\frac{f(0 + h) - f(0)}{h} = h \sin(1/h),$$

does have a limit as $h \to 0$ (use the squeeze limit law), and, consequently, that $f′(0)$ exists and is zero.

Thus, f is an everywhere differentiable function whose derivative $f′$ is not a continuous function.

6. Earlier we came to the shape principle by combining the zero-crossing property for f' with Theorems 2' and 2''. Prove it now directly from the extreme-value property of a continuous function and Theorem 5. Here is what you are to prove:

> *If f is continuous on an interval I and has no critical points in the interior of I, then f is monotone on I.*

[*Hint*: Suppose f is not monotone. Then there must be three points $a < b < c$ in I such that $f(b)$ is either the maximum or the minimum of the values of f at these three points. Then $f'(x) = 0$ at some point between a and c.]

We know from algebra that a polynomial $p(x)$ of degree n,

$$p(x) = x^n + a_{n-1}x^{n-1} + \cdots + a_0,$$

has at most n real roots. That is, there are *at most n* distinct real numbers that are solutions of the equation $p(x) = 0$. There may very well be *fewer* than n. Thus, $p(x) = x^2 + 1$ is a polynomial of degree 2 having *no* real roots.

7. Prove that a polynomial of *odd degree* has at least one real root.

8. Call a polynomial p of degree n *simple* if it does have n distinct real roots. Prove that if p is simple, then so is its derivative $q = p'$.

CHAPTER 6

THE INTEGRAL

**1
THE
ANTIDERIVATIVE**

A major discovery of Newton and Leibniz was that many problems in geometry and physics can be solved by "backwards differentiation," or *antidifferentiation*. We shall see how this comes about later in this chapter, after we have taken up the definite integral. We begin here with an introduction to the antiderivative. The technique of antidifferentiation is developed in Chapter 8.

To get started, consider what we know about the motion of a particle traveling along a coordinate line with constant velocity 5. In Chapter 3 we decided that if s is the position coordinate of the particle, then its velocity v is given by the derivative ds/dt. So here we are assuming that s is a function of t that has the constant derivative 5. The question is: What does this tell us about the function itself?

You undoubtedly can visualize *one* function of t having the constant derivative 5, namely $5t$. Moreover, adding a constant term doesn't change the derivative, so if s is any one of

$$5t + 3, \quad 5t - 10, \quad 5t + \sqrt{2}, \quad 5t + C,$$

then s is a function of t for which $ds/dt = 5$. Are there others? The answer is *no*, because of the following fact.

THEOREM 1 *If h_1 and h_2 are differentiable functions having the same derivative on an interval I, then there is a constant C such that*

$$h_2(x) = h_1(x) + C$$

for all x in I.

Proof If $h'_1 = h'_2$ on I, then $(h_2 - h_1)' = 0$ on I. But any function whose derivative is identically zero must be a constant. (This is Theorem 1 of Chapter 5.) So $h_2 - h_1 = C$ and $h_2 = h_1 + C$. ∎

Using this principle we see that if $ds/dt = 5$, then s must differ from $5t$ by a constant, so $s = 5t + C$ for some constant C. This is all we can conclude without further information.

The problem above involved finding a function whose derivative is known. A differentiable function F such that $F' = f$ is called an *antiderivative* of f, and the process of finding F from f is *antidifferentiation*. We saw above that

A given function f does not have a uniquely determined antiderivative, but if we can find one antiderivative F, then every other antiderivative is of the form F + constant.

For example, $x^3/3$ is one antiderivative of x^2, and every other antiderivative is then necessarily of the form

$$\frac{x^3}{3} + C.$$

An antiderivative F of f is also called an *integral* of f, and antidifferentiation is also called *integration*. The most general integral $F(x) + C$ is called the *indefinite integral* of f, because of the arbitrary constant C. This constant is called the *constant of integration*. The Leibniz notation for the indefinite integral of a function f is

$$\int f(x)\, dx,$$

a notation that will prove to be especially useful in Chapter 8. For example,

$$\int x^2 \, dx = \frac{x^3}{3} + C \quad \text{and} \quad \int f(x) \, dx = F(x) + C.$$

We also write $\int f = F + C$. In this context, x^2 (or $f(x)$, or f) is called the *integrand*.

The polynomial rules of Chapter 3, read backward, now become the following integration rules:

$$\int x^k \, dx = \frac{x^{k+1}}{k+1} + C \quad \text{(if } k \neq -1\text{)},$$

$$\int (af(x) + bg(x)) \, dx = a \int f(x) \, dx + b \int g(x) \, dx.$$

We'll check them shortly, but consider first an example of how they are used.

Example 1

$$\int (3x + 2x^3) \, dx = 3 \int x \, dx + 2 \int x^3 \, dx \qquad \text{(By the second rule)}$$

$$= 3\left(\frac{x^2}{2} + C_1\right) + 2\left(\frac{x^4}{4} + C_2\right) \qquad \text{(By the first rule)}$$

$$= \frac{3x^2 + x^4}{2} + (3C_1 + 2C_2).$$

But we can write this as

$$\frac{3x^2 + x^4}{2} + C,$$

because $3C_1 + 2C_2$ is itself just an arbitrary constant. $\qquad \square$

In practice, when we have to compute the indefinite integral of a sum of functions, we just add up the particular integrals (antiderivatives) that we find and then add an arbitrary constant at the end. For example, the above calculation would be done like this:

$$\int (3x + 2x^3) \, dx = 3 \cdot \frac{x^2}{2} + 2 \cdot \frac{x^4}{4} + C$$

$$= \frac{3x^2 + x^4}{2} + C.$$

Example 2
$$\int (2x^2 + \cos x)\, dx = 2\frac{x^3}{3} + \sin x + C. \qquad \square$$

Proof of the integration rules. Since

$$\frac{d}{dx}\left(\frac{x^{k+1}}{k+1}\right) = \frac{(k+1)x^k}{k+1} = x^k,$$

we see that $x^{k+1}/(k+1)$ is a particular antiderivative of x^k, and the first rule follows.

For the second rule we suppose that F and G are antiderivatives of f and g, respectively, and then have

$$(aF + bG)' = aF' + bG' = af + bg,$$

so

$$\int (af + bg) = (aF + bG) + C$$

$$= a\int f + b\int g.$$

The Initial-Value Problem   The graphs of the antiderivatives of x^2 are the graphs of the functions

$$y = \int x^2\, dx = \frac{x^3}{3} + C$$

for the various values of the constant C, as shown below. They form a family of curves filling up the plane, exactly one curve going through any given point (x_0, y_0).

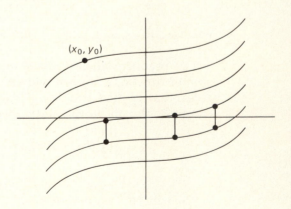

This means that if we ask for an antiderivative $F(x)$ of x^2 having a given value at a given point, say $F(-1) = 2$, then there is a unique solution: The function F must be of the form

$$F(x) = \int x^2 \, dx = \frac{x^3}{3} + C,$$

for some value of the constant C, and C is determined by the extra requirement that $F(-1) = 2$. Thus

$$2 = F(-1) = -\frac{1}{3} + C,$$

so that $C = 2 + 1/3 = 7/3$, giving the unique solution

$$F(x) = \frac{x^3 + 7}{3}.$$

The requirement that $F(-1) = 2$ is called an *initial condition*, and the total problem is called an *initial-value problem*. The above initial-value problem, then, is to determine the unique function F such that

$$F'(x) = x^2, \qquad F(-1) = 2.$$

In pure Leibniz notation it is the problem of determining y as a function of x, given that

$$\frac{dy}{dx} = x^2,$$

and that

$$y = 2 \qquad \text{when } x = -1.$$

In this notation the solution would be written out as follows. First,

$$y = \int x^2 \, dx = \frac{x^3}{3} + C,$$

where the constant C is to be determined by the initial condition. Substituting the initial values $(x, y) = (-1, 2)$ gives

$$2 = \frac{(-1)^3}{3} + C,$$

so $C = 7/3$, as before.

We now turn to initial-value problems in which we determine the motion of an object from its known velocity or acceleration.

Example 3 A particle travels along the number line with constant velocity 5. Find its position s as a function of t if $s = -10$ when $t = 1$.

This is an initial-value problem. From $ds/dt = 5$ we obtain $s = 5t + C$, as before, although we might now write this

$$s = \int 5 \, dt = 5t + C.$$

Substituting the initial values $(t, s) = (1, -10)$ gives

$$-10 = 5 \cdot 1 + C,$$

so that $C = -15$ and

$$s = 5t - 15. \qquad \square$$

It is said that in the seventeenth century Galileo experimented by dropping objects from the tower in Pisa and concluded that, neglecting air resistance that would slow down light objects more, all falling objects drop with a constant acceleration of 32 ft per $\sec^2$.

If the position of the falling body is measured along a vertical coordinate system with the positive direction upward, then its constant acceleration is -32, because its velocity is becoming increasingly negative. The acceleration due to gravity is usually designated g, so we have $g = -32$ in the present axis system. Thus

$$-32 = \frac{dv}{dt} = \frac{d^2 s}{dt^2}.$$

Example 4 A stone is thrown vertically upward with an initial velocity of 50 ft/sec. What is its velocity two seconds later? How high will it rise?

Solution Since $dv/dt = g = -32$, we have

$$v = \int g \, dt = \int (-32) \, dt = -32t + C.$$

If we start measuring time at the instant the stone is thrown, then the initial condition is that $v = 50$ when $t = 0$. Thus $50 = -32 \cdot 0 + C$ and $C = 50$, giving the velocity equation

$$v = 50 - 32t.$$

At $t = 2$ we have $v = 50 - 64 = -14$ ft/sec. The stone is already falling.

The highest point in its trajectory will be reached when the stone just stops rising and starts to fall, i.e., when $v = 0$. This occurs when

$$0 = 50 - 32t$$

or $t = 25/16$ seconds. But in order to find how far the stone has risen we have to integrate the equation

$$v = \frac{ds}{dt} = 50 - 32t.$$

We get

$$s = 50t - 16t^2 + C.$$

If we measure s from the point at which the stone is thrown, then the initial condition is $s = 0$ at $t = 0$. This gives $C = 0$, and at $t = 25/16$ we have

$$s = \frac{50 \cdot 25}{16} - 16\left(\frac{25}{16}\right)^2 = \frac{625}{16} \text{ ft,}$$

or approximately 39 ft.

(The highest point is the maximum value of s and can be determined by finding where $ds/dt = 0$. But this is just the condition $v = 0$ that we arrived at intuitively above.) ☐

PROBLEMS FOR SECTION 1

Compute the following integrals. Be sure to include the constant of integration in each answer.

1. $\displaystyle\int x^6 \, dx$

2. $\displaystyle\int \sqrt{x} \, dx$

3. $\displaystyle\int x \, dx$

4. $\displaystyle\int dx$

5. $\displaystyle\int x^{1/3} \, dx$

6. $\displaystyle\int (x^2 + x + 1) \, dx$

7. $\displaystyle\int \left(3x^3 - \frac{1}{3x^3}\right) dx$

8. $\displaystyle\int (t^2 - 2t + 3) \, dt$

9. $\displaystyle\int (2x + 3)(x - 2) \, dx$

10. $\displaystyle\int x^{2/3} \, dx$

11. $\displaystyle\int \frac{dx}{\sqrt{x}}$

12. $\displaystyle\int \frac{dx}{\sqrt[3]{x}}$

13. $\displaystyle\int 3ay^2 \, dy$

14. $\displaystyle\int (x^{3/2} - 2x^{2/3} + 5\sqrt{x} - 3) \, dx$

15. $\displaystyle\int \frac{4x^2 - 2\sqrt{x}}{x}\, dx$

16. $\displaystyle\int \sqrt{x}(3x - 2)\, dx$

17. $\displaystyle\int (9t^2 + 25t + 14)\, dt$

18. $\displaystyle\int \frac{dx}{x^2}$

19. $\displaystyle\int \frac{dx}{\sqrt[3]{x^2}}$

20. $\displaystyle\int (x^3 + 2)^2 3x^2\, dx$

21. Prove that

$$\int (ax + b)^k\, dx = \frac{(ax + b)^{k+1}}{a(k + 1)} + C$$

(where the exponent k is rational, and $k \neq -1$).

22. $\displaystyle\int \sqrt{x + 1}\, dx$

23. $\displaystyle\int \frac{1}{\sqrt{x + 1}}\, dx$

24. $\displaystyle\int \sqrt{2x + 1}\, dx$

25. $\displaystyle\int (3x + 2)^{-2/3}\, dx$

26. $\displaystyle\int \sqrt{1 - 4x}\, dx$

27. $\displaystyle\int \sin t\, dt$

28. $\displaystyle\int \cos 2x\, dx$

29. $\displaystyle\int (\sin x + x + C)\, dx$

30. $\displaystyle\int \sec^2 x\, dx$

31. $\displaystyle\int (\sec^2 3\theta + \sec \theta \tan \theta)\, d\theta$

32. $\displaystyle\int \cos^2 x\, dx$ [*Hint:* $\cos^2 x = (1 + \cos 2x)/2$.]

Solve the following initial-value problems.

33. $f'(x) = x - 3, \quad f(2) = 9$

34. $g'(x) = x + 3 - 5x^2, \quad g(6) = -20$

35. $f'(y) = y^3 - b^2 y, \quad f(2) = 0$

36. $f'(x) = bx^3 + ax + 4, \quad f(b) = 10$

37. $f'(t) = \sqrt{t} + \dfrac{1}{\sqrt{t}}, \quad f(4) = 0$

38. Given $dy/dx = (2x + 1)$, $y = 7$ when $x = 1$. Find the value of y when $x = 3$.

39. Given $dA/dx = \sqrt{2px}$, $A = p^2/3$ when $x = p/2$. Find the value of A when $x = 2p$.

40. Given the expressions below for acceleration, find the relation between s (displacement) and t if $s = 0$, $v = 20$, when $t = 0$.

  a) $a = 32$

  b) $a = 4 - t$

41. With what velocity will a stone strike the ground if dropped from the top of a building 100 ft high? ($g = 32$ if distance is measured positively in the downward direction.)

42. If the stone in Problem 41 is thrown from the top of the building with an initial velocity of 100 ft/sec downward, what will be its velocity upon hitting the ground? What if it has an *upward* initial velocity of 100 ft/sec?

43. A train leaving a railroad station has an acceleration of

$$\frac{1}{2} + \frac{2}{100}\, t \text{ ft/sec}^2.$$

How far will the train move in the first 20 sec of motion?

44. What constant acceleration is required to

a) Move a particle 50 ft in 5 sec?

b) Slow a particle from a velocity of 45 ft/sec to a stop in 15 ft?

46. A car traveling v_0 ft/sec hits its brakes and starts decelerating at a ft/sec². How far will it travel before stopping?

45. Neglecting air resistance, show that an object dropped from a height h will strike the ground with a velocity $\sqrt{2gh}$, where g is the acceleration due to gravity.

2
CONVERGENT SEQUENCES

The principal topic of the present chapter is the definite integral, with a few applications to show how it works. Since a definite integral will be specified as the limit of a *sequence* of Riemann sums, we give here a brief introduction to sequences and sequential convergence.

Example 1

Let a_n be the area of the regular polygon of n sides inscribed in a circle of radius 1. Then a_n is an approximation to the area of the circle that improves as n is taken larger, and approaches the circle area as its limit as n tends to infinity. We thus have an *infinite sequence*

$$a_3, a_4, \ldots, a_n, \ldots$$

of numbers, each the area of a polygon, such that

$$a_n \to \pi \qquad \text{as } n \to \infty. \qquad \square$$

In this chapter and in Chapter 9 we are going to be concerned with similar approximating sequences that approach areas, lengths, volumes, and other quantities as their limits.

The notion of an infinite sequence arises very naturally from examples like the one above. Some procedure or formula determines a number a_n for every integer n from a certain point on. It might seem desirable always to start with a "first" term a_1, but the sequence above starts naturally at $n = 3$ because a polygon presumably has at least three sides. On the other hand, in the sequence

$$1, 2, 4, 8, \ldots, 2^n, \ldots$$

of powers of 2, we would probably like to think of $8 = 2^3$ as the third term a_3, and 2^n as the nth term a_n, in which case the beginning number $1 = 2^0$ is the zeroth term a_0. The three dots $\cdots$ can be loosely read "and so on."

The infinite sequence in Example 1 is designated

$$\{a_n\}_{n=3}^{\infty} \qquad \text{or} \qquad \{a_n\}_{3}^{\infty}.$$

When it is not important to specify the starting index we often write just $\{a_n\}$.

We shall be especially interested in sequences that approach limits, in the way that $a_n \to \pi$ in the first example above. For a much simpler example, consider the sequence

$$1, \frac{1}{2}, \frac{1}{3}, \frac{1}{4}, \dots,$$

where the nth term is $1/n$ for every positive integer n. Then

$$\frac{1}{n} \to 0 \qquad \text{as } n \to \infty,$$

in exactly the same way that

$$\frac{1}{x} \to 0 \qquad \text{as } x \to \infty.$$

(The integers n *do* go all the way out to ∞. That is, for any positive real number A there is an integer n that is larger than A. This is a basic fact about numbers.)

Sequence limits are thus much like function limits, the analogy being with the limit of $f(x)$ as x tends to infinity.

DEFINITION   If $\{a_n\}$ is an infinite sequence of numbers, then we say that a_n *approaches the limit l as n approaches infinity* if we can make the difference $l - a_n$ as small as we wish merely by taking n large enough.

We then write

$$\boxed{a_n \to l \qquad \text{as } n \to \infty,}$$

or

$$\boxed{\lim_{n \to \infty} a_n = l.}$$

Example 2   If $a_n = 1/n$, then

$$a_n \to 0 \qquad \text{as } n \to \infty,$$

as noted above. Technically, if we want a_n closer to 0 than a given small positive number ϵ, that is, if we want

$$\frac{1}{n} < \epsilon,$$

then we only have to take n larger than $1/\epsilon$. This proves that we can cause $1/n$ to be as close to 0 as we wish by taking n suitably large.

□

Example 3 Later in the chapter we will meet a type of sequence $\{a_n\}$ with the property that

$$|a_n - l| < \frac{K}{n}$$

for all n (where K is a constant). So if we want a_n to be closer to its limit l than $1/100$, that is, if we want

$$|a_n - l| < \frac{1}{100},$$

it is sufficient to take n large enough so that

$$\frac{K}{n} \leq \frac{1}{100},$$

or

$$n \geq 100\,K.$$

□

A sequence having the limit l is said to *converge* to l. A sequence that doesn't converge to any l is said to *diverge*. Divergence can occur in two "pure" forms: *oscillatory divergence*, as in

$$1, \quad -1, \quad 1, \quad -1, \quad 1, \quad \ldots,$$

where $a_n = (-1)^n$; and *divergence to* ∞, as in

$$1, \quad 4, \quad 9, \quad 16, \quad 25, \quad \ldots,$$

where $a_n = n^2$. Often a divergent sequence will exhibit a mixture of these two types of divergent behavior.

The computation of sequential limits is governed by the same algebraic limit laws that control function limits:

A. *If $a_n \to a$ and $b_n \to b$ as $n \to \infty$, then*

$$a_n + b_n \to a + b,$$

$$a_n b_n \to ab,$$

$$\frac{a_n}{b_n} \to \frac{a}{b} \qquad \left(\begin{matrix}\textit{if the denominators} \\ \textit{are nonzero}\end{matrix}\right).$$

B. *If f is continuous at $x = a$ and if $x_n \to a$ as $n \to \infty$, then*

$$f(x_n) \to f(a) \qquad \textit{as } n \to \infty.$$

C. (Squeeze limit law) *If $0 \le a_n \le b_n$ for all n, and if $b_n \to 0$ as $n \to \infty$, then $a_n \to 0$ as $n \to \infty$.*

D. *If $a_n \to a$ as $n \to \infty$ and if $a > 0$, then $a_n > 0$ for all n from some n_0 on.*

All of these limit laws should seem to be correct. For example, if a_n is very close to a and b_n is very close to b, then surely $a_n + b_n$ is close to $a + b$.

Example 4   Let $\{a_n\}$ be the sequence from Example 1. We know that $a_n \to \pi$ as $n \to \infty$. Then (A) tells us that

$$a_n^2 \to \pi^2 \qquad \text{and} \qquad \frac{a_n}{n} = a_n \cdot \frac{1}{n} \to \pi \cdot 0 = 0.$$

Also, by (B), $\cos a_n \to -1$ (since $\cos \pi = -1$).   □

Example 5   Starting with the fact that $1/n \to 0$ as $n \to \infty$, and applying rules from (A), we see that

$$\lim_{n \to \infty} \frac{4 - n}{1 + 3n} = \lim_{n \to \infty} \frac{\dfrac{4}{n} - 1}{\dfrac{1}{n} + 3} = \frac{0 - 1}{0 + 3} = -\frac{1}{3}.$$

Similarly,

$$\lim_{n \to \infty} \frac{n^2 - 5n}{4 + n^3} = \lim_{n \to \infty} \frac{\dfrac{1}{n} - \dfrac{5}{n^2}}{\dfrac{4}{n^3} + 1} = \frac{0 + 0}{0 + 1} = 0.$$

Note that in both cases we first divided by the highest power of n in the denominator. This is the way rational functions of x were treated in Chapter 2, Section 4. $\square$

Example 6 Show that $\sin \dfrac{1}{n} \to 0$ as $n \to \infty$.

Solution We can use the squeeze limit law (C), because

$$0 < \sin \frac{1}{n} < \frac{1}{n}$$

and $1/n \to 0$. Or we can use (B), since $\sin x$ is continuous at $x = 0$ and $\sin(0) = 0$. $\square$

We use the words *increasing* and *decreasing* in a slightly weaker sense when we are talking about sequences. A sequence $\{a_n\}$ is said to be *increasing* if

$$a_{n+1} \geq a_n \qquad \text{for all } n.$$

If $a_{n+1} > a_n$ for all n, then we say that $\{a_n\}$ is *strictly increasing*. The meaning of decreasing is similarly modified.

Example 7 $a_n = 1/n$ is strictly decreasing. $\square$

Example 8 If a_n is the n-place decimal truncation of the expansion of π, so that $a_1 = 3.1$ and $a_5 = 3.14159$, then $\{a_n\}$ is an increasing sequence. However, we can't claim that it is strictly increasing, because presumably there will be a zero somewhere along the infinite-decimal expansion of π, and if the zero occurs in the jth place we will have

$$a_{j-1} = a_j.$$

(Decimal expansions are discussed at the beginning of Chapter 11.) $\square$

Example 9 In the first section of Chapter 7, a number e will be defined having the property that

$$\left(1 + \frac{1}{n}\right)^n < e < \left(1 + \frac{1}{n}\right)^{n+1}$$

for all n. Show that

$$\left(1 + \frac{1}{n}\right)^n \to e \qquad \text{as} \qquad n \to \infty.$$

Solution   From the given inequality we get

$$0 < e - \left(1 + \frac{1}{n}\right)^n < \frac{1}{n}\left(1 + \frac{1}{n}\right)^n < \frac{e}{n}.$$

Therefore,

$$e - \left(1 + \frac{1}{n}\right)^n \to 0 \qquad \text{as} \qquad n \to \infty,$$

by the squeeze limit law.   □

Example 10   If $0 < t < 1$, show that $t^n \to 0$ as $n \to \infty$.

Solution   We can write

$$t = \frac{1}{1 + b},$$

where $b > 0$. Then

$$0 < t^n = \frac{1}{(1 + b)^n} = \frac{1}{1 + nb + \cdots} < \frac{1}{nb},$$

for all n. Since $1/n \to 0$ as $n \to \infty$, by Example 3, it follows
from the squeeze limit law that $t^n \to 0$ as $n \to \infty$.   □

**Sigma
Notation**   We also introduce here some notation that is frequently helpful.
This is the *summation* notation, involving $\sum$, the Greek letter capital
sigma.

Suppose we want to indicate the sum of the first 99 terms of the
sequence $\{a_n\}$. Up till now we would have written the sum schematically as

$$a_1 + a_2 + \cdots + a_{99}.$$

Here, as before, the three dots are read "and so on," or "and so on up
to"; they instruct us to continue as we have started. The sigma
notation gives us another way of describing such finite sums. For the
above sum it is

$$\sum_{i=1}^{99} a_i,$$

which is read "the sum of the terms a_i from $i = 1$ to $i = 99$." Here are some other examples:

$$\sum_{i=13}^{25} a_i = a_{13} + a_{14} + \cdots + a_{25};$$

$$\sum_{j=1}^{N} a_j = a_1 + a_2 + \cdots + a_N;$$

$$\sum_{k=2}^{5} k^2 = 4 + 9 + 16 + 25 = 54;$$

$$\sum_{k=1}^{n} k = 1 + 2 + \cdots + n = \frac{n(n+1)}{2}.$$

The subscript i (or j or k) is used as a variable that takes on integer values, as in sequences, and its values run consecutively from a given beginning value to a given final value.

The last formula above is true but needs proof. It is one of a string of similar formulas, the next one being

$$\sum_{k=1}^{n} k^2 = \frac{n(n+1)(2n+1)}{6}.$$

The leading terms of these formulas will be useful to us in the next section. Assuming the above two formulas, we see that

$$\sum_{k=1}^{n} k^0 = \sum_{k=1}^{n} 1 = n,$$

$$\sum_{k=1}^{n} k^1 = \frac{n(n+1)}{2} = \frac{n^2}{2} + \text{a term of degree 1},$$

$$\sum_{k=1}^{n} k^2 = \frac{n(n+1)(2n+1)}{6} = \frac{n^3}{3} + \text{terms of degree} \leq 2.$$

These equations suggest that for each fixed positive integer p,

$$\sum_{k=1}^{n} k^p = \frac{n^{p+1}}{p+1} + \text{a polynomial in } n \text{ of degree } p,$$

the equation holding for all values of n. This is true. It can be proved for $p = 1, 2, \ldots$, in order, each proof depending on the earlier results.

Example 10   To illustrate this we shall derive the formula for $p = 3$, assuming it true for $p = 1, 2$. We start with the binomial expansion

$$(k - 1)^4 = k^4 - 4k^3 + 6k^2 - 4k + 1,$$

which we rewrite

$$k^4 - (k-1)^4 = 4k^3 - 6k^2 + 4k - 1.$$

If we give k the values $1, 2, \ldots, n$, and add these n equations, then the sum of the left collapses to n^4. We thus have

$$n^4 = 4 \sum_{k=1}^{n} k^3 - 6 \sum_{k=1}^{n} k^2 + 4 \sum_{k=1}^{n} k - n$$

$$= 4 \sum_{k=1}^{n} k^3 + \text{a polynomial in } n \text{ of degree } 3$$

(by the assumed results for the cases $p = 1$ and $p = 2$). Solving this equation for $\sum k^3$ gives the formula we are after.   □

PROBLEMS FOR SECTION 2

Evaluate the limit as $n \to \infty$ of the following sequences.

1. $2 - \dfrac{3}{n} + \dfrac{4}{n^2}$

2. $\dfrac{n+1}{n-1}$

3. $\dfrac{n}{n^2+1}$

4. $\dfrac{100n - n^2}{2n^2 + 1}$

5. $\dfrac{5n+2}{3-n}$

6. $\dfrac{(2-n)(3-n)}{(1+2n)(1+3n)}$

7. $\dfrac{2n}{\sqrt{4+n^2}}$

8. $\sqrt{n+1} - \sqrt{n}$

9. $\sqrt{n^2+n} - n$

10. $\sqrt{n^3+n^2} - \sqrt{n^3}$

Here is a principle related to (B). (It can be proved from (B).)

$$\text{If } \lim_{x \to 0} f(x) = l, \quad \text{then} \quad \lim_{n \to \infty} f\left(\frac{1}{n}\right) = l.$$

11. Use the new principle stated above to prove that $\lim n \sin(1/n) = 1$.

More limits.

12. $n^2\left(1 - \cos\dfrac{1}{n}\right)$

13. $(n+1)^{1/3} - n^{1/3}$

14. $(n^2+n)^{1/3} - n^{2/3}$

15. $(n^3+n^2)^{1/3} - n$

16. Use an algebraic limit law from (A) and the squeeze limit law (C) to prove the more general squeeze limit law:

If $a_n \le b_n \le c_n$ for all n, and if $a_n \to l$ and $c_n \to l$ as $n \to \infty$, then $b_n \to l$ as $n \to \infty$.

▤ **17.** The inequality in Example 9 gives a way to estimate e. Using a pocket (scientific) calculator, try a few values of n to get a feel for the approximation, and then determine the value of e with an error less than one in the second decimal place.

18. Show from Example 9 that

$$\left(\frac{n+1}{n-1}\right)^{n/2} \to e \qquad \text{as } n \to \infty,$$

Use the summation notation to describe each of the following sums.

20. $1 + \dfrac{1}{2} + \dfrac{1}{3} + \cdots + \dfrac{1}{50}$

22. $2 + 4 + 6 + \cdots + 18$

24. $2 + 4 + 8 + \cdots + 1024$

26. $x_1 y_2 + x_2 y_3 + \cdots + x_n y_{n+1}$

19. Compute the sequence terms in Problem 18 for $n = 50, 100, 150, 200$. What does e seem to be through three decimal places?

21. $2 + 3 + 4 + \cdots + 17$

23. $\dfrac{1}{2} + \dfrac{2}{3} + \dfrac{3}{4} + \cdots + \dfrac{99}{100}$

25. $a_1 + a_3 + a_5 + \cdots + a_{37}$

27. Prove that

$$\sum_{k=1}^{n} k^4 = \frac{n^5}{5} + \text{a polynomial in } n \text{ of degree } 4.$$

3

THE DEFINITE INTEGRAL

As an introduction to the definite integral we shall consider two questions:

What is the average value of a continuous function over a given interval?

How do we determine the area of a region under a function graph?

Suppose that $T = f(t)$ is the temperature at time t recorded at a weather station on a certain day. We want to find the average temperature for that day. The station uses a 24-hour clock, so the domain of f is the time interval $[0, 24]$. And we assume f to be continuous.

In order to find the average temperature for that day, we might take six temperature readings, one at the end of each 4-hour interval, starting from midnight: $T_1 = f(4)$, $T_2 = f(8)$, ..., $T_6 = f(24)$. The average reading would then be the sum of these six readings divided by 6:

$$A_6 = \frac{T_1 + T_2 + T_3 + T_4 + T_5 + T_6}{6}$$

$$= \frac{f(4) + f(8) + f(12) + f(16) + f(20) + f(24)}{6}$$

$$= \frac{1}{6} \sum_{k=1}^{6} f(4k).$$

This computation of the average temperature probably does not give the right answer. For example, suppose it is a hot summer day, and that at 2:00 in the afternoon (hour 14 on the 24-hour clock) there is a short thunderstorm that cools the air for an hour. This temporary dip in temperature wouldn't even show in the average computed above.

Clearly we would do better by taking 24 readings on the hour and using their average

$$A_{24} = \frac{f(1) + f(2) + \cdots + f(24)}{24} = \frac{1}{24} \sum_{k=1}^{24} f(k).$$

But this probably wouldn't be exactly right either, for the same reason that small temporary fluctuations might not be adequately represented in the average. Still better would be the average of 48 readings taken on the half-hour,

$$A_{48} = \frac{f(1/2) + \cdots + f(24)}{48} = \frac{1}{48} \sum_{k=1}^{48} f\left(\frac{k}{2}\right),$$

and so on.

We thus have various ways of computing an average daily temperature, none of them exactly right. Our intuition tells us that there is a *true average temperature* $\bar{T}$ that these approximating averages come close to, and that by taking a sufficiently large number of equally spaced temperature readings, we could, in principle, make the approximation as good as we wished.

Here is a general description of the above line of thought. We are given a continuous function f on a closed interval $[a, b]$, and we want to determine the number A that is the average of the values of f over the interval. We start with a finite sample of the values, say $f(x_1), f(x_2), \ldots, f(x_N)$, and compute the ordinary arithmetic average of the sample:

$$A_N = \frac{1}{N}[f(x_1) + f(x_2) + \cdots + f(x_N)] = \frac{1}{N} \sum_{i=1}^{N} f(x_i).$$

We require the N sampling points $x_1, x_2, \ldots, x_N$ to be evenly spread out over the interval, so that the sample will represent the overall behavior of $f(x)$ as fairly as possible.

To get the particular sampling set that we used earlier, we first divide the interval $[a, b]$ into N equal subintervals, each of length

$$\Delta x = \frac{b - a}{N}.$$

Then we choose their right-hand endpoints:

$$x_1 = a + \Delta x,$$
$$x_2 = x_1 + \Delta x = a + 2\,\Delta x,$$
$$\cdot$$
$$\cdot$$
$$\cdot$$
$$x_N = x_{N-1} + \Delta x = a + N\,\Delta x = b,$$

as the sample evaluation points. (It is convenient also to set $x_0 = a$. This makes the labeling scheme more symmetric, and it has various other little advantages. See Fig. 1.) This set of right-hand endpoints will be our standard sampling set.

Figure 1

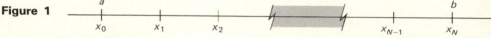

With such an evenly distributed sampling set, it seems that the sample average A_N should be close to the true average A. And the larger the sample the closer A_N should come to A. We therefore conjecture the theorem:

THEOREM A *Let f be continuous on $[a, b]$. For each positive integer N, let A_N be the average of the values of f taken over the standard sampling set of size N. Then the sequence $\{A_N\}$ is convergent.*

We also concluded that this limit is the average value of f over $[a, b]$. But this is an *interpretation* of the result and not part of the purely mathematical theorem. The theorem simply says that sequences formed in a certain way necessarily converge.

Area When we look at areas we find that a similar theorem seems to be true. Let f be continuous and positive over $[a, b]$, and let G be the region under the graph of f and over $[a, b]$. Figure 2 shows how G can be approximated by covering rectangles. To get these rectangles

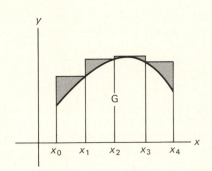
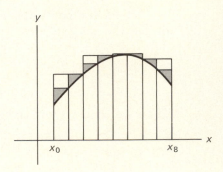

Figure 2

we first subdivide $[a, b]$ into N equal subintervals, each of length $\Delta x = (b - a)/N$. Then over each subinterval as base we erect the vertical rectangle whose altitude is the *maximum value* of f on the base subinterval.

The area of the jth rectangle is $\Delta x\, M_j$, where M_j is the maximum value of f on the jth subinterval, for $j = 1, 2, \ldots, N$. The sum of these N rectangular areas is thus

$$\Delta x[M_1 + M_2 + \cdots + M_N] = \Delta x \sum_{j=1}^{N} M_j.$$

We designate this sum U_N, and call it the Nth *upper sum* for f over $[a, b]$ (because it involves the *maximum* values of f on the N subintervals). Since U_N is the area of a configuration that covers G, we have

$$U_N \geq A,$$

where A is the area of G.

The difference $U_N - A$ is the amount of shaded area lying between the graph and the tops of the rectangles. This shaded area is greatly reduced by doubling the number of rectangles. In fact, if we examine what happens when a typical rectangle is replaced by two, we see that the error is roughly cut in half. A roughly triangular error area is replaced by two out of four roughly equal triangular quarters.

Since the difference $U_N - A$ is roughly halved each time N doubles, it appears that $U_N - A \to 0$ as $N \to \infty$, that is, that

$$\lim_{N \to \infty} U_N = A.$$

In particular, we conjecture:

THEOREM B *Let f be positive and continuous on $[a, b]$, and, for each positive integer N, let U_N be the Nth upper sum for f over $[a, b]$. Then the sequence $\{U_N\}$ converges.*

The Definite Integral We now have a pair of similar looking theorems that seem to be true because of two different interpretations of their terms. In fact, *the two theorems are instances of the same mathematical theorem.* This underlying theorem will have many interpretations and applications, so it is best stated in neutral terms that do not suggest any particular interpretation.

To set this up we first describe how to form an *evaluation set* for f over $[a, b]$. Given an equally spaced subdivision

$$a = x_0 < x_1 < \cdots < x_N = b,$$

as introduced earlier, we choose an *arbitrary* point from each of the subintervals: we choose *any* point c_1 in the first subinterval, *any* point c_2 in the second subinterval, etc., as shown in Fig. 3. Thus,

$$x_0 \le c_1 \le x_1,$$
$$x_1 \le c_2 \le x_2,$$
$$\vdots$$
$$x_{N-1} \le c_N \le x_N,$$

and these inequalities are the *only* restrictions on the evaluation set $c_1, c_2, \ldots, c_N$. Then the sum

$$S = \Delta x[f(c_1) + \cdots + f(c_N)] = \Delta x \sum_{i=1}^{N} f(c_i)$$

is called an *Nth Riemann sum*, or simply a *Riemann sum*, for f over $[a, b]$. The *standard Nth sum* uses the standard evaluation set (the right-hand endpoints) and is thus

$$S_N = \Delta x[f(x_1) + \cdots + f(x_N)] = \Delta x \sum_{i=1}^{N} f(x_i).$$

Figure 3

We can now state the theorem.

THEOREM 2   *Let f be continuous on [a, b]. For each positive integer N let S_N be an Nth Riemann sum for f over [a, b]. Then the sequence $\{S_N\}$ is convergent. Furthermore, any other sequence formed in the same way has the same limit.*

There is one final version that is worth having.

DEFINITION   Let f be *any* function defined on [a, b], continuous or not. Suppose that the conclusions of Theorem 2 hold for f. That is, suppose that every sequence of Riemann sums for f over [a, b] is convergent, and all to the same limit. Then f is said to be *integrable* over [a, b].

Then Theorem 2 becomes, loosely,

THEOREM 2′   *Continuous functions are integrable.*

We shall simply assume Theorem 2. Its proof is very technical and normally is not taken up until the later years of the calculus sequence. However, the theorem will appear to be true in practically every application that we make of the Riemann sum approximations.

DEFINITION   If f is integrable on [a, b] then the common limit of all the Riemann sum sequences $\{S_N\}$ is called the *definite integral* of f over [a, b]. It is designated

$$\int_a^b f \qquad \text{(function notation)},$$

or

$$\int_a^b f(x)\, dx \qquad \text{(Leibniz notation)}.$$

The b and a above and below the integral sign are the upper and lower *limits of integration*, respectively. The x in the Leibniz form is a "dummy variable" and can be replaced by any other variable. Thus,

$$\int_a^b f(x)\, dx = \int_a^b f(t)\, dt.$$

In a few simple cases we can evaluate a definite integral $\int_a^b f$ by explicitly calculating the limit $\lim_{N \to \infty} S_N$, where S_N is the Nth standard sum.

Example 1   Prove that

$$\int_0^b x^2 \, dx = \frac{b^3}{3}.$$

Solution   We are asked to prove that

$$\lim_{N \to \infty} \Delta x [f(x_1) + \cdots + f(x_N)] = \frac{b^3}{3},$$

where $\Delta x = b/N$, and the subdividing points are $x_1 = \Delta x$, $x_2 = 2 \Delta x$, $x_3 = 3 \Delta x$, etc. Thus, $f(x_1) = x_1^2 = (\Delta x)^2$, $f(x_2) = x_2^2 = (2 \Delta x)^2$, etc., so

$$\begin{aligned}
S_N &= \Delta x [f(x_1) + f(x_2) + \cdots + f(x_N)] \\
&= \Delta x [(\Delta x)^2 + (2 \Delta x)^2 + \cdots + (N \Delta x)^2] \\
&= (\Delta x)^3 [1^2 + 2^2 + \cdots + N^2] = (\Delta x)^3 \sum_{k=1}^{N} k^2.
\end{aligned}$$

We know from the last section that

$$\sum_{k=1}^{N} k^2 = \frac{N^3}{3} + p(N),$$

where p is a polynomial of degree 2. Therefore

$$\begin{aligned}
S_N &= \left(\frac{b}{N}\right)^3 \left[\frac{N^3}{3} + p(N)\right] \\
&= \frac{b^3}{3} \left[1 + \frac{3p(N)}{N^3}\right].
\end{aligned}$$

Now $p(N)/N^3 \to 0$ as $N \to \infty$, because p has degree <3. (See Chapter 2, Section 4.) So $S_N \to b^3/3$, as claimed. $\square$

Let us now return to the two questions we asked at the beginning.

It appeared that the average value AV of a continuous function f over $[a, b]$ should be given by $AV = \lim A_N$, where, for each N, A_N is the average of the values of f over the Nth standard evaluation set. Let us abbreviate $\sum_1^N f(x_j)$ by Σ_N. Then

$$A_N = \frac{1}{N} \Sigma_N \quad \text{and} \quad S_N = \Delta x \, \Sigma_N,$$

where S_N is the Nth standard Riemann sum. So

$$A_N = \frac{1}{N \, \Delta x} \, S_N = \frac{1}{b - a} \, S_N.$$

Therefore Theorem 2 guarantees that the limit exists, and

$$AV = \lim_{N \to \infty} A_N = \frac{1}{(b - a)} \lim_{N \to \infty} S_N = \frac{1}{(b - a)} \int_a^b f.$$

Thus,

$$\boxed{AV = \frac{1}{b - a} \int_a^b f}$$

is the formula for the average value of f over $[a, b]$.

Example 2   In Example 1 we computed that

$$\int_0^b x^2 \, dx = \frac{b^3}{3}.$$

Therefore the average value of $f(x) = x^2$ over $[0, b]$ is

$$AV = \frac{1}{b - 0} \int_0^b x^2 \, dx = \frac{1}{b} \cdot \frac{b^3}{3} = \frac{b^2}{3}. \qquad \square$$

The other question asked about the area A under a function graph. Here we can give a geometric proof. First we form an Nth *lower sum* L_N that is analogous to the upper sum U_N. Let m_j be the *minimum* value of f on the jth subinterval of the subdivision for $j = 1, 2, \ldots, N$, and set

$$L_N = \Delta x[m_1 + m_2 + \cdots + m_N] = \Delta x \sum_{i=1}^{N} m_i.$$

Then L_N can be interpreted as a sum of areas of *nonoverlapping* rectangles that are *inscribed in* the region G, as shown in Fig. 4. Therefore,

$$L_N \leq \text{area of } G.$$

Altogether we have

$$L_N \leq A \leq U_N$$

for each N, where A is the area of the region G.

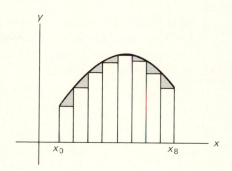

Figure 4

Now the Nth upper sum

$$U_N = \Delta x[M_1 + \cdots + M_N] = \Delta x \sum_{i=1}^{N} M_i$$

is a Riemann sum, because each *maximum* value M_j is in particular a *value:* $M_j = f(c_j)$ for some point c_j in the jth subinterval. The Nth lower sum L_N is also a Riemann sum, by the same observation.

Therefore, by Theorem 2, the sequences $\{L_N\}$ and $\{U_N\}$ both converge, and to the same limit. So A must equal this common limit, by the squeeze limit law. And this limit is $\int_a^b f$, by definition. The following theorem is thus a corollary of Theorem 2 and properties of area.

THEOREM 3   *If f is a positive continuous function on $[a, b]$, then $\int_a^b f$ is equal to the area of the region under the graph of f from $x = a$ to $x = b$.*

Example 3   The area (of the region) under $y = x^2$ from $x = 0$ to $x = 1$ is

$$\int_0^1 x^2 \, dx = \frac{1}{3},$$

again by the result in Example 1.   □

One final point. Any Riemann sum is an approximation to the corresponding integral, and the upper and lower sums approximate the integral from above and below. We can use these approximations to estimate integrals that are difficult or impossible to calculate exactly.

Example 4   Estimate $\int_1^2 \sqrt{x} \, dx$ by the upper and lower sums for a fourfold subdivision of $[1, 2]$.

Solution Since $f(x) = \sqrt{x}$ is increasing, the minimum values m_j occur at the left-hand endpoints of the subintervals. Using a pocket calculator, or the square root table (Table 4 in Appendix 6), we get

$$L_4 = \frac{1}{4}[\sqrt{1} + \sqrt{1.25} + \sqrt{1.5} + \sqrt{1.75}]$$

$$= \frac{1}{8}[\sqrt{4} + \sqrt{5} + \sqrt{6} + \sqrt{7}] = 1.166.$$

Similarly,

$$U_4 = \frac{1}{8}[\sqrt{5} + \sqrt{6} + \sqrt{7} + \sqrt{8}] = 1.270.$$

Thus

$$1.166 < \int_1^2 \sqrt{x}\,dx < 1.270.$$

The correct value is 1.219 to the nearest three decimal places.   □

When we visualize the area interpretations of these upper and lower Riemann sums, we might feel that the integral must be approximately halfway between them, in which case the average of these two Riemann sums should be a better estimate than either is by itself. In the above example, this average is

$$\frac{L_4 + U_4}{2} = \frac{1.166 + 1.270}{2} = 1.218,$$

which is exceedingly close to the true value. Why this is so will be discussed in Chapter 9, Section 8.

We make one more experiment, computing the sum of just the odd-numbered function values and doubling the result:

$$2\,\Delta x[f(x_1) + f(x_3)] = \frac{1}{2}[\sqrt{1.25} + \sqrt{1.75}] = 1.220.$$

This is much the simplest calculation, yet it gives a result as good as the average above. Moreover, this is no accident. It illustrates that by working less hard, but with more sophistication, we can often greatly improve our performance. See Section 8.

PROBLEMS FOR SECTION 3

Given any positive integer p, there is a polynomial f of degree at most p such that

$$1^p + 2^p + 3^p + \cdots + n^p = \frac{n^{p+1}}{p+1} + f(n)$$

for all positive integers n. This was discussed in the last section. Assuming this fact, and using Example 1 as a model, establish each of the following identities.

1. $\int_0^b x\,dx = \frac{b^2}{2}$ **2.** $\int_0^b x^3\,dx = \frac{b^4}{4}$ **3.** $\int_0^b x^4\,dx = \frac{b^5}{5}$ **4.** $\int_0^b x^n\,dx = \frac{b^{n+1}}{n+1}$

5. $\int_a^b x\,dx = \frac{b^2}{2} - \frac{a^2}{2}$ **6.** $\int_a^b x^2\,dx = \frac{b^3}{3} - \frac{a^3}{3}$ (expand the squares in the Riemann sum)

7. $\int_a^b x^3\,dx = \frac{b^4}{4} - \frac{a^4}{4}$

8. Derive the results in Problems 5 through 7 by combining a fact about area with the earlier results and the geometric interpretation of the integral.

Estimate the following integrals by calculating the upper and lower Riemann sums for the indicated subdivisions:

9. $\int_3^4 \sqrt{x}\,dx; n = 6$ **10.** $\int_1^2 \sqrt{x}\,dx; n = 10$ **11.** $\int_1^2 \frac{dx}{x}; n = 5$

12. $\int_1^2 \frac{dx}{x}; n = 10$ **13.** $\int_0^1 \frac{dx}{1+x^2}; n = 5$ **14.** $\int_0^1 \frac{dx}{1+x^2}; n = 10$

4 THE FUNDAMENTAL THEOREM OF CALCULUS

The calculation of areas by the definite integral process in the last section descends from a method used by the ancient Greeks. This method, called exhaustion, enabled Archimedes to compute the area of a parabolic segment (an ancient version of our Example 1), and to compute a few other particular geometric magnitudes. For nearly two thousand years this cluster of computations by Archimedes stood as an isolated achievement. Then, during the seventeenth century, the method of exhaustion was successfully applied to a few more individual problems, and in the nineteenth century it evolved into a general theory of integration.

But meanwhile Newton and Leibniz discovered that if a quantity could be calculated exactly by exhaustion, then it could be computed *much more easily* from antiderivatives. Having looked at the modern version of exhaustion in the last section, we turn now to a modern version of Newton's important discovery: the calculation of definite integrals by antidifferentiation. This remarkable result is so basic that it is called the Fundamental Theorem of Calculus. The evaluation of a definite integral is the final calculation in all sorts of problems, both applied and theoretical. Without the Fundamental Theorem we would hardly ever be able to get the exact answer. The direct evaluation of $\lim_{N \to \infty} S_N$ is simply too hard.

THEOREM 4   **First Fundamental Theorem of Calculus** *Suppose that f is integrable on* $[a, b]$ *and suppose that F is an antiderivative of f over* $[a, b]$. *Then*

$$\int_a^b f = F(b) - F(a).$$

Proof We will show that for every N there is an Nth Riemann sum R_N having the value

$$R_N = F(b) - F(a).$$

Of course, this constant sequence will then be its own limit. On the other hand, $\lim R_N = \int_a^b f$, by definition. So

$$\int_a^b f = \lim_{N \to \infty} R_N = F(b) - F(a),$$

and we have the theorem. It remains to find R_N.

We start from the equally spaced subdividing points

$$a = x_0 < x_1 < \cdots < x_N = b.$$

By the mean-value principle there is a point c_1 lying strictly between x_0 and x_1 such that $F(x_1) - F(x_0) = F'(c_1) \Delta x = f(c_1) \Delta x$. There is a corresponding equation for each subinterval of the subdivision, so we have the following N equations:

$$F(x_1) - F(a) = f(c_1) \Delta x$$
$$F(x_2) - F(x_1) = f(c_2) \Delta x$$
$$F(x_3) - F(x_2) = f(c_3) \Delta x$$
$$\vdots$$
$$F(b) - F(x_{N-1}) = f(c_N) \Delta x.$$

When we add this stack of N equations, all the terms on the left cancel in pairs except for two, and we have

$$F(b) - F(a) = \Delta x \sum_{j=1}^{N} f(c_j).$$

The sum on the right is a Riemann sum R_N, so we are done. ∎

Example 1 Since $F(x) = x^3/3$ is an antiderivative of x^2, the Fundamental Theorem tells us that

$$\int_a^b x^2 \, dx = F(b) - F(a) = \frac{1}{3}(b^3 - a^3).$$

Comparing this with the calculation in Example 1 of the last section will give some idea of the power of the Fundamental Theorem. □

There is an intermediate step that is helpful in more complicated problems. Instead of the above, we would write

$$\int_a^b x^2 \, dx = \frac{x^3}{3}\Big]_a^b = \frac{b^3}{3} - \frac{a^3}{3}.$$

That is, we first write down an antiderivative, together with a bracket labeled as above to show how it is to be evaluated. The convention is that $f(x)]_a^b = f(b) - f(a)$. Then we carry out the evaluation as a second step.

Note that *any* antiderivative can be used. For a direct check on this, suppose we change to another by adding a constant: $G(x) = F(x) + C$. Then C cancels out when we evaluate between limits:

$$G(x)]_a^b = G(b) - G(a) = (F(b) + C) - (F(a) + C)$$
$$= F(b) - F(a).$$

Example 2
$$\int_1^2 (x^2 - x) \, dx = \frac{x^3}{3} - \frac{x^2}{2}\Big]_1^2$$
$$= \left(\frac{8}{3} - \frac{4}{2}\right) - \left(\frac{1}{3} - \frac{1}{2}\right)$$
$$= \frac{7}{3} - \frac{3}{2} = \frac{5}{6}.$$
□

Example 3

$$\int_1^2 \sqrt{x}\, dx = \frac{2}{3} x^{3/2} \Big]_1^2$$

$$= \frac{2}{3}(2\sqrt{2} - 1) \approx 1.219.$$

This is the integral that we estimated at the end of the last section.   □

Example 4

$$\int_0^1 x^{100}\, dx = \frac{x^{101}}{101} \Big]_0^1 = \frac{1}{101};$$

$$\int_0^\pi \sin x\, dx = -\cos x \Big]_0^\pi = 1 - (-1) = 2;$$

$$\int_0^\pi \cos x\, dx = \sin x \Big]_0^\pi = 0 - 0 = 0;$$

$$\int_1^2 \frac{dt}{t^2} = -\frac{1}{t} \Big]_1^2 = -\frac{1}{2} - (-1) = \frac{1}{2}.$$ □

Example 5   Find the average value of $\sin x$ over the interval $[0, \pi]$.

Solution   By the formula from Section 1 and the Fundamental Theorem,

$$AV = \frac{1}{\pi} \int_0^\pi \sin x\, dx = \frac{2}{\pi}.$$

This is about 2/3, which may seem surprising, since the values of $\sin x$ may seem pretty evenly distributed over the range $[0, 1]$. However, a careful look at the graph of $\sin x$ shows that actually

$$\sin x \geq \frac{1}{2}$$

on 2/3 of the interval.   □

Example 6   Find the area under the graph of $\cos x$ from $x = 0$ to $x = \pi/2$.

Solution

$$A = \int_0^{\pi/2} \cos x\, dx = \sin x \Big]_0^{\pi/2} = 1 - 0 = 1.$$ □

Example 7 Find the area of the region between the x-axis and the graph of $y = 2x - x^2$.

Solution A rough sketch will often help to get a feel for a problem. In particular, it may help to find the correct limits of integration. Here we have a downward opening parabola that lies above the x-axis between its x-intercepts $x = 0, 2$. See Fig. 1. Then

$$A = \int_0^2 (2x - x^2)\, dx = \left[x^2 - \frac{x^3}{3} \right]_0^2 = \frac{4}{3}.$$ □

Figure 1

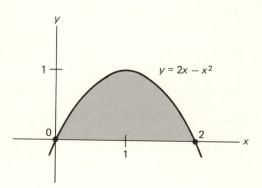

$y = 2x - x^2$

PROBLEMS FOR SECTION 4

Evaluate the following definite integrals.

1. $\int_0^1 (x - x^2)\, dx$

2. $\int_1^2 (x - x^2)\, dx$

3. $\int_{-1}^1 (x^4 - x^2)\, dx$

4. $\int_0^a (ax - x^2)\, dx$

5. $\int_x^y 3t^2\, dt$

6. $\int_0^{\pi/2} \cos t\, dt$

7. $\int_0^1 (x^3 - x^{1/3})\, dx$

8. $\dfrac{1}{b - a} \int_a^b x\, dx$

In the following area problems, the first step should normally be drawing a sketch.

9. Find, by integration, the area of the triangle bounded by the line $y = 2x$, the x-axis, and the line $x = 4$. Verify your answer by using the formula $A = \frac{1}{2}bh$.

10. Find, by integration, the area of the trapezoid bounded by the line $x + y = 15$, the x-axis, and the lines $x = 3$ and $x = 10$. Verify your answer by use of the formula $A = \frac{1}{2}(a + b)h$.

Find the area bounded by the given curve, the x-axis, and the given vertical lines.

11. $y = x^3$; $x = 0, x = 4$

12. $y = x^2 + x + 1$; $x = 2, x = 3$

13. $y = x^2 + 4x$; $x = -4, x = -2$

14. $y^2 + 4x = 0$; $x = -1, x = 0$

15. $y = 2x + \dfrac{1}{x^2}$; $x = 1, x = 4$

16. $y = x^2$; $x = 2, x = 5$

17. Find the area of the region bounded by the coordinate axes and the line $x + y = 2$.

18. Verify the formula for the area of a triangle in the case of the right triangle with vertices at the origin, the point (b, h) and the point $(b, 0)$.

19. Verify the formula for the area of a trapezoid in the case of the vertices $(0, b_1)$, (h, b_2), and the points on the x-axis under these two points.

20. Find the area of the region between the parabola $y = x^2$ and the line $y = 4$.

21. Find the area under one arch of the curve $y = \cos x$.

22. Find the area of the region under the graph of $y = \sec^2 x$, from $x = 0$ to $x = \pi/4$.

Find the average value of the following functions over the specified interval.

23. $y = 2x^3$; $[-1, 1]$

24. $y = 4 - x^2$; $[-2, 2]$

25. $y = x^2 - x + 1$; $[0, 2]$

26. $y = (x/2) + 1$; $[2, 6]$

27. $y = 2x + 1$; $[-1, 3]$

28. $y = x^n$; $[0, 1]$ (Why is the answer reasonable?)

29. A car travels 20 miles an hour for $\frac{1}{2}$ hour, 30 miles per hour for 2 hours, and 40 miles per hour for $\frac{1}{2}$ hour. What is its average velocity?

30. A typist's speed over a four-hour interval increases as he warms up and decreases as he tires. His speed in words per minute can be approximated by $w(t) = 6[4^2 - (t - 1)^2]$. Find his speed at the beginning of the interval, the end of the interval, his maximum speed, and his average speed over the 4-hour period.

31. If a particle is moving along a coordinate line, and if its position at time t is given by $s = f(t)$, then we saw in Chapter 3 that its average velocity over the interval $[t_0, t_1]$ is given by

$$\frac{s_1 - s_0}{t_1 - t_0} = \frac{f(t_1) - f(t_0)}{t_1 - t_0},$$

and that its instantaneous velocity at time t is

$$v = \frac{ds}{dt} = f'(t).$$

Show now that the average velocity really is the average value of the instantaneous velocity, according to the notion of average value developed in this section.

32. If $y = f(x)$ and if $f'(x_0)$ is interpreted as the rate of change of y with respect to x at x_0, show that

$$\frac{\Delta y}{\Delta x} = \frac{f(x + \Delta x) - f(x)}{\Delta x}$$

is the average rate of change of y with respect to x over the interval $[x, x + \Delta x]$.

33. A particle moves along a straight line with varying velocity $v = g(t)$. Give a direct justification, in terms of finite-sum approximations, for the fact that the distance traveled during the time interval $[a, b]$ is given by

$$s = \int_a^b v \, dt = \int_a^b g(t) \, dt.$$

Start with the basic idea that over a very short time interval from t_0 to $t_0 + \Delta t$ the velocity is nearly constant, with value $v_0 = g(t_0)$, so the distance traveled is approximately $v_0 \, \Delta t = g(t_0) \, \Delta t$.

5

PROPERTIES OF THE INTEGRAL; MORE AREAS.

Besides the Fundamental Theorem, there are some other properties of the definite integral that help greatly in making computations. Here is a short list. All functions mentioned are assumed to be integrable over the intervals in question.

1. $\displaystyle\int_a^b (f + g) = \int_a^b f + \int_a^b g;$

2. $\displaystyle\int_a^b kf = k \int_a^b f,$ where k is a constant.

3. If $a < b < c$, then $\displaystyle\int_a^c f = \int_a^b f + \int_b^c f.$

4. If $f \geq g$ on $[a, b]$, then $\displaystyle\int_a^b f \geq \int_a^b g.$

4'. If $m \leq f(x) \leq M$ for all x in $[a, b]$, then

$$m(b - a) \leq \int_a^b f \leq M(b - a).$$

These laws have straightforward proofs based on the sequence limit laws and the definition of the integral. Only (3) is a little tricky. On the other hand, (3) is geometrically obvious. The area under the graph of f from a to b, plus the area from b to c, equals the area from a to c (Fig. 1). We shall discuss the proofs at the end of the section.

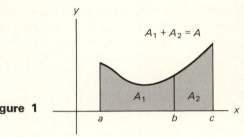
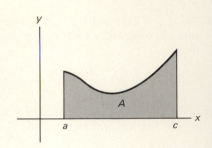

Figure 1

Example 1   Compute

$$\int_0^2 |x^2 - 1|\, dx.$$

Solution   Since $x^2 - 1$ is negative on $(0, 1)$ and positive on $(1, 2)$, we have, by (3),

$$\int_0^2 |x^2 - 1|\, dx = \int_0^1 (1 - x^2)\, dx + \int_1^2 (x^2 - 1)\, dx$$

$$= \left[x - \frac{x^3}{3} \right]_0^1 + \left[\frac{x^3}{3} - x \right]_1^2$$

$$= \frac{2}{3} + \left[\left(\frac{8}{3} - 2 \right) - \left(-\frac{2}{3} \right) \right] = 2. \qquad \square$$

Example 2   By (4),

$$\int_0^1 \sqrt{1 + x^2}\, dx \le \int_0^1 \sqrt{1 + x}\, dx = \frac{2}{3} (1 + x)^{3/2} \Big]_0^1$$

$$= \frac{2}{3} (2\sqrt{2} - 1)$$

$$\le \frac{2}{3} (2.83 - 1) = 1.22. \qquad \square$$

The first three laws have more general formulations that are often easier to use. Thus, (1) and (2) together say that the definite integral is *linear*:

$$\boxed{\int_a^b (c_1 f_1 + c_2 f_2) = c_1 \int_a^b f_1 + c_2 \int_a^b f_2,}$$

where c_1 and c_2 are constants. This can be read. "The integral of a linear combination is that same linear combination of the separate integrals."

We can also add further terms and apply (1) and/or (2) again for each new term, to get

$$\int_a^b (c_1f_1 + c_2f_2 + \cdots + c_nf_n) = c_1 \int_a^b f_1 + c_2 \int_a^b f_2 + \cdots + c_n \int_a^b f_n.$$

We regard these generalizations as being an understood part of the properties (1) and (2).

Areas Between Graphs We consider now the area *between* two graphs. This slightly generalizes the earlier situation where we considered the area between the x-axis and a function graph.

THEOREM 5 *Let f and g be continuous functions and suppose that $g(x) \leq f(x)$ over the interval $[a, b]$. Then the area of the region between the two graphs from $x = a$ to $x = b$ is $\int(f - g)]_a^b$.*

Proof Choose a constant C such that $g(x) + C$ is positive over $[a, b]$. The graph of $g + C$ is simply the graph of g translated upward a distance C (Fig. 2). The same holds for $f + C$. The area (of the region) between the given graphs equals the translated area between the translated graphs. (This is a property of area: congruent regions always have the same area.) And this translated area can be evaluated as the area

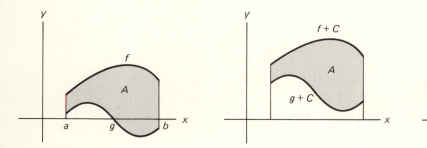

Figure 2

under $f + C$ minus the area under $g + C$. So, by Theorem 3,

$$A = \int_a^b (f + C) - \int_a^b (g + C) = \int_a^b [(f + C) - (g + C)]$$

$$= \int_a^b (f - g). \qquad \blacksquare$$

Example 3 Find the area of the finite region bounded by the graphs of $y = x^2$ and $y = \sqrt{x}$.

Solution You will generally want to draw a reasonably accurate sketch in a problem of this sort, to ensure that you have the right configuration in mind. Here, for example, we note that $y = \sqrt{x}$ is the *upper* graph between $x = 0$ and $x = 1$, the points of intersection of the two curves (Fig. 3). (These are found algebraically by solving the two equations simultaneously. This reduces to solving the equation $x^2 = \sqrt{x}$, or $x^4 = x$, from which we find the solutions $x = 0$ and $x = 1$.) These are the left and right edges of the region in question, so our area is

$$A = \int_0^1 (\sqrt{x} - x^2)\, dx = \left[\frac{2}{3} x^{3/2} - \frac{x^3}{3} \right]_0^1 = \frac{2}{3} - \frac{1}{3} = \frac{1}{3}.$$

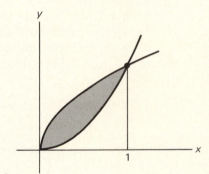

Figure 3

In Theorem 5 there is no requirement that the functions be positive.

Example 4 Find the area between the parabola $x = y^2$ and the vertical line $x = 1$.

Solution   Solving the first equation for y, we have

$$y = \pm\sqrt{x},$$

and we see that the area lies between the lower function graph $y = -\sqrt{x}$ and the upper function graph $y = \sqrt{x}$ (Fig. 4). So

$$A = \int_0^1 [\sqrt{x} - (-\sqrt{x})]\, dx = 2 \cdot \frac{2}{3} x^{3/2} \bigg]_0^1 = \frac{4}{3}.$$

Figure 4

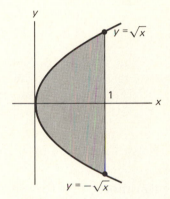

The arguments leading to Theorems 3 and 5 can be carried out just as well with the roles of the axes interchanged. Thus,

THEOREM 5′   *If $g(y) \le f(y)$ for all y in the interval $[a, b]$, then the area between the graphs $x = g(y)$ and $x = f(y)$, from $y = a$ to $y = b$, is*

$$\int_a^b [f(y) - g(y)]\, dy.$$

Proof   Instead of running through the same proof in this new configuration, we can reduce the new situation directly to the old as follows. The standard function graphs for the functions of Theorem 5′ would have the y-axis horizontal, and then the geometric configurations are exactly the same as for Theorem 5. So the area is

$$\int_a^b (f - g) = \int_a^b [f(y) - g(y)]\, dy$$

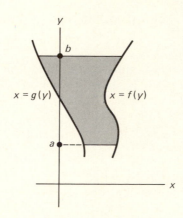

Figure 5

by Theorem 5. Now rotate the whole configuration over the 45° line $y = x$, obtaining the configuration of Theorem 5′ (Fig. 5). The area remains unchanged, so it is still given by $\int_a^b (f - g)$. ∎

Example 5   Solve Example 4 by integration with respect to y.

Solution   Now the region is viewed as extending from $x = y^2$ up to $x = 1$, between the limits -1 and 1. Thus,

$$A = \int_{-1}^{1} (1 - y^2)\, dy = y - \frac{y^3}{3}\Bigg]_{-1}^{1} = \left(1 - \frac{1}{3}\right) - \left(-1 + \frac{1}{3}\right) = \frac{4}{3}. \quad \square$$

Proofs of Integral Properties   Finally, we shall sketch the proofs of the laws (1) through (4). Let us temporarily adopt the notation $S_N(g)$ for the Nth standard Riemann sum for the function g over $[a, b]$. We then observe that

$$S_N(f + g) = S_N(f) + S_N(g),$$

because the $f + g$ sum can simply be rearranged into the right side by grouping together all the terms involving f and then all the terms involving g. This equation holds for all N and hence holds in the limit, giving (1).

The proof of (2) is similar.

For (4), we start with the fact that $f(x_j) \geq g(x_j)$ for $j = 1, \ldots, N$, so $S_N(f) \geq S_N(g)$. This holds for all N and hence in the limit. The inequality (4′) is a corollary of (4) and the fact that $\int_a^b c\, dx = c(b - a)$.

We come now to (3). The proof will omit details.

We start with the Nth standard Riemann sum for f over $[a, c]$,

$$S = \Delta x[f(x_1) + f(x_2) + \cdots + f(x_N)],$$

where $\Delta x = (c - a)/N$. Suppose that b lies in the Mth subinterval,

$$x_M \leq b \leq x_{M+1},$$

as shown in Fig. 6. The idea now is straightforward. We break S into two parts, the first M terms plus the last $N - M$ terms,

$$S = U + V.$$

We then show that the first part U is practically a Riemann sum for f over $[a, b]$, etc.

Figure 6

In fact, the endpoints $x_1, \ldots, x_M$ form an evaluation set for $[a, b]$, though not the standard set (this must be checked). So U becomes a Riemann sum when the interval width is adjusted. Specifically, if $\Delta x' = (b - a)/M$ then

$$\frac{\Delta x'}{\Delta x} U = \Delta x'[f(x_1) + \cdots + f(x_M)]$$

is an Mth Riemann sum for f over $[a, b]$. Therefore,

$$\frac{\Delta x'}{\Delta x} U \to \int_a^b f$$

as $M \to \infty$, by Theorem 2. Now it is not hard to show from the way M was chosen above that

$$\frac{M}{N} \to \frac{b - a}{c - a}$$

as $N \to \infty$. From this in turn it follows that $M \to \infty$ and that $\Delta x'/\Delta x \to 1$, both as $N \to \infty$. Therefore,

$$U \to \int_a^b f$$

as $N \to \infty$, which is what we wanted.

By exactly the same argument,

$$V \to \int_b^c f.$$

And of course S has the limit $\int_a^c f$, by definition. So in the limit the equation $S = U + V$ turns into

$$\int_a^c f = \int_a^b f + \int_b^c f.$$

PROBLEMS FOR SECTION 5

In each of the following examples, find the area of the finite region bounded by the indicated graphs. Always draw a sketch.

1. $y = 4x$; $y = 2x^2$

2. $y^2 = x$; $y = 4$; $x = 0$

3. $y^2 = 2x$; $x - y = 4$

4. $y^2 = 6x$; $x^2 = 6y$

5. $y^2 = 4x$; $x^2 = 6y$

6. $y^2 = 4x$; $2x - y = 4$

7. $y = 4 - x^2$; $y = 4 - 4x$

8. $y = 6x - x^2$; $y = x$

9. $y = x^3 - 3x$; $y = x$.

10. $y^2 = 4x$; $x = 12 + 2y - y^2$

11. $x^2 y = x^2 - 1$; $y = 1, x = 1, x = 4$

12. $y = x^2$; $y = x^3, x = 1, x = 2$

13. $y = x^2$; $y = x^3$

14. $w = 2 - x^2$; $w = x$

15. $s = 1 - t^2$; $s = t^2$

16. $t = y^4$; $t = 0, y = 1$

17. $y = x^{1/3}$; $y = x/4$.

18. $y = \sin x$; $y = 2x/\pi$

19. Find the area of the region between the graph of $y = x^3$ and its tangent line at $x = 1$.

20. Find the area of the region between $y = 1$ and $y = \cos x$, between two successive points of contact of these two graphs.

21. $x = 9y - y^3$; $y = 0, y = 3$

22. $y^2 = 4x$; $y = 0, y = 4$

23. $x = -(y^2 + 4y)$; $x = 0$

24. $y = 4 - x^2$; $y = 0, y = 3$

25. Find the area of the region bounded by the parabola $y = x^2$ and the line $y = x + 2$.

26. The same problem for $y = x^2$ and $y = 2 - x^2$.

27. Prove that

$$L_n \le \int_a^b f \le U_n,$$

where U_n and L_n are the nth upper and lower Riemann sums for f over $[a, b]$. (Write down property (4′) for each subinterval of a subdivision and then use a generalization of property (3).)

6
THE VOLUME
OF A SOLID
OF REVOLUTION

So far we have applied the definite integral to the computation of *average values* and *areas*. In this section a certain integral will be interpreted as a *volume*. Further interpretations and applications will be given in Chapter 9.

Suppose that f is continuous and positive on the closed interval $[a, b]$. If the upper half-plane is rotated about the x-axis, then each point on the graph $y = f(x)$ has a circular path, and the whole graph sweeps out a certain surface, called a *surface of revolution*. The plane region between the graph, the x-axis, and the vertical lines $x = a$ and $x = b$ sweeps out a *solid of revolution*. The boundary of the solid S lies partly on the surface of revolution and partly on the vertical planes through $x = a$ and $x = b$ (Fig. 1). We propose to calculate the volume of this solid.

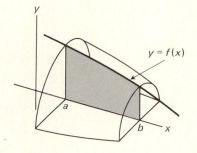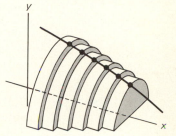

Figure 1

If we similarly rotate the inscribed rectangles of the Nth lower Riemann sum, we get a collection of nonoverlapping thin circular cylinders (disks, or wafers) that lie *inside* the solid S. The jth disk has altitude Δx and radius m_j, where m_j is the *minimum* value of f on the jth subinterval. Its volume is therefore

$$\pi r^2 h = \pi m_j^2 \, \Delta x.$$

The sum of these disk volumes is less than the volume V of S,

$$\Delta x \sum_{j=1}^{N} \pi m_j^2 \le V.$$

Finally, if we rotate the rectangles of the Nth *upper* sum, we get a collection of disks that *cover* S, and whose total volume is therefore *greater* than V. That is,

$$V \le \Delta x \sum_{j=1}^{N} \pi M_j^2,$$

where M_j is the *maximum* value of f on the jth subinterval, for $j = 1, \ldots, N$.

Now πm_j^2 and πM_j^2 are the minimum and maximum values on the jth subinterval of the function

$$g(x) = \pi[f(x)]^2,$$

so the two sums shown above are the Nth lower and upper sums for $g(x)$ over $[a, b]$. Thus V lies between two Nth Riemann sums for $g(x)$. This holds for every N so in the limit $V = \int_a^b g = \int_a^b \pi f^2$ (by the squeeze limit law). We have proved

THEOREM 6   *If f is a positive continuous function over the interval $[a, b]$ and if V is the volume of the solid of revolution generated by rotating about the x-axis the region under the graph of $y = f(x)$ between $x = a$ and $x = b$, then*

$$V = \int_a^b \pi(f(x))^2 \, dx.$$

Example   Find the volume of the solid of revolution generated by rotating the region under the graph of $y = x^2$, from $x = 1$ to $x = 3$.

Solution

$$V = \int_1^3 \pi(x^2)^2 \, dx = \int_1^3 \pi x^4 \, dx = \left. \frac{\pi x^5}{5} \right]_1^3$$

$$= \pi \frac{243}{5} - \pi \frac{1}{5} = (48\tfrac{2}{5})\pi. \qquad \square$$

PROBLEMS FOR SECTION 6

1. Prove, by integration, the formula

$$V = \frac{4}{3} \pi r^3$$

for the volume of a sphere. (The sphere is generated by revolving the circle $x^2 + y^2 = r^2$ about a diameter.)

2. Find, by integration, the volume of the truncated cone generated by revolving the area bounded by $y = 6 - x$, $y = 0$, $x = 0$, and $x = 4$ about the x-axis.

Find the volume generated by revolving about the x-axis the regions bounded by the graphs of the following equations.

3. $y = x^3$; $\quad y = 0, x = 1$ **4.** $9x^2 + 16y^2 = 144$ **5.** $y = x^2 - 6x$; $\quad y = 0$

6. $y^2 = (2 - x)^3$; $\quad y = 0, x = 0, x = 1$ **7.** $(x - 1)y = 2$; $\quad y = 0, x = 2, x = 5$

8. $y = x^{1/3}$; $\quad y = 0, x = 0, x = 8$ **9.** $y = \sqrt{1 - x^4}$; $\quad y = 0$

10. $y = (1 + x)^{1/4}, y = 0, x = 0, x = 1$ **11.** $y = x(1 + x^3)^{1/4}, y = 0, x = 0, x = 1$

12. $y = \sin x$; $\quad y = 0, x = 0, x = \pi$
[*Hint*: $\sin^2 x = (1 - \cos 2x)/2$]

13. $y = \sec x$; $\quad y = 0, x = 0, x = \pi/4$ **14.** $y = \sqrt{\cos(\pi x)}$; $\quad x = -\frac{1}{2}, x = \frac{1}{2}$

15. Show that the volume generated by rotating the ellipse

$$\frac{x^2}{a^2} + \frac{y^2}{b^2} = 1$$

about the x-axis is

$$V = \frac{4}{3} \pi b^2 a.$$

(Note that this generalizes the formula for the volume of a sphere.)

16. Find the limit as $n \to \infty$ of the volume obtained by rotating about the x-axis the region between the x-axis and the graph of $y = 1 + x^n$, from $x = 0$ to $x = 1$. What is the geometric interpretation of this limit?

17. Prove the formula

$$V = \frac{1}{3} \pi r^2 h$$

for the volume of a cone. (Rotate the region bounded by the straight line $y = rx/h$, the x-axis, and the vertical line $x = h$, about the x-axis.)

18. Find the formula for the volume of a spherical cap.

7

THE SECOND FUNDAMENTAL THEOREM

The Fundamental Theorem and its use in computing integrals raises the following question: Does every continuous function f *have* an antiderivative? Whether or not we succeed in explicitly finding an antiderivative F, can we count on its *existence*?

Suppose the answer is yes. Then notice what happens when we replace the upper limit of integration b in the Fundamental Theorem by a variable s. The evaluated integral

$$F(s) - F(a) = \int_a^s f$$

is now a function of s of the form $F(s)$ + constant and hence is an antiderivative of f. That is, if f has *any* antiderivative at all, then the function

$$H(s) = \int_a^s f$$

is one. So we start with H and attempt to prove directly that it is an antiderivative of f. This is the second Fundamental Theorem.

THEOREM 7   **Second Fundamental Theorem of Calculus**   *Let f be continuous on $[a, b]$ and let H be defined on $[a, b]$ by*

$$H(x) = \int_a^x f$$

for all x in $[a, b]$. Then H is differentiable and $H' = f$.

It was this theorem that Newton proved, except that he used areas instead of integrals. (The definite integral didn't appear on the scene until about 1825.) We shall give Newton's area proof.

Suppose that f is positive, and let A be the area under the graph of f from a to x. The figure will be clearer if we use a different variable, say t, for the horizontal axis, as shown in Fig. 1.

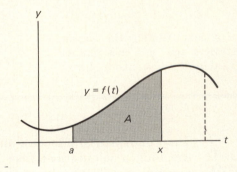

Figure 1

The varying area A is a function of the varying right-hand edge coordinate x, and we propose to prove:

THEOREM 8   *A is a differentiable function of x, and*

$$\frac{dA}{dx} = f(x).$$

Proof   To start with, we give x a positive increment Δx. The area A then increases by an increment ΔA, shown in the middle in Fig. 2, and ΔA is squeezed between two rectangular areas:

$$m \, \Delta x \leq \Delta A \leq M \, \Delta x,$$

where m and M are the minimum and maximum values of f on the incremental interval $[x, x + \Delta x]$. Then

$$m \leq \frac{\Delta A}{\Delta x} \leq M.$$

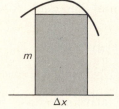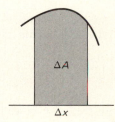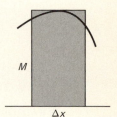

Figure 2

Now m is a function value $f(c)$ for some c between x and $x + \Delta x$. Since $c \to x$ as $\Delta x \to 0$ and since f is continuous, it follows that $m \to f(x)$ as $\Delta x \to 0$. Similarly for M. Then $\Delta A / \Delta x$ is squeezed to the same limit, so

$$\frac{dA}{dx} = \lim_{\Delta x \to 0} \frac{\Delta A}{\Delta x} = f(x).$$

If Δx is negative, the first inequality is reversed. But it reverses back on dividing by Δx, with the same final result. ∎

This area proof turns into a proof of the Second Fundamental Theorem when properties of integrals are substituted for properties of area. It will be left to the reader to verify this.

Example 1   Find the derivative of the function

$$f(x) = \int_0^x \sin \sqrt{t} \, dt.$$

Solution   By the Second Fundamental Theorem

$$f'(x) = \sin \sqrt{x}.$$

□

Example 2   Find the derivative of the function

$$f(x) = \int_0^{x^2} \sqrt{1 + t^3}\, dt.$$

Solution   If we set

$$g(x) = \int_0^x \sqrt{1 + t^3}\, dt$$

then

$$g'(x) = \sqrt{1 + x^3}$$

by the Second Fundamental Theorem. Also, $f(x) = g(x^2)$, so the chain rule gives

$$\begin{aligned} f'(x) &= g'(x^2) \cdot 2x \\ &= 2x\sqrt{1 + x^6}. \end{aligned}$$ □

Example 3   Show that $\int_x^b f$ is an antiderivative of $-f(x)$.

Solution   If F is an antiderivative of f, then

$$\int_x^b f = F(b) - F(x),$$

by the First Fundamental Theorem. This is an antiderivative of $-f(x)$.   □

Example 4   Find the derivative of

$$f(x) = \int_x^{x^2} \sqrt{1 + t^3}\, dt.$$

Solution   This combines Examples 2 and 3. We rewrite the expression for $f(x)$ as

$$f(x) = \int_x^a \sqrt{1 + t^3}\, dt + \int_a^{x^2} \sqrt{1 + t^3}\, dt,$$

and then have

$$f'(x) = -\sqrt{1 + x^3} + 2x\sqrt{1 + x^6}$$
$$\text{(by Ex. 3)} \quad \text{(by Ex. 2)}$$ □

It is useful to define $\int_b^a f$ when $b > a$ by

$$\int_b^a f = - \int_a^b f.$$

The First Fundamental Theorem then remains true:

$$\int_b^a f = - \int_a^b f = -[F(b) - F(a)]$$

$$= F(a) - F(b).$$

Moreover, the additivity property (3) then holds with complete generality:

3′. For *any* three numbers a, b, and c:

$$\int_a^c f = \int_a^b f + \int_b^c f.$$

We check this for a continuous function f. Let F be any antiderivative of f (on some interval containing a, b, and c). Then

$$\int_a^c f = F(c) - F(a) = (F(c) - F(b)) + (F(b) - F(a))$$

$$= \int_a^b f + \int_b^c f.$$

New Functions The area (integral) antiderivatives given by the Second Fundamental Theorem constitute an important source of new functions. For example, according to the integration formula

$$\int x^k \, dx = \frac{x^{k+1}}{k + 1} + C \qquad (k \neq -1),$$

every power of x has an antiderivative that is another power of x (times a constant), *except for the power* $k = -1$. Yet $1/x$ has a perfectly well-defined antiderivative, namely,

$$F(x) = \int_1^x \frac{dt}{t}.$$

This is a wholly new function, and it turns out to be an important one. Later on, it will be shown to be a logarithm function. It is called the *natural* logarithm function and is designated $\ln x$. Thus, by definition,

$$\ln x = \int_1^x \frac{dt}{t}$$

for all $x > 0$. When $x > 1$, $\ln x$ is the area of the shaded region in Fig. 3.

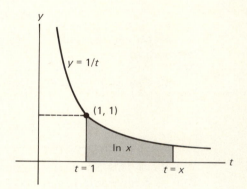

Figure 3

PROBLEMS FOR SECTION 7

Use the Second Fundamental Theorem to differentiate the following functions.

1. $f(x) = \displaystyle\int_0^x \frac{dt}{1 + t^2}$

2. $f(x) = \displaystyle\int_0^{x^2} \frac{dt}{1 + t}$

3. $f(x) = \displaystyle\int_x^{x^2} \frac{dt}{t}$

4. $f(x) = \displaystyle\int_1^{\sqrt{x}} \sqrt{1 + t^4}\, dt$

5. $f(x) = \displaystyle\int_x^a \sqrt{a^2 - t^2}\, dt$

6. $f(x) = \displaystyle\int_0^{\sqrt{x}} \sin(t^2)\, dt$

7. $f(x) = \displaystyle\int_0^{\sin x} \sqrt{1 - t^2}\, dt$

8. $f(x) = \displaystyle\int_0^{2\tan x} \sqrt{4 + t^2}\, dt$

9. $f(x) = \displaystyle\int_2^{\sec x} \frac{dt}{\sqrt{t^2 - 1}}$

10. $f(x) = \displaystyle\int_0^{1/x} \sqrt{\cos t}\, dt$

11. $f(x) = \displaystyle\int_0^{x^2} \sin(\sqrt{t})\, dt$

12. $f(x) = \displaystyle\int_0^{h(x)} g(t)\, dt$

13. $\displaystyle\int_0^{\sqrt{x}} \sec t\, dt$

14. $f(x) = \displaystyle\int_0^{x+a} \sin^2 t\, dt$

15. $f(x) = \displaystyle\int_0^{\tan^2 x} \frac{dt}{1 + t}$

16. $f(x) = \displaystyle\int_x^{x^2} g(t)\, dt$

17. $f(x) = \displaystyle\int_{1/x}^x \frac{dt}{1 + t^2}$

18. Show that

$$\int_{1/x}^{1} \frac{dt}{t} = \int_{1}^{x} \frac{dt}{t}$$

(compare derivatives).

20. Show that

$$\int_{0}^{1} \frac{dx}{1 + x^2} = \frac{\pi}{4}$$

(look at the problem above).

22. Show, from Problem 21, that

$$\frac{\pi}{6} = \int_{0}^{1/2} \frac{dx}{\sqrt{1 - x^2}}.$$

19. Let $f(x)$ be defined by

$$f(x) = \int_{0}^{x} \frac{dt}{1 + t^2}.$$

Prove that $f(\tan x) = x$ for all x in $(-\pi/2, \pi/2)$.

21. Let $f(x)$ be defined by

$$f(x) = \int_{0}^{x} \frac{dt}{\sqrt{1 - t^2}}.$$

Prove that $f(\sin x) = x$ for all x in $(-\pi/2, \pi/2)$.

23. Write out the proof of Theorem 7 by imitating the area proof given for Theorem 8.

**8
RIEMANN SUM
APPROXI-
MATIONS**

How close is a particular Riemann sum S_N to $\int_a^b f$? This question may seem pointless if we know how to evaluate $\int_a^b f$ exactly, but suppose we don't. Then the only way to calculate the integral may be through its Riemann sums, or modifications of these sums specifically tailored for a close fit.

Suppose, for example, that we want to calculate values of the function

$$\ln x = \int_{1}^{x} \frac{dt}{t}.$$

This is a wholly new function, as we noted in the last section. It is an antiderivative of $f(x) = 1/x$, and there is no such antiderivative among the functions we know about so far. So, the only way we have to calculate $\ln x$ at the moment is by its Riemann sums. (We shall study $\ln x$ in the next section.)

We shall now reason with areas, for intuition's sake. However, every result obtained this way can subsequently be made independent of area. We just replace each property of area that was used by the corresponding property of the definite integral.

To start with, let f be any monotone function, say increasing. Then the nth lower and upper sums for f over $[a, b]$, L_n and U_n, involve its values at the left-hand and right-hand endpoints of the n-fold subdivision.

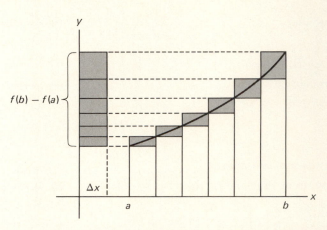

Figure 1

When we combine the diagrams for U_n and L_n a remarkable fact appears. The difference $U_n - L_n$, represented as the sum of the areas of the small shaded rectangles in Fig. 1, is exactly $\Delta x[f(b) - f(a)]$. This is because these difference rectangles can be slid left and stacked to form exactly a single rectangle of base-width Δx and altitude $f(b) - f(a)$. Note that since $\Delta x = (b - a)/n$, we can also write $U_n - L_n = K/n$, where K is the constant $(b - a)[f(b) - f(a)]$.

Altogether, we have established the following estimate:

THEOREM 9   *Let f be continuous, positive, and increasing over the interval $[a, b]$, and let U_n and L_n be the nth upper and lower Riemann sums for f over $[a, b]$. Then*

$$L_n < \int_a^b f < U_n,$$

and

$$U_n - L_n = \Delta x[f(b) - f(a)] = \frac{K}{n},$$

where $K = (b - a)[f(b) - f(a)]$. Thus each of U_n and L_n approximates $\int_a^b f$ with an error less than K/n in magnitude.

REMARK   The identity

$$U_n - L_n = \Delta x[f(b) - f(a)] = K/n$$

is a purely algebraic fact that is true for *any* function defined over

$[a, b]$: we just note that if the Riemann sums are written out, then the terms in $U_n - L_n$ cancel in pairs except for the last term of U_n and the first term of L_n, leaving

$$\Delta x\, f(x_n) - \Delta x\, f(x_0) = \Delta x[f(b) - f(a)].$$

Since $\int_a^b f$ is probably not exactly halfway between L_n and U_n, we can expect it to be closer to one of these numbers than half the distance $U_n - L_n$ between them. In general we can't tell which sum is closer, but we can if f has constant concavity over $[a, b]$. Suppose, for example, that f is increasing and concave down. The error rectangle over a single subdivision interval is divided by the graph of f into the upper error area for the upper Riemann sum U_n (shaded in Fig. 2), and the lower error area for L_n (unshaded).

Figure 2

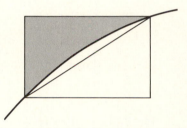

Since the graph lies above its chord when f is concave down, the upper error area is less than half the total rectangle area. This holds for all the individual errors, and therefore for their sum. So we have the sharper estimate:

If f is positive, increasing, and concave down, then the upper Riemann sum U_n approximates $\int_a^b f$ from above with an error E satisfying

$$\boxed{|E| \le \frac{1}{2} \Delta x[f(b) - f(a)].}$$

Similarly,

If f is decreasing and concave up, then L_n approximates $\int_a^b f$ from below with an error less than $\frac{1}{2}|f(b) - f(a)|\,\Delta x$.

Example 1   Use a fivefold subdivision of $[1, 2]$ to estimate $\ln 2$.

Solution   Since

$$\ln 2 = \int_1^2 \frac{dx}{x},$$

and since $1/x$ is decreasing and concave up, we use the lower Riemann sum S_5 for $f(x) = 1/x$ over $[1, 2]$. We have

$$S_5 = \frac{1}{5}\left[\frac{1}{1.2} + \frac{1}{1.4} + \frac{1}{1.6} + \frac{1}{1.8} + \frac{1}{2}\right]$$

$$= 0.6456 \cdots$$

(by calculation). We know that S_5 approximates $\ln 2$ from below with an error less than

$$\frac{1}{2}|f(b) - f(a)|\,\Delta x = \frac{1}{2} \cdot \frac{1}{2} \cdot \frac{1}{5} = 0.05.$$

Therefore,

$$0.645 < \ln 2 < 0.696.$$

This shows that the decimal expansion of $\ln 2$ starts

$$\ln 2 = 0.6 \cdots. \qquad \square$$

The squeeze inequality above also suggests that the one-place decimal closest to $\ln 2$ is 0.7. The next example will show that this is correct.

The Midpoint Riemann Sum   Now let us take the evaluation points $c_1, c_2, \ldots, c_n$ to be the *midpoints* of the subintervals of the n-fold subdivision, and let $\bar{S}_n$ be the resulting Riemann sum. The following result will be established in the problems.

THEOREM 10   *Suppose that the second derivative f'' exists and is bounded by B on $[a, b]$ ($|f''(x)| \leq B$ for all x in $[a, b]$). Then $\bar{S}_n$ approximates $\int_a^b f$ with an error E satisfying*

$$\boxed{|E| \leq \frac{B}{24}(b - a)(\Delta x)^2.}$$

Furthermore, the error is positive if f is concave up over $[a, b]$, and negative if f is concave down.

Example 2   Use a fivefold subdivision of $[1, 2]$ and Theorem 10 to estimate $\ln 2$.

Solution

$$\ln 2 = \int_1^2 \frac{dx}{x} \approx \bar{S}_5 = \frac{1}{5}\left[\frac{1}{1.1} + \frac{1}{1.3} + \frac{1}{1.5} + \frac{1}{1.7} + \frac{1}{1.9}\right]$$
$$= 0.6919\cdots.$$

Since $f(x) = 1/x$ has the second derivative $2/x^3$, which is positive and bounded by 2 on $[1,2]$, the error in this estimate is positive and less than

$$\frac{2}{24} \cdot 1 \cdot \left(\frac{1}{5}\right)^2 = \frac{1}{300} = 0.0033\cdots$$

by Theorem 10. Thus

$$0.6919 < \ln 2 < 0.6953$$

and $\ln 2 = 0.69\cdots$. □

Note that the midpoint-sum estimate in Example 2 is 15 times better than the lower sum estimate in Example 1.

At the end of Section 3, when we took the odd-numbered terms in a fourfold sum, we were actually taking the *midpoints* of a twofold sum, and the above error formula partially explains the accuracy we obtained.

PROBLEMS FOR SECTION 8

1. Show that if $[1, \frac{3}{2}]$ is divided into eight equal subintervals, then the upper Riemann sum for $1/x$ is

$$\sum_{k=0}^{7} \frac{1}{16 + k}.$$

2. In the manner of the above problem, write down a formula for the lower Riemann sum for $1/x$ obtained by dividing $[1, 2]$ into 1000 equal subintervals.

3. a) Write down the fourth standard Riemann sum S_4 for $f(x) = x$, over the interval $[1, \frac{3}{2}]$, leaving its terms as fractions.

b) Compute $\int_1^{3/2} x\, dx$ and show that the error in using S_4 as an estimate of the integral is exactly 1/32.

c) Show therefore that the sharper error estimate for a concave function given in the text cannot be improved.

4. a) Write down the fourth lower Riemann sum s_4 for $f(x) = x^2$ over the interval $[1, \frac{3}{2}]$, leaving its terms as fractions.

b) Compute $\int_1^{3/2} x^2\, dx$ and show that the actual error is less than the sharper error bound for a concave function.

5. We wish to use a standard Riemann sum S_n for $f(x) = 1/(1 + x^2)$ over the interval $[0, 1]$ to compute

$$\frac{\pi}{4} = \int_0^1 \frac{dx}{1 + x^2}.$$

(See Problem 20 in Section 7.) Show that S_n will approximate $\pi/4$ accurately to the nearest two decimal places if $n = 100$ by using an appropriate estimate from the text.

7. Suppose we wish to compute $\pi = 4\int_0^1 dx/(1 + x^2)$ to the nearest four decimal places, i.e., with an error less than $5(10)^{-5}$, by using a standard Riemann sum. How fine a subdivision would have to be used?

6. Compute the fourth standard Riemann sums for $f(x) = 1/(1 + x^2)$ over the interval $[0, 1]$. show that

$$2.88 < \pi < 3.39.$$

8. The same question for $\ln 2 = \int_1^2 dx/x$.

The following exercises center around Theorem 10. First we use it; then we prove it.

9. Show that

$$\ln 2 \approx 2\left[\frac{1}{17} + \frac{1}{19} + \frac{1}{21} + \frac{1}{23} + \frac{1}{25} + \frac{1}{27} + \frac{1}{29} + \frac{1}{31}\right]$$

with a positive error less than $(1/768)$. Calculate the decimal estimate this gives.

11. Show geometrically that $\bar{S}_n < \int_a^b f$ if f is decreasing and concave up. (Draw a typical subinterval and compare the errors in the area estimate on the two halves of the interval.)

10. If we wish to use the midpoint-evaluation Riemann sum $\bar{S}_n$ to approximate $\log 2 = \int_1^2 dx/x$ accurately to the nearest two decimal places, show that it is sufficient to take $n = 5$.

12. Show that

$$\ln \frac{3}{2} \approx \left[\frac{2}{17} + \frac{2}{19} + \frac{2}{21} + \frac{2}{23}\right]$$

with a positive error less than $(1/1536)$. Divide out the fractions to four decimal places and hence show that

$$\ln \frac{3}{2} \approx 0.405$$

correct to three decimal places.

13. Show that

$$\frac{\pi}{4} \approx \frac{1}{5}\left[\frac{1}{1.01} + \frac{1}{1.09} + \frac{1}{1.25} + \frac{1}{1.49} + \frac{1}{1.81}\right]$$

with an error less than $(1/300)$. (Assume that $|f''(x)| \le 2$ on $[0, 1]$, where $f(x) = 1/(1 + x^2)$.) Convert this approximation to a decimal estimate.

14. Note that the sum in Problem 12 is the first half of the sum in Problem 9. The second half of this latter sum is a midpoint-sum estimate of

$$\int_{3/2}^{2} \frac{dt}{t}.$$

Use the error estimate separately on this second half of the sum and then combine with the result in Problem 12 to get a better result than in Problem 9.

16. Show that the rectangular area $f(\bar{x}) \Delta x$ is the same as the trapezoidal area under the tangent line at $\bar{x}$, as suggested by the figure, when $\bar{x}$ is the midpoint of the increment integral.

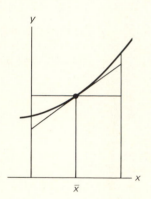

Show therefore that

$$\left(\int_{x}^{x+\Delta x} f \right) - f(\bar{x}) \, \Delta x$$

is equal to the integral

$$\int_{x}^{x+\Delta x} [f(t) - f(\bar{x}) - f'(\bar{x})(t - \bar{x})] \, dt.$$

15. Use the same device to improve the error bound in Problem 13. Assume that $|f''(x)| \leq 2/3$ on $[0.4, 1]$.

17. Use the tangent-line error formula (Theorem 8, Chapter 5) to show that if f'' is bounded by B between x and $x + \Delta x$, then the above integral is at most

$$\frac{B}{24} (\Delta x)^3$$

in magnitude. This bounds the error over each subdivision interval. Now add these errors up and hence prove Theorem 10.

CHAPTER 7

THE NATURAL LOGARITHM AND EXPONENTIAL FUNCTIONS; INVERSE FUNCTIONS

In this chapter we complete our list of basic functions by adding four new functions: $\ln x$, e^x, arctan x, and arcsin x. These functions are defined from known functions either by integration or by inversion. We have already seen how integration can be a source of new functions, by the Second Fundamental Theorem. Inverse functions will be introduced in this chapter, in Section 2.

1
THE NATURAL LOGARITHM FUNCTION

In Chapter 6 we looked briefly at the function $\ln x$ defined by

$$\ln x = \int_1^x \frac{dt}{t}.$$

We noted there that $\ln x$ is a wholly new function; the integrand x^{-1} is the one case for which the power rule cannot be reversed to give an antiderivative. In fact, $\ln x$ is a *logarithm* function, as will be shown below. The integral formula defines $\ln$ for all $x > 0$. When $x > 1$ we can visualize $\ln x$ as the area of the shaded region in Fig. 1. When $0 < x < 1$, $\ln x$ is the negative of the corresponding area.

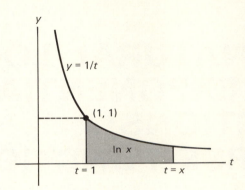

Figure 1

The principal things that we know about ln so far are

Basic Properties

$$\ln'(x) = \frac{1}{x}$$

for all positive x (by the Second Fundamental Theorem), and

$$\ln 1 = 0.$$

One consequence is that $\ln x$ is increasing over its whole domain $(0, \infty)$, since it has an everywhere positive derivative. Also, $\ln x$ is positive when $x > 1$ and negative when $x < 1$. Now we shall prove, by using calculus, that

$$\ln ab = \ln a + \ln b$$

for any two positive numbers a and b. This identity is the characteristic property of a logarithm; any function having this property is called a *logarithm* function.

THEOREM 1   *The function* $\ln x$ *is a logarithm function.*

Proof   Let a be any positive constant. Then

$$\frac{d}{dx}\ln(ax) = \ln'(ax)\frac{d}{dx}(ax)$$

$$= \frac{1}{ax} \cdot a = \frac{1}{x},$$

by the chain rule. Thus $\ln ax$ has the same derivative as $\ln x$ and hence differs from $\ln x$ by a constant:

$$\ln ax = \ln x + C.$$

To evaluate C we put $x = 1$ and get

$$\ln a = \ln 1 + C = 0 + C = C.$$

Thus $C = \ln a$ and

$$\ln ax = \ln x + \ln a.$$

This holds for any positive a and any positive x, so we are done. ∎

We call $\ln x$ the *natural* logarithm of x, because its derivative formula and other properties are simpler than those of other logarithms.

The logarithm law holds for any number of factors. For example, if a, b, and c are positive, then two applications of the law give

$$\ln(abc) = \ln(ab) + \ln c = \ln a + \ln b + \ln c.$$

We can continue to extend the law one factor at a time in this way. The general form is

If $a_1, a_2, \ldots, a_n$ are positive real numbers, then

$$\boxed{\ln(a_1 a_2 \cdots a_n) = \sum_{i=1}^{n} \ln a_i.}$$

If

$$a_1 = a_2 = \cdots = a_n = a,$$

then the boxed equation becomes $\ln(a^n) = n \ln a$. And by arguing a little further, in this same algebraic manner, it can be shown that

COROLLARY

$$\boxed{\ln a^r = r \ln a}$$

for all rational numbers r.

This identity can also be proved by calculus, as follows. If $x > 0$, then

$$\frac{d}{dx} \ln(x^r) = \frac{1}{x^r} \frac{d}{dx} x^r = \frac{rx^{r-1}}{x^r} = \frac{r}{x}.$$

But this is also the derivative of $r \ln x$. The two functions therefore differ only by a constant:

$$\ln(x^r) = r \ln x + C,$$

and we find that $C = 0$ upon setting $x = 1$.

We can get a very good idea of the graph of $\ln x$ by plotting the approximate values $\ln 2 \approx 0.7$ and

$$\ln 2^n = n \ln 2 \approx (0.7)n.$$

Figure 2 shows the resulting graph.

Figure 2

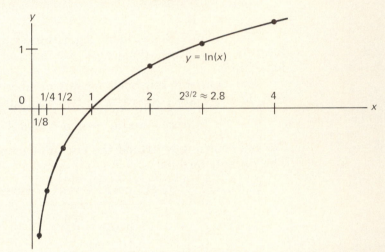

The *base* of a logarithm function is the number whose logarithm is 1. So the base of the natural logarithm, designated e, is defined by

$$1 = \ln e = \int_1^e \frac{dt}{t}.$$

We can visualize the base e as the value of x for which the shaded region in Fig. 1 has area equal to 1. This leads to easy, but mostly crude, estimates of e. Figure 3 shows three such estimates.

The first is the simplest (but also the least informative). The inscribed and circumscribed rectangular areas show that

$$\frac{1}{2} < \ln 2 < 1.$$

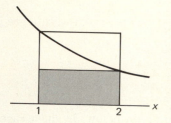
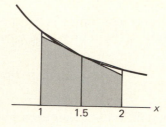
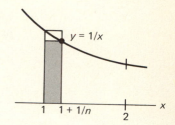

Figure 3

(This is actually a squeeze between upper and lower Riemann sums.) Multiplying by 2, and remembering that $2\ln 2 = \ln 4$, we get

$$1 < \ln 4 < 2.$$

Since $1 = \ln e$, we have, in particular, that

$$\ln 2 < \ln e < \ln 4,$$

and hence that

$$2 < e < 4,$$

which tells us something about where e is.

The areas indicated in the center figure lead, in a similar way, to the inequality

$$2.5 < e < 2.9.$$

(See Problem 49.) Actually,

$$e = 2.718$$

to three decimal places, as we shall see in Chapter 12.

The third figure is like the first and shows that

$$\frac{1}{n+1} < \ln\left(1 + \frac{1}{n}\right) < \frac{1}{n}$$

for any positive integer n. Multiplying first by n and then by $n + 1$, and using the above corollary, will lead this time to the inequality

$$\left(1 + \frac{1}{n}\right)^n < e < \left(1 + \frac{1}{n}\right)^{n+1}.$$

The difference between the outside terms is $1/n$ times the left term and hence is less than e/n. Since this $\to 0$ as $n \to \infty$, the squeeze limit law gives us the interesting formula

$$e = \lim_{n \to \infty} \left(1 + \frac{1}{n}\right)^n.$$

Some differentiation examples are given below. In most cases we apply the chain rule to the basic formula

$$\frac{d}{dx}\ln x = \frac{1}{x}.$$

With the built-in chain rule, this formula becomes

$$\frac{d}{dx}\ln u = \frac{1}{u}\frac{du}{dx},$$

$$d(\ln u) = \frac{du}{u}.$$

Examples 1 $\quad \dfrac{d}{dx}\ln(1 + x^2) = \dfrac{1}{1 + x^2}\dfrac{d}{dx}(1 + x^2) = \dfrac{2x}{1 + x^2}$

2 $\quad \dfrac{d}{dx}\ln kx = \dfrac{k}{kx} = \dfrac{1}{x}$

3 $\quad \dfrac{d}{dx}\ln(\cos x) = \dfrac{1}{\cos x} \cdot (-\sin x) = -\tan x$

4 $\quad \dfrac{d}{dx}\ln(\ln x) = \dfrac{1}{\ln x}\left(\dfrac{1}{x}\right) = \dfrac{1}{x \ln x}$

5 $\quad \dfrac{d}{dx}(x \ln x - x) = x \cdot \dfrac{1}{x} + \ln x \cdot 1 - 1 = \ln x$

6 $\quad \dfrac{d}{dx}\ln\left(\dfrac{1-x}{1+x}\right) = \dfrac{1}{\left(\dfrac{1-x}{1+x}\right)} \cdot \dfrac{(1+x)(-1) - (1-x)\cdot 1}{(1+x)^2}$

$$= \frac{-2}{(1-x)(1+x)} = -\frac{2}{1-x^2}$$

7   Since $\ln[(1 - x)/(1 + x)] = \ln(1 - x) - \ln(1 + x)$,

$$\frac{d}{dx} \ln\left(\frac{1 - x}{1 + x}\right) = \frac{1}{(1 - x)}(-1) - \frac{1}{(1 + x)} = \frac{-2}{(1 - x)(1 + x)} \qquad \square$$

PROBLEMS FOR SECTION 1

Reformulate the following expressions by using the logarithm law.

1. $\ln(x - 1) + \ln(x + 1)$ **2.** $2 \ln x + 3 \ln(x - 2)$ **3.** $\ln(x^2 - 2x + 1)$

4. $\ln(x - 1) - \ln(x + 1) + \ln(x^2 + x + 1) - \ln(x^2 - x + 1)$

5. Prove the more general formula

$$\frac{d}{dx} \ln |x| = \frac{1}{x} \qquad (\text{if } x \neq 0).$$

Differentiate the following functions.

6. $\ln(x + 1)$ **7.** $\ln(x^2 + 1)$ **8.** $\ln\sqrt{x^2 + 4}$

9. $x \ln x$ **10.** $\ln(1 + x^{1/3})$ **11.** $(\ln x)/x$

12. $\ln(\sec x)$ **13.** $\ln\left(x + \frac{1}{x}\right)$ **14.** $x^2 \ln x - x^2/2$

15. $\dfrac{x^{n+1}}{n + 1}\left(\ln x - \dfrac{1}{n + 1}\right)$ **16.** $(\ln x)^2$

Prove the following formulas.

17. $\dfrac{d}{dx} \ln(\sin x) = \cot x$ **18.** $\dfrac{d}{dx} \ln(\sec x) = \tan x$

19. $\dfrac{d}{dx} \ln(\sec x + \tan x) = \sec x$ **20.** $\dfrac{d}{dx} \ln\left(\dfrac{1 + \sin x}{1 - \sin x}\right) = 2 \sec x$

21. $\dfrac{d}{dx} \ln(x + \sqrt{1 + x^2}) = \dfrac{1}{\sqrt{1 + x^2}}$ **22.** $\dfrac{d}{dx} \ln(x + \sqrt{x^2 - 1}) = \dfrac{1}{\sqrt{x^2 - 1}}$

23. $\dfrac{d}{dx} \ln\left(\dfrac{1 + \sqrt{1 + x^2}}{x}\right) = -\dfrac{1}{x\sqrt{1 + x^2}}$

24. Using the notation $u' = du/dx$, etc., show that

$$\frac{d}{dx} \ln(uv) = \frac{u'}{u} + \frac{v'}{v},$$

$$\frac{d}{dx} \ln(uvw) = \frac{u'}{u} + \frac{v'}{v} + \frac{w'}{w}.$$

25. Prove that $a \ln x < x^a - 1$ when $x > 1$ for any positive rational number a. (Use the mean-value theorem.)

26. Using the above inequality, show that for any positive rational number b,

$$\ln x < x^b$$

when x is large enough. (Take $a = b/2$ and manipulate a little.) Show therefore that $\ln x/x^k \to 0$ as $x \to \infty$, for any positive rational number k.

27. Find the minimum value of $f(x) = x \ln x$.

28. Find the minimum value of $f(x) = x^2 \ln x$.

More derivatives.

29. $\ln(ax^2 + bx + c)$

30. $\sin(\ln x)$

31. $\ln x/(1 + x^2)$

32. $\sqrt{\ln(1 + x)}$

33. $\ln \tan x$

34. $(\sin 2x)(\ln 3x)$

35. $1/[1 + \ln(ax + b)]$

36. $\ln(\sin x + \cos x)$

37. $\cos(\ln^2 x)$

38. $\ln(\ln(\ln x))$

39. $\ln(1 - x)\ln(1 + x)$

40. $\dfrac{\ln x}{1 + (\ln x)^2}$

41. $\sec^2(x + \ln x)$

42. $\sqrt{1 + 2(\ln x)^3}$

43. $\ln^2(\cos^2(x^2))$

44. $[\ln(1 + \sin^2 4x)]^k$

45. $\cos\sqrt{\ln(\ln x)}$

46. $\frac{1}{2}\sec^2 x + \ln(\sin x)$

47. Graph $\ln x/x$ in the spirit of Chapter 5, Section 3, showing any critical point(s) and inflection point(s). (Note that $\ln x/x \to 0$ as $x \to \infty$, from Problem 26.)

48. Graph $\ln x/x^2$.

49. Show that $f(x) = \ln x/x^k$ assumes a maximum value, for any positive rational number k. Use this result to show that $f(x) \to 0$ as $x \to \infty$. (Boundedness for one k implies convergence to zero for another k.)

50. Show that the middle diagram in Fig. 3 leads to the inequality

$$\frac{2}{3} < \ln 2 < \frac{17}{24}$$

and therefore to

$$2^{4/3} < 2^{24/17} < e < 2^{3/2}.$$

Use a pocket calculator to show that $2.5 < e < 2.9$.

2
INVERSE
FUNCTIONS

The remaining new functions to be introduced in this chapter are all characterized as being inverses of known functions. We develop the general framework for inverse functions in this section and then take up the individual functions one at a time in succeeding sections.

DEFINITION We say that g *inverts f on the set I* if

$$\boxed{g(f(x)) = x}$$

for all x in I.

Example 1 Since $(x^3)^{1/3} = x$ for all x, $g(x) = x^{1/3}$ inverts $f(x) = x^3$ on $(-\infty, \infty)$.
Since $\sqrt{x^2} = x$ if $x \geq 0$, $g(x) = \sqrt{x}$ inverts $f(x) = x^2$ on $I = [0, \infty)$.

□

The inverting identity says that g *undoes what f has done on I.*
Thus, f takes x to $f(x)$ and g takes $f(x)$ back to x. We can think of g
as being "f backwards," as shown schematically in Fig. 1.

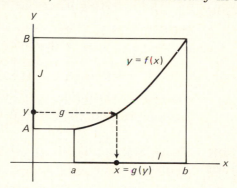

Figure 1

It is implicit in the above definition that the domain of f includes
I and the domain of g includes $f(I)$. We now show that the inverting
relationship is symmetrical.

THEOREM 2 *If g inverts f on I, then f inverts g on $J = f(I)$.*

Proof If we apply f to both sides of the equation

$$g(f(x)) = x,$$

we get

$$f(g(f(x))) = f(x),$$

for all x in I. Setting $y = f(x)$ turns this into

$$f(g(y)) = y,$$

for all y in J. That is, f inverts g on J. ■

In this situation we call f and g a pair of *mutually inverse*
functions; each function is the *inverse* of the other. More precisely,
their restrictions to I and J are mutually inverse functions.

Example 2 The functions $f(x) = x^3$ and $g(x) = x^{1/3}$ are mutually inverse (with no
domain restrictions).

If g is the restriction of the squaring function to the interval $I = [0, \infty)$,

$$g(x) = x^2 \qquad \text{on } [0, \infty),$$

then g is the inverse of $f(x) = \sqrt{x}$. □

The inverting identities

$$
\boxed{
\begin{aligned}
g(f(x)) &= x && \text{on } I, \\
f(g(y)) &= y && \text{on } J = f(I)
\end{aligned}
}
$$

are also called the *cancellation identities*, since they state that each function cancels the effect of the other.

Invertible Functions

There are two practical procedures for showing that a function f is invertible (on a given set).

(I) We may be able to solve the equation

$$y = f(x)$$

for x in terms of y, obtaining an *equivalent* equation

$$x = g(y).$$

Then g is the inverse of f. To see this, just note that substituting from either equation into the other gives the inverting identities.

Example 3

We show that $g(x) = x/(1 - x)$ has an inverse by this method. Solving for x in

$$y = \frac{x}{1 - x},$$

we have, first cross-multiplying,

$$y - yx = x,$$
$$y = x(y + 1),$$
$$x = \frac{y}{1 + y}.$$

Thus $f(y) = y/(1 + y)$ is the inverse of $g(x) = x/(1 - x)$.

We know that these functions have to cancel each other, but it is interesting to check it directly. We have

$$f(g(x)) = f\left(\frac{x}{1-x}\right) = \frac{\dfrac{x}{1-x}}{1 + \dfrac{x}{1-x}} = \frac{x}{(1-x)+x} = x.$$

The other identity $g(f(y)) = y$ works out similarly. $\square$

However, we will not normally be able to find an explicit expression for the inverse function. In fact, inverse functions are a prime source for new functions, as we shall see shortly.

(II) We may be able to apply one of the following two tests.

THEOREM 3 *If f is continuous and monotone on an interval I, then f is invertible on I, and its inverse g is also continuous and monotone, on the interval $J = f(I)$.*

THEOREM 4 **Inverse-function theorem** *If f is differentiable and f' is never zero on an interval I, then f is invertible on I, and its inverse is differentiable (on its domain $J = f(I)$).*

Theorem 3 is simply a restatement of the final remarks in Chapter 2. Theorem 4 combines the shape principle (to get f monotone) with Theorem 3 (to get f invertible) and the inverse function rule from Chapter 4.

Example 4 Show that

$$f(x) = 3x^3 - 3x^2 + 2x + 1$$

is invertible (on its whole domain $(-\infty, \infty)$).

Solution We have

$$f'(x) = 9x^2 - 6x + 2 = (3x - 1)^2 + 1,$$

which is everywhere positive. So f is invertible, by Theorem 4. $\square$

In simpler language, Example 4 shows that the equation

$$y = 3x^3 - 3x^2 + 2x + 1$$

determines x as a function of y.

Example 5   We know that $\ln x$ has the everywhere positive derivative $1/x$ on $(0, \infty)$. It therefore has a differentiable inverse, by Theorem 4. That is, there is a positive, differentiable function $E(x)$, defined on $(-\infty, \infty)$, such that

$$\ln(E(x)) = x \qquad \text{for all } x,$$
$$E(\ln x) = x \qquad \text{for all positive } x. \qquad \square$$

REMARK   Theorem 4 remains true when f has a finite number of critical points, provided that f' is otherwise of constant sign. For instance, if $f' > 0$ except at the critical points, then f is increasing on each subinterval of I determined by the critical points, and hence on the whole of I. In this situation the inverse function will (in general) also have a finite number of critical points.

Example 6   Show that $f(x) = x - \sin x$ is invertible on $[-3\pi, 3\pi]$.

Solution   The derivative

$$f'(x) = 1 - \cos x$$

is positive except at $x = -2\pi$, 0 and 2π, so f is increasing and hence invertible, by the remark above. $\qquad \square$

Inverse functions can be recognized from their graphs.

THEOREM 5   *The function g is the inverse of f if and only if the graph of g is the mirror image of the graph of f in the line $y = x$.*

Proof   If f and g are inverse, then, for any two numbers x and y, $x = g(y)$ if and only if $y = f(x)$. So these two equations are equivalent and have the same graph, which is the graph of f. The graph of g, on the other hand, is the graph of the equation $y = g(x)$, and this is the mirror image of the above graph in the line $y = x$, as we saw in Chapter 1, Sections 5 and 9.

      This reasoning can be reversed, showing that if the graphs are mirror images then the functions are mutually inverse. $\qquad \blacksquare$

Example 7   If

$$y = f(x) = (x - 1)^3,$$

then we can solve for x and get

$$x = g(y) = y^{1/3} + 1.$$

So the functions

$$f(x) = (x - 1)^3,$$
$$g(x) = x^{1/3} + 1$$

are mutually inverse. Their graphs are shown in Fig. 2.

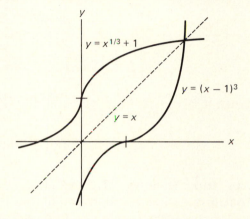

Figure 2

Example 8 The graphs of $\ln x$ and its inverse $E(x)$ are shown in Fig. 3.

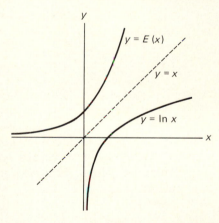

Figure 3

Remark (I) Theorems 3 and 4 have very simple partial converses, which explains why we need to look only at monotone functions in our search for inverses.

(A)   *If f is differentiable on the interval I and if f has a differentiable inverse g,*

$$g(f(x)) = x \text{ for every } x \text{ in } I,$$

then f is necessarily monotone.

Proof   By the chain rule

$$g'(f(x)) \cdot f'(x) = 1.$$

Therefore f' is never zero, and must have constant sign. Therefore f is monotone.   ■

In fact, differentiability can be replaced by continuity in (A).

(B)   *If f is continuous and invertible on an interval I, then f is necessarily monotone on I.*

See Problem 42 for hints about the proof.

Concavity and Inverse Functions

Remark (II)   The graphs below suggest what happens to concavity when a graph is reflected in, or rotated over, the 45° line $y = x$. It appears that *an increasing graph reverses its concavity when reflected,* and that *a decreasing graph preserves concavity.*

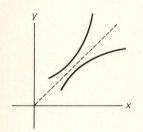 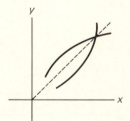 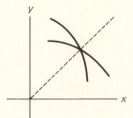 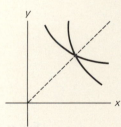

These conjectures are readily verified. We saw earlier that two functions related this way are mutually inverse and, in particular, satisfy

$$f(g(x)) = x.$$

Then $f'(g(x))g'(x) = 1$ and

$$g'(x) = \frac{1}{f'(g(x))},$$

by the chain rule. We have been through all of this before. Differentiating once more, and then substituting from the above equation, we have

$$g''(x) = -\frac{1}{[f'(g(x))]^2} \cdot f''(g(x)) \cdot g'(x)$$

$$= -\frac{f''(g(x))}{[f'(g(x))]^3} = -\frac{f''(y)}{[f'(y)]^3},$$

where $y = g(x)$. We are supposing that f' is never 0. Thus if f is increasing, then f' is everywhere positive and the formula above shows that g'' and f'' have opposite signs, so that concavity reverses. Similarly, if f is decreasing, then f' and $(f')^3$ are everywhere negative, and the formula shows that concavity is preserved.

PROBLEMS FOR SECTION 2

Some of the functions in Problems 1 through 15 are already known to be invertible. But give a (new) proof in each case by applying Theorem 4 or the remark that followed.

1. $f(x) = x^3$

2. $f(x) = x^2$, on the interval $[0, \infty)$

3. $f(x) = x^5$

4. $f(x) = x^3 + x$

5. $f(x) = 2x - x^2$, on $[-1, 1]$

6. $f(x) = x^3 + x^2$, on $[0, \infty)$

7. $f(x) = x^3 - 3x^2 + 3x$

8. $f(x) = \dfrac{x^5}{5} - \dfrac{x^4}{2} + \dfrac{x^3}{3}$

9. $f(x) = x^3 - x^2 + x$

10. $f(x) = x\sqrt{1 + x^2}$

11. $f(x) = \dfrac{x}{\sqrt{1 + x^2}}$

12. $f(x) = \sin x$, on $[-\pi/2, \pi/2]$

13. $f(x) = \tan x$, on $(-\pi/2, \pi/2)$

14. $f(x) = x \cos x$, on $[-\pi/4, \pi/4]$

15. $f(x) = \sin^3 x$, on $[-\pi/2, \pi/2]$

16. Prove that $f(x) = x^3/(1 + x^2)$ is invertible, and show that its inverse function is everywhere defined, i.e., its domain is $(-\infty, \infty)$. (Apply Theorem 4.)

17. Show that $f(x) = (x + 4)/3$ and $g(x) = 3x - 4$ are mutually inverse functions by computing $f(g(x))$ and $g(f(x))$.

18. Show that $f(x) = (4 - x)/(1 + x)$ is its own inverse by computing $f(f(x))$.

19. Compute $f(g(x))$ and $g(f(x))$, where

$$f(x) = \frac{x}{\sqrt{1 + x^2}} \quad \text{and} \quad g(x) = \frac{x}{\sqrt{1 - x^2}}.$$

What is your conclusion?

20. Find the values of a and b such that the linear function $f(x) = ax + b$ is self-inverse. There is more than one solution. Graph each type.

Show that each of the following functions f has an inverse function g by solving the equation for x in terms of y. Give $g(x)$ in each case, and sketch the two graphs.

21. $y = f(x) = 3x - 5$

22. $y = f(x) = 1/x$

23. $y = f(x) = x^{3/5}$

24. $y = f(x) = 1 + x^{1/7}$

25. $y = f(x) = \dfrac{1 - x}{1 + x}$

26. $y = f(x) = x^3 + 3x^2 + 3x + 1$

27. $y = f(x) = \dfrac{x}{\sqrt{1 + x^2}}$

28. $y = f(x) = \sqrt{x}$ (Be careful here.)

29. Compute $f(g(x))$ and $g(f(x))$, where

$$f(x) = x^2 - 2x + 1, \qquad g(x) = 1 + \sqrt{x}.$$

Show that f and g are mutually inverse if the domain of f is suitably restricted.

30. Show that

$$f(x) = (x^2 + 1)^{1/4}$$

and

$$g(x) = \sqrt{(x^2 + 1)(x^2 - 1)}$$

are mutually inverse if their domains are suitably restricted.

31. Show that $f(x) = \sqrt{1 - x^2}$ is its own inverse, provided that its domain is suitably restricted.

32. a) Show that if $g(f(x)) = x$ for every x in the domain of f, then the domain of f is included in the range of g and the range of f is included in the domain of g.

b) Show therefore that if f and g are mutually inverse, then

$$\text{domain } f = \text{range } g,$$
$$\text{range } f = \text{domain } g.$$

33. Prove:

If an equation in x and y determines each of x and y as a function of the other, then these two functions are mutually inverse.

(Argue in any manner, using function symbols only if you want to.)

34. Show that if f and g are both invertible, then so is their composition product $f \circ g$, and

$$\text{inv}(f \circ g) = (\text{inv}) g \circ (\text{inv}) f.$$

35. Show that

$$f(x) = \frac{ax + b}{cx + d}$$

is self-inverse if and only if

a) $d = -a$, or

b) $d = a, c = b = 0.$

36. Show that if f is increasing and equal to its own inverse, then $f(x) = x$.

37. Suppose that an equation E in x and y determines each of x and y as a function of the other. Show that these mutually inverse functions are the same function if and only if interchanging x and y in the equation E yields an equivalent equation.

39. Draw a graph and find an equation for a decreasing self-inverse function f that is concave down.

41. Show that if f and g are mutually inverse, then so are $f(ax)$ and $g(x)/a$, for any nonzero constant a.

38. Keeping in mind the principle stated in Problem 37, draw the graph of a self-inverse function f that is decreasing and concave up. Find an equation for such a function.

40. What can you say about the function solution of the equation

$$x^{1/3} + y^{1/3} = 2?$$

Sketch its graph.

42. Prove Remark IB. (Show that if f is continuous and *not* monotone on an interval I then there must be three points on the graph with the middle one higher than both the others (or lower). Show therefore that there are distinct values c_1 and c_2 in I such that $f(c_1) = f(c_2)$, and hence that the equation $y = f(x)$ does *not* determine x as a function of y.)

**3
THE
EXPONENTIAL
FUNCTION**

The exponential function $f(x) = e^x$ is one of the most important functions in mathematics and in the applications of mathematics. For example, we shall see in Section 5 that a "population" grows exponentially, whether it be a population of bacteria growing in a culture, or a bank account growing at continuously compounded interest. In this section the exponential function will be defined and its principal properties established.

In elementary algebra one learns how to calculate with powers a^m and roots $a^{1/n}$ of a positive "base" number a. Such calculations are governed by the following *laws of exponents*:

$$a^{x+y} = a^x a^y,$$
$$(a^x)^y = a^{xy},$$
$$a^1 = a,$$
$$a^x b^x = (ab)^x.$$

These laws are proved for all rational numbers x and y by purely algebraic manipulations. And by using these laws we can handle numbers like $5^{1/3}$, $2^{8/7}$, $3^{-1/2}$, etc.

In calculus we need irrational exponents, too. For example, we need the function $f(x) = 2^x$ to be defined for all real numbers x, so that expressions like

$$f(\pi) = 2^\pi, \qquad f(\sqrt{3}) = 2^{\sqrt{3}},$$

etc., are meaningful. But how should we define 2^x when x is an irrational number?

In order to see how to do this, we first restrict our attention to the base e. Recall from Section 1 that e is the base of the natural logarithm function ($\ln e = 1$; $e \approx 2.718$), and that if x is a rational number, then

$$\ln e^x = x \ln e = x.$$

In the language of Section 2, this equation says that e^x is inverted by the natural logarithm function when x is rational. We thus have a new description of e^x, and it can be used to define e^x when x is irrational. Accordingly, we define the extended exponential function e^x to be the inverse $E(x)$ of the natural logarithm function. That is, for every number x,

> e^x is the number whose natural logarithm is x.

It follows that

> e^x has domain $(-\infty, \infty)$ and range $(0, \infty)$;
>
> $\ln(e^x) = x$ for all real numbers x;
>
> $e^{\ln y} = y$ for all $y > 0$.

It also follows that e^x is differentiable, by Theorem 4. To find what its derivative is, we only have to differentiate the identity

$$\ln(e^x) = x$$

by the chain rule. This gives

$$\frac{1}{e^x} \frac{d}{dx} e^x = 1,$$

and

$$\boxed{\frac{d}{dx} e^x = e^x.}$$

Thus the exponential function is its own derivative! So is ke^x, for any constant k, but we shall see shortly that no other function has this property.

With built-in chain rule the formula is

$$\frac{d}{dx} e^u = e^u \frac{du}{dx}$$

or

$$de^u = e^u \, du.$$

Example 1

$$\frac{d}{dx} e^{x^2} = e^{x^2} \cdot 2x = 2xe^{x^2},$$

$$\frac{d}{dx} \sin(e^x) = e^x \cos(e^x),$$

$$\frac{d}{dt} e^t \sin t = e^t \cos t + e^t \sin t$$

$$= e^t(\sin t + \cos t). \qquad \Box$$

Example 2 If $y = xe^x$, show that

$$\frac{d^2y}{dx^2} - 2\frac{dy}{dx} + y = 0.$$

Solution We have

$$\frac{dy}{dx} = xe^x + e^x,$$

$$\frac{d^2y}{dx^2} = xe^x + e^x + e^x$$

$$= xe^x + 2e^x,$$

$$\frac{d^2y}{dx^2} - 2\frac{dy}{dx} + y = (xe^x + 2e^x) - 2(xe^x + e^x) + xe^x = 0. \qquad \Box$$

Example 3 Show that a differentiable function $y = f(x)$ satisfies the equation

$$\frac{dy}{dx} = y \qquad (1)$$

over an interval I if and only if $y = ce^x$ on I, for a suitable constant c.

Solution That the function ce^x satisfies the equation (1) is exactly what the derivative rule $de^x/dx = e^x$ says. Now suppose that f is some other function satisfying (1) on an interval I. Then the quotient derivative

$$\frac{d}{dx}\frac{f(x)}{e^x}$$

has the numerator

$$e^x f'(x) - f(x)e^x = e^x f(x) - f(x)e^x = 0.$$

Thus

$$\frac{f(x)}{e^x} = \text{constant},$$

and $f(x) = ce^x$, as claimed.   □

We still have to prove that e^x satisfies the law of exponents $e^{x+y} = e^x e^y$ for all real numbers x and y. This can be done in three completely different ways: (1) as an exercise in algebraic symbol manipulation; (2) by a continuity argument; (3) by calculus. We naturally choose calculus.

THEOREM 6 *The exponential function e^x satisfies the law of exponents*

$$e^{x+y} = e^x \cdot e^y$$

for all real numbers x and y.

Proof For any constant a,

$$\frac{d}{dx}e^{x+a} = e^{x+a}\frac{d}{dx}(x + a) = e^{x+a}.$$

Therefore,

$$e^{x+a} = ce^x,$$

by Example 3. Setting $x = 0$ shows that $c = e^a$. Thus

$$e^{x+a} = e^a e^x$$

for all real numbers x and a.   ■

The graph of the exponential function is shown in Fig. 1. It can be obtained by reflecting the graph of its inverse function $y = \ln x$ in

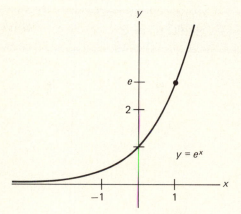

Figure 1

the line $y = x$. (See p. 309.) It can also be sketched quickly and quite accurately by using the approximation

$$e^{0.7n} \approx 2^n.$$

We conclude with the following remarkable identity for the exponential function: for every number x,

$$e^x = \lim_{n \to \infty} \left(1 + \frac{x}{n}\right)^n.$$

Proof   We know that

$$\frac{\ln(1 + h)}{h} \to 1$$

as $h \to 0$, because the left side is just the difference quotient for $\ln x$ at $x = 1$. If we multiply by a fixed number x and set $r = x/h$ (so that $h = x/r$), we see that

$$r \ln\left(1 + \frac{x}{r}\right) \to x$$

as $r \to \infty$. This holds in particular as $r \to \infty$ through integer values n, and since $n \ln u = \ln u^n$, we have

$$\ln\left(1 + \frac{x}{n}\right)^n \to x$$

as $n \to \infty$. Now just apply the exponential function to both sides, and use the identity $e^{\ln t} = t$ on the left.   ■

PROBLEMS FOR SECTION 3

Differentiate.

1. $xe^x - e^x$

2. e^x/x

3. $x^2 e^x - 2xe^x + 2e^x$

4. e^{x^2}

5. $e^{1/x}$

6. $e^{2x} - e^{-x}$

7. $\sqrt{e^x - 1}$

8. $(e^x - e^{-x})/(e^x + e^{-x})$

9. $e^{x/2}\sqrt{x - 1}$

10. $e^x + e^{2x} + e^{3x}$

11. $(e^x)^a$

12. $e^{\sqrt{x}}$

13. $e^{k \ln x}$

14. x/e^x

15. $(e^{3x} + x)^{1/3}$

16. $e^x/(1 - e^x)$

17. $(e^{2x} + 2 + e^{-2x})^{1/2}$

18. $\left(\dfrac{1 - e^x}{1 + e^x}\right)^{1/2}$

19. $\dfrac{xe^{-x}}{1 + x^2}$

20. $e^{ax} \cos bx$

21. $\dfrac{1}{2} e^x(\sin x + \cos x)$

22. $e^x \ln x$

23. $e^{\cos x}$

24. $\ln(e^x + 1)$

25. e^{e^x}

26. $\sin(e^{ax})$

27. $e^{\sqrt{1-x^2}}$

28. $\dfrac{e^{u^2}}{u^2}$

29. $e^{-x^2}(x + 1)$

30. e^{-1/x^2}

31. $\sin(\sqrt{1 - e^{-ax}})$

32. $xe^{\sqrt{x}}$

33. $e^{3x}\sqrt{1 + \cos^2 5x}$

34. $\ln \cos e^{\sqrt{x}}$

35. $y = \dfrac{e^x(1 + \cos^2 2x)}{\ln(1 + x^2)}$

36. $e^{-1/\ln x}$

In Problems 37 through 39 determine the value or values of k for which $y = e^{kx}$ satisfies the given equation.

37. $\dfrac{d^2 y}{dx^2} + \dfrac{dy}{dx} - 2y = 0$

38. $\dfrac{d^2 y}{dx^2} + 2\dfrac{dy}{dx} + y = 0$

39. $\dfrac{d^2 y}{dx^2} + 2\dfrac{dy}{dx} - 3y = 0$

40. Show that $y = e^{kx}$ satisfies the differential equation

$$y'' + ay' + by = 0$$

if and only if

$$k^2 + ak + b = 0.$$

(Here $y' = dy/dx$ and $y'' = d^2y/dx^2$.)

41. Show that $y = xe^{kx}$ satisfies the above equation if and only if $a^2 = 4b$ and $k = -a/2$.

42. Show that $y = e^{mx} \sin rx$ satisfies the second-order differential equation

$$\dfrac{d^2 y}{dx^2} - 2m\dfrac{dy}{dx} + (m^2 + r^2)y = 0.$$

43. If $y = xe^{-2x}$, show that

$$\dfrac{d^2 y}{dx^2} + 4\dfrac{dy}{dx} + 4y = 0.$$

44. Suppose that the tangent line to the graph of $y = e^x$ at the point x_0 intersects the x-axis at $x = x_1$. Show that $x_0 - x_1 = 1$.

46. Find the maximum value of $x^2 e^{-x}$ on $[0, 4]$.

48. Graph $y = x^k e^{-x}$ on $[0, \infty)$, where k is a positive rational number.

50. A particle moves along the x-axis from the origin toward $x = 1$, its position at time t being given by $x = 1 - e^{-t}$. Show that its acceleration is the negative of its distance from 1.

45. Show that $f(x) = x^k e^{-x}$ has a maximum value on $[0, \infty)$, for any positive exponent k. Show therefore that $f(x) \to 0$ as $x \to \infty$. (Use the maximum for a different value of k.)

47. Graph $y = xe^{1-x}$ on $[0, \infty)$.

49. Find the minimum value of xe^x on $[-2, 0]$.

51. A particle moves along the x-axis, its position at time t being given by $x = te^{-t}$.

  a) How far to the right does it get?

  b) Show that its position x, velocity v, and acceleration a are related by

$$a + 2v + x = 0.$$

4

THE GENERAL EXPONENTIAL FUNCTION

In Section 3 we briefly discussed the exponential function a^x with a general positive base a but then concentrated on the base e. We now return to the base a. In Section 1 we saw that

$$\ln a^x = x \ln a$$

for every rational number x. It follows that

$$\boxed{a^x = e^{x \ln a}}$$

for all rational x, by the cancellation identity $e^{\ln y} = y$. For irrational x, we make this equation the *definition* of a^x.

We therefore always use this identity to reduce problems involving other bases to the known rules for the base e.

Example 1 We start by proving the "mixed" law

$$\ln a^x = x \ln a.$$

Proof $\ln(a^x) = \ln(e^{x \ln a}) = x \ln a$, by the cancellation law $\ln e^y = y$. ■

Example 2

$$\frac{d}{dx}(2^x) = \frac{d}{dx} e^{(x \ln 2)}$$

$$= e^{x \ln 2} \cdot \frac{d}{dx}(x \ln 2)$$

$$= 2^x \cdot \ln 2.$$ □

Example 3 At last we can prove the power rule for an *arbitrary* exponent a. We have

$$\frac{d}{dx}x^a = \frac{d}{dx}e^{a\ln x}$$

$$= e^{a\ln x}\frac{d}{dx}(a\ln x)$$

$$= x^a \cdot \frac{a}{x}$$

$$= ax^{a-1}. \qquad \square$$

Example 4
$$\frac{d}{dx}x^{\sqrt{5}} = \sqrt{5}\,x^{\sqrt{5}-1}. \qquad \square$$

Example 5
$$\frac{d}{dx}x^x = \frac{d}{dx}e^{x\ln x}$$

$$= e^{x\ln x}\frac{d}{dx}(x\ln x)$$

$$= x^x[1 + \ln x]. \qquad \square$$

The function

$$[f(x)]^{g(x)} = e^{g(x)\ln f(x)}$$

is defined only when $f(x)$ is positive; it is then differentiable whenever f and g are differentiable. One way of computing its derivative is illustrated in the examples above. There is another procedure, called *logarithmic differentiation,* where we compute the derivative of the logarithm of the function in question.

Example 6 Find the derivative of $y = x^x$.

Solution Since

$$\ln y = \ln x^x = x\ln x,$$

differentiating with respect to x gives

$$\frac{1}{y}\frac{dy}{dx} = x \cdot \frac{1}{x} + \ln x \cdot 1 = 1 + \ln x,$$

so

$$\frac{dy}{dx} = y[1 + \ln x] = x^x[1 + \ln x],$$

as before. □

Example 7 Products and quotients can be attacked by logarithmic differentiation. We shall illustrate this by finding the rule for the differential of a product with three factors. If

$$y = uvw,$$

then

$$\ln |y| = \ln |u| + \ln |v| + \ln |w|.$$

Applying the differential rules and Problem 5 in Section 1, we see that

$$\frac{dy}{y} = \frac{du}{u} + \frac{dv}{v} + \frac{dw}{w},$$

or

$$dy = uvw\left(\frac{du}{u} + \frac{dv}{v} + \frac{dw}{w}\right).$$

This scheme is not new, however. It was a version of the product rule in Chapter 4. □

PROBLEMS FOR SECTION 4

Differentiate the following.

1. $x^{\sqrt{x}}$

2. 3^{2x}

3. 9^x

4. $x^{\ln x}$

5. $(x^a)^x$

6. $x^{(a^x)}$

7. $(\sin x)^x$

8. $(\ln x)^{\ln x}$

9. $10^{(ax^2+bx+c)}$

10. $(e^e)^x$

11. $x^{\tan x}$

12. $(x^2)^{x^2}$

13. $x^{1/x}$

14. $(1 + a^2)^{x^2}$

15. $(\ln x)^x$

16. $(1 + \cos^2 x)^{\ln x}$

17. $x^{(e^x)}$

18. $x^{(e^{-x^2})}$

19. We define $\log_a x$ as the function inverse to a^x. Show that

$$\log_a x = \frac{\ln x}{\ln a}.$$

20. Show that

$$a^{x+y} = a^x a^y, \qquad (a^x)^y = a^{xy}, \qquad \text{and} \qquad a^1 = a.$$

In the following problems, compute dy/dx by logarithmic differentiation

21. $y = \sqrt{1 + x^2}$

22. $y^3 = x(x - 1)$

23. $y^n = x^m$

24. $y = x^{(e^x)}$

25. $(xy)^x = 4$

26. $y = (\sin x)^x$

27. $y = x^{\tan x}$

28. $y^3 + y = x^3$

29. $y = \sqrt{\left(\dfrac{1 - x}{1 + x}\right)}$

30. $y = \dfrac{(1 + x)^{1/3}(1 - x)^{2/3}}{(1 + x^2)^{1/6}}$

31. $y = [(1 - x)(1 - 2x)(1 - 3x)(1 - 4x)]^{1/4}$

32. $y^{3/5} = \dfrac{\sqrt{(2x - 1)^3(x + 2)^5}}{(3x + 2)^{2/3}(2x - 3)^{1/7}}$

**5
GROWTH AND
DECAY**

We saw in Section 3 that "up to a multiplicative constant," $y = e^x$ is the unique function satisfying the differential equation

$$\frac{dy}{dx} = y.$$

That is, if f is a function such that

$$f' = f,$$

then $f(x) = ce^x$ for some constant c.

   If we carry through the same argument, using e^{kx} instead of e^x, we find that

THEOREM 7   *A function* $y = f(x)$ *satisfies the equation*

$$\boxed{\frac{dy}{dx} = ky}$$

if and only if $y = ce^{kx}$ *for some constant c.*

   The equation

$$\frac{dy}{dt} = ky$$

is the basic law of population growth. Neglecting special inhibiting or stimulating factors, a population normally reproduces itself at a rate proportional to its size. This is exactly what the above equation says. Bacteria colonies grow this way as long as they have normal environment, and so does a bank balance with a fixed rate of continuously compounded interest. Theorem 7 shows that a population normally grows exponentially, and if you will look back at the shape of an exponential graph (Fig. 1 in Section 3) you will see what "population explosion" is all about.

The constant k is sometimes called the *rate of exponential growth*. This is not the rate of change of the population size, which is

$$\frac{dy}{dt} = ky,$$

but the constant that y must be multiplied by to get its rate of change. It is thus a different use of the word *rate*. It is like the *interest rate* paid by a bank. If the interest rate is 0.07, we do not mean that your bank balance y is growing at the rate of 0.07 dollars per year but at the rate of $0.07y$ dollar per year. We therefore express the rate as 7% per year, rather than 0.07 dollars per year. We could say that the rate is 0.07 dollars *per dollar* per year. (Where the interest is compounded continuously, the interest rate is a true exponential growth rate.)

When a quantity Q is growing exponentially,

$$Q = ce^{kt},$$

the time T it takes to double depends only on the exponential growth rate k. Conversely, the growth rate k can be found from the doubling time T.

Example 1   Suppose that a culture of bacteria doubles its size from 0.15 grams to 0.3 grams during the seven-hour period from $t = 15$ hrs to $t = 22$ hrs. What is its exponential growth rate k?

Solution   Assuming normal exponential growth

$$y = ce^{kt},$$

our data are

$$0.15 = ce^{k \cdot 15},$$
$$0.30 = ce^{k \cdot 22}.$$

Dividing the second equation by the first gives us

$$2 = e^{k(22-15)} = e^{k \cdot 7}.$$

Thus,

$$k \cdot 7 = \ln 2,$$

$$k = \frac{\ln 2}{7}.$$

A calculator, or Table 1 in Appendix 6, gives $\log 2 = 0.6931$ to four decimal places. So the growth rate k is approximately 0.1. □

In general the exponential growth rate k and the doubling time T are related by

$$kT = \ln 2 \approx 0.6931,$$

so that either determines the other. The proof is like the example above.

If the exponential growth rate is *negative*, say $-k$ where k is positive, then the equation $dy/dt = -ky$ shows y to be *decreasing* as a function of time, and the solution $y = ce^{-kt}$ shows it to be decreasing exponentially. This is called exponential *decay*. A radioactive element behaves this way; it disintegrates at a rate proportional to the amount present. In this context, we have the *half-life* of the element instead of the doubling time. This is the time it takes for $y = ce^{-kt}$ to decrease to half of an original value. That is, if

$$y_0 = ce^{-kt_0},$$

$$\frac{y_0}{2} = ce^{-kt_1},$$

then $T = t_1 - t_0$ is the half-life. Again we find that

$$kT = \ln 2,$$

so that the half-life T depends only on the decay rate k. In particular, it is independent of the initial population size y_0 and the initial time t_0.

If we know the exponential growth (or decay) rate k, then the remaining constant c is determined by a single initial condition (t_0, y_0), as before.

Carbon 14 is a radioactive isotope of carbon 12 (the normal form) with a half-life of roughly 5700 years. It is constantly being produced from other elements by the action of cosmic rays, and as a result it seems to be in equilibrium in the environment, holding at a constant percentage of the total amount of carbon. In particular, a living organism contains the standard percentage of carbon 14 during its lifetime. Once it dies, however, its carbon 14 content decreases through radioactive decay. This means that the age of an organic remnant can be estimated by measuring its present content of carbon 14.

Example 2   Charcoal from an ancient tree that was burned during a volcanic eruption is found to have only 45% of the standard amount of carbon 14. When did the volcano erupt?

Solution   The half-life $T = 5700$ determines the exponential decay rate k according to

$$kT = k5700 = \ln 2 = 0.6931.$$

Thus,

$$k = \frac{0.6931}{5700}.$$

If the volcano erupted t years ago, then our information is that

$$0.45 = e^{-kt}.$$

Taking natural logarithms, we have

$$kt = -\ln 0.45 \approx 0.799$$

(by pocket calculator, or Table 1 in Appendix 6). Therefore,

$$t = \frac{0.799}{k} = \frac{0.799}{0.693} 5700 \approx 6570 \text{ years ago.} \qquad \square$$

PROBLEMS FOR SECTION 5

1. A colony of bacteria has an exponential growth rate of 1% per hour. What is its doubling time?

2. A culture of bacteria is found to increase by 41% in 12 hours. What is its exponential growth rate per hour? (Note that $1.41 \simeq \sqrt{2}$.)

▦ **3.** A growth rate of 100% per day corresponds to what exponential growth rate per hour?

▦ **4.** A bank advertises that it compounds interest continuously and that it will double your money in 10 years. What is its annual interest rate?

▦ **5.** A certain radioactive material has a half-life of 1 year. When will it be 99% gone? ($\ln 10 \approx 2.3$)

▦ **6.** A quantity Q_1 grows exponentially with a doubling time of one week. Q_2 grows exponentially with a doubling time of three weeks. If the initial amounts of Q_1 and Q_2 are the same, when will Q_1 be twice the size of Q_2?

▦ **7.** There is nothing sacred about doubling time. Any two measurements of an exponentially growing population $y = ce^{kt}$ will determine both parameters k and c Show that if y has the values y_0 and y_1 at times t_0 and t_1, then

$$k = \frac{\ln(y_1/y_0)}{t_1 - t_0}.$$

▦ **8.** In a chemical reaction a substance A decomposes at a rate proportional to the amount of A present. It is found that 8 pounds of A will reduce to 4 pounds in 3 hours. At what time will there be only 1 pound left?

▦ **9.** In a chemical reaction a substance B decomposes at a rate proportional to the amount of B present. It is found that 8 pounds of B will reduce to 7 pounds in one hour. Make a rough calculation to show that more than four hours are needed for 8 pounds of B to reduce to 4 pounds.

▦ **10.** The temperature T of a cooling body drops at a rate that is proportional to the difference $T - C$, where C is the constant temperature of the surrounding medium. Use Theorem 7 to show that

$$T = ae^{-kt} + C.$$

▦ **11.** The temperature of a cup of freshly poured coffee is 200° and the room temperature is 70°. The coffee cools 10° in 5 minutes. How much longer will it take to cool 20° more? (Use the above formula.)

▦ **12.** If a cold object is placed in a warm medium whose temperature is held constant at C°, the object warms up according to the same general principle. State a general law of temperature change that includes both the warming and cooling situations.

▦ **13.** Grain seeds from a burial mound are found to contain one fourth of the standard amount of carbon 14. About how old are they?

▦ **14.** A bone fragment found in Africa contains 10% of the standard amount of carbon 14. How old is it?

6
THE ARCSINE FUNCTION

The next new function, the *arcsine* function, is the inverse of sin x on $[-\pi/2, \pi/2]$. Its value at x is designated arcsin x, which can be thought of as an abbreviation for "the angle whose sine is x." Thus, for each number x between -1 and 1,

arcsin x is the number between $-\pi/2$ and $\pi/2$ whose sine is x.

The restriction of $\sin x$ to $[-\pi/2, \pi/2]$, or to another similar interval, is necessary in order to get an inverse. According to Section 2, a continuous function has an inverse (on an interval) if and only if it is monotone. So $\sin x$ has to be restricted to an interval on which it is monotone—it is increasing on $[-\pi/2, \pi/2]$—and then Theorem 3 guarantees an inverse function.

The domain of $\arcsin x$ is $[-1, 1]$ and its range is $[-\pi/2, \pi/2]$. The inverting (cancellation) identities are

$$\sin(\arcsin x) = x \qquad \text{(for } x \text{ in } [-1, 1]),$$
$$\arcsin(\sin x) = x \qquad \text{(for } x \text{ in } [-\pi/2, \pi/2]).$$

The graph of $\arcsin x$ is the reflection in the line $y = x$ of the graph of $\sin x$ (restricted to $[-\pi/2, \pi/2]$), as shown in Fig. 1.

Figure 1

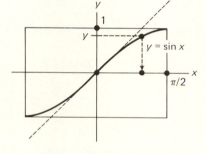
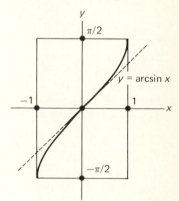

According to Theorem 4, $y = \arcsin x$ is everywhere differentiable in $[-1, 1]$ except at the two endpoints. To compute its derivative, we can differentiate the first cancellation identity. The algebra is a little simpler if instead we differentiate the equation

$$x = \sin y$$

implicitly, knowing that it determines y as a differentiable function of x. This gives

$$1 = \frac{d}{dx}\sin y = \cos y \cdot \frac{dy}{dx}.$$

Therefore

$$\frac{dy}{dx} = \frac{1}{\cos y} = \frac{1}{\pm\sqrt{1 - \sin^2 y}} = \frac{1}{\pm\sqrt{1 - x^2}}.$$

Moreover, $\cos y > 0$ when $-\pi/2 < y < \pi/2$ so the plus sign before the radical is correct. Thus

$$
\frac{d}{dx}\arcsin x = \frac{1}{\sqrt{1-x^2}},
$$

$$
\frac{d}{dx}\arcsin u = \frac{1}{\sqrt{1-u^2}}\cdot\frac{du}{dx},
$$

$$
d(\arcsin u) = \frac{du}{\sqrt{1-u^2}}.
$$

Examples 1   If a is positive, then

$$
\frac{d}{dx}\arcsin\frac{x}{a} = \frac{1}{\sqrt{1-\left(\dfrac{x}{a}\right)^2}}\cdot\frac{1}{a} = \frac{1}{\sqrt{a^2-x^2}}.
$$

2 $\dfrac{d}{dx}(\sqrt{1-x^2}\arcsin x - x) = \dfrac{\sqrt{1-x^2}}{\sqrt{1-x^2}} + \arcsin x\cdot\dfrac{-x}{\sqrt{1-x^2}} - 1$

$$
= -\frac{x\arcsin x}{\sqrt{1-x^2}}.
$$

3 $\dfrac{d}{dx}\arcsin\dfrac{1}{x} = \dfrac{1}{\sqrt{1-(1/x)^2}}\cdot\dfrac{-1}{x^2} = \dfrac{-\sqrt{x^2}}{x^2\sqrt{x^2-1}} = \dfrac{-1}{|x|\sqrt{x^2-1}}.$

The domain of this function consists of two separate intervals: $(-\infty, -1]$ and $[1, \infty)$. $\square$

The inverse of $\cos x$, called $\arccos x$, is similar to $\arcsin x$. We have to choose an interval on which $\cos x$ is monotone, and the usual choice is $[0, \pi]$, so that this time we are dealing with *decreasing* functions (Fig. 2).

Proceeding as before, we find that

$$
\frac{d}{dx}\arccos x = -\frac{1}{\sqrt{1-x^2}}.
$$

With built in chain rule, the formula is

$$
\frac{d}{dx}\arccos u = -\frac{1}{\sqrt{1-u^2}}\frac{du}{dx}.
$$

Figure 2

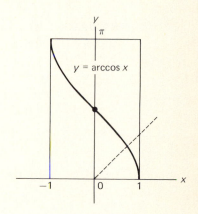

Example 4   Find y' if $y = \arccos x^2$.

Solution

$$y' = -\frac{1}{\sqrt{1-x^4}} \cdot 2x.$$ □

REMARK   The derivative of $\arcsin x + \arccos x$ is zero, and the sum is therefore a constant! Once our attention has been directed to this fact, we find that we can establish it directly from the definition of the two functions. It depends on the identity

$$\cos(\pi/2 - y) = \sin y$$

for all y, and the fact that $\pi/2 - y$ runs from π to 0 as y runs from $-\pi/2$ to $\pi/2$. Thus, if x is the common value in the above identity, then

$$\arcsin x = y,$$

$$\arccos x = \frac{\pi}{2} - y,$$

and

$$\arcsin x + \arccos x = \frac{\pi}{2}.$$

The derivative formula for $\arccos x$ can now be obtained by differentiating $\pi/2 - \arcsin x$.

REMARK The derivative formula suggests another approach to arcsin x. We define a new function $As(x)$ with domain $(-1, 1)$ by the integral

$$As(x) = \int_0^x \frac{dt}{\sqrt{1-t^2}}.$$

Then $As(x)$ is differentiable and

$$As'(x) = \frac{1}{\sqrt{1-x^2}},$$

by the Second Fundamental Theorem. But then

$$\frac{d}{dx} As(\sin x) = \frac{1}{\sqrt{1-\sin^2 x}} \cdot \cos x = 1$$

for x in $(-\pi/2, \pi/2)$, by the chain rule. Therefore,

$$As(\sin x) = x + C.$$

Putting $x = 0$ shows that $C = 0$, at which point we know that $As(x)$ is the inverse of $\sin x$.

PROBLEMS FOR SECTION 6

Find the derivatives of the following functions.

1. $\arcsin \sqrt{x}$
2. $\arcsin kx$
3. $\arcsin x^2$
4. $\arcsin(\sin^2 x)$
5. $\arcsin(\cos x)$
6. $\arcsin(1 - 2x)$
7. $x \arcsin x + \sqrt{1-x^2}$
8. $\arcsin x + \sqrt{1-x^2}$
9. $\ln\left(\arcsin \frac{1}{x}\right)$
10. $\arcsin(e^{-x^2})$
11. $\arcsin\left(\frac{x+a}{1+ax}\right)$
12. $\sqrt{\arcsin(2\sin x)}$
13. $\arcsin x + x\sqrt{1-x^2}$
14. $\arcsin\left(\frac{1-x}{1+x}\right)$
15. $\arcsin\left(\frac{\cos x}{1+\sin x}\right)$
16. $\cos^2(\arcsin x)$
17. $\cos(\arcsin x) - \sqrt{1-x^2}$
18. $\arcsin \frac{x}{\sqrt{1+x^2}}$

19. If we don't restrict the domain of $g(y) = y^2$, then g fails to be the inverse of $f(x) = \sqrt{x}$. But g is a *left inverse* of f, in the sense that $g(f(x)) = x$ for all x in the domain of f. What breaks down is that g fails to be a right inverse of f. Show, similarly, that $\sin y$ is a left inverse of $\arcsin x$ but not a right inverse.

20. Show that $\pi - \arcsin x$ is another right inverse of $\sin x$.

21. Describe the most general differentiable right inverse of $\sin x$ with domain $[-1, 1]$.

23. Do Problems 1 and 6, and compare their answers. Try to figure out what the explanation is.

25. Derive the formula for the derivative of $\arccos x$ by differentiating the right side of the above identity.

22. Draw the graph of $\arcsin(\sin x)$ for $-2\pi \le x \le 2\pi$.

24. Show that

$$\arccos x = \arcsin \sqrt{1 - x^2}$$

for all x in the interval $[0, 1]$.

7
THE
ARCTANGENT
FUNCTION

To invert the tangent function we use the same restricted domain as for the sine function, except that now the endpoints are missing. Thus, if x is restricted to the open interval $(-\pi/2, \pi/2)$, then $f(x) = \tan x$ is an increasing function with an everywhere positive derivative, and so has a differentiable inverse, by Theorem 4. This new inverse function is called the *arctangent* function, and its value at x is designated $\arctan x$, suggesting "the angle whose tangent is x." Thus, for each number x,

$$\arctan x = \text{the number in } (-\pi/2, \pi/2) \text{ whose tangent is } x.$$

Figure 1 shows its graph.

Figure 1

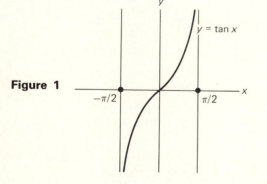
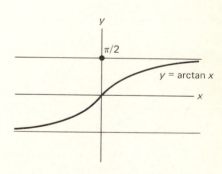

The domain of $\arctan x$ is the whole real-number system $(-\infty, \infty)$. This fact makes $\arctan x$ a more generally useful function than $\arcsin x$. The *range* of $\arctan x$ is $(-\pi/2, \pi/2)$.

The cancellation identities are

$$\boxed{\begin{array}{ll} \tan(\arctan x) = x & \text{for every } x, \\ \arctan(\tan x) = x & \text{for } x \text{ in } (-\pi/2, \pi/2). \end{array}}$$

We can find the derivative of arctan x by differentiating the equation

$$x = \tan y$$

implicitly, knowing that it determines y as a differentiable function of x. We get

$$1 = \frac{d}{dx}(\tan y) = \sec^2 y \frac{dy}{dx},$$

so

$$\frac{dy}{dx} = \frac{1}{\sec^2 y} = \frac{1}{1 + \tan^2 y} = \frac{1}{1 + x^2}.$$

Thus,

$$\boxed{\begin{array}{l} \dfrac{d}{dx} \arctan x = \dfrac{1}{1 + x^2}, \\[2mm] \dfrac{d}{dx} \arctan u = \dfrac{1}{1 + u^2} \cdot \dfrac{du}{dx}, \\[2mm] d \arctan u = \dfrac{du}{1 + u^2}. \end{array}}$$

Examples 1 $\dfrac{d}{dx} \arctan \dfrac{x}{a} = \dfrac{1}{1 + \left(\dfrac{x}{a}\right)^2} \cdot \dfrac{1}{a} = \dfrac{a}{a^2 + x^2}$

2 $\dfrac{d}{dx}(1 + x^2)\arctan x = \dfrac{1 + x^2}{1 + x^2} + (\arctan x)2x$

$$= 2x \arctan x + 1$$

3 If $x \neq -1$, then

$$\frac{d}{dx} \arctan \frac{x-1}{x+1} = \frac{1}{1 + \frac{(x-1)^2}{(x+1)^2}} \cdot \frac{(x+1) - (x-1)}{(x+1)^2}$$

$$= \frac{2}{(x+1)^2 + (x-1)^2} = \frac{1}{x^2+1}.$$

See Problem 26 for what is behind this interesting result.   □

PROBLEMS FOR SECTION 7

Differentiate the following functions.

1. $\arctan x^2$

2. $\arctan 3x$

3. $\arctan ax$

4. $\arctan \sqrt{x}$

5. $\arctan \sqrt{x^2 - 1}$

6. $\arctan\left(\dfrac{\sin x}{1 + \cos x}\right)$

7. $\arctan x + \ln \sqrt{1 + x^2}$

Prove the following identities.

8. $\dfrac{d}{dx} [x \arctan x - \ln \sqrt{1 + x^2}] = \arctan x$

9. $\dfrac{d}{dx} [\sqrt{x} - \arctan \sqrt{x}] = \dfrac{\sqrt{x}}{2(1 + x)}$

10. $\dfrac{d}{dx} \arctan\left(\dfrac{e^x - e^{-x}}{2}\right) = \dfrac{2}{e^x + e^{-x}}$

11. If we invert $\cot x$ on the interval $(0, \pi)$, show that

$$\text{arccot } x = \frac{\pi}{2} - \arctan x,$$

and hence that

$$\frac{d}{dx} \text{arccot } x = -\frac{1}{x^2 + 1}.$$

12. Show that for positive values of x,

$$\text{arccot } x = \arctan \frac{1}{x}.$$

Then compute the derivative of arccot x from this identity, and compare with the above problem.

13. If we invert $x = \sec y$ on the interval $[0, \pi]$, excluding $\pi/2$, show that

$$\frac{d}{dx} \text{arcsec } x = \frac{1}{|x| \sqrt{x^2 - 1}}.$$

14. Prove similarly that

$$\frac{d}{dx} \operatorname{arccsc} x = -\frac{1}{|x|\sqrt{x^2 - 1}}.$$

Differentiate (using the formulas in Problems 11 through 14 when necessary).

15. $\operatorname{arccot}(\tan x)$

16. $\operatorname{arcsec} \sqrt{x^2 + 1}$

17. $\arcsin \dfrac{x}{\sqrt{1 + x^2}}$

18. $\operatorname{arsec} \dfrac{1}{x}$

19. $\operatorname{arccot}(\ln x)$

20. $\tan(2 \arctan x)$

21. $(1 + x^2)\arctan x$

22. $\ln[\arctan(e^x)]$

23. $\arctan \sqrt{x^2/(1 - x^2)}$

24. $\operatorname{arcsec}\left[\dfrac{e^x + e^{-x}}{2}\right]$

25. $\ln \cos \arctan x$

26. It appeared from Example 3 in the text that

$$\arctan \frac{x - 1}{x + 1} = \arctan x + C.$$

Show that this is so and that $C = -\pi/4$.

28. Show that

$$\arctan x = \arcsin \frac{x}{\sqrt{1 + x^2}}.$$

30. Find the maximum value of

$$f(x) = \arctan(x + 1) - \arctan(x - 1).$$

32. Sketch the graph of

$$y = 2 \arctan \frac{x}{2} - \arctan x$$

(critical points, concavity, behavior at infinity). You will have to guess at a couple of y values.

34. A statue a feet tall stands on a pedestal whose top is b feet above an observer's eye level. If the observer stands x feet away from the pedestal, show that the angle θ subtended by the statue at the observer's eyes is given by

$$\theta = \arctan \frac{a + b}{x} - \arctan \frac{b}{x}.$$

27. Draw the graph of $\arctan(\tan x)$ over $[-2\pi, 2\pi]$.

29. Show that

$$\arcsin x = \arctan \frac{x}{\sqrt{1 - x^2}}.$$

31. Find the value of x where

$$f(x) = \arctan 2x - \arctan x$$

has its maximum value on the interval $[0, \infty)$.

33. Sketch the graph of

$$y = \arctan 2x - \arctan x.$$

You will have to guess at a couple of y values.

35. Find where the observer must stand in order to maximize the angle of vision θ given by the above formula.

CHAPTER 8
THE TECHNIQUE OF INTEGRATION

If we start with the five functions x, e^x, $\ln x$, $\sin x$, and $\arcsin x$, and build new functions by using the algebraic operations and composition, possibly repeatedly and in combination with each other, we generate the class of *elementary functions*. Such functions are said to have *closed form* because they can be explicitly written down.

The functions $\cos x$ and $\arctan x$ were omitted from the basic list because they arise as simple compositions:

$$\cos x = \sin(x + \pi/2),$$

$$\arctan x = \arcsin \frac{x}{\sqrt{1 + x^2}}.$$

The derivative of an elementary function is always an elementary function, and it can be found in a systematic way by applying the differentiation rules. When we first turn to integration, we probably expect the same pattern; that is, we expect every elementary function to have an elementary antiderivative that we can find in an explicit and systematic manner. We saw in Chapter 6 that polynomials behave this way; in fact, polynomials can be integrated as easily as they can be differentiated.

This is not so, however, in general. There is, in general, no systematic integration procedure that can be followed step by step to a guaranteed answer. There may not even *be* an answer, at least until we invent a new function for the purpose. For example, such a simple-looking function as $f(x) = e^{-x^2}$ has no antiderivative at all within the class of elementary functions. It does have an anti-derivative F, by the Second Fundamental Theorem, and we shall learn how to calculate $F(x)$ in Chapter 12. But it can be proved that there is no way of representing F in terms of functions we already know about. That is, $F(x)$ is not an elementary function.

Despite such uncertainties, it is very useful to find anti-derivatives explicitly in terms of known functions when it is possible and feasible. That this process must be viewed as an art rather than a systematic routine should not be discouraging. Many students find, after they get the hang of the three or four basic tricks of the trade, that integrating is more interesting than differentiating because it is more like solving puzzles.

We recall that systematic differentiation involves

a) The derivative formulas for the functions x^a, e^x, $\sin x$, $\cos x$, $\ln x$, arcsin x, and arctan x;

b) The basic differentiation rules:
    1. linearity, $(af + bg)' = af' + bg'$;
    2. the product rule, $(fg)' = fg' + gf'$;
    3. the chain rule, $(f \circ g)' = (f' \circ g) \cdot g'$.

The quotient rule is omitted because it really is a secondary rule derived from the above: f/g is the product $f(1/g)$, and $1/g$ is the composition g^{-1}, a special case of the general power function g^a.

Integration involves these same formulas and rules, but applied in the reverse direction. We start with the following formulas:

$$\int x^a \, dx = \frac{x^{a+1}}{a + 1} + C, \qquad \text{provided } a \neq -1;$$

$$\int \frac{dx}{x} = \ln x + C \qquad \text{(the case } a = -1\text{)};$$

$$\int e^x \, dx = e^x + C;$$

$$\int \sin x \, dx = -\cos x + C;$$

$$\int \cos x \, dx = \sin x + C;$$

$$\int \frac{dx}{1 + x^2} = \arctan x + C;$$

$$\int \frac{dx}{\sqrt{1 - x^2}} = \arcsin x + C.$$

Then more complicated functions are attacked by the reversed rules. We have already considered linearity in Chapter 6, and this leaves only the chain rule and the product rule. The reversed chain rule is called *integration by substitution* or *change of variable*. The reversed product rule is called *integration by parts*.

In addition, there is one integration procedure that does not correspond to a differentiation rule: Occasionally we can subdue a balky integrand by algebraically manipulating it into a more docile form. The systematic integration of rational functions is accomplished this way (in conjunction with the rules), and there are a couple of other situations in which we can benefit by such tactics.

However, no matter how clever we are, there will always be more functions that we cannot integrate than ones we can, and some technique of numerical integration is necessary. This *estimation* approach to integration was started in Chapter 6 and will be continued in Chapter 9.

The logarithm integral has a complication that we must discuss. Although the integrand $1/x$ is defined for all x except $x = 0$, the formula $\int dx/x = \ln x + C$ is valid only for $x > 0$, since negative numbers don't have logarithms. However, if $x < 0$, then $\ln(-x)$ exists, and

$$\frac{d}{dx} \ln(-x) = \frac{1}{-x} \cdot (-1) = \frac{1}{x}.$$

Altogether, we have

$$\frac{d}{dx} \ln |x| = \frac{1}{x}, \qquad x \neq 0.$$

It is customary, therefore, to write

$$\int \frac{dx}{x} = \ln |x| + C, \qquad x \neq 0.$$

This is true over any *interval* in the domain of $\ln |x|$, but it is imprecise as stated since the complete domain is not an interval. For example, the function

$$f(x) = \begin{cases} \ln |x|, & \text{when } x < 0, \\ \ln |x| + 2, & \text{when } x > 0, \end{cases}$$

is an antiderivative that is not in the above form. The theorem guaranteeing that $F(x) + C$ is the most general antiderivative was only established over an *interval* lying in the domains of both F and $f = F'$.

The same considerations apply to

$$\int dx/x^2 = -1/x + C,$$

and to other negative powers of x.

1 CHANGE OF VARIABLE: DIRECT SUBSTITUTION

We consider first the reversed chain rule.

So far the dx in Leibniz's notation $\int f(x) \, dx$ has played no role, but now we shall find it useful to take dx at its face value as the differential of x.

Consider the integral

$$\int (\sin x)^2 \cos x \, dx.$$

If we set $u = \sin x$, then $du = \cos x \, dx$, and the integral takes on the new form

$$\int u^2 \, du.$$

Since

$$\int u^2 \, du = \frac{u^3}{3} + C,$$

and since $u = \sin x$, the answer we are looking for ought to be

$$\int (\sin x)^2 \cos x \, dx = \frac{(\sin x)^3}{3} + C.$$

We see that it is, by differentiating,

$$\frac{d}{dx}\left(\frac{1}{3}\sin^3 x + C\right) = \frac{1}{3} \cdot 3 \sin^2 x \cos x$$

$$= \sin^2 x \cos x.$$

This is a chain-rule differentiation, and our change of variables from x to u can thus be considered as a backward application of the chain rule.

The general pattern of the above calculation is this. We have a complicated function $f(x)$ that can be seen to be in the form $f(x) = g(h(x))h'(x)$, and we want to compute the integral

$$\int f(x)\, dx = \int g(h(x))h'(x)\, dx.$$

If we set $u = h(x)$, then $du = h'(x)\, dx$, and the integral takes on the new form

$$\int g(u)\, du.$$

We suppose that we can find an antiderivative G of g, so

$$\int g(u)\, du = G(u) + C.$$

Then, since $u = h(x)$, the answer we are looking for ought to be

$$\boxed{\int g(h(x))h'(x)\, dx = G(h(x)) + C.}$$

We see that it is, by the chain rule:

$$\frac{d}{dx} G(h(x)) = G'(h(x))h'(x) = g(h(x))h'(x) = f(x).$$

Thus, if we take Leibniz's notation at its face value, as involving a *differential*, then it automatically gives us the correct integration procedure that corresponds to the chain rule.

One might think that this device requires an integrand of such special form that it would hardly ever be applicable, but it is surprising how often it can be used. Here are some further examples:

Example 1
$$\int \frac{2x\, dx}{1 + x^2} = \int \frac{du}{u} \qquad \left(\begin{array}{l} \text{set } u = 1 + x^2; \\ \text{then } du = 2x\, dx \end{array} \right)$$
$$= \ln u + C = \ln(1 + x^2) + C. \qquad \square$$

Example 2
$$\int \frac{2x\, dx}{(1 + x^2)^2} = \int \frac{du}{u^2} = -\frac{1}{u} + C = -\frac{1}{1 + x^2} + C. \qquad \square$$

Example 3

$$\int \frac{\ln x}{x}\, dx = \int u\, du \qquad \left(\begin{aligned} u &= \ln x, \\ du &= \frac{dx}{x} \end{aligned} \right)$$

$$= \frac{u^2}{2} + C = \frac{(\ln x)^2}{2} + C. \qquad \square$$

Example 4   Frequently we will have to multiply by a constant to get $h'(x)\, dx$. Thus, in

$$\int x e^{x^2}\, dx,$$

we try setting $u = x^2$. Then $du = 2x\, dx$, which we don't quite have. However, we do have

$$x\, dx = \frac{du}{2},$$

and this is good enough because a constant can be factored out:

$$\int x e^{x^2}\, dx = \int e^{x^2} (x\, dx) = \int \frac{e^u\, du}{2} = \frac{1}{2} \int e^u\, du$$

$$= \frac{1}{2} e^u + C = \frac{1}{2} e^{x^2} + C. \qquad \square$$

The simplest substitution of all occurs when we have the derivative of a known function except for an added constant inside.

Example 5

$$\int \cos(x + 3)\, dx = \int \cos u\, du \qquad \left(\begin{aligned} u &= x + 3, \\ du &= dx \end{aligned} \right)$$

$$= \sin u + C = \sin(x + 3) + C. \qquad \square$$

Example 6

$$\int \frac{dx}{x + a} = \int \frac{du}{u} \qquad \left(\begin{aligned} u &= x + a, \\ du &= dx \end{aligned} \right)$$

$$= \ln |u| + C = \ln |x + a| + C. \qquad \square$$

Example 7   A multiplicative constant inside is about as simple,

$$\int \cos 2x\, dx = \frac{1}{2} \int \cos u\, du \qquad \left(\begin{aligned} u &= 2x, \\ du &= 2\, dx, \\ dx &= \frac{du}{2} \end{aligned} \right)$$

$$= \frac{1}{2} \sin 2x + C. \qquad \square$$

Note that we could have viewed this substitution as setting $x = u/2$, $dx = du/2$. In the next example this is the *natural* procedure. (See Section 6.)

Example 8
$$\int \frac{dx}{\sqrt{a^2 - x^2}} = \int \frac{a\,du}{\sqrt{a^2 - a^2u^2}} = \int \frac{du}{\sqrt{1 - u^2}} \qquad \begin{pmatrix} x = au, \\ dx = a\,du \\ a > 0 \end{pmatrix}$$
$$= \arcsin u + C = \arcsin \frac{x}{a} + C. \qquad \square$$

Example 9
$$\int \frac{dx}{x^2 + a^2} = \int \frac{a\,du}{a^2u^2 + a^2} = \frac{1}{a} \int \frac{du}{u^2 + 1} \qquad \begin{pmatrix} x = au, \\ dx = a\,du \end{pmatrix}$$
$$= \frac{1}{a} \arctan u + C = \frac{1}{a} \arctan \frac{x}{a} + C. \qquad \square$$

The last two examples come up so frequently that it may be worth remembering these slightly more general forms of the basic integral formulas. Here they are without the details:

$$\boxed{\begin{aligned} \int \frac{dx}{\sqrt{a^2 - x^2}} &= \arcsin \frac{x}{a} + C, \\ \int \frac{dx}{x^2 + a^2} &= \frac{1}{a} \arctan \frac{x}{a} + C. \end{aligned}}$$

Example 10
$$\int \frac{dx}{\sqrt{4 - x^2}} = \arcsin \frac{x}{2} + C. \qquad \square$$

PROBLEMS FOR SECTION 1

Integrate the following.

1. $\int \sin 2x \, dx$

2. $\int e^{x/2} \, dx$

3. $\int \sin^2 x \cos x \, dx$

4. $\int x \cos x^2 \, dx$

5. $\int t^2 e^{-t^3} \, dt$

6. $\int \sqrt{ax + b} \, dx$

7. $\int \cos(2 - x)\, dx$

8. $\int e^x \cos(e^x)\, dx$

9. $\int \dfrac{\sin x}{1 + \cos x}\, dx$

10. $\int \tan x\, dx$

11. $\int \dfrac{\ln x\, dx}{x}$

12. $\int \dfrac{dx}{x \ln x}$

13. $\int \dfrac{2x + 1}{x^2 + x - 1}\, dx$

14. $\int \dfrac{\sin \theta}{\cos^2 \theta}\, d\theta$

15. $\int y^3 \sqrt{a + y}\, dy$

16. $\int \dfrac{x^{1/2}\, dx}{1 + x^{3/4}}$

17. $\int \dfrac{(x + 2)\, dx}{x \sqrt{x - 3}}$

18. $\int (x^3 + 2)^{1/3} x^2\, dx$

19. $\int \dfrac{8x^2\, dx}{(x^3 + 2)^3}$

20. $\int \dfrac{dx}{4 - 2x}$

21. $\int \dfrac{e^{\sqrt{x}}}{\sqrt{x}}\, dx$

22. $\int \dfrac{dx}{(1 + x)^{3/2} + (1 + x)^{1/2}}$

23. $\int (e^x + 1)^3 e^x\, dx$

24. $\int \dfrac{x\, dx}{x^2 - 1}$

25. $\int \dfrac{(\ln x)^2\, dx}{x}$

26. $\int \dfrac{x\, dx}{1 + x^4}$

27. $\int \dfrac{\arctan t}{1 + t^2}\, dt$

28. $\int \dfrac{dx}{x^{1/2}(1 + x)}$

29. $\int \dfrac{\sin \sqrt{x}}{\sqrt{x}}\, dx$

30. $\int \dfrac{\cos(\ln x)\, dx}{x}$

31. $\int \dfrac{\cos t}{\sqrt{1 - \sin t}}\, dt$

32. $\int \dfrac{x^2}{\sqrt{1 - x}}\, dx$

33. $\int \sqrt{x^4 + x^2}\, dx$

34. $\int \dfrac{e^{2t}\, dt}{1 + e^{2t}}$

35. $\int \dfrac{e^t}{1 + e^{2t}}\, dt$

36. $\int \tan \theta \sec^2 \theta\, d\theta$

37. $\int \tan^n \theta \sec^2 \theta\, d\theta$

38. $\int \sec^n \theta \tan \theta\, d\theta$

39. $\int \sin^n \theta \cos \theta\, d\theta$

40. $\int \dfrac{\sin t\, dt}{(a + b \cos t)^n}$

41. $\int \dfrac{dx}{x(\ln x)^a}, \quad a > 1$

42. $\int \dfrac{dx}{x \ln x \ln(\ln x)}$

43. $\int \dfrac{dx}{x \ln x [\ln(\ln x)]^a}, \quad a > 1$

2
INTEGRATION BY PARTS

We make the product rule

$$\frac{d}{dx}\, uv = u\frac{dv}{dx} + v\frac{du}{dx}$$

into an integration rule by solving for the term $u\, dv/dx$ and integrating:

$$\boxed{\int u\frac{dv}{dx}\, dx = uv - \int v\frac{du}{dx}\, dx.}$$

Although the indefinite integral of $d(uv)/dx$ is $uv + C$, we have suppressed the constant C since a constant of integration is contained implicitly in the remaining integrals. This rule is called *integration by parts*. We can use it whenever an integrand is a product of two functions, *provided that we already know how to integrate one of them*.

Example 1   The integral

$$\int x \cos x \, dx$$

can be considered to be in the form

$$\int u \frac{dv}{dx} \, dx$$

with $u = x$, $dv/dx = \cos x$. We can then use the formula because we know how to integrate $dv/dx = \cos x$ to find v. Any antiderivative is suitable, and we generally (but not always) choose the simplest. So here we set $v = \sin x$. Since $du/dx = 1$, the formula gives

$$\int x \cos x \, dx = x \sin x - \int \sin x \cdot 1 \cdot dx$$

$$= x \sin x + \cos x + C. \qquad \square$$

In terms of differentials, the integration-by-parts formula takes the neater form

$$\boxed{\int u \, dv = uv - \int v \, du.}$$

For instance, in the above example we have $x = u$ and $\cos x \, dx = dv$, from which we can conclude that $du = dx$ and $v = \sin x$.

Normally, this procedure is useful only when the resulting new integral is simpler than the original, or at least more amenable. This doesn't necessarily happen. For example, if we had reversed the roles of x and $\cos x$ in the above example, we would get

$$\int (\cos x)x \cdot dx = (\cos x)\frac{x^2}{2} + \int \frac{x^2}{2} \sin x \, dx,$$

which has made things worse rather than better, since $\int x^2 \sin x \, dx$ is more complicated than $\int x \cos x \, dx$. Thus, when we know how to integrate *both* factors of the integrand, and accordingly have a choice as to how to apply the integration-by-parts formula, we can expect that only one of the possibilities will be useful (and maybe neither will).

Example 2

$$\int \ln x \, dx = x \ln x - \int x \frac{dx}{x} \qquad \left(\begin{aligned} u &= \ln x, \, dv = dx \\ du &= \frac{dx}{x}, \, v = x \end{aligned} \right)$$

$$= x \ln x - \int dx$$

$$= x \ln x - x + C. \qquad \square$$

Example 3 Even powers of $\sin x$ and $\cos x$ can successfully be tackled by parts.

$$\int \cos^2 x \, dx = \int \cos x \cdot \cos x \, dx \qquad \left(\begin{aligned} u &= \cos x, \\ dv &= \cos x \, dx \end{aligned} \right)$$

$$= \cos x \sin x - \int \sin x (-\sin x \, dx)$$

$$= \cos x \sin x + \int \sin^2 x \, dx$$

$$= \cos x \sin x + \int (1 - \cos^2 x) \, dx$$

$$= \cos x \sin x + x - \int \cos^2 x \, dx.$$

The two functions labeled $\int \cos^2 x \, dx$ may differ by a constant, so

$$\int \cos^2 x \, dx = \frac{\cos x \sin x + x}{2} + C. \qquad \square$$

Example 4

$$\int \tan^2 x \, dx = \int \frac{\sin^2 x}{\cos^2 x} \, dx = \int \sin x \cdot \frac{\sin x \, dx}{\cos^2 x}$$

$$= \sin x \left(\frac{1}{\cos x} \right) - \int \cos x \left(\frac{1}{\cos x} \right) dx$$

$$= \tan x - x + C. \qquad \square$$

Example 5 In order to integrate $\int e^x \cos x \, dx$ we have to apply the integration-by-parts process twice.

$$\int e^x \cos x \, dx = e^x \sin x - \int e^x \sin x \, dx \qquad \begin{pmatrix} u = e^x, \\ dv = \cos x \, dx \end{pmatrix}$$

$$= e^x \sin x - \left[-e^x \cos x + \int e^x \cos x \, dx \right]. \quad \begin{pmatrix} u = e^x, \\ dv = \sin x \, dx \end{pmatrix}$$

Here the two occurrences of $\int e^x \cos x \, dx$ designate functions that may differ by a constant. Solving the equation gives

$$\int e^x \cos x \, dx = \frac{1}{2} e^x (\sin x + \cos x) + C. \qquad \square$$

In a problem involving two or more stages, the two factors *must* be treated consistently. Above we differentiated e^x at the first integration, and therefore we were required to differentiate e^x again in the second integration. If we hadn't, then the second stage would simply have undone the work of the first stage, and we would have ended up with the useless equation

$$\int e^x \cos x \, dx = \int e^x \cos x \, dx.$$

Example 6 Higher even powers of $\sin x$ and $\cos x$ can be approached as in Example 3, but more than one step will be needed.

$$\int \cos^4 x \, dx = \int \cos^3 x \cos x \, dx$$

$$= \cos^3 x \sin x - \int \sin x \cdot 3 \cos^2 x (-\sin x) \, dx$$

$$= \cos^3 x \sin x + \int 3 \cos^2 x \sin^2 x \, dx$$

$$= \cos^3 x \sin x + \int 3 \cos^2 x (1 - \cos^2 x) \, dx$$

$$= \cos^3 x \sin x - 3 \int \cos^4 x \, dx + 3 \int \cos^2 x \, dx.$$

Solving in this equation for $\int \cos^4 x \, dx$, we get

$$\int \cos^4 x \, dx = \frac{\cos^3 x \sin x}{4} + \frac{3}{4} \int \cos^2 x \, dx,$$

which reduces to the integral formula in Example 3. Similarly, $\int \cos^6 x \, dx$ would involve two such reducing steps.   □

Example 7   Such degree-reducing formulas can be found for all trigonometric powers.

$$\int \frac{dx}{\cos^5 x} = \int \sec^5 x \, dx = \int \sec^3 x \cdot \sec^2 x \, dx$$

$$= \sec^3 x \tan x - \int \tan x \cdot 3 \sec^2 x \cdot \sec x \tan x \, dx$$

$$= \sec^3 x \tan x - 3 \int \sec^5 x \, dx + 3 \int \sec^3 x \, dx.$$

Therefore

$$\int \sec^5 x \, dx = \frac{1}{4} \sec^3 x \tan x + \frac{3}{4} \int \sec^3 x \, dx.$$ □

PROBLEMS FOR SECTION 2

Integrate.

1. $\displaystyle\int x \sin x \, dx$ 2. $\displaystyle\int e^x \cos x \, dx$ 3. $\displaystyle\int x\sqrt{x+1} \, dx$ 4. $\displaystyle\int x \ln x \, dx$

5. $\displaystyle\int y^2 \sin y \, dy$ 6. $\displaystyle\int x e^{ax} \, dx$ 7. $\displaystyle\int x^2 e^{-x} \, dx$ 8. $\displaystyle\int x \cos ax \, dx$

9. $\displaystyle\int \sin^2 x \, dx$ 10. $\displaystyle\int \arctan x \, dx$ 11. $\displaystyle\int \arcsin x \, dx$ 12. $\displaystyle\int x^n \ln x \, dx$

13. $\displaystyle\int \frac{\ln x \, dx}{x}$ 14. $\displaystyle\int \sec^3 x \, dx$ 15. $\displaystyle\int \sin \sqrt{x} \, dx$ 16. $\displaystyle\int \sin(\ln x) \, dx$

17. $\displaystyle\int \ln(1 + x^2) \, dx$ 18. $\displaystyle\int \frac{x e^x \, dx}{(1+x)^2}$ 19. $\displaystyle\int x \arctan x \, dx$ 20. $\displaystyle\int x \arcsin x \, dx$

21. $\displaystyle\int \frac{\ln x}{(x+1)^2} \, dx$ 22. $\displaystyle\int x^3 \sin x \, dx$ 23. $\displaystyle\int \arctan \sqrt{x} \, dx$ 24. $\displaystyle\int x^2 \ln(x+1) \, dx$

25. $\displaystyle\int e^{ax} \cos bx \, dx$ 26. $\displaystyle\int (\ln x)^2 \, dx$

In the following examples find degree-reducing formulas analogous to the special cases worked out in Examples 6 and 7.

27. $\displaystyle\int \sec^n x \, dx$ **28.** $\displaystyle\int \cos^n x \, dx$ **29.** $\displaystyle\int (\ln x)^n \, dx$

3
RATIONAL FUNCTIONS

In this section and the next two, we shall look at two general situations in which an integrand can be algebraically manipulated into a form that can be integrated.

First, we look at some simple rational functions. The basic device employed here—called a *partial fraction decomposition*—can be used to integrate any rational function whatsoever. We shall discuss this systematic process further in Section 4.

Example 1

$$\int \frac{x \, dx}{(x - 1)(x + 3)}$$

The crucial fact is that there are constants a and b such that

$$\frac{x}{(x - 1)(x + 3)} = \frac{a}{x - 1} + \frac{b}{x + 3}.$$

In order to determine these constants, first combine the terms on the right:

$$\frac{x}{(x - 1)(x + 3)} = \frac{a(x + 3) + b(x - 1)}{(x - 1)(x + 3)}.$$

Then the two numerators have to be equal:

$$x = (a + b)x + (3a - b).$$

But two polynomials are equal only if their corresponding coefficients are equal, so

$$1 = a + b, \qquad 0 = 3a - b.$$

Solving these equations simultaneously, we find that $a = 1/4$ and $b = 3/4$. Therefore,

$$\int \frac{x \, dx}{(x - 1)(x + 3)} = \frac{1}{4} \int \frac{dx}{x - 1} + \frac{3}{4} \int \frac{dx}{x + 3}$$

$$= \frac{1}{4} \ln |x - 1| + \frac{3}{4} \ln |x + 3| + C$$

$$= \ln |x - 1|^{1/4} |x + 3|^{3/4} + C. \qquad \square$$

Example 2 The above scheme will not work for

$$\int \frac{x\,dx}{(x-1)^2}.$$

(Try it.) When the factors in the denominator are the same, as they are here, we have to modify the scheme slightly. This time we look for constants a and b such that

$$\frac{x}{(x-1)^2} = \frac{a}{(x-1)} + \frac{b}{(x-1)^2}.$$

Proceeding as in the first example we find that $a = b = 1$, so

$$\int \frac{x\,dx}{(x-1)^2} = \int \frac{dx}{(x-1)} + \int \frac{dx}{(x-1)^2}$$

$$= \ln|x-1| - \frac{1}{(x-1)} + C. \qquad \square$$

An alternative procedure here is to write $x = (x-1) + 1$ in the numerator:

$$\frac{x}{(x-1)^2} = \frac{(x-1)+1}{(x-1)^2} = \frac{1}{(x-1)} + \frac{1}{(x-1)^2}.$$

Example 3 The method of "judicious substitution."

In Example 1, after combining terms and equating numerators, we have

$$x = a(x+3) + b(x-1).$$

At this point we can obtain a by eliminating the b term, as follows. Setting $x = 1$ turns the equation into $1 = a \cdot 4$, so $a = 1/4$. Similarly, setting $x = -3$ gives $-3 = b(-4)$ and $b = 3/4$. These substitutions can be done mentally and we can just write the answers down.

Note that we combine and equate numerators but do *not* collect coefficients. If we do, going on to

$$x = (a+b)x + (3a-b),$$

then the substitutions aren't so obvious.

Now consider Example 2. There we get

$$x = a(x-1) + b.$$

Setting $x = 1$ gives us $b = 1$ at once. Any other value of x will then give a. Thus $x = 2$ gives $2 = a + 1$, $a = 1$. But this second calculation is not quite in the same spirit.

This method works best when the denominator consists of two or more distinct linear factors. In more complicated situations it may not help at all. □

Example 4 In order for these devices to work, the integrand must be a *proper* rational function; that is, the degree of its numerator must be less than the degree of its denominator. If it isn't proper, we use polynomial long division to reduce to the proper case. For example, we find by long division that

$$\frac{x^3 - x + 2}{x^2 - x} = x + 1 + \frac{2}{x^2 - x}.$$

Therefore,

$$\int \frac{x^3 - x + 2}{x^2 - x}\, dx = \frac{x^2}{2} + x + 2\int \frac{dx}{x(x-1)},$$

and the latter integrand is then computed as above. Thus

$$\frac{1}{x(x-1)} = -\frac{1}{x} + \frac{1}{(x-1)},$$

and the final answer is

$$\frac{x^2}{2} + x + 2\ln\left|\frac{x-1}{x}\right| + C = \frac{x^2}{2} + x + \ln\left(1 - \frac{1}{x}\right)^2 + C. \qquad \square$$

Example 5

$$\int \frac{x\, dx}{x^2 + 2x - 3} = \int \frac{x\, dx}{(x-1)(x+3)},$$

which is the same as the first example. □

A quadratic denominator $ax^2 + bx + c$ may have *distinct* linear factors (Example 1), a *single* repeated linear factor (Example 2), or *no* real linear factors (Example 6). Normally we need to know which case we have in order to integrate. If we can see how to factor we usually do so (Example 5). If we can't see how to factor we usually proceed by *completing the square*.

Example 6

$$\int \frac{dx}{x^2 + 2x + 3}$$

Here we can't factor the denominator, and in fact this is an arctangent integral rather than a logarithm. In order to see what is going on, we complete the square in the denominator and have

$$\int \frac{dx}{x^2 + 2x + 3} = \int \frac{dx}{(x+1)^2 + 2} = \frac{1}{\sqrt{2}} \int \frac{du}{u^2 + 1}$$

(where we have set $x + 1 = \sqrt{2}\, u, dx = \sqrt{2}\, du$). This is the arctan integral

$$\frac{1}{\sqrt{2}} \arctan u + C = \frac{1}{\sqrt{2}} \arctan\left(\frac{x+1}{\sqrt{2}}\right) + C. \qquad \square$$

Note that if the numerator in the above example were $(2x + 2)$, then the answer would be $\ln|x^2 + 2x + 3|$, by simple substitution. If the numerator were $2x$, we would rewrite it as $(2x + 2) - 2$, and so have a combination of a logarithm and arctangent.

The device of completing the square could be used even for a quadratic that can be factored, such as in Example 4:

$$\begin{aligned}
x^2 + 2x - 3 &= (x+1)^2 - 4 \\
&= [(x+1) - 2][(x+1) + 2] \\
&= (x-1)(x+3).
\end{aligned}$$

Example 7   We have just seen that there are three distinct possibilities for a proper rational function with a quadratic denominator, depending on how the denominator factors. If the degree of the denominator is three, then there will be three unknown coefficients to determine in the partial-fraction decomposition, and four different possible setups for them. Here are two.

1. *Three different linear factors:*

$$\frac{1}{(x-1)(x+2)(x+3)} = \frac{a}{x-1} + \frac{b}{x+2} + \frac{c}{x+3}.$$

2. *One linear factor* and *one irreducible quadratic factor:*

$$\frac{1}{(x-1)(x^2+1)} = \frac{a}{(x-1)} + \frac{bx+c}{(x^2+1)}.$$

The coefficients are computed in the same way as before. Thus (2) goes on:

$$\frac{1}{(x-1)(x^2+1)} = \frac{a(x^2+1)+(bx+c)(x-1)}{(x-1)(x^2+1)};$$

$$1 = (a+b)x^2 + (c-b)x + (a-c);$$

$$\left.\begin{array}{l} a+b=0 \\[2mm] c-b=0 \\[2mm] a-c=1 \end{array}\right\} \quad \begin{array}{l} a=\dfrac{1}{2}, \\[2mm] b=-\dfrac{1}{2}, \\[2mm] c=-\dfrac{1}{2}. \end{array}$$

$$\int \frac{dx}{(x-1)(x^2+1)} = \frac{1}{2}\int \frac{dx}{(x-1)} - \frac{1}{2}\int \frac{x\,dx}{x^2+1} - \frac{1}{2}\int \frac{dx}{x^2+1}$$

$$= \frac{1}{2}\ln|x-1| - \frac{1}{4}\ln(x^2+1) - \frac{1}{2}\arctan x + C$$

$$= \frac{1}{2}\left[\ln\frac{|x-1|}{\sqrt{x^2+1}} - \arctan x\right] + C.$$

In both of the above two cases, the numerator 1 can be replaced by any polynomial of degree less than 3 (the quotient must be proper) without changing the procedure. For example, if we set

$$\frac{x^2+2}{(x-1)(x^2+1)} = \frac{a}{x-1} + \frac{bx+c}{x^2+1},$$

we just change the final set of three equations in the three unknowns a, b, and c. When we equate numerators this time, we have

$$x^2+2 = (a+b)x^2 + (c-b)x + (a-c),$$

so

$$\left.\begin{array}{l} a+b=1 \\[2mm] c-b=0 \\[2mm] a-c=2 \end{array}\right\} \quad \begin{array}{l} a=\dfrac{3}{2}, \\[2mm] b=-\dfrac{1}{2}, \\[2mm] c=-\dfrac{1}{2}. \end{array}$$

□

Example 8 In the first case above we can use judicious substitution. On combining terms and equating numerators, we have

$$1 = a(x + 2)(x + 3) + b(x - 1)(x + 3) + c(x - 1)(x + 2).$$

We see that

$$a = 1/12 \quad \text{on setting } x = 1,$$
$$b = -1/3 \quad \text{on setting } x = -2,$$
$$c = 1/4 \quad \text{on setting } x = -3. \qquad \square$$

The general rational function is treated in essentially the same way, but the complications can be greater. Further discussion will be found in the next section.

PROBLEMS FOR SECTION 3

Integrate

1. $\int \dfrac{dx}{x^2 - 9}$

2. $\int \dfrac{x^3 \, dx}{x^2 + 9}$

3. $\int \dfrac{dx}{x^2 - 9x}$

4. $\int \dfrac{x \, dx}{x^2 + x - 2}$

5. $\int \dfrac{dx}{x^2 + 2x - 2}$

6. $\int \dfrac{dx}{x^2 + 2x + 2}$

7. $\int \dfrac{dx}{x^2 + 3x + 2}$

8. $\int \dfrac{x^2 \, dx}{x^2 + 2x + 1}$

9. $\int \dfrac{dx}{4x^2 + 9}$

10. $\int \dfrac{dx}{2x^2 + x - 3}$

11. $\int \dfrac{x \, dx}{x^2 + 4x - 5}$

12. $\int \dfrac{4 \, dx}{x^3 + 4x}$

13. $\int \dfrac{x^2 \, dx}{x^2 + 2x + 1}$

14. $\int \dfrac{y^2 + 1}{y^2 + y + 1} \, dy$

15. $\int \dfrac{(x + 1) \, dx}{x^2 + 4x - 5}$

16. $\int \dfrac{2x^3 + x + 3}{x^2 + 1} \, dx$

17. $\int \dfrac{2x^2 + x + 2}{x(x - 1)^2} \, dx$

18. $\int \dfrac{dx}{x^3 + 8}$

19. $\int \dfrac{\cos \theta \, d\theta}{1 - \sin^2\theta}$

20. $\int \dfrac{\sin \theta \, d\theta}{\cos^2\theta + \cos \theta - 2}$

21. $\int \dfrac{e^t \, dt}{e^{2t} + 3e^t + 2}$

22. $\int \dfrac{dx}{x^3 + x^2 + 5x}$

23. $\int \dfrac{dx}{x^3 - 4x}$

24. $\int \dfrac{dx}{x^3 - 4x^2}$

25. $\int \dfrac{x^2 + 1}{x^3 - 4x^2} \, dx$

26. $\int \dfrac{dx}{x^{1/2}(x - 1)}$

27. $\int \dfrac{dx}{x^{1/2}(x^{1/3} + 1)}$

28. $\int \dfrac{x^3 \, dx}{x^2 - 1}$

29. $\int \dfrac{2x^3 + x}{x^2 + 2} \, dx$

30. $\int \dfrac{x^4}{x^2 - x + 1} \, dx$

31. $\int \dfrac{dx}{(1-x)(x^2+4x+5)}$ **32.** $\int \dfrac{2x^2+3x+4}{x^2+2x+3}\,dx$ **33.** $\int \dfrac{x^5}{(x+1)^3}\,dx$

34. $\int \dfrac{2x+1}{x^2+x-6}\,dx$ **35.** $\int \dfrac{x\,dx}{x^3-x^2+x-1}$ **36.** $\int \dfrac{x^3+x^2+x+1}{x^3-x^2-x+1}\,dx$

37. $\int \dfrac{2x+1}{x^3+x^2-6x}\,dx$ **38.** $\int \dfrac{x^4+1}{x^3+x}\,dx$ **39.** $\int \dfrac{x^2-x}{x^3-1}\,dx$

40. $\int \dfrac{x^6-x^3+1}{x^3+1}\,dx$

4
RATIONAL FUNCTIONS (CONCLUDED)

Partial-fraction decompositions depend on a lemma from polynomial algebra that we shall assume.

LEMMA   *If* $r(x) = p(x)/q(x)$ *is a proper rational function, and if its denominator* $q(x)$ *can be written in the form*

$$q(x) = q_1(x)q_2(x),$$

where $q_1(x)$ *and* $q_2(x)$ *are polynomials that have no factor in common (i.e., are relatively prime), then* $r(x)$ *can be written*

$$\frac{p(x)}{q(x)} = \frac{p_1(x)}{q_1(x)} + \frac{p_2(x)}{q_2(x)},$$

where each quotient is a proper rational function.

The lemma doesn't tell us how to find the new numerators $p_1(x)$ and $p_2(x)$, but this is easy once we know that the decomposition is possible.

Consider, for example, the rational function $r(x) = x/(x-1)(x+3)$, where $q_1(x)$ and $q_2(x)$ are each of the form $x + c$. A rational function with such a denominator will be proper only if its numerator is a constant, i.e., a polynomial of degree zero. The lemma therefore guarantees that there are constants a and b such that

$$\frac{x}{(x-1)(x+3)} = \frac{a}{x-1} + \frac{b}{x+3},$$

and we proceed as in Section 3.

Consider next

$$r(x) = \frac{x^2 - 3}{(x - 1)(x^2 + 1)}.$$

A rational function with $x^2 + 1$ in the denominator is proper only if its numerator has the form $bx + c$, this being the most general polynomial of degree less than 2. The lemma therefore guarantees constants a, b, and c, such that

$$\frac{x^2 - 3}{(x - 1)(x^2 + 1)} = \frac{a}{x - 1} + \frac{bx + c}{x^2 + 1},$$

again a situation that we worked out in Section 3.

If $q_2(x)$ in turn could be factored into relatively prime factors, then the lemma could be applied again to the rational function $p_2(x)/q_2(x)$. Proceeding in this way, the lemma extends to denominators having many relatively prime factors $q_1(x)$, $q_2(x)$, ..., $q_n(x)$ and shows that we can always find polynomials $p_1(x)$, ..., $p_n(x)$ such that

$$\frac{p(x)}{q(x)} = \frac{p_1(x)}{q_1(x)} + \frac{p_2(x)}{q_2(x)} + \cdots + \frac{p_n(x)}{q_n(x)},$$

each quotient being a proper rational function.

For example, we know that there are constants a, b, and c, such that

$$\frac{x - 2}{x(x + 1)(x + 2)} = \frac{a}{x} + \frac{b}{x + 1} + \frac{c}{x + 2},$$

another type considered in Section 3.

Determining the constants in the above examples falls under the general heading of methods involving *undetermined coefficients*. In the first example, the answer we are looking for is a *linear combination* of $1/(x - 1)$ and $1/(x + 3)$, that is, a sum of constants times these two functions. We start off with a linear combination having undetermined coefficients a and b, and then determine the coefficients by algebra.

In every case, *the number of undetermined coefficients is equal to the degree of the original denominator.*

The second example also involves a linear combination having undetermined coefficients, this time a linear combination

$$a\left(\frac{1}{x - 1}\right) + b\left(\frac{x}{x^2 + 1}\right) + c\left(\frac{1}{x^2 + 1}\right)$$

of the three functions

$$\frac{1}{x-1}, \qquad \frac{x}{x^2+1}, \qquad \frac{1}{x^2+1}.$$

If a denominator factor is repeated there is more to do. For example, the lemma guarantees constants a, b, and c, such that

$$\frac{3x^2+2}{x(x-1)^2} = \frac{a}{x} + \frac{bx+c}{(x-1)^2},$$

because x and $(x-1)^2$ are relatively prime. We thus end up with a linear combination of the three functions

$$\frac{1}{x}, \qquad \frac{x}{(x-1)^2}, \qquad \frac{1}{(x-1)^2}.$$

We can write down the integrals of the first and third of these functions, but not the second. However,

$$\frac{x}{(x-1)^2} = \frac{(x-1)+1}{(x-1)^2} = \frac{1}{x-1} + \frac{1}{(x-1)^2},$$

and now we can integrate.

Instead of resorting to this second change of form, it is customary to incorporate it into the original linear combination decomposition. That is, we look for a, b, and c, such that

$$\frac{3x^2+2}{x(x-1)^2} = \frac{a}{x} + \frac{b}{x-1} + \frac{c}{(x-1)^2},$$

again as in Section 3.

Similarly, if $(x^2+1)^2$ is a denominator factor, we look for two terms

$$\frac{ax+b}{x^2+1} + \frac{cx+d}{(x^2+1)^2},$$

instead of one term

$$\frac{ax^3+bx^2+cx+d}{(x^2+1)^2}.$$

We then have a linear combination of the following four functions:

$$\frac{x}{x^2+1}, \qquad \frac{1}{x^2+1}, \qquad \frac{x}{(x^2+1)^2}, \qquad \frac{1}{(x^2+1)^2}.$$

The first three can be directly integrated, but the last falls into the one category of elementary partial fractions that we don't yet know how to integrate:

$$\int \frac{dx}{(x^2 + 1)^n},$$

where n is an integer ≥ 2. However, the substitution $x = \tan \theta$ reduces the problem to the treatment of the even powers of $\cos x$ that we discussed at the end of Section 2.

There is a theorem, called the Fundamental Theorem of Algebra, that asserts that every polynomial with real coefficients can be expressed as a product of linear polynomials, such as $x - 1$, and quadratic polynomials, such as $x^2 + 1$ and $x^2 + 2x + 2$, which cannot be factored into linear polynomials with real coefficients. Therefore, in theory, every rational function has a denominator like those in the examples we have been working, except that the number of factors might be large. This is why, in theory, every rational function can be integrated.

In practice, finding the linear and irreducible quadratic factors of a polynomial may be very difficult. For example,

$$x^4 + 4 = (x^2 + 2x + 2)(x^2 - 2x + 2),$$

but this factorization is by no means obvious, and a more complicated fourth-degree polynomial could be even more difficult to factor.

PROBLEMS FOR SECTION 4

Integrate.

1. $\int \dfrac{dx}{x^4 - x^3}$

2. $\int \dfrac{(x^2 + 1)\,dx}{x^4 - x^3}$

3. $\int \dfrac{dx}{x^4 - x^2}$

4. $\int \dfrac{x^2 + x + 1}{x^4 - x^2}\,dx$

5. $\int \dfrac{dx}{x^4 + x^2}$

6. $\int \dfrac{x^3 + 1}{x^4 + x^2}\,dx$

7. $\int \dfrac{dx}{(1 - x^2)^2}$

8. $\int \dfrac{x^3 + x}{(1 - x^2)^2}\,dx$

9. $\int \dfrac{dx}{(x^2 - 1)(x^2 - 4)}$

10. $\int \dfrac{x^2 + 2}{(x^2 - 1)(x^2 - 4)}\,dx$

11. $\int \dfrac{dx}{(x^2 + 1)(x^2 + 4)}$

12. $\int \dfrac{x^2 + 2}{(x^2 + 1)(x^2 + 4)}\,dx$

13. $\int \dfrac{x + 2}{x^4 - 1}\,dx$

14. $\int \dfrac{x^5 + 1}{x^4 - x}\,dx$

15. $\int \dfrac{x^4 + x^2 + 2}{x^4 - x^2 - 2}\,dx$

16. $\displaystyle\int \frac{x^6 - x^5 + 1}{x^4 + 3x^2 + 2}\, dx$ **17.** $\displaystyle\int \frac{dx}{x^5 - x^3 - x^2 + 1}$ **18.** $\displaystyle\int \frac{dx}{x^6 - 1}$

19. $\displaystyle\int \frac{dx}{x^6 + 3x^4 + 2x^2}$ **20.** $\displaystyle\int \frac{dx}{(x^2 - 1)^3}$

21. a) Show that

$$\frac{x^3}{(x^2 + 1)^2} = \frac{ax + b}{(x^2 + 1)} + \frac{cx + d}{(x^2 + 1)^2}$$

for suitable constants a, b, c, and d.

b) Using the result in (a), compute

$$\int \frac{x^3}{(x^2 + 1)^2}\, dx.$$

22. Show that

$$\frac{Ax^3 + Bx^2 + Cx + D}{(x^2 + 1)^2}$$

can always be written in the form

$$\frac{ax + b}{x^2 + 1} + \frac{cx + d}{(x^2 + 1)^2}.$$

23. Find the degree-reducing formula for

$$\int \frac{dx}{(x^2 + a^2)^n},$$

by using the substitution $x = a \tan \theta$.

**5
TRIGONOMETRIC
POLYNOMIALS**

Trigonometric polynomials can also be integrated in a systematic way. A trigonometric polynomial is a sum of products of trigonometric functions and, by using the definitions of these functions in terms of sine and cosine, any such product can be reduced to the form

$$\sin^m x \cos^n x,$$

where m and n are integers (positive, zero, or negative). The problem is therefore to integrate any such reduced product. We consider three cases.

I. If $\cos x$ occurs to an *odd* power, then the integration can be performed by simple substitution, by the following device:

$$\cos^{2p+1} x\, dx = \cos^{2p} x \cos x\, dx = (1 - \sin^2 x)^p \cos x\, dx$$
$$= (1 - u^2)^p\, du \qquad\qquad (u = \sin x).$$

Example 1

$$\int \cos^3 x\, dx = \int (1 - \sin^2 x)\cos x\, dx = \int \cos x\, dx - \int \sin^2 x \cos x\, dx$$

$$= \sin x - \frac{\sin^3 x}{3} + C. \qquad\qquad \square$$

The same idea lets us integrate any product $\sin^m x \cos^n x$ where m and n are integers and at least one is odd. But when the odd integer is *negative*, the final integrand is a rational function, as in the following example.

Example 2

$$\int \frac{dx}{\cos x} = \int \frac{\cos x\, dx}{\cos^2 x} \qquad \left(\begin{matrix}\text{set } u = \sin x,\\ du = \cos x\, dx\end{matrix}\right)$$

$$= \int \frac{du}{1 - u^2} = \frac{1}{2}\int\left[\frac{1}{1-u} + \frac{1}{1+u}\right] du$$

$$= \frac{1}{2}\ln\left[\frac{1+u}{1-u}\right] + C = \ln\left[\frac{1+\sin x}{1-\sin x}\right]^{1/2} + C.$$

This answer can be simplified by multiplying top and bottom inside the brackets by $(1 + \sin x)$. It becomes

$$\ln\left[\frac{1+\sin x}{|\cos x|}\right] + C = \ln|\sec x + \tan x| + C. \qquad \square$$

Example 3

$$\int \frac{dx}{\cos^5 x} = \int \frac{\cos x\, dx}{\cos^6 x} = \int \frac{du}{(1-u^2)^3}$$

Here the partial-fraction decomposition is

$$\frac{1}{(1-u)^3(1+u)^3} = \frac{a}{1-u} + \frac{b}{(1-u)^2} + \frac{c}{(1-u)^3}$$

$$+ \frac{d}{1+u} + \frac{e}{(1+u)^2} + \frac{f}{(1+u)^3},$$

and a lot of algebra is required. $\qquad \square$

Example 4 If there is a *positive* odd power, then the other exponent does not have to be an integer.

$$\int \sqrt{\sin x}\, \cos^3 x\, dx = \int \sqrt{\sin x}(1 - \sin^2 x)\cos x\, dx$$

$$= \int (u^{1/2} - u^{5/2})\, du = \frac{2}{3} u^{3/2} - \frac{2}{7} u^{7/2} + C$$

$$= \frac{2}{3}(\sin x)^{3/2} - \frac{2}{7}(\sin x)^{7/2} + C. \qquad \square$$

II. If m and n are *even positive* integers, we can reduce the total power $m + n$ in either one of two ways:

a) integration by parts, as in Section 2, Example 5;

b) use of the half-angle formulas

$$\cos^2 x = \frac{1 + \cos 2x}{2} \quad \text{and} \quad \sin^2 x = \frac{1 - \cos 2x}{2}.$$

Example 5

$$\int \cos^2 x \, dx = \int \frac{(1 + \cos 2x)}{2} \, dx = \frac{x}{2} + \frac{\sin 2x}{4} + C.$$

Since $\sin 2x = 2 \sin x \cos x$ (by the addition law for $\sin(x + x)$), this answer can be rewritten

$$\frac{x + \sin x \cos x}{2} + C. \qquad \square$$

Example 6 Two applications of the power-reducing identity yield

$$\cos^4 x = \left(\frac{1 + \cos 2x}{2} \right)^2 = \frac{1}{4} + \frac{1}{2} \cos 2x + \frac{\cos^2 2x}{4}$$

$$= \frac{1}{4} + \frac{\cos 2x}{2} + \frac{1}{4} \left(\frac{1 + \cos 4x}{2} \right) = \frac{3}{8} + \frac{\cos 2x}{2} + \frac{\cos 4x}{8}.$$

Therefore

$$\int \cos^4 x \, dx = \frac{3x}{8} + \frac{\sin 2x}{4} + \frac{\sin 4x}{32} + C. \qquad \square$$

Example 7 The above two examples combine to give

$$\int \sin^2 x \cos^2 x \, dx = \int (1 - \cos^2 x) \cos^2 x \, dx$$

$$= \int \cos^2 x \, dx - \int \cos^4 x \, dx$$

$$= \frac{x}{2} + \frac{\sin 2x}{4} - \frac{3x}{8} - \frac{\sin 2x}{4} - \frac{\sin 4x}{32} + C$$

$$= \frac{x}{8} - \frac{\sin 4x}{32} + C. \qquad \square$$

In this way we can attack any product $\sin^m x \cos^n x$, where m and n are *even positive integers*.

III.   *Both m and n are even and at least one of them negative.* In order to handle such integrals we add the formulas

$$\int \sec^2 x \, dx = \tan x + C,$$

$$\int \csc^2 x \, dx = -\cot x + C$$

to our basic list, and we convert the identity $\sin^2 x + \cos^2 x = 1$ to the equivalent forms

$$\tan^2 x + 1 = \sec^2 x,$$
$$1 + \cot^2 x = \csc^2 x,$$

in order to apply these additional integrals.

The following examples show how to handle the various special types.

1. *An even power of* sec *x.* We can always bring about degree reductions through a succession of integrations by parts, but the method illustrated below is much easier.

Example 8

$$\int \sec^6 x \, dx = \int (\tan^2 x + 1)^2 \sec^2 x \, dx \qquad \begin{pmatrix} u = \tan x, \\ du = \sec^2 x \, dx \end{pmatrix}$$

$$= \int (u^2 + 1)^2 \, du$$

$$= \frac{u^5}{5} + \frac{2u^3}{3} + u + C$$

$$= \frac{\tan^5 x}{5} + \frac{2 \tan^3 x}{3} + \tan x + C. \qquad \square$$

Example 9   The same procedure works for an even power of sec *x* multiplied by *any* power of tan *x*. Thus,

$$\int \tan^k x \sec^4 x \, dx = \int u^k (u^2 + 1) \, du \qquad \begin{pmatrix} u = \tan x, \\ du = \sec^2 x \, dx \end{pmatrix}$$

$$= \frac{u^{k+3}}{k+3} + \frac{u^{k+1}}{k+1} + C$$

$$= \frac{\tan^{k+3} x}{k+3} + \frac{\tan^{k+1} x}{k+1} + C. \qquad \square$$

2. *An even power of* tan x. We proceed in a number of steps, each reducing the power by 2.

Example 10

$$
\begin{aligned}
\int \tan^6 x \, dx &= \int \tan^4 x (\sec^2 x - 1) \, dx \\
&= \frac{\tan^5 x}{5} - \int \tan^4 x \, dx, \\
&= \frac{\tan^5 x}{5} - \int \tan^2 x (\sec^2 x - 1) \, dx \\
&= \frac{\tan^5 x}{5} - \frac{\tan^3 x}{3} + \int \tan^2 x \, dx \\
&= \frac{\tan^5 x}{5} - \frac{\tan^3 x}{3} + \int (\sec^2 x - 1) \, dx \\
&= \frac{\tan^5 x}{5} - \frac{\tan^3 x}{3} + \tan x - x + C. \qquad \square
\end{aligned}
$$

The co-functions csc x and cot x are handled in the same way, the only difference being a minus sign when we integrate.

3. $\dfrac{\sin^m x}{\cos^n x}$, *where m and n are positive, even integers and $m > n$.* We write the numerator as

$$
\sin^m x = (1 - \cos^2 x)^{m/2},
$$

expand, and cancel the terms of the resulting sum against the denominator. The new terms can be integrated by earlier methods. Although this procedure always works, it can be tedious and sometimes inefficient. If $m \le n$, it is much better to apply (1) or (2) directly.

The same procedure applies if sin x and cox x are interchanged.

Example 11 For a very simple example, we have

$$
\begin{aligned}
\int \frac{\sin^4 x}{\cos^2 x} \, dx &= \int \frac{(1 - \cos^2 x)^2}{\cos^2 x} \, dx \\
&= \int (\sec^2 x - 2 + \cos^2 x) \, dx \\
&= \tan x - 2x + \int \cos^2 x \, dx,
\end{aligned}
$$

where we now compute the latter integral by earlier methods. $\qquad \square$

4. The only case that remains is $\sin^m x \cos^n x$ where m and n are both negative even integers. We can rewrite this in the form

$$\sec^{2j} x \, \csc^{2k} x.$$

Now use the identity

$$\sec^2 x \, \csc^2 x = \sec^2 x + \csc^2 x$$

repeatedly, and as long as possible. The final result will be a sum of terms that can be integrated by earlier devices. So again, we finally get an answer.

Example 12

$$\begin{aligned}
\sec^4 x \, \csc^2 x &= \sec^2 x (\sec^2 x + \csc^2 x) \\
&= \sec^4 x + (\sec^2 x + \csc^2 x) \\
&= \tan^2 x \, \sec^2 x + 2 \sec^2 x + \csc^2 x.
\end{aligned}$$

Therefore,

$$\begin{aligned}
\int \frac{dx}{\cos^4 x \, \sin^2 x} &= \int \sec^4 x \, \csc^2 x \, dx \\
&= \frac{\tan^3 x}{3} + 2 \tan x - \cot x + C.
\end{aligned}$$ □

PROBLEMS FOR SECTION 5

Integrate the following.

1. $\displaystyle\int \sin^2 x \, dx$

2. $\displaystyle\int \sin^3 x \, dx$

3. $\displaystyle\int \sin^4 x \, dx$

4. $\displaystyle\int \cos^2 3x \, dx$

5. $\displaystyle\int \sin^3 6x \cos 6x \, dx$

6. $\displaystyle\int \cos^3 2\theta \sin 2\theta \, d\theta$

7. $\displaystyle\int \cos^3 x \sin^{-4} x \, dx$

8. $\displaystyle\int \frac{\sin^5 y \, dy}{\sqrt{\cos y}}$

9. $\displaystyle\int \cos^6 \theta \, d\theta$

10. $\displaystyle\int \cos^3 \left(\frac{x}{2}\right) \sin^2 \left(\frac{x}{2}\right) dx$

11. $\displaystyle\int \frac{\sin^3 2x}{\sqrt[3]{\cos 2x}} \, dx$

12. $\displaystyle\int \sin^3 x \cos^3 x \, dx$

13. $\displaystyle\int \frac{dx}{\sin x \cos x}$

14. $\displaystyle\int \sin 3x \cot 3x \, dx$

15. $\displaystyle\int \csc^4 x \cot^2 x \, dx$

16. $\displaystyle\int \tan^4 x \, dx$

17. $\displaystyle\int \csc^4 \frac{x}{4} \, dx$

18. $\displaystyle\int \sec^3 x \, dx$

19. $\displaystyle\int \tan^3 x \sec^{5/2} x \, dx$

20. $\displaystyle\int \tan x \sqrt{\sec x} \, dx$

21. $\int \tan^3 x \, dx$. Do this in two ways:

a) By substitution, since $\sin x$ (and $\cos x$ also) occurs to an odd power;

b) By using $\tan^2 x = \sec^2 x - 1$.

22. $\int \sec x \, dx$. Do this in two ways:

a) By substitution, since $\cos x$ occurs to an odd power;

b) By tricky substitution, multiplying top and bottom by $(\sec x + \tan x)$.

23. $\int \tan^5 x \, dx$ **24.** $\int \dfrac{\cos^5 x}{\sin^5 x} \, dx$

25. $\int \tan^6 x \, dx$ **26.** $\int \sec^6 x \, dx$

27. In Example 2 the absolute-value sign was not introduced inside the logarithm until the very end. Why was this permissible?

More integrals.

28. $\int \dfrac{\cos^4 x}{\sin^2 x} \, dx$

29. $\int \dfrac{\sin^6 3x}{\cos^2 3x} \, dx$

30. $\int \dfrac{dx}{\sin^4 x \cos^2 x}$

31. $\int \sec^4 x \csc^4 x \, dx$

32. $\int \csc^4 ax \, dx$

33. $\int x \cot^4(x^2) \, dx$

34. $\int \sin^3 x \ln \cos x \, dx$

35. $\int x \sec^2 x \, dx$

36. $\int x \sec^4 x \, dx$

37. $\int \sec^4 x \ln \sin x \, dx$

38. $\int \csc^2 x \ln \sin x \, dx$

39. $\int \sec^4 x \ln \cos x \, dx$

40. $\int \dfrac{\cos^3(\ln x) \, dx}{x}$

41. $\int \dfrac{\sin^4 \sqrt{2x} \, dx}{\sqrt{x}}$

42. $\int e^x \sin^2 x \, dx$

43. $\int e^x \sin^n x \, dx$ (reduction)

6
TRIGONOMETRIC
SUBSTITUTIONS

Here we change variables again, but in a different way. The most common examples involve the trigonometric functions.

Example 1 In order to integrate

$$\int \sqrt{4 - x^2} \, dx,$$

we try setting $x = 2 \sin u$, which will at least eliminate the radical.

Then

$$dx = 2 \cos u\, du,$$

$$\sqrt{4 - x^2} = 2\sqrt{1 - \sin^2 u} = 2 \cos u,$$

(See comment after Example 5.)

$$\int \sqrt{4 - x^2}\, dx = 4 \int \cos^2 u\, du = 4 \int \frac{1 + \cos 2u}{2}\, du \qquad \text{(as in Section 5)}$$

$$= 2u + \sin 2u + C$$

$$= 2 \sin u \cos u + 2u + C.$$

Now we have to write the answer in terms of the original variable x. We have $\sin u$ and $\cos u$ from above, but for u itself we have to know the inverse of the substitution function. In this case $u = \arcsin x/2$, so

$$\int \sqrt{4 - x^2}\, dx = \frac{1}{2}\, x\sqrt{4 - x^2} + 2 \arcsin x/2 + C$$

is the final answer. Note that $x = 2 \sin u$ must be restricted to the domain $[-\pi/2, \pi/2]$ to make it invertible. $\square$

This process is called *inverse substitution*. Instead of setting $h(x) = u$ as we did in Section 1, we now set $x = k(u)$, $dx = k'(u)\, du$. This makes for a very easy transformation of the integral, but we have to know an inverse function $u = h(x)$ in order to express the u answer in terms of the original variable x.

There are three such useful trigonometric substitutions, including the one used in Example 1:

a) *to rationalize* $\sqrt{a^2 - x^2}$, *set* $x = a \sin u$;

b) *to rationalize* $\sqrt{a^2 + x^2}$, *set* $x = a \tan u$,

c) *to rationalize* $\sqrt{x^2 - a^2}$, *set* $x = a \sec u$.

$$\left(\begin{array}{c} \text{with} \\ a > 0 \\ \text{in each} \\ \text{case} \end{array} \right)$$

Example 2 Integrate

$$\int \frac{dx}{\sqrt{a^2 + x^2}}.$$

Solution We set $x = a \tan u$, $dx = a \sec^2 u\, du$ (for u in

$(-\pi/2, \pi/2))$. Then

$$\int \frac{dx}{\sqrt{a^2 + x^2}} = \int \sec u \, dw = \int \frac{\cos u \, du}{\cos^2 u}$$

(See comment after Example 5.)

$$= \int \frac{dv}{1 - v^2} = \frac{1}{2} \int \left[\frac{1}{1 - v} + \frac{1}{1 + v} \right] dv$$

$$= \ln\left[\frac{1 + v}{1 - v} \right]^{1/2} + C = \ln\left[\frac{1 + \sin u}{1 - \sin u} \right]^{1/2} + C$$

$$= \ln |\sec u + \tan u| + C \qquad \text{(as in Section 5).}$$

There remains the question, What is $\sec u$ when $\tan u = x/a$? We could write

$$\sec u = \sec(\arctan x/a)$$

and have an answer. But this is more complicated than necessary, since any such composition of a trigonometric function with an inverse trigonometric function can always be simplified. In Fig. 1 we label a right triangle in the simplest way possible consistent with $\tan u = x/a$, and see that when $\tan u = x/a$, then

$$\sec u = \sqrt{a^2 + x^2}/a.$$

Figure 1

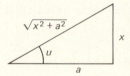

This could also be worked out directly,

$$\sec u = \sqrt{\sec^2 u} = \sqrt{1 + \tan^2 u} = \sqrt{1 + \left(\frac{x}{a}\right)^2},$$

but the right-triangle device is easier. In any case, the final answer is

$$\int \frac{dx}{\sqrt{a^2 + x^2}} = \ln\left[\frac{\sqrt{a^2 + x^2} + x}{a} \right] + C = \ln[\sqrt{a^2 + x^2} + x] + C. \qquad \square$$

Example 3 Integrate

$$\int \frac{dx}{\sqrt{x^2 - a^2}}.$$

Solution Assuming that $x > a > 0$, we set $x = a \sec u$, $dx = a \sec u \tan u \, du$, where u is in $[0, \pi/2)$. Then

$$\int \frac{dx}{\sqrt{x^2 - a^2}} = \int \frac{a \sec u \tan u \, du}{a \tan u} = \int \sec u \, du$$

$$= \ln |\sec u + \tan u| + C$$

$$= \ln \left| \frac{x + \sqrt{x^2 - a^2}}{a} \right| + C \quad \text{(using Fig. 2)}$$

$$= \ln |x + \sqrt{x^2 - a^2}| + C.$$

The answer can be checked to be correct for negative x, too.

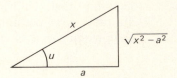

Figure 2

Example 4 The same substitution and diagram is used here:

$$\int \frac{\sqrt{x^2 - a^2} \, dx}{x} = \int \frac{a^2 \sec u \tan^2 u}{a \sec u} \, du = a \int \tan^2 u \, du$$

$$= a \int (\sec^2 u - 1) \, du = a \tan u - au + C$$

$$= \sqrt{x^2 - a^2} - a \arctan \frac{\sqrt{x^2 - a^2}}{a} + C.$$

Example 5 Integrate

$$\int \sqrt{2x - x^2} \, dx.$$

Solution We complete the square under the radical, and then proceed as in Example 1.

$$\int \sqrt{2x - x^2}\, dx = \int \sqrt{1 - (1 - x)^2}\, dx \qquad \left(\begin{array}{l} \text{set } 1 - x = \sin u, \\ dx = -\cos u\, du \end{array} \right)$$

$$= -\int \cos^2 u\, du = -\frac{u + \sin u \cos u}{2} + C$$

$$= -\frac{\arcsin(1 - x) + (1 - x)\sqrt{2x - x^2}}{2} + C. \qquad \square$$

The computations above have an important feature that needs to be pointed out. It was careless to write

$$\cos u = \sqrt{1 - \sin^2 u}$$

without checking signs, because $\cos u$ is negative as often as it is positive. However, $u = \arcsin x$ here, so u is restricted to the interval $[-\pi/2, \pi/2]$, and $\cos u$ is positive, as was assumed.

Similarly, $\sec u$ is positive when $u = \arctan x$, so it is correct to write

$$\sqrt{1 + x^2} = \sqrt{1 + \tan^2 u} = \sqrt{\sec^2 u} = \sec u.$$

Here are a few rationalizing substitutions that are not trigonometric.

Example 6   Integrate

$$\int \frac{x\, dx}{(1 + x)^{1/3}}.$$

Solution   We would like to set $u = (1 + x)^{1/3}$ in order to eliminate the radical. Fortunately, we can easily solve the substitution equation for x in terms of u,

$$x = u^3 - 1,$$

$$dx = 3u^2\, du,$$

so

$$\int \frac{x\, dx}{(1 + x)^{1/3}} = \int \frac{(u^3 - 1)3u^2\, du}{u} = 3 \int (u^4 - u)\, du$$

$$= 3\left(\frac{u^5}{5} - \frac{u^2}{2}\right) + C = \frac{3u^2}{10}(2u^3 - 5) + C$$

$$= \frac{3}{10}(1 + x)^{2/3}(2x - 3) + C. \qquad \square$$

Example 7

$$\int \frac{dx}{x^{1/2}(1 + x^{1/3})} = \int \frac{5u^5 \, du}{u^3(1 + u^2)} \qquad \left(\begin{array}{l} x = u^6, \\ dx = 5u^5 \, du \end{array} \right)$$

$$= 5 \int \frac{u^2 \, du}{1 + u^2} = 5 \int \left[1 - \frac{1}{1 + u^2} \right] du$$

$$= 5[u - \arctan u] + C$$

$$= 5[x^{1/6} - \arctan x^{1/6}] + C. \qquad \square$$

Example 8   We can do Example 4 in this way, too.

$$\int \frac{\sqrt{x^2 - a^2} \, dx}{x} = \int \frac{\sqrt{x^2 - a^2} \, x \, dx}{x^2} \qquad \left(\begin{array}{l} \text{set } u^2 = x^2 - a^2, \\ u \, du = x \, dx \end{array} \right)$$

$$= \int \frac{u \cdot u \, du}{u^2 + a^2}$$

$$= \int \left[1 - \frac{a^2}{u^2 + a^2} \right] du$$

$$= u - a \arctan \frac{u}{a} + C$$

$$= \sqrt{x^2 - a^2} - a \arctan \frac{\sqrt{x^2 - a^2}}{a} + C. \qquad \square$$

This device can be used whenever one of the radicals

$$\sqrt{x^2 - a^2}, \qquad \sqrt{a^2 - x^2}, \qquad \sqrt{a^2 + x^2}$$

is combined with an *odd* power of x.

We have not yet justified the inverse substitution process. Here is a general description of what we have been doing. Setting $x = k(u)$ and $dx = k'(u) \, du$, we write

$$\int f(x) \, dx = \int f(k(u))k'(u) \, du = \int g(u) \, du,$$

where $g(u) = f(k(u))k'(u)$. We now find an antiderivative $G(u)$ of $g(u)$, and claim that then

$$\int f(x) \, dx = G(h(x)) + C.$$

where h is the inverse of the function k.

In order to see that this is correct we simply differentiate:

$$\frac{d}{dx} G(h(x)) = G'(h(x))h'(x)$$

$$= g(h(x))h'(x)$$

$$= f(k(h(x)))k'(h(x))h'(x),$$

where the final complicated expression comes from the definition of g. Since k and h are inverse functions,

$$k(h(x)) = x \qquad \text{and} \qquad k'(h(x))h'(x) = 1.$$

The complicated answer above therefore collapses to just $f(x)$, so

$$\frac{d}{dx} G(h(x)) = f(x),$$

as claimed.

PROBLEMS FOR SECTION 6

In the following integrations a labeled right triangle will often be helpful when substituting back, at the end.

1. $\int \dfrac{x^3\, dx}{\sqrt{4 - x^2}}$

2. $\int \dfrac{dx}{x\sqrt{x^2 - a^2}}$

3. $\int \dfrac{dx}{(x^2 + a^2)^{3/2}}$

4. $\int \dfrac{dx}{(x^2 - 9)^{3/2}}$

5. $\int x\sqrt{16 - x^2}\, dx$

6. $\int \dfrac{x\, dx}{\sqrt{(x^2 + a^2)^3}}$

7. $\int \dfrac{dx}{x\sqrt{x^2 + 1}}$

8. $\int \dfrac{\sqrt{9 - 4x^2}}{x}\, dx$

9. $\int \dfrac{dx}{x\sqrt{25 - x^2}}$

10. $\int \dfrac{dy}{y^2\sqrt{y^2 - 7}}$

11. $\int \dfrac{dx}{x\sqrt{9 + 4x^2}}$

12. $\int \dfrac{\sqrt{16 - t^2}}{t^2}$

13. $\int \dfrac{x^2\, dx}{(a^2 - x^2)^{3/2}}$

14. $\int \dfrac{x^2\, dx}{\sqrt{2x - x^2}}$

15. $\int \dfrac{dx}{(x^2 - 2x - 3)^{3/2}}$

The following problems would normally be tackled in a different way, but this time do them using the trigonometric rationalizing substitutions.

16. $\int \dfrac{dy}{\sqrt{a^2 - y^2}}$

17. $\int \dfrac{x\, dx}{\sqrt{1 - x^2}}$

18. $\int \dfrac{dx}{x^2 + 4}$

19. $\int \dfrac{x\, dx}{\sqrt{1 + x^2}}$

20. $\int \dfrac{dx}{x^2 - a^2}$

21. $\int \dfrac{x\, dx}{\sqrt{x^2 - a^2}}$

22. $\int \dfrac{dx}{(x^2 + a^2)^2}$

Integrate by finding a suitable substitution.

23. $\displaystyle\int \frac{dx}{1 + \sqrt{x}}$ **24.** $\displaystyle\int \frac{dx}{1 + x^{1/3}}$ **25.** $\displaystyle\int \frac{1 + x^{1/2}}{1 + x^{1/3}}\,dx$ **26.** $\displaystyle\int \frac{1 + x}{\sqrt{1 - x}}\,dx$

27. $\displaystyle\int \frac{\sqrt{1 - x}}{1 + x}\,dx$ **28.** $\displaystyle\int \frac{\cos t}{2 - \sin^2 t}\,dt$ **29.** $\displaystyle\int \frac{\sin t\,dt}{1 + \cos^2 t}$ **30.** $\displaystyle\int \sqrt{\frac{1 - x}{1 + x}}\,dx$

31. $\displaystyle\int \frac{x^3\,dx}{a^2 - x^2}$ **32.** $\displaystyle\int \frac{x^3\,dx}{\sqrt{a^2 - x^2}}$ **33.** $\displaystyle\int \frac{\sqrt{x}}{1 + x}\,dx$ **34.** $\displaystyle\int \frac{x^{1/3}}{1 + x}\,dx$

35. $\displaystyle\int \frac{\sqrt{1 + x^4}}{x^3}\,dx$ **36.** $\displaystyle\int \frac{\sqrt{1 + x^{4/3}}}{x^{1/3}}\,dx$

37. Show that a right triangle as labeled in the text for reading off the values of the trigonometric functions when $x = a\tan u$ will give the right signs for these functions when x is negative $(-\pi/2 < u < 0)$.

38. Show that a right triangle labeled to read off the values of the trigonometric functions when $x = a\sin u$ will give the right signs of these functions when x is negative $(-\pi/2 \le u < 0)$.

39. The right triangle labeled to show the substitution $x = a\sec u$ does not give the correct signs for the other trigonometric functions when x is negative $(\pi/2 < u < \pi)$. Show that the signs come out right for x negative if one leg is labeled $-\sqrt{x^2 - a^2}$.

40. a) Example 4 in the text was worked under the assumption that x is positive. Show that the answer is also correct for x negative.

b) Work Example 4 out again for the case of negative x. [*Hint*: The proper procedure is related to the conclusion in Problem 39.]

41. a) Example 3 in the text was worked out under the assumption that x is positive. Show by differentiation that the conclusion is correct for negative x.

b) Work Example 3 out again for the case that x is negative.

7 EXTRA INTEGRATION PROBLEMS

1. $\displaystyle\int (e^x)^2\,dx$ **2.** $\displaystyle\int \frac{x^3 + 1}{x - 1}\,dx$ **3.** $\displaystyle\int \frac{\sin x}{\cos^3 x}\,dx$

4. $\displaystyle\int \frac{\sin x^{2/3}}{x^{1/3}}\,dx$ **5.** $\displaystyle\int e^{(ex)}e^x\,dx$ **6.** $\displaystyle\int \frac{\ln \sqrt{x}}{x}\,dx$

7. $\displaystyle\int \frac{\ln x}{\sqrt{x}}\,dx$ **8.** $\displaystyle\int \sqrt{x}\,\ln x\,dx$ **9.** $\displaystyle\int \frac{x^2\,dx}{\sqrt{1 - x^2}}$

10. $\displaystyle\int \frac{dx}{\sqrt{x}\sqrt{x+1}}$

11. $\displaystyle\int x^{n-1}\sqrt{a+bx^n}\,dx$

12. $\displaystyle\int \frac{x^{n-1}\,dx}{\sqrt{a+bx^n}}$

13. $\displaystyle\int \frac{dx}{x^2\sqrt{x^2+4}}$

14. $\displaystyle\int \frac{\arcsin x\,dx}{\sqrt{1-x^2}}$

15. $\displaystyle\int \frac{dx}{x(\ln x)^3}$

16. $\displaystyle\int e^{2x}\sin(e^x)\,dx$

17. $\displaystyle\int \frac{x^3}{\sqrt{x^2+1}}\,dx$

18. $\displaystyle\int xe^{\sqrt{x}}\,dx$

19. $\displaystyle\int \frac{dx}{\sqrt{x}\cos\sqrt{x}}$

20. $\displaystyle\int \frac{x\,dx}{(a^2-x^2)^{3/2}}$

21. $\displaystyle\int \sin x\,\ln(\cos x)\,dx$

22. $\displaystyle\int \frac{dx}{x\cos(\ln x)}$

23. $\displaystyle\int \frac{x^2+1}{x^3-4x}\,dx$

24. $\displaystyle\int \frac{dx}{x^{1/3}(x-1)}$

25. $\displaystyle\int \frac{dx}{x^{1/2}(x^{1/2}+1)}$

26. $\displaystyle\int \tan^4x\sec^4x\,dx$

27. $\displaystyle\int \tan^6x\sec^2x\,dx$

28. $\displaystyle\int \sec^6x\tan^2x\,dx$

29. $\displaystyle\int \frac{\arcsin x}{\sqrt{1-x^2}}\,dx$

30. $\displaystyle\int \frac{dx}{x^{1/2}(2+x)}$

31. $\displaystyle\int \frac{\tan\sqrt{x}}{\sqrt{x}}\,dx$

32. $\displaystyle\int \frac{\sin(\ln x)\,dx}{x}$

33. $\displaystyle\int \frac{\cos^3t}{\sqrt{1-\sin t}}\,dt$

34. $\displaystyle\int \frac{x}{\sqrt{1-2x}}\,dx$

35. $\displaystyle\int \sqrt{x^4-x^2}\,dx$

36. $\displaystyle\int \frac{e^{2t}\,dt}{1-e^{2t}}$

37. $\displaystyle\int \frac{e^t}{1-e^{2t}}\,dt$

38. $\displaystyle\int \tan\theta\sin^2\theta\,d\theta$

39. $\displaystyle\int \tan^n\theta\sec^4\theta\,d\theta$

40. $\displaystyle\int \sec^n\theta\tan^3\theta\,d\theta$

41. $\displaystyle\int \sin^n\theta\cos^3\theta\,d\theta$

42. $\displaystyle\int \frac{t\,dt}{(a+bt)^n}$

43. $\displaystyle\int \frac{dx}{x^2\sqrt{9+4x^2}}$

44. $\displaystyle\int \frac{\sqrt{16-t^2}}{t^3}\,dt$

45. $\displaystyle\int \frac{x\,dx}{(a^2-x^2)^{3/2}}$

46. $\displaystyle\int \frac{x\,dx}{\sqrt{2x-x^2}}$

47. $\displaystyle\int \frac{dx}{(x^2-2x+3)^{3/2}}$

48. $\displaystyle\int \frac{\sqrt{1-x^2}}{1+x}\,dx$

49. $\displaystyle\int \frac{\cos^3t}{2-\sin^2t}\,dt$

50. $\displaystyle\int \frac{\sin^3t\,dt}{1+\cos^2t}$

51. $\displaystyle\int \sqrt{\frac{1-2x}{1+x}}\,dx$

52. $\displaystyle\int \frac{x^5\,dx}{a^2-x^2}$

53. $\displaystyle\int \frac{x^5\,dx}{\sqrt{a^2-x^2}}$

54. $\displaystyle\int \frac{\sqrt{x}}{1+x^3}\,dx$

55. $\displaystyle\int \frac{x^{1/3}}{1+x^3}\,dx$

56. $\displaystyle\int x^2\arctan x\,dx$

57. $\displaystyle\int x^2\arcsin x\,dx$

58. $\displaystyle\int \frac{\ln(x+1)}{x^2}\,dx$

59. $\displaystyle\int x^3\cos x\,dx$

60. $\displaystyle\int \arctan\sqrt{x^{1/3}}\,dx$

61. $\displaystyle\int x^a\ln x\,dx$

62. $\displaystyle\int e^{ax}\sin bx\,dx$

63. $\displaystyle\int (\ln x)^3\,dx$

64. Prove that the area of the ellipse

$$\frac{x^2}{a^2} + \frac{y^2}{b^2} = 1$$

is πab.

66. Compute the volume of the solid generated by rotating about the x-axis the region under the graph of $y = \ln x$, from $x = 1$ to $x = 2$.

65. One arch of the curve $y = \sin x$ is revolved about the x-axis. Compute the volume of the region thus generated.

67. Calculate the area under the graph of $f(x) = \arctan x$ from $x = 0$ to $x = 1$.

8
SIMPLE
DIFFERENTIAL
EQUATIONS

The laws governing natural phenomena are often expressed by differential equations. For example, we saw in Chapter 7 that the differential equation

$$\frac{dy}{dt} = ky$$

can be interpreted as the law of *normal population growth*. In the absence of inhibiting or stimulating factors, a population reproduces itself at a rate proportional to its size, which is exactly what the above equation says. Then, solving the equation, we are startled to learn that a population grows exponentially: The solution of the population equation is

$$y = y_0 e^{kt},$$

where the constant y_0 is the size of the initial population, i.e., the size at time $t = 0$. Moreover, we can now answer such quantitative questions as, How frequently will the population double in size? And when will it reach a magnitude that we have chosen, perhaps arbitrarily, as being critical?

Thus the differential equation states our understanding of the growth process as a *rate-of-change* phenomenon, and its solution shows us the growth pattern and predicts the future.

The differential equation

$$\frac{d^2 y}{dx^2} + k^2 y = 0$$

is another example. We shall see in Chapter 19 that it is the law governing the motion of a free, frictionless, elastic system.

As these examples suggest, differential equations lead to important applications of calculus.

In general, a differential equation is an equation connecting an unknown function with one or more of its derivatives. Here are some further examples.

$$3\frac{dy}{dx} + \frac{x}{y^2} = 0,$$

$$(x - y)\frac{dy}{dx} = x + y,$$

$$\left(\frac{dy}{dx}\right)^2 + y^2 = 1,$$

$$\frac{d^2y}{dx^2} + 3\frac{dy}{dx} + 2 = \sin x.$$

Each states a relationship between an unknown function $y = f(x)$, its derivative $y = f'(x)$, and, in the last example, its second derivative, $d^2y/dx^2 = f''(x)$. The first three are differential equations of the *first order* and the last is of the *second order*. A *solution* of a differential equation is a function $y = f(x)$ making the equation true. Thus, for the first example, it is a function f such that

$$3f'(x) + \frac{x}{[f(x)]^2} \equiv 0.$$

Solving a differential equation means finding all of its solution functions. We shall consider here only first-order equations, and only one technique for solving them, called *separation of variables*. Chapter 19 carries the subject further.

To start with, a first-order equation is normally written in *differential* form. For the first two examples above, this means simply multiplying across by dx, giving

$$3\,dy + \frac{x}{y^2}\,dx = 0,$$

$$(x - y)\,dy = (x + y)\,dx.$$

We now try to separate the variables, i.e., write the equation in the form

$$g(y)\,dy + h(x)\,dx = 0.$$

This cannot be done in the second equation, and other techniques must be used there. But the first equation can be multiplied across by

y^2, giving the equation

$$3y^2\, dy + x\, dx = 0,$$

which does have its variables separated. We now just integrate termwise, obtaining

$$\int 3y^2\, dy + \int x\, dx = C,$$

or

$$y^3 + \frac{x^2}{2} = C.$$

The general solution is thus

$$y = \left(C - \frac{x^2}{2}\right)^{1/3}.$$

Note the constant of integration. In general,

THEOREM 1   *The general solution to the differential equation*

$$\boxed{g(y)\, dy + h(x)\, dx = 0}$$

is

$$\boxed{\int g(y)\, dy + \int h(x)\, dx = C.}$$

That is, if $G(y)$ and $H(x)$ are antiderivatives of $g(y)$ and $h(x)$, respectively, then the solutions of the differential equation are exactly the differentiable functions defined implicitly by the equation

$$G(y) + H(x) = C,$$

for all values of the constant C.

Proof   We argue backward. The function

$$G(f(x)) + H(x)$$

is constant if and only if its derivative is 0, that is, if and only if

$$G'(f(x))f'(x) + H'(x) = 0.$$

That is, a differentiable function $y = f(x)$ satisfies the equation

$$G(y) + H(x) = \text{constant}$$

if and only if

$$G'(y)\frac{dy}{dx} + H'(x) = 0$$

or

$$g(y)\frac{dy}{dx} + h(x) = 0,$$

$$(\text{or} \quad g(y)\,dy + h(x)\,dx = 0).$$

Thus the solutions of the differential equation are exactly the differentiable functions defined implicitly by the equations $G(y) + H(x) = C$ for various constants C. ∎

Example 1 The equation

$$\frac{dy}{dx} + \frac{x}{y} = 0$$

can be separated by multiplying by $y\,dx$, the equation then becoming

$$y\,dy + x\,dx = 0.$$

By Theorem 1, this differential equation has the general solution

$$\int y\,dy + \int x\,dx = C,$$

or

$$\frac{y^2}{2} + \frac{x^2}{2} = C,$$

or

$$x^2 + y^2 = C$$

(replacing $2C$ by C). As curves in the coordinate plane, the solutions are the *circles about the origin*. A *function* is a solution if and only if its graph lies along one of these curves. The function solutions divide up into the upper half-circle functions

$$y = \sqrt{C - x^2},$$

and the lower half-circle functions

$$y = -\sqrt{C - x^2}.$$

□

Viewing a circle as the union of upper and lower half circles is geometrically unappealing, and in this sense it may be more satisfying to view the solutions of

$$h(x)\,dx + g(y)\,dy = 0$$

as the curves

$$H(x) + G(y) = C,$$

rather than as the functions whose graphs lie along these curves.

Example 2   Solve the differential equation

$$\frac{dy}{dx} + xy = 0.$$

Solution   First separate the variables,

$$\frac{dy}{y} + x\,dx = 0,$$

and then integrate, obtaining

$$\ln|y| + \frac{x^2}{2} = c.$$

Therefore,

$$\ln|y| = c - \frac{x^2}{2},$$

$$|y| = e^{c-x^2/2} = e^c e^{-x^2/2} = Ce^{-x^2/2}.$$

The right side is never 0, so y can never change sign. Therefore,

$$y = Ce^{-x^2/2},$$

where $C > 0$ if y is always positive and $C < 0$ if y is always negative. Note that the constant of integration appears here as the multiplicative constant C.   □

It may happen that the process of separating the variables loses a solution of the original equation. For example, we divided by y to separate variables in the above differential equation. This introduces the new requirement in the separated equation that $y \neq 0$ and warns us to check back to see whether the added restriction loses a solution.

We see by inspection that the constant function $y = 0$ is indeed such a lost solution.

In this example, the lost solution reappears in the final form of the general solution. The solutions of the separated equation correspond to values of C different from 0, but we can try putting $C = 0$, and we find that we recover the lost solution of the original equation.

One-parameter Families The general solution of a first-order differential equation always contains a constant of integration. For example, the general solution of

$$x\frac{dy}{dx} = 2y$$

is

$$y = Cx^2.$$

Each value of the constant C determines a solution curve. As C varies the solution curves vary, and in their totality they fill up some region G in the plane. In the above example the solution curves fill up the whole plane except for the y-axis (Fig. 1). In such situations we call the constant of integration C a *parameter*, and we say that the solutions of the differential equation form a *one-parameter* family of curves filling up G.

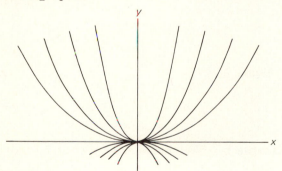

Figure 1

We can often recapture a differential equation from its general solution by differentiating and eliminating the parameter.

Example 3 Along each curve of the one-parameter family

$$y = Cx^2,$$

we have

$$\frac{dy}{dx} = 2Cx.$$

Since y/x^2 has the constant value C along the curve, we can set $C = y/x^2$ in the second equation and thus recapture the differential equation

$$\frac{dy}{dx} = \frac{2y}{x}.$$ □

Example 4   The parameter C in

$$y = x^2 + C$$

disappears upon differentiation. The differential equation of this one-parameter family is thus

$$\frac{dy}{dx} = 2x.$$

The solution curves fill up the whole plane (Fig. 2).

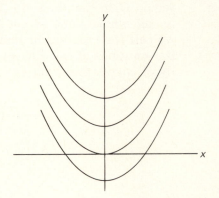

Figure 2

Example 5   Sketch the one-parameter family

$$y = \sin(x + C).$$

Find its differential equation. Then solve the differential equation and recover the one-parameter family.

Solution   The sine curves $y = \sin(x + C)$ fill up the strip $-1 < y < 1$ with exactly *two* curves passing through each point, as shown in Fig. 3. The differential equation must therefore assign *two* slopes to each point (x, y) in the strip, and this suggests an equation that is *quadratic* in dy/dx. In fact, it is

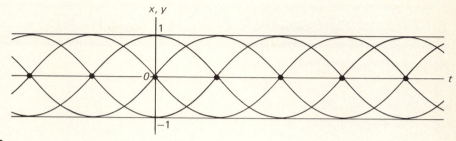

Figure 3

a version of the identity

$$\sin^2 x + \cos^2 x = 1.$$

We have

$$y = \sin(x + C),$$

$$\frac{dy}{dx} = \cos(x + C).$$

Here we can eliminate the parameter by squaring these two equations and adding, which gives the differential equation

$$y^2 + \left(\frac{dy}{dx}\right)^2 = 1.$$

The differential equation is solved by separating variables:

$$\frac{dy}{dx} = \pm\sqrt{1 - y^2}$$

$$\frac{dy}{\sqrt{1 - y^2}} = \pm dx$$

$$\arcsin y = \pm x + C$$

$$y = \sin(\pm x + C).$$

The minus sign can be disregarded since it simply has the effect of a different value of C: $\sin(-x + C) = \sin(x - C + \pi)$. We thus have recovered the one-parameter family we began with. (We have disregarded complications that can arise when the various arcsin solution curves are patched together.) □

PROBLEMS FOR SECTION 8

Solve the following differential equations.

1. $\dfrac{dy}{dx} = 2y$ **2.** $y\dfrac{dy}{dx} = 1$ **3.** $\dfrac{dy}{dx} = ky$ **4.** $\dfrac{dy}{dx} = y^2$

5. $\dfrac{dy}{dx} = y^2 x$ **6.** $\dfrac{dy}{dx} = e^y$ **7.** $\dfrac{dy}{dx} = xe^y$ **8.** $\dfrac{dy}{dx} = ye^x$

9. $\dfrac{dy}{dx} = \sqrt{1 - y^2}$ **10.** $\dfrac{dy}{dx} = x^2$ **11.** $\dfrac{dy}{dx} = \dfrac{2x}{y}$ **12.** $\dfrac{dy}{dx} = \dfrac{2y}{x}$

13. $x^3 \dfrac{dy}{dx} = y^2(x - 4)$ **14.** $4xy\,dx + (x^2 + 1)\,dy = 0$

15. $dy/dx = f(x)$ (Show that separation of variables amounts merely to integrating $f(x)$ for this type of equation.)

16. $\dfrac{dy}{dx} = y^2 \cos x$ **17.** $x\,dy + \cos^2 y\,dx = 0$

18. $\sqrt{1 - x^2}\,dy + \sqrt{1 - y^2}\,dx = 0$ **19.** $\dfrac{dy}{dx} = \cos^2 y \sin x$

20. $dy/dx = (y/x) + (y/x)^2$.   Here the variables cannot be separated. Show, however, that the substitution $y = vx$ reduces the given equation to one in x and v in which the variable *can* be separated, and hence solve the equation. (Note that dy/dx becomes $d(xv)\,dx = v + x\,dv/dx$.)

21. Show that the equation

$$\frac{dy}{dx} = f\left(\frac{y}{x}\right)$$

reduces to a variables-separable equation in v and x under the substitution $y = vx$.

22. $(x^2 + y^2)\,dx - 2xy\,dy = 0$.

In each of the following problems, find the differential equation of the given one-parameter family of curves. Sketch a few curves of the family, and indicate the region of the plane that they cover.

23. $y = x + C$ **24.** $y = Cx$ **25.** $x^2 + y^2 = C$

26. $y = \sqrt{x + C}$ **27.** $y = \sqrt{1 - x^2} + C$ **28.** $y = Ce^x/(1 + Ce^x)$

29. The one-parameter family of curves

$$y = (x + C)^2$$

has the added complication that *two* curves of the family pass through each point of its region G. Show that this is so, and sketch a few of the curves. Also find a differential equation of the family, and show from the differential equation why one would expect two solutions through each point.

30. Do the same for the one-parameter family

$$(y + C)^2 + x^2 = 1.$$

9
SOME
APPLICATIONS
OF DIFFERENTIAL
EQUATIONS

Differential equations frequently arise from word problems. Such problems have the added complication—and this may be the harder part of the solution—that first the appropriate differential equation has to be extracted from the statement of the problem. We start with geometric examples.

An *orthogonal trajectory* of a family of curves is a curve that crosses each curve of the family at right angles. The orthogonal trajectories of a one-parameter family of curves will themselves form a second one-parameter family, and they can be found by solving a differential equation.

Example 1   Find the orthogonal trajectories of the family of parabolas $y = Cx^2$.

Solution   We have seen (Example 3, Section 8) that the differential equation of this family of curves is

$$\frac{dy}{dx} = \frac{2y}{x}.$$

A curve cutting each of these parabolas orthogonally (i.e., perpendicularly) will have a slope that is everywhere the *negative reciprocal* of the above slope, so it will be a solution of the equation

$$\frac{dy}{dx} = -\frac{x}{2y},$$

or

$$2y\,dy + x\,dx = 0.$$

Solving this equation, we see that the orthogonal trajectories are the curves

$$y^2 + \frac{x^2}{2} = C,$$

a one-parameter family of *ellipses* (Fig. 1).

Figure 1

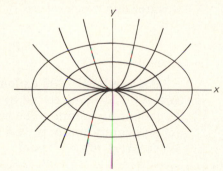

$\square$

Example 2   A curve has the property that its normal at a point (x_0, y_0) intersects the x-axis in a point $(x_1, 0)$ in such a way that $x_1 - x_0$ has the constant value a. Find all such curves.

Solution   We use the notation $(dy/dx)_0$ for the value of dy/dx at the point (x_0, y_0) on the curve. The normal then has the slope $-1/(dy/dx)_0$ and equation

$$y - y_0 = \frac{-1}{(dy/dx)_0}(x - x_0).$$

We obtain its x-intercept by setting $y = 0$, so

$$x_1 = x_0 + y_0\left(\frac{dy}{dx}\right)_0.$$

Therefore, by hypothesis,

$$a = x_1 - x_0 = y_0\left(\frac{dy}{dx}\right)_0.$$

This holds at every point (x_0, y_0) on the curve, so the differential equation of the curve is

$$y\frac{dy}{dx} = a.$$

Separating variables and integrating gives the solution

$$y^2 = 2ax + C.$$

The curves having the given property thus form a family of parabolas opening to the right.   □

Example 3   If x_1 gallons of water at a temperature T_1 are mixed with x_2 gallons of water at a temperature T_2, the resulting uniform temperature T is

$$T = \frac{x_1 T_1 + x_2 T_2}{x_1 + x_2}.$$

An insulated 1000-gallon tank of water initially at the temperature 200°F is being cooled by running in water at 40°F, at the rate of 30 gallons per minute. The water is stirred, so that its temperature is approximately uniform, and the excess runs off. What is the water temperature after 1 hour?

Solution Let $T = T(t)$ be the temperature at time t in minutes, so that $T(0) = 200$. In Δt minutes, $30 \Delta t$ gallons of water are added at the temperature 40°, so

$$T + \Delta T = \frac{40(30 \Delta t) + T(1000 - 30 \Delta t)}{1000},$$

$$\Delta T = \frac{1200 \Delta t - 30T \Delta t}{1000},$$

and

$$\frac{dT}{dt} = \frac{1}{100}(120 - 3T).$$

Solving this equation by separating the variables, and using the initial condition $T(0) = 200$, gives

$$T = T(t) = 40 + 160e^{-3t/100}.$$

So

$$T(60) = 40 + 160e^{-1.8} \approx 40 + 160(0.165)$$
$$\approx 66.4°. \qquad \square$$

Example 4 A more realistic view (model) of population growth will take into account other factors besides the size of the population: For example, there may be some compelling reason why the population x can never grow beyond a limiting size L, due perhaps to a limitation on food or living space. In that case we would expect the growth rate to lessen as the population size increases, and to approach 0 as x approaches L. This hypothesis does not by itself determine the new growth law. The simplest way for such an inhibition to express itself is for the growth rate dx/dt to be proportional both to the population size x and to its remaining possible room for growth $L - x$. Then

$$\frac{dx}{dt} = kx(L - x).$$

There might, on the other hand, be a reason why the limiting size L is "doubly depressing" on the growth rate, so that

$$\frac{dx}{dt} = kx(L - x)^2.$$

In order to solve the first equation we separate variables and integrate:

$$\int \frac{dx}{x(L-x)} = \int k\,dt,$$

$$\frac{1}{L}\ln\frac{x}{L-x} = kt + c,$$

$$\frac{x}{L-x} = Ce^{Lkt},$$

where the integration constant $C = e^c$ is positive. This equation can be solved for x, and the solutions are the functions

$$x = \frac{CLe^{Lkt}}{1+Ce^{Lkt}} = \frac{L}{1+(e^{-Lkt}/C)}.$$

Note that $x \to L$ as $t \to \infty$ for each value of C.

The solution curves fill up the strip $0 < x < L$, as indicated in Fig. 2. If we don't project these growth curves into the past, i.e., if we don't use *negative time*, then the solution curves are confined to the half strip $0 < x < L$, where $t \geq 0$. For any x_0 between 0 and L, there is a growth curve emanating from $(0, x_0)$, showing the evolution of a population that starts at size x_0. That is, $x = x_0$ at $t = 0$ is an *initial condition* that picks out a unique solution curve.

Figure 2

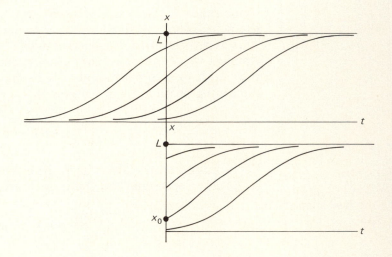

At $t = 0$,

$$x_0 = \frac{CL}{1 + C},$$

or, solving for C,

$$C = \frac{x_0}{L - x_0}.$$

We lose the solutions $x = 0$ and $x = L$ when we separate variables in the above equation. The first lost solution is recovered in the general solution by giving the integration constant C the value of 0, but the solution $x = L$ is not part of the general solution. (In a sense, it corresponds to setting $C = \infty$, as shown by the above equation for C.)

Note also that the lost solutions form the boundary of the region that is filled up by the solution curves of the general solution.   □

PROBLEMS FOR SECTION 9

Find the orthogonal trajectories of each of the following families of curves. Sketch a few curves of the given family and a few of the orthogonal trajectories.

1. The straight lines $y = Cx$

2. The hyperbolas $y = (1/x) + C$

3. The parabolas $y^2 = x + C$

4. The parabolas $y = C/x$

5. The circles $x^2 + y^2 = C$

6. The curves $y = C - \arctan x$

7. The circles $x^2 + Cx + y^2 = 0$

8. The ellipses $a^2 y^2 + x^2 = C$

9. If $y = (x + C)/(1 - Cx)$, show that

$$\frac{dy}{dx} = \frac{1 + y^2}{1 + x^2}.$$

10. Solve the above differential equation, and show that its general solution can be expressed in the above form.

11. The Gompertz growth law has the equation $y' = -ky \ln y$. Find the law.

12. A population grows at a rate proportional to its size, but, because of a steadily deteriorating environment, its growth rate is also *inversely proportional to time*. Find the law of population growth.

13. The growth rate of a population of size P is proportional to P, but it is also inversely proportional to $P^{1/2}$, because of difficulties in maintaining the food supply. Find the law of population growth.

14. The velocity of water flowing from a pipe at the bottom of a cylindrical tank is proportional to the square root of the depth of the water in the tank. If the water level drops from 10 feet to 9 feet in the first hour, how long will it take to reach the level of x feet?

15. A chemical reaction converts a substance A into a substance B at a rate that is proportional to the amount of A present and inversely proportional to the amount of B. Find a formula relating the amount of B to the time t. (If x is the amount of A and y is the amount of B, assume that x and y have the initial values $x = c$, $y = 0$.)

16. Show directly from the differential equation of Example 3 that the solution curves have points of inflection at $x = L/2$.

17. An insulated 500-gallon tank is full of water at 210°F. Water at 60°F is running in at the rate of 10 gallons per minute, the excess running off through an overflow pipe. The water is stirred, so that its temperature can be assumed to be uniform. What is its temperature after a half hour?

18. The tangent to a curve at a point P intersects the x-axis at Q. The normal at P intersects the x-axis at R. The curve is such that the segments RP and PQ always have equal lengths. Find all such curves.

19. A plane curve has the property that each of its tangents intersects the axis in two points whose midpoint is the point of tangency. Find all such curves.

20. The normal to a curve at a point P intersects the x-axis at Q. The curve is such that the segments OP and PQ always have equal lengths. Find all such curves.

21. The tangent to a plane curve at the point P on the curve intersects the x-axis at Q. The curve has the property that for every point of tangency P, the segment PQ is bisected by the y-axis. Find all such curves.

22. The tangent to a plane curve at the point P on the curve intersects the x-axis at the point Q. The curve has the property that the segments OP and PQ are always the same length. Find all such curves.

23. The growth rate of a population is the difference between the birth rate and the death rate. The birth rate is proportional to the population size, but the death rate is proportional to the square of its size. Find the law of population growth.

24. Solve the above problem if the death rate is proportional to $P^{3/2}$, where P is the population size.

25. Fifty pounds of salt are dissolved in a tank holding 100 gallons of water. Pure water is added at the rate of 1 gallon per minute and the solution is drained off at same rate. Assuming that the solution is agitated sufficiently so that mixture can be considered to be always uniform, how much salt is left after two hours?

26. A chemical dissolves in water at a rate that is jointly proportional to the amount undissolved and to the further amount that can be dissolved before saturation occurs.

Suppose that x_0 pounds of the chemical are introduced into G gallons of pure water, and that the solution will become saturated after P pounds have dissolved. Find a formula for the amount of dissolved chemical as a function of time.

27. A mixture consists initially of equal parts of chemicals x and y, which combine in a reaction in the ratio of two parts of x to one part of y to produce a chemical z. The reaction rate is jointly proportional to the amount of x and y present. Find the amount of z that has been produced, as a function of time, supposing there is none to begin with.

29. Suppose that water is being added to the tank in the above problem at the constant rate of 2 cubic feet per second, the excess normally flowing off the top. How long will it take for the water level to fall 5 feet when the outlet valve at the bottom is opened?

31. A cynical economist conjectures that the total money supply M grows at a rate that is directly proportional both to M and to the time measured from a fixed base time t_0. Find the formula for M as a function of t that this conjecture leads to.

28. It is experimentally determined that water flows out from a tank container through a hole in the bottom with a velocity v given by

$$v = 8\sqrt{h}\ \text{ft/sec,}$$

where h is the water depth in feet (assuming the hole to be large enough so that "edge effects" are negligible).

Suppose that the tank is a vertical circular cylinder 10 feet in diameter and 15 feet high, and that the hole is 6 inches in diameter. How long will it take to empty a full tank?

30. A tank is a solid of revolution, with axis vertical, of such a shape that its water level falls at a constant rate when the valve at the bottom is opened. What is its shape? (Use the formula in Problem 28.)

CHAPTER 9

THE DEFINITE INTEGRAL: FURTHER DEVELOPMENT AND APPLICATIONS

1 PROPERTIES OF THE DEFINITE INTEGRAL

When we work with the integral, in either computation or theory, it is essential to know its principal properties. So far we have been using the following properties:

1. $\displaystyle\int_a^b (f + g) = \int_a^b f + \int_a^b g$;

2. $\displaystyle\int_a^b kf = k\int_a^b f$, where k is a constant;

3. $\displaystyle\int_a^c f = \int_a^b f + \int_b^c f$, where a, b, and c are any three points in an interval I on which f is integrable;

4. If $f \geq g$ on $[a, b]$, then $\displaystyle\int_a^b f \geq \int_a^b g$.

4′. If $m \leq f(x) \leq M$ on $[a, b]$, then

$$m(b - a) \leq \int_a^b f \leq M(b - a).$$

5. The two forms of the fundamental theorem.

Now consider substitution. When we are looking for an antiderivative $\int f(x)\,dx$, we can substitute $x = k(u)$, $dx = k'(u)\,du$, but it is necessary that k be invertible (over the x interval in question), because we have to substitute back the inverse function $u = h(x)$ to get the final answer. This was the gist of Section 6 in Chapter 8. But, in the computation of a definite integral $\int_a^b f$, the final step is unnecessary, and we can substitute with almost no worries at all.

THEOREM 1   *If the integral $\int f(x)\,dx$ transforms into the integral $\int g(u)\,du$ under the substitution $x = k(u)$, $dx = k'(u)\,du$, all functions being continuous over the intervals in question, then*

$$6. \quad \int_a^b f(x)\,dx = \int_A^B g(u)\,du$$

whenever $a = k(A)$ and $b = k(B)$.

Of course, we have to be able to find the new limits A and B, but this is frequently easier than finding the whole inverse function. In fact, the transformation function k isn't required to be invertible here.

Example 1   In Chapter 8 we found that

$$\int \frac{dx}{\sqrt{a^2 + x^2}} = \int \sec u\,dx = \int \frac{dv}{1 - v^2} = \ln\left[\frac{1 + v}{1 - v}\right]^{1/2},$$

where the substitutions were

$$x = a \tan u, \qquad v = \sin u.$$

We then had to substitute back for the answer. However,

$$\int_0^a \frac{dx}{\sqrt{a^2 + x^2}} = \int_0^{\pi/4} \sec u\,du = \int_0^{\sqrt{2}/2} \frac{dv}{1 - v^2} = \ln\left[\frac{1 + \sqrt{2}/2}{1 - \sqrt{2}/2}\right]^{1/2},$$

and we are done, except possibly for algebraic simplification. We have used the fact that $a \tan(\pi/4) = a$ and that $\sin(\pi/4) = \sqrt{2}/2$. The limits of integration here are obviously rather special ones in relation to the integrand, but such special limits occur frequently in applications.

Incidentally, the answer given above can be simplified by "rationalizing the denominator" to give, finally,

$$\int_0^a \frac{dx}{\sqrt{a^2 + x^2}} = \ln(\sqrt{2} + 1). \qquad \square$$

Proof of Theorem By hypothesis,

$$g(u) = f(k(u))k'(u).$$

If F is an antiderivative of f, then $F(k(u))$ is an antiderivative of $g(u)$, by the chain rule. (This is just *direct* substitution.) Therefore,

$$\int_A^B g = F(k(B)) - F(k(A)).$$

But if $a = k(A)$ and $b = k(B)$, then

$$F(k(B)) - F(k(A)) = F(b) - F(a) = \int_a^b f.$$

Therefore, $\int_a^b f = \int_A^B g$, as claimed. $\qquad\blacksquare$

The integration-by-parts formula, in its definite-integral setting, offers no surprises. It is

---

7. $\displaystyle\int_a^b f(x)g'(x)\,dx = f(x)g(x)\bigg]_a^b - \int_a^b f'(x)g(x)\,dx,$

or

$\displaystyle\int_a^b fg' = f(b)g(b) - f(a)g(a) - \int_a^b f'g.$

The power of this formula may be suprising, though. In many advanced areas of mathematics it is a crucial tool. Examples will be found in Section 8.

The ordinary mean-value theorem can be given the integral form:

$$\int_a^b f = f(X)(b - a)$$

for some X strictly between a and b. [In terms of an antiderivative F

of f, this just says that

$$F(b) - F(a) = F'(X)(b - a).]$$

But the "weighted" mean-value theorem below requires an inequality argument depending on Property (4).

THEOREM 2   *Suppose that f and g are continuous on $[a, b]$, and that $g(x)$ is never zero on the open interval (a, b). Then there is a number X in $[a, b]$ such that*

$$\textbf{8.} \quad \int_a^b fg = f(X) \int_a^b g.$$

Proof   By hypothesis, g cannot change sign on $[a, b]$, and we can suppose that $g > 0$ on (a, b). Let m and M be the minimum and maximum values of f on $[a, b]$. The inequality

$$m \leq f(x) \leq M$$

can be multiplied through by the nonnegative number $g(x)$, and becomes

$$mg(x) \leq f(x)g(x) \leq Mg(x).$$

This holds for all x in $[a, b]$, so we can integrate the inequality, by (4):

$$m \int_a^b g \leq \int_a^b fg \leq M \int_a^b g.$$

The outside terms here are the minimum and maximum values of the continuous function $f(x) \int_a^b g$. The middle term is therefore a value, by the intermediate-value principle. But this is all we had to show.   ■

PROBLEMS FOR SECTION 1

Integrate.

1. $\displaystyle\int_0^{\pi/2} \cos^3\theta \, d\theta$ **2.** $\displaystyle\int_0^{\pi} \cos^2 x \, dx$ **3.** $\displaystyle\int_0^{\pi/3} \tan^3 x \sec^2 x \, dx$

4. $\displaystyle\int_0^{\pi/4} \tan^2 x \sec^2 x \, dx$

5. $\displaystyle\int_0^{\pi/6} \tan^3 \theta \sec \theta \, d\theta$

6. $\displaystyle\int_0^{\pi} \frac{dx}{1 + \sin x}$

7. $\displaystyle\int_{-1}^{1} \frac{dx}{4 - x^2}$

8. $\displaystyle\int_{-1}^{1} \frac{dx}{x^2 + 2x + 5}$

9. $\displaystyle\int_0^{\ln 2} \frac{dx}{1 + e^x}$

10. $\displaystyle\int_0^{1} \arctan x \, dx$

11. $\displaystyle\int_0^{1/2} \arctan \sqrt{x} \, dx$

12. $\displaystyle\int_0^{\pi/3} x \sec^2 x \, dx$

13. $\displaystyle\int_a^{2a} x^3 \sqrt{x^2 - a^2} \, dx$

14. $\displaystyle\int_0^{1} \frac{dx}{(1 + x^2)^{3/2}}$

15. $\displaystyle\int_a^{\sqrt{3}a} \frac{dx}{x^2\sqrt{a^2 + x^2}}$

16. $\displaystyle\int_0^{1} \frac{dx}{(2 - x^2)^{3/2}}$

17. $\displaystyle\int_0^{a} \frac{x \, dx}{\sqrt{a^2 - x^2}}$

18. $\displaystyle\int_0^{r} \sqrt{r^2 - x^2} \, dx$

19. $\displaystyle\int_0^{1} \sqrt{1 + x^2} \, dx$

20. $\displaystyle\int_1^{2} \sqrt{x^2 - 1} \, dx$

21. $\displaystyle\int_0^{\infty} \frac{dx}{1 + x^2}$ $\left(\text{Compute } \displaystyle\int_0^{r} \text{ and let } r \to \infty.\right)$

22. $\displaystyle\int_0^{\infty} \frac{dx}{4 + x^2}$

23. $\displaystyle\int_0^{\infty} \frac{dx}{(1 + x^2)^{3/2}}$

24. a) Show that

$$\int_0^{a} f(-x) \, dx = \int_{-a}^{0} f(x) \, dx$$

for any function f.

b) Show that therefore

$$\int_{-a}^{a} f(x) \, dx = 0$$

if f is an odd function on the interval $[-a, a]$, and that

$$\int_{-a}^{a} f(x) \, dx = 2\int_0^{a} f(x) \, dx$$

if f is even on $[-a, a]$.

25. Show that

$$\int_A^{B} f(cx + d) \, dx = \frac{1}{c} \int_{cA+d}^{cB+d} f(x) \, dx.$$

26. Show that

$$\int_0^{1} x(1 - x) \, dx = \frac{1}{6}.$$

27. Show that

$$\int_a^{b} (x - a)(b - x) \, dx = \frac{(b - a)^3}{6}.$$

28. Show that Problem 27 follows from Problem 26 under a change of variable.

29. Show that

$$\int_0^{1} x^2(1 - x)^2 \, dx = \frac{1}{30}.$$

30. Show that

$$\int_a^b (x - a)^2 (b - x)^2 \, dx = \frac{(b - a)^5}{30},$$

a) by a direct computation;
b) from Problem 29, by a change of variable.

2
VOLUMES BY
SLICING

The volume formula derived below is basic. It will be applied both here and in Chapter 17.

Suppose we put a coordinate line along beside a solid body S (or, possibly, skewering S). We shall call the coordinate line the t-axis. We want to consider how the solid intersects various planes perpendicular to the t-axis (Fig. 1). The cross section of the solid in the plane $t = t_0$ will be a plane region R_{t_0}. Nearby cross sections can be projected onto the same plane (by moving them parallel to the t-axis),

Figure 1

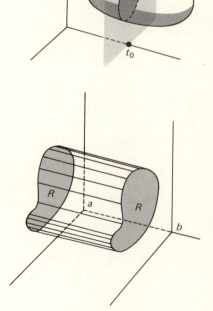

Figure 2

so we can compare how the cross section R_t changes as we move the slicing plane. In general, the cross section R_t will vary with t. If it does not, then we have a *cylinder* (Fig. 2).

DEFINITION If the solid S runs from the plane at $t = a$ to the plane at $t = b$, and if all its cross sections R_t at points t between a and b are the same plane region R, then we say that S is a *cylinder*, with base R and altitude $h = b - a$.

The volume of a cylinder is the product of its base area and its altitude:

$$V = Ah,$$

where A is the area of the base and h is the altitude. We shall assume this formula, and use it to derive a very general formula for the volume V of a solid body S.

Figure 3

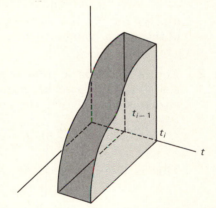

Suppose the first and the last planes to touch the solid S are at $t = a$ and $t = b$, and consider a subdivision $a = t_0 < t_1 < \cdots < t_n = b$ of the interval $[a, b]$ into n equal subintervals of common length $\Delta t = (b - a)/n$. Then the planes $t = t_i$ slice the solid into incremental slabs ΔS_i. (Imagine slicing a loaf of French bread. Figure 3 shows a quarter of a slab.) Each slab ΔS_i is approximately a cylinder of base area $A(t_i)$ and altitude Δt, where $A(t)$ is the cross-sectional area. So

$$\Delta V_i \approx A(t_i) \Delta t$$

is its approximate volume, by the cylinder volume formula. Therefore

$$V = \sum_{i=1}^{n} \Delta V_i \approx \Delta t \sum_{i=1}^{n} A(t_i).$$

It appears that this approximation to V improves with finer subdivision (finer slicing). It is also a Riemann sum for $A(t)$. Thus,

THEOREM 3   *Let S be a solid whose cross section R_t varies continuously as t runs from a to b. Let $A(t)$ be the area of the cross section R_t. Then the volume of S between $t = a$ and $t = b$ is given by*

$$V = \int_a^b A(t)\, dt.$$

Example 1   The solid of revolution generated by rotating the graph of f about the x-axis has the cross-sectional area $A(x) = \pi[f(x)]^2$. Therefore,

$$V = \int_a^b A(x)\, dx = \int_a^b \pi[f(x)]^2\, dx,$$

and we recover the formula of Section 6, Chapter 6 as a special case of the above "volumes-by-slicing" formula.   □

Example 2   Figure 4 shows a solid whose cross sections perpendicular to the x-axis are all isosceles right triangles with one vertex on the half-parabola $y = \sqrt{1 - x}$. The area of the triangle at x is

$$\frac{1}{2} bh = \frac{1}{2}\sqrt{1 - x}\sqrt{1 - x} = \frac{1 - x}{2}.$$

This is the cross-sectional area $A(x)$, and the volume of the solid

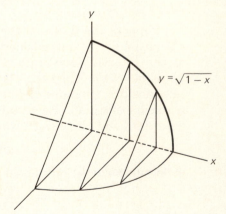

Figure 4

between $x = 0$ and $x = 1$ is therefore

$$V = \int_0^1 A(x)\, dx$$

$$= \int_0^1 \frac{1-x}{2}\, dx = \frac{x}{2} - \frac{x^2}{4}\Big]_0^1 = \frac{1}{4}.$$ □

Example 3 The region between $y = e^x$, the y-axis and the line $y = e$, is rotated about the y-axis. Find the volume swept out.

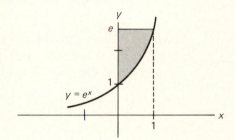

Figure 5

Solution We naturally try to use the cross sections perpendicular to the y-axis, since they are circles. At $y_0 = e^{x_0}$ the cross-section radius is $x_0 = \ln y_0$, so

$$V = \int_a^b A(y)\, dy$$

$$= \int_1^e \pi(\ln y)^2\, dy.$$

The integration can be done by parts, giving

$$\pi y(\ln y)^2\big]_1^e - \pi\int_1^e y \cdot \frac{2\ln y}{y}\, dy = \pi e - 2\pi\int_1^e \ln y\, dy$$

$$= \pi e - 2\pi y \ln y\big]_1^e + 2\pi\int_1^e y\frac{1}{y}\, dy$$

$$= \pi e - 2\pi e + 2\pi[e - 1]$$

$$= \pi(e - 2).$$ □

Example 4 The parabola $y = x^2$ is rotated about the line $y = 1$. The resulting surface bounds one finite solid and two infinite solids. Find the volume of the finite piece.

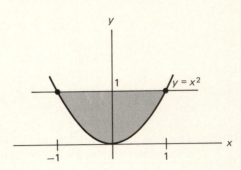

Figure 6

Solution   The parabola $y = x^2$ intersects the line $y = 1$ over the
points $x = \pm 1$. At any point x between them, the cross
section of the solid has area $\pi r^2 = \pi(1 - x^2)^2$. Therefore

$$V = \int_a^b A(x)\, dx$$

$$= \int_{-1}^{1} \pi(1 - x^2)^2\, dx = \pi \int_{-1}^{1} (1 - 2x^2 + x^4)\, dx$$

$$= \pi \left[x - \frac{2x^3}{3} + \frac{x^5}{5} \right]_{-1}^{1} = \frac{16}{15}\, \pi.$$ □

The formula $V = \int_a^b A(t)\, dt$ is important and deserves additional
justification. We go back to the examination of a typical increment
(see Fig. 7). Let ΔS be the incremental slab of width Δt that is sliced
from the solid by the planes $t = t_0$ and $t = t_0 + \Delta t$, and let ΔV be the
volume of ΔS. We assume that the cross section R_t varies continuously with t, in the following sense. The incremental slab ΔS completely includes a cylinder C' whose base area A' is only slightly less
than $A(t_0)$, and is completely included in a cylinder C'' whose base
area A'' is only slightly greater than $A(t_0)$. That is, the incremental
slab ΔS is "squeezed between" two cylinders C' and C'', whose base
areas A' and A'' can be made as close to $A(t_0)$ as desired by making Δt
sufficiently small.

Consequently,

Volume of $C' \leq$ Volume of $\Delta S \leq$ Volume of C'',

or

$$A'\, \Delta t \leq \Delta V \leq A''\, \Delta t,$$

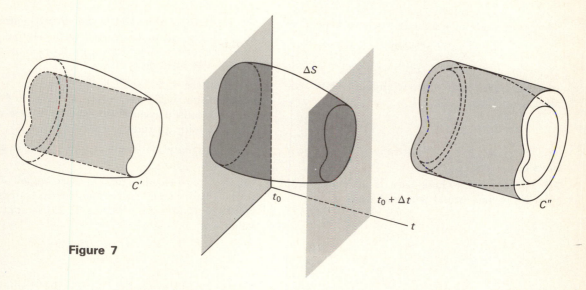

Figure 7

or

$$A' \le \frac{\Delta V}{\Delta t} \le A''.$$

Then, since A' and A'' both approach $A(t_0)$ as Δt approaches 0, $\Delta V/\Delta t$ is squeezed to the same limit. If $V(t)$ is the volume between a and t, then what we have shown is that $dV(t)/dt = A(t)$. Therefore

$$V = V(b) - V(a)$$

$$= \int_a^b A(t)\, dt,$$

by the First Fundamental Theorem.

PROBLEMS FOR SECTION 2

Find the volume of each of the solids described below.

In the first four of these problems, the base of the solid is the unit circular disk $x^2 + y^2 \le 1$, and its cross section perpendicular to the x-axis is the given figure:

1. A semicircle.   **2.** A square.   **3.** An equilateral triangle.

4. A rectangle of height $|y^3|$ (where $x^2 + y^2 = 1$).

5. The base of the solid is the square centered at the origin with one vertex at $(1, 1)$, and its cross sections perpendicular to the x-axis are triangles of altitude $(1 - x^2)$.

6. The same base, with triangular cross sections of altitude $1 + x$.

7. The base of the solid is the square centered at the origin with one vertex at $(1, 0)$, and its cross sections perpendicular to the x-axis are triangles with altitude $1 - x^4$.

8. The same base, with triangular cross sections of altitude 1.

9. The pyramid with square base of side a, and altitude h.

10. The pyramid whose base is an equilateral triangle of side a and whose altitude is h.

11. A cone with an irregular base region R is formed by drawing line segments from boundary points of R to a fixed vertex P. Prove that its volume is

$$V = \frac{1}{3} Ah,$$

where A is the area of R and h is the altitude of the cone. (Assume that similar plane figures have areas proportional to the squares of their diameters. The diameter of a plane figure is the length of the longest segment that can be drawn between two of its points.)

12. The base is the region between $y = 1 - e^x$ and $y = e^x - 1$, from $x = 0$ to $x = \pi$. The cross sections perpendicular to the x-axis are triangles of altitude $\sin x$.

13. The same base region, with square cross sections.

14. The base is the region between the x-axis and the first arch of $y = \sin x$. The cross section is an equilateral triangle.

15. The same base, with rectangular cross sections whose altitudes are the squares of their bases.

16. The same base, with triangular cross sections of altitude $x(\pi - x)$.

17. The "wood chopper's wedge" pictured in the left-hand column is cut from a cylinder of radius r by a plane through a diameter of the base circle, making an angle of $30° = \pi/6$ radians with the base plane. Compute the volume using plane sections perpendicular to the wedge edge.

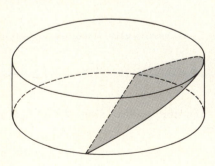

18. Solve the above problem by sections parallel to the wedge edge.

20. Show that the volume of the ellipsoid

$$\frac{x^2}{a^2} + \frac{y^2}{b^2} + \frac{z^2}{c^2} = 1$$

is $(4/3)\pi abc$. [*Hint*: The cross section of this solid perpendicular to the z-axis at the point $z = z_0$ is the ellipse

$$\frac{x^2}{a^2} + \frac{y^2}{b^2} = \left(1 - \frac{z_0^2}{c^2}\right).$$

The area of the standard ellipse

$$\frac{x^2}{a^2} + \frac{y^2}{b^2} = 1$$

is πab, by Extra Problem 64, Chapter 8.]

19. A tent is made by stretching canvas from a circular base of radius r to a semicircular rib, erected at right angles to the base at the ends of a diameter. Compute the volume of the tent.

21. Two circular cylinders intersect to form a solid S. The cylinders have the same radius a and have perpendicular intersecting axes. Find the volume of S.

3
VOLUMES BY SHELLS

There is a second formula for the volume of a solid of revolution, expressing as an x-integral the volume which is generated by revolving a region about the y-axis. (See Fig. 1.) This time the incremental volume ΔV associated with the Δx interval $[x, x + \Delta x]$ is approximately the volume of a thin cylindrical shell, obtained by rotating the thin vertical Δx strip about the y-axis, as shown in Fig. 1.

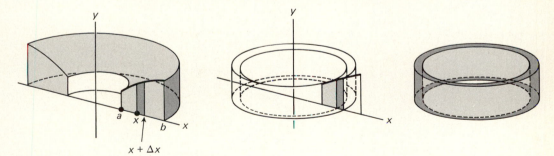

Figure 1

This shell has inner circumference $2\pi x$, altitude $f(x)$, and thickness Δx. Its volume is approximately the product of these three dimensions:

$$\Delta V \approx 2\pi x f(x) \, \Delta x.$$

The total volume is approximately a sum of such incremental volumes, and hence approximately a Riemann sum for the function $2\pi x f(x)$. This gives the formula of the theorem below.

THEOREM 4 *If the region under the graph of f, from $x = a$ to $x = b$, is revolved around the y-axis, then the volume swept out is*

$$V = \int_a^b 2\pi x f(x) \, dx = 2\pi \int_a^b x f(x) \, dx.$$

It is understood that $0 \le a < b$ and that $f \ge 0$ on $[a, b]$, so the back half of the solid can be pictured as in the middle figure.

We omit a more careful treatment of the theorem here because the proof comes up naturally in the discussion of polar coordinates in Chapter 17.

Example 1 Suppose that $f(x) = 2x - x^2$. Then $f(x)$ is above the x-axis between $x = 0$ and $x = 2$, and

$$V = 2\pi \int_0^2 x(2x - x^2) \, dx$$

$$= 2\pi \left[\frac{2x^3}{3} - \frac{x^4}{4} \right]_0^2$$

$$= 2\pi \left[\frac{16}{3} - \frac{16}{4} \right] = \frac{8}{3} \pi.$$

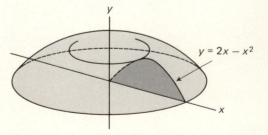

Figure 2

If $g(x) < f(x)$ over the interval $[a, b]$, and if R is the region between the graphs over this interval, then the volume swept out by rotating R about the y-axis is given by

$$V = \int_a^b 2\pi x[f(x) - g(x)]\,dx = 2\pi \int_a^b x[f(x) - g(x)]\,dx.$$

This generalizes our first situation. Now the incremental shell (Fig. 3) has volume ΔV given approximately by

$$\Delta V \approx 2\pi x[f(x) - g(x)]\,\Delta x,$$

which leads to the above integral formula.

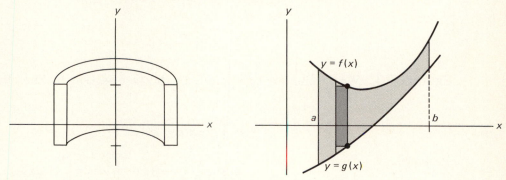

Figure 3

We can also obtain the general formula from the first case in the way we did earlier for the area between two graphs, by subtracting the two integrals.

Example 2 The region in Fig. 4 between the line $y = x$ and the parabola $y = x^2 - x$ is rotated about the y-axis. Find the volume generated.

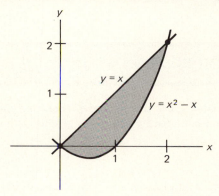

Figure 4

Solution   The graphs intersect at $x = 0$ and $x = 2$. Therefore,

$$V = 2\pi \int_0^2 x[f(x) - g(x)]\, dx$$

$$= 2\pi \int_0^2 x[x - (x^2 - x)]\, dx$$

$$= 2\pi \left[\frac{2x^3}{3} - \frac{x^4}{4}\right]_0^2$$

$$= 2\pi \left(\frac{16}{3} - \frac{16}{4}\right)$$

$$= \frac{8}{3}\pi. \qquad \qquad \square$$

Example 3   We recompute Example 3 in the last section by the shell formula. Thus

$$V = \int_0^1 2\pi x(e - e^x)\, dx$$

$$= 2\pi e \frac{x^2}{2}\Big]_0^1 - 2\pi x e^x\Big]_0^1 + 2\pi \int_0^1 e^x\, dx$$

$$= \pi e - 2\pi e + 2\pi(e - 1)$$

$$= \pi(e - 2). \qquad \qquad \square$$

Note that the integration needed for the shell method was simpler than that required for the disk method. In general, one cannot anticipate which method will be easier.

PROBLEMS FOR SECTION 3

In the following problems, use the method of shells to compute the volume generated by rotating about the y-axis the region bounded by the given graphs. Draw the plane region in each case.

1. $y = x;\quad y = 0, x = a$

2. $y = x^2;\ y = 1, x = 0$ (Compute this volume also by the earlier formula.)

3. $y^2 = x^3;\quad x = 4$

4. $y = x^2 - 2x;\quad y = 2x$

5. $y = x - x^3$; $\quad y = 0$

6. The region to the right of the line $x = a$ and inside the circle $x^2 + y^2 = r^2$ (where $a < r$).

7. $y = \sin x$; $\quad y = 0$, $\quad x = 0$, $\quad x = \pi$

8. $y = \ln x$; $\quad y = 0$, $\quad x = 1$, $\quad x = 2$

9. $y = e^{-x^2}$; $\quad y = 0$, $\quad x = 0$, $\quad x = \infty$

10. $y = e^x$; $\quad y = -e^x$, $\quad x = 0$, $\quad x = a$

11. The region outside the hyperbola $x^2 - y^2 = 4$ and inside the line $x = 5$.

12. $y = e^{kx}$; $\quad x = 0, x = a$

13. $y = \sin 2x$; $\quad x = \pi/4, x = \pi/2$

14. $y = \sin x^2$; $\quad x = a, x = b$
$\quad (0 \le a < b \le \sqrt{\pi})$

15. $y = \sqrt{1 + x^4}$; $\quad x = 0, x = 1$

16. $y = \ln x$; $\quad x = 1, x = e$

17. Calculate the area of the region between the graph of $y = \ln x$, its tangent line at $x = 1$, and the line $x = 2$.

18. Find the volume swept out when the above region is rotated about the y-axis.

19. Same problem when the region is rotated about the x-axis.

20. Find the formula for the volume of a sphere by the shell method.

21. A torus is a doughnut-like solid formed by rotating a circular disk about an axis not touching it. If the radius of the circle is r and the distance from the center of the circle to the axis is R $(R > r)$, show that the volume of the torus is

$$V = 2\pi^2 r^2 R.$$

22. A cylindrical core of radius a is removed from a spherical apple of radius r. Find the volume of the cored apple.

4

ARC LENGTH

In high school geometry the length s of a circular arc is characterized roughly as follows: We inscribe a polygonal arc composed of n segments (Fig. 1), compute its length s_n, and then obtain s as the limit of s_n as n tends to infinity.

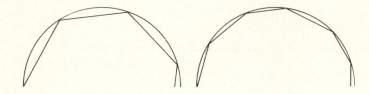

Figure 1

We shall now show that this procedure works for the graph of any function that is *smooth* in the sense of having a *continuous* derivative.

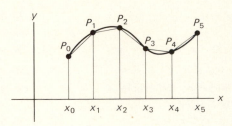

Figure 2

Consider, then, the graph of a smooth function f from $x = a$ to $x = b$, and inscribe the polygonal arc, with vertices $P_0, P_1, \ldots, P_n$, determined by subdividing the interval $[a, b]$ into n equal subintervals (Fig. 2).

Consider, to begin with, the kth polygon side. We isolate and magnify it in Fig. 3.

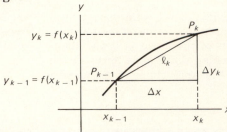

Figure 3

Its length l_k is the distance between P_{k-1} and P_k,

$$l_k = \sqrt{(\Delta x)^2 + (\Delta y_k)^2}.$$

By the mean-value principle, there is a point c_k between x_{k-1} and x_k such that

$$\Delta y_k = f'(c_k)\, \Delta x,$$

and, when this is substituted in the radical above, we end up with

$$l_k = \sqrt{1 + [f'(c_k)]^2}\, \Delta x$$

as the length of the segment $P_{k-1}P_k$.

The length of the inscribed polygon is thus

$$l = \sum_{k=1}^{n} l_k = \Delta x \sum_{k=1}^{n} \sqrt{1 + [f'(c_k)]^2}.$$

But this is simply a Riemann sum for the function

$$g(x) = \sqrt{1 + [f'(x)]^2},$$

and therefore approaches the limit

$$\int_a^b \sqrt{1 + [f'(x)]^2} \, dx$$

as $n \to \infty$. We have thus derived the formula

$$\boxed{\text{length} = \int_a^b \sqrt{1 + [f'(x)]^2} \, dx}$$

for the length of the graph of f over the interval $[a, b]$. If $y = f(x)$, this can be written

$$\int_a^b \sqrt{1 + \left(\frac{dy}{dx}\right)^2} \, dx.$$

This formula is equivalent to a very intuitive property of arc length. First note that if s is the length from $t = a$ to $t = x$,

$$s = \int_a^x \sqrt{1 + [f'(t)]^2} \, dt,$$

then

$$\frac{ds}{dx} = \sqrt{1 + \left(\frac{dy}{dx}\right)^2}$$

by the Second Fundamental Theorem. And this is the limit of $\Delta s/\Delta x$ as $\Delta x \to 0$. But $l/\Delta x$ has the same limit, since

$$\frac{l}{\Delta x} = \frac{\sqrt{(\Delta x)^2 + (\Delta y)^2}}{\Delta x} = \sqrt{1 + \left(\frac{\Delta y}{\Delta x}\right)^2}.$$

So their quotient has the limit 1. Thus

$$\lim \frac{\Delta s}{l} = \lim \frac{\Delta s/\Delta x}{l/\Delta x} = \frac{\lim \Delta s/\Delta x}{\lim l/\Delta x} = 1.$$

That is,

$$\boxed{\frac{\text{arc length}}{\text{chord length}} \to 1,}$$

as chord length $\to 0$. This is the intuitive property referred to above. If we assume this property of arc length, then the integral formula can be *proved*, reversing the above steps and using the First Fundamental Theorem.

The arc length formula leads to complicated integrands and there are very few functions for which we can work out the integral explicitly. Even simple changes of scale can affect the complexity of the calculation. For example, we can easily find the length of the graph of $f(x) = (e^x + e^{-x})/2$, but not $g(x) = e^x + e^{-x}$!

Example 1   Find the length of the graph of

$$f(x) = \frac{e^x + e^{-x}}{2}$$

from $x = 0$ to $x = a$.

Solution   Here $f'(x) = (e^x - e^{-x})/2$, and

$$1 + [f'(x)]^2 = 1 + \frac{e^{2x} - 2 + e^{-2x}}{4} = \frac{e^{2x} + 2 + e^{-2x}}{4}$$

$$= \left[\frac{e^x + e^{-x}}{2}\right]^2.$$

The arc length is therefore

$$\int_0^a \sqrt{1 + [f'(x)]^2}\, dx = \int_0^a \frac{e^x + e^{-x}}{2}\, dx = \left.\frac{e^x - e^{-x}}{2}\right]_0^a$$

$$= \frac{e^a - e^{-a}}{2}.$$ □

Example 2   Find the length of the graph of $\ln x$ from $x = 1$ to $x = 2$.

Solution   We have to integrate

$$\int \sqrt{1 + [f'(x)]^2}\, dx = \int \sqrt{1 + \left(\frac{1}{x}\right)^2}\, dx = \int \frac{\sqrt{x^2 + 1}}{x}\, dx.$$

Setting first $x = \tan \theta$, $dx = \sec^2\theta\, d\theta$, and then $u = \cos \theta$, $du = -\sin \theta\, d\theta$, this becomes

$$\int \frac{\sec^3\theta\, d\theta}{\tan \theta} = \int \frac{\sin \theta\, d\theta}{\sin^2\theta \cos^2\theta} = -\int \frac{du}{(1 - u^2)u^2}$$

$$= -\int \left[\frac{1}{2(1 + u)} + \frac{1}{2(1 - u)} + \frac{1}{u^2}\right] du$$

$$= \frac{1}{2}\ln\frac{1 - u}{1 + u} + \frac{1}{u}.$$

Retracing the substitutions $u = \cos\theta$, $x = \tan\theta$, we obtain $u = 1/\sqrt{x^2 + 1}$. The arc length is then

$$\left[\frac{1}{2}\ln\left(\frac{\sqrt{x^2 + 1} - 1}{\sqrt{x^2 + 1} + 1}\right) + \sqrt{x^2 + 1}\right]_1^2$$

$$= \left[\ln\left(\frac{\sqrt{x^2 + 1} - 1}{x}\right) + \sqrt{x^2 + 1}\right]_1^2$$

$$= \ln\frac{\sqrt{5} - 1}{2(\sqrt{2} - 1)} + \sqrt{5} - \sqrt{2}. \qquad \Box$$

Problem 14 asks for this computation to be redone using the definite integral substitution rule from Section 1. Keep this rule in mind for all the problems.

Example 3 Find the length of the graph of

$$f(x) = \int_1^x \sqrt{t^3 - 1}\, dt$$

over $[1, 4]$.

Solution Here $f'(x) = \sqrt{x^3 - 1}$, by the Second Fundamental Theorem, so

$$\sqrt{1 + [f'(x)]^2} = \sqrt{1 + (x^3 - 1)} = x^{3/2}.$$

The length is thus

$$\int_1^4 x^{3/2}\, dx = \frac{2}{5}x^{5/2}\bigg]_1^4 = \frac{2}{5}(32 - 1) = \frac{62}{5}. \qquad \Box$$

PROBLEMS FOR SECTION 4

Find the lengths of the graphs of the following equations (or functions), over the given intervals.

1. $y = \ln\cos x$, $[0, \pi/4]$

2. $y = x^{1/2} - x^{3/2}/3$, $[1, 2]$

3. $y = ax^{3/2}$, $[0, 1]$

4. $3y = 2(x^2 + 1)^{3/2}$, $[0, a]$

5. $y^3 = x^2$, $[0, 8]$

6. $y = (1 - x^{2/3})^{3/2}$, $[0, 1]$

7. $y = e^x$, $[0, 1]$

8. $y = ax^2$, $[0, 1]$

9. $y = mx + k$, $[a, b]$

10. $y = \sqrt{x}$, $[1, 4]$

11. $8y = x^4 + \dfrac{2}{x^2}$, $[1, 2]$

12. $y = \ln(1 - x^2)$, $[0, 1/2]$

13. $y = \dfrac{x^2}{8} - \ln x$, $[1, 2]$

14. Redo Example 2 using the definite integral substitution rule.

15. $f(x) = \displaystyle\int_1^x \sqrt{t^4 - 1}\, dt$, $[1, 2]$

16. $f(x) = \displaystyle\int_0^x \sqrt{e^t - 1}\, dt$, $[0, a]$

17. $f(x) = \int_a^x \sqrt{[g'(t)]^2 - 1}\, dt$, $[a, b]$, where $|g'(t)| \geq 1$ on $[a, b]$.

18. If $f(x) = (e^x + e^{-x})/2$, show that the length of the graph of f from $x = 0$ to $x = a$ is $f'(a)$.

19. Let f be a smooth decreasing function whose graph runs from $(0, 1)$ to $(1, 0)$. Prove that the length of the graph of f necessarily lies between $\sqrt{2}$ and 2. [*Hint*: Show that the length of any inscribed polygonal arc necessarily lies between these numbers.]

20. In the same way, show, for any two positive numbers a and b, that if the graph of f decreases from $(0, b)$ to $(a, 0)$, then the length of the graph necessarily lies between $\sqrt{a^2 + b^2}$ and $a + b$. What would be the corresponding statement for an increasing graph?

21. Over the interval $[0, 1]$, the graph of $f(x) = x^n$ increases smoothly from 0 to 1. Let a_n be the point where f has the value $\sqrt{1/n}$: $f(a_n) = \sqrt{1/n}$.

a) Prove that $a_n > 1 - \sqrt{1/n}$. [*Hint*: Show that if $a_n \leq 1 - \sqrt{1/n}$, then the area under the graph must be larger than $1/n$. But compute the area.]

b) Show that the length of the graph of f over $[0, 1]$ is greater than $2 - 2\sqrt{1/n}$ (because there is a simple inscribed polygon with length greater than this).

22. Show, partly on the basis of the above result, that the bounds 2 and $\sqrt{2}$ in Problem 9 cannot be improved upon.

23. Let f and g be mutually inverse functions with domain I and J, respectively. Show that the length of the graph of g over J equals the length of the graph of f over I.

5

THE AREA OF A SURFACE OF REVOLUTION

We consider now the surfaces that are generated when the graphs in Fig. 1 are rotated about the x-axis. The graph of f sweeps out a surface S, and the approximating polygonal arc sweeps out an approximating surface made up of pieces of cones. The area of S is evidently approximated by the sum of the areas of these conical segments. The kth conical piece has the width l_k, where

$$l_k = \sqrt{1 + [f'(c_k)]^2}\, \Delta x.$$

Its points trace out circles having the approximate circumference $2\pi f(c_k)$. Its area is therefore approximately the product of these two

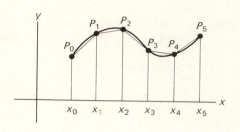

Figure 1

dimensions. So for the area of the kth slice of S we have the approximation

$$\Delta A_k \approx 2\pi f(c_k)\sqrt{1 + [f'(c_k)]^2}\,\Delta x.$$

Summing over k gives a Riemann sum approximation to A, and hence the integral formula

$$A = 2\pi \int_a^b f(x)\sqrt{1 + [f'(x)]^2}\,dx.$$

The sketchy derivation outlined above can be improved later on (Chapter 17) when the general notion of surface area is discussed. See Problems 19 and 20 in Section 6 of that chapter.

Example 1 Find the area of the surface generated by rotating the semi-parabola $y = \sqrt{x}$ about the x-axis, from $x = 0$ to $x = a$.

Solution

$$A = 2\pi \int_0^a \sqrt{x} \cdot \sqrt{1 + 1/4x}\,dx$$

$$= 2\pi \int_0^a \frac{1}{2}\sqrt{4x + 1}\,dx$$

$$= \pi \cdot \frac{1}{4} \cdot \frac{2}{3}(4x + 1)^{3/2}\Big]_0^a = \frac{\pi}{6}[(4a + 1)^{3/2} - 1]. \qquad \square$$

It can be shown (Problem 13) that if f is monotone (and hence invertible) over $[a, b]$, and if the graph of f is rotated about the y-axis, then the area of the surface swept out is

$$A = 2\pi \int_a^b x\sqrt{1 + [f'(x)]^2}\,dx.$$

Example 2   Find the surface area generated by rotating about the y-axis the arc of the parabola $y = x^2$ that lies over $[0, 1]$.

Solution

$$A = 2\pi \int_0^1 x\sqrt{1 + 4x^2}\, dx$$

$$= \frac{\pi}{4} \cdot \frac{2}{3} (1 + 4x^2)^{3/2} \Big]_0^1$$

$$= \frac{\pi}{6} (5^{3/2} - 1). \qquad \square$$

PROBLEMS FOR SECTION 5

The following graphs are to be rotated about the x-axis, over the given interval. In each case find the area of the surface swept out.

1. $3y = x^3$; $[0, 1]$ **2.** $8y = x^4 + 2/x^2$; $[1, 2]$ **3.** $y = x^3/3 + 1/4x$; $[1, 2]$

4. $y = x^{1/2} - x^{3/2}/3$; $[1, 2]$ **5.** $y = e^x$; $[0, \ln 2]$ **6.** $y = \sin x$; $[0, \pi]$

7. $y = 3x^{1/3}$; $[0, 8]$ **8.** $y = 1/x$; $[1, 2]$ **9.** $2y = 3x^{2/3}$; $[0, 8]$

10. $y = x^{3/2}$; $[1, 4]$

For each of the following equations, the graph over the given interval is to be rotated about the y-axis. Find the area of the surface thus generated.

11. $3y = x^{3/2}$; $[0, 5]$ **12.** $y = kx^2$; $[0, a]$

13. $8y = x^4 + 2/x^2$; $[1, 2]$ **14.** $y = x^3/3 - 1/4x$; $[1, 2]$

15. $y = x^{1/2} - x^{3/2}/3$; $[1, 2]$ **16.** $y = x^2/8 - \ln x$; $[1, 2]$

17. $y = \ln(1 - x^2)$; $[0, 1/2]$ **18.** $y = 3x^{1/3}$; $[0, 8]$

19. $y = (1 - x^{2/3})^{3/2}$; $[0, 1]$ **20.** $3y = 2(x^2 + 1)^{3/2}$; $[0, a]$

21. Show that the surface area of a sphere of radius r is $4\pi r^2$.

22. Show that the surface area of the spherical strip cut off from a sphere of radius r by two parallel planes a distance d apart $(d < 2r)$ is $2\pi rd$.

23. Prove the second surface area formula from the first. (Start with the first formula for the function g inverse to f, over the corresponding interval $[A, B]$. Then make the change of variables $x = f(y)$, $dx = f'(y)\, dy$.)

6
FLUID PRESSURE

In a tank of water the water pressure increases with depth. A cubic foot of water weighs 62.4 pounds, and at a point x feet below the surface the pressure is $62.4x$ pounds per square foot (lb/ft^2).

If the bottom of the container is horizontal, and water is d feet deep, then the pressure on the bottom is $62.4d$ lb/ft^2 at all points of the bottom. The total force exerted by the water on the bottom is then

$$F = \text{constant pressure times area}$$
$$= 62.4d\ A \text{ pounds},$$

where A is the area of the bottom in square feet.

Pressure in a fluid is exerted equally in all directions. Thus, at a point 5 ft below the surface, water at the left is pressing to the right with a pressure of 312 lb/ft$^2 = 2.17$ lb/in^2, and water at the right is pressing to the left with the same pressure. Of course these pressures balance and there is no resulting movement of the water.

However, if a balloon filled with air were submerged to a depth of 5 ft, then its volume would be greatly diminished, to the point where the air pressure in the balloon balanced the water pressure outside the balloon.

Related to this omni-directional property of fluid pressure is a feature that at first may seem paradoxical. In any connected channel of standing water, the pressure at any point is determined entirely by how far it is below the level of the water surface. It is *wholly independent* of the *quantity* of water at higher elevations. For example, in the configurations outlined in Fig. 1 only a tiny fraction of the water is high, and yet this tiny fraction determines the pressure that is experienced throughout the great bulk of the water.

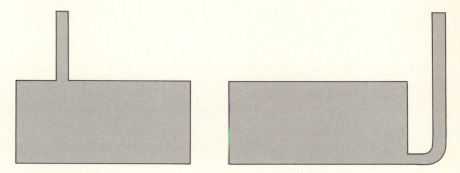

Figure 1

Now suppose that a vertical flat plate is immersed in water and let $f(x)$ be its width at the depth x (see Fig. 2). Along an incremental strip between the depths x and $x + \Delta x$, the pressure will have the approximately constant value $62.4x$, and the force ΔF exerted by the

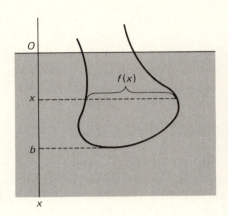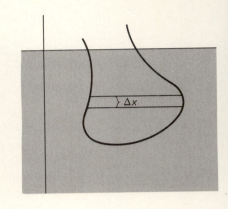

Figure 2

water on one side of this strip will be approximately this constant value times the area of the strip. Since the strip area is approximately $f(x)\,\Delta x$, we have

$$\Delta F \approx p\,\Delta A \approx 62.4x \cdot f(x)\,\Delta x.$$

The total force exerted by the water on one side of the plate is given approximately by a sum of such incremental strip forces:

$$F \approx \Delta x \sum 62.4x_j f(x_j).$$

Therefore F should be given exactly by the integral formula:

$$F = 62.4 \int_0^b x f(x)\,dx.$$

Example 1   A cistern has square ends, 6 ft × 6 ft. Find the total force against one end when the cistern is full of water.

Solution   Here $f(x)$ has the constant value 6, so

$$F = 62.4 \int_0^b x f(x)\,dx = 62.4 \int_0^6 6x\,dx \approx 6739\,\text{lb}. \qquad \square$$

Example 2   A trough with vertical ends in the shape of equilateral triangles is filled with water to a depth of 4 ft. What is the force of the water against one end?

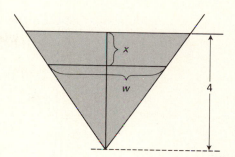

Figure 3

Solution The altitude h and base b of an equilateral triangle are related by

$$h = \frac{\sqrt{3}}{2}\,b.$$

Therefore the trough width $f(x)$ at the depth of x feet (below the water surface) is

$$f(x) = \frac{2}{\sqrt{3}}\,(4 - x).$$

Therefore

$$F = 62.4 \int_0^b xf(x)\,dx$$

$$= 62.4 \int_0^4 x\frac{2}{\sqrt{3}}(4 - x)\,dx$$

$$= \frac{124.8}{\sqrt{3}} \left(2x^2 - \frac{x^3}{3}\right)_0^4$$

$$= \frac{124.8}{\sqrt{3}} \cdot \frac{32}{3} \approx 768.57 \text{ lb.} \qquad \square$$

Example 3 Suppose the trough in Example 2 is 10 ft long. Find the total force on the *sides* of the trough.

Solution The altitude h and side s of an equilateral triangle are related by $h = \sqrt{3}\,s/2$. Therefore, the area of the trough's side strips between the depths of x and $x + \Delta x$ is

(total side length)(slant height between those depths)

$$= (2 \cdot 10)\frac{2}{\sqrt{3}}\,\Delta x = \frac{40}{\sqrt{3}}\,\Delta x.$$

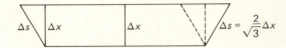

Figure 4

This replaces the strip area $f(x)\,\Delta x$ of our earlier discussion, and the integral for the total force is now

$$F = \int_0^4 (62.4x)\left(\frac{40}{\sqrt{3}}\,dx\right) = \frac{2000}{\sqrt{3}}\,62.4 \approx 72{,}000\ \text{lb.} \qquad \square$$

PROBLEMS FOR SECTION 6

1. A vat is in the shape of a trough with triangular ends. The triangles are equilateral, with altitude 6 feet. What is the force on one end when the vat is full of water?

2. The vat in Problem 1 is 12 feet long. What is the total force on its sides?

3. A conical tank filled with water is 10 feet deep and 10 feet in diameter across the top. What is the total force exerted by the water on the tank?

4. The inner face of a dam is inclined 30° from the vertical and is in the shape of a trapezoid. It is 900 feet across the top, 100 feet across the bottom, and 100 vertical feet from top (water level) to bottom. What is the total force of the water on the dam?

5. A tank is in the form of a right circular cylinder lying horizontally. Its ends have diameter 4 feet. The tank is full of alcohol weighing $50\ \text{lb/ft}^3$. What is the force on one end of the tank?

6. In Problem 5 what is the total force on the curving side of the tank if the tank is 8 feet long?

7. In the same problem, what is the force against one end if the tank is half full?

8. A spherical tank of diameter 16 feet is half full of a heavy liquid weighing $70\ \text{lb/ft}^3$. What is the total force exerted by the liquid on the tank?

9. Find the total force in Problem 8 when the liquid is 12 feet deep.

7
IMPROPER
INTEGRALS

We approach the notion of an improper integral through questions about area.

Example 1 Find the total area under the graph of $y = 1/x^2$ from $x = 1$ to $x = \infty$.

Solution As worded, the problem doesn't make sense, since the only area formula we know is for a finite region. However, the area from $x = 1$ to $x = b$ is

$$\int_1^b \frac{dx}{x^2} = -\frac{1}{x}\bigg]_1^b = 1 - \frac{1}{b}.$$

Then

$$\lim_{b \to \infty} [\text{area from 1 to } b] = \lim_{b \to \infty}\left(1 - \frac{1}{b}\right) = 1.$$

This limit is what we *mean* by the area from 1 to ∞. So here we have an example of an infinitely long region with a finite area. See Fig. 1.

Figure 1

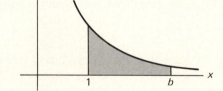

Example 2 Show that the region under the graph of $y = 1/\sqrt{x}$ from $x = 1$ to ∞ has infinite area.

Solution As in Example 1, the area A from 1 to ∞ is computed as the limit as $b \to \infty$ of the area from 1 to b (Fig. 2). So

$$A = \lim_{b \to \infty}\int_1^b \frac{dx}{\sqrt{x}} = \lim_{b \to \infty} 2\sqrt{x}\bigg]_1^b$$

$$= \lim_{b \to \infty}(2\sqrt{b} - 2) = \infty.$$

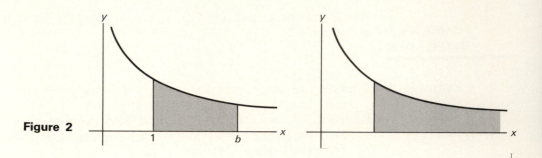

Figure 2

The limit $\lim_{b\to\infty} \int_a^b f$ is called an *improper integral* and is designated $\int_a^\infty f$. Thus, by definition,

$$\int_a^\infty f(x)\,dx = \lim_{b\to\infty} \int_a^b f(x)\,dx.$$

If the limit exists (as a finite number l) we say that the improper integral *converges*, and that its value is l. If the limit does not exist, we say that the improper integral *diverges*.

In the above two examples we reasoned that areas of regions extending to infinity are given by improper integrals.

Another type of improper integral is illustrated in the next example.

Example 3   Find the area under the graph of $y = 1/\sqrt{x}$ from $x = 0$ to $x = 1$.

Solution   This region is infinitely long along the y-axis, rather than along the x-axis (Fig. 3). However, we can follow essentially the same scheme. We first compute the area over the interval $[a, 1]$, where a is a small positive number. Over this interval the integrand $1/\sqrt{x}$ is continuous and there is no problem. Then we can let a tend to zero and see what happens "in the limit." We have

$$A = \lim_{a\to 0^+} \int_a^1 \frac{dx}{\sqrt{x}} = \lim_{a\to 0^+} 2\sqrt{x}\,\Big]_a^1$$

$$= \lim_{a\to 0^+} (2 - 2\sqrt{a}) = 2.$$

The area we were looking for is thus 2.

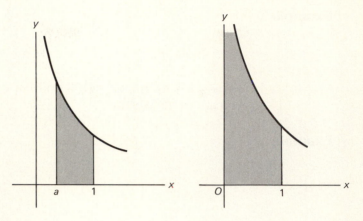

Figure 3

In general:

If f is continuous on $(a, c]$, but $f(x) \to \infty$ as $x \downarrow a$, then the improper integral $\int_a^c f$ is defined as the limit

$$\int_a^c f(x)\, dx = \lim_{b \downarrow a} \int_b^c f(x)\, dx.$$

Thus the area in Example 3 was given by the improper integral $\int_0^1 dx/\sqrt{x}$.

Example 4 The improper integral $\int_0^1 dx/x^2$ is divergent:

$$\int_0^1 \frac{dx}{x^2} = \lim_{a \to 0^+} \int_a^1 \frac{dx}{x^2} = \lim_{a \to 0} \left. -\frac{1}{x} \right]_a^1$$

$$= \lim_{a \to 0^+} \left(\frac{1}{a} - 1 \right) = +\infty.$$

It is useful to keep in mind what kind of improper integral we get when the integrand is *any* power of x. We have (assuming $a > 0$)

$$\int_a^\infty x^k\, dx = \begin{cases} -\dfrac{a^{k+1}}{k+1} & \text{if} \quad k < -1, \\[2mm] \infty & \text{if} \quad k \geq -1. \end{cases}$$

$$\int_0^a x^k\, dx = \begin{cases} \infty & \text{if} \quad k \leq -1, \\[2mm] \dfrac{a^{k+1}}{k+1} & \text{if} \quad k > -1. \end{cases}$$

Changing variables is permitted in improper integrals.

Example 5
$$\int_2^\infty \frac{dx}{x(\ln x)^2} = \int_{\ln 2}^\infty \frac{du}{u^2} \qquad \left(\begin{aligned} u &= \ln x, \\ du &= dx/x \end{aligned} \right)$$
$$= 1/\ln 2. \qquad \square$$

Sometimes an improper integral becomes a definite integral under a change of variables.

Example 6
$$\int_0^\infty \frac{dx}{1 + x^2} = \int_0^{\pi/2} d\theta \qquad \left(\begin{aligned} \theta &= \arctan x, \\ d\theta &= dx/(1 + x^2) \end{aligned} \right)$$
$$= \theta \Big]_0^{\pi/2} = \pi/2. \qquad \square$$

Comparison Tests   We can compare the size of two improper integrals without computing them:

> If $0 \le f(x) \le g(x)$ on $[a, \infty)$, then
> $$\int_a^\infty f(x) \, dx \le \int_a^\infty g(x) \, dx.$$

This is obvious on the basis of areas. The hypothesis says that the region under the graph of f is part of the region under the graph of g; we conclude that the f region has the smaller area. The formal proof is an application of property (4) from the first section.

This remark gives us a way of showing that $\int_a^\infty f(x) \, dx$ is convergent even when we can't evaluate the integral. We look for a larger (simpler) function g that we know has a convergent integral. Then

$$\int_a^\infty f \le \int_a^\infty g < \infty.$$

Or, if we have reason to think that $\int_a^\infty f$ is probably divergent, then we interchange the roles of f and g in the above inequalities and look for a *smaller* function g that we know has a *divergent* integral. Then

$$\infty = \int_a^\infty g \le \int_a^\infty f,$$

so $\int_a^\infty f = \infty$.

Example 7   We could compute

$$\int_a^\infty \frac{x\,dx}{x^3 + x^2 + x + 1}$$

because the integrand is a rational function, but it would be complicated. However, we can easily see that the integral converges, because the integrand is less than $x/x^3 = 1/x^2$, and

$$\int_a^\infty \frac{dx}{x^2} = \frac{1}{a}.$$

The integrals of x^k, listed earlier, are useful for these comparisons.

Example 8   Determine whether the improper integral

$$\int_1^\infty \sqrt{\frac{1 + x^2}{1 + x^4}}\,dx$$

is convergent or divergent.

Solution   When x is very large the function under the radical is approximately $x^2/x^4 = 1/x^2$, so the integrand is approximately $\sqrt{1/x^2} = 1/x$. We therefore expect divergence, in comparison with $\int_1^\infty dx/x = \infty$.

In fact, the rough approximations made above are actually inequalities in the right direction:

$$\frac{1 + x^2}{1 + x^4} \geq \frac{x^2}{x^4} = \frac{1}{x^2}$$

when $1 \leq x$. So the integrand is $\geq 1/x$ on $[1, \infty)$ and the improper integral is $\geq \int_1^\infty dx/x = \infty$.

Example 9   The integral

$$\int_0^\infty e^{-x^2/2}\,dx$$

cannot be computed by the techniques we know so far. However, we can show that it is finite quite easily. We just write

$$\int_0^\infty e^{-x^2/2}\,dx = \int_0^2 e^{-x^2/2}\,dx + \int_2^\infty e^{-x^2/2}\,dx,$$

where we know that the first integral on the right-hand side is finite. Then we note

$$e^{-x^2/2} \le e^{-x}$$

on $[2, \infty)$, because if $x \ge 2$ then $x^2/2 \ge x$. Thus $-x^2/2 \le -x$ and

$$\int_2^\infty e^{-x^2/2} \, dx \le \int_2^\infty e^{-x} \, dx = e^{-2}.$$

Therefore the given integral is convergent.   □

PROBLEMS FOR SECTION 7

Determine whether each of the following improper integrals is convergent or divergent, and calculate its value if convergent.

1. $\displaystyle\int_1^\infty \frac{dx}{x^3}$

2. $\displaystyle\int_1^\infty \frac{dx}{x^{3/2}}$

3. $\displaystyle\int_1^\infty \frac{dx}{x^{2/3}}$

4. $\displaystyle\int_0^\infty \frac{dx}{1+x^2}$

5. $\displaystyle\int_0^\infty \frac{dx}{1+x}$

6. $\displaystyle\int_0^1 \frac{dx}{x^{2/3}}$

7. $\displaystyle\int_0^1 \frac{dx}{x^3}$

8. $\displaystyle\int_0^1 \frac{dx}{\sqrt{1-x^2}}$

9. $\displaystyle\int_0^{\pi/2} \sec x \tan x \, dx$

10. $\displaystyle\int_0^{\pi/2} \sec^2 x \, dx$

11. $\displaystyle\int_{-1}^1 \frac{dx}{x^2}$

12. $\displaystyle\int_{-1}^1 \frac{dx}{x^{2/3}}$

13. We know (Problem 3) that the region under the graph of $y = x^{-2/3}$ from $x = 1$ to ∞ has infinite area. Yet the volume generated by revolving this region about the x-axis is finite. Prove this, by calculating the volume.

14. The region under the graph of $y = 1/x$, from $x = 1$ to $x = \infty$, is rotated about the x-axis. Show that the resulting solid has finite volume but infinite surface area.

Prove the following inequalities by comparing integrals with integrals.

15. $\displaystyle\int_0^\infty \frac{dx}{x^2 + x + 1} < \frac{\pi}{2}$

16. $\displaystyle\int_2^\infty \frac{x \, dx}{x^3 + 1} < \frac{1}{2}$

17. $\displaystyle\int_2^\infty \frac{\sqrt{x} \, dx}{x^2 + 1} < \sqrt{2}$

18. $\displaystyle\int_1^\infty e^{-x^2} \, dx < \frac{1}{2e}$

Determine whether or not the following improper integrals are convergent or divergent.

19. $\displaystyle\int_1^\infty \frac{dx}{\sqrt{1+x^3}}$

20. $\displaystyle\int_1^\infty \left(\frac{\sin \theta}{\theta}\right)^2 d\theta$

21. $\displaystyle\int_0^\infty e^{-x^3} \, dx$

22. $\displaystyle\int_1^\infty \frac{\ln t}{t^2}\, dt$

23. $\displaystyle\int_1^\infty \frac{dx}{\ln x\sqrt{1+x^2}}$

24. $\displaystyle\int_2^\infty \frac{dx}{(\ln x)^3}$

25. $\displaystyle\int_1^\infty \sin^2\!\Big(\frac{1}{x}\Big)\, dx$

26. $\displaystyle\int_1^\infty \frac{\sin(u^{-1/2})}{u}\, du$

27. $\displaystyle\int_0^\infty e^x e^{-x^2}\, dx$

Evaluate the following improper integrals.

28. $\displaystyle\int_a^\infty e^{-\sqrt{x}}\, dx$

29. $\displaystyle\int_2^\infty \frac{dx}{x\ln x}$

30. $\displaystyle\int_2^\infty \frac{dx}{x(\ln x)^2}$

31. $\displaystyle\int_1^\infty \frac{e^{-\sqrt{x}}}{\sqrt{x}}\, dx$

32. $\displaystyle\int_1^\infty \sqrt{x}\, e^{-\sqrt{x}}\, dx$

33. $\displaystyle\int_2^\infty \frac{dx}{x(\ln x)^a}$

34. $\displaystyle\int_2^\infty \frac{dx}{x\ln x[\ln(\ln x)]^2}$

8
THE
TRAPEZOIDAL
APPROXIMATION

By now the reader has experienced some of the difficulties one can meet in evaluating a definite integral $\int_a^b f$, and can better appreciate the need for numerical estimates. If f is a function, like $f(x) = e^{-x^2}$, for which no elementary antiderivative exists, then a direct numerical evaluation is our only recourse. But even if we can integrate f, the antiderivative F that we find may be very complicated, and the evaluation

$$\int_a^b f = F(b) - F(a)$$

may be more difficult than using some direct numerical estimate.

The only numerical method that we have considered so far is the approximation of $\int_a^b f$ by a Riemann sum. We found that the midpoint sum, in particular, is quite accurate. We can do better still, however, by slightly modifying the standard Riemann sum.

The most obvious change to make is to replace each rectangle associated with a Riemann sum by a corresponding trapezoid, as shown in Fig. 1. The area of a trapezoid is $h(l_1 + l_2)/2$, where l_1 and l_2 are the lengths of the parallel sides, i.e., the two base lengths.

Thus, the sum of the two trapezoidal areas on the right gives the estimate

$$\int_a^b f \approx \Delta x\left[\frac{f(a)}{2} + f(m) + \frac{f(b)}{2}\right].$$

Figure 1

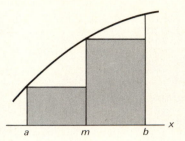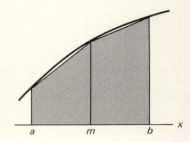

We expect from the look of the geometric figures that this estimate will be much better than the original Riemann-sum estimate.

In general, we subdivide the interval $[a, b]$ into n equal subintervals of common length $\Delta x = (b - a)/n$ the subdividing points being $a = x_0, x_1, \ldots, x_n = b$ (Fig. 2).

Figure 2

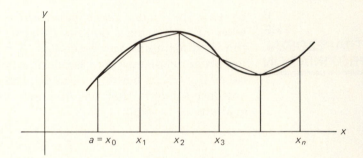

Then we add up all the individual trapezoidal estimates

$$\int_{x_{i-1}}^{x_i} f \approx \Delta x \frac{f(x_{i-1}) + f(x_i)}{2},$$

from $i = 1$ to $i = n$, and end up with the approximation

$$\int_a^b f \approx \Delta x \left[\frac{f(x_0)}{2} + f(x_1) + f(x_2) + \cdots + f(x_{n-1}) + \frac{f(x_n)}{2} \right]$$

$$= \Delta x \left[\frac{f(a) + f(b)}{2} + \sum_{k=1}^{n-1} f(x_k) \right].$$

This formula is called the *trapezoidal rule*. It is in a class with the midpoint Riemann-sum estimate, as the following error formula shows.

THEOREM 5 **Error Formula for Trapezoidal Rule** *If the second derivative of f exists and is bounded by B over the interval* $[a, b]$, *then the error in using the trapezoidal rule to estimate* $\int_a^b f$ *is at most*

$$\frac{B(b-a)}{12}(\Delta x)^2$$

in magnitude. Furthermore, the trapezoidal sum is larger (smaller) than $\int_a^b f$ *if f is concave up (down) on* $[a, b]$.

We shall sketch the proof at the end of the section.

Example 1 We use the same fivefold subdivision of $[1, 2]$ as in the last section of Chapter 6 to estimate $\ln 2$ by the trapezoidal rule. We have

$$\ln 2 = \int_1^2 \frac{dx}{x} \approx \frac{1}{5}\left[\frac{1}{2} + \frac{1}{1.2} + \frac{1}{1.4} + \frac{1}{1.6} + \frac{1}{1.8} + \frac{1}{4}\right]$$
$$= 0.69563\cdots$$

with a negative error less than

$$\frac{2}{12}\cdot 1 \cdot \left(\frac{1}{5}\right)^2 = \frac{1}{150} = 0.00666\cdots$$

in magnitude. Thus

$$0.68896 < \ln 2 < 0.69564.$$

This suggests that $\ln 2 = 0.69$ to the nearest two decimal places but doesn't prove it. We *can* say (see Chapter 5, Section 7) that

$$\ln 2 \approx 0.692$$

with an error less than 0.004 in magnitude. $\qquad\square$

The trapezoidal rule can be vastly improved by including a correction that takes into account the exact directions of the graph at the two endpoints. Here is the result.

THEOREM 6 **The Modified Trapezoidal Rule** *If the fourth derivative* $f^{(4)}$ *is continuous and bounded by K on the interval* $[a, b]$, *then*

$$\int_a^b f \approx \text{trapezoidal estimate} - (\Delta x)^2 \frac{f'(b) - f'(a)}{12},$$

with an error at most

$$\boxed{\frac{K(b-a)}{720}(\Delta x)^4.}$$

If the sign of $f^{(4)}$ is constant on $[a, b]$, then the error has the same sign.

Example 2   Once again we calculate $\ln 2$ using a fivefold subdivision. We have (using Example 1)

$$\ln 2 = \int_1^2 \frac{dx}{x} \approx \text{trapezoidal approximation} - \left(\frac{1}{5}\right)^2 \frac{1 - 1/4}{12}$$

$$= 0.69563 \cdots - 0.0025$$

$$= 0.69313 \cdots.$$

Here $f^{(4)}(x) = 24/x^5$, which is positive and bounded by 24 over the interval $[1, 2]$. The error is therefore positive and less than

$$\frac{K(b-a)}{720}(\Delta x)^4 = \frac{24}{720}\left(\frac{1}{5}\right)^4 = 0.00005 \cdots.$$

Thus

$$0.69313 < \ln 2 < 0.69319,$$

and

$$\ln 2 = 0.6931 \cdots. \qquad \square$$

Discussion of the Proofs   The proof of the trapezoidal error formula depends on the following identity:

$$\int_a^b f = (b - a)\frac{f(a) + f(b)}{2} + \frac{1}{2}\int_a^b (x - a)(x - b)f''(x)\, dx.$$

To check this equation, start with the integral on the right and integrate by parts twice. Now $(x - a)(x - b)$ has constant sign over (a, b), so

$$\int_a^b (x - a)(x - b)f''(x)\, dx = f''(X)\int_a^b (x - a)(x - b)\, dx$$

for some X in (a, b), by the weighted mean-value theorem for integrals. (See Section 1.) Furthermore, we find upon integration that

$$\int_a^b (x - a)(x - b)\, dx = -\frac{(b - a)^3}{6}.$$

Therefore

$$\int_a^b f = (b - a)\frac{f(a) + f(b)}{2} - \frac{f''(X)}{12}(b - a)^3.$$

This is the basic error formula. When we write it down for all the subintervals of a subdivision and add up the resulting n equations, we get

$$\int_a^b f = \text{trapezoidal sum} - \frac{(\Delta x)^3}{12}\sum_1^n f''(X_j).$$

Since $|f''(X_j)| \le K$ for each j, and since $n\,\Delta x = b - a$, the term on the right is bounded by

$$\frac{(\Delta x)^2}{12}K(b - a),$$

and we are done.

The last displayed equation above also gives a clue to the modified formula. The right-hand term contains a Riemann sum that approximates $\int_a^b f'' = f'(b) - f'(a)$. Replacing the above Riemann sum by the integral, we have the modified formula.

The proof of the modified formula goes in exactly the same way as before, starting this time from

$$\int_a^b [(x - a)(x - b)]^2 f^{(4)}(x)\, dx$$

and integrating by parts *four* times.

PROBLEMS FOR SECTION 8

1. Calculate the trapezoidal approximation to ln 2 arising from a tenfold subdivision of $[1, 2]$. Show that this sum approximates ln 2 with a negative error less than 0.002 in magnitude.

2. How fine a subdivision do we need to take in order to ensure that the trapezoidal sum will approximate ln 2 to the nearest four decimal places (i.e., with an error less than $0.00005 = 5(10)^{-5}$)?

▦ **3.** Calculate the trapezoidal approximation to

$$\frac{\pi}{4} = \int_0^1 dx/(1 + x^2)$$

arising from a tenfold subdivision of $[0, 1]$. Show that the error is less than 0.002 in magnitude.

5. Compute the trapezoidal approximation to $\int_0^b x^2\, dx$ for the trivial subdivision of $[0, b]$ (no subdividing points). Compare with the value of the integral, and hence show that the error formula cannot be improved.

7. What would the error be in the estimate of $\ln 2$ provided by the modified trapezoidal rule if $[1, 2]$ were divided into 10 subintervals? 20 subintervals?

9. Show that the modified trapezoidal estimate gives the exact answer if f is a polynomial of degree at most 3.

11. Assume the above problem, and apply the modified trapezoidal rule to a fourth-degree polynomial using the trivial subdivision of $[a, b]$ (no subdividing points). State the result as a lemma about fifth-degree polynomials.

The identity

$$\pi = 4 \arctan 1 = 4 \int_0^1 \frac{dx}{1 + x^2}$$

can be used to calculate π. In the following problems we consider the tenfold subdivision of $[0, 1]$ by the one-place decimals $0.0, 0.1, \ldots, 0.9, 1.0$. The calculation of the higher derivatives of $f(x) = 1/(1 + x^2)$ becomes rather complicated, and you may assume that $|f^{(4)}(x)| \le 24$ on the interval $[0, 1]$. The correct beginning of the decimal expansion of π is

$$\pi = 3.141592653589 \cdots.$$

▦ **12.** Show that the error in the approximation of $\pi = 4 \int_0^1 dx/(1 + x^2)$ is

a) less than 0.0067 when the unmodified trapezoidal sum is used;

b) less than $0.000014 = 1.4(10)^{-5}$ when the modified trapezoidal rule is used.

4. Show that the trapezoidal approximation is exact (i.e., has zero error) for a linear integrand $f(x) = Ax + B$.

6. Compute the trapezoidal approximation to $\int_0^1 x^3\, dx$ for the fivefold subdivision of $[0, 1]$. Show that the actual error is exactly one half the error bound provided by the error formula.

▦ **8.** Estimate $\ln 2$ by the modified trapezoidal rule, using a sixfold subdivision of $[1, 2]$. Compute the error estimate as the first step, and use it as a guide in deciding how far to carry out the decimal expansions.

10. Show that the error is exactly $(\Delta x)^4(b - a)/720$ times the constant value of $f^{(4)}(x)$ if f is a fourth-degree polynomial. Show this first for $f(x) = x^4$. Then extend the result to the general case by virtue of Problem 9.

A desk calculator gave the following values for $1/(1 + x^2)$ at the subdividing points:

1/1.01	0 . 9 9 0 0 9 9 0 0 9 9 0 0
1/1.04	0 . 9 6 1 5 3 8 4 6 1 5 3 8
1/1.09	0 . 9 1 7 4 3 1 1 9 2 6 6 0
1/1.16	0 . 8 6 2 0 6 8 9 6 5 5 1 7
1/1.2	0 . 8 0 0 0 0 0 0 0 0 0 0 0
1/1.36	0 . 7 3 5 2 9 4 1 1 7 6 4 7
1/1.49	0 . 6 7 1 1 4 0 9 3 9 5 9 7
1/1.64	0 . 6 0 9 7 5 6 0 9 7 5 6 0
1/1.81	0 . 5 5 2 4 8 6 1 8 7 8 4 5

It is clear from the top line that the calculator did not round off at the last place, and the above values are therefore the twelfth-place truncations in the decimal expansions. So each approximates its entry with an error lying between 0 and 10^{-12}. Note that one row is exact.

13. Compute the sum of the above twelfth-place decimals, and modify it to correspond to the trapezoidal sum. (Two entries are missing, as well as the factor Δx.) Then multiply by 4. The result is an approximation of π. Show that the error is approximately 0.0017. Show also that this error is due almost entirely to the trapezoidal sum, and not to the above approximation of this sum. That is, show that the error in your approximation of the exact trapezoidal sum is insignificant as compared to the error in the trapezoidal sum itself.

14. Now modify the approximation by the endpoint-derivative correction term, and compare with the expansion of π. Your answer should be correct through eight decimal places.

15. The actual estimate of π obtained above is much better than theoretically predicted error (Problem 12). Go over the proof of the modified trapezoidal rule with the present case in mind and try to see why this might be.

16. Carry out the proof of Theorem 5 that is sketched in the text.

17. Prove that

$$\int_a^b f = (b - a)\frac{f(a) + f(b)}{2}$$

$$+ (b - a)^2 \frac{f'(b) - f'(a)}{2}$$

$$+ \frac{1}{24}\int_a^b (x - a)^2(x - b)^2 f^{(4)}(x)\, dx.$$

(Start with the integral on the right and integrate by parts four times.)

18. Show that

$$\int_a^b (x - a)^2(x - b)^2\, dx = \frac{(b - a)^5}{30}.$$

19. Combine the above two results to get an equation for $\int_a^b f$ in which the integral on the right is replaced by

$$\frac{f^{(4)}}{720}(b-a)^5$$

for some X between a and b. Then prove Theorem 6.

9
SIMPSON'S RULE

Simpson's rule is a good compromise between accuracy and simplicity. It is not as accurate as the modified trapezoidal rule, by a factor of 4, but it avoids the two endpoint-derivative evaluations. The basic estimate depends on a rather complicated integration by parts similar to the derivations outlined in the last section but less straightforward. We shall omit the proof.

The Basic Estimate   *The "weighted" Riemann-sum estimate*

$$\int_{a-h}^{a+h} f(x)\,dx \approx \frac{h}{3}[f(a-h)+4f(a)+f(a+h)]$$

is accurate to within an error of

$$\frac{K}{90}h^5,$$

if the fourth derivative of f exists and is bounded by K over the interval $[a-h, a+h]$.

THEOREM 7   **Simpson's Rule**   *If the points $a = x_0, x_1, \ldots, x_n = b$ subdivide the interval $[a, b]$ into an even number of equal subintervals of length $\Delta x = (b-a)/n$ each, then*

$$\int_a^b f \approx \frac{\Delta x}{3}[f(x_0) + 4f(x_1) + 2f(x_2) + 4f(x_3)$$
$$+ \cdots + 4f(x_{n-1}) + f(x_n)].$$

If $f^{(4)}(x)$ is bounded by K over the interval $[a, b]$, then the error in the approximation is at most

$$\frac{K(b-a)}{180}(\Delta x)^4$$

in magnitude. When $f^{(4)}$ has constant sign over $[a, b]$, the error has the opposite sign.

Note that the coefficient pattern in this estimating sum, including the outside factor $(1/3)$, is

$$\frac{1}{3}, \frac{4}{3}, \frac{2}{3}, \frac{4}{3}, \frac{2}{3}, \frac{4}{3}, \ldots, \frac{2}{3}, \frac{4}{3}, \frac{1}{3}.$$

Since there are $n + 1$ terms, the sum of these coefficients is n, the number of intervals in the subdivision. The right side of Simpson's rule is thus Δx times a "weighted" average of the values $f(x_i)$, where $i = 0, 1, \ldots, n$.

The proof of Simpson's rule simply adds up the three-point basic estimates for the $n/2$ successive blocks of two subintervals each in the given subdivision.

Example   We consider our familiar $\ln 2$ problem for the last time. However, we have to change the number of subintervals, since the subdivision must be even, so we shall use six. The estimate is

$$\ln 2 = \int_1^2 \frac{dx}{x} \approx \frac{1}{18}\left[\frac{1}{1} + \frac{4}{7/6} + \frac{2}{8/6} + \frac{4}{9/6} + \frac{2}{10/6} + \frac{4}{11/6} + \frac{1}{2}\right]$$

$$= \frac{1}{3}\left[\frac{1}{6} + \frac{4}{7} + \frac{2}{8} + \frac{4}{9} + \frac{2}{10} + \frac{4}{11} + \frac{1}{12}\right]$$

$$= 0.693169\cdots.$$

Here $f^{(4)}(x) = 24/x^5$, so the error is negative and less than

$$\frac{K(b - a)}{180}(\Delta x)^4 = \frac{24}{180}\left(\frac{1}{6}\right)^4 = 0.000102\cdots$$

in magnitude. Therefore

$$0.69306 < \ln 2 < 0.69317. \qquad \square$$

Simpson's rule may seem, at first, very arbitrary. How would one ever guess that it would give an improved approximation? Here is one way.

According to the error formulas at the end of Chapter 6 and in Section 8, we can expect the midpoint Riemann sum to be twice as accurate as the corresponding trapezoidal sum. Figure 1 shows a geometric reason for anticipating this.

The midpoint tangent line fits the graph more closely than the chord, and on the opposite side (when f has constant concavity).

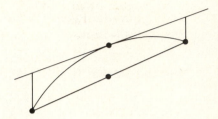

Figure 1

These considerations suggest that a weighted average of the midpoint sum and the trapezoidal sum, in the ratio of 2 to 1, might hit the nail almost exactly on the head. It does, and it is Simpson's rule.

There is another way of relating the trapezoidal rule to Simpson's rule. To obtain the trapezoidal rule we approximate an arc by its chord. That is, we approximate a function f between two of its values by the *linear* function that agrees with f at those two values. We can obtain Simpson's rule by going one step further and approximating f over the range of *three* equally spaced values by the *quadratic* function that agrees with f at those three values.

When f is approximated in this way at x_0, x_1, and x_2, and the approximating quadratic function is integrated from x_0 to x_2, the result is

$$\frac{\Delta x}{3}[f(x_0) + 4f(x_1) + f(x_2)].$$

Simpson's rule just patches together a succession of these quadratic approximations.

PROBLEMS FOR SECTION 9

1. Show that Simpson's rule is exact for a polynomial of degree 3 or less.

2. Apply Simpson's rule to $f(x) = x^4$ over the interval $[-a, a]$ with $\Delta x = a$. Show that the error is exactly

$$\frac{K(b-a)}{180}(\Delta x)^4$$

in magnitude, where K is the constant value of $f^{(4)}$. This shows that the error formula cannot be improved.

▦ **3.** Estimate $\pi = 4 \int_0^1 dx/(1 + x^2)$ by applying Simpson's rule to the data provided for Problem 13, Section 8. This is just a heavy dose of arithmetic, but the use of calculus in applied problems generally ends this way. Your error should be approximately $2.6(10)^{-5}$.

4. Now compute the expected error from the statement of Simpson's rule and compare with the actual error obtained above.

CHAPTER 10
POLAR COORDINATES; PARAMETRIC EQUATIONS; L'HÔPITAL'S RULE

Sometimes a graph seems to have a special affinity for the origin. It loops about and seems to be constantly pulled inward toward the origin (Fig. 1).

Figure 1

Such a curve will not be a function graph in a Cartesian coordinate system, and its equation may be complicated and unrevealing of what is going on. When a point traces such a curve, the most natural description may be in terms of its distance from and orientation to the origin, i.e., in terms of its *polar coordinates*.

Recall that any number θ, considered as an angular coordinate, determines a unique ray, or half-line, drawn from the origin (Fig. 2). Then a nonnegative number r determines the unique point P on the ray at the distance r from the origin. The numbers r and θ together

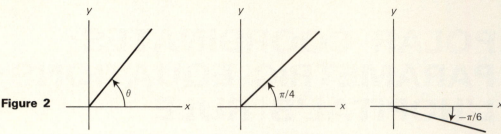

Figure 2

determine the point P uniquely, and are called *polar coordinates* of P (Fig. 3). The rectangular coordinates of P are obtained from the polar coordinates r, θ by the change-of-coordinates equations

$$x = r \cos \theta, \qquad y = r \sin \theta.$$

Note that $x^2 + y^2 = r^2$.

The polar-coordinate correspondence

$$(r, \theta) \to P$$

is not one-to-one. That is, it is not true, conversely, that P determines a unique pair of polar coordinates. An angle θ can always be modified by adding a multiple of 2π without changing the ray it determines, so P determines its polar coordinates only "up to" the addition of a multiple of 2π to its angular coordinate θ.

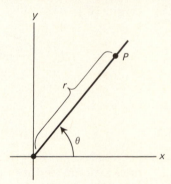

Figure 3

So far it has been assumed that r is positive or zero. But the equations $x = r \cos \theta$, $y = r \sin \theta$ define a point P for *any* numbers r and θ. If r is negative, then P is obtained geometrically by first measuring off the angle θ to obtain a ray and then proceeding along this *ray backward across the origin* a distance $|r|$ on the other side, as shown in Fig. 4. Such negative values of r are not normally used, but we shall see that some graphs seem incomplete unless negative r is allowed.

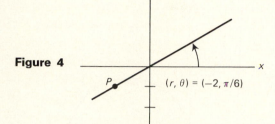

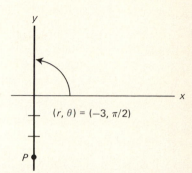

Figure 4

$(r, \theta) = (-2, \pi/6)$ $(r, \theta) = (-3, \pi/2)$

Any equation in r and θ has a *polar graph*, consisting of all points in the plane whose polar coordinates satisfy the equation. The polar graph of a function f is the polar graph of the equation

$$r = f(\theta).$$

Example 1 The polar graph of $r = 1$ is the unit circle about the origin (Fig. 5).

Figure 5

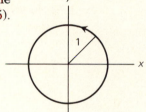

□

Example 2 The polar graph of $r = \theta$ is a spiral, as shown in Fig. 6.

Figure 6

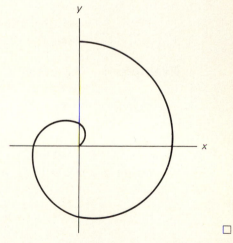

□

Example 3 The graph of $r = \cos \theta$ is another circle, but this is not so obvious. We can try to get an idea of the shape of an unknown polar graph by

θ	$r = \cos\theta$
0	1
$\pi/6$	$\sqrt{3}/2 \approx 0.86$
$\pi/4$	$\sqrt{2}/2 \approx 0.70$
$\pi/3$	$1/2$
$\pi/2$	0

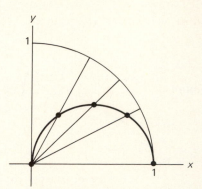

Figure 7

computing and plotting a few selected points. We do this in Fig. 7 for half the graph, and see that the graph is probably some kind of *oval*.

A better procedure, when it works, is to find and recognize the Cartesian equation of the graph. Here we use the change-of-coordinate equation $x = r\cos\theta$ together with the given equation $r = \cos\theta$, to get, in order,

$$r = \cos\theta = x/r,$$
$$r^2 = x,$$
$$x^2 + y^2 = x,$$
$$x^2 - x + y^2 = 0,$$
$$\left(x - \frac{1}{2}\right)^2 + y^2 = \frac{1}{4},$$

and finally we recognize that the graph is the circle of radius 1/2 about the center $(1/2, 0)$. □

Example 4 The graph of $r = 1 + \cos\theta$, shown in Fig. 8, is called a *cardioid*.

Figure 8

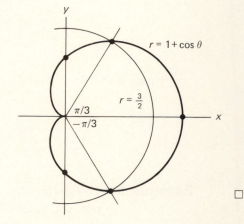

$r = 1 + \cos\theta$

$r = \dfrac{3}{2}$

$\pi/3$

$-\pi/3$

□

Example 5 Find the points of intersection of the cardioid $r = 1 + \cos\theta$ and the circle $r = 3/2$.

> **Solution** The curves will intersect at the points obtained by solving the two equations simultaneously. Eliminating r, we have
>
> $$3/2 = 1 + \cos\theta$$
>
> $$\cos\theta = 1/2$$
>
> $$\theta = \pm\pi/3. \qquad \square$$

Example 6 If e is a positive number less than 1, show that the polar graph of

$$r = \frac{1}{1 - e\cos\theta}$$

is an ellipse. Do this by finding the Cartesian equation of the graph, as in Example 3. (The number e is called the *eccentricity* of the ellipse.)

> **Solution** Setting $\cos\theta = x/r$ and cross-multiplying, the equation becomes, in turn,
>
> $$r\left(1 - e\frac{x}{r}\right) = 1,$$
>
> $$r = 1 + ex,$$
>
> $$x^2 + y^2 = r^2 = (1 + ex)^2$$
>
> $$= 1 + 2ex + e^2x^2,$$
>
> $$x^2(1 - e^2) - 2ex + y^2 = 1,$$
>
> $$(1 - e^2)\left(x - \frac{e}{1 - e^2}\right)^2 + y^2 = 1 + \frac{e^2}{1 - e^2}$$
>
> $$= \frac{1}{1 - e^2};$$
>
> $$\frac{(x - h)^2}{a^2} + \frac{y^2}{b^2} = 1,$$
>
> where $a = 1/(1 - e^2)$, $b = 1/\sqrt{1 - e^2}$, $h = e/(1 - e^2)$.

If $e = 4/5$, then $a = 25/9$, $b = 5/3$, $h = 20/9$; the ellipse is shown in Fig. 9.

Figure 9

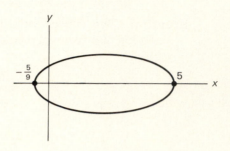

Example 7 If negative r is not allowed, then the graph of $r = \cos 2\theta$ is as shown at the left in Fig. 10. At the right is the "full" graph, using negative r.

Figure 10

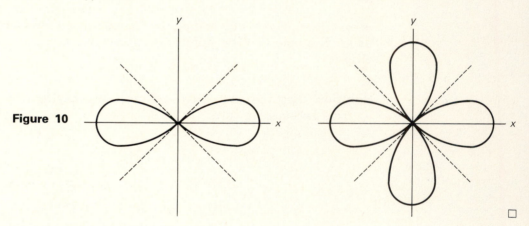

PROBLEMS FOR SECTION 1

Sketch the polar graph of each of the following equations.

1. $r = \cos 3\theta$
2. $r = \sin 2\theta$
3. $r = 2\cos 5\theta$
4. $r = 2\sin 4\theta$
5. $r^2 = \cos 2\theta$
6. $r^2 = \cos \theta$
7. $r = 2\sin \theta$ (Find the x, y equation first.)
8. $r = 1 + \sin \theta$
9. $r = 1 - \cos \theta$
10. $r = 1 - 2\cos \theta$

11. Show that $r \sin \theta = 2$ is the equation of the horizontal straight line two units above the x-axis.

12. In view of the above problem, consider the equation

$$r \sin(\theta - \theta_0) = 2.$$

It has the above form if we set $\phi = \theta - \theta_0$. On this basis, show by a geometric argument that it must be the equation of the line that is at a distance 2 from the origin and that makes the angle θ_0 with the direction of the x-axis.

13. Show that the polar graph of the equation

$$r = \frac{1}{1 - \cos \theta}$$

is a parabola. (Follow Example 6.)

14. Show that the polar graph of the equation

$$r = \frac{1}{1 - \epsilon \cos \theta}$$

is a hyperbola if $\epsilon > 1$.

15. Show more generally that for any positive constants e and d the polar graph of

$$r = \frac{d}{1 - e \cos \theta}$$

is an ellipse if $e < 1$, a parabola if $e = 1$, and a hyperbola if $e > 1$.

16. Consider the locus traced by a point whose distance from the origin is a constant e times its distance from the line $x = -k$. Assuming the above problem, show that this locus is an ellipse if $e < 1$, a parabola if $e = 1$, and a hyperbola if $e > 1$.

Find the points of intersection of the following pairs of polar graphs.

17. The cardioid $r = 1 + \cos \theta$ and the circle $r = 1/2$

18. $r = 1 + 2 \cos \theta$, $r = 1$

19. The two cardioids $r = 1 + \cos \theta$, $r = 1 + \sin \theta$

20. The cardioid $r = 1 + \cos \theta$ and the circle $r = 3 \cos \theta$

It seems reasonable to specify the direction of a polar graph $r = f(\theta)$ be the angle ψ (Greek psi) it makes with the polar ray, as in Fig. 11. That this is the natural angle to use is shown by the simplicity

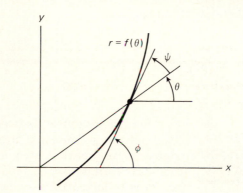

Figure 11

of the formula:

$$\cot \psi = \frac{f'(\theta)}{f(\theta)} = \frac{1}{r} \frac{dr}{d\theta}.$$

The following problems center around this formula.

21. Derive the identity

$$\cot(\alpha - \beta) = \frac{1 + \tan \alpha \tan \beta}{\tan \alpha - \tan \beta}$$

from the trigonometric addition formulas (Chapter 4, Section 3).

Referring to the above figure again, the slope of the polar graph is $\tan \phi$, where $\phi = \psi + \theta$. Thus,

$$\tan \phi = \frac{dy}{dx} = \frac{dy/d\theta}{dx/d\theta}$$

$$= \frac{f'(\theta)\sin \theta + f(\theta)\cos \theta}{f'(\theta)\cos \theta - f(\theta)\sin \theta},$$

where the values on the right come from $x = r \cos \theta = f(\theta)\cos \theta$, etc. Now expand

$$\cot \psi = \cot(\phi - \theta)$$

by the formula above, substitute the above value for $\tan \phi$ and simplify. Your answer should be the simple formula for $\cot \psi$ stated earlier (after Problem 20).

22. Prove the identities

$$\tan x = \frac{\sin 2x}{1 + \cos 2x}$$

and

$$\cot x = \tan\left(\frac{\pi}{2} - x\right)$$

from the trigonometric addition formulas.

Compute $\cot \psi$ for each of the following polar graphs. In each case, try to interpret the answer geometrically. (Problem 22 can help.)

23. $r = e^{\theta}$ **24.** $r = \sec \theta$ **25.** $r = \cos \theta$ **26.** $r = 1 + \cos \theta$

2
AREA IN POLAR COORDINATES

Polar area is measured between given values of the *angular* coordinate, and the geometric picture is thus different (Fig. 1). The incremental approximating area is now in the form of a circular sector with vertex angle $\Delta\theta$, instead of a rectangle with base Δx, and this change is reflected in the new formula below.

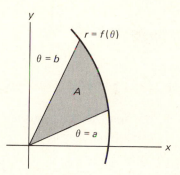

Figure 1

THEOREM 1 *Let f be a positive continuous function and let A be the area of the region bounded by the polar graph of the function f and the two rays θ = a and θ = b. Then*

$$A = \frac{1}{2} \int_a^b f^2(\theta) \, d\theta.$$

Example 1 Find the area swept over in the first revolution of the spiral $r = \theta$ (Fig. 2).

Solution Here $r = f(\theta) = \theta$, and θ runs from 0 to 2π. Therefore,

$$A = \frac{1}{2} \int_0^{2\pi} f(\theta)^2 \, d\theta = \frac{1}{2} \int_0^{2\pi} \theta^2 \, d\theta = \frac{1}{2} \left[\frac{\theta^3}{3} \right]_0^{2\pi} = \frac{4}{3} \pi^3,$$

by the theorem.

Figure 2

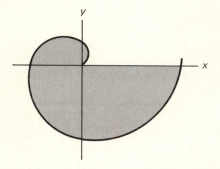

□

The form of the integrand in the theorem stems from the formula for the area of a circular sector (Fig. 3). If the sector central angle is

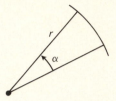

Figure 3

α, then the sector area A is the fraction $\alpha/2\pi$ of the area of the whole circle, so

$$A = \left(\frac{\alpha}{2\pi}\right)\pi r^2 = \frac{1}{2}\,r^2\alpha.$$

We prove Theorem 1 in the manner of Chapter 6.

Proof of Theorem We start as usual with an equally spaced subdivision of the angular interval $[a, b]$, $a = \theta_0 < \theta_1 < \cdots < \theta_n = b$. For each k, let m_k and M_k be the minimum and maximum values of $f(\theta)$ on the kth subinterval $[\theta_{k-1}, \theta_k]$. Also, let ΔA_k be the area lying "over" this incremental subinterval, i.e., the area of the incremental region bounded by the polar graph of $r = f(\theta)$ and the rays $\theta = \theta_{k-1}$ and $\theta = \theta_k$. We have pictured one of these incremental regions in Fig. 4. It includes an "inscribed" circular sector of radius $r = m_k$ and is included in a "circumscribed" circular sector of radius $r = M_k$. Therefore,

area of inside sector $\leq \Delta A_k \leq$ area of outside sector,

or

$$\frac{1}{2}\,m_k^2\,\Delta\theta \leq \Delta A_k \leq \frac{1}{2}\,M_k^2\,\Delta\theta.$$

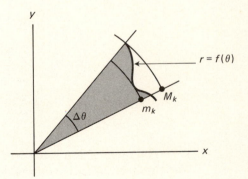

Figure 4

Summing these inequalities from $k = 1$ to $k = n$, we have the inequality

$$\Delta\theta \sum_{k=1}^{n} \frac{1}{2} m_k^2 \le A \le \Delta\theta \sum_{k=1}^{n} \frac{1}{2} M_k^2.$$

Each of these two sums is a Riemann sum for the function $(1/2)f^2(\theta)$, so each of them approaches the limit $(1/2)\int_a^b f^2(\theta)\,d\theta$ as $n \to \infty$. Finally, since A is squeezed between these varying sums, it must equal their limit. ∎

Example 2   Find the area enclosed by the cardioid $r = 1 + \cos\theta$ (see Fig. 5).

Figure 5

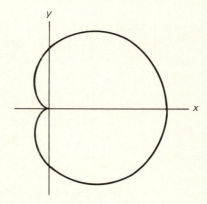

Solution

$$A = \frac{1}{2}\int_0^{2\pi} r^2\,d\theta = \frac{1}{2}\int_0^{2\pi} (1 + \cos\theta)^2\,d\theta$$

$$= \frac{1}{2}\int_0^{2\pi} (1 + 2\cos\theta + \cos^2\theta]\,d\theta$$

$$= \frac{1}{2}\int_0^{2\pi} \left[1 + 2\cos\theta + \frac{1}{2} + \frac{\cos 2\theta}{2}\right]\,d\theta$$

$$= \left[\frac{3}{4}\theta + \sin\theta + \frac{\sin 2\theta}{8}\right]_0^{2\pi}$$

$$= \frac{3\pi}{2}.$$
□

If $0 < f(\theta) < g(\theta)$ over the interval $[\alpha, \beta]$, then the area between the two polar graphs from $\theta = \alpha$ to $\theta = \beta$ is given by

$$A = \frac{1}{2} \int_\alpha^\beta (g^2 - f^2).$$

Example 3 Find the area of the region bounded on the outside by the cardioid $r = 1 + \sin \theta$ and on the inside by the circle $r = 1$.

Solution The graphs intersect when $1 + \sin \theta = 1$, or $\sin \theta = 0$. The intersection points are thus at $\theta = 0$ and $\theta = \pi$, as shown in Fig. 6.

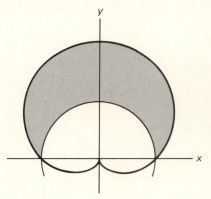

Figure 6

The area is therefore

$$\frac{1}{2} \int_0^\pi [(1 + \sin \theta)^2 - 1] \, d\theta = \frac{1}{2} \int_0^\pi (2 \sin \theta + \sin^2 \theta) \, d\theta$$

$$= \frac{1}{2} \int_0^\pi \left(2 \sin \theta + \frac{1 - \cos 2\theta}{2}\right) d\theta$$

$$= \frac{1}{2} \left[-2 \cos \theta + \frac{\theta}{2} - \frac{\sin 2\theta}{4}\right]_0^\pi$$

$$= \frac{1}{2} \left(4 + \frac{\pi}{2}\right) = 2 + \frac{\pi}{4}. \qquad \square$$

In general, the polar graphs $r = f(\theta)$ and $r = g(\theta)$ will intersect when $f(\theta) = g(\theta)$, so this equation must be solved for θ to find the angles of intersection. A sketch of the graphs helps to guard against overlooking one or more solutions. For example, the equation

$\sin 3\theta = 1$ has *three* distinct solutions between 0 and 2π; namely, $\theta = \pi/6, 5\pi/6, 3\pi/2$.

PROBLEMS FOR SECTION 2

1. Find the area swept out in the first revolution of the spiral $r = \theta^2$.

2. The area formula of Theorem 1 should be consistent with the formula for the area of a circle. Show that it is. (Use Theorem 1 to retrieve the value πr^2 as the area of a circle of radius r.)

3. Find the area of the region bounded by the polar graph $r = \cos\theta$ and the rays $\theta = 0$, $\theta = \pi/2$.

4. Find the area bounded by the polar graph $r = \sec\theta$ and the rays $\theta = 0$, $\theta = \pi/4$.

5. Show that the area cut off by the spiral $r = e^\theta$ in successive quadrants increases by the constant factor e^π.

6. Find the area of the region bounded on the outside by the cardioid $r = 1 + \cos\theta$ and on the inside by $r = 2\cos\theta$.

In each of the following problems, calculate the area of the region bounded by the given polar graph or graphs.

7. $r = 1 - \sin\theta$ **8.** $r = 2 + \cos\theta$ **9.** $r = 3 - 2\sin\theta$ **10.** $r^2 = 2\cos 2\theta$

11. $r^2 = a\cos n\theta$ (one loop) **12.** $r = a\cos 3\theta$ (one loop) **13.** $r = a\cos n\theta$ (one loop)

14. $r = 1 + 2\cos\theta$ (outside loop) **15.** $r = 1 + 2\cos\theta$ (inside loop) **16.** $r = 1/\sqrt{1 + \theta}$, $\theta = 0$, $\theta = \pi$

17. $r = 1/(1 + \theta)$, $\theta = 0$, $\theta = \pi$ **18.** $r = 3\sec\theta$, $r = 4\cos\theta$ **19.** $r = \sec\theta$, $r = 2\cos\theta$

20. $r = 1 + \cos\theta$, $r = 3/2$

For each pair of polar graphs find the area of the region lying:

21. Outside the circle $r = 1/2$ and inside the cardioid $r = 1 - \cos\theta$

22. Outside the cardioid $r = 1 + \cos\theta$; inside the circle $r = 1$

23. Outside the cardioid $r = 1 + \cos\theta$; inside the circle $r = 3\cos\theta$

24. Inside both cardioids $r = 1 + \cos\theta$, $r = 1 + \sin\theta$.

3
PATHS

This section introduces a new way of describing curves in the plane.

Example 1 Consider the two equations

$$x = t^3 - 3t,$$
$$y = t^2.$$

For any given value of t, the corresponding values of x and y, taken together, give a point in the xy-coordinate plane. For instance, when $t = -2$, we have $(x, y) = (-2, 4)$. As t varies, the point

$$(x, y) = (t^3 - 3t, t^2)$$

varies and traces out the curve in the xy-plane shown in Fig. 1. That is, the collection of all points (x, y) of the form $(x, y) = (t^3 - 3t, t^2)$ makes up this curve. We shall see later how we know the graph looks like this.

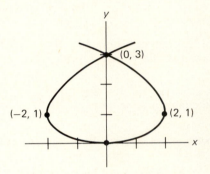

Figure 1

Note that we don't plot the variable t, but plot coordinates x and y that are determined by t. Thus, t is the independent variable, but it doesn't appear in the geometric picture. We call the equations $x = t^3 - 3t$, $y = t^2$ a *parametric representation* of the curve, and t is the *parameter*.   □

Any pair of functions f and g having a common domain interval can be used together to define a curve in the above way. The parametric equations are

$$x = g(t),$$
$$y = h(t),$$

and the curve is made up of the points $(x, y) = (g(t), h(t))$. We imagine the parameter t changing, and visualize the varying point $(x, y) = (g(t), h(t))$ tracing the curve. Often we think of t as time and $(x, y) = (g(t), h(t))$ as the position at time t of a moving point or particle. Then the path of this moving particle lies along the curve. Because of this interpretation, a parametric representation of a curve is often called a *path*.

The example above shows that a path may not lie along the graph of a function. In fact, one reason for introducing parametric representations is to obtain a new way of discussing curves that are more general than function graphs.

Sometimes we can eliminate the parameter t from the parametric equations

$$x = g(t),$$
$$y = h(t),$$

and obtain a single equation in x and y for the graph of the path. In the above example,

$$x = t^3 - 3t,$$
$$y = t^2,$$

if we square x, then the right side involves only powers of $t^2 = y$, and we obtain the equation

$$x^2 = y^3 - 6y^2 + 9y.$$

Thus, the point $(x, y) = (t^3 - 3t, t^2)$ moves along the graph of this equation.

Often we can recognize a path by eliminating the parameter in the above way. In each of the following examples, the parameter is easily eliminated to get an equation in x and y, and in each case we can easily sketch the graph of the equation. Then the path of the point moving along this known curve can be indicated by adding arrows.

Example 2 $x = t^3$, $y = t^2$. The path is shown in Fig. 2. This example is closely related to the first example.

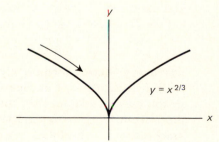

Figure 2

$y = x^{2/3}$

Example 3 $x = \cos t$; $y = \sin t$. The path runs along the unit circle (Fig. 3).

Figure 3

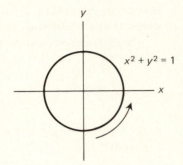

$x^2 + y^2 = 1$

☐

Example 4 $x = \cos t^3$; $y = \sin t^3$. Again the path traces the unit circle (Fig. 4).

Figure 4

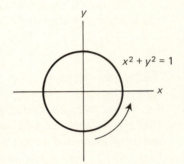

$x^2 + y^2 = 1$

☐

The point of this last example is the use of the word *path*. Some people would say that Examples 3 and 4 involve the same path, although it is traced somewhat differently in the two situations. Others feel that the idea of a path includes the way it is traced, and that these examples are, therefore, *different* paths. This ambiguity in the meaning of "path" causes no difficulty.

In Example 2, even though the coordinate functions

$$x = g(t) = t^3 \qquad \text{and} \qquad y = h(t) = t^2$$

are everywhere continuously differentiable, the graph itself has a kink (called a *cusp*) at the origin, and the path direction along the graph takes a sudden 180° turn there. It will be seen later that such a discontinuity in path direction can be ruled out by requiring path *smoothness* in the following sense.

DEFINITION A path $(x, y) = (g(t), h(t))$ is *smooth* if g and h have continuous derivatives and if g' and h' are never simultaneously zero.

We shall need a formula for the slope of the path in terms of g' and h'. In order to find such a formula, we consider a general path point $(x, y) = (g(t), h(t))$ and give the parameter t a (nonzero) increment Δt, obtaining a new path point

$$(x + \Delta x, y + \Delta y) = (g(t + \Delta t), h(t + \Delta t)).$$

The secant line between these two path points has the slope $\Delta y / \Delta x$. The tangent slope m is the limit of the secant slope as the second point approaches the first point along the path and, therefore, as Δt approaches 0. Thus

$$m = \lim_{\Delta t \to 0} \frac{\Delta y}{\Delta x} = \lim_{\Delta t \to 0} \frac{\dfrac{\Delta y}{\Delta t}}{\dfrac{\Delta x}{\Delta t}} = \frac{h'(t)}{g'(t)}.$$

This calculation is valid only if $g'(t) \neq 0$. Then $\Delta x \neq 0$ when Δt is sufficiently small (Theorem 4, Chapter 5), and there is no division by zero. However, the formula is still correct in the case $g'(t) = 0$, $h'(t) \neq 0$, provided it is interpreted as saying that the slope is infinite and the tangent line is vertical. For in this case the above calculation can be repeated to show that the *reciprocal* slope is zero,

$$\lim_{\Delta t \to 0} \frac{\Delta x}{\Delta y} = \frac{g'(t)}{h'(t)} = 0.$$

And the fact that $\Delta x / \Delta y$ approaches zero means geometrically that the secant line approaches a vertical line as $\Delta t \to 0$. We have proved the following theorem.

THEOREM 2 *A smooth path $(x, y) = (g(t), h(t))$ has a tangent line at every point, with slope m given by*

$$\boxed{m = \frac{h'(t)}{g'(t)}.}$$

If $g'(t) = 0$, the formula is interpreted as saying that the slope is infinite and the tangent line is vertical.

Since the slope h'/g' and the reciprocal slope g'/h' are both continuous functions of t wherever defined, it follows that the tangent line varies continuously.

Sketching Paths

We can sketch smooth paths much as we sketched graphs in Chapter 5. When we first considered function graphs, we decided that a graph could change from rising to falling (or vice versa) only by going through a point where the tangent is *horizontal*. A path like that in Fig. 1 has a complementary property: It can change from moving forward to moving backward (or vice versa) only by going through a point where the tangent is *vertical*, that is, where $g'(t) = 0$. Figure 5 illustrates this.

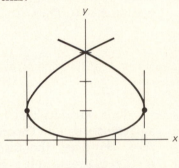

Figure 5

So the points where $g'(t) = 0$ divide the path into pieces, each of which is traced steadily in one direction over the x-axis, either to the right, like a function graph, or to the left, like a function graph traced backward. The proof of this observation (Theorem 3) depends on showing that each of these pieces actually does lie along the graph of a function.

Each function-graph section of the path is divided in turn into *monotone* pieces by its *horizontal* tangents, as in Chapter 5. A smooth path thus has two types of critical points.

DEFINITION

A *critical point* of a smooth path is a point where the tangent line is either vertical or horizontal, i.e., a point where either $g'(t) = 0$ or $h'(t) = 0$.

We saw above that the critical points of a smooth path divide it up into monotone-function-graph pieces, traced forward or backward. Such pieces are qualitatively characterized by their four possible "quadrant directions": to the right and up, to the right and down, to the left and up, to the left and down.

Example 5 We know that the path

$$x = \cos t,$$
$$y = \sin t$$

traces the unit circle $x^2 + y^2 = 1$ counterclockwise. Let us see, however, what the critical point analysis by itself would tell us about the path.

The path critical points are the points where either derivative is zero. We find

$$\frac{dx}{dt} = -\sin t = 0 \qquad \text{at } t = 0, \pi, 2\pi, \text{etc.,}$$

$$\frac{dy}{dt} = \cos t = 0 \qquad \text{at } t = \pi/2, 3\pi/2, \text{etc.}$$

We compute the corresponding path points, and get the following table.

t	0	$\pi/2$	π	$3\pi/2$	2π	$\cdots$
x	1	0	-1	0	1	$\cdots$
y	0	1	0	-1	0	$\cdots$

These points are in consecutive order on the path, which runs along a monotone graph between each pair. The general configuration must therefore be like that shown in Fig. 6.

The slope formula $m = g'(t)/h'(t)$ shows whether the critical point tangent is vertical or horizontal. However, this can also be decided geometrically. For example, the tangent at the critical point $(1, 0)$ must be vertical because the path doubles back from that point.

Figure 6

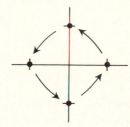

The simplest curve that we can draw that will meet these requirements is some kind of oval, and we thus get the right general shape from these path considerations (Fig. 7).

Figure 7

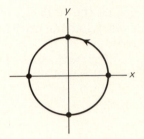

$\square$

Example 6   We now show that we have been using the correct figure for our first path

$$x = g(t) = t^3 - 3t,$$
$$y = h(t) = t^2.$$

First,

$$\frac{dx}{dt} = 3(t^2 - 1) = 0 \quad \text{at } t = \pm 1,$$

$$\frac{dy}{dt} = 2t = 0 \quad \text{at } t = 0,$$

so that altogether these critical parameter values break the t-axis up into four intervals as shown in Fig. 8.

Figure 8

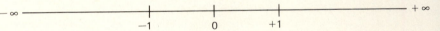

We then make up the corresponding critical point table.

t	$-\infty$	-1	0	1	∞
x	$-\infty$	2	0	-2	∞
y	$+\infty$	1	0	1	∞

Note also that $x = t^3 - 3t = 0$ when $t = 0, \pm\sqrt{3}$, so we can add the y-intercept at $y = (\sqrt{3})^2 = 3$. Plotting these points gives the framework shown at the left in Fig. 9, and we draw as smooth a path as we can, following these directions.

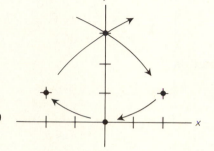 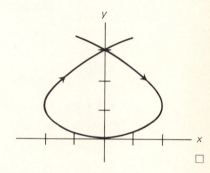

Figure 9

When a curve loops around the origin, it is frequently convenient to express it as a path with an angle θ as the parameter, as we already have done in a couple of cases. However, the angle θ may or may not be the angular coordinate of the path point.

Example 7 The ellipse

$$\frac{x^2}{a^2} + \frac{y^2}{b^2} = 1$$

has the important parametric representation

$$x = a \cos \theta, \qquad y = b \sin \theta.$$

But θ is not the angular coordinate of the point $(x, y) = (a \cos \theta, b \sin \theta)$ on this ellipse. The geometric meaning of θ here is shown in Fig. 10.

Figure 10

Angle as Parameter

The angular coordinate θ is a natural parameter when the polar graph of a function f is interpreted as a path. Along the polar graph the x and y coordinates,

$$x = r \cos \theta,$$
$$y = r \sin \theta,$$

satisfy the equation

$$r = f(\theta),$$

so that

$$x = f(\theta)\cos \theta,$$
$$y = f(\theta)\sin \theta$$

along the graph. But these are the equations of a path with θ as parameter.

Example 8

The spiral $r = \theta$ is traced by the path

$$x = \theta \cos \theta,$$
$$y = \theta \sin \theta.$$ □

Example 9

The cardioid $r = 1 + \cos \theta$ is traced by the path

$$x = \cos \theta + \cos^2\theta,$$
$$y = \sin \theta + \sin \theta \cos \theta.$$ □

Paths along Function Graphs

We conclude with a look at when and how a path runs along a function graph.

Suppose that a smooth path $(x, y) = (g(t), h(t))$ runs for a while along the graph of a differentiable function F, say while t is in an interval I. Then the coordinate-parameter functions

$$x = g(t) \quad \text{and} \quad y = h(t)$$

are related by the equation

$$y = F(x),$$

or

$$h(t) = F(g(t)),$$

for t in I, and

$$\frac{dy}{dt} = F'(x)\frac{dx}{dt},$$

by the chain rule. It follows that dx/dt can never be zero. For if $dx/dt = 0$, then the formula shows that $dy/dt = 0$ also, and this contradicts the definition of a smooth path. Therefore,

$$F'(x) = \frac{dy/dt}{dx/dt} = \frac{h'(t)}{g'(t)},$$

on I. Since $F'(x)$ is the slope m of the graph of F, we thus recover the formula

$$m = \frac{h'(t)}{g'(t)}$$

for the slope of a path in terms of the parameter functions.

Theorem 3 is a partial converse of the above remarks.

THEOREM 3 *If f and g are differentiable functions on an interval I and if $g'(t)$ is never zero in I, then as t runs across I the path $(x, y) = (g(t), h(t))$ runs along the graph of a differentiable function F, forward if $x = g(t)$ is increasing, and backward if g is decreasing.*

Proof Since g' is never zero on I, we know that g has a differentiable inverse function $t = k(x)$, by the inverse function theorem (Chapter 7, Theorem 4). Then

$$y = h(t) = h(k(x))$$

while t is in I. So if we set $F(x) = h(k(x))$, then F is differentiable and the equation

$$y = F(x)$$

holds along the path while t is crossing I. If $x = g(t)$ is increasing, then the path runs along the graph of F in the normal left-to-right direction. If $x = g(t)$ is decreasing, then the path traces the graph of F backward. ∎

Along an interval I containing no path critical points both derivatives $g'(t)$ and $h'(t)$ are different from zero. Since $g' \neq 0$, the path runs along the graph of a function F. Since $h' \neq 0$, the formula $F'(x) = h'(t)/g'(t)$ shows that $F' \neq 0$, and therefore that F is monotone. Thus, between successive critical points, a smooth path runs along a monotone function graph, as claimed earlier.

PROBLEMS FOR SECTION 3

In each of the following problems eliminate the parameter to obtain an equation in x and y for the graph along which the path runs. Sketch and identify (if possible) the graph, and indicate by one or more arrows how it is being traced. Compute the path slope dy/dx as a function of t and determine the path critical points (points where $dy/dx = 0$ or ∞, or is undefined).

1. $x = t, y = t^2$
2. $x = t - 1, y = 2t + 3$
3. $x = t - 1, y = t^3$

4. $x = 3 \sec t, y = 2 \tan t$
5. $x = 2t^{1/3}, y = 3t^{1/3}$
6. $x = t^2 - 1, y = t + 1$

7. $x = 2 \sin t, y = 3 \cos t$
8. $x = t + 1, y + t(t + 4)$
9. $x = 2 \cos t, y = 4 \sin t$

10. $x = a \cos^3 t, y = a \sin^3 t$
11. $x = 2 \cos t, y = 2 \sin t$
12. $x = a \cos t, y = b \sin t$

13. The function f has the graph shown below.

14. Sketch the path $x = -t, y = f(t)$, where f is the function of Problem 13.

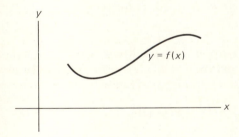

$y = f(x)$

Sketch the following paths.

a) $x = t, y = f(t)$.

b) $x = -t, y = f(-t)$.

c) $x = f(t), y = t$.

15. Sketch the path $x = t^2$, $y = t^4$, including some arrows to show how it is traced.

16. Do the same for the path $x = t^3 - 3t$, $y = (t^3 - 3t)^2$.

Parametric equations can be used to great advantage in a practical way when analyzing the motion of projectiles. In this case, the motion may be represented by

$$x = v_0 t \cos \theta,$$

$$y = v_0 t \sin \theta - \frac{1}{2} g t^2,$$

where v_0 is the initial velocity, θ is the initial angle of inclination from the horizontal, and g is the acceleration of gravity (32 ft/sec^2).

17. A shell is fired with an initial velocity of 100 ft/sec and $\cos \theta = 3/5$ and $\sin \theta = 4/5$. Find the time t when the shell hits the ground, and the distance between the gun and the spot where the shell lands; also, find the maximum height reached by the shell.

18. A ball is thrown off the top of a building 100 ft high with a velocity of 60 ft/sec and at an angle of 45° above the horizontal. Find

a) The maximum height attained by the ball.

b) The distance from the building to the point where the ball hits the ground.

c) The velocity of the ball when it strikes the ground.

Analyze each of the following paths by Theorem 3, following the scheme used in Example 6. Include the x and y intercepts in your table if you can determine them, and indicate any singular points that you find (points where the path is not smooth).

19. $x = t^2, y = t^2$ **20.** $x = t^2, y = t^2 - 2t$ **21.** $x = t^3 - 3t, y = t^3 - 3t^2$

22. $x = t^2, y = 2t^3 - 9t^2 + 12t$ **23.** $x = t^3 - 3t^2 + 3t, y = t^3 - 3t^2$

24. $x = t^2 - t, y = t^4 - 2t^2$ **25.** $x = 4t^3 - 3t^2, y = t^4 - 2t^2$

26. Elsewhere we have called a function f smooth if it is continuously differentiable. Show that a function is smooth if and only if its graph is a smooth path when it is traced in the standard way (with the independent variable as parameter).

27. It is reasonable to call a *curve* smooth if there is a smooth path running along it. It is always possible to trace a smooth curve by a different path that fails to meet the smoothness requirement at one or more points. (Later we shall see that this amounts to tracing the curve by a particle that comes to a stop one or more times.) Show that Examples 3 and 4 in the text illustrate this possibility.

28. Show, as claimed in the text, that the path $x = a \cos \theta, y = b \sin \theta$ does lie along the ellipse $x^2/a^2 + y^2/b^2 = 1$.

29. Use the parametric representation

$$x = a \cos \theta,$$
$$y = b \sin \theta$$

for the ellipse

$$x^2/a^2 + y^2/b^2 = 1,$$

to prove the following theorem.

29. (*Continued*)

> **Theorem**   *Let F and F′ be the points* $(c, 0)$ *and* $(-c, 0)$ *on the x-axis, where* $c = \sqrt{a^2 - b^2}$. *Let* $P = (x, y)$ *be any point on the ellipse. Then the segments PF and PF′ have the lengths*
>
> $$\overline{PF} = a - ex \qquad and \qquad \overline{PF'} = a + ex,$$
>
> *where* $e = c/a$.

This shows that $\overline{PF} + \overline{PF'} = 2a$; the ellipse is the locus of a point P moving in such a way that the sum of its distance from the two fixed points F and F' is constant and equal to $2a$. See Fig. 11.

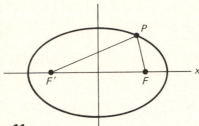

Figure 11

31. Let F and F' be the points $(c, 0)$ and $(-c, 0)$, where $c = \sqrt{a^2 + b^2}$. Let P be any point on the hyperbola of Problem 30. Using the above parametric representation, show that the segments PF and PF' have the lengths

$$PF = |ex - a|, \quad PF' = |ex + a|,$$

where $e = c/a$. [*Hint*: Set $\tan^2\theta = \sec^2\theta - 1$ at an appropriate moment.]

30. Show that the hyperbola

$$\frac{x^2}{a^2} - \frac{y^2}{b^2} = 1$$

has the parametric representation

$$x = a \sec \theta,$$
$$y = b \tan \theta.$$

32. Assuming the above equations show that

$$|PF - PF'| = 2a.$$

(Consider separately the cases $x > 0$, $x < 0$. Note that $e > 1$ and that $|x| \geq 1$ on the graph.) Restate this equation in words as a locus characterization of the hyperbola, similar to that for the ellipse in Problem 29.

4
PATH LENGTH

We saw in the last chapter that we could define the length of a smooth function graph as the limit of lengths of inscribed polygons, and that this limit had a natural formulation as a definite integral (Fig. 1). There is a similar, but more general, formula for paths.

If we rotate the plane over the $y = x$ diagonal, such a function graph rotates to a nonstandard function graph along which x is a function of y. But the inscribed polygonal lengths remain the same and they necessarily have the same limit, so the *length* of a piece of a

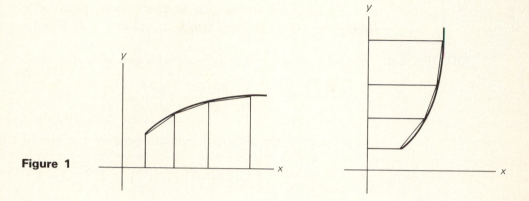

Figure 1

graph remains unchanged by such a rotation. However, the description of the length calculation for the new graph proceeds with the roles of x and y interchanged, so the integral formula for the length is now

$$\int_a^b \sqrt{1 + \left(\frac{dx}{dy}\right)^2}\, dy.$$

A smooth path can be broken up into a succession of arcs of the above two types. The length of a smooth path is therefore a sum of lengths, each of which can be computed by one of these two formulas. But this is a piecemeal way to deal with path length, and we clearly would like a single integral formula with the parameter as variable.

The most straightforward way to approach such a path length formula would be to start over again with inscribed polygons and express their lengths in terms of the parameter t. This works all right, but it involves an extra complication, namely, that the resulting finite t sums are not quite Riemann sums and a new integration theorem has to be proved.

Instead, we shall adopt the antiderivative approach. We know that for small arcs and their chords,

$$\frac{\text{arc length}}{\text{chord length}} \to 1$$

as chord length $\to 0$. (See Section 4 in Chapter 9.)

So let s be the length of the smooth path $(x, y) = (g(t), h(t))$ measured from the fixed point (x_0, y_0) at $t = t_0$. Then,

THEOREM 4   *The path length s is a differentiable function of t, and*

$$\left(\frac{ds}{dt}\right)^2 = \left(\frac{dx}{dt}\right)^2 + \left(\frac{dy}{dt}\right)^2.$$

In particular, the length of the path between $t = t_1$ and $t = t_2$ is

$$\text{length} = \int_{t_1}^{t_2} \sqrt{\left(\frac{dx}{dt}\right)^2 + \left(\frac{dy}{dt}\right)^2}\, dt.$$

Proof   We assume that $t > t_0$, so that s is an increasing function of t and $\Delta s / \Delta t$ is positive. If we give t a positive increment Δt, then the chord length l for the corresponding arc increment is

$$l = \sqrt{(\Delta x)^2 + (\Delta y)^2}.$$

See Fig. 2. Therefore,

$$\frac{\Delta s}{\Delta t} = \frac{\Delta s}{l} \cdot \frac{l}{\Delta t} = \frac{\Delta s}{l} \sqrt{\left(\frac{\Delta x}{\Delta t}\right)^2 + \left(\frac{\Delta y}{\Delta t}\right)^2}.$$

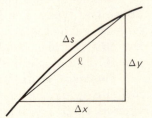

Figure 2

Now let $\Delta t \to 0$. Then Δx and Δy both approach 0 (differentiable functions being continuous), so $l \to 0$ and $\Delta s / l \to 1$. Thus $ds/dt = \lim_{\Delta t \to 0}(\Delta s / \Delta t)$ exists and

$$\frac{ds}{dt} = \sqrt{\left(\frac{dx}{dt}\right)^2 + \left(\frac{dy}{dt}\right)^2}.$$

The integral formula follows.

If Δt is negative, then we have to keep track of minus signs in the above calculation, but the result is the same.

However, if $t < t_0$, so that s is a *decreasing* function of t, then the length of the incremental arc is $-\Delta s$ for positive Δt, and this extra minus sign leads to

$$\frac{ds}{dt} = \lim_{\Delta t \to 0} \frac{\Delta s}{\Delta t} = -\sqrt{\left(\frac{dx}{dt}\right)^2 + \left(\frac{dy}{dt}\right)^2}.$$

In every case, however,

$$\left(\frac{ds}{dt}\right)^2 = \left(\frac{dx}{dt}\right)^2 + \left(\frac{dv}{dt}\right)^2.$$

∎

REMARK   In terms of differentials, this equation becomes

$$\boxed{(ds)^2 = (dx)^2 + (dy)^2,}$$

which is easy to remember by thinking of the Pythagorean theorem for an "infinitesimal" right triangle.

REMARK   A function graph traced from left to right is a particular path with parameter $t = x$:

$$x = x,$$
$$y = f(x).$$

For this special path, our new path-length formula reduces to the original function formula

$$\frac{ds}{dx} = \sqrt{1 + \left(\frac{dy}{dx}\right)^2}.$$

Example 1   The path $(x, y) = (\cos \theta, \sin \theta)$ traces the unit circle once as θ runs from 0 to 2π. The length formula should therefore give the answer 2π. We check that it does:

$$\int_0^{2\pi} \sqrt{\left(\frac{dx}{d\theta}\right)^2 + \left(\frac{dy}{d\theta}\right)^2}\, d\theta = \int_0^{2\pi} \sqrt{\sin^2\theta + \cos^2\theta}\, d\theta$$

$$= \int_0^{2\pi} 1 \cdot d\theta$$

$$= \theta\Big]_0^{2\pi}$$

$$= 2\pi.$$

□

Example 2   The path $(x, y) = (\cos(\theta^2), \sin(\theta^2))$ traces the unit circle once as θ runs from 0 to $\sqrt{2\pi}$. The answer should again be 2π.

$$\int_0^{\sqrt{2\pi}} \sqrt{\left(\frac{dx}{d\theta}\right)^2 + \left(\frac{dy}{d\theta}\right)^2} \, d\theta = \int_0^{\sqrt{2\pi}} \sqrt{4\theta^2 \sin^2\theta^2 + 4\theta^2 \cos^2\theta^2} \, d\theta$$

$$= \int_0^{\sqrt{2\pi}} 2\theta \, d\theta = \theta^2 \Big]_0^{\sqrt{2\pi}} = 2\pi. \qquad \square$$

Example 3   The path

$$x = t^3 - 3t, \qquad y = 3t^2$$

is the same as our original example of a path except that the y-coordinates are multipled by 3 (see Fig. 3). This makes a big difference in the length computation. There is no obvious way of doing it for the original path, but it is very easy for this modified path. Here it is:

$$s = \int \sqrt{\left(\frac{dx}{dt}\right)^2 + \left(\frac{dy}{dt}\right)^2} \, dt = \int \sqrt{9(t^2 - 1)^2 + 36t^2} \, dt$$

$$= \int \sqrt{9(t^2 + 1)^2} \, dt = \int 3(t^2 + 1) \, dt = t^3 + 3t + C.$$

The closed loop in the path runs from $t = -\sqrt{3}$ to $t = \sqrt{3}$, and its length is

$$s \Big]_{-\sqrt{3}}^{\sqrt{3}} = (t^3 + 3t) \Big]_{-\sqrt{3}}^{\sqrt{3}} = 12\sqrt{3}.$$

Figure 3

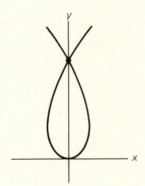

$\square$

Length in Polar Coordinates

The expression of the polar graph

$$r = f(\theta)$$

as the path

$$x = f(\theta)\cos \theta$$

and

$$y = f(\theta)\sin \theta$$

gives us an easy derivation of the arc-length formula in polar coordinates. We have

$$\frac{dx}{d\theta} = f'(\theta)\cos \theta - f(\theta)\sin \theta,$$

$$\frac{dy}{d\theta} = f'(\theta)\sin \theta + f(\theta)\cos \theta,$$

and when we square each equation and add, we find that

$$\left(\frac{dx}{d\theta}\right)^2 + \left(\frac{dy}{d\theta}\right)^2 = [f'(\theta)]^2 + [f(\theta)]^2$$

$$= r^2 + \left(\frac{dr}{d\theta}\right)^2.$$

The parametric arc-length formula

$$s = \int_a^b \sqrt{\left(\frac{dx}{d\theta}\right)^2 + \left(\frac{dy}{d\theta}\right)^2}\, d\theta$$

thus becomes the formula

$$s = \int_a^b \sqrt{[f(\theta)]^2 + [f'(\theta)]^2}\, d\theta$$

$$= \int_a^b \sqrt{r^2 + \left(\frac{dr}{d\theta}\right)^2}\, d\theta$$

for the length of the polar graph of f from $\theta = a$ to $\theta = b$.

Example 4 Find the length of the cardioid $r = 1 + \cos\theta = 2\cos^2(\theta/2)$ shown in Fig. 4.

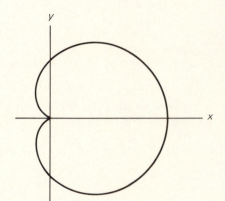

Figure 4

Solution

$$s = \int_0^{2\pi} \sqrt{r^2 + \left(\frac{dr}{d\theta}\right)^2}\, d\theta$$

$$= \int_0^{2\pi} \sqrt{4\cos^4(\theta/2) + 4\cos^2(\theta/2)\sin^2(\theta/2)}\, d\theta$$

$$= 2\int_0^{2\pi} \sqrt{\cos^2\theta/2}\, d\theta = 2\int_0^{2\pi} |\cos(\theta/2)|\, d\theta$$

$$= 4\int_0^{\pi} \cos(\theta/2)\, d\theta = 8\sin(\theta/2)\Big]_0^{\pi}$$

$$= 8.$$

$\square$

Example 5 Find the length of the first revolution of the spiral $r = \theta$.

Solution By the formula above, this length is

$$s = \int_0^{2\pi} \sqrt{r^2 + \left(\frac{dr}{d\theta}\right)^2}\, d\theta = \int_0^{2\pi} \sqrt{\theta^2 + 1}\, d\theta.$$

This is a difficult integral. Substituting first $\theta = \tan u$ and then $\sin u = v$, we get

$$\int \sqrt{\theta^2 + 1}\, d\theta = \int \sec^3 u\, du = \int \frac{\cos u\, du}{(1 - \sin^2 u)^2}$$

$$= \int \frac{dv}{(1 - v^2)^2}.$$

We then express $1/(1 - v^2)^2$ as the sum of its partial fractions. Skipping the details, the answer is

$$\frac{1}{(1 - v^2)^2} = \frac{1}{4}\left[\frac{1}{(1 - v)^2} + \frac{1}{(1 + v)^2} + \frac{1}{1 - v} + \frac{1}{1 + v}\right].$$

Therefore,

$$\int \frac{dv}{(1 - v^2)^2} = \frac{1}{4}\left[\frac{1}{1 - v} - \frac{1}{1 + v} - \ln(1 - v) + \ln(1 + v)\right]$$

$$= \frac{1}{4}\left[\frac{2v}{1 - v^2} + \ln\left(\frac{1 + v}{1 - v}\right)\right].$$

The v limits of integration are found to be 0 and $2\pi/\sqrt{1 + 4\pi^2}$, and when these are substituted the answer simplifies to

$$s = \int_0^{2\pi} \sqrt{\theta^2 + 1}\, d\theta = \pi\sqrt{4\pi^2 + 1} + \frac{1}{2}\ln[\sqrt{4\pi^2 + 1} + 2\pi]. \qquad \square$$

PROBLEMS FOR SECTION 4

Find the derivative of the path length (ds/dt) for each of the following paths.

1. $x = t, y = t^2$

2. $x = \cos t, y = \sin t$

3. $x = 3\sin t, y = \cos t$

4. $x = t^3, y = t^2$

Compute the lengths of the following paths over the given parameter intervals.

5. $3x = 4t^{3/2}, 2y = t^2 - 2t$; $[0, a]$

6. $x = a\cos^3 t, y = a\sin^3 t$; $[0, \pi/2]$

7. $x = t^3 + 3t^2, y = t^3 - 3t^2$; a) $[0, 1]$
b) $[0, 2]$

8. $x = t^3, y = 2t^2$; $[0, 1]$

9. $x = e^t\cos t, y = e^t\sin t$; $[0, 1]$

10. $x = t^2\cos t, y = t^2\sin t$; $[0, 1]$

11. $x = \cos t + t \sin t$, $y = \sin t - t \cos t$; $[0, 1]$

12. A smooth path $(x, y) = (g(t), h(t))$ runs along the graph of a function $y = f(x)$. Show that if x is increasing with t, then the formulas

$$\frac{ds}{dx} = \sqrt{1 + \left(\frac{dy}{dx}\right)^2}, \quad \frac{ds}{dt} = \sqrt{\left(\frac{dx}{dt}\right)^2 + \left(\frac{dy}{dt}\right)^2}$$

are equivalent, by the chain rule.

13. Continue the line of reasoning begun in the above problem, and thus give an alternative proof of the parametric path-length formula

$$s = \int_a^b \sqrt{\left(\frac{dx}{dt}\right)^2 + \left(\frac{dy}{dt}\right)^2} \, dt.$$

14. Show that the formula

$$s = \int_c^d \sqrt{1 + \left(\frac{dx}{dy}\right)^2} \, dy$$

for the length of a graph $x = g(y)$ is a special case of the parametric formula for path length.

Find the lengths of the following polar graphs.

15. The first revolution of the spiral $r = \theta^2$

16. $r = \cos \theta$, from $\theta = 0$ to $\theta = \pi/2$

17. $r = \sec \theta$, from $\theta = 0$ to $\theta = \pi/4$

18. $r = e^\theta$, from $\theta = 0$ to $\theta = a$

19. $r = 1/\theta$, from $\theta = \pi/2$ to $\theta = \pi$

20. Carry out the computation showing that

$$\left(\frac{dx}{d\theta}\right)^2 + \left(\frac{dy}{d\theta}\right)^2 = r^2 + \left(\frac{dr}{d\theta}\right)^2,$$

where $x = r \cos \theta$, $y = r \sin \theta$, and r is a function of θ.

5
THE
PARAMETRIC
MEAN-VALUE
PRINCIPLE

There is a generalization of the mean-value principle that has the same geometric interpretation. Suppose we have a smooth path $(x, y) = (g(t), h(t))$ running from (x_0, y_0) to (x_1, y_1). Then the secant line through these two points must be parallel to the tangent line at some in-between point (Fig. 1). If the lines are not vertical, this is

(x_1, y_1)

(x_0, y_0)

Figure 1

equivalent to their slopes being equal:

$$\frac{y_1 - y_0}{x_1 - x_0} = \text{slope of tangent line.}$$

And we saw in Section 1 that the tangent-line slope is $h'(t)/g'(t)$. We thus arrive at the following theorem. (As before, differentiability is unnecessary at the endpoints.)

THEOREM 5 *Let the functions g and h be continuous on* $[a, b]$ *and differentiable on* (a, b). *Suppose that* $g(a) \neq g(b)$, *and that* $g'(t)$ *and* $h'(t)$ *are never simultaneously zero. Then there is at least one point T strictly between a and b at which*

$$\boxed{\frac{h(b) - h(a)}{g(b) - g(a)} = \frac{h'(T)}{g'(T)}.}$$

Proof The proof is essentially the same as for the mean-value theorem. We try to determine the constant m so that the function

$$k(t) = h(t) - mg(t)$$

has equal values at a and b:

$$h(a) - mg(a) = h(b) - mg(b).$$

When we solve this equation for m, we get

$$m = \frac{h(b) - h(a)}{g(b) - g(a)}$$

as the required value. This is possible because $g(a) \neq g(b)$ by hypothesis. Since now $k(a) = k(b)$, Rolle's Theorem guarantees the existence of a number T lying strictly between a and b such that $k'(T) = 0$. That is,

$$h'(T) = mg'(T).$$

If $g'(T) = 0$, then also $h'(T) = 0$, contradicting the hypothesis of the theorem. Therefore $g'(T) \neq 0$ and we can divide by it to get the equation of the theorem. ∎

The following special case of the theorem better fits some of the applications.

THEOREM 6 *Let g and h be differentiable functions on an interval I containing*
x = a, and suppose that g(a) = h(a) = 0. Suppose also that
g'(x) ≠ 0 when x ≠ a. Then for each x different from a, there is a point
X lying strictly between a and x such that

$$\boxed{\frac{h(x)}{g(x)} = \frac{h'(X)}{g'(X)}.}$$

As a first application of Theorem 6, we let $h(x)$ be the error $E(x)$
in the tangent-line approximation, as discussed in Chapter 5:

$$E(x) = f(x) - f(a) - f'(a)(x - a).$$

Note that $E(a)$ and $E'(a)$ are both zero, and that $E''(x) = f''(x)$. Let
$g(x) = (x - a)^2$. Then $g(a)$ and $g'(a)$ are also both zero. We can then
apply Theorem 6 twice in a row, and conclude that

$$\frac{E(x)}{(x - a)^2} = \frac{E'(y)}{2(y - a)} = \frac{E''(X)}{2} = \frac{f''(X)}{2},$$

where y lies between a and x, and X lies between a and y. Thus,

COROLLARY *Let f be twice differentiable on an interval I containing the point a.*
Then for every other point x in I, the error

$$E = f(x) - [f(a) + f'(a)(x - a)]$$

in the tangent-line approximation to f at a can be written in the form

$$\boxed{E = \frac{f''(X)}{2}(x - a)^2,}$$

where X is some number lying strictly between a and x.

It follows from the above formula that if f'' is bounded by K on
an interval I about a, then

$$|E(x)| \leq \frac{K}{2}(x - a)^2$$

for each x in I.

6
L'HÔPITAL'S
RULES;
INDETERMINATE
FORMS

The new mean-value theorem lets us apply calculus to the evaluation of certain limits.

If $f(x) \to A$ and $g(x) \to B$ as $x \to a$, then the quotient limit law says that

$$\lim_{x \to a} \frac{f(x)}{g(x)} \begin{cases} = \dfrac{A}{B} & \text{if } B \neq 0, \\[2mm] = +\infty & \text{if } B = 0 \text{ and } A > 0, \\[2mm] \text{is undetermined if } A = B = 0; \text{ the law does not apply.} \end{cases}$$

In the third case, we say that $f(x)/g(x)$ has the *indeterminate form* 0/0 at $x = 0$. Sometimes we say simply that *the limit is indeterminate*. This is all right provided it is understood that we are referring to the whole limit process, and not to the answer.

The limit may very well exist. For example,

$$\lim_{x \to 0} \frac{x}{x} = 1,$$

even though the limit is indeterminate.

Some of the earliest limits we looked at had the indeterminate form 0/0. These were the difference quotient limits in our various explicit derivative calculations, such as

$$\lim_{x \to a} \frac{x^3 - a^3}{x - a}, \qquad \lim_{h \to 0} \frac{(a + h)^3 - a^3}{h}.$$

Each such limit required a special device for its evaluation, to get around the 0/0 indeterminacy. But now that this derivative machinery has all been set up, it can be used in the reverse direction to evaluate 0/0 indeterminate forms, according to the following theorem.

THEOREM 7   *Suppose that $f(x) \to 0$ and $g(x) \to 0$ as $x \to a$. Suppose also that*

$$\frac{f'(x)}{g'(x)} \to L$$

as $x \to a$ (where L may be infinite). Then

$$\frac{f(x)}{g(x)} \to L$$

as $x \to a$. This is true whether a is finite or infinite, and whether the limits are one-sided or two-sided.

The theorem hypotheses include the implicit assumptions that $f'(x)$ and $g'(x)$ exist and $g'(x) \neq 0$ for all $x \neq a$ in some open interval I abutting or containing a. The proof will be given after some examples.

In working out a problem, we first check that the desired limit is indeed indeterminate. If both functions are continuous at a, this means checking that $f(a) = g(a) = 0$. Then we write

$$\lim_{x \to a} \frac{f(x)}{g(x)} = \lim_{x \to a} \frac{f'(x)}{g'(x)}.$$

This is understood as meaning that the limit on the left exists and equals the limit on the right, *provided* the right limit is known to exist.

Example 1   Since $1 - x$ and $\ln x$ are both 0 at $x = 1$, their quotient is indeterminate there and we can apply the theorem, with $x \to 1$. Thus

$$\lim_{x \to 1} \frac{1 - x}{\ln x} = \lim_{x \to 1} \frac{-1}{1/x} = -1.$$

Here the second limit had a trivial evaluation since $-1/(1/x) = -x$ is continuous at $x = 1$. But it may happen that the second limit is also indeterminate, and a second application of l'Hôpital's rule is needed to evaluate it. $\square$

Example 2   Evaluate the limit of $(1 - \cos x)/x^2$ as $x \to 0$.

Solution

$$\lim_{x \to 0} \frac{1 - \cos x}{x^2} = \lim_{x \to 0} \frac{\sin x}{2x} = \lim_{x \to 0} \frac{\cos x}{2} = \frac{1}{2}. \qquad \square$$

Before applying l'Hôpital's rule, be sure that its conditions are met. If not, the rule will probably give the wrong answer.

Example 3   Evaluate $\lim_{x \to 1}(x^2 - 1)/(x^2 + 1)$.

Solution   By the quotient rule,

$$\lim_{x \to 1} \frac{x^2 - 1}{x^2 + 1} = \frac{0}{2} = 0.$$

L'Hôpital's rule would give the wrong answer:

$$\lim_{x \to 1} \frac{x^2 - 1}{x^2 + 1} = \lim_{x \to 1} \frac{2x}{2x} = 1.$$

But of course the rule cannot be applied, because the limit at 1 is not indeterminate. □

Example 4 Evaluate $\lim_{x \to 0} (1 + x)^{1/x}$.

Solution The limit

$$\lim_{x \to 0} (1 + x)^{1/x}$$

has the indeterminate form 1^∞ and seems to be outside the scope of the theorem. But we *can* compute the limit of $\ln(1 + x)^{1/x}$ as $x \to 0$, because it can be written in the indeterminate form 0/0. Thus

$$\lim_{x \to 0} \ln(1 + x)^{1/x} = \lim_{x \to 0} \frac{\ln(1 + x)}{x} = \lim_{x \to 0} \frac{1/(1 + x)}{1} = 1.$$

But if we know that $\ln y \to 1$, then $y = e^{\ln y} \to e^1 = e$, since e^x is continuous at $x = 1$. Thus

$$\lim_{x \to 0} (1 + x)^{1/x} = e.$$ □

Example 5

$$\lim_{x \to +\infty} \frac{\ln(1 + 1/x)}{\ln(1 - 1/x)} = \lim_{x \to +\infty} \frac{(-1/x^2)[1/(1 + 1/x)]}{(1/x^2)[1/(1 - 1/x)]}$$

$$= \lim_{x \to +\infty} -\frac{x - 1}{x + 1} = -1.$$ □

Proof of Theorem We start with the case where a is finite. Then f and g become continuous at a if we define $f(a) = g(a) = 0$, and we can therefore apply Theorem 6. Thus for any $x \neq a$ in the domain interval I, there exists a number X lying strictly between x and a such that

$$\frac{f(x)}{g(x)} = \frac{f'(X)}{g'(X)}.$$

Now let $x \to a$. Then $X \to a$, and $f'(X)/g'(X) \to L$, by hypothesis. Since the two quotients are equal, we conclude that $f(x)/g(x) \to L$ as $x \to a$.

The case $x \to +\infty$ is reduced to the case $x \to 0^+$ by a change of variable. See Problem 43. ∎

The Form L'Hôpital's procedure can also be used to evaluate limits having
∞/∞ the indeterminate form ∞/∞.

THEOREM 8 *Suppose that $|f(x)| \to +\infty$ and $|g(x)| \to +\infty$, in each case as $x \to a$. Suppose also that*

$$\frac{f'(x)}{g'(x)} \to L$$

as $x \to a$ (where L may be infinite). Then

$$\frac{f(x)}{g(x)} \to L$$

as $x \to a$.

The proof will be sketched after some examples. Here, also, a can be finite or infinite, and the limits can be two-sided or one-sided.

Example 6 Since x and $\ln x$ both approach ∞ as $x \to +\infty$, we can apply the theorem. Thus

$$\lim_{x \to +\infty} \frac{\ln x}{x} = \lim_{x \to +\infty} \frac{1/x}{1} = \frac{0}{1} = 0. \qquad \square$$

Example 7 The limit

$$\lim_{x \to 0^+} x \ln x$$

has the indeterminate form $0 \cdot \infty$. But we can rewrite it so that Theorem 8 applies, as follows:

$$\lim_{x \downarrow 0} x \ln x = \lim_{x \downarrow 0} \frac{\ln x}{(1/x)} = \lim_{x \downarrow 0} \frac{1/x}{-1/x^2}$$

$$= \lim_{x \downarrow 0} -x = 0. \qquad \square$$

Example 8 Two applications of Theorem 8 give

$$\lim_{x \to +\infty} \frac{x^2}{e^x} = \lim_{x \to +\infty} \frac{2x}{e^x} = \lim_{x \to +\infty} \frac{2}{e^x} = 0. \qquad \square$$

Partial Proof of Theorem 8   We consider the case $x \to +\infty$. If $f(x) \to +\infty$ as $x \to +\infty$, then we can find a function $h(x)$ such that

1. $h(x) \to +\infty$ as $x \to +\infty$;

2. $\dfrac{f(h(x))}{f(x)} \to 0$ as $x \to +\infty$.

We skip showing that $h(x)$ exists and go on. It follows from (2) that $h(x) \to \infty$ much more slowly than does x, in particular, $h(x) < x$ when x is large.

So now suppose that f and g satisfy the hypotheses of the theorem, and let the function h be related in the above way to both f and g. Then

$$\lim_{x \to +\infty} \frac{f(x)}{g(x)} = \lim_{x \to +\infty} \frac{f(x)[1 - f(h(x))/f(x)]}{g(x)[1 - g(h(x))/g(x)]}$$

$$\left(\begin{matrix} \text{because both bracketed} \\ \text{functions} \to 1 \text{ as } x \to \infty \end{matrix} \right)$$

$$= \lim_{x \to +\infty} \frac{f(x) - f(h(x))}{g(x) - g(h(x))}$$

$$= \lim_{x \to +\infty} \frac{f'(X)}{g'(X)} \qquad \text{(where } h(x) < X < x \text{)}$$

by the parametric mean-value theorem. Now $X \to \infty$ as $x \to \infty$ because $X > h(x)$, so the last limit can be rewritten

$$\lim_{X \to \infty} \frac{f'(X)}{g'(X)}.$$

And this limit does exist, with the value L, by hypothesis. So, altogether, we have proved that

$$\frac{f(x)}{g(x)} \to L$$

as $x \to +\infty$.   ■

PROBLEMS FOR SECTION 6

Evaluate the following limits, mostly by l'Hôpital's rule.

1. $\displaystyle \lim_{t \to 0} \frac{\cos t - 1}{t}$ **2.** $\displaystyle \lim_{x \to 0} \frac{e^x - 1}{x}$ **3.** $\displaystyle \lim_{\theta \to 0} \theta \csc \theta$ **4.** $\displaystyle \lim_{x \to 0} x \cot x$

5. $\lim\limits_{x\to 0}\dfrac{x-\ln(1+x)}{x^2}$ **6.** $\lim\limits_{t\to 0}\dfrac{1-\cos t}{t^2}$ **7.** $\lim\limits_{\theta\to 0}\dfrac{1-\cos\theta}{\sin^2\theta}$ **8.** $\lim\limits_{t\to 0}\dfrac{1-\cos^2 t}{t^2}$

9. $\lim\limits_{x\to 0}\dfrac{x-\sin x}{x-\tan x}$ **10.** $\lim\limits_{x\to 0}\dfrac{e^x-x-1}{x^2}$ **11.** $\lim\limits_{t\to 0}\dfrac{2\sqrt{1+t}-t-2}{t^2}$

12. $\lim\limits_{x\to 0}\dfrac{2\cos x+x^2-2}{x^4}$ **13.** $\lim\limits_{x\to 0}\dfrac{\ln(1+x)}{x}$ **14.** $\lim\limits_{x\to 0}\ln(1+x)^{1/x}$

15. $\lim\limits_{x\to 0}x\ln x$ **16.** $\lim\limits_{x\to 0}x\ln(\sin x)$ **17.** $\lim\limits_{t\to 0}\dfrac{\cos t}{1+t^2}$

18. $\lim\limits_{x\to 0}\dfrac{\cos x}{x^2}$ **19.** $\lim\limits_{x\to 0}x\sin\dfrac{1}{x}$

20. Prove the weighted-integral mean-value theorem (Theorem 2 in Chapter 9, Section 1) from the parametric mean-value theorem. (Let F be an antiderivative of fg and let G be an antiderivative of g. Consider the quotient $[F(b)-F(a)]/[G(b)-G(a)]$.)

21. Show that
$$\lim\limits_{x\to 0}\dfrac{f(x)-f(-x)}{x}=2f'(0).$$

22. Show that
$$\lim\limits_{x\to 0}\dfrac{f(x)+f(-x)-2f(0)}{x^2}=f''(0).$$

23. Prove that
$$\lim\limits_{x\to 0}(1+x)^{1/x}=e.$$

More limits.

24. $\lim\limits_{x\to\infty}x^{1/x}$ **25.** $\lim\limits_{x\to\infty}x\sin\dfrac{1}{x}$ **26.** $\lim\limits_{x\to\infty}x(\pi-\arctan x)$

27. $\lim\limits_{x\to\infty}\dfrac{e^{-\sqrt{x}}}{e^{-x}}$ **28.** $\lim\limits_{x\to 0^+}x^x$ **29.** $\lim\limits_{x\to\infty}\dfrac{\ln x}{\sqrt{x}}$

30. $\lim\limits_{x\to\infty}\dfrac{\ln\ln x}{\ln x}$ **31.** $\lim\limits_{x\to\infty}\dfrac{e^x}{x^{100}}$ **32.** $\lim\limits_{x\to\infty}\dfrac{\sqrt{1+x^2}}{x}$

33. $\lim\limits_{x\to 0}\dfrac{e^x+e^{-x}-\sin^2 x+2}{x^4}$ **34.** $\lim\limits_{x\to 1}\dfrac{x^3-x^2+x-1}{x^3+x^2-x-1}$ **35.** $\lim\limits_{x\to 2}\dfrac{x^3-x^2-x-2}{x^3-2x^2+x-2}$

36. $\lim\limits_{x\to 0}\dfrac{3^x-2^x}{\ln(1+x)}$ **37.** $\lim\limits_{x\to e}\dfrac{\ln(\ln x)}{\ln x-1}$ **38.** $\lim\limits_{x\to 0}\dfrac{\ln(\cos x)}{\ln(\cos 2x)}$

39. $\lim\limits_{x\to\infty}\left(1+\dfrac{1}{x}\right)^x$ **40.** $\lim\limits_{x\to +\infty}(1+e^{-x})^x$ **41.** $\lim\limits_{x\to +\infty}\left(1+\dfrac{1}{x}\right)^{\ln x}$ **42.** $\lim\limits_{x\to 0}\dfrac{x-\sin x}{x^3}$

43. Finish the proof of Theorem 7.

If $\lim\limits_{x\to\infty}g(x)=\lim\limits_{x\to\infty}h(x)=0$, and if $g'(x)\neq 0$ for all sufficiently large x, then
$$\lim\limits_{x\to\infty}\dfrac{h(x)}{g(x)}=\lim\limits_{x\to\infty}\dfrac{h'(x)}{g'(x)},$$
in the same sense as before.

[*Hint*: Set $H(y)=h(1/y)$, $G(y)=g(1/y)$, and try to apply the first part of Theorem 7 to H/G.]

CHAPTER 11
INFINITE SERIES

It will be shown in the next chapter that the derivative sum rule remains true for certain "infinitely long polynomials":

$$\text{If}\quad f(x) = \sum_{n=0}^{\infty} a_n x^n, \quad \text{then}\quad f'(x) = \sum_{n=1}^{\infty} n a_n x^{n-1}.$$

Before considering the calculus of such infinite polynomials, however, it is necessary to learn something about the nature of the infinite sum operation. This chapter lays the groundwork.

We start with the infinite decimal representation of a positive real number. This is important in itself and connects with infinite series.

**1
DECIMAL
REPRE-
SENTATIONS**

Ordinary arithmetic is concerned with *rational* numbers, i.e., with quotients of integers, such as 5/3, −2/7, 1/100, 4 = 4/1, and 3.1 = 31/10. A real number that is *not* rational is called *irrational*. For example it can be shown that $\sqrt{2}$, $\sqrt{3}$, π, and e are all irrational. (If p is a positive prime integer, such as 2, 3, 5, or 7, then it is relatively easy to show that $\sqrt{p}$ is irrational. See Problem 17. But the proofs that π and e are irrational are much harder.)

A rational number that can be expressed as a fraction with a power of 10 as denominator is called a *decimal fraction,* and is normally written as a *finite decimal.* For example,

$$\frac{31}{10} = 3.1,$$

$$\frac{141}{100} = 1.41,$$

$$\frac{2}{1000} = 0.002.$$

Finite decimals are simpler to handle than general rational numbers. The rules for adding and multiplying fractions reduce, in this case, to the simpler rules for integers (plus keeping track of the decimal point).

The crucial fact we wish to discuss is that *every real number can be represented by an infinite decimal expansion.* In order to see how this goes, suppose that x is some particular number in the interval $[1, 2]$ and suppose that x is not itself a finite decimal. Figure 1 shows how the one-place decimals divide the number line into intervals of length 1/10. In particular, $[1, 2]$ is divided into ten such intervals. Note that we have written the integers 1 and 2 as the one-place decimals 1.0 and 2.0. The number x must lie in the interior of one of these ten subintervals of $[1, 2]$, that is, between a uniquely determined pair of adjacent one-place decimals. In the figure, x is shown as lying in the subinterval $[1.6, 1.7]$. *The decimal expansion of x records this location by starting off $1.6 \cdots$.*

Figure 1

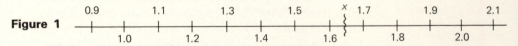

The two-place decimals divide the interval $[1.6, 1.7]$ into ten subintervals, each of length 1/100. for example, $[1.60, 1.61]$ is the first of these and $[1.64, 1.65]$ is the fifth. Again, exactly one of these ten subintervals contains x. It appears from the figure that this might be $[1.64, 1.65]$, in which case the next digit in the decimal expansion of x is 4 and the expansion starts off $1.64 \cdots$. Then the three-place decimals divide $[1.64, 1.65]$ into ten subintervals of length $1/1000 = 10^{-3}$, and the one containing x is recorded by the digit in the third decimal place in the expansion of x. And so on. Thus, *the first n places in the decimal expansion of x are given by the unique n-place decimal*

a_n such that

$$a_n < x < a_n + 10^{-n}.$$

In this way we determine the unique infinite sequence of digits that make up the decimal expansion of x.

The decimal expansion of x can be viewed as the base ten "address" of x on the number line. We have seen that x must *have* such a base ten address, by thinking in a general way about how a particular number x must sit amongst the finite decimals, but *finding* the expansion for x is another matter.

We can find the decimal expansion of a rational number by continued long division. For example, to find the expansion of 27/88, we start dividing 88 into 27, and get

$$\frac{27}{88} = 0.306818181\cdots,$$

where the three dots indicate that the sequence of digits goes on forever. This is a very simple infinite decimal expansion because the two-digit block 81 repeats forever.

The decimal expansion of $\sqrt{2}$ can be obtained by a process (algorithm) for extracting square roots that is often taught in school. It is more complicated than long division, but similar in being nonterminating (unless the beginning number happens to be a finite decimal that is a perfect square, such as $\sqrt{1.44} = 1.2$). If we apply this process to compute $\sqrt{2}$, we get

$$\sqrt{2} = 1.414213\cdots,$$

where again the three dots indicate that the sequence of digits goes on forever. This time, though, there isn't any block of digits that forever repeats.

The decimal expansion representing π begins

$$\pi = 3.14159\cdots.$$

The calculation of this expansion is harder still, but there are various ways of going about it.

At the opposite extreme, an n-place decimal can be viewed as already being an infinite decimal, with all zeros from the nth place on. It is then a special case of a repeating decimal.

Whether a number is rational or irrational shows up in its infinite decimal expansion. An expansion of a rational number is *always* a repeating decimal, as in the example above. After a certain point it consists entirely of repetitions of a certain block of digits. Here is a brief indication of why this is so.

We obtain the decimal expansion of m/n by continued long division. Suppose we have gone far enough so that only zeros are being brought down. Suppose also that no subsequent remainder is ever zero; this rules out m/n being equal to a decimal fraction. The only possible remainders are then the $n-1$ numbers $1, \ldots, n-1$, so during the next n steps of the long division there must be a repetition of a remainder r. Then this remainder r and the block of steps between the two occurrences of r will repeat forever, as will the block of quotient digits arising from these steps.

It is convenient to indicate the repeated block by an overhead bar, and in this notation the first example is

$$\frac{27}{88} = 0.306\overline{81}.$$

Conversely, we shall see in the next section that any repeating decimal represents a rational number. Thus, *the irrational numbers are exactly those numbers having nonrepeating infinite decimal expansions.*

Since the finite decimal 3.14 can be obtained by cutting off the decimal expansion of π after two places, we call 3.14 the two-place *truncation* of the expansion of π. What it says about the location of π was discussed above (and is shown in Fig. 2).

Figure 2
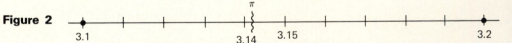

The following examples illustrate the "positioning" role of such truncations. Their solutions depend on the laws for inequalities, which are reviewed in Appendix 1.

Example 1 A classical approximation to π is 22/7. Assuming that the expansion of π begins $3.141\cdots$, show that $\pi < 22/7$ and that $22/7 - \pi < 0.002$.

Solution Dividing out 22/7, we see that its decimal expansion begins $3.142\cdots$. Therefore,

$$3.141 < \pi < 3.142 < \frac{22}{7} < 3.143.$$

This shows that π is less than 22/7 and that

$$\frac{22}{7} - \pi < 3.143 - 3.141 = 0.002.$$ □

Example 2 The decimal expansion of e begins $2.71 \cdots$. Show from this that $\sqrt{e} < 5/3$.

Solution Since $(5/3)^2 = 25/9 = 2.77 \cdots$, we have

$$\left(\frac{5}{3}\right)^2 > 2.77 > 2.72 > e,$$

and hence

$$\frac{5}{3} > \sqrt{e}.$$ □

Example 3 Knowing that the decimal expansions of π and $\sqrt{2}$ start 3.14 and 1.41 respectively, what can we say about $\sqrt{2}\pi$?

Solution We are assuming that

$$3.14 < \pi < 3.15,$$
$$1.41 < \sqrt{2} < 1.42.$$

Multiplying these two inequalities together, we find that

$$4.4274 < \sqrt{2}\,\pi < 4.4730.$$

In particular,

$$4.4 < \sqrt{2}\,\pi < 4.5,$$

so the expansion of $\sqrt{2}\,\pi$ is 4.4, to one decimal place. (But we don't know from this calculation which of 4.4 and 4.5 is the one-place decimal *closest* to $\sqrt{2}\,\pi$.) □

Passing from the decimal expansion of x to the *closest n*-place decimal is called *rounding off* to n places. Thus 3.14 is both the two-place truncation of the expansion of π,

$$3.14 < \pi < 3.15,$$

and also the two-place round-off

$$|3.14 - \pi| < 0.005.$$

On the other hand, the expansion of π through five places is $3.14159\cdots$, so the four-place truncation is 3.1415 while the four-place round-off is 3.1416.

As this example suggests, we can generally obtain the n-place round-off if we know the expansion through $n + 1$ places.

The two-place round-off of the number 0.325 represents a rarely occurring special case, since 0.325 lies exactly halfway between 0.32 and 0.33, and therefore does not determine a unique closest two-place decimal. Here we adopt the artificial convention of rounding to the even digit. Thus 0.325 rounds to 0.32, while 0.335 rounds to 0.34.

Tables contain rounded-off values. Table 1 in Appendix 6 gives 0.6931 as the value for $\ln 2$, which means that

$$|\ln 2 - 0.6931| < 5(10)^{-5},$$

or

$$\ln 2 \approx 0.6931$$

with an error less than $5(10)^{-5}$ in magnitude. Calculator values may be either rounded off or truncated.

Completeness of $\mathbb{R}$

We saw earlier how each positive number x determines an infinite decimal expansion that records the exact location of x on the number line. We now proceed in the opposite direction. Each infinite decimal defines a sequence of intervals in the manner discussed above, each being one-tenth of the preceding one, and altogether squeezing down on an exact location on the number line. And each such exact location is marked by a point on the number line, that is, by a real number. Thus *every infinite decimal is the expansion of a real number*. This property is called the *completeness* of the real number system. It says that every "position" amongst the finite decimals is occupied by a number.

For flexibility, we need a more general formulation of this property. For example, if we compute in base two arithmetic, then we end up specifying a real number by a different type of interval sequence, where each interval is one of two *halves* of the preceding interval. So we need the fact that *any* squeezing-down sequence of closed intervals determines a number.

DEFINITION

A sequence of closed intervals $I_n = [a_n, b_n]$ is *nested* if each interval I_n includes the next interval I_{n+1} (and hence includes all later intervals I_{n+j}).

This just means that $a_n \leq a_{n+1}$ and $b_n \geq b_{n+1}$ for each n (and, of course, $a_n \leq b_n$), so it is a simple way of saying simultaneously that $\{a_n\}$ is an increasing sequence, that $\{b_n\}$ is a decreasing sequence, and that $a_n \leq b_n$ for all n.

Example 4   The intervals $[a_n, b_n] = [-1/n, 1/n]$ form a nested sequence.   □

> **The Nested-interval Principle**   *Suppose that $\{I_n\}$ is a nested sequence of closed intervals whose lengths approach 0 as n tends to infinity. Then there is a unique real number x that lies in them all.*

If $I_n = [a_n, b_n]$ in the nested-interval principle, then $a_n \to x$ as $n \to \infty$. Thus, in Example 4, $x = 0$ and $a_n = -1/n \to 0$ as $n \to \infty$. For another example, if a_n is the n-place truncation of the decimal expansion of x, then $a_n \to x$ as $n \to \infty$.

The formal proof first converts the inequality $a_n \leq x \leq b_n$ to the form

$$0 \leq x - a_n \leq b_n - a_n.$$

Then, since $b_n - a_n \to 0$, it follows that $x - a_n \to 0$ by the squeeze limit law. Therefore, $a_n \to x$. Similarly, $b_n \to x$ as $n \to \infty$.

Also useful is the following "one-sided" formulation of completeness:

> **The Monotone-limit Principle**   *If the sequence $\{a_n\}$ is increasing and bounded above, then it necessarily converges.*

This version also is very plausible: if the numbers a_n increase with n, and are blocked from above by a number b, then it seems that they must pile up at some point $l \leq b$, in which case l must be the limit of the sequence. (This is the same way we reasoned in Chapter 2 when we were talking about an increasing function having one-sided limits.) Furthermore,

$$a_n \leq l$$

for all n. This is to be expected: if a_n increases to the limit l, then l sits over all the sequence terms a_n. The formal proof is an application of the limit law D for sequences. (See Chapter 6, Section 2.)

The monotone-limit principle can be proved from the nested-interval principle, and conversely. See Problems 18 through 24 below, and also Appendix 5.

The completeness of the real numbers is the basic fact underlying all the theory needed for calculus. Appendix 5 discusses another form of this property and derives some theoretical consequences.

PROBLEMS FOR SECTION 1

1. Archimedes proved that

$$3\frac{10}{71} < \pi < 3\frac{1}{7}.$$

Show from this that 22/7 approximates π with an error less than 1/497.

3. From a classical partial-fraction expansion of e, we find that

$$2\frac{334}{465} < e < 2\frac{385}{536}.$$

Show therefore that either of these rational numbers approximates e with an error less than 1/249,000.

2. Find what the above double inequality tells us about the decimal expansion of π.

4. Compute e from the above data to the nearest five decimal places.

Here are some decimal expansions through two decimal places:

$$\pi = 3.14\cdots, \qquad \sqrt{2} = 1.41\cdots, \qquad e = 2.71\cdots,$$
$$\ln 3 = 1.09\cdots, \qquad \sin\sqrt{2} = 0.98\cdots.$$

Determine what can be said about the one-place decimal approximations to each of the following products. In each case the answer should be *either* the decimal expansion through one decimal place, *or* the closest one-place decimal approximation (or both).

5. $e\sqrt{2}$ **6.** $\pi \ln 3$ **7.** $\sqrt{2}\sin\sqrt{2}$ **8.** e^2

9. $\pi \sin\sqrt{2}$ **10.** $e \ln 3$ **11.** $\ln 3 \sin\sqrt{2}$

12. If $|x - 1.4849| < 0.0001$, how close must x be to 1.485?

13. If $|x - 1.099| < 0.005$, how close must x be to 1.1?

14. If $|x - 4.473| < 0.001$, how close must x be to 4.47?

15. If $|x - 2.152| < 0.003$, how close must x be to 2.15? 2.16?

16. If $|x - 2.289| < 0.001$, how close must x be to 2.28? 2.29?

17. Prove that $\sqrt{2}$ is irrational, as follows. If $\sqrt{2} = m/n$, then $m^2 = 2n^2$. Assume that each integer k can be written in a unique way as a product of a power of 2 and an odd integer. Then derive a contradiction.

19. Let $I_n = [a_n, b_n]$ be a nested sequence of closed intervals. Use the monotone-limit property and its alternative form above, to prove that both endpoint sequences $\{a_n\}$ and $\{b_n\}$ converge. If their limits are a and b, respectively, prove that $a \leq b$.

18. Use the monotone-limit property (and the limit laws) to prove the following variant:

A decreasing sequence that is bounded below necessarily converges.

20. Prove the nested-interval property from the monotone-limit property. (The problem above is part of this proof, and your job here is to finish it.)

The following four problems outline a proof of the monotone-limit property from the nested interval property. You are supposed to finish off each step, assuming the earlier results.

21. Suppose that $\{a_n\}$ is an increasing sequence lying entirely in $[0, 1]$. Fix the positive integer m and consider the m-place decimals in $[0, 1]$. Show that among these m-place decimals, there is a smallest one, b_m, that is an upper bound of the sequence $\{a_n\}$.

22. We now have an upper bound b_m for each m. Show that this new sequence $\{b_m\}$ is decreasing.

23. Show next that
$$b_n - a_n \to 0 \quad \text{as } n \to \infty.$$
[*Hint*: Fix N and show that $b_N - 1/10^N$ is less than some a_n, with $n > N$. Then picture the four numbers
$$b_N, \quad b_N - \frac{1}{10^N}, \quad a_n, \quad \text{and} \quad b_n$$
on the line.]

24. Finally, prove that $\{a_n\}$ converges, by the nested-interval property.

25. Suppose that $f(x)$ is an increasing function on $(0, \infty)$, and that $f(x) < B$ for all x. Arguing informally, show that $\lim_{x \to \infty} f(x)$ exists and is at most B. (First consider the sequence $a_n = f(n)$.)

2
THE SUM OF AN INFINITE SERIES

We shall now see that we can sometimes calculate what the sum of an infinite collection of numbers ought to be even though we can't actually carry out the infinite number of additions that seem to be required. For example, we shall see below that there is a sense in which the sum of all the powers of 1/2, starting from $1 = (1/2)^0$, is exactly equal to 2,

$$1 + \frac{1}{2} + \frac{1}{4} + \frac{1}{8} + \cdots = 2,$$

even though the actual calculation of the infinitely many additions indicated on the left is impossible.

This is an example of a *geometric series*. For the general geometric series, we are given a number t between -1 and 1, and we want to calculate the infinite sum of all the powers of t,

$$1 + t + t^2 + \cdots + t^n + \cdots.$$

We clearly can't perform all the additions that are indicated above, but we *can* start adding and see what happens as we keep on going. The sum of the terms from 1 through t^n is called the nth *partial sum*, and is designated s_n. Thus,

$$s_1 = 1 + t,$$
$$s_2 = 1 + t + t^2,$$
$$\vdots \qquad \vdots$$
$$s_n = 1 + t + t^2 + \cdots + t^n.$$

Suppose there is an exact infinite sum s. Then it should seem likely, as we keep on adding, that this accumulating partial sum s_n comes closer and closer to s. That is, we ought to be able to compute the exact infinite sum s as the limit:

$$s = \lim_{n \to \infty} s_n.$$

This computation actually is possible for the geometric series, because of the factorization formula

$$1 - t^{n+1} = (1 - t)(1 + t + t^2 + \cdots + t^n).$$

We can write this as

$$1 - t^{n+1} = (1 - t)s_n,$$

so

$$s_n = \frac{1 - t^{n+1}}{1 - t}.$$

But we know that $t^{n+1} \to 0$ as $n \to \infty$, because $|t| < 1$. Therefore,

$$s_n = \frac{1 - t^{n+1}}{1 - t} \to \frac{1}{1 - t}$$

as $n \to \infty$. Thus, supposing that there is an exact infinite sum s of the geometric series $1 + t + \cdots + t^n + \cdots$, we can calculate it as the limit of the accumulating partial sum s_n:

$$s = \lim_{n \to \infty} s_n = \frac{1}{1-t}.$$

Example 1

$$1 + \frac{1}{2} + \frac{1}{4} + \cdots + \left(\frac{1}{2}\right)^n + \cdots = \frac{1}{1 - \frac{1}{2}} = 2. \qquad \square$$

Example 2

$$1 + \frac{3}{7} + \frac{9}{49} + \cdots + \left(\frac{3}{7}\right)^n + \cdots = \frac{1}{1 - (3/7)} = \frac{7}{4}. \qquad \square$$

Example 3 If $t = -1/5$, then we have

$$1 - \frac{1}{5} + \frac{1}{25} + \cdots + \left(-\frac{1}{5}\right)^n + \cdots = \frac{1}{1 - (-1/5)} = \frac{5}{6}. \qquad \square$$

Now comes a slight shift in point of view. Further investigation will not turn up any other way to calculate the exact infinite sum than as the limit $s = \lim s_n$, and we therefore *define* the infinite sum to be the number calculated in this way. We also say that the series *converges*, and that it *converges to s*. Therefore,

THEOREM 1 *If $|t| < 1$, then the geometric series $1 + t + t^2 + \cdots + t^n + \cdots$ converges, and its sum is $1/(1 - t)$.*

The sum of any other infinite series

$$a_1 + a_2 + \cdots + a_n + \cdots$$

is investigated in the same way. We form the sum s_n of the terms through a_n, and then see what happens as we keep on adding, i.e., as n increases.

DEFINITION If the partial sums s_n approach a limit s as $n \to \infty$ then we define s to be the exact *sum* of the infinite series, and we say that *the series converges to s*.

Example 4 The finite decimals $s_1 = 3.1$, $s_2 = 3.14$, $s_3 = 3.141, \cdots$, obtained by truncating the infinite decimal expansion of π, are just the partial sums of the infinite series

$$3 + \frac{1}{10} + \frac{4}{(10)^2} + \frac{1}{(10)^3} + \frac{5}{(10)^4} + \frac{9}{(10)^5} + \cdots.$$

Therefore, the fact that $s_n \to \pi$ as $n \to \infty$ can now be reinterpreted as the fact that this series converges to π.   □

When an infinite series fails to have a sum we say that it *diverges*. This can happen in two quite different ways. Suppose that we consider the infinite series

$$1 + 1 + \cdots + 1 + \cdots,$$

that is, the series

$$a_1 + a_2 + \cdots + a_n + \cdots$$

for which all the terms a_n have the value 1. For this series, the partial sum s_n is just the integer n, which becomes arbitrarily large and doesn't approach any number s. We say that s_n tends to $+\infty$ (plus infinity), and that the series *diverges* to $+\infty$.

A more interesting example of a series that diverges to $+\infty$ is the so-called *harmonic* series,

$$1 + \frac{1}{2} + \frac{1}{3} + \frac{1}{4} + \cdots + \frac{1}{n} + \cdots.$$

Here, the fact that the series diverges isn't so obvious, principally because we don't have a simple formula for the partial sum s_n that we can examine. However, by gathering the terms into groups we see that

$$1 + \frac{1}{2} + \overline{\frac{1}{3} + \frac{1}{4}} + \overline{\frac{1}{5} + \frac{1}{6} + \frac{1}{7} + \frac{1}{8}} + \cdots$$

$$\geq 1 + \frac{1}{2} + \underbrace{\frac{1}{4} + \frac{1}{4}} + \underbrace{\frac{1}{8} + \frac{1}{8} + \frac{1}{8} + \frac{1}{8}} + \cdots$$

$$= 1 + \frac{1}{2} \quad + \frac{1}{2} \quad\quad + \frac{1}{2} \quad\quad + \cdots.$$

Thus, by going out along the harmonic series far enough, we can make the partial sum s_n larger than $1 + m(1/2)$, no matter how large we have taken the integer m. Therefore, the harmonic series diverges to $+\infty$.

The other kind of divergent behavior is illustrated by taking $a_n = (-1)^n$ so that the series (starting with a_0) is

$$1 + (-1) + 1 + (-1) + \cdots + (-1)^n + \cdots.$$

After we have added n terms, we have either 0 or 1, depending on whether n is even or odd. The partial sum doesn't grow arbitrarily large, but neither does it settle down near one number. Instead it oscillates forever.

These two "pure" forms of divergence can be combined, as in

$$1 + (-2) + 3 + (-4) + \cdots + n(-1)^{n+1} + \cdots .$$

The following condition is frequently useful.

THEOREM 2   *If the series $a_0 + a_1 + \cdots$ converges, then $a_n \to 0$ as $n \to \infty$. Therefore, if the nth term a_n does not approach the limit zero as $n \to \infty$, then the series $a_0 + a_1 + \cdots$ diverges.*

Proof   Suppose that the series converges, with sum s. That is, the partial sum s_n approaches the limit s as $n \to \infty$. Arguing informally, we can see that since s_n and s_{n-1} are both very close to s, their difference, which is a_n, must be close to zero.

The formal argument would be that since we have both
$$s_n \to s \quad \text{and} \quad s_{n-1} \to s \quad \text{as} \quad n \to \infty, \quad \text{then} \quad a_n = s_n - s_{n-1} \to s - s = 0. \quad\blacksquare$$

The harmonic series shows that this theorem cannot be reversed: it is a divergent series and yet the general term $1/n$ has the limit zero. The condition that $a_n \to 0$ is thus *necessary*, but not *sufficient*, for the convergence of the series $a_1 + a_2 + \cdots + a_n + \cdots$.

Properties of Series Sums   In working with the sums of convergent series we are helped by the fact that they behave in many ways like finite sums. For example, we shall see later on that

$$e = 1 + 1 + \frac{1}{2} + \frac{1}{3!} + \frac{1}{4!} + \cdots + \frac{1}{n!} + \cdots ,$$

and that

$$1/e = 1 - 1 + \frac{1}{2} - \frac{1}{3!} + \frac{1}{4!} + \cdots + (-1)^n \frac{1}{n!} + \cdots .$$

(Here 4!, read "4 factorial," is $4 \cdot 3 \cdot 2 \cdot 1$, and

$$n! = n(n - 1)(n - 2) \cdots 3 \cdot 2 \cdot 1.)$$

Assuming these two equations, we would certainly expect to be able to add the two series term by term, and to conclude that

$$e + 1/e = 2 + 2\left(\frac{1}{2}\right) + 2\left(\frac{1}{4!}\right) + \cdots + 2\left(\frac{1}{(2m)!}\right) + \cdots.$$

Then we would like to be able to multiply throughout by 1/2, ending up with

$$\frac{e + 1/e}{2} = 1 + \frac{1}{2!} + \frac{1}{4!} + \cdots + \frac{1}{(2m)!} + \cdots.$$

We would also like to be able to prove that $e < 3$ by the following argument:

$$e = 1 + 1 + \frac{1}{2} + \frac{1}{3!} + \frac{1}{4!} + \cdots + \frac{1}{n!} + \cdots$$

$$= 1 + 1 + \frac{1}{2} + \frac{1}{2 \cdot 3} + \frac{1}{2 \cdot 3 \cdot 4} + \cdots$$

$$< 1 + 1 + \frac{1}{2} + \frac{1}{2 \cdot 2} + \frac{1}{2 \cdot 2 \cdot 2} + \cdots$$

$$= 1 + \left[1 + \frac{1}{2} + \left(\frac{1}{2}\right)^2 + \left(\frac{1}{2}\right)^3 + \cdots + \left(\frac{1}{2}\right)^n + \cdots\right]$$

$$= 1 + \frac{1}{1 - 1/2} = 1 + 2 = 3.$$

The rules that let us do things like this have to be derived from the rules about sequential limits, because a series sum *is* a sequence limit.

However, before looking at these laws we introduce the sigma ($\sum$) notation for infinite series. Remember that $\sum_{j=1}^{n} a_j$ is the sum of the terms a_j from $j = 1$ to $j = n$:

$$\sum_{j=1}^{n} a_j = a_1 + a_2 + \cdots + a_n.$$

This notation was discussed in Chapter 6, Section 2. It is then natural to denote the sum of a convergent infinite series by

$$\sum_{j=1}^{\infty} a_j \qquad \text{or} \qquad \sum_{j=0}^{\infty} a_j.$$

If we start at the mth term, the sum is $\sum_{j=m}^{\infty} a_j$. We sometimes designate the series itself by

$$\sum a_j \quad \text{or} \quad \sum_{j=1} a_j,$$

the rationale being that since we are not indicating the sum to any particular point we are looking at the "unsummed" series. Also, we shall often suppress the index under $\sum$ when it is obvious what the index should be, as in

$$\sum_1^\infty a_j,$$

The general laws for combining series can be discussed conveniently in the $\sum$ notation. In the statements of these laws, any equation of the form $\sum_1^\infty x_j = X$ is to be read:

The infinite series $\sum_1^\infty x_j$ is convergent and its sum is X.

THEOREM 3 *If $\sum_1^\infty a_j = A$ and $\sum_1^\infty b_j = B$, then*

$$\sum_1^\infty (a_j + b_j) = A + B$$

and

$$\sum_1^\infty ca_j = cA,$$

where c is any constant. Furthermore, if $a_j \le b_j$ for all j, then

$$A \le B.$$

Proof We are assuming that the partial sums $\sum_1^n a_j$ and $\sum_1^n b_j$ have the limits A and B, respectively, as n tends to ∞. Since we can rearrange a finite sum in any way, it then follows that

$$\sum_1^n (a_j + b_j) = \sum_1^n a_j + \sum_1^n b_j \to A + B$$

as $n \to \infty$, by the law for the limit of the sum of two sequences. This shows that

$$\sum_{j=1}^\infty (a_j + b_j) = A + B,$$

by the definition of the infinite sum. Similarly,

$$\sum_{1}^{n} ca_j = c\sum_{1}^{n} a_j \to cA$$

as $n \to \infty$, so that

$$\sum_{j=1}^{\infty} ca_j = cA.$$

Finally, if $a_j \leq b_j$ for all j, then

$$\sum_{j=1}^{n} a_j \leq \sum_{j=1}^{n} b_j,$$

for all n. Therefore, in the limit, $A \leq B$. This completes the proof of the theorem. ∎

Whether a series converges or not is independent of any given initial block of terms, because we always have the identity

$$\sum_{1}^{\infty} a_j = \sum_{1}^{N} a_j + \sum_{N+1}^{\infty} a_j,$$

and any finite sum is always defined. This means that if either of the two infinite series converges, then so does the other, with equality holding between the sums. It can be considered a special case of Theorem 3 if each of the two incomplete sums on the right is considered to be a full infinite series with zeros in all the missing positions.

The first term on the right above is the partial sum s_N for the given series. The second term, the series sum from the $(N + 1)$st term on, is called the Nth *remainder series*, and is frequently designated r_N. Thus, if the series converges, then the equation

$$s = s_N + r_N$$

expresses the infinite sum s as the sum of the first N terms plus the remainder. Note that then

$$r_N \to 0$$

as $N \to \infty$, since $r_N = s - s_N$ and $s_N \to s$.

Example 5   If we assume the formula

$$1 + t + \cdots + t^n + \cdots = \frac{1}{1 - t},$$

then we can recover the partial-sum formula algebraically, as follows:

$$\frac{1}{1-t} = s_n + r_n = s_n + [t^{n+1} + t^{n+2} + \cdots]$$

$$= s_n + t^{n+1}[1 + t + \cdots]$$

$$= s_n + t^{n+1} \cdot \frac{1}{1-t}.$$

Therefore

$$s_n = \frac{1}{1-t} - \frac{t^{n+1}}{1-t} = \frac{1 - t^{n+1}}{1-t}. \qquad \square$$

Example 6 We can now see why a repeating decimal always represents a rational number: it is essentially a geometric series from a certain point on. Consider our early example, $0.306\overline{81}$. We group together the terms in the repeating block, and write the infinite decimal as the series sum,

$$\frac{306}{1000} + \frac{1}{1000}\left[\frac{81}{100} + \frac{81}{(100)^2} + \cdots\right]$$

$$= \frac{306}{1000} + \frac{1}{1000} \cdot \frac{81}{100} \cdot \sum_{k=0}^{\infty} \frac{1}{100^k}$$

$$= \frac{306}{1000} + \frac{1}{1000} \cdot \frac{81}{100} \cdot \frac{100}{99} = \frac{306}{1000} + \frac{1}{1000} \cdot \frac{9}{11}$$

$$= \frac{306 \cdot 11 + 9}{11,000} = \frac{3375}{11,000} = \frac{27}{88}. \qquad \square$$

Example 7 This geometric series interpretation also shows that each finite decimal has two decimal expansions. Thus,

$$0.1999\cdots = 0.1\overline{9} = \frac{1}{10} + \frac{9}{(10)^2} + \frac{9}{(10)^3} + \cdots$$

$$= \frac{1}{10} + \frac{9}{100} \sum_{0}^{\infty} \frac{1}{10^j}$$

$$= \frac{1}{10} + \frac{9}{100} \cdot \frac{10}{9} = \frac{2}{10} = 0.2\overline{0}.$$

Any other type of number has a unique decimal expansion. This was shown at the beginning of the last section. $\qquad \square$

PROBLEMS FOR SECTION 2

Find the sum of each of the following series.

1. $1 + \frac{1}{4} + \frac{1}{16} + \frac{1}{64} + \cdots + \left(\frac{1}{4}\right)^n + \cdots$

2. $1 - \frac{1}{6} + \frac{1}{36} - \frac{1}{216} + \cdots + \left(-\frac{1}{6}\right)^n + \cdots$

3. $1 - \frac{2}{3} + \frac{4}{9} - \frac{16}{27} + \cdots + \left(-\frac{2}{3}\right)^n \cdots$

4. $1 + \frac{3}{4} + \frac{9}{16} + \frac{27}{64} + \cdots + \left(\frac{3}{4}\right)^n + \cdots$

5. $2 + \frac{2}{3} + \frac{2}{9} + \frac{2}{27} + \cdots + 2\left(\frac{1}{3}\right)^n + \cdots$

6. $3 + \frac{6}{5} + \frac{12}{25} + \frac{24}{125} + \cdots + 3\left(\frac{2}{5}\right)^n + \cdots$

7. $\frac{1}{4} - \frac{1}{8} + \frac{1}{16} + \cdots + \frac{1}{4}\left(-\frac{1}{2}\right)^n + \cdots$

8. $\frac{7}{6} + \frac{35}{36} + \frac{175}{216} + \frac{875}{1296} + \cdots + \frac{7}{6}\left(\frac{5}{6}\right)^n + \cdots$

9. $\frac{1}{4} + \frac{1}{6} + \frac{1}{9} + \frac{2}{27} + \cdots + \frac{1}{4}\left(\frac{2}{3}\right)^n + \cdots$

10. $\sum_{n=0}^{\infty} \frac{2^n}{3^n}$

11. $\sum_{n=3}^{\infty} \frac{3^n}{4^n}$

12. $\sum_{0}^{\infty} \frac{2^n}{5^{n/2}}$

Suppose that each of the following partial sums continues as a geometric series, possibly multiplied by a constant k, so that the nth term is kr^n. Find its sum.

13. $3 + \frac{9}{4} + \cdots$

14. $\frac{1}{9} + \frac{2}{27} + \cdots$

15. $\frac{1}{5} - \frac{1}{25} + \cdots$

16. $\frac{1}{2} + \frac{1}{3} + \cdots$

17. Show that a geometric series $\sum r^n$ diverges if $|r| \geq 1$. (Apply Theorem 2.)

18. Suppose that $a_n \geq 1/n$ for every n. Assuming that the harmonic series diverges to $+\infty$, show that the series $\sum a_n$ diverges to $+\infty$.

Assuming the above fact (and possibly using Theorem 3), show that each of the following series is divergent to $+\infty$.

19. $\frac{1}{2} + \frac{1}{4} + \cdots + \frac{1}{2n} + \cdots$
(Suppose it converges and multiply by 2.)

20. $1 + \frac{1}{3} + \frac{1}{5} + \cdots + \frac{1}{2n-1} + \cdots$

21. $\frac{1}{4} + \frac{1}{7} + \frac{1}{10} + \cdots + \frac{1}{3n-1} + \cdots$

22. Prove the following divergence criteria from Theorem 3.

    a) If $\sum a_n$ diverges and $c \neq 0$, then $\sum ca_n$ diverges.

    b) If $\sum (a_n + b_n)$ diverges, then either $\sum a_n$ diverges or $\sum b_n$ diverges.

Express each of the following repeating decimals as an infinite series and compute its sum.

23. $0.232323\cdots$

24. $0.\overline{63}$

25. $0.\overline{315}$

26. $0.012012012\cdots$ **27.** $2.\overline{01}$ **28.** $4.162162162\cdots$

29. Show that $s_n = 1 - 1/(n + 1)$ if s_n is the nth partial sum of the following series:

$$\frac{1}{2} + \frac{1}{6} + \frac{1}{12} + \cdots + \frac{1}{n(n + 1)} + \cdots.$$

[*Hint*: $(1/6) = (1/2) - (1/3)$.]

30. Consider the series

$$1 + \frac{1}{4} + \frac{1}{9} + \frac{1}{16} + \cdots + \frac{1}{n^2} + \cdots.$$

a) Show that its sequence of partial sums s_n is strictly increasing.

b) Show that $t_n = s_n + 1/n$ is decreasing sequence.

c) Conclude that the present series is convergent and that its sum is less than 2.

Show that each of the following series converges, and find its sum.

31. $\dfrac{1}{1\cdot 3} + \dfrac{1}{3\cdot 5} + \dfrac{1}{5\cdot 7} + \cdots$

$$+ \frac{1}{(2n - 1)(2n + 1)} + \cdots$$

[*Hint*: $2/(3\cdot 5) = (1/3) - (1/5)$.]

32. $\dfrac{1}{2\cdot 4} + \dfrac{1}{4\cdot 6} + \dfrac{1}{6\cdot 8} + \cdots$

33. $\dfrac{1}{1\cdot 3} + \dfrac{1}{2\cdot 4} + \dfrac{1}{3\cdot 5} + \cdots + \dfrac{1}{n(n + 2)}$

34. Find the sum of the series

$$\frac{1}{1\cdot 4} + \frac{1}{2\cdot 5} + \frac{1}{3\cdot 6} + \cdots + \frac{1}{n(n + 3)}.$$

35. A ball bearing is dropped from 10 feet onto a heavy metal plate. Being very elastic, but not *perfectly* elastic, the ball bounces each time to a height that is 19/20 of its preceding height. How far does the bounding ball travel altogether?

36. A man walks a mile, from A to B, at a steady pace of 3 miles per hour. According to an ancient paradox attributed to the Greek Zeno, he will never get to B, because he first must pass the halfway point, then the 3/4 point, then the 7/8 point, and so on, so that he must pass an infinite number of distance markers on the way, which is clearly impossible. Our modern answer to Zeno is that the man does get to B, in 20 minutes, because two different geometric series each have a finite sum. Explain.

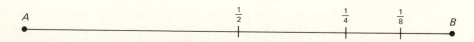

3
THE
COMPARISON
TEST

The next theorem is the crux of the whole subject. It allows us to establish the convergence of a series by comparing its terms with those of a known convergent series, and is called the *comparison test for convergence.*

THEOREM 4   **The Comparison Test**   *Suppose that* $0 \leq a_n \leq b_n$ *for all n. Then,*

(i) *if* $\sum b_j$ *converges then* $\sum a_j$ *converges and*

$$\sum_1^\infty a_j \leq \sum_1^\infty b_j;$$

(ii) *if* $\sum a_j$ *diverges then* $\sum b_j$ *diverges.*

Example 1   We saw in the last section that the series

$$1 + \frac{1}{2!} + \frac{1}{3!} + \cdots + \frac{1}{n!} \cdots$$

is dominated, in the sense of the theorem, by a geometric series whose sum is 2. Therefore, the displayed series is convergent and its sum is at most 2.

The series

$$1 + \frac{1}{\sqrt{2}} + \cdots + \frac{1}{\sqrt{n}} + \cdots$$

diverges, because $1/\sqrt{n} \geq 1/n$ for all n and $\sum 1/n$ is known to diverge. $\square$

Proof of Theorem   We prove (i). Let s_n be the nth partial sum of the series $\sum a_j$, and set $B = \sum_1^\infty b_j$. Then:

1. the sequence $\{s_n\}$ is increasing, because

$$s_n - s_{n-1} = a_n \geq 0;$$

2. the sequence $\{s_n\}$ is bounded above by B, since

$$s_n = \sum_1^n a_j \leq \sum_1^n b_j \leq B.$$

Therefore, the sequence $\{s_n\}$ converge, by the monotone-limit principle. That is, the series $\sum a_n$ is convergent. The inequality was part of Theorem 3. The second statement in the theorem follows from the first. ∎

If a number s is the sum of an infinite series, we may be able to use the partial sums s_n to estimate s. The error $s - s_n$ in the approximation is the sum of the nth remainder series

$$r_n = a_{n+1} + a_{n+2} + \cdots,$$

and we may be able to show that $s - s_n$ is small by comparing this remainder series with a convergent series whose sum is known. This type of comparison doesn't use the full force of Theorem 4, but only the inequality between the sums. It really is an application of Theorem 3.

Example 2 This is the ultimate explanation of our original interpretation of an infinite decimal expansion. We noted in Example 2 in the last section that the decimal expansion

$$\pi = 3.14159\cdots$$

means that π is the sum of the infinite series

$$\pi = 3 + \frac{1}{10} + \frac{4}{10^2} + \frac{1}{10^3} + \frac{5}{10^4} + \frac{9}{10^5}\cdots.$$

The second remainder series,

$$r_2 = \frac{1}{10^3} + \frac{5}{10^4} + \frac{9}{10^5} + \cdots,$$

is term-by-term dominated by the series

$$\frac{9}{10^3} + \frac{9}{10^4} + \frac{9}{10^5} + \cdots = \frac{9}{10^3}\left[\sum_{k=0}^{\infty}\frac{1}{10^k}\right] = \frac{9}{10^3}\cdot\frac{10}{9} = \frac{1}{10^2}.$$

Therefore $r_2 < 0.01$, and

$$3.14 = s_2 < \pi = s_2 + r_2 < 3.14 + 0.01 = 3.15.$$

That is, the bracketing inequality

$$3.14 < \pi < 3.15$$

is now seen to be a consequence of the comparison inequality for series sums, applied to the second remainder series r_2. $\quad\square$

Example 3 Assuming that e is the sum of the series

$$e = 1 + 1 + \frac{1}{2!} + \cdots + \frac{1}{n!} + \cdots,$$

we can compute e by comparing its remainder series with a geometric series. Suppose, for example, that we compare the remainder series r_4 term-by-term with the geometric series that its first terms suggest. We have

$$r_4 = \frac{1}{5!} + \frac{1}{6!} + \frac{1}{7!} + \cdots$$

$$= \frac{1}{5!}\left[1 + \frac{1}{6} + \frac{1}{6 \cdot 7} + \cdots\right]$$

$$< \frac{1}{5!}\left[1 + \frac{1}{6} + \frac{1}{6^2} + \cdots\right] = \frac{1}{5!}\left[\frac{1}{1 - 1/6}\right] = 0.01.$$

Thus s_4 approximates e with an error at most 1 in the second decimal place. We have

$$s_4 = 1 + 1 + \frac{1}{2} + \frac{1}{6} + \frac{1}{24} = 2 + \frac{17}{24} = 2.708\bar{3},$$

and

$$2.708\bar{3} < e < 2.718\bar{3}.$$

Actually, the upper bound $s_4 + 0.01$ is *much* closer to e than is s_4. This is because the geometric series fits the remainder r_4 so closely. Using the first four terms of r_4, we find that

$$0.01 - r_4 = \frac{1}{5!}\frac{6}{5} - r_4 < \frac{1}{5!}\left[\frac{1}{5} - \frac{1}{6} - \frac{1}{6 \cdot 7} - \frac{1}{6 \cdot 7 \cdot 8}\right] = \frac{1}{5!}\left(\frac{11}{5 \cdot 6 \cdot 7 \cdot 8}\right)$$

$$< 6 \cdot 10^{-5}.$$

So $r_4 > 0.01 - 6 \cdot 10^{15}$, and

$$e = s_4 + r_4 > s_4 + 0.01 - 6 \cdot 10^{-5} = 2.718273.$$

Thus

$$2.718273 < e < 2.718333,$$

and

$$e = 2.7183$$

to the nearest four places.   □

In many cases, the comparison between $\sum a_n$ and $\sum b_n$ can be made most easily by showing that the sequence a_n/b_n has a limit.

THEOREM 5 **The Limit Comparison Test** *Let* $\sum a_n$ *and* $\sum b_n$ *be series with positive terms, and suppose that* $\lim_{n\to\infty}(a_n/b_n)$ *exists. Then* $\sum a_n$ *converges if* $\sum b_n$ *converges. (So* $\sum b_n$ *diverges if* $\sum a_n$ *diverges.)*

Proof The difference between a_n/b_n and its limit l can be made as small as desired by taking n large enough. In particular, the difference is less than 1 for all n from some particular value N on

$$\frac{a_n}{b_n} < l + 1 \qquad \text{for all} \qquad n \geq N.$$

Thus

$$a_n < (l + 1)b_n$$

for $n \geq N$, and $\sum a_n$ converges by comparison with the series $\sum (l + 1)b_n = (l + 1)\sum b_n$. ∎

Example 4 Another series that can be shown to be convergent by directly evaluating its partial sum s_n is $\sum 1/n(n + 1)$. (See Problem 29, Section 2.) Show from this that

$$\sum \frac{1}{n^2}$$

is convergent. (This is a convenient series to use for further comparisons.)

Solution We apply Theorem 5. Since

$$\frac{1/n^2}{1/n(n + 1)} = \frac{n(n + 1)}{n^2} = 1 + \frac{1}{n} \to 1$$

as $n \to \infty$, the convergence of $\sum 1/n^2$ follows from the theorem. □

Example 5 We could apply the theorem in the same way for $\sum 1/n(n + 2)$, but here it is just as easy to make the direct comparison

$$\frac{1}{n(n + 2)} < \frac{1}{n^2}$$

and use Example 4 and the comparison test. □

PROBLEMS FOR SECTION 3

Use the comparison test, including Theorem 5, to determine whether the following series converge or diverge. For convergence, compare with a geometric series or the series $\sum 1/n^2$. For divergence, compare with the harmonic series $\sum 1/n$. In some instances, you may have to multiply by a constant before comparing.

1. $\sum \dfrac{1}{n+1}$

2. $\sum \dfrac{1}{n^3}$

3. $\sum \dfrac{1}{2\sqrt{n}}$

4. $\sum \dfrac{1}{n^2+1}$

5. $\sum \dfrac{1}{n^2-1}$

6. $\sum \dfrac{n-1}{n^2+1}$

7. $\sum \dfrac{1}{3n+5}$

8. $\sum \dfrac{3n+5}{n^3}$

9. $\sum \dfrac{\ln n}{n^3}$

10. $\sum \dfrac{1}{2^n-1}$

11. $\sum \sin \dfrac{1}{n^2}$

12. $\sum \dfrac{1}{n!}$

13. $\sum \dfrac{n!}{n^n}$

14. $\sum \dfrac{1}{n(n-2)}$

15. $\sum \dfrac{1}{n+\sqrt{n}}$

16. $\sum \dfrac{1}{\sqrt{n^2+n}}$

17. $\sum \dfrac{1}{n}\sin\dfrac{1}{n}$

18. $\sum \dfrac{1}{n}\cos\dfrac{1}{n}$

19. $\sum \ln\left(1+\dfrac{1}{n}\right)$

20. $\sum \ln\left(1+\dfrac{1}{n^2}\right)$

21. $\sum \dfrac{1}{(n^7-n^6+1)^{1/3}}$

22. $\sum \left(1-\cos\dfrac{1}{n}\right)$

23. $\sum \sin\left[\left(\dfrac{\pi}{4}\right)^n\right]$

24. $\sum \dfrac{1}{\sqrt{e^n}}$

25. Carry out the proof of the error formula

$$E < \frac{1}{(n+3)!}\left(1+\frac{6}{n}\right)$$

for the approximation of e by $s_n + 1/(n \cdot n!)$, following the calculation given in the text for the special case $n = 4$.

27. Using the infinite-series interpretation of the decimal expansion $\pi = 3.1415\cdots$, prove that

$$3.141 < \pi < 3.142.$$

26. Carry out the computation of e given in Example 3 through five more factorials (i.e., through 1/9!). Compute $S_9 + 1/(9 \cdot 9!)$ through ten decimal places, and use the error formula given in Problem 25 to show that this estimation of e is accurate through eight decimal places.

4
THE RATIO TEST:
THE INTEGRAL
TEST

We shall need other corollaries of the comparison test. Most important is the *ratio test*, which concerns a class of series that can be shown to be covergent by comparison with the geometric series. First an example.

Example 1   In the series

$$\frac{1}{2} + \frac{2}{2^2} + \frac{3}{2^3} + \cdots \frac{n}{2^n} + \cdots,$$

a pair of successive terms a_n and a_{n+1} have the ratio

$$\frac{a_{n+1}}{a_n} = \frac{(n+1)/2^{n+1}}{n/2^n} = \left(\frac{n+1}{n}\right)\left(\frac{1}{2}\right) = \left(1 + \frac{1}{n}\right)\frac{1}{2}.$$

This "test ratio" a_{n+1}/a_n varies with n, but it has the limit 1/2 as $n \to \infty$, Therefore, if we choose any number larger than 1/2, say 3/4, then we will have

$$\frac{a_{n+1}}{a_n} < \frac{3}{4}$$

for all integers n beyond some value N. We don't have to know what N is, but we see by inspection that the inequality holds for all $n \geq 3$. So

$$a_4 < \frac{3}{4} a_3,$$

$$a_5 < \frac{3}{4} a_4 < \left(\frac{3}{4}\right)^2 a_3,$$

$$a_6 < \frac{3}{4} a_5 < \left(\frac{3}{4}\right)^3 a_3,$$

$$\vdots$$

$$a_n < \frac{3}{4} a_{n-1} < \left(\frac{3}{4}\right)^{n-3} a_3.$$

Therefore, if we sum starting with the third term, the comparison test shows that $\sum a_j$ is convergent and that

$$\sum_{n=3}^{\infty} a_n < a_3 \sum_{k=0}^{\infty} \left(\frac{3}{4}\right)^k = a_3\left(\frac{1}{1 - 3/4}\right) = 4a_3. \qquad \square$$

There was a further restriction on our choice of 3/4 above. We needed a number larger than 1/2 *and less than* 1, so that the final comparison series is a *convergent* geometric series. Here is the theorem.

THEOREM 6   **The Ratio Test**   *Let $\sum a_n$ be a series of positive terms, and suppose that*

$$\frac{a_{n+1}}{a_n} \to l$$

as $n \to \infty$. Then $\sum a_n$ converges if $l < 1$ and diverges if $l > 1$.

Proof   Suppose that $l < 1$ and choose a number t between l and 1: $l < t < 1$. Since $a_{n+1}/a_n \to l$, we will have $a_{n+1}/a_n < t$, i.e.,

$$a_{n+1} < ta_n,$$

for all indices n starting at some value N. Thus

$$a_{N+1} < ta_N,$$
$$a_{N+2} < ta_{N+1} < t^2 a_N,$$
$$a_{N+3} < ta_{N+2} < t^3 a_N,$$
$$\vdots$$
$$a_{N+m} < ta_{n+m-1} < t^m a_N.$$

Therefore, if we start at the Nth term, the series $\sum a_n$ converges by comparison with the geometric series $a_N \sum t^m$.

If $l > 1$, then it follows as above that $a_{n+1} > a_n$ for all n beyond some value N. Thus the series terms a_j form an increasing sequence from N on. In particular the sequence $\{a_j\}$ does not converge to 0, so $\sum a_j$ diverges, by Theorem 2.   ∎

Example 2   Consider the series

$$\frac{1}{2} + \frac{2^2}{2^2} + \frac{3^2}{2^3} + \cdots + \frac{n^2}{2^n} + \cdots.$$

Then

$$\frac{a_{n+1}}{a_n} = \frac{(n+1)^2/2^{n+1}}{n^2/2^n}$$

$$= \left(\frac{n+1}{n}\right)^2 \frac{1}{2}$$

$$= \left(1 + \frac{1}{n}\right)^2 \Big/ 2 \to \frac{1}{2} \quad \text{as } n \to \infty.$$

The series therefore converges, by the theorem.   □

The theorem says nothing about the case $l = 1$. This is because nothing *can* be said: the harmonic series is a *divergent* series for which the test ratio has the limit 1, whereas the series $\sum 1/n^2$ is a *convergent* series for which $l = 1$.

The Integral Test Another useful offshoot of the comparison test is obtained by comparing the terms of a series with blocks of area under a graph. For example, at the moment we cannot say anything about the convergence of the series

$$1 + \frac{1}{2\sqrt{2}} + \frac{1}{3\sqrt{3}} + \cdots + \frac{1}{n\sqrt{n}} + \cdots.$$

However, look at Fig. 1. We see that the terms of the above series (from 2 on) are the areas of rectangles under the graph of $y = x^{-3/2}$.

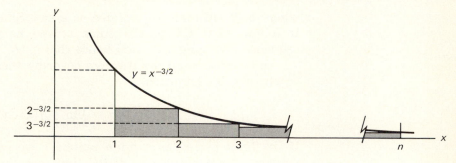

Figure 1

So if A_n is the area under this graph from $n - 1$ to n, then

$$\frac{1}{n\sqrt{n}} < A_n$$

for $n \geq 2$. But

$$\sum_2^n A_k = \int_1^n x^{-3/2} = -2x^{-1/2}\Big]_1^n = 2 - \frac{2}{\sqrt{n}},$$

which has the limit 2 as $n \to \infty$. The series $\sum A_k$ thus converges, with sum 2. Therefore the series $\sum_2^\infty 1/n\sqrt{n}$ converges, by comparison, with sum ≤ 2. When we add on the first term that was missing in the above comparison we end up with

$$\sum_{k=1}^\infty k^{-3/2} \leq 3.$$

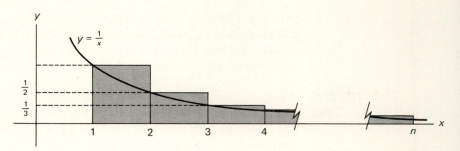

Figure 2

In the same way, but using upper rectangles rather than lower rectangles, we obtain a new proof that the harmonic series $\sum 1/n$ diverges. Figure 2 shows that

$$\ln n = \int_1^n \frac{dx}{x} < \sum_{k=1}^{n-1} \frac{1}{k},$$

where both sides are interpreted as areas. Since $\ln n \to \infty$ as $n \to \infty$, it follows that the partial sums of the harmonic series form a sequence diverging to ∞, and hence that $\sum 1/n$ is divergent.

In order to state the theorem covering these two situations, we recall that, by definition,

$$\int_a^\infty f(x)\,dx = \lim_{b \to \infty} \int_a^b f(x)\,dx.$$

If the limit exists (as a finite number), we say that $\int_a^\infty f$ is a convergent improper integral. If the limit does not exist the improper integral is divergent.

THEOREM 7   **The Integral Test**   *Let f be a positive decreasing continuous function defined on the interval* $[1, \infty)$. *Then,*

  a) *if the improper integral* $\int_1^\infty f$ *converges, then the series* $\sum f(n)$ *converges, and*

$$\sum_2^\infty f(n) \le \int_1^\infty f(x)\,dx \le \sum_1^\infty f(n);$$

  b) *if the improper integral* $\int_1^\infty f(x)\,dx$ *diverges then the series* $\sum f(n)$ *diverges.*

  Proof   Left to the reader. Just write down in general terms the arguments from the two examples. The second inequality of part (a) follows from the same argument used to prove part (b). ∎

REMARK Since a series converges if and only if a remainder series $\sum_N a_k$ converges, we can restate the theorem in terms of the improper integral $\int_N^\infty f$.

Example 3 Show that the series

$$\sum_{n=1}^{\infty} \frac{1}{n^p}$$

converges if $p > 1$ and diverges if $0 < p \leq 1$.

Solution The series is of the form $\sum f(n)$, where $f(x) = x^{-p}$. Now

$$\int_1^\infty \frac{dx}{x^p} = \begin{cases} 1/(p-1), & \text{if } p > 1; \\ \infty, & \text{if } 0 < p \leq 1 \end{cases}$$

(see Chapter 9, Section 7). since $f(x) = x^{-p}$ is positive and decreasing on $[1, \infty)$, we can apply the theorem, and conclude that $\sum n^{-p}$ converges if $p > 1$ and diverges if $0 < p \leq 1$. ☐

Example 3 reestablishes both the known divergence of the harmonic series $\sum 1/n$ and the known convergence of the series $\sum 1/n^2$. These two series have been useful standards for comparison testing, and the other series in this family are similarly useful.

Example 4 Test the series

$$\sum \frac{\sqrt{n-1}}{n(n+1)}$$

for convergence.

Solution Since

$$\frac{\sqrt{n-1}}{n(n+1)} < \frac{\sqrt{n}}{n^2} = n^{-3/2},$$

for all n, and since $\sum n^{-p}$ converges when $p = 3/2$, by Example 3, we can conclude that the given series is convergent by the comparison test. ☐

PROBLEMS FOR SECTION 4

Use the ratio test to conclude what you can about the convergence or divergence of each of the following series.

1. $\sum \dfrac{n+4}{n2^n}$

2. $\sum nx^n$

3. $\sum \dfrac{n^3}{2^n}$

4. $\sum n(n-1)x^{n-2}$

5. $\sum \dfrac{1}{n^3}$

6. $\sum \dfrac{n^4}{n!}$

7. $\sum \dfrac{1}{\sqrt{n}}$

8. $\sum \dfrac{n!}{n^n}$ $\left(\text{Use: } \left(1+\dfrac{1}{n}\right)^n \to e.\right)$

9. $\sum \dfrac{2^n+n}{3^n}$

Determine the convergence or divergence of the series whose general term is

10. ne^{-n^2}

11. ne^{-n}

12. $\ln n / n^2$

13. $\dfrac{1}{n \ln n}$

14. $\dfrac{1}{n(\ln n)^p}$ $(p > 1)$

15. Show that $\sum_1^\infty 1/(n^2) < 2$. (Apply Theorem 6.)

16. Show that $\sum_1^\infty 1/(n^2+1) < \dfrac{\pi}{2}$.

17. Show that $\sum_1^\infty 1/(n^3) < \dfrac{3}{2}$.

Test for convergence or divergence the series whose general term (from a certain point on) is

18. $\dfrac{1}{\sqrt{n^4-1}}$

19. $\dfrac{1}{n+\sqrt{n}}$

20. $\dfrac{1}{\sqrt{n!}}$

21. $\dfrac{n^{10}}{10^n}$

22. $\dfrac{10^n}{n!}$

23. $\sin\dfrac{1}{n}$

24. $\dfrac{1}{100n+10^6}$

25. $\dfrac{1}{\ln n}$

26. $\dfrac{1}{n^{1/3}}$

27. $\dfrac{1}{\sqrt{n(n-1)}}$

28. $\dfrac{n-1}{(n+1)\sqrt{n}}$

29. $\dfrac{n^2 2^n}{3^n}$

30. $\dfrac{\ln n}{n\sqrt{n}}$

31. $\ln\left(\cos\dfrac{1}{n}\right)$

32. $\dfrac{\sqrt{n+1}}{n^2}$

33. $\dfrac{1}{n\ln(\ln n)}$

34. $\dfrac{1}{n \ln n \ln(\ln n)}$

35. $\dfrac{1}{n \ln n(\ln \ln n)^p}$

36. $\left(1+\dfrac{1}{n}\right)^{-n^2}$

37. $\dfrac{1}{2^{\sqrt{n}}}$

38. $\dfrac{(n!)^2}{(2n)!}$

39. $\dfrac{2^n(n!)^2}{(2n)!}$

40. Prove that $n!/n^n \to 0$ as $n \to \infty$.

41. Prove that $2^n/n! \to 0$ as $n \to \infty$.

42. The finite sum

$$\ln(n!) = \ln 2 + \ln 3 + \cdots + \ln n$$

can be compared with $\int \ln x\, dx$ between appropriate limits, as in the proof of the integral test. Show in this way that

$$\ln(n-1)! < n \ln n - n + 1 < \ln(n!)$$

for $n \geq 2$. Conclude that

$$(n-1)! < en^n e^{-n} < n!$$

44. Finally, interpret $\ln k$ as the area of a trapezoid below the tangent line to the graph of $\ln x$ at $x = k$, the trapezoid extending from $x = k - 1/2$ to $x = k + 1/2$. Thus $\ln(n!)$ is a sum of trapezoidal areas that *include* the area under the graph of $\ln x$, except for sticking out too far by 1/2 on the right, and not reaching quite far enough (by 1/2) on the left. Show therefore that

$$\ln n! + \frac{1}{4} - \frac{1}{2} \ln n > n \ln n - n + 1,$$

and conclude that

$$n! > e^{7/8} n^n e^{-n} n^{1/2}.$$

43. Using the trapezoidal approximation in a similar way, show that

$$\sqrt{n}(n-1)! < en^n e^{-n},$$

or

$$n! < en^n e^{-n} n^{1/2}.$$

5
ABSOLUTE CONVERGENCE AND CONDITIONAL CONVERGENCE

We shall have to know something about the convergence of a series $\sum a_n$ whose terms have both signs. If the sum $\sum |a_n|$ of their absolute values converges, it should seem probable that the series $\sum a_n$ itself must converge, and to a smaller sum, because of the cancelling that goes on as we add together the positive and negative terms. In this section we shall prove two theorems about such "mixed sign" convergence.

Example 1
Consider the geometric series with ratio $t = -(1/2)$. Theorem 1 shows that

$$1 - \frac{1}{2} + \frac{1}{4} - \frac{1}{8} + \cdots + \left(-\frac{1}{2}\right)^n + \cdots$$

has the sum

$$\frac{1}{1 - \left(-\dfrac{1}{2}\right)} = \frac{1}{1 + 1/2} = \frac{2}{3}.$$

This is much smaller than the sum of the powers of 1/2,

$$1 + \frac{1}{2} + \frac{1}{4} + \cdots = 2,$$

and the reason is the cancelling that goes on as we add. We can see exactly the cancelling that occurs by dividing up the series into its positive and negative parts:

$$P = 1 + \frac{1}{4} + \frac{1}{16} + \cdots = \frac{1}{1 - (1/4)} = \frac{4}{3},$$

$$N = -\left(\frac{1}{2} + \frac{1}{8} \cdots \right)$$

$$= \left(-\frac{1}{2}\right)\left[1 + \frac{1}{4} + \cdots\right]$$

$$= \left(-\frac{1}{2}\right)P = -\frac{2}{3},$$

$$\sum = P + N = \frac{4}{3} + \left(-\frac{2}{3}\right) = \frac{2}{3}. \qquad \square$$

The crux of the matter is a kind of "squeeze" lemma for series that may have terms of both signs.

LEMMA   *If $a_n \le b_n \le c_n$ for all n, and if the series $\sum a_n$ and $\sum c_n$ both converge, then $\sum b_n$ converges, and*

$$\sum_{1}^{\infty} a_n \le \sum_{1}^{\infty} b_n \le \sum_{1}^{\infty} c_n.$$

Proof   Since

$$0 \le b_n - a_n \le c_n - a_n$$

for all n by hypothesis, and since $\sum (c_n - a_n)$ is convergent by Theorem 3, it follows by comparison that $\sum (b_n - a_n)$ is convergent. But $b_n = (b_n - a_n) + a_n$, so $\sum b_n$ also converges, again by Theorem 3. The inequality for the sums was also part of Theorem 3.   ∎

DEFINITION   A series $\sum a_n$ is said to *converge absolutely* if $\sum |a_n|$ converges.

THEOREM 8   *If a series $\sum a_n$ converges absolutely, then it converges, and*

$$\left| \sum_1^\infty a_n \right| \le \sum_1^\infty |a_n|.$$

Proof   Apply the lemma to the inequality

$$- |a_n| \le a_n \le |a_n|.$$

The details are omitted.   ∎

Example 2   Consider the series

$$\sum_{n=1}^\infty (-1)^{[\ln n]} 2^{-n}.$$

Remember that $[\ln n]$ is the greatest integer $\le \ln n$. There is no simple rule telling us when this integer is even or when it is odd, so we have no knowledge of when the changes of sign occur in this series. Nevertheless we know that the series converges, by Theorem 8, since the series of absolute values is just the geometric series $\sum 2^{-n}$.

   The only thing we know about the sum of the given series is that it lies between -1 and $+1$ (from the inequality in Theorem 8).   □

Theorems 6 and 8 have the following useful corollary:

COROLLARY   **The Ratio Test**   *Suppose $|a_{n+1}/a_n| \to l$ as $n \to \infty$. Then $\sum a_n$ converges absolutely if $l < 1$, and diverges if $l > 1$.*

Example 3   Discuss the convergence of the series

$$\sum (-1)^n n \pi^{-n}.$$

Solution   We apply the ratio test, and get

$$\left| \frac{a_{n+1}}{a_n} \right| = \frac{(n+1)\pi^{-n-1}}{n\pi^{-n}} = \left(1 + \frac{1}{n}\right) \frac{1}{\pi} \to \frac{1}{\pi}$$

as $n \to \infty$. Since $1/\pi < 1$, the series is absolutely convergent, by the ratio test.   □

The comparison tests also have useful formulations in terms of absolute convergence, as follows:

If $\sum b_n$ converges absolutely, and if $|a_n| \leq |b_n|$ for all n, then $\sum a_n$ converges absolutely.

If $\sum b_n$ converges absolutely, and if $|a_n/b_n| \to l$ as $n \to \infty$, then $\sum a_n$ converges absolutely.

These versions will have applications in the next chapter.

When the terms of a series *alternate* in sign, the following special test is available.

THEOREM 9   **The Alternating Series Test**   *Let $\{a_n\}$ be a decreasing sequence of positive numbers such that $a_n \to 0$ as $n \to \infty$. Then the series*

$$a_0 - a_1 + a_2 + \cdots + (-1)^n a_n + \cdots$$

is convergent. Moreover, its sum s is approximated by its nth partial sum s_n with an error bounded by the next term a_{n+1}:

$$|s - s_n| \leq a_{n+1}.$$

The proof is given below.

Example 4   The series

$$1 - \frac{1}{2} + \frac{1}{3} - \frac{1}{4} + \cdots + (-1)^{n+1}\frac{1}{n} + \cdots$$

is convergent, by the theorem. However it is not absolutely convergent because the harmonic series

$$1 + \frac{1}{2} + \frac{1}{3} + \cdots + \frac{1}{n} + \cdots$$

diverges.   □

A series that is convergent but *not* absolutely convergent is said to converge *conditionally*. Conditionally convergent series are of little interest to us here. They converge so slowly that their partial sums are of little use in estimating their sums. Moreover, we are not allowed to handle them algebraically with freedom. For example, we cannot express the above sum as the difference of the sums of the positive and negative parts.

$$s = \left(1 + \frac{1}{3} + \frac{1}{5} + \cdots\right) - \left(\frac{1}{2} + \frac{1}{4} + \frac{1}{6} + \cdots\right),$$

because each of these series diverges to infinity and we have the nonsensical

$$s = \infty - \infty.$$

So for us the importance of an alternating series is the error estimate $|s - s_n| \leq a_{n+1}$ in situations where we do have rapid, absolute convergence.

Proof of Theorem   We first note that the odd partial sums form an increasing sequence. This is because every odd sum but the first is obtained from the preceding one by adding a nonnegative number:

$$s_3 = s_1 + (a_2 - a_3), \qquad s_5 = s_3 + (a_4 - a_5), \qquad \text{etc.}$$

So $s_1 \leq s_3 \leq s_5 \leq \cdots$.

In a similar way we can see that the even partial sums form a decreasing sequence, because each one (but the first) is obtained from the preceding one by subtracting a nonnegative number:

$$s_2 = s_0 - a_1 + a_2 = s_0 - (a_1 - a_2),$$

$$s_4 = s_2 - (a_3 - a_4), \quad \text{etc.}$$

So $s_0 \geq s_2 \geq s_4 \geq \cdots$.

Finally, each odd sum is less than the next even sum because

$$s_{2m-1} + a_{2m} = s_{2m}.$$

In order to organize all this information we form closed intervals with the numbers s_n as endpoints, as follows:

$$I_1 = [s_1, s_2], \qquad I_2 = [s_3, s_4], \ldots, I_m = [s_{2m-1}, s_{2m}], \ldots.$$

The three conclusions above then translate exactly into the fact that the intervals I_m form a nested sequence of closed intervals with lengths a_{2m} tending to 0.

It follows from the nested-interval principle that the two endpoint sequences converge as $m \to \infty$, and to the same limit s. That is, $s_n \to s$ as n runs to ∞ through the even integers, and also as n runs to ∞ through the odd integers. Therefore, $s_n \to s$ as $n \to \infty$.

Finally, since

$$s_{2m-1} \leq s \leq s_{2m},$$

we have

$$s - s_{2m-1} \leq s_{2m} - s_{2m-1} = a_{2m}.$$

Also $s_{2m+1} \leq s \leq s_{2m}$, so

$$s_{2m} - s \leq s_{2m} - s_{2m+1} = a_{2m+1}.$$

Thus in all cases

$$|s - s_n| \leq a_{n+1}. \qquad \blacksquare$$

Example 5 *All* the hypotheses of the theorem have to be met to ensure convergence. For example, the series

$$3 - \frac{1}{2} + \frac{3}{3} - \frac{1}{4} + \frac{3}{5} - \cdots + \frac{1 + (-1)^{n+1}2}{n} + \cdots$$

diverges, even though $a_n \to 0$ and the signs alternate. What allows divergence here is that a_n is not a *decreasing* sequence. The proof of divergence is left as a problem. $\qquad \square$

REMARK If $\{a_n\}$ is strictly decreasing, then all inequalities in the proof are strict, including the error bound.

PROBLEMS FOR SECTION 5

The series in Problems 1 through 13 alternate in sign. Determine in each case whether the series converges absolutely, conditionally, or not at all.

1. $\sum (-1)^n/n^2$

2. $\sum (-1)^n/\sqrt{n}$

3. $\sum (-1)^n 2^n/n^2$

4. $\sum (-1)^n/\ln n$

5. $\sum (-1)^n (n-1)/n$

6. $\sum (-1)^n (\sqrt{n} - 1)/n$

7. $\sum (-1)^n (n-1)/n^2$

8. $\sum (-1)^n (\sqrt{n} - 1)/n^2$

9. $\sum (-1)^n \ln n/n$

10. $\sum (-1)^n \ln n/\ln 2n$

11. $\sum (-1)^n \ln n/\ln n^2$

12. $\sum (-1)^n \ln n/\ln(n^n)$

13. $\sum (-1)^n \sin(n\pi/6)/n$

14. Show that the series in Example 5 is divergent.

15. Construct a divergent alternating series with terms that decrease in absolute value.

EXTRA PROBLEMS FOR CHAPTER 11

Test for convergence the series having the following general terms, distinguishing between absolute convergence and conditional convergence.

1. $\dfrac{1}{n^2 + n}$ **2.** $\dfrac{1}{n^3 - 50}$ **3.** $\dfrac{1}{n^2 - n}$

4. $\dfrac{(-1)^n + \sqrt{n}}{n}$ **5.** $\dfrac{n^2}{2^n}$ **6.** $\dfrac{1 + (-1)^n \sqrt{n}}{n^2}$

7. $\dfrac{n^2}{(n-2)(n-1)(n+1)(n+2)}$ **8.** $\dfrac{n}{1 + (-1)^n n^2}$ **9.** $\dfrac{(-1)^n}{n \ln n}$

10. $\dfrac{1}{n\sqrt{n-1}}$ **11.** $\dfrac{2^n}{n^2}$ **12.** $\dfrac{(-1)^n}{1 + \sqrt{n}}$

13. $\dfrac{1}{\sqrt{n^3 - 5}}$ **14.** $(-1)^n \sin \dfrac{1}{n}$ **15.** $\dfrac{\ln n}{n^2}$

16. $r^{(-1)^n}$ **17.** $\dfrac{e^n}{n!}$ **18.** $\dfrac{1}{(n-2)^2}$

19. $\sin \dfrac{(-1)^n}{n}$ **20.** $\cos \dfrac{(-1)^n}{n}$ **21.** $\dfrac{1}{1 + (-1)^n n^{3/2}}$

22. $\dfrac{n}{1 + (-1)^n n^{3/2}}$ **23.** $\sin\left(\dfrac{1}{n^{3/2}}\right)$ **24.** $\dfrac{1 + (-1)^n \sqrt{n}}{n}$

25. $\dfrac{1}{(n \ln n)^2}$ **26.** $\dfrac{(-1)^n n}{5n + 10}$ **27.** $\dfrac{\sin(x + n\pi)}{n}$

28. $\dfrac{(-1)^n n!}{n^n}$ **29.** $\dfrac{(-1)^{n(n+1)/2}}{n}$ **30.** $\left(1 - \dfrac{1}{n}\right)^{n^2}$

31. $r^{(n^2)}$ **32.** $e^{(-1)^n n}$ **33.** $\dfrac{5^n (n!)^2}{(2n)!}$

34. $\dfrac{x^{(n^2)}}{2n}$ **35.** $\dfrac{n!}{\sqrt{(2n)!}}$ **36.** $\dfrac{3^n (n!)^2}{(2n)!}$

37. $n^{(-1)^n}$ **38.** e^{-n^2} **39.** $r^{(n^{3/2})}$

40. $\dfrac{(\ln n)^a}{n^b}$ **41.** $e^{-\sqrt{n}}$ **42.** $\dfrac{(n!)^2 x^n}{(2n)!}$

43. $1 - \dfrac{1}{2} + \dfrac{1}{4} - \dfrac{1}{5} + \dfrac{1}{7} - \dfrac{1}{8} + \cdots$

(harmonic series, omitting every third term, and alternating)

44. $1 - \dfrac{1}{2} - \dfrac{1}{3} + \dfrac{1}{4} - \dfrac{1}{5} - \dfrac{1}{6} + \dfrac{1}{7} - \dfrac{1}{8} - \dfrac{1}{9} + \cdots$

(harmonic series, alternating unevenly in the pattern one positive term followed by two negative terms)

45. $1 + \dfrac{1}{2} - \dfrac{1}{3} - \dfrac{1}{4} + \dfrac{1}{5} + \dfrac{1}{6} - \dfrac{1}{7} - \dfrac{1}{8} + \cdots$

(harmonic series, with signs changing in blocks of two)

CHAPTER 12
POWER SERIES

1
POWER SERIES

We can now show that the elementary transcendental functions e^x, $\sin x$, etc., can be expressed as "infinite polynomials," in the sense that each is an infinite sum of terms $a_n x^n$. For example, the series that we have been using for e will be established by showing more generally that

$$e^x = 1 + x + \frac{x^2}{2!} + \frac{x^3}{3!} + \cdots + \frac{x^n}{n!} + \cdots$$

for all x. Such a series is called a *power series* because its nth term is a constant times x^n. This remarkable equation for e^x, and the others like it that we shall establish, all depend on a few simple facts about power series that we shall derive in this section and the next.

This whole subject is important for computation. For example, most of the numerical tables for transcendental functions, such as the tables for the natural logarithm, sine and cosine, were computed from the power-series expansions of these functions. A calculator computes these functions in essentially the same way, and *very* quickly, every time the button is pushed.

In general, any series of the form

$$a_0 + a_1 x + \cdots + a_n x^n + \cdots$$

is called a power series. Here $a_0, a_1, \cdots$ are constants, called the *coefficients* of the power series. Such a series cannot be said to converge or diverge as it stands, since it is not an infinite series of numbers. However, it becomes a numerical series if we give x a numerical value, and we can expect that it may converge for some values of x and diverge for other values. Here is the basic fact.

THEOREM 1   *If a power series $\sum a_n x^n$ converges for a particular nonzero value of x, say $x = c$, then the series converges absolutely for every x that is smaller in absolute value, i.e., for every x in the interval $(-|c|, |c|)$.*

Proof   We are assuming that $\sum a_n c^n$ converges. Then $a_n c^n \to 0$ as $n \to \infty$, by Theorem 2 in the last chapter, so $|a_n c^n| \leq 1$ from some point on, say from the integer N on. We can now compare with the geometric series. For suppose that $|x| < |c|$ and set $t = |x/c|$. Then $0 \leq t < 1$, and

$$|a_n x^n| = |a_n c^n t^n| = |a_n c^n| \cdot t^n \leq t^n,$$

when $n \geq N$. Therefore the remainder series

$$\sum_N^\infty a_n x^n$$

converges absolutely by comparison with the geometric series $\sum_N t^n$. This proves that $\sum a_n x^n$ converges absolutely for every x in the interval $(-|c|, |c|)$.   ∎

Interval of Convergence   It can be proved (from Theorem 1 and the completeness of the real numbers) that a power series $\sum a_n x^n$ converges exactly on an interval centered at the origin. This is a theoretical result and does not provide any way of determining the interval. However, the ratio test often provides a direct computation of the open interval of convergence and thus bypasses such theoretical considerations.

This will not settle the question of convergence at the interval endpoints, and other tests have to be used for this.

Example 1   In order to determine whether the series

$$x + 2x^2 + 3x^3 + \cdots + n x^n + \cdots$$

converges for a given number x we try the ratio test. The test ratio for $\sum |nx^n|$ is

$$\left| \frac{(n + 1)x^{n+1}}{nx^n} \right| = \left(\frac{n + 1}{n} \right) |x| = \left(1 + \frac{1}{n} \right) |x|,$$

which approaches $|x|$ as n approaches infinity. Therefore, we know from the ratio test that $\sum nx^n$ will converge absolutely if $|x| < 1$ and diverge if $|x| > 1$. The domain of convergence is thus the interval $(-1, 1)$, with the possible addition of one or both endpoints. Since the ratio test fails when the limit $|x|$ has the value 1, the series behavior at the endpoints $x = \pm 1$ has to be determined by other means. Here the nth terms in the endpoint series are $\pm n$, so these two series both diverge. The open interval $(-1, 1)$ is thus the exact domain of convergence. We say that the radius of convergence of the power series is 1. □

Example 2 The series

$$(x - 3) + 2(x - 3)^2 + \cdots + n(x - 3)^n + \cdots$$

is an example of a more general type of power series. We apply the ratio test exactly as in the above example:

$$\left| \frac{(n + 1)(x - 3)^{n+1}}{n(x - 3)^n} \right| = \left(1 + \frac{1}{n} \right) |x - 3| \to |x - 3|.$$

We conclude as above that the series converges absolutely if $|x - 3| < 1$ and diverges if $|x - 3| \geq 1$, so the interval of convergence is defined by the inequality

$$-1 < x - 3 < 1, \qquad \text{or} \qquad 2 < x < 4.$$

It is thus the interval $(2, 4)$, centered about the point $x = 3$. The radius of convergence is again 1. This series can be viewed as the series of Example 1 shifted 3 units to the right. □

Example 3 Consider the series $\sum x^n/n$. The ratio test shows that it converges if $|x| < 1$ and diverges if $|x| > 1$, but it leaves up in the air the behavior of the series at $x = \pm 1$.

At $x = 1$ the series becomes the harmonic series, which we know to be divergent.

At $x = -1$ it is the *alternating harmonic* series

$$-1 + \frac{1}{2} - \frac{1}{3} + \frac{1}{4} + \cdots + \frac{(-1)^n}{n} + \cdots,$$

which converges, by the alternating series test.

The domain of convergence is thus the set of numbers x such that $-1 \leq x < 1$, that is, the interval $[-1, 1)$, closed on the left and open on the right. The radius of convergence is 1. □

Example 4   For the series

$$1 + x + \frac{x^2}{2!} + \frac{x^3}{3!} + \cdots + \frac{x^n}{n!} + \cdots,$$

we have the test ratio

$$\left| \frac{x^{n+1}/(n+1)!}{x^n/n!} \right| = \frac{|x|}{n+1},$$

which approaches 0 as $n \to \infty$, no matter what value x has. The series therefore converges absolutely for all x, and the interval of convergence is $(-\infty, \infty)$. The radius of convergence is ∞. □

Example 5   The series $\sum n! \, x^n$ converges only for $x = 0$. Its radius of convergence is 0. □

Differentiability of Power Series   If the power series $\sum a_n x^n$ converges on the interval $(-r, r)$, then its sum depends on x and hence is a function f of x,

$$f(x) = a_0 + a_1 x + a_2 x^2 + \cdots + a_n x^n + \cdots.$$

The major question we have to answer is this: Does the law that the derivative of a sum is equal to the sum of the derivatives hold for such *infinite* sums? In other words, must the function f above necessarily have a derivative f' that is given by

$$f'(x) = a_1 + 2a_2 x + \cdots + na_n x^{n-1} + \cdots$$

on the interval $(-r, r)$?

It is really remarkable that the answer to this question is *yes*, and that we can therefore apply calculus to the study of such infinite-sum functions. We first show that the sum of the derivatives exists.

THEOREM 2 *If the power series $\sum a_n x^n$ converges for each x in the open interval $(-r, r)$, then it and its term-by-term differentiated series $\sum n a_n x^{n-1}$ both converge absolutely for every x in the interval $(-r, r)$.*

>   **Proof** Consider any x in $(-r, r)$. Then choose a positive number u such that $|x| < u < r$. This number u will now play the role that c played in Theorem 1. Thus, since $\sum a_n u^n$ converges by hypothesis, it follows from Theorem 1 that $\sum a_n x^n$ is absolutely convergent.
>
>   For the differentiated series we have to go back to the line of reasoning we used in Theorem 1. Since $\sum a_n u^n$ converges, it follows that $|a_n u^n| \le 1$ for all n from some integer N on. Set $t = |x/u|$ and assume $x \ne 0$. Then
>
> $$|n a_n x^n| = n |a_n u^n| \cdot t^n \le n t^n$$
>
>   for $n \le N$. Since $0 < t < 1$, the series $\sum n t^n$ converges by the ratio test. Then $\sum n a_n x^n$ converges absolutely by comparison. Finally, we can multiply this series by $1/x$ and conclude that $\sum n a_n x^{n-1}$ is absolutely convergent (Theorem 3 in the last chapter). Of course, every power series converges trivially at $x = 0$, so the proof is complete. ∎

This brings us to the major theorem.

THEOREM 3 *If $\sum a_n x^n$ converges on the interval $(-r, r)$, then its sum function*

$$f(x) = a_0 + a_1 x + a_2 x^2 + \cdots + a_n x^n + \cdots$$

is differentiable and

$$f'(x) = a_1 + 2a_2 x + \cdots + n a_n x^{n-1} + \cdots.$$

We shall give the proof of this theorem later on.

Example 6 As a special case of the geometric series we have

$$\frac{1}{1 - x^2} = 1 + x^2 + x^4 + \cdots = \sum_0^\infty x^{2n}.$$

Therefore

$$\frac{2x}{(1 - x^2)^2} = 2x + 4x^3 + \cdots = \sum_1^\infty 2n x^{2n-1}$$

by Theorem 3, so

$$\frac{1}{(1-x^2)^2} = 1 + 2x^2 + 3x^4 + \cdots = \sum_{1}^{\infty} nx^{2n-2}$$

$$= \sum_{0}^{\infty} (n+1)x^{2n}.$$

The final series can also be obtained by squaring the first series, but we have not discussed how to do this.   □

Theorem 3 implies that we also can integrate a power series term by term. For, if

$$f(x) = a_0 + a_1 x + \cdots + a_n x^n + \cdots$$

on $(-r, r)$, then the term-by-term integrated series

$$a_0 x + \frac{a_1 x^2}{2} + \cdots + \frac{a_n x^{n+1}}{n+1} + \cdots$$

also converges on $(-r, r)$, because, after we factor x out, its terms are smaller and we can apply the comparison test. Thus

$$F(x) = a_0 x + \frac{a_1 x^2}{2} \cdots + \frac{a_n x^{n+1}}{n+1} + \cdots$$

is defined on $(-r, r)$, and $F'(x) = f(x)$, by Theorem 3. Since obviously $F(0) = 0$, we have proved:

THEOREM 4   *If the power series*

$$f(x) = a_0 + a_1 x + \cdots + a_n x^n + \cdots$$

converges on $(-r, r)$, then so does its term-by-term integrated series, and the sum $F(x)$ of the integrated series is the antiderivative of $f(x)$ having the value 0 at 0.

COROLLARY 1   *For every x in the interval $(-1, 1)$,*

$$\ln(1 + x) = x - \frac{x^2}{2} + \frac{x^2}{3} + \cdots + (-1)^{n+1} \frac{x^n}{n} + \cdots.$$

Proof   Since

$$1 - x + x^2 + \cdots + (-1)^n x^n + \cdots = \frac{1}{1+x} \left(= \frac{1}{1-(-x)} \right)$$

on $(-1, 1)$, the sum of the integrated series

$$x - \frac{x^2}{2} + \frac{x^3}{3} + \cdots + (-1)^n \frac{x^{n+1}}{n+1} + \cdots$$

is the antiderivative of $1/(1 + x)$ which has the value 0 at $x = 0$, and hence is just $\ln(1 + x)$, without any constant of integration. ∎

COROLLARY 2   *For every x in the interval $(-1, 1)$,*

$$\arctan x = x - \frac{x^3}{3} + \frac{x^5}{5} - \frac{x^7}{7} + \cdots + (-1)^m \frac{x^{2m+1}}{2m+1} + \cdots.$$

Proof   Left to the reader. [*Hint*: Expand $1/(1 + x^2)$ in a geometric series.] ∎

REMARK   The corollaries suggest that

$$\ln 2 = 1 - \frac{1}{2} + \frac{1}{3} - \frac{1}{4} + \cdots + (-1)^{n+1} \frac{1}{n} + \cdots,$$

$$\frac{\pi}{4} = 1 - \frac{1}{3} + \frac{1}{5} - \frac{1}{7} + \cdots + (-1)^m \frac{1}{2m+1} + \cdots.$$

Both identities are true, although we have not proved them.

If we set $x = t - 1$ in Corollary 1, write the answer out, and then replace t by x (not the original x), we have

COROLLARY 1′   *For every x in the interval $(0, 2)$*

$$\ln x = (x - 1) - \frac{(x - 1)^2}{2} + \frac{(x - 1)^3}{3} + \cdots + (-1)^{n+1} \frac{(x - 1)^n}{n} + \cdots$$

$$= \sum_{1}^{\infty} (-1)^{n+1} \frac{(x - 1)^n}{n}.$$

Power Series about x_0   This is an example of a power series *centered about $x = 1$*. The series in Example 2 is centered about $x = 3$. Every other power series we have looked at has been centered about $x = 0$. In general, a power series centered about $x = x_0$ is a series of the form

$$a_0 + a_1(x - x_0) + a_2(x - x_0)^2 + \cdots + a_n(x - x_0)^n + \cdots.$$

Setting $u = x - x_0$ converts this to a power series in u centered about the origin

$$a_0 + a_1 u + a_2 u^2 + \cdots + a_n u^n + \cdots.$$

The series in $x - x_0$ converges for a given value of x if and only if the u series converges for the value $u = x - x_0$. (They are the same numerical series.) This means that every question about a general power series can be transformed into a question concerning a related power series about the origin.

By this device we can easily show

THEOREMS 1′–3′   *If the power series $\sum a_n(x - x_0)^n$ converges for some value of x different from x_0, say $x = x_0 + c$, then it converges absolutely on the interval $(x_0 - r, x_0 + r)$, where $r = |c|$.*

If the series converges on an open interval I centered at x_0, and if f is its sum on I, then f is differentiable on I and

$$f'(x) = \sum_{1}^{\infty} n a_n (x - x_0)^{n-1}$$

for all x in I.

When we attempt to represent a function $f(x)$ as the sum of a power series there is sometimes no hope for a series about the origin. For example, $f(x) = \ln x$ has the domain $(0, \infty)$ and blows up as $x \to 0$, whereas any function represented by a series about 0 is differentiable at 0. So $\ln x$ cannot be the sum of a power series about the origin. The more general power series of Theorems 1′–3′ are thus needed.

However, a power series will be assumed to be about 0 unless otherwise stated.

PROBLEMS FOR SECTION 1

In each problem below determine the *open* interval of convergence by the ratio test. The test will fail at every endpoint. Then determine endpoint behavior by other tests.

1. $1 + x + \dfrac{x^2}{2} + \dfrac{x^3}{3} + \cdots + \dfrac{x^n}{n} + \cdots$

2. $1 + \dfrac{x}{2} + \dfrac{x^2}{4} + \cdots + \dfrac{x^n}{2^n} + \cdots$

3. $1 + x + \dfrac{x^2}{2} + \dfrac{x^3}{2 \cdot 3} + \dfrac{x^4}{2 \cdot 3 \cdot 4} + \cdots + \dfrac{x^n}{n!} + \cdots$

4. $1 + 3x + 9x^2 + \cdots + 3^n x^n + \cdots$

5. $1 + 3x + 5x^2 + \cdots + (2n + 1)x^n + \cdots$

6. $1 + \dfrac{x^2}{2} + \dfrac{x^4}{4} + \dfrac{x^6}{8} + \cdots + \dfrac{x^{2n}}{2^n} + \cdots$

7. $x - \dfrac{x^3}{3} + \dfrac{x^5}{5} - \dfrac{x^7}{7} + \cdots + (-1)^n \dfrac{x^{2n+1}}{2n + 1} \cdots$

8. $1 - \dfrac{x}{1 \cdot 2} + \dfrac{x^2}{2 \cdot 4} - \dfrac{x^3}{3 \cdot 8} + \cdots + (-1)^n \dfrac{x^n}{n\,2^n} + \cdots$

9. $\dfrac{(-1)^n x^n}{n}$

10. $\dfrac{x^n}{n\,2^n}$

11. $\dfrac{x^n}{\sqrt{n}}$

12. $\dfrac{(x - 3)^n}{n^2}$

13. $\dfrac{(-1)^n (x - 2)}{n\,2^n}$

14. $\dfrac{n(x - 1)^n}{2^n}$

15. $\dfrac{(x - 1)^n}{n\,3^n}$

16. $\dfrac{(x - 1)^n}{n^2 3^n}$

17. $\sum \dfrac{(x - 2)^n}{2^n}$

18. $\sum \dfrac{(x - 1)^n}{2^n}$

19. $\sum 2^n (x - 1)^n$

20. $\sum \dfrac{(x - 3)^n}{n\,2^n}$

21. $\sum \dfrac{(x - 3)^n}{n^2 3^n}$

The following problems are based on the geometric series

$$\frac{1}{1 - x} = 1 + x + x^2 + \cdots + x^n + \cdots.$$

22. a) Using Theorem 3, find a power series whose sum is $1/(1 - x)^2$.

b) Compute the exact sum of the series $\sum_1^\infty n/2^n$.

23. Compute the exact sum of the series $\sum_1^\infty n^2/2^n$ by some device like that used in the problem above.

Use Theorems 3 and 4 to identify the following functions.

24. $f(x) = 1 + \dfrac{x}{2} + \dfrac{x^2}{3} + \cdots + \dfrac{x^n}{n + 1} + \cdots$

25. $f(x) = 1 + 2x + 3x^2 + \cdots + (n + 1)x^n + \cdots$

26. $f(x) = 2 + 3x + 4x^2 + 5x^3 + \cdots$
$\qquad\qquad + (n + 2)x^n + \cdots$

27. $f(x) = 1 + 4x + 9x^2 + \cdots + (n + 1)^2 x^n$
$\qquad\qquad\qquad\qquad + \cdots$

28. $f(x) = 1 + x + \dfrac{x^2}{1 \cdot 2} + \dfrac{x^3}{2 \cdot 3} + \cdots$
$\qquad\qquad + \dfrac{x^n}{(n - 1)n} + \cdots$

2
MACLAURIN
AND TAYLOR
SERIES

It is difficult to overemphasize the importance of Theorem 3. We saw above how it determines power series expansions for $\ln(1 + x)$ and $\arctan x$. We show next how it determines the series coefficients a_n in terms of the derivatives of the sum function. This will lead to the series formulas for e^x, $\sin x$, $\cos x$, and $(1 + x)^\alpha$.

THEOREM 5 *If*

$$f(x) = a_0 + a_1 x + a_2 x^2 + \cdots + a_n x^n + \cdots$$

over an interval $(-r, r)$, *then* f *has derivatives of all orders in this interval, and*

$$a_n = \frac{f^{(n)}(0)}{n!}$$

for every n, *where* $f^{(n)}$ *is the nth derivative of* f. *The series can thus be rewritten*

$$f(x) = f(0) + f'(0)x + \frac{f''(0)x^2}{2!} + \cdots + \frac{f^{(n)}(0)x^n}{n!} + \cdots.$$

REMARK In order to interpret $f(0)$ as $f^{(0)}(0)/(0)!$ we adopt the conventions that

$$0! = 1 \quad \text{and} \quad f^{(0)}(x) = f(x).$$

Proof of Theorem By Theorems 2 and 3, we can differentiate the above series as many times as we wish. Thus

$$f(x) = a_0 + a_1 x + a_2 x^2 + a_3 x^3 + \cdots + a_n x^n + \cdots$$
$$f'(x) = a_1 + 2a_2 x + 3a_3 x^2 + \cdots + na_n x^{n-1} + \cdots$$
$$f''(x) = 2a_2 + 3 \cdot 2a_3 x + \cdots + n(n-1)a_n x^{n-2} + \cdots$$
$$f'''(x) = 3!\, a_3 + \cdots + n(n-1)(n-2)a_n x^{n-3} + \cdots$$
$$\vdots \qquad\qquad \vdots$$
$$f^{(n)}(x) = n!\, a_n + \cdots$$

Setting $x = 0$ in these equations reduces them to

$$f(0) = a_0$$
$$f'(0) = a_1$$
$$f''(0) = 2a_2$$
$$f'''(0) = 3!\, a_3$$
$$\vdots$$
$$f^{(n)}(0) = n!\, a_n.$$

Thus $a_j = f^{(j)}(0)/j!$ for all j. ∎

DEFINITION   Let f be any function that is infinitely differentiable on some interval about the origin. Then the series

$$\sum \frac{f^{(n)}(0)}{n!} x^n$$

is called the *Maclaurin series* for f.

Theorem 5 says that *if f is the sum of a power series, then* that series has to be the Maclaurin series for f. It does *not* say that if we can form the Maclaurin series for f, then f has to be its sum. Sometimes f is the sum of its Maclaurin series, sometimes not.

Example 1   The Maclaurin series for e^x is easy to find since e^x is equal to all of its derivatives, so

$$f^{(n)}(0) = e^0 = 1$$

for all n. The series is thus

$$1 + x + \frac{x^2}{2} + \frac{x^3}{3!} + \cdots + \frac{x^n}{n!} + \cdots.$$

Therefore, according to the theorem, if *any* power series has e^x as its sum, then this has to be the one. The theorem does not say that e^x *is* the sum of the above series. This does happen to be true; it just doesn't follow from Theorem 5.   □

We shall give two proofs of this important fact, one here using the growth equation $dy/dx = y$, and the other in Section 5 using the remainder formula.

THEOREM 6   *For every real number x,*

$$e^x = 1 + x + \frac{x^2}{2!} + \cdots + \frac{x^n}{n!} + \cdots = \sum_0^\infty \frac{x^n}{n!}.$$

Proof   The ratio test shows that the series on the right converges for all x. Let $g(x)$ be its sum. We have to prove that $g(x) = e^x$. We do this by differentiating the series and discovering a relationship between g' and g. In fact (by Theorem 3),

$$g'(x) = 0 + 1 + \frac{2x}{2!} + \frac{3x^2}{3!} + \frac{4x^3}{4!} + \cdots$$

$$= 1 + x + \frac{x^2}{2!} + \frac{x^3}{3!} + \cdots = g(x).$$

Thus $g' = g$. But we know (or can work out again by separation of variables) that any function satisfying the differential equation $dy/dx = y$ is of the form $y = Ce^x$. So $g(x) = Ce^x$ for some constant C. Setting $x = 0$ in this equation shows that $C = 1$, and we are done. ∎

We go on with our catalogue of the major power series expansions.

THEOREM 7   *For every real number x,*

$$\cos x = 1 - \frac{x^2}{2!} + \frac{x^4}{4!} - \frac{x^6}{6!} + \cdots + (-1)^m \frac{x^{2m}}{(2m)!} + \cdots.$$

Proof   This proof goes exactly like the one above, and we shall only list the steps.

1. The series is the Maclaurin series for $\cos x$ and so is the only power series that can possibly represent $\cos x$.

2. The series converges for all x (by the ratio test) and so defines a function g.

3. The function g satisfies the differential equation $g'' = -g$ and the initial conditions $g(0) = 1$, $g'(0) = 0$.

4. But $\cos x$ is the unique function satisfying this differential equation and these initial conditions (Chapter 4, Section 3). ∎

THEOREM 8   *For every real number x,*

$$\sin x = x - \frac{x^3}{3!} + \frac{x^5}{5!} - \frac{x^7}{7!} + \cdots + (-1)^m \frac{x^{2m+1}}{(2m + 1)!} + \cdots.$$

Proof   Differentiate the equation in Theorem 7 (or prove from scratch in the manner of Theorem 7). ∎

THEOREM 9   **The Binomial Series.**   *For any fixed real number α,*

$$(1 + x)^\alpha = 1 + \alpha x + \frac{\alpha(\alpha - 1)}{2!} x^2 + \cdots$$
$$+ \frac{\alpha(\alpha - 1) \cdots (\alpha - n + 1)}{n!} x^n + \cdots.$$

on the interval $(-1, 1)$.

This proof is the hardest yet, but the pattern is the same.

Proof

1. Computing the Maclaurin series for $(1 + x)^\alpha$ gives the series shown. It is the only power series (about the origin) that can possibly represent $(1 + x)^\alpha$.

2. The series converges on the interval $(-1, 1)$, by the ratio test, and thus represents some function g. We have to show that $g(x) = (1 + x)^\alpha$.

3. The coefficients in this series are the famous binomial coefficients. If we introduce the standard notation

$$\binom{\alpha}{n} = \frac{\alpha(\alpha - 1) \cdots (\alpha - n + 1)}{n!}$$

for the coefficient of x^n, then it can be checked that

$$\binom{\alpha + 1}{n + 1} = \binom{\alpha}{n} \cdot \frac{\alpha + 1}{n + 1}, \qquad \binom{\alpha}{n + 1} = \binom{\alpha}{n} \cdot \frac{\alpha - n}{n + 1},$$

and therefore (with a speck of algebra)

$$\binom{\alpha + 1}{n + 1} = \binom{\alpha}{n + 1} + \binom{\alpha}{n}$$

(Pascal's identity).

4. If the series sum is called $g_\alpha(x)$, then

$$g_\alpha' = \alpha g_{\alpha-1},$$

and

$$(1 + x)g_{\alpha-1}(x) = g_\alpha(x),$$

both equations holding because of the above coefficient identities. If follows that g_α satisfies the differential equation

$$(1 + x)\frac{dy}{dx} = \alpha y.$$

5. Solving this differential equation by separating the variables, and evaluating the constant of integration from the initial condition $g_\alpha(0) = 1$, shows that

$$g_\alpha(x) = (1 + x)^\alpha.$$

The pitfall warned about earlier really does exist. There *is* a function h that *has* a Maclaurin series that *does* converge to a function g *different* from h.

Example 2   If we define h by

$$h(x) = e^{-1/|x|} \quad \text{if } x \neq 0,$$
$$h(0) = 0,$$

then it can be proved that h is infinitely differentiable and that all the derivatives of h are 0 at $x = 0$. Yet the graph of h is like this:

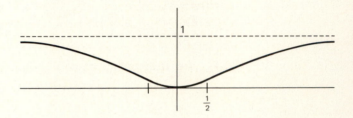

It is decreasing on $(-\infty, 0]$, increasing on $[0, \infty)$, concave up on $[-\frac{1}{2}, \frac{1}{2}]$ and concave down elsewhere.

The Maclaurin series for h has only 0 coefficients and its sum is the 0 function, not h. (See Problem 25–27, which outline the proof for a function similar to h.)   □

Taylor Series   The proof of Theorem 5 could just as well have been given for a power series centered at x_0. It takes more space to write down the terms of such a series, and the page would have been more crowded, that is all. The result is the general form of Theorem 5.

THEOREM 5′   *If*

$$f(x) = a_0 + a_1(x - x_0) + \cdots + a_n(x - x_0)^n + \cdots = \sum_0^\infty a_n(x - x_0)^n$$

on an open interval I centered about x_0, then f has derivatives of all orders in I, and

$$a_n = \frac{f^{(n)}(x_0)}{n!}$$

for every n. The series can thus be rewritten

$$f(x) = f(x_0) + f'(x_0)(x - x_0) + \cdots + \frac{f^{(n)}(x_0)}{n!}(x - x_0)^n + \cdots$$

$$= \sum_0^\infty \frac{f^{(n)}(x_0)}{n!}(x - x_0)^n.$$

REMARK Theorem 5′ can also be obtained from Theorem 5 by a change of variable.

DEFINITION If f is infinitely differentiable on some interval about x_0, then the power series

$$\sum_0^\infty \frac{f^{(n)}(x_0)}{n!}(x - x_0)^n$$

is called the *Taylor series* for f about x_0.

Thus a Maclaurin series is the special case of a Taylor series in which $x_0 = 0$.

Any function f that is infinitely differentiable on an interval about x_0 *has* a Taylor series at x_0, but in general f will not be the sum of its Taylor series. Many such functions *are* given by their Taylor series, but a proof has to be supplied in each case.

Example 3 The Taylor series for $\cos x$ about $x_0 = \pi/4$ is easily found to be

$$\frac{\sqrt{2}}{2}\left[1 - (x - \pi/4) - \frac{(x - \pi/4)^2}{2!} + \frac{(x - \pi/4)^3}{3!} + \frac{(x - \pi/4)^4}{4!} - - + + \cdots\right].$$

The signs alternate in *blocks of two*.

The proof sketched for Theorem 7 works just as well here and shows that $\cos x$ is the sum of this series. □

PROBLEMS FOR SECTION 2

Compute the Maclaurin series for each of the following functions f by evaluating its derivatives $f^{(n)}(0)$.

1. $1/(1 - x)$ **2.** $\ln(1 - x)$ **3.** x^3

4. $1/(1 + x)$ **5.** $x^3 - 2x + 5$ **6.** $\sqrt{1 + x}$

Assuming Theorems 4 through 9, express each of the following functions as a sum of a power series. Also give the interval of convergence in each case.

7. e^{-x^2} **8.** $\cos \sqrt{x}$ **9.** $\arctan x^3$ **10.** $\displaystyle\int_0^x \cos \sqrt{t}\, dt$

11. $\displaystyle\int_0^x e^{-t^2/2}$ **12.** $\displaystyle\int_0^x \frac{\sin t}{t}\, dt$ **13.** $\displaystyle\int_0^x \frac{1 - \cos t}{t^2}\, dt$

14. Complete the proof of Theorem 7 by giving the details for steps (a), (b), and (c).

15. Prove Theorem 8.

16. Show directly that if $g'(x) = \alpha g(x)/(1 + x)$, then $g(x) = c(1 + x)^\alpha$ for some constant c. (How can you show that $g(x)/(1 + x)^\alpha$ is a constant?)

17. a) Show that the infinite series defined in Theorem 10 converges on $(-1, 1)$.

b) Show that its sum $g(x)$ has the property that $g'(x)(1 + x) = \alpha g(x)$.

c) Assuming the result of Problem 16, show that $g(x) = (1 + x)^\alpha$.

Write out the Taylor series for

18. e^x about $x = a$ **19.** $\sqrt{x}$ about $x = 1$ **20.** $1/x$ about $x = 1$

21. $\sin x$ about $x = \pi/2$ **22.** $\ln x$ about $x = 1$ **23.** $\sin x$ about $x = \pi$

24. $\sin x$ about $x = \pi/4$

25. We know that $x^n e^{-x} \to 0$ as $x \to \infty$ for any positive integer n (see Chapter 7, Section 3, Problem 45). Show therefore that

$$\frac{e^{-1/x^2}}{x^m} \to 0$$

as $x \to 0$, for any positive integer m. It follows then that

$$p\!\left(\frac{1}{x}\right)e^{-1/x^2} \to 0$$

as $x \to 0$ for any polynomial in $1/x$:

$$p\!\left(\frac{1}{x}\right) = \frac{a_n}{x^n} + \frac{a_{n-1}}{x^{n-1}} + \cdots + \frac{a_1}{x} + a_0.$$

26. Show that the derivative of a function of the form

$$f(x) = p\!\left(\frac{1}{x}\right)e^{-1/x^2}$$

is again of this form, with a different polynomial. It is sufficient to show this for one of the terms of $f(x)$, that is, for a "monomial" term $a e^{-1/x^2}/x^m$.

27. Now define $F(x)$ by

$$F(x) = e^{-1/x^2} \quad \text{if } x \neq 0,$$

$$F(0) = 0.$$

a) Show that F is everywhere differentiable, with $F'(0) = 0$ and with $F'(x)$ of the form $p(1/x)e^{-1/x^2}$, when $x \neq 0$, where p is a polynomial. (Use Problems 25 and 26.)

b) Then show, continuing to use Problems 25 and 26, that all the derivatives of F exist and that $F^{(n)}(0) = 0$ for every n.

The function F is thus an infinitely differentiable function such that $F(x) \neq 0$ when $x \neq 0$, and yet its Maclaurin-series coefficients are all 0.

Express each of the following as the sum of a power series.

28. arcsin x (Start with the binomial series with exponent $\alpha = -1/2$.)

29. $\sin^2 x$ (Use a trigonometric identity.)

30. $\displaystyle\int_0^x \sqrt{t}\, e^t\, dt$. This will be a series in fractional powers of x.

31. $\displaystyle\int_0^x \frac{\arctan t}{t}\, dt$

3
PROOF OF THE DERIVATIVE SUM RULE FOR POWER SERIES

We come finally to the postponed proof of the basic theorem on differentiating power series. It does not involve any new principles but is a little more complicated than anything we have done so far in this chapter.

During the proof we shall say that a series $\sum b_n$ dominates a series $\sum a_n$. This means that $|b_n| \geq |a_n|$ for all n.

THEOREM 3   *If*

$$f(x) = a_0 + a_1 x + a_2 x^2 + \cdots + a_n x^n + \cdots$$

on an open interval $I = (-r, r)$, then f is differentiable and

$$f'(x) = a_1 + 2a_2 x + \cdots + n a_n x^{n-1} + \cdots$$

on I.

Proof   We know from Theorem 2 that the term-by-term differentiated series $\sum n a_n x^{n-1}$ converges on the same interval $(-r, r)$. Call its sum $g(x)$. We want to prove, for any given x in this interval, that $f'(x)$ exists and equals $g(x)$. In other words, we want to prove that

$$\frac{f(x + \Delta x) - f(x)}{\Delta x} - g(x)$$

approaches 0 as Δx approaches 0. According to Theorem 3 in the last chapter, this algebraic combination is the sum of the series

$$\sum_{n=1}^{\infty} a_n \left[\frac{(x + \Delta x)^n - x^n}{\Delta x} - nx^{n-1} \right].$$

We shall show below that this series is dominated by Δx times the series $\sum n(n-1)a_n b^{n-2}$ (for some positive b less than r and for all sufficiently small Δx). This b series is simply the twice-differentiated series for f evaluated at b and hence is absolutely convergent, by Theorem 2 again. So if we set

$$s = \sum_{n=2}^{\infty} n(n-1) |a_n| b^{n-2},$$

then it will follow directly from Theorems 3 and 8 in the last chapter that the two series sums satisfy the inequality

$$\left| \frac{f(x + \Delta x) - f(x)}{\Delta x} - g(x) \right| \leq s |\Delta x|.$$

The left side therefore has the limit zero at $\Delta x \to 0$. That is,

$$g(x) = \lim_{\Delta x \to 0} \frac{f(x + \Delta x) - f(x)}{\Delta x},$$

or $g(x) = f'(x)$, as we claimed.

Our proof will thus be complete when we have defined b and have shown that

$$\left| \frac{(x + \Delta x)^n - x^n}{\Delta x} - nx^{n-1} \right| \leq n(n-1)b^{n-2} \cdot |\Delta x|,$$

for all $n \geq 2$ and all sufficiently small Δx.

We start by choosing a constant b such that $|x| < b < r$. We can also suppose that $|x + \Delta x| < b$, since this holds for all sufficiently small Δx. Now recall the tangent-line error formula (Chapter 10, Section 6), which tells us that

$$\left|(x + \Delta x)^n - x^n - nx^{n-1}\,\Delta x\right| = \left|\frac{n(n-1)X^{n-2}(\Delta x)^2}{2}\right|$$

for some number X between x and $x + \Delta x$. Then $|X| < b$, and we obtain the needed inequality upon replacing X by b and dividing by $|\Delta x|$. ∎

4
TAYLOR AND
MACLAURIN
POLYNOMIALS

Suppose we know the derivatives of f up through the order n in some interval about x_0. Then we can write down the polynomial that *would* be the nth partial sum of the Taylor series for f about x_0 *if* f had such an expansion. This polynomial,

$$p_n(x) = f(x_0) + f'(x_0)(x - x_0) + \cdots + \frac{f^{(n)}}{n!}(x - x_0)^n$$

$$= \sum_{k=0}^{n} \frac{f^{(k)}(x_0)}{k!}(x - x_0)^k,$$

is called the nth *Taylor polynomial for f about x_0*. If $x_0 = 0$, it is called the nth *Maclaurin polynomial for f*.

These polynomials are extremely powerful and flexible tools for learning about the behavior of f near x_0. The point is that they approximate f near x_0 as though they *were* the partial sums of an infinite series representing f, even when they are *not*.

Before looking into these approximations, we give some examples of Taylor and Maclaurin polynomials. The error will be taken up in the next section.

Example 1   The fourth Maclaurin polynomial of e^x is

$$p(x) = 1 + x + \frac{x^2}{2} + \frac{x^3}{6} + \frac{x^4}{24}.$$ □

Example 2   The fifth Maclaurin polynomial of $\cos x$ is

$$p(x) = 1 - \frac{x^2}{2!} + \frac{x^4}{4!}.$$ □

In a power series the nth partial sum is generally understood to be the sum through the term $a_n x^n$, regardless of how many preceding coefficients a_n are zero.

As noted above, we can also form Taylor and Maclaurin polynomials for functions that are *not* sums of power series. All that is required is that the function in question have derivatives up through some order n around x_0.

Example 3   The function

$$f(x) = x^{7/2} + x^3 - 2x^2 + x + 3$$

has derivatives through the order 3 around the origin, and you can check that its second Maclaurin polynomial is

$$f(0) + \frac{f'(0)}{1}x + \frac{f''(0)}{2!}x^2 = 3 + x - 2x^2.$$

We could also form its third Maclaurin polynomial. Note, however, that $f^{(4)}(0)$ does not exist, because

$$\frac{d^4 x^{7/2}}{dx^4} = \frac{7}{2} \cdot \frac{5}{2} \cdot \frac{3}{2} \cdot \frac{1}{2} x^{-1/2},$$

which is undefined at $x = 0$. Therefore, f does not have any further Maclaurin polynomials.   □

Example 4   The function $\tan x$ can be proved to be the sum of its Maclaurin series about the origin, but we don't know any way of obtaining the series coefficients except to compute them by the Maclaurin formula. We can obtain the first few coefficients in this manner, but the differentiations required become more and more cumbersome, and after a while we give up. Here we shall compute the third Maclaurin polynomial. We have

$$f(0) = 0;$$

$$\frac{d}{dx}\tan x = \sec^2 x = \tan^2 x + 1, \qquad f'(0) = 1;$$

$$\frac{d^2}{dx^2}\tan x = 2\tan x \sec^2 x$$

$$= 2\tan^3 x + 2\tan x, \qquad f''(0) = 0;$$

$$\frac{d^3}{dx^3}\tan x = 6\tan^2 x \sec^2 x + 2\sec^2 x$$

$$= 6\tan^4 x + 8\tan^2 x + 2, \qquad f'''(0) = 2.$$

The third Maclaurin polynomial of $\tan x$ is thus

$$p(x) = x + \frac{x^3}{3}.$$

It is also the fourth Maclaurin polynomial. (This can be proved without calculation from simple facts about odd and even functions.)

□

Example 5   The third Taylor polynomial for $f(x) = \ln x$ at $x = 1$ is

$$p(x) = f(1) + \frac{f'(1)}{1}(x - 1) + \frac{f''(1)}{2!}(x - 1)^2 + \frac{f'''(1)}{3!}(x - 1)^3$$

$$= 0 + \frac{1}{1}(x - 1) + \frac{(-1)}{2!}(x - 1)^2 + \frac{2}{3!}(x - 1)^3$$

$$= (x - 1) - \frac{1}{2}(x - 1)^2 + \frac{1}{3}(x - 1)^3.$$

□

Example 6   The second Taylor polynomial for $f(x) = \sin x$ around $a = \pi/4$ is

$$f(\pi/4) + \frac{f'(\pi/4)}{1}(x - \pi/4) + \frac{f''(\pi/4)}{2!}(x - \pi/4)^2$$

$$= \sin(\pi/4) + [\cos(\pi/4)](x - \pi/4) - [\sin(\pi/4)](x - \pi/4)^2/2$$

$$= \frac{\sqrt{2}}{2}\left[1 + \left(x - \frac{\pi}{4}\right) - \frac{(x - \pi/4)^2}{2}\right].$$

□

A polynomial of degree n is its own nth Maclaurin polynomial. More generally,

LEMMA   *Any polynomial of degree n,*

$$p(x) = a_0 + a_1 x + \cdots + a_n x^n$$

is equal to its own nth Taylor polynomial about any point a.

Proof   If we write each power x^k as the binominal power

$$x^k = [(x - a) + a]^k$$

and expand, then the result, after collecting terms, is an expression for $p(x)$ in the form

$$p(x) = c_0 + c_1(x - a) + \cdots + c_n(x - a)^n.$$

This is an infinite sum in a trivial sense, and Theorem 5′ can be applied. If it seems clumsy to use an infinite series theorem for a finite calculation, then look at the proof and rethink it for a polynomial. One way or the other, we get

$$c_k = \frac{p^{(k)}(a)}{k!}$$

as before, and the polynomial on the right above is the nth Taylor polynomial for $p(x)$. So we are done.   ∎

Example 6   Expand $p(x) = x^3 - x$ about $x = 2$.

Solution   We could write

$$x^3 = ((x - 2) + 2)^3, \text{ etc.,}$$

and expand these binomials. Or we can write down the Taylor expansion about $x = 2$, as follows:

$$p'(x) = 3x^2 - 1, \quad p''(x) = 6x, \quad p'''(x) = 6;$$
$$p(2) = 6, \quad p'(2) = 11, \quad p''(2) = 12, \quad p'''(2) = 6;$$
$$p(x) = 6 + \frac{11}{1}(x - 2) + \frac{12}{2!}(x - 2)^2 + \frac{6}{3!}(x - 2)^3.$$

Therefore,

$$x^3 - x = 6 + 11(x - 2) + 6(x - 2)^2 + (x - 2)^3.$$ □

COROLLARY   *If $p(x)$ is the nth Taylor polynomial for $f(x)$ about $x = a$, then*

$$p^{(k)}(a) = f^{(k)}(a)$$

for $k = 0, 1, \ldots, n$.

Proof   By the lemma, the kth coefficient of $p(x)$,

$$c_k = \frac{f^{(k)}(a)}{k!},$$

is also equal to $p^{(k)}(a)/k!$. This holds for $k = 0, 1, \ldots, n$, and proves the corollary.   ∎

PROBLEMS FOR SECTION 4

Find the nth Maclaurin polynomial $p_n(x)$ of the function $f(x)$, where

1. $f(x) = \arcsin x, \quad n = 3$

2. $f(x) = \ln(1 + x), \quad n = 3$

3. $f(x) = \sec x, \quad n = 3$

4. $f(x) = \ln(\cos x), \quad n = 4$

5. $f(x) = \sin\left(x + \dfrac{\pi}{4}\right), \quad n = 4$

6. $f(x) = (1 + x)^{\alpha}, \quad n = 4$

7. $f(x) = \tan x, \quad n = 4$

8. $f(x) = x/\sin x, \quad n = 3$

Find the nth Taylor polynomial of the function f around the point a, where

9. $f(x) = \ln x, \quad n = 3, a = 1$

10. $f(x) = \sin x, \quad n = 4, a = \pi/4$

11. $f(x) = \sin x, \quad n = 4, a = \pi/6$

12. $f(x) = \sin x, \quad n = 4, a = \pi/3$

13. $f(x) = e^x, \quad n = 3, a = a$

14. $f(x) = \tan x, \quad n = 3, a = \pi/4$

15. Express $p(x) = x^3 + 2x^2 + 1$ as its third Taylor polynomial about $a = 1$.

16. Express $p(x) = x^4 - 5x^3 + 10x$ as its fourth Taylor polynomial about $a = 2$.

17. Prove that every Maclaurin polynomial of an odd (even) function is odd (even).

18. The second Maclaurin polynomial of $f(x)$ is $2 + (x/2) - (x^2/8)$. Show that the second Maclaurin polynomial of $1/f(x)$ is

$$\frac{1}{2} - \frac{x}{8} + \frac{x^2}{16},$$

by computing the derivatives of $g = 1/f$ and evaluating them at the origin.

19. Let the function $g(x)$ in the above problem have the second Maclaurin polynomial $A + Bx + Cx^2$. Supposing that f and g both have continuous third derivatives, we can write

$$f(x) = 2 + \frac{x}{2} - \frac{x^2}{8} + x^3 F(x),$$

$$g(x) = A + Bx + Cx^2 + x^3 G(x),$$

where F and G are continuous. Use the fact that $f(x)g(x) = 1$ to solve for the coefficients A, B, and C.

20. The second Maclaurin polynomial of $f(x)$ is $1 + x + (x^2/6)$. Find the second Maclaurin polynomial of $g = 1/f$:

    a) by the method of Problem 18;

    b) by the method of Problem 19.

21. Find the third Maclaurin polynomial for $\tan x$ by the method of Problem 19, using this time the equation $\cos x \tan x = \sin x$ and the known polynomials for sine and cosine.

22. Suppose that f has derivatives up through order $n - 1$ on an interval I about the origin, with

$$f(0) = f'(0) = \cdots = f^{(n-1)}(0) = 0.$$

Suppose furthermore that $f^{(n-1)}(x)$ is differentiable at $x = 0$, so that $f^{(n)}(0)$ exists. Show that $f(x)/x^n \to f^{(n)}(0)/n!$ as $x \to 0$. (Apply l'Hôpital's rule.)

24. Let p and q be two polynomials of degree n at most, and suppose each of them is an nth-order approximation of f around the origin, in the sense that

$$\lim_{x \to 0} \frac{f(x) - p(x)}{x^n} = 0,$$

$$\lim_{x \to 0} \frac{f(x) - q(x)}{x^n} = 0.$$

Show that the two polynomials p and q must necessarily be the same. In view of the above problem, we see that this property uniquely characterizes the nth Maclaurin polynomial of f.

23. If f has derivatives of order $n - 1$ on an interval about the origin and if $f^{(n-1)}(x)$ is differentiable at $x = 0$, then

$$f(0), \qquad f'(0), \qquad \cdots, \qquad f^{(n)}(0)$$

all exists, and we can write down the nth Maclaurin polynomial $p_n(x)$ for f. Show that

$$\lim_{x \to 0} \frac{f(x) - p_n(x)}{x^n} = 0.$$

(Assume and apply the conclusion in the preceding problem.)

25. Use Problem 24 to prove that every Maclaurin polynomial of an even (odd) function is even (odd).

**5
THE
REMAINDER;
COMPUTATIONS**

We now look into how well a function is approximated by its Taylor and Maclaurin polynomials.

Let f be the sum of a power series

$$f(x) = a_0 + a_1 x + \cdots + a_n x^n + \cdots + = \sum_0^\infty a_j x^j$$

on $(-r, r)$. Then the error when $f(x)$ is approximated by its nth Maclaurin polynomial

$$p(x) = a_0 + a_1 x + \cdots + a_n x^n$$

is given by the nth remainder series $R(x) = f(x) - p(x)$. This can be rewritten

$$R(x) = a_{n+1} x^{n+1} + \cdots + a_{n+j} x^{n+j} + \cdots$$
$$= x^{n+1}[a_{n+1} + a_{n+2} x + \cdots + a_{n+j} x^{j-1} + \cdots]$$
$$= x^{n+1} h(x),$$

where h is the sum of a power series converging on $(-r, r)$. In particular, h is continuous, and on any *closed* subinterval I, say $I = [-r/2, r/2]$, $|h|$ has a maximum value K. Thus

$$|f(x) - p(x)| \le K|x|^{n+1}$$

on I. This shows the general nature of the approximation of f by its partial-sum polynomial $s(x)$; as x approaches zero, the error approaches zero like a multiple of x^{n+1}.

The same thing is true if f is only known to have $n + 1$ continuous derivatives on an interval about the origin. This is the key result. We state it below more generally for approximations about $x = a$.

If $p_n(x)$ is the nth Taylor polynomial for $f(x)$ about a, then the difference

$$f(x) - p_n(x)$$

is called the nth Taylor *remainder* for f at a and is designated $R_n(x)$. It is the error in the approximation of f by p_n, and the equation

$$f = p_n + R_n$$

thus expresses f as the approximating polynomial plus the error. If n is being held fixed, we shall sometimes drop the subscripts and write

$$f = p + R.$$

THEOREM 10   *Let f have derivatives up through at least the order $n + 1$ on an interval I about $x = a$. Then the nth Taylor remainder for f at a can be expressed in the form*

$$R_n(x) = \frac{f^{(n+1)}(X)}{(n + 1)!}(x - a)^{n+1}$$

for any x in I and for some number X lying between a and x.

Proof   We note first that the remainder $R(x)$ and its first n derivatives all vanish at a,

$$R^{(k)}(a) = f^{(k)}(a) - p^{(k)}(a) = 0$$

for $k = 0, 1, \ldots, n$, by the corollary in the last section.

In the rest of the proof we really should use a string of subscripted variables such as $x, x_1, x_2, \ldots, x_n$. But the subscripts would cause the formulas to appear more complicated than they actually are and we shall instead resort to the fiction that $x, y, z, \ldots, w$ stand for an arbitrarily long string of variables.

We now apply the second version of the parametric mean-value theorem (Chapter 10, Section 5) repeatedly to the quotient $R(x)/(x - a)^{n+1}$. First

$$\frac{R(x)}{(x - a)^{n+1}} = \frac{R'(y)}{(n + 1)(y - a)^n}$$

for some number y between a and x. Then

$$\frac{R'(y)}{(n + 1)(y - a)^n} = \frac{R''(z)}{(n + 1)n(z - a)^{n-1}}$$

for some point z between a and y. We can continue in this way right up through the nth derivative, because all the derivatives $R^{(k)}(a)$ are 0 for $k \le n$. Thus

$$\frac{R(x)}{(x - a)^{n+1}} = \frac{R'(y)}{(n + 1)(y - a)^n} = \cdots = \frac{R^{(n)}(w)}{(n + 1)!\,(w - a)}$$

$$= \frac{R^{(n+1)}(X)}{(n + 1)!},$$

where X lies between a and x. Now

$$R^{(n+1)} = (f - p)^{(n+1)} = f^{(n+1)} - p^{(n+1)} = f^{(n+1)}$$

because p is a polynomial of degree at most n, so that $p^{(n+1)} = 0$. Making this substitution in the term on the very right above, and multiplying through by $(x - a)^{n+1}$, we have the formula of the theorem.   ∎

The formula for $R_n(x)$ given in the theorem is called Lagrange's form of the remainder.

The expression of $f(x)$ in the form $f(x) = p(x) + R(x)$ now becomes a formula for $f(x)$ made up of its nth Taylor polynomial at a plus Lagrange's form of the remainder:

$$f(x) = f(a) + \frac{f'(a)}{1}(x - a) + \frac{f''(a)}{2!}(x - a)^2 + \cdots + \frac{f^{(n)}(a)}{n!}(x - a)^n$$

$$+ \frac{f^{(n+1)}(X)}{(n + 1)!}(x - a)^{n+1},$$

for some number X between a and x. This is called *Taylor's formula with remainder*.

The remainder formula is very useful. For example, if we can determine that $|f^{(n+1)}(x)|$ assumes the maximum value M on the interval I, then

$$|f(x) - p_n(x)| \leq \frac{M}{(n+1)!} |x - a|^{n+1}$$

for all x in I, and we have the error-estimating inequality we were looking for.

Beyond this, if f is infinitely differentiable, then we can fix x and vary n, and possibly show that the nth remainder $R_n(x) = f(x) - p_n(x)$ approaches 0. We would then have another way of proving that f is the sum of its Taylor series.

Example 1   Show that $f(x) = e^x$ is the sum of its Maclaurin series.

Solution   The Maclaurin formula with remainder is

$$e^x = p_n(x) + R_n(x) = 1 + \frac{x}{1} + \frac{x^2}{2!} + \cdots + \frac{x^n}{n!} + e^X \frac{x^{n+1}}{(n+1)!}$$

for some X between 0 and x. If x is positive then $e^X \leq e^x$, and

$$R_n(x) \leq \frac{e^x x^{n+1}}{(n+1)!}.$$

(If x is negative, then $e^X \leq 1$ and we omit the e^x factor.) We know that $x^n/n! \to 0$ as $n \to \infty$. Therefore $R_n(x) \to 0$ for all x. That is

$$p_n(x) = e^x - R_n(x) \to e^x,$$

so

$$e^x = \lim_{n \to \infty} p_n(x) = \lim_{n \to \infty} \sum_{k=0}^{n} \frac{x^k}{k!} = \sum_{k=0}^{\infty} \frac{x^k}{k!}.$$

We thus have a second proof that e^x is the sum of its Maclaurin series. However when x is positive the above estimate of the remainder $R_n(x)$ is not as good as the one we obtained in Chapter 11 by comparing the remainder series with the geometric series.   □

Computations   It was mentioned earlier that function values have generally been computed from power series expansions. That is, a function value $f(x)$ is approximated by the value of a Taylor or Maclaurin polynomial $p_n(x)$, and the answer is then shown to be good enough by estimating the error, which is the remainder

$$R_n(x) = f(x) - p_n(x).$$

We shall illustrate by examples three ways of doing this:

1. If $f(x)$ is the sum of an alternating series then the error is less than the next term:

$$|R_n(x)| < |a_{n+1}x^{n+1}|.$$

2. Assuming, again, that $f(x)$ is the sum of a power series, then $R_n(x)$ is the sum of the nth remainder series,

$$R_n(x) = a_{n+1}x^{n+1} + a_{n+2}x^{n+2} + \cdots$$

and we may be able to compare this series with a known and easily computed series, such as the geometric series.

3. We may be able to estimate $R_n(x)$ from a formula for the remainder, such as the Lagrange formula proved earlier in this section. This method can be used even when we don't know a complete series expansion for f.

Example 2   The nonzero terms of the Maclaurin series for $\sin x$ form an alternating series if $|x| \leq 2$. Therefore $x - x^3/3!$ approximates $\sin x$ with an error less than $x^5/5!$. For example,

$$\sin 0.2 \approx 0.2 - 0.008/6$$
$$= 0.2 - 0.001333 \cdots$$
$$= 0.198666 \cdots$$

with an error less than

$$\frac{(0.2)^5}{120} = \frac{0.00032}{120} < 0.000003.$$

The partial sums of an alternating series are alternately too large and too small, so here

$$0.198666 < \sin 0.2 < 0.198670,$$

and

$$\sin 0.2 = 0.19867$$

to five decimal places.   □

Example 3   Determine a range of x over which the first three nonzero terms in the Maclaurin series for $\sin x$ are accurate as an approximation to $\sin x$ to within an error of 1 in the fifth decimal place. Consider only positive x.

Solution The error is less than $x^7/7!$ so it will be sufficient to restrict x so that

$$\frac{x^7}{7!} < 10^{-5},$$

or

$$x^7 < (7!)10^{-5} = 5040 \cdot 10^{-5} = 0.0504$$

or

$$x < (0.0504)^{1/7}.$$

A calculator gives

$$(0.0504)^{1/7} = 0.65 \cdots.$$

So if $x < 0.65$, then

$$x - \frac{x^3}{3!} + \frac{x^5}{5!}$$

approximates $\sin x$ with an error less than 10^{-5}. □

In Section 3 of the last chapter we gave a computation of e in which the remainder series was compared to a geometric series. This was sufficient to illustrate the second procedure mentioned above, so we turn finally to examples that depend on the Lagrange remainder formula.

Example 4 The Taylor series for $\sin x$ about $x = \pi/4$ is not an alternating series, and the remainder is best tackled by the third method. We computed the second Taylor polynomial in Example 6 of the last section. By Theorem 10, this polynomial approximates $\sin x$ with the error (remainder)

$$\frac{-\cos(X)}{3!}\left(x - \frac{\pi}{4}\right)^3,$$

where X is some number between x and $\pi/4$. Since $|\cos X| \le 1$, the error is at most

$$\frac{\left|x - \frac{\pi}{4}\right|^3}{3!}.$$

For values of x near $\pi/4$, this approximation to $\sin x$ will have an accuracy that could be obtained from the Maclaurin series only by using many more terms. However, this advantage is offset in part by the fact that here we may have to use decimal approximations to $\pi/4$.

□

Example 5 Suppose that, for some reason, we need to know the value of

$$\int_0^{1/2} \tan(x^2)\,dx$$

to several decimal places. How would we compute it?

Solution We saw in Example 4 in the last section that the third Maclaurin polynomial for $\tan t$ is

$$\tan t \approx t + \frac{t^3}{3}.$$

The error in this approximation can be written

$$R_3(t) = \frac{\tan^{(4)}(T)}{4!}\,t^4,$$

by Theorem 10. Here $t = x^2$ lies between 0 and 1/4, and hence between 0 and $\pi/12$, and over this range we find (after further calculation) that

$$0 < \tan^{(4)}T < 6.$$

We therefore have, first,

$$\int_0^{1/2} \tan(x^2)\,dx \approx \int_0^{1/2}\left(x^2 + \frac{x^6}{3}\right)dx = \left(\frac{x^3}{3} + \frac{x^7}{21}\right)_0^{1/2}$$

$$= \frac{1}{24}\cdot\frac{113}{112}$$

$$= 0.04203869\cdots$$

(by pocket calculator). Second, the error in this calculation is

$$\int_0^{1/2} R_3(x^2)\,dx.$$

By the above inequality on $\tan^{(4)}T$, this integral is positive and less than

$$\int_0^{1/2} \frac{6}{4!} x^8 \, dx = \frac{1}{4} \frac{x^9}{9} \Bigg]_0^{1/2}$$

$$= \frac{1}{9 \cdot 2^{11}} \le 5.5(10)^{-5}.$$

We have thus shown that

$$\int_0^{1/2} \tan(x^2) \, dx = 0.0420$$

to four decimal places.

☐

PROBLEMS FOR SECTION 5

Find the nth Maclaurin formula with remainder, $f(x) = p_n(x) + r_n(x)$, for each function f below. Also determine a constant K such that

$$|f(x) - p_n(x)| \le Kx^{n+1}$$

on the interval $I = [-\frac{1}{2}, \frac{1}{2}]$.

1. $f(x) = \ln \cos x, \quad n = 3$ **2.** $f(x) = \arcsin x, \quad n = 2$ **3.** $f(x) = \sec x, \quad n = 2$

4. $f(x) = \tan x, \quad n = 2$ **5.** $f(x) = e^{-x}, \quad n = 4$ **6.** $f(x) = \sin\left(x - \frac{\pi}{4}\right), \quad n = 4$

7. $f(x) = (1 + x)^{3/2}, \quad n = 3$ **8.** $f(x) = \int_0^x e^{-t^2} \, dt, \quad n = 3$

Find the nth Taylor formula with remainder, $f(x) = p_n(x) + r_n(x)$, about the point $x = a$, where

9. $f(x) = \cos x, \quad n = 4, a = \pi/4$ **10.** $f(x) = \cos x, \quad n = 4, a = \pi/3$

11. $f(x) = \sin x, \quad n = 4, a = \pi/6$ **12.** $f(x) = e^x, \quad n = n, a = a$

13. $f(x) = x^{1/3}, \quad n = 3, a = 1$ **14.** $f(x) = x^{-1/4}, \quad n = 3, a = 1$

15. $f(x) = x^r, \quad n = 3, a = 1$ **16.** $f(x) = \tan x, \quad n = 2, a = \pi/4$

17. $f(x) = \ln \cos x, n = 3, a = \pi/4$ **18.** Show that

$$\frac{\sqrt{2}}{2} \left[1 - \frac{\pi}{20} - \frac{\pi^2}{800} \right]$$

approximates $\sin \pi/5$ with an error less than 0.001.

19. Prove that $\sin x$ is the sum of its Maclaurin series by using the Maclaurin formula with remainder, as in Example 1 in the text.

21. Prove that $\sin x$ is the sum of its Taylor series about $\pi/2$ by using the Taylor formula with remainder and proving that the remainder goes to zero.

23. Prove that e^x is the sum of its Taylor series about $x = a$, by evaluating the remainder and showing that it goes to zero.

⊞ 25. Compute $\sin 0.2$ to eight decimal places. (First show that the error after the first three terms is less than 3×10^{-9}.)

⊞ 27. Prove that

$$\int_0^x e^{-t^2/2}\, dt = x - \frac{x^3}{3 \cdot 2} + \frac{x^5}{5 \cdot 2^2 \cdot 2!} - \frac{x^7}{7 \cdot 2^3 \cdot 3!}$$
$$+ \cdots + \frac{(-1)^n x^{2n+1}}{(2n+1)2^n n!} + \cdots$$

Use the above series to compute $\int_0^1 e^{-t^2/2}\, dt$ to three decimal places.

⊞ 29. Compute $\int_0^1 \cos \sqrt{x}\, dx$ to four decimal places.

31. In the above series, show that the error after n nonzero terms is at most

$$\frac{2}{2n+1} x^{2n+1} \left(\frac{1}{1-x^2} \right).$$

(Compare the remainder with a geometric series.)

⊞ 33. Compute $\ln 3/2$ to eight decimal places.

20. Prove that $1/(1-x)$ is the sum of its Maclaurin series over the interval $(-1, 1/2)$ by using the Maclaurin formula with remainder.

22. Show that the Taylor series for $\sin x$ about $a = \pi/6$ expresses the identity

$$\sin x = \sin\left(x - \frac{\pi}{6}\right)\cos\frac{\pi}{6} + \cos\left(x - \frac{\pi}{6}\right)\sin\frac{\pi}{6}.$$

24. Show that for any value of x the Maclaurin series for $\sin x$ is alternating from some term on.

⊞ 26. Compute $\sin 0.5$ to three decimal places.

⊞ 28. Express

$$\int_0^1 \frac{\sin x}{x}\, dx$$

as the sum of an infinite series. Then use the series to compute the integral to four decimal places.

30. The series for $\ln(1+x)$ in Theorem 4, Corollary 1 is not useful for computation because it converges too slowly. Show that it can be modified to give the following improved version:

$$\ln\frac{1+x}{1-x} = 2\left[x + \frac{x^3}{3} + \frac{x^5}{5} + \cdots + \frac{x^{2n-1}}{2n-1} + \cdots\right].$$

(First write down the $\ln(1-x)$ series and then combine.)

⊞ 32. Compute $\ln 3/2$ to four decimal places by using the above series. (First show that the error after the first three series terms, i.e., through the x^5 term, is less than 10^{-5}.) Note how much more efficient this computation is than the Riemann-sum method of Chapter 9.

⊞ 34. Compute $\ln 2$ to five decimal places.

35. The arctan x series

$$\arctan x = x - \frac{x^3}{3} + \frac{x^5}{5}$$
$$+ \cdots + (-1)^n \frac{x^{2n+1}}{2n+1} + \cdots$$

provides the standard method of computing π, but here again a little artistic modification pays enormous dividends. Prove first that, if $\tan \theta = 1/5$, then

$$\tan\left(4\theta - \frac{\pi}{4}\right) = \frac{1}{239}.$$

(Use the formulas for $\tan(x + y)$ and $\tan(x - y)$ from Appendix 2. Compute $\tan 2\theta$, $\tan 4\theta$, and $\tan(4\theta - \pi/4)$, in that order.) Conclude that

$$\pi = 16 \arctan \frac{1}{5} - 4 \arctan \frac{1}{239}.$$

This remarkable identity has been known for nearly 300 years.

37. The decimal expansion of $1/239$ (and hence that of $4/239$) repeats in blocks of 7. Carry out the long division far enough to determine this expansion. Then use the result from the above problem, and the inequality $\arctan x < x$, to show that

$$0.01673620 < 4 \arctan \frac{1}{239} < 0.01673641.$$

39. Combine the above two results using Problem 14, and so prove that

$$3.1415925 < \pi < 3.1415928.$$

Thus $\pi = 3.141593$ to the nearest six decimal places, and the expansion of π begins $3.141592 \cdots$

36. We start by seeing what we can do with only the first term of the series for arctan $1/239$. Using just the crude estimate $239 > 200$ in the error term, show that $4/239$ approximates 4 arctan $1/239$ with an error less than $2(10)^{-7}$.

38. Now compute $16 \arctan 0.2$ through the $x^9/9$ term, and show that

$$3.15832895 < 16 \arctan 0.2 < 3.15832899.$$

40. Show that if arctan $1/239$ is replaced by its series expansion through the second term (the $x^3/3$ term) in the formula

$$\pi = 16 \arctan \frac{1}{5} - 4 \arctan \frac{1}{239},$$

and if arctan $1/5$ is replaced by its expansion through the eighth term (the $x^{15}/15$ term), then the right side approximates π with an error less than 5×10^{-12}. (Use the crude estimates $239 > 200$ and $2^{13} = 8192 < 10^4$.)

CHAPTER 13
VECTORS IN THE PLANE AND IN SPACE

The interaction between vectors and calculus leads to a huge increase in the range of problems to which calculus can be successfully applied, and such vector calculus is one of the principal topics of advanced courses in calculus. In this book we can only introduce the subject and give a few applications.

This chapter develops the algebra of vectors in the plane and in space, along with its geometric interpretations. Chapter 14 then treats the differential calculus for vector functions of a parameter t, and Chapter 16 takes up vector interpretations of differentiation that arise in the study of functions of two or three independent variables. Chapter 15 introduces such functions and has no vector material.

1
VECTORS IN THE PLANE

This section introduces vectors and vector operations in the plane and sets up the laws of vector algebra.

A *vector* is any entity that is specified by a magnitude and a direction.

Example 1   The velocity of a particle moving in the plane (or in space) is a vector. It is specified by the direction in which the particle is moving and by the speed of the particle (the magnitude of its velocity).   □

Example 2   Forces are vectors. For example, the gravitational force exerted by the moon on a circling rocket is directed toward the center of the moon, and its magnitude is proportional to $1/d^2$, where d is the distance from the rocket to the center of the moon.   □

Example 3

Arrows   Any vector can be represented geometrically by an *arrow* (directed line segment). Thus a force applied to an object at a point P can be represented by an arrow stemming from P in the direction of the force, the magnitude of the force being given by the arrow length. The same force, applied to an object at a point R, would then be represented by an arrow having the same length and direction but stemming from R (Fig. 1). In other words, two *different* arrows represent the *same* vector force if they have the same direction and length.

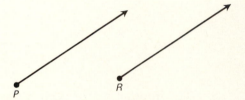

Figure 1

The velocity of a moving object can be represented by an arrow in the same way, and in the same way two different objects having the same vector velocity call for two different arrows having the same direction and length.

We use the notation $\overrightarrow{PQ}$ for the directed line segment (arrow) *from P to Q*. Two arrows that represent the same vector are said to be equal. Thus the equation

$$\overrightarrow{PQ} = \overrightarrow{RS}$$

means that the two arrows have the same length and direction; although different as directed line segments, they are the same as vectors.

The situation is exactly like the representation of rational numbers by fractions. We write

$$\frac{4}{6} = \frac{2}{3},$$

meaning that the two fractions represent the same rational number;

although different as fractions they are the same as rational numbers. And just as we refer to the *number* 2/3, meaning the number represented by the fraction, so shall we refer to the *vector* $\overrightarrow{PQ}$, meaning the vector represented by the arrow. □

Vector Addition Two vector forces applied to a point P act in combination as though they were a single force, called their *resultant*, or *sum*. The sum vector is found by the parallelogram rule. Vector velocities also combine by the parallelogram rule; in fact, any two vectors of the same kind combine in this way.

The Parallelogram Rule The sum (resultant) of the vectors $\overrightarrow{PQ}$ and $\overrightarrow{PR}$ is the vector $\overrightarrow{PS}$, where S is the fourth vertex of the parallelogram having PQ and PR as adjacent sides (Fig. 2).

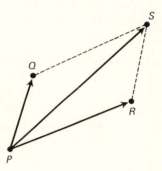

Figure 2

A better description of this sum vector is indicated in Fig. 3. Its terminal point S is obtained by "laying off" from Q the arrow $\overrightarrow{QS}$ that is equal to $\overrightarrow{PR}$.

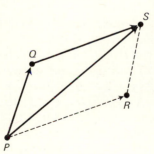

Figure 3

This procedure covers a special case that is missed by the parallelogram rule, namely, the case where $\overrightarrow{PQ}$ and $\overrightarrow{PR}$ lie along the same line and no genuine parallelogram exists. We shall therefore adopt the "laying off" characterization as the definition of vector addition.

DEFINITION   The *sum* of the vectors $\overrightarrow{PQ}$ and $\overrightarrow{PR}$ is the vector $\overrightarrow{PS}$, where the point S is determined so that

$$\overrightarrow{QS} = \overrightarrow{PR}$$

When we want to refer to a vector without pinning ourselves down to one of its representing arrows, we shall generally use a lower case boldface letter such as **u**, **v**, or **x**.

The sum of two vectors **u** and **v** was defined above in terms of arrow representatives. However, it is independent of which *particular* arrow was chosen to represent **u** as shown in Fig. 4. We can therefore write the sum

$$\mathbf{u} + \mathbf{v},$$

without referring to arrows.

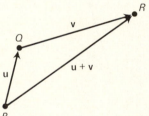

Figure 4

The equation

$$\mathbf{u} = \overrightarrow{PQ}$$

is read, "**u** is the vector (represented by the arrow) $\overrightarrow{PQ}$."

The following two examples illustrate the parallelogram rule. They are not essential.

Example 4   A bow exerts a force of twenty pounds on an arrow just before release. That force is the vector sum of the two forces exerted by the bow ends along the string of the bow.

On the right in Fig. 5, this vector sum is the vertical diagonal of the parallelogram. If the pulled string makes an angle of 15° with

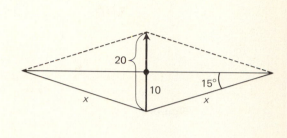

Figure 5

the line through the bow tips, then the common magnitude x of these two string-tension forces is given by

$$\frac{10}{x} = \sin 15° \approx 0.26,$$

$$x \approx \frac{10}{0.26} \approx 40 \text{ pounds.} \qquad \square$$

Example 5   A canoeist wishes to paddle across a river to a point directly opposite on the other bank. The canoeist can paddle at 4 miles per hour and the river current runs 2 miles per hour. In what direction must the canoe be aimed?

        Solution   The actual velocity of the canoe is the vector sum of the water velocity and its velocity relative to the water. That vector sum will be perpendicular to the river bank if the canoe is aimed upstream at such an angle θ with the perpendicular that $\sin \theta = 2/4$ (see Fig. 6). Thus $\theta = 30°$.

Figure 6

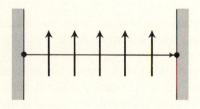 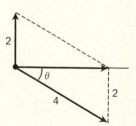

Components   So far we have been discussing vectors without assuming them to be confined to a particular plane, and without making any use of a

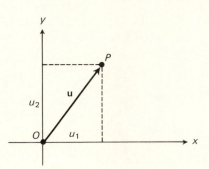

Figure 7

coordinate system. We turn now to the coordinate plane. Until further notice, all vectors will be assumed to lie in the standard x, y plane. Then each vector $\mathbf{u}$ is represented by a unique arrow $\overrightarrow{OP}$ stemming from the origin (Fig. 7), and hence $\mathbf{u}$ acquires *components*, defined as follows.

DEFINITION　If $\mathbf{u} = \overrightarrow{OP}$, then the coordinates of P are called the *components* of $\mathbf{u}$, and $\mathbf{u}$ is called the *position vector* of P.

A vector in the coordinate plane can now be identified with an ordered pair of numbers, its components, just as earlier we identified a point in the coordinate plane with its pair of coordinates. Thus we write $\mathbf{u} = (u_1, u_2)$, where (u_1, u_2) are the components of $\mathbf{u}$. It turns out that vector operations have particularly simple characterizations as operations on components.

THEOREM 1　*If* $\mathbf{u} = (u_1, u_2)$ *and* $\mathbf{v} = (v_1, v_2)$ *then*

$$\mathbf{u} + \mathbf{v} = (u_1 + v_1, u_2 + v_2).$$

That is, vector addition is given by ordinary arithmetic addition of corresponding vector components.

Partial Proof　Figure 8 shows (in one case) that laying off the vector $\mathbf{v} = (v_1, v_2)$ from the point $P = (x, y)$ has the effect of adding the components of $\mathbf{v}$ to the coordinates of P. Any other special case can be verified in the same way. So if $\mathbf{u}$ is the position vector of P, then the sum $\mathbf{u} + \mathbf{v}$ has components $(x + v_1, y + v_2)$.

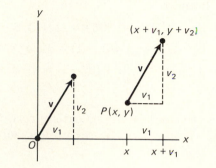 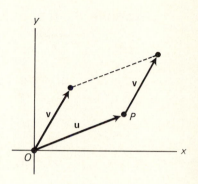

Figure 8

Another important vector operation is multiplying a vector by a number.

DEFINITION The *product* of a vector **u** and a positive number t, designated $t\mathbf{u}$, is the vector **v** having the same direction as **u**, but t times the magnitude of **u**. If t is negative, then $\mathbf{v} = t\mathbf{u}$ has the direction opposite to that of **u** and $|t|$ times the magnitude of **u**.

Since similar right triangles have corresponding sides proportional, we have (Fig. 9):

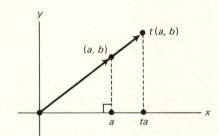

Figure 9

THEOREM 2 *If* $\mathbf{u} = (a, b)$, *then*

$$t\mathbf{u} = (ta, tb).$$

In vector discussions, numbers and numerical quantities are called *scalars*. Thus $t\mathbf{u}$ is the product of a scalar and a vector.

Example 6 If $\mathbf{a} = (1, 2)$ and $\mathbf{b} = (3, -1)$, then

$$\mathbf{a} + \mathbf{b} = (1, 2) + (3, -1) = (1 + 3, 2 - 1) = (4, 1)$$
$$2\mathbf{a} + 4\mathbf{b} = 2(1, 2) + 4(3, -1) = (2, 4) + (12, -4) = (14, 0)$$
$$(-3)\mathbf{a} + \mathbf{b} = (-3)(1, 2) + (3, -1) = (-3, -6) + (3, -1) = (0, -7).$$

$\square$

The two vector operations satisfy certain laws of algebra. These laws permit problems to be set up and solved much as in ordinary algebra.

The Vector Laws For any vectors $\mathbf{u}$, $\mathbf{v}$, and $\mathbf{w}$,

> **V1.** $\mathbf{u} + \mathbf{v} = \mathbf{v} + \mathbf{u}$,
> **V2.** $(\mathbf{u} + \mathbf{v}) + \mathbf{w} = \mathbf{u} + (\mathbf{v} + \mathbf{w})$.

There is a special vector, designated $\mathbf{0}$ and called the zero vector, such that

> **V3.** $\mathbf{u} + \mathbf{0} = \mathbf{0} + \mathbf{u} = \mathbf{u}$

for every vector $\mathbf{u}$. Moreover, for each vector $\mathbf{u}$ there is a uniquely determined "opposite" vector, designated $-\mathbf{u}$ and called the *negative of* $\mathbf{u}$, such that

> **V4.** $\mathbf{u} + (-\mathbf{u}) = \mathbf{0}$.

The above laws concern vector addition only. Multiplication by scalars satisfies the further laws:

> **V5.** $t(\mathbf{u} + \mathbf{v}) = t\mathbf{u} + t\mathbf{v}$,
> **V6.** $(s + t)\mathbf{u} = s\mathbf{u} + t\mathbf{u}$,
> **V7.** $s(t\mathbf{u}) = (st)\mathbf{u}$,
> **V8.** $1\mathbf{u} = \mathbf{u}$.

The first two laws are the commutative and associative laws for vector addition. The fifth and sixth are distributive laws: they state that the operation of multiplying a vector by a scalar distributes over both addition of vectors and addition of numbers. The seventh law is a mixed associative law relating multiplication of numbers to the multiplication of a vector by a number.

The zero vector has the components $(0, 0)$ and is represented geometrically by "degenerate" arrows $\overrightarrow{PP}$ that go nowhere; they have zero length and no direction. The negative of $\mathbf{u} = (u_1, u_2) = \overrightarrow{PQ}$ is given algebraically by $-\mathbf{u} = (-u_1, -u_2)$, and geometrically by the reversed arrow: $-\mathbf{u} = \overrightarrow{QP}$. In terms of these arrows and the "laying off" definition of addition, the equation $\mathbf{u} + (-\mathbf{u}) = \mathbf{0}$ is

$$\overrightarrow{PQ} + \overrightarrow{QP} = \overrightarrow{PP}.$$

Figure 10 shows how the commutative and associative laws for vector addition can be verified geometrically.

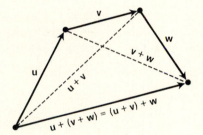 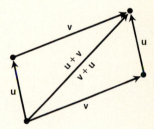

Figure 10

The trouble with such geometric proofs is that it is usually impossible to draw figures in a sufficiently "general position" to represent all cases. It is much easier to give purely algebraic proofs on the basis of Theorems 1 and 2 and the corresponding laws for numbers. For example, if $\mathbf{u} = (u_1, u_2)$ and $\mathbf{v} = (v_1, v_2)$, then

$$
\begin{aligned}
t[\mathbf{u} + \mathbf{v}] &= t[(u_1, u_2) + (v_1, v_2)] \\
&= t(u_1 + v_1, u_2 + v_2) && \text{(Theorem 1)} \\
&= (t(u_1 + v_1), t(u_2 + v_2)) && \text{(Theorem 2)} \\
&= (tu_1 + tv_1, tu_2 + tv_2) && \text{(distributive law for numbers)} \\
&= (tu_1, tu_2) + (tv_1, tv_2) && \text{(Theorem 1)} \\
&= t(u_1, u_2) + t(v_1, v_2) && \text{(Theorem 2)} \\
&= t\mathbf{u} + t\mathbf{v}.
\end{aligned}
$$

Subtracting $\mathbf{u}$ is defined as adding $-\mathbf{u}$:

$$\mathbf{v} - \mathbf{u} = \mathbf{v} + (-\mathbf{u}).$$

Numerically, this amounts simply to subtracting corresponding components, since

$$(v_1, v_2) - (u_1, u_2) = (v_1, v_2) + (-u_1, -u_2)$$
$$= (v_1 - u_1, v_2 - u_2).$$

Notation Component notation is traditionally inconsistent. The components of a vector **u** are normally indicated by subscripts on the corresponding lightface letter, as in

$$\mathbf{u} = (u_1, u_2).$$

On the other hand, the position vector of a variable point is usually indicated

$$\mathbf{x} = (x, y),$$

in order to keep the standard x, y usage and also in order to avoid subscripts, since a subscripted letter may suggest a constant. The position vector of a *fixed* point is treated in either manner, as in

$$\mathbf{x}_0 = (x_0, y_0), \qquad \text{or} \qquad \mathbf{a} = (a, b), \qquad \text{or} \qquad \mathbf{a} = (a_1, a_2).$$

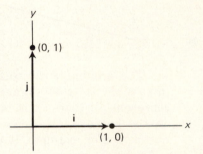

Figure 11

Finally, we introduce a traditional method for expressing a vector in terms of its components. The position vectors of the unit points on the coordinate axes are designated **i** and **j**,

$$\boxed{\mathbf{i} = (1, 0), \qquad \mathbf{j} = (0, 1),}$$

and are called the *standard basis vectors* (Fig. 11). Then,

LEMMA *Any vector* $\mathbf{u} = (u_1, u_2)$ *is given by*

$$\boxed{\mathbf{u} = u_1\mathbf{i} + u_2\mathbf{j}.}$$

That is, any vector $\mathbf{u}$ is a *linear combination* of the standard basis vectors $\mathbf{i}$ and $\mathbf{j}$, with coefficients equal to the components of $\mathbf{u}$.

Proof By Theorems 1 and 2, we have

$$\mathbf{u} = (u_1, u_2) = (u_1, 0) + (0, u_2)$$
$$= u_1(1, 0) + u_2(0, 1)$$
$$= u_1\mathbf{i} + u_2\mathbf{j}. \qquad \blacksquare$$

When vectors are written this way, their manipulation involves constant use of the laws of vector algebra. For example, the computation of the coefficients (components) of $\mathbf{u} + \mathbf{v}$,

$$\mathbf{u} + \mathbf{v} = (u_1\mathbf{i} + u_2\mathbf{j}) + (v_1\mathbf{i} + v_2\mathbf{j})$$
$$= (u_1 + v_1)\mathbf{i} + (u_2 + v_2)\mathbf{j},$$

would require, if written out in complete detail, several applications of the associative law for vector addition, one application of the commutative law, and two applications of the distributive law $(s + t)\mathbf{u} = s\mathbf{u} + t\mathbf{u}$.

PROBLEMS FOR SECTION 1

1. Draw a figure for the geometric verification of the vector law **V5**.

2. For each pair of vectors $\mathbf{x}$ and $\mathbf{a}$ given below, find the requested algebraic combination.

    a) $\mathbf{x} = (1, 4)$, $\mathbf{a} = (2, 1)$; find $\mathbf{x} - \mathbf{a}$.

    b) $\mathbf{x} = (3/2, 1/4)$, $\mathbf{a} = (7/3, 1/6)$; find $\mathbf{x} + 6\mathbf{a}$.

    c) $\mathbf{x} = (3/2, 1/4)$, $\mathbf{a} = (7/3, 1/6)$; find $6(\mathbf{x} + \mathbf{a})$.

    d) $\mathbf{x} = (1, 4)$, $\mathbf{a} = (2, 7)$; find $\frac{1}{2}\mathbf{x} - 2\mathbf{a}$.

3. Solve the bow problem if the pulled bow string exerts a force of 20 pounds on the arrow when the angle between the bow tips and the deflected string is 10°.

4. Solve the bow problem when an angle of 30° produces a force of 30 pounds.

5. Solve the canoe problem when the canoe and stream speeds are $2.8 \approx 2\sqrt{2}$ and 2 miles per hour, respectively.

6. Solve the canoe problem when the two speeds are $2.3 \approx 4/\sqrt{3}$ and 2.

7. A small plane flies 200 mph. The pilot wishes to fly due north at a time when a westerly wind of 40 mph is blowing. How should he set his course? What will be his true speed (relative to ground)?

8. Do the same problem for plane and wind speeds of 130 and 50 mph, respectively.

Prove the following laws from the component characterizations of the vector operations and the laws of algebra for numbers.

9. $\mathbf{a} + \mathbf{x} = \mathbf{a}$ if and only if $\mathbf{x} = \mathbf{0}$

10. $\mathbf{a} + \mathbf{u} = \mathbf{0}$ if and only if $\mathbf{u} = -\mathbf{a}$

11. $(\mathbf{u} + \mathbf{x}) + \mathbf{a} = \mathbf{u} + (\mathbf{x} + \mathbf{a})$

12. $(s + t)\mathbf{x} = s\mathbf{x} + t\mathbf{x}$

13. $0\mathbf{a} = \mathbf{0}$

14. $(-1)\mathbf{a} = -\mathbf{a}$

15. A theorem of geometry says that if two sides of a quadrilateral are parallel and equal, then the other two sides are also parallel and equal, so the figure is a parallelogram. Formulate and prove this as a theorem about arrows and the vector operations.

2
PARAMETRIC
EQUATIONS
FOR A LINE

When we refer to "the point $\mathbf{x}$" we mean the point P having the position vector $\mathbf{x}$, so that $\mathbf{x} = \overrightarrow{OP}$. Algebraically, $\mathbf{x}$ and P are indistinguishable: the coordinates of P are the components of $\mathbf{x}$. Two nonzero vectors $\mathbf{a}$ and $\mathbf{b}$ are *parallel* if they have the same or opposite directions, that is, if $\mathbf{b} = t\mathbf{a}$ for some scalar t.

A line l is determined if we know a point $\mathbf{x}_0$ on l and a nonzero vector $\mathbf{u}$ parallel to l. For then a point $\mathbf{x}$ lies on l if and only if the arrow from $\mathbf{x}_0$ to $\mathbf{x}$ is parallel to $\mathbf{u}$, that is, if and only if

$$\mathbf{x} - \mathbf{x}_0 = t\mathbf{u},$$

or

$$\mathbf{x} = \mathbf{x}_0 + t\mathbf{u},$$

from some scalar t. See Fig. 1.

As the real number t runs from $-\infty$ to $+\infty$, the varying point $\mathbf{x}$ sweeps out the line. In this situation, t is called a *parameter*, and the

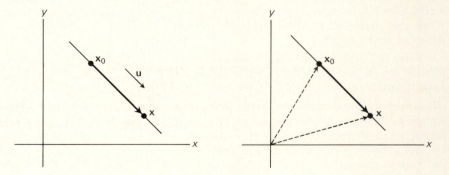

Figure 1

above equation is a *parametric equation* for the line. This is consistent with the use of the word *parameter* in Chapter 10. The parameter t is not "part of" the point $\mathbf{x}$, but t determines $\mathbf{x}$.

The result above is important, and we restate it as a theorem.

THEOREM 3 *The vector equation*

$$\boxed{\mathbf{x} = \mathbf{x}_0 + t\mathbf{u}}$$

is a parametric equation for the line through the point $\mathbf{x}_0$ in the direction of the nonzero vector $\mathbf{u}$.

Example 1 The line through the point $\mathbf{x}_0 = (1, 2)$ in the direction of $\mathbf{u} = (3, -1)$ has the parametric equation

$$\begin{aligned} \mathbf{x} &= \mathbf{x}_0 + t\mathbf{u} \\ &= (1, 2) + t(3, -1). \end{aligned}$$

This is the equation

$$(x, y) = (1 + 3t, 2 - t),$$

and is equivalent to the pair of numerical (scalar) equations

$$\begin{aligned} x &= 1 + 3t, \\ y &= 2 - t. \end{aligned}$$

Note that these are parametric equations for the line in exactly the sense of Section 3 in Chapter 10. The parameter can easily be eliminated to obtain an x, y-equation for the line. For instance, $t = 2 - y$ from the second equation, so

$$x = 1 + 3t = 1 + 3(2 - y) = 7 - 3y.$$

The line has the equation

$$x + 3y = 7. \qquad \square$$

Two points determine a line. If $\mathbf{a}$ and $\mathbf{b}$ are (the position vectors of) two distinct points on l, then $\mathbf{u} = \mathbf{b} - \mathbf{a}$ gives the direction of l, so the parametric equation $\mathbf{x} = \mathbf{a} + t\mathbf{u}$ can be written

$$\boxed{\mathbf{x} = \mathbf{a} + t(\mathbf{b} - \mathbf{a}).}$$

Example 2   The line through the points $\mathbf{a} = (1, 2)$ and $\mathbf{b} = (4, 1)$ has the parametric equation

$$\mathbf{x} = (1, 2) + t[(4, 1) - (1, 2)]$$
$$= (1, 2) + t(3, -1)$$

and is the line of Example 1.   □

Proportional Division of a Segment

When we write the two-point equation $\mathbf{x} = \mathbf{a} + t(\mathbf{b} - \mathbf{a})$ in the form

$$\mathbf{x} - \mathbf{a} = t(\mathbf{b} - \mathbf{a}),$$

it says that $\mathbf{x}$ is t times as far from $\mathbf{a}$ as $\mathbf{b}$ is. For example, $\mathbf{x}$ will be the midpoint of the segment $\mathbf{ab}$ if $t = 1/2$, and solving for $\mathbf{x}$ with this value of t yields the midpoint formula

$$\boxed{\mathbf{x} = \frac{\mathbf{a} + \mathbf{b}}{2}.}$$

When $t = 2/3$, we get the point 2/3 of the way from $\mathbf{a}$ to $\mathbf{b}$, and its formula comes out to be

$$\mathbf{x} = \frac{1}{3}\mathbf{a} + \frac{2}{3}\mathbf{b}.$$

It divides the segment ab in the ratio 2 to 1. In general,

THEOREM 4   *If $s + t = 1$ and both numbers are positive, then*

$$\boxed{\mathbf{x} = s\mathbf{a} + t\mathbf{b}}$$

is the point on the segment $\mathbf{ab}$ that divides it in the ratio t/s.

Note that the dividing point is nearer the endpoint having the *larger coefficient.*

Example 3   The midpoint of the segment from $\mathbf{a} = (1, 2)$ to $\mathbf{b} = (4, 4)$ is

$$\mathbf{x} = \frac{\mathbf{a} + \mathbf{b}}{2} = \frac{(1, 2) + (4, 4)}{2}$$
$$= \frac{1}{2}(5, 6) = \left(\frac{5}{2}, 3\right). \quad\quad □$$

Example 4 The point (3/4) of the way from **a** to **b** on the above segment is

$$\mathbf{x} = \frac{1}{4}\mathbf{a} + \frac{3}{4}\mathbf{b}$$

$$= \frac{(1,2) + 3(4,4)}{4}$$

$$= \frac{(13,14)}{4} = \left(\frac{13}{4}, \frac{7}{2}\right).$$ □

We can already solve some problems in geometry by this new algebra.

Example 5 Prove that the medians of a triangle are concurrent.

Solution We start off by computing the point **x** where the two medians shown in Fig. 2 intersect. Since **x** lies on the segment **mc**, it can be written

$$\mathbf{x} = t\mathbf{m} + (1 - t)\mathbf{c}$$

for some number t between 0 and 1. Moreover, $\mathbf{m} = (\mathbf{a} + \mathbf{b})/2$. Thus,

$$\mathbf{x} = \frac{t}{2}\mathbf{a} + \frac{t}{2}\mathbf{b} + (1 - t)\mathbf{c}.$$

But **x** also lies on the segment **an**, so that, similarly,

$$\mathbf{x} = (1 - s)\mathbf{a} + \frac{s}{2}\mathbf{b} + \frac{s}{2}\mathbf{c}$$

for some number s between 0 and 1. These two representations must give the same point **x**, and this will happen if $s = t$ and $(t/2) = (1 - t)$. This gives $s = t = (2/3)$, and

$$\mathbf{x} = \frac{\mathbf{a} + \mathbf{b} + \mathbf{c}}{3}.$$

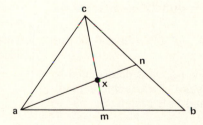

Figure 2

Since this representation is symmetric in the letters **a**, **b**, and **c**, the point **x** lies on all three medians. (In more detail, if we had started with another pair of medians, the calculation of their point of intersection would be the same as the above calculation except that the letters **a**, **b**, and **c** would be permuted. The resulting formula for the point of intersection would thus be the same as the above formula except for this permutation of the letters, and the two points of intersection are thus the same point.)   □

PROBLEMS FOR SECTION 2

Write out the following examples of the parametric equation of Theorem 3, and reduce each equation to the equivalent pair of scalar equations.

1. $x_0 = (1, 2), u = (-3, 0)$

2. $x_0 = (-2, 3), u = (1, 1)$

3. $x_0 = (0, 0), u = (3, 4)$

4. $x_0 = (0, 0), u = (a, b)$

5. $x_0 = (x_0, y_0), u = (1, 1)$

6. $x_0 = (2, -1), u = (1, 4)$

7. Find the point of intersection of the lines $x = (1, 2) + s(1, 1)$ and $x = (2, 2) + t(3, -1)$.

In each of the following problems, write the parametric equation of the line through the given points. Also give the equivalent pair of scalar equations.

8. $x_0 = (1, 4), x_1 = (2, -2)$

9. $x_0 = (1, 1), x_1 = (3, 0)$

10. $x_0 = (1, 1), x_1 = (4, 1)$

11. $x_0 = (2, 0), x_1 = (2, 1)$

12. $x_0 = (-1, 1/2), x_1 = (1/2, 3/4)$

In each of the following problems, find the point **x** that divides the line segment into the given ratio.

13. $x_0 = (1, 1), x_1 = (-6, 0)$; **x** is 1/3 of the way from x_1 to x_0

14. $x_0 = (0, 1), x_1 = (10, 4)$; **x** is 1/4 of the way from x_0 to x_1

15. $x_0 = (1/2, 1/3), x_1 = (4/3, 1/2)$; **x** is 2/5 of the way from x_0 to x_1

16. $x_0 = (-1, -4), x_1 = (-6, -2)$; **x** is 7/8 of the way from x_1 to x_0

17. $x_0 = (4, -3), x_1 = (-1, 6)$; **x** is 5/6 of the way from x_0 to x_1

18. Prove that the diagonals of a parallelogram bisect one another.

19. Prove that the line segment determined by the midpoints of two sides of a triangle is parallel to the third side and half its length.

20. In Example 5 it was noted that if $s = t$ and $1 - t = t/2$, so that $s = t = 2/3$, then the two representations of $\mathbf{x}$ are the same. It was implicit that $s = t = 2/3$ was the *only* solution, but this was never proved. Show this now, by showing that if the two representations of $\mathbf{x}$ are subtracted, then the resulting single equation can be cast into the form

$$k_1(\mathbf{b} - \mathbf{a}) = k_2(\mathbf{c} - \mathbf{a}),$$

where the coefficients k_1 and k_2 involve s and t. Conclude that $k_1 = k_2 = 0$, etc.

22. A theorem of geometry says that if the opposite sides of a quadrilateral are parallel, then the opposite sides are also equal, and the figure is a parallelogram. However, this clearly can be false if all four points lie on a straight line.

Prove the following correct algebraic version of the theorem.

> **Theorem** If $(\mathbf{a}_2 - \mathbf{x}_2) = s(\mathbf{a}_1 - \mathbf{x}_1)$ *and* $(\mathbf{a}_2 - \mathbf{a}_1) = t(\mathbf{x}_2 - \mathbf{x}_1)$, *then either* $s = t = 1$ *or the four points are collinear.*

21. Let $\mathbf{x}_1$, $\mathbf{x}_2$, $\mathbf{x}_3$, and $\mathbf{x}_4$ be the vertices of an arbitrary quadrilateral. Show that the midpoints of the sides are the vertices of a parallelogram.

23. Prove the following geometric theorem.

> **Theorem** *A line parallel to one side of a triangle cuts the other two sides proportionately. (Take one vertex at the origin.)*

3
THE DOT PRODUCT

In this section we define a way of multiplying two vectors together and getting a number. This new product operation has to do with measurements of lengths and angles in the plane. We consider first the unit vector $\mathbf{u}$ in the direction of a nonzero vector $\mathbf{x}$.

The ordinary absolute value symbol $|\mathbf{x}|$ will be used for the magnitude of the vector $\mathbf{x} = (x, y)$:

$$\boxed{|\mathbf{x}| = \sqrt{x^2 + y^2}.}$$

Geometrically, $|\mathbf{x}|$ is the *distance* from the origin to the point $\mathbf{x}$ and also the *length* of any arrow representing $\mathbf{x}$. According to the definition of $t\mathbf{a}$,

$$\boxed{|t\mathbf{a}| = |t| \cdot |\mathbf{a}|.}$$

Unit Vectors A *unit vector* is a vector of length one. Any nonzero vector $\mathbf{x}$ determines a unique unit vector $\mathbf{u}$ in the same direction, that is, a unit vector **u** such that

$$\mathbf{u} = t\mathbf{x}, \qquad t > 0.$$

For these requirements lead to

$$1 = |\mathbf{u}| = |t\mathbf{x}| = |t| \cdot |\mathbf{x}|, \quad \text{and} \quad |t| = \frac{1}{|\mathbf{x}|}.$$

But t is positive, so $t = 1/|\mathbf{x}|$. Thus **u** is uniquely determined as

$$\boxed{\mathbf{u} = \frac{\mathbf{x}}{|\mathbf{x}|}.}$$

The fact that $\mathbf{x}/|\mathbf{x}|$ is a unit vector can of course be checked directly. It is called the *normalization* of **x**.

Example 1   The unit vector in the direction of $\mathbf{x} = (1, 2)$ is $\mathbf{u} = \mathbf{x}/|\mathbf{x}| = (1, 2)/\sqrt{5} = (1/\sqrt{5}, 2/\sqrt{5})$.   □

Since there is exactly one unit vector in a given direction, we sometimes identify a direction with its unit vector.

The trigonometric identity

$$\cos^2\theta + \sin^2\theta = 1$$

can be interpreted as saying that

$$(\cos\theta, \sin\theta)$$

is a unit vector. Moreover, every unit vector **u** can be expressed this way: since **u** is a point on the unit circle, it follows that

$$\boxed{\mathbf{u} = (\cos\theta, \sin\theta),}$$

where θ is the angle from the direction of the positive x-axis to the direction of **u**.

If **u** is the unit vector (direction) of **x**, then

$$(\cos\theta, \sin\theta) = \mathbf{u} = \frac{\mathbf{x}}{|\mathbf{x}|} = \left(\frac{x}{r}, \frac{y}{r}\right),$$

where $r = |\mathbf{x}| = \sqrt{x^2 + y^2}$. The cross-multiplied equation

$$\boxed{\mathbf{x} = |\mathbf{x}|\,\mathbf{u},}$$

or

$$(x, y) = r(\cos \theta, \sin \theta),$$

can be called the polar-coordinate representation of the vector $\mathbf{x} = (x, y)$. It expresses $\mathbf{x}$ as the product of its length times its direction.

The Dot Product Now let $\mathbf{a}$ be a second nonzero vector, with unit vector

$$(\cos \alpha, \sin \alpha) = \frac{\mathbf{a}}{|\mathbf{a}|} = \left(\frac{a}{l}, \frac{b}{l}\right),$$

where $l = |\mathbf{a}| = \sqrt{a^2 + b^2}$.

Then $\phi = \theta - \alpha$ is the angle from $\mathbf{a}$ to $\mathbf{x}$ (see Fig. 1), and

$$\cos \phi = \cos(\theta - \alpha) = \cos \theta \cos \alpha + \sin \theta \sin \alpha$$

$$= \frac{x}{r} \cdot \frac{a}{l} + \frac{y}{r} \cdot \frac{b}{l} = \frac{ax + by}{rl}.$$

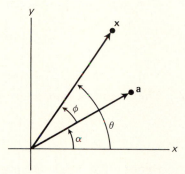

Figure 1

We could also compute $\sin \phi$. But it turns out that the $\sin \phi$ formula doesn't generalize to three dimensions, whereas the $\cos \phi$ formula does. This reflects the fact that there is no natural way to define a *signed* angle *from* one direction *to* another in three dimensions. We can only look at the *undirected* angle *between* two vectors, and such an angle is specified by its cosine (since $\cos \phi = \cos(-\phi)$).

The numerator in the above formula for $\cos \phi$ comes up so often that we give it a special name. It is called the *dot product* of the vectors $\mathbf{a}$ and $\mathbf{x}$ and is designated $\mathbf{a} \cdot \mathbf{x}$.

DEFINITION   The *dot product* $\mathbf{a} \cdot \mathbf{x}$ of the vectors $\mathbf{a} = (a, b)$ and $\mathbf{x} = (x, y)$ is defined by

$$\mathbf{a} \cdot \mathbf{x} = ax + by.$$

Example 2   The dot product of the vectors $(1, 4)$ and $(-3, 2)$ is

$$(1, 4) \cdot (-3, 2) = 1(-3) + 4 \cdot 2 = -3 + 8 = 5. \qquad \square$$

Example 3   The length $|\mathbf{x}|$ of the vector $\mathbf{x}$ is

$$|\mathbf{x}| = (\mathbf{x} \cdot \mathbf{x})^{1/2},$$

since $|\mathbf{x}|^2 = x^2 + y^2 = x\,x + y\,y = \mathbf{x} \cdot \mathbf{x}.$ $\qquad \square$

The formula for $\cos \phi$ that we computed above is an important property of the dot product.

THEOREM 5   *If ϕ is the angle between two nonzero vectors $\mathbf{a}$ and $\mathbf{x}$, then*

$$\cos \phi = \frac{\mathbf{a} \cdot \mathbf{x}}{|\mathbf{a}|\,|\mathbf{x}|}.$$

COROLLARY   *The vectors $\mathbf{a}$ and $\mathbf{x}$ are perpendicular if and only if*

$$\mathbf{a} \cdot \mathbf{x} = 0.$$

Proof   The angle ϕ between the vectors is $90°$ $(\pi/2)$ if and only if $\cos \phi = 0$, and this is equivalent to $\mathbf{a} \cdot \mathbf{x} = 0$, by the formula of the theorem. $\qquad \blacksquare$

Example 4   In order to check the vectors $\mathbf{a} = (1, 2)$ and $\mathbf{x} = (4, -2)$ for perpendicularity, we compute

$$\mathbf{a} \cdot \mathbf{x} = ax + by = 1 \cdot 4 + 2(-2) = 0.$$

Thus, they are perpendicular ($\mathbf{a} \perp \mathbf{x}$). $\qquad \square$

Example 5   Find the angle ϕ between the vectors $(1, -1)$ and $(3, 2)$.

Solution

$$\cos \phi = \frac{\mathbf{a} \cdot \mathbf{x}}{|\mathbf{a}| \, |\mathbf{x}|} = \frac{1 \cdot 3 + (-1)2}{\sqrt{1 + 1}\sqrt{9 + 4}} = \frac{1}{\sqrt{26}},$$

and $\phi = \arccos 1/\sqrt{26}$. A calculator gives $\phi \approx 1.37$ rad. ☐

Example 6 Show that the straight line

$$ax + by + c = 0$$

is perpendicular to the vector $\mathbf{a} = (a, b)$.

Solution Let $\mathbf{x}_1$ and $\mathbf{x}_0$ be any two distinct points on the line. Then

$$ax_1 + by_1 + c = 0, \qquad ax_0 + by_0 + c = 0,$$

and hence (subtracting),

$$a(x_1 - x_0) + b(y_1 - y_0) = 0.$$

But this just says that

$$\mathbf{a} \cdot (\mathbf{x}_1 - \mathbf{x}_0) = 0,$$

and hence that $\mathbf{a} \perp (\mathbf{x}_1 - \mathbf{x}_0)$. Since the direction of $\mathbf{x}_1 - \mathbf{x}_0$ is the direction of the arrow $\overrightarrow{\mathbf{x}_0 \mathbf{x}_1}$ along the line, we see that the vector $\mathbf{a}$ is perpendicular to the line. ☐

Reversing the above steps gives a new derivation of the equation of a line. Given a line l in the coordinate plane, choose a fixed point $\mathbf{x}_0 = (x_0, y_0)$ on l and a fixed vector $\mathbf{a} = (a, b)$ perpendicular to the direction of l. Then a point $\mathbf{x}$ will be on l if and only if the arrow $\overrightarrow{\mathbf{x}_0 \mathbf{x}}$ is perpendicular to the vector $\mathbf{a}$ (see Fig. 2). In view of Theorem 5,

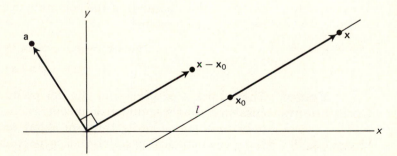

Figure 2

this is exactly the condition

$$\mathbf{a} \cdot (\mathbf{x} - \mathbf{x}_0) = 0.$$

So

$$a(x - x_0) + b(y - y_0) = 0,$$

and

$$ax + by + c = 0 \qquad (c = -ax_0 - by_0),$$

are equations of the line l.

Properties of the Dot Product

We now do for the dot product what we did in the beginning for vector addition and multiplication by scalars: We list the algebraic properties that let us use $\mathbf{x} \cdot \mathbf{a}$ directly, without recourse to its component formula. They are

(s1)
(s2)
(s3)
(s4)

$$\mathbf{x} \cdot \mathbf{x} = |\mathbf{x}|^2,$$
$$\mathbf{x} \cdot \mathbf{a} = \mathbf{a} \cdot \mathbf{x},$$
$$(\mathbf{x}_1 + \mathbf{x}_2) \cdot \mathbf{a} = \mathbf{x}_1 \cdot \mathbf{a} + \mathbf{x}_2 \cdot \mathbf{a},$$
$$(t\mathbf{x}) \cdot \mathbf{a} = t(\mathbf{x} \cdot \mathbf{a}).$$

As before, the proofs follow from properties of numbers when vectors are expressed in terms of components. Thus,

$$
\begin{aligned}
(\mathbf{x}_1 + \mathbf{x}_2) \cdot \mathbf{a} &= ((x_1, y_1) + (x_2, y_2)) \cdot (a, b) \\
&= (x_1 + x_2, y_2 + y_2) \cdot (a, b) \\
&= (x_1 + x_2)a + (y_1 + y_2)b \\
&= (x_1 a + y_1 b) + (x_2 a + y_2 b) \\
&= (x_1, y_1) \cdot (a, b) + (x_2, y_2) \cdot (a, b) \\
&= \mathbf{x}_1 \cdot \mathbf{a} + \mathbf{x}_2 \cdot \mathbf{a},
\end{aligned}
$$

which is (s3).

Note that, because of the commutative law (s2), both (s3) and (s4) can be turned around:

$$\mathbf{a} \cdot (\mathbf{x}_1 + \mathbf{x}_2) = \mathbf{a} \cdot \mathbf{x}_1 + \mathbf{a} \cdot \mathbf{x}_2,$$
$$\mathbf{a} \cdot (t\mathbf{x}) = t(\mathbf{a} \cdot \mathbf{x}).$$

Vector Components

These formal properties of the dot product will be basic in the discussion of the three-dimensional dot product in Section 5. Here we make only one application: a theorem about resolving an arbitrary vector into "components" parallel and perpendicular to a given vector.

THEOREM 6 *Let* **a** *be a fixed nonzero vector. Then every vector* **x** *can be expressed in a unique way as a sum*

$$\mathbf{x} = c\mathbf{a} + \mathbf{n}$$

of a vector $c\mathbf{a}$ *parallel to* **a** *and a vector* **n** *perpendicular to* **a**.

Proof Supposing that **x** has an expression of the above form, we can find what the constant c must be by taking the dot product with **a**:

$$\mathbf{x} \cdot \mathbf{a} = c\mathbf{a} \cdot \mathbf{a} + \mathbf{n} \cdot \mathbf{a}$$
$$= c(\mathbf{a} \cdot \mathbf{a}),$$

since $\mathbf{n} \cdot \mathbf{a} = 0$ because $\mathbf{n} \perp \mathbf{a}$. Thus c must have the value

$$c = \frac{\mathbf{x} \cdot \mathbf{a}}{\mathbf{a} \cdot \mathbf{a}}.$$

Conversely, if c is defined this way then $\mathbf{x} - c\mathbf{a}$ is perpendicular to **a**, for then

$$(\mathbf{x} - c\mathbf{a}) \cdot \mathbf{a} = \mathbf{x} \cdot \mathbf{a} - c(\mathbf{a} \cdot \mathbf{a}) = 0.$$

So we set $\mathbf{n} = \mathbf{x} - c\mathbf{a}$ and have the decomposition of the theorem. ∎

The vector $c\mathbf{a}$ is called the *vector component of* **x** *parallel to* **a**. If **a** is a unit vector then the coefficient c is just the dot product

$$c = \mathbf{x} \cdot \mathbf{a}.$$

Also, $|c| = |c\mathbf{a}|$, since $|\mathbf{a}|$ is 1. In this case the coefficient c is called the *scalar component of* **x** *in the direction* **a**. The word *component* by itself is used, ambiguously, in both senses. It may be helpful to look back at the original context of the word. If

$$\mathbf{u} = (u_1, u_2) = u_1\mathbf{i} + u_2\mathbf{j}$$

then $u_1\mathbf{i}$ is the vector component of **u** along the x-axis, and u_1 is the scalar component of **u** in the direction **i** (that is, in the direction of the positive x-axis).

Example 7 Find the vector component of $\mathbf{x} = (2, -3)$ parallel to $\mathbf{a} = (4, 1)$.

Solution By definition, it is $c\mathbf{a} = c(4, 1)$, where

$$c = \frac{\mathbf{x} \cdot \mathbf{a}}{\mathbf{a} \cdot \mathbf{a}} = \frac{(2, -3) \cdot (4, 1)}{(4, 1) \cdot (4, 1)} = \frac{5}{17}.$$

Thus

$$\frac{5}{17}(4, 1) = \left(\frac{20}{17}, \frac{5}{17}\right)$$

is the desired vector component. $\quad\square$

PROBLEMS FOR SECTION 3

Determine the length of each of the following vectors, and give the unit vector in the same direction. Draw each vector as an arrow from the origin and also as an arrow from the point $(-1, 2)$.

1. $\mathbf{v} = (3, 4)$ **2.** $\mathbf{x} = (4, -1)$ **3.** $\mathbf{a} = (-3, 1)$

4. $\mathbf{x} = (1/2, 1/2)$ **5.** $\mathbf{v} = (1, -2)$

Find $\mathbf{a} \cdot \mathbf{x}$ and the angle ϕ between $\mathbf{a}$ and $\mathbf{x}$. Draw $\mathbf{a}$ and $\mathbf{x}$ as arrows from the origin, and label ϕ.

▦ 6. $\mathbf{a} = (2, 1), \mathbf{x} = (1, 4)$ **▦ 7.** $\mathbf{a} = (1/2, -2), \mathbf{x} = (1, -1)$ **▦ 8.** $\mathbf{a} = (3, -1/2), \mathbf{x} = (-1/2, 2/3)$

9. $\mathbf{a} = (1/4, -1/3), \mathbf{x} = (1/3, 1/4)$ **▦ 10.** $\mathbf{a} = .(2, -1), \mathbf{x} = (2, 2)$

11. $\mathbf{u}$ is a unit vector perpendicular to $\mathbf{a} = (1, 2)$. Determine $\mathbf{u}$ (two answers).

12. $\mathbf{u}$ is a unit vector making the angle $\pi/3$ with $\mathbf{a} = (1, 0)$. Determine $\mathbf{u}$ (two answers).

13. $\mathbf{u}$ is a unit vector making the angle $\pi/3$ with $\mathbf{a} = (1, 1)$. Determine $\mathbf{u}$.

14. The same question when $\mathbf{a} = (1, 2)$.

15. Determine the collection of all vectors $\mathbf{x}$ making the angle $\pi/3$ with $\mathbf{a} = (1, 0)$.

16. Find the equation of the line through $\mathbf{x}_0 = (1, -5)$ perpendicular to the vector $\mathbf{a} = (-2, 1)$.

17. Find the equation of the line through the point $\mathbf{x}_0 = (3, 2)$ perpendicular to the vector $(-2, 3)$.

18. Let l be the line through the point $\mathbf{x}_0$ perpendicular to the nonzero vector $\mathbf{a}$. Show that l goes through the origin if and only if $\mathbf{x}_0$ is perpendicular to $\mathbf{a}$.

19. Let l be the line through the point $\mathbf{x}_0$ perpendicular to the nonzero vector $\mathbf{a}$. Show that its distance from the origin is $|\mathbf{a} \cdot \mathbf{x}_0|/|\mathbf{a}|$.

20. Prove the law (s4): $t(\mathbf{x} \cdot \mathbf{a}) = (t\mathbf{x}) \cdot \mathbf{a}$.

For each pair $\mathbf{a}$ and $\mathbf{x}$ given below, find the (scalar) component of $\mathbf{x}$ in the direction of $\mathbf{a}$.

21. $\mathbf{a} = (-3, 2), \mathbf{x} = (1, 0)$ **22.** $\mathbf{a} = (1, 1), \mathbf{x} = (2, 1)$ **23.** $\mathbf{a} = (1, -2), \mathbf{x} = (2, 1)$

24. $\mathbf{a} = (1, -2), \mathbf{x} = (1, 2)$ **25.** $\mathbf{a} = (10, 1), \mathbf{x} = (1, 0)$ **26.** $\mathbf{a} = (1, 0), \mathbf{x} = (10, 1)$

27. $\mathbf{a} = (10, 1), \mathbf{x} = (0, 1)$

28. If **a** is a given nonzero vector and k is a given constant, we know that $\mathbf{a} \cdot \mathbf{x} = k$ is the equation of a line perpendicular to **a**. Now reinterpret this locus as the collection of all vectors **x** having a certain component property.

4
COORDINATES
IN SPACE

A vector in space has *three* components. They are determined in the same manner as in the plane, except that now a *three*-dimensional coordinate system is involved, having three mutually perpendicular lines in space through a common origin O as axes. We generally draw and label such space axes as shown in Fig. 1. This configuration is called a *right-handed* coordinate system, because if the thumb of the right hand points in the direction of the positive z-axis, then the curl of the fingers gives the positive direction of rotation in the x,y-plane, from the positive x-axis to the positive y-axis.

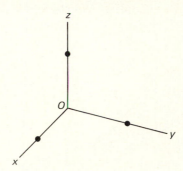

Figure 1

Each axis is given a coordinate system with common origin at O and a common unit of distance. The plane through the y- and z-axes is then a coordinate plane, called the yz-coordinate plane. It is viewed as a vertical "back wall" plane. The xy-plane is a horizontal "floor" plane, and it has the normal xy-plane appearance when viewed from a point on the positive z-axis. The xz-plane is a vertical "left wall" plane perpendicular to the yz-plane. Figures in these coordinate planes have to be distorted when they are drawn on the page. In particular, the fact that their axes are perpendicular cannot be shown and must be visualized. See Fig. 2.

The three coordinate planes divide space into eight regions, called *octants*, coming together at O. Imagine a room corner inside a building. Then the floor extends into other rooms and the walls

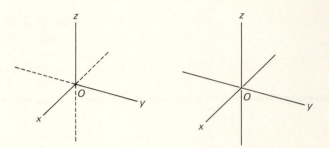

Figure 2

extend down to the floor below. We can indicate the extended edges that we can't see by dotted lines or light lines. There are eight rooms sharing the corner point O, four on our floor having O as a floor point, and four on the floor below having O as a ceiling corner point.

Consider now an arbitrary point p in space. Its coordinates can be obtained by dropping perpendiculars to the axes, just as in the plane. However, the resulting figure, shown at the left in Fig. 3, is hard to visualize. We therefore describe the process differently.

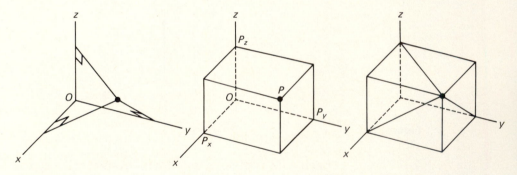

Figure 3

First, pass the plane through p perpendicular to the x-axis. Then the x-coordinate of p is determined by where this plane intersects the x-axis. If the other two coordinates of p are defined in the same way, then the three planes through p and the three coordinate planes determine altogether a rectangular "coordinate box" for p. It can be described as the rectangular box having three of its edges along the coordinate axes and having p as the vertex diagonally opposite to the origin O. The middle figure above shows this box, and it should appear to be three-dimensional. In the right figure the three original perpendicular lines are now seen as lying on faces of the box.

Just as we did in one and two dimensions, we identify the geometric point p with its coordinate triple (x, y, z) and speak of the point (x, y, z).

If the line segment joining the points

$$p_1 = (x_1, y_1, z_1) \qquad \text{and} \qquad p_2 = (x_2, y_2, z_2)$$

is not parallel to any coordinate plane, then p_1 and p_2 are diagonally opposite vertices of a certain rectangular box, just as were p and the origin O in the above coordinate box.

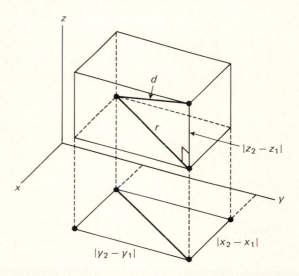

Figure 4

The edge lengths are now given by the coordinate differences, as shown in Fig. 4, and the distance formula is again a consequence of the Pythagorean theorem:

$$d^2 = r^2 + (z_2 - z_1)^2 = (x_2 - x_1)^2 + (y_2 - y_1)^2 + (z_2 - z_1)^2.$$

Thus,

The distance d between the points (x_1, y_1, z_1) and (x_2, y_2, z_2) in coordinate three-space is given by

$$d = \sqrt{(x_2 - x_1)^2 + (y_2 - y_1)^2 + (z_2 - z_1)^2}.$$

There is a new problem in space figures. If we indicate a space point P by a dot in the usual way, then we don't really know where P is, because there is a whole line in space that looks "end on" like a

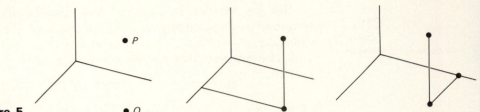

Figure 5

single point. The simplest way to fix the location of P is to show also
the point Q directly below it in the xy-plane. If you can visualize the
point Q, shown in Fig. 5, as a point in the floor plane, and P as a
point directly over it, then you should be able to "see" P in space.
We can also convey the space location of P by other auxiliary lines,
the coordinate box giving probably the strongest impression of where
P is.

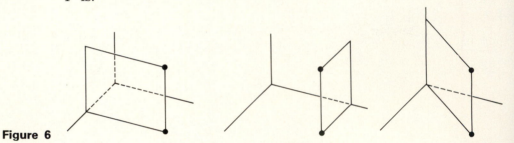

Figure 6

PROBLEMS FOR SECTION 4

Draw a three-dimensional coordinate system and locate the following points on it.

1. $P_1(6, 2, 1)$ **2.** $P_2(-6, 3, 1)$ **3.** $P_3(1, -2, 5)$

4. $P_4(2, 3, 4)$ **5.** $P_5(2, 1, -3)$ **6.** $P_6(-1, -2, 3)$

7. $P_7(2, -1, -4)$ **8.** $P_8(-2, 3, -1)$ **9.** $P_9(-6, -2, -1)$

10. $P_{10}(2, 1, -4)$ **11.** Find the distance between P_1 and P_4.

12. Find the distance between P_4 and P_9.   **13.** Find the distance between P_{10} and P_1.

14. Find the distance between P_6 and P_5.   **15.** Find the distance between P_3 and P_2.

What is the locus of a point

16. Whose x-coordinate is always 0?   **17.** Whose y-coordinate is always 3?

18. Whose x- and z-coordinates are always 0?

5
VECTORS IN
SPACE

At this point the reader is asked to go over the discussion in Section 1 and note that everything said there is valid in space, the sole difference being that a space vector has three components instead of two. In particular, the component characterization of the sum $\mathbf{x} + \mathbf{a}$ and the product $t\mathbf{x}$ are now

$$(x, y, z) + (a, b, c) = (x + a, y + b, z + c),$$
$$t(x, y, z) = (tx, ty, tz).$$

The equation

$$\mathbf{x} = \mathbf{x}_0 + t(\mathbf{x}_1 - \mathbf{x}_0)$$

is a parametric equation for the line through the distinct points $\mathbf{x}_0$ and $\mathbf{x}_1$, just as before.

Example 1 The line through $\mathbf{x}_0 = (1, 2, 3)$ and $\mathbf{x}_1 = (3, -2, 1)$ has the parametric equation

$$\mathbf{x} = (1, 2, 3) + t(2, -4, -2).$$

This is the equation

$$(x, y, z) = (1 + 2t, 2 - 4t, 3 - 2t),$$

and is equivalent to the three scalar equations

$$x = 1 + 2t, \qquad y = 2 - 4t, \qquad z = 3 - 2t.$$

Here the parameter t cannot be eliminated, and the parametric equations are *not* equivalent to a single linear equation in x, y, and z. In fact, we shall see in the next section that such an equation has a *plane* as its graph. □

Example 2 The midpoint of the segment from $\mathbf{x}_1 = (1, 2, 3)$ to $\mathbf{x}_2 = (3, -4, 5)$ is

$$\frac{\mathbf{x}_1 + \mathbf{x}_2}{2} = \frac{(1, 2, 3) + (3, -4, 5)}{2} = \frac{(4, -2, 8)}{2} = (2, -1, 4). \qquad □$$

Example 3 We saw earlier that the three medians of the triangle $\mathbf{a}_1\mathbf{a}_2\mathbf{a}_3$ are concurrent in the point $(\mathbf{a}_1 + \mathbf{a}_2 + \mathbf{a}_3)/3$. This point is called the *centroid* of the triangle. It is algebraically the average value of the three vectors $\mathbf{a}_1$, $\mathbf{a}_2$, and $\mathbf{a}_3$. The proof given for this median result is valid for a triangle in space. Now consider four space points $\mathbf{a}_1$, $\mathbf{a}_2$, $\mathbf{a}_3$, and $\mathbf{a}_4$ not lying in any one plane (Fig. 1). They are the vertices of *tetrahedron*. Through each vertex we draw a line to the centroid of the opposite face, and we ask whether these four lines are concurrent.

Here we shall just guess the answer and then verify that it is correct. We guess that

$$\frac{\mathbf{a}_1 + \mathbf{a}_2 + \mathbf{a}_3 + \mathbf{a}_4}{4},$$

the *average* of the four vertex vectors, is the point we want. Is it on the segment joining the vertex $\mathbf{a}_1$ to the centroid $(\mathbf{a}_2 + \mathbf{a}_3 + \mathbf{a}_4)/3$? Yes, because the equation

$$\frac{\mathbf{a}_1 + \mathbf{a}_2 + \mathbf{a}_3 + \mathbf{a}_4}{4} = \frac{1}{4}\mathbf{a}_1 + \frac{3}{4}\left(\frac{\mathbf{a}_2 + \mathbf{a}_3 + \mathbf{a}_4}{3}\right)$$

shows it to be the point 3/4 of the way along the segment from the vertex to the centroid.

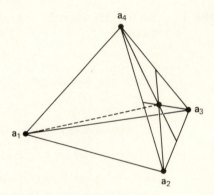

Figure 1

The magnitude $|\mathbf{x}|$ of the vector $\mathbf{x} = (x, y, z)$ is the distance from the origin to $\mathbf{x}$, and its formula is

$$|\mathbf{x}| = \sqrt{x^2 + y^2 + z^2}.$$

Then the formula for the distance d between the space points $\mathbf{x}_1$ and $\mathbf{x}_2$ can be rewritten, in terms of vector subtraction, as $d = |\mathbf{x}_1 - \mathbf{x}_2|$.

Unit Vectors   The *direction* of $\mathbf{a} = (a, b, c)$ is again presented by its unit vector

$$\mathbf{u} = \frac{\mathbf{a}}{|\mathbf{a}|} = \left(\frac{a}{r}, \frac{b}{r}, \frac{c}{r}\right),$$

where $r = |\mathbf{a}| = \sqrt{a^2 + b^2 + c^2}$. But the geometric interpretation of the components of $\mathbf{u}$ is different.

If a coordinate box is drawn with vertex at $\mathbf{a}$, and if the *face* diagonals are drawn from $\mathbf{a}$ (two are shown in Fig. 2), then it is apparent that

$$\frac{a}{r} = \cos \alpha, \qquad \frac{b}{r} = \cos \beta, \qquad \frac{c}{r} = \cos \gamma,$$

where α, β, and γ are the angles between the direction of $\mathbf{a}$ and the positive x-, y-, and z-axes. Thus,

$$\mathbf{u} = \frac{\mathbf{a}}{|\mathbf{a}|} = (\cos \alpha, \cos \beta, \cos \gamma),$$

and $\mathbf{u}$ is the triple of so-called "direction cosines" of $\mathbf{a}$.

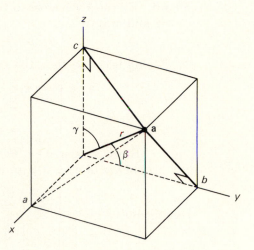

Figure 2

The Dot Product

The dot product in three dimensions has essentially the same definition as before,

$$(x, y, z) \cdot (r, s, t) = xr + ys + zt,$$

and it has the same properties (s1) through (s4), for the same reasons. In view of this second example of a dot product, we can take these properties as the *axioms* for a dot product. Here they are again, for ready reference:

(s1)	$\mathbf{x} \cdot \mathbf{x} = |\mathbf{x}|^2,$
(s2)	$\mathbf{x} \cdot \mathbf{a} = \mathbf{a} \cdot \mathbf{x},$
(s3)	$(\mathbf{x}_1 + \mathbf{x}_2) \cdot \mathbf{a} = \mathbf{x}_1 \cdot \mathbf{a} + \mathbf{x}_2 \cdot \mathbf{a},$
(s4)	$(t\mathbf{x}) \cdot \mathbf{a} = t(\mathbf{x} \cdot \mathbf{a}).$

The formula relating the dot product to $\cos \phi$ cannot be proved as before. Instead, we shall see that it is a consequence of the dot-product axioms. Consider first the special case of perpendicularity.

THEOREM 6   *The vectors* $\mathbf{a}$ *and* $\mathbf{x}$ *are perpendicular if and only if*

$$\mathbf{a} \cdot \mathbf{x} = 0.$$

Proof   By the Pythagorean theorem, $\mathbf{a}$ and $\mathbf{x}$ are perpendicular if and only if

$$|\mathbf{a}|^2 + |\mathbf{x}|^2 = |\mathbf{a} - \mathbf{x}|^2.$$

Using (s1) and then the two versions of (s3), this becomes

$$\mathbf{a} \cdot \mathbf{a} + \mathbf{x} \cdot \mathbf{x} = (\mathbf{a} - \mathbf{x}) \cdot (\mathbf{a} - \mathbf{x})$$
$$= \mathbf{a} \cdot (\mathbf{a} - \mathbf{x}) - \mathbf{x} \cdot (\mathbf{a} - \mathbf{x})$$
$$= \mathbf{a} \cdot \mathbf{a} - \mathbf{a} \cdot \mathbf{x} - \mathbf{x} \cdot \mathbf{a} + \mathbf{x} \cdot \mathbf{x}.$$

Since $\mathbf{x} \cdot \mathbf{a} = \mathbf{a} \cdot \mathbf{x}$, this equation reduces to $2(\mathbf{a} \cdot \mathbf{x}) = 0$, and hence to $\mathbf{a} \cdot \mathbf{x} = 0$, proving the theorem.   ∎

Example 4   We see that $\mathbf{a} = (-1, 2, 3)$ and $\mathbf{x} = (3, 3, -1)$ are perpendicular, by computing

$$\mathbf{a} \cdot \mathbf{x} = (-1)3 + 2 \cdot 3 + 3 \cdot (-1) = -3 + 6 - 3 = 0. \qquad \square$$

Example 5   Find the most general vector $\mathbf{x}$ perpendicular to

$$\mathbf{a}_1 = (1, 2, 3) \qquad \text{and} \qquad \mathbf{a}_2 = (2, 3, 1).$$

Solution $\mathbf{x} = (x, y, z)$ will be perpendicular to these two vectors if and only if $\mathbf{a}_1 \cdot \mathbf{x} = 0$ and $\mathbf{a}_2 \cdot \mathbf{x} = 0$, that is, if and only if

$$x + 2y + 3z = 0,$$
$$2x + 3y + z = 0.$$

Eliminating z (by multiplying the second equation by 3 and subtracting), we see that

$$5x + 7y = 0 \quad \text{and} \quad y = -(5/7)x.$$

Similarly, we find that $z = x/7$. Thus a vector $\mathbf{x}$ is perpendicular to both $\mathbf{a}_1 = (1, 2, 3)$ and $\mathbf{a}_2 = (2, 3, 1)$ if and only if it can be written in the form

$$\mathbf{x} = \left(x, -\frac{5}{7}x, \frac{1}{7}x\right) = x\left(1, -\frac{5}{7}, \frac{1}{7}\right). \qquad \square$$

Next we repeat an argument given in Section 3.

THEOREM 7 *If $\mathbf{x}$ and $\mathbf{a}$ are any two vectors, with $\mathbf{a} \neq 0$, then there is a uniquely determined scalar c such that $\mathbf{x} - c\mathbf{a}$ is perpendicular to $\mathbf{a}$.*

Proof This is obvious geometrically from a diagram, but we want to deduce it just from the axioms. If $\mathbf{x} - c\mathbf{a} \perp \mathbf{a}$, then

$$0 = (\mathbf{x} - c\mathbf{a}) \cdot \mathbf{a} = \mathbf{x} \cdot \mathbf{a} - c(\mathbf{a} \cdot \mathbf{a})$$
$$= \mathbf{x} \cdot \mathbf{a} - c\,|\mathbf{a}|^2,$$

by (s3), (s4), and (s1). Therefore,

$$c = \frac{\mathbf{x} \cdot \mathbf{a}}{|\mathbf{a}|^2}.$$

Conversely, with this value of c, the calculation above can be reversed and $\mathbf{x} - c\mathbf{a}$ is perpendicular to $\mathbf{a}$. ■

The $\cos \phi$ law is implicit in this result, for we now have a right triangle as shown in Fig. 3 at the left if $c > 0$, and at the right if $c < 0$. In particular, the sign of c is the sign of $\cos \phi$, and since

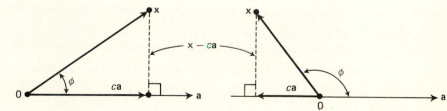

Figure 3

$\cos \phi = $ adjacent/hypotenuse *except* for sign, it follows that

$$\cos \phi = \frac{c\,|\mathbf{a}|}{|\mathbf{x}|} = \frac{\mathbf{x} \cdot \mathbf{a}}{|\mathbf{a}|^2} \cdot \frac{|\mathbf{a}|}{\mathbf{x}} = \frac{\mathbf{x} \cdot \mathbf{a}}{|\mathbf{x}|\,|\mathbf{a}|}.$$

We have proved:

THEOREM 8   *If ϕ is the angle between the nonzero vectors $\mathbf{a}$ and $\mathbf{x}$, then*

$$\cos \phi = \frac{\mathbf{a} \cdot \mathbf{x}}{|\mathbf{a}|\,|\mathbf{x}|}.$$

REMARK   This argument could have been used in the plane as well.

Example 6   Find the angle between $(1, 2, 1)$ and $(1, -1, 2)$.

Solution

$$\cos \phi = \frac{\mathbf{a} \cdot \mathbf{x}}{|\mathbf{a}|\,|\mathbf{x}|} = \frac{1 \cdot 1 + 2(-1) + 1 \cdot 2}{\sqrt{6}\sqrt{6}} = \frac{1}{6},$$

so

$$\phi = \arccos\left(\frac{1}{6}\right).$$

A calculator gives $\phi \approx 1.4$ radians.   □

Just as in the plane, the unit vectors along the coordinate axes are given special designations and are called the standard basic vectors. Now there are three,

$$\mathbf{i} = (1, 0, 0) \qquad \mathbf{j} = (0, 1, 0), \qquad \mathbf{k} = (0, 0, 1).$$

As before,

$$\mathbf{x} = (x, y, z) = x\mathbf{i} + y\mathbf{j} + z\mathbf{k}$$

for every vector $\mathbf{x}$.

PROBLEMS FOR SECTION 5

In the following, find the unit vector in the same direction as the given vector.

1. $(2, 1, 2)$ **2.** $(-3, 6, -6)$ **3.** $(7, -6, -6)$

4. $(-3, 4, 12)$ **5.** $(1/2, 1/2, 1/4)$

Given $\mathbf{x}$ and $\mathbf{a}$, find $\mathbf{x} \cdot \mathbf{a}$ and the angle ϕ between $\mathbf{x}$ and $\mathbf{a}$.

6. $\mathbf{x} = (2, 1, 2), \mathbf{a} = (1, 1, 0)$

7. $\mathbf{x} = (1/2, 0, -1/2), \mathbf{a} = (2, 1, 2)$

8. $\mathbf{x} = (3, 1, \sqrt{2}), \mathbf{a} = (1, 0, 0)$

9. $\mathbf{x} = (1, 1/2, -1), \mathbf{a} = (0, 1, 1/2)$

10. $\mathbf{x} = (1, 1, 1), \mathbf{a} = (2, -2, 1)$ $(\phi = \arctan(\ ))$

Write the parametric equation of the line through the two given points.

11. $(1, -2, 1), (2, 1, -1)$

12. $(1/2, 0, 1), (1/2, -1, 2)$

13. $(-2, -1, 2), (1, 3, 4)$

14. $(4, 1/4, -1), (-1, 1/2, 1)$

15. $(-1, -2, 1), (2, 0, 2)$

Write the scalar equations of the line through the two given points.

16. $(-3, 1, 2), (2, -1, -1)$

17. $(1, -1, 4), (1, 1/2, 0)$

18. $(-1, 1/2, -1/3), (-2, -2, 1)$

19. $(-2, -1, 0), (1/2, -1, -2)$

20. $(1/4, 1/2, 1/3), (1, 0, 1)$

Write the parametric equation of the line through $\mathbf{x}_0$ in the direction of the vector $\mathbf{a}$, where

21. $\mathbf{x}_0 + (0, 0, 0), \mathbf{a} = (1, -2, -1)$

22. $\mathbf{x}_0 = (4, -2, 2), \mathbf{a} = (-2, 1, 1)$

23. $\mathbf{x}_0 = (1, 1, 1), \mathbf{a} = (2, 0, -1)$

24. $\mathbf{x}_0 = (-3, 1, 2), \mathbf{a} = (5, -2, -3)$

25. Let $\mathbf{a}$ and $\mathbf{l}$ be two unit vectors, making angles, α, β, γ, and λ, μ, ν, respectively, with the coordinate axes. Interpret the formula

$$\cos \phi = \cos \alpha \cos \lambda + \cos \beta \cos \mu + \cos \gamma \cos \nu.$$

Find the most general vector $\mathbf{x}$ perpendicular to $\mathbf{a}_1$ and $\mathbf{a}_2$ in the following problems.

26. $\mathbf{a}_1 = (1, 2, 3), \mathbf{a}_2 = (0, 1, 2)$

27. $\mathbf{a}_1 = (1, 0, 1), \mathbf{a}_2 = (1, 1, 0)$

28. $\mathbf{a}_1 = (2, -1, 1), \mathbf{a}_2 = (0, 2, 2)$

29. Show that direction cosines in the plane would be equivalent to the $(\sin \theta, \cos \theta)$ characterization of a unit vector.

6
PLANES;
THE CROSS
PRODUCT

We saw earlier that the dot-product perpendicularity condition led to an easy derivation of the equation of a line. In three dimensions, the same proof leads to the equation of a *plane*.

Let $\mathbf{x}_0 = (x_0, y_0, z_0)$ be any fixed point and let $\mathbf{a} = (a, b, c)$ be a fixed nonzero vector. Let W be the plane that passes through the point $\mathbf{x}_0$ and is perpendicular to the direction of $\mathbf{a}$. A point $\mathbf{x}$ will lie on the plane W if and only if the arrow $\overrightarrow{\mathbf{x}_0 \mathbf{x}}$ is perpendicular to $\mathbf{a}$ (see Fig. 1). Thus, $\mathbf{x}$ is on W if and only if

$$\mathbf{a} \cdot (\mathbf{x} - \mathbf{x}_0) = 0,$$

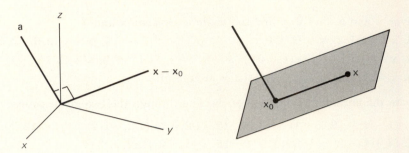

Figure 1

or

$$a(x - x_0) + b(y - y_0) + c(z - z_0) = 0.$$

This equation is therefore the "point-direction" equation of the plane.
    If we set

$$(ax_0 + by_0 + cz_0) = d,$$

the equation takes the general linear form

$$ax + by + cz = d.$$

Conversely, any such first-degree equation is the equation of a plane, provided that at least one of the coefficients a, b, and c is not zero. For, if $\mathbf{x}_0 = (x_0, y_0, z_0)$ is any fixed triple satisfying the equation

$$ax_0 + by_0 + cz_0 = d,$$

then subtracting the two equations gives the equivalent form

$$a(x - x_0) + b(y - y_0) + c(z - z_0) = 0,$$

or

$$\mathbf{a} \cdot (\mathbf{x} - \mathbf{x}_0) = 0.$$

Thus, the points $\mathbf{x} = (x, y, z)$ whose coordinates satisfy the equation $ax + by + cz - d = 0$ are exactly the points $\mathbf{x}$ such that $\mathbf{x} - \mathbf{x}_0$ is perpendicular to $\mathbf{a}$, that is, the points lying on the plane W through $\mathbf{x}_0$ and perpendicular to $\mathbf{a}$. Altogether we have proved

THEOREM 9   *The plane through the point (x_0, y_0, z_0) and perpendicular to the nonzero vector $\mathbf{a} = (a, b, c)$ has the equation*

$$\mathbf{a} \cdot (\mathbf{x} - \mathbf{x}_0) = 0,$$

or

$$ax + by + cz = d,$$

where $d = \mathbf{a} \cdot \mathbf{x}_0$. Conversely, the graph of the equation

$$ax + by + cz = d$$

is a plane perpendicular to $\mathbf{a} = (a, b, c)$, provided that $\mathbf{a}$ is not the zero vector.

Example 1   Find the plane through the point $\mathbf{x}_0 = (2, -1, 4)$ and perpendicular to the vector $\mathbf{n} = (-3, 5, 1)$.

Solution   Substituting in the equation

$$\mathbf{n} \cdot (\mathbf{x} - \mathbf{x}_0) = 0$$

gives

$$-3(x - 2) + 5(y + 1) + 1(z - 4) = 0,$$

or

$$-3x + 5y + z = -7$$

as the equation of the plane.   □

The Cross Product   There is another kind of product, where the product of two vectors is a *vector*. To see how it comes up, consider two nonzero vectors $\mathbf{a} = (a_1, a_2, a_3)$ and $\mathbf{b} = (b_1, b_2, b_3)$ that are not parallel. Then there is an essentially unique (unique up to a scalar multiple) nonzero vector $\mathbf{x}$ that is perpendicular to them both. The conditions on $\mathbf{x}$, namely $\mathbf{a} \cdot \mathbf{x} = 0$ and $\mathbf{b} \cdot \mathbf{x} = 0$, are the following two equations in three unknowns:

$$a_1 x + a_2 y + a_3 z = 0,$$
$$b_1 x + b_2 y + b_3 z = 0.$$

Solving them by the usual method of elimination leads to a particular solution vector $\mathbf{x}$ that is called the *cross product* of $\mathbf{a}$ and $\mathbf{b}$ and designated $\mathbf{a} \times \mathbf{b}$: namely;

DEFINITION

$$\mathbf{a} \times \mathbf{b} = (a_2 b_3 - b_2 a_3,\ a_3 b_1 - b_3 a_1,\ a_1 b_2 - b_1 a_2).$$

At some point it must be shown that $\mathbf{a} \times \mathbf{b}$ is not zero. This can be checked directly from the formula, and will be left as an exercise. In summary;

THEOREM 10   *If* **a** *and* **b** *are two nonparallel (and nonzero) space vectors then a vector* **x** *is perpendicular to them both if and only if* **x** $= t(\mathbf{a} \times \mathbf{b})$ *for some scalar t.*

Example 2   Find the plane that contains the three points $\mathbf{x}_0 = (1, 2, 1), \mathbf{x}_1 = (2, -2, 0),$ and $\mathbf{x}_2 = (1, 0, -3)$.

Solution   The most elementary procedure is to substitute the above values into the general equation of a plane and thus obtain three equations for the unknown coefficients.

A more efficient procedure is to note that the vectors $\mathbf{x}_1 - \mathbf{x}_0 = (1, -4, -1)$ and $\mathbf{x}_2 - \mathbf{x}_0 = (0, -2, -4)$ are parallel to the plane, so their cross product

$$(16 - 2, 0 + 4, -2 - 0) = (14, 4, -2)$$

must be normal to the plane. The plane therefore has the equation

$$14(x - 1) + 4(y - 2) - 2(z - 1) = 0,$$

or

$$7x + 2y - z = 10. \qquad \square$$

Example 3   Find the direction of the line of intersection of the two planes

$$2x + 3y - z = 4 \qquad \text{and} \qquad x + y + z = 0.$$

Solution   The coefficient vectors $(2, 3, -1)$ and $(1, 1, 1)$ are normal to the two planes and hence to the line of intersection, so their cross product gives the desired direction. It is

$$(3 + 1, -1 - 2, 2 - 3) = (4, -3, -1). \qquad \square$$

Readers familiar with 3×3 determinants will find the following remark useful:

REMARK   The cross product $\mathbf{a} \times \mathbf{b}$ is simply the formal expansion of the determinant

$$\mathbf{a} \times \mathbf{b} = \begin{vmatrix} \mathbf{i} & \mathbf{j} & \mathbf{k} \\ a_1 & a_2 & a_3 \\ b_1 & b_2 & b_3 \end{vmatrix}.$$

For by following the usual expansion procedure (using minors of the top row), we get

$$\mathbf{i}\begin{vmatrix} a_2 & a_3 \\ b_2 & b_3 \end{vmatrix} - \mathbf{j}\begin{vmatrix} a_1 & a_3 \\ b_1 & b_3 \end{vmatrix} + \mathbf{k}\begin{vmatrix} a_1 & a_2 \\ b_1 & b_2 \end{vmatrix}$$

$$= (a_2 b_3 - b_2 a_3)\mathbf{i} + (a_3 b_1 - b_3 a_1)\mathbf{j} + (a_1 b_2 - b_1 a_2)\mathbf{k}.$$

This may be the best way of computing cross products. Thus in Example 3 above we write down the determinant

$$\begin{vmatrix} \mathbf{i} & \mathbf{j} & \mathbf{k} \\ 2 & 3 & -1 \\ 1 & 1 & 1 \end{vmatrix}$$

and read the answer $4\mathbf{i} - 3\mathbf{j} - \mathbf{k} = (4, -3, -1)$.

Like the dot product, the cross product has properties that let us work with $\mathbf{a} \times \mathbf{b}$ directly, without recourse to its components. They are listed in Problems 16 through 24.

PROBLEMS FOR SECTION 6

Write the equation of the plane through the given point and perpendicular to the given vector.

1. Point $(0, 0, 0)$; vector $(1, 2, 1)$

2. Point $(-1, 1, 1/2)$; vector $(3, -1, -2)$

3. Point $(4, 1, 1)$; vector $(1/2, 1, 0)$

4. Point $(1, -1, 4)$; vector $(-1, -2, 1)$

5. Point $(6, 4, -6)$; vector $(1/3, 1/4, -1/6)$

In Problems 6 through 9 find an independent pair of vectors $\mathbf{a}$ and $\mathbf{x}$ each perpendicular to the given vector.

6. $(1, 1, 0)$ **7.** $(1, 2, -1)$ **8.** $(2, 2, 2)$ **9.** $(5, -1, 3)$

10. Find the point of intersection of the line

$$\mathbf{x} = (1, 2, -1) + t(2, -2, 3)$$

and the plane

$$3x + y - z = 2.$$

11. Let W be the plane through the origin and the two points $\mathbf{a}_1(2, -1, 1)$ and $\mathbf{a}_2(1, 3, 1)$. Prove that its intersection with the xy-coordinate plane is the line

$$\mathbf{x} = t(1, -4, 0).$$

Eliminate the parameter and find the xy-equation of the line.

12. Find the plane W through the two lines

$$\mathbf{x} = (1, 2, -1) + t(2, -2, 3),$$
$$\mathbf{x}' = (1, 2, -1) + s(5, 1, 0).$$

13. Find the line of intersection of the planes

$$x + 2y - z = 0 \quad \text{and} \quad 3x - y + 2z = 0.$$

14. Find the plane through the origin perpendicular to each of the planes

$$x + y + z = 3 \qquad \text{and} \qquad x - 2y - z = 1.$$

15. Find the plane through $(1, 2, -1)$ perpendicular to each of the planes

$$2x - y + z = 0 \qquad \text{and} \qquad x + 3y - 4z = 1.$$

Like the dot product, the cross product has a collection of algebraic properties that we can establish from its definition, and which let us then use the cross product without necessarily referring back to its components. Prove the following properties of the cross product.

16. $\mathbf{a}_1 \times \mathbf{a}_2 = 0$ if and only if $\mathbf{a}_1$ and $\mathbf{a}_2$ are dependent. [*Hint:* With $\mathbf{a}_1 = (a_1, b_1)$, etc., suppose $a_1 \neq 0$ and set $k = (a_2/a_1)$. Show that if $\mathbf{a}_1 \times \mathbf{a}_2 = 0$, then $\mathbf{a}_2 = k\mathbf{a}_1$.]

17. $\mathbf{a}_1 \times \mathbf{a}_2 = -(\mathbf{a}_2 \times \mathbf{a}_1)$

18. $(\mathbf{a}_1 \times \mathbf{a}_2) \cdot \mathbf{a}_3 = \mathbf{a}_1 \cdot (\mathbf{a}_2 \times \mathbf{a}_3)$

19. $\mathbf{a}_1 \times \mathbf{a}_2$ is perpendicular to both $\mathbf{a}_1$ and $\mathbf{a}_2$. (Use Problems 18 and 16.)

20. $\mathbf{a} \times (\mathbf{u}_1 + \mathbf{u}_2) = (\mathbf{a} \times \mathbf{u}_1) + (\mathbf{a} \times \mathbf{u}_2)$

21. $|\mathbf{a}_1 \times \mathbf{a}_2|^2 = |\mathbf{a}_1|^2 |\mathbf{a}_2|^2 - |\mathbf{a}_1 \cdot \mathbf{a}_2|^2$

22. $|\mathbf{a}_1 \times \mathbf{a}_2| = |\mathbf{a}_1| |\mathbf{a}_2| |\sin \phi|$, where ϕ is the angle between the directions of $\mathbf{a}_1$ and $\mathbf{a}_2$ (from Problem 21 and the dot-product cosine formula).

23. $|\mathbf{a}_1 \times \mathbf{a}_2|$ is the area of the parallelogram having $\mathbf{a}_1$ and $\mathbf{a}_2$ for sides.

24. $|(\mathbf{a}_1 \times \mathbf{a}_2) \cdot \mathbf{a}_3|$ is the volume of the parallelopiped having edges $\mathbf{a}_1$, $\mathbf{a}_2$, and $\mathbf{a}_3$.

25. Let the vectors $\mathbf{a}$ and $\mathbf{b}$ be independent (nonparallel and nonzero). Show that the triple $\mathbf{a}$, $\mathbf{b}$, $\mathbf{a} \times \mathbf{b}$ has right-handed orientation. (Let W be the plane through the origin containing $\mathbf{a}$ and $\mathbf{b}$, and hence perpendicular to $\mathbf{a} \times \mathbf{b}$. Imagine W rotating continuously to the position of the x,y-plane, carrying $\mathbf{a}$ and $\mathbf{b}$ along with it to a pair of vectors in the x,y-plane. Then $\mathbf{a} \times \mathbf{b}$ rotates continuously to a vector pointing along the z-axis, and the magnitude of $\mathbf{a} \times \mathbf{b}$ remains constant, by Problem 22, etc.)

26. We have seen, in the examples and problems, that two nonparallel vectors $\mathbf{a}_1$ and $\mathbf{a}_2$ determine a nonzero vector $\mathbf{a}$ perpendicular to them both, and uniquely up to scalar multiples. That is, if $\mathbf{x}$ is any other vector perpendicular to both $\mathbf{a}_1$ and $\mathbf{a}_2$, then $\mathbf{x} = t\mathbf{a}$ for some scalar t. Use this fact to prove the following result:

If $\mathbf{x}$ *is a vector perpendicular to* $\mathbf{a}_1$, $\mathbf{a}_2$, *and* $\mathbf{a}$, *then* $\mathbf{x} = 0$.

In particular, if $\mathbf{a}_1$, $\mathbf{a}_2$, and $\mathbf{a}_3$ are any three mutually perpendicular nonzero vectors, then the only vector perpendicular to them all is 0.

CHAPTER 14
VECTOR-VALUED FUNCTIONS

This chapter is concerned with vector aspects of the calculus of paths. The first four sections treat plane paths; Section 5 recapitulates this vector calculus for space paths and briefly discusses a new phenomenon (torsion) that the third dimension makes possible. Finally, Section 6 presents Newton's astonishing derivation of the universal law of gravitation from Kepler's observations about planetary motion.

1
THE TANGENT VECTOR TO A PATH

In this section we discuss tangents to curves from the vector point of view. First we introduce vector-valued functions.

In the same way that a pair of numbers x and y combine to form a single vector

$$\mathbf{x} = (x, y),$$

Vector-valued Functions

a pair of functions

$$x = g(t),$$

$$y = h(t),$$

can be combined to form a single *vector-valued* function

$$\boxed{(x, y) = (g(t), h(t))}$$

or

$$\boxed{\mathbf{x} = \mathbf{g}(t).}$$

The notions of limit, continuity, etc., carry over to such functions with little change. For example, in order to discuss the continuity of a vector-valued function we fix a value,

$$\mathbf{x}_0 = (x_0, y_0) = (g(t_0), h(t_0)),$$

and then give t an increment Δt to get a new value

$$\mathbf{x} = (x, y) = (g(t_0 + \Delta t), h(t_0 + \Delta t)).$$

The *vector increment* is then

$$\boxed{\Delta \mathbf{x} = \mathbf{x} - \mathbf{x}_0 = (x - x_0, y - y_0) = (\Delta x, \Delta y).}$$

It is the vector $\overrightarrow{\mathbf{x}_0\mathbf{x}}$ from the fixed point $\mathbf{x}_0$ to the varying point $\mathbf{x}$ (see Fig. 1). Then the vector-valued function is *continuous* at $\mathbf{x}_0$ if $\Delta \mathbf{x} \to 0$ as $\Delta t \to 0$. This means that the magnitude of $\Delta \mathbf{x}$ approaches 0,

$$|\Delta \mathbf{x}| = \sqrt{(\Delta x)^2 + (\Delta y)^2} \to 0,$$

and this occurs just when both Δx and Δy approach 0 as Δt approaches 0. That is,

> *A vector-valued function* $\mathbf{x} = (g(t), h(t))$ *is continuous at* $t = t_0$ *if and only if both component functions* g *and* h *are continuous there.*

In the same way we find that a vector function $\mathbf{x} = (g(t), h(t))$ has the limit $\mathbf{l} = (l, m)$ as t approaches t_0 if and only if $g(t) \to l$ *and* $h(t) \to m$ *as* $t \to t_0$.

That is,

$$\lim_{t \to t_0} (g(t), h(t)) = \left(\lim_{t \to t_0} g(t), \lim_{t \to t_0} h(t) \right),$$

where the vector limit exists if and only if both scalar limits exist.

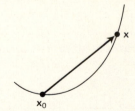

Figure 1

Vector-valued Derivatives

Our main concern is the *derivative* of a vector-valued function $\mathbf{x}$ of a parameter t at a point $t = t_0$. There is nothing new in the definition except the use of vector limits. We form the difference quotient

$$\frac{\Delta \mathbf{x}}{\Delta t} = \frac{1}{\Delta t}(\Delta x, \Delta y) = \left(\frac{\Delta x}{\Delta t}, \frac{\Delta y}{\Delta t}\right),$$

and take the limit as $\Delta t \to 0$. Then (omitting zero subscripts)

$$\frac{d\mathbf{x}}{dt} = \lim_{\Delta t \to 0} \frac{\Delta \mathbf{x}}{\Delta t} = \left(\lim_{\Delta t \to 0} \frac{\Delta x}{\Delta t}, \lim_{\Delta t \to 0} \frac{\Delta y}{\Delta t}\right) = \left(\frac{dx}{dt}, \frac{dy}{dt}\right),$$

where the left limit exists if and only if both right limits exist. Thus,

THEOREM 1   *The vector function*

$$\mathbf{x} = (x, y) = (g(t), h(t))$$

is differentiable at a given value of t if and only if its component functions g and h are both differentiable there, in each case

$$\boxed{\frac{d\mathbf{x}}{dt} = \left(\frac{dx}{dt}, \frac{dy}{dt}\right) = (g'(t), h'(t)).}$$

We can also write

$$\boxed{\mathbf{x} = \mathbf{g}(t) \qquad \text{and} \qquad \frac{d\mathbf{x}}{dt} = \mathbf{g}'(t).}$$

This gives us a new way to look at paths. A smooth path is simply a continuously differentiable vector-valued function $\mathbf{g}$ with $\mathbf{g}'$ nowhere zero. We will always assume path smoothness here.

Tangent Vectors

Now consider the geometric meaning of the above limit. If $\Delta t > 0$, then the vector $\Delta \mathbf{x}/\Delta t$ has the same direction as the increment vector $\Delta \mathbf{x} = \mathbf{x} - \mathbf{x}_0$. So if we locate the vector $\Delta \mathbf{x}/\Delta t$ at $\mathbf{x}_0$, then it will lie along the arrow $\overrightarrow{\mathbf{x}_0\mathbf{x}}$. It will be longer than $\overrightarrow{\mathbf{x}_0\mathbf{x}}$ if $\Delta t < 1$, the multiplying scalar $1/\Delta t$ then being greater than 1. See Fig. 2. The direction of the limit vector

$$\frac{d\mathbf{x}}{dt} = \lim_{\Delta t \to 0} \frac{\Delta \mathbf{x}}{\Delta t}$$

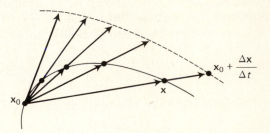

Figure 2

is thus the limit of the direction of the secant arrow $\overrightarrow{\mathbf{x}_0\mathbf{x}_1}$ as $\mathbf{x}_1$ approaches $\mathbf{x}_0$ along the path. But the limit of the direction of the secant arrow is what we *mean* by the tangent direction. Therefore,

> *The path derivative $d\mathbf{x}/dt$ can be interpreted geometrically as a vector tangent to the path.* It is called *the path tangent vector.*

We normally draw it as an arrow stemming from the path point $\mathbf{x}$, as in Fig. 3.

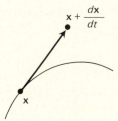

Figure 3

Example 1   Find the tangent vector to the path

$$\mathbf{x} = (x, y) = (t^3 - 3t, t^2)$$

at the point corresponding to $t = -1$. The same for $t = 0$, $t = \frac{3}{2}$, and $t = 2$.

Solution   At the general point $\mathbf{x} = (x, y) = (t^3 - 3t, t^2)$, the path has the tangent vector

$$\frac{d\mathbf{x}}{dt} = \left(\frac{dx}{dt}, \frac{dy}{dt}\right) = (3t^2 - 3, 2t).$$

At $t = -1$, the path point $\mathbf{x}$ is $(2, 1)$ and the tangent vector is

$$\left.\frac{d\mathbf{x}}{dt}\right|_{t=-1} = (3(-1)^2 - 3, 2(-1)) = (0, -2).$$

When $t = 0$,

$$\mathbf{x} = (0,0), \qquad \frac{d\mathbf{x}}{dt} = (-3,0).$$

When $t = \frac{3}{2}$,

$$\mathbf{x} = \left(-\frac{9}{8}, \frac{9}{4}\right), \qquad \frac{d\mathbf{x}}{dt} = \left(\frac{15}{4}, 3\right).$$

When $t = 2$,

$$\mathbf{x} = (2,4), \qquad \frac{d\mathbf{x}}{dt} = (9,4).$$

This path is the same one we began our path discussion with in Chapter 10. Figure 4 shows the four tangent vectors as tangent arrows at the corresponding path points.

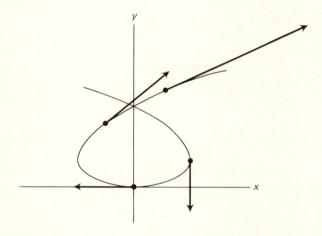

Figure 4

When the vector $\mathbf{a}$ is constant and the scalar t is variable, we shall feel free to write the product $t\mathbf{a}$ in the more traditional order, $\mathbf{a}t$.

Example 2   Show that a path whose tangent vector is constant and not zero must necessarily be a straight line.

Solution   Let $\mathbf{x} = (g(t), h(t))$ be the path and let $\mathbf{a}$ be the nonzero constant tangent vector. Our hypothesis is that

$$\frac{d\mathbf{x}}{dt} = (g'(t), h'(t)) = \mathbf{a} = (a, b),$$

for all t. That is,

$$g'(t) = a,$$
$$h'(t) = b,$$

for all t. Integrating these two equations, we get

$$x = g(t) = at + c,$$
$$y = h(t) = bt + d,$$

or

$$\mathbf{x} = \mathbf{a}t + \mathbf{c}.$$

This is a parametric equation for a line, as we saw in the last chapter.   □

Example 3   Find the line tangent to the curve $\mathbf{x} = (x, y) = (2\cos t, \sin t)$ at the point corresponding to $t_0 = \pi/6$.

Solution   At $t_0 = \pi/6$,

$$\mathbf{x}_0 = (2\cos \pi/6, \sin \pi/6)$$
$$= (\sqrt{3}, 1/2),$$
$$\left.\frac{d\mathbf{x}}{dt}\right|_{\pi/6} = (-2\sin t, \cos t)_{\pi/6}$$
$$= (-1, \sqrt{3}/2).$$

The line through $\mathbf{x}_0$ in the direction of the vector $\mathbf{a}$ has the parametric equation

$$\mathbf{x} = \mathbf{x}_0 + \mathbf{a}t.$$

Here the direction is given by the tangent vector, and the equation of the line is

$$\mathbf{x} = (\sqrt{3}, 1/2) + (-1, \sqrt{3}/2)t$$
$$= (\sqrt{3} - t, 1/2 + (\sqrt{3}/2)t).$$

The scalar equations are

$$x = \sqrt{3} - t,$$
$$y = \frac{1 + \sqrt{3}\,t}{2}.$$
                                                                            □

Example 4 Show that if the vector variable $\mathbf{x} = (x, y)$ is a differentiable function of the parameter t, then $|\mathbf{x}| = \sqrt{x^2 + y^2}$ is a constant if and only if $\mathbf{x}$ is always perpendicular to $d\mathbf{x}/dt$.

Solution The assumption of constant magnitude can be written

$$x^2 + y^2 = |\mathbf{x}|^2 = \text{constant},$$

where x and y are differentiable functions of t. Differentiating with respect to t gives

$$2x\frac{dx}{dt} + 2y\frac{dy}{dt} = 0,$$

or

$$2\left(\mathbf{x} \cdot \frac{d\mathbf{x}}{dt}\right) = 0.$$

Thus $\mathbf{x}$ is perpendicular to $d\mathbf{x}/dt$. Conversely, if these vectors are perpendicular, then the above argument can be reversed, to give

$$\frac{d}{dt}(x^2 + y^2) = 0,$$

$$x^2 + y^2 = \text{constant}.$$

We have shown the following: a path $\mathbf{x} = \mathbf{g}(t)$ runs along a circle about the origin if and only if the path tangent vector $d\mathbf{x}/dt$ is always perpendicular to the position vector $\mathbf{x}$. □

Example 5 If $\mathbf{x}_0 = (g(t_0), h(t_0))$ is the point closest to the origin on the path $\mathbf{x} = (g(t), h(t))$, then the tangent vector to the path at $\mathbf{x}_0$ is necessarily perpendicular to the vector $\mathbf{x}_0$.

This is because $|\mathbf{x}|^2 = g^2(t) + h^2(t)$ has its minimum value at $\mathbf{x}_0$, so its derivative is zero there. But

$$\frac{d\,|\mathbf{x}|^2}{dt} = 2\left(\mathbf{x} \cdot \frac{d\mathbf{x}}{dt}\right),$$

as we saw in Example 4, so

$$\mathbf{x}_0 \cdot \frac{d\mathbf{x}}{dt}\bigg|_0 = 0,$$

as claimed. □

A tangent vector

$$\frac{d\mathbf{x}}{dt} = (g'(t), h'(t))$$

can perfectly well be the zero vector $(0, 0)$, in which case it does not specify a unique direction. However, this is exactly what we ruled out when we defined a smooth path, so a smooth path has a continuously varying, nowhere-zero, tangent vector. In particular, a smooth path has a continuously changing direction.

PROBLEMS FOR SECTION 1

Find the derivative $d\mathbf{x}/dt$ for each of the following vector functions. Also, determine the points where $d\mathbf{x}/dt$ is perpendicular to $\mathbf{x}$.

1. $\mathbf{f}(t) = (t, 1/t)$
2. $\mathbf{x} = (e^{2t}, e^{-t})$
3. $\Phi(t) = (t^2, t + 3)$

4. $\mathbf{x} = (2t, t + 1)$
5. $\mathbf{x} = (\cos t, \sin t)$
6. $\mathbf{u} = (\sqrt{1 - t^2}, t)$

7. $\mathbf{g}(t) = (3 \cos t, 2 \sin t)$
8. $\mathbf{x} = (t^2, (1 + t)^2)$

Find the vector tangent to each of the paths below at the given point. Then write down a parametric equation for the tangent line at the point.

9. $\mathbf{x} = (40t, 30t - 16t^2)$; when $t = 1$
10. $\mathbf{x} = (4 \cos t, 3 \sin t)$; when $t = \pi/3$

11. $\mathbf{x} = (e^t \cos t, e^t \sin t)$; when $t = 0$
12. $\mathbf{x} = (t^2, t^3)$; when $t = 2$

13. $\mathbf{x} = (e^t + e^{-t}, e^t - e^{-t})$; when $t = 0$
14. $\mathbf{x} = (1/t, 1/(t^2 + 1))$; when $t = -1$

15. $\mathbf{x} = (\log t, 2/t)$; when $t = 1$
16. $\mathbf{x} = ((t^2 + 1)/(t^2 + 2), 2t)$; when $t = 0$

17. The ellipse

$$\frac{x^2}{a^2} + \frac{y^2}{b^2} = 1$$

has the parametric equation

$$\mathbf{x} = (a \cos \theta, b \sin \theta).$$

Use this representation to find a parametric equation of the tangent line to the ellipse at $(x_0, y_0) = (a \cos \theta_0, b \sin \theta_0)$.

18. The path $\mathbf{x} = (a \sec \theta, b \tan \theta)$ runs along the hyperbola $(x/a)^2 - (y/b)^2 = 1$. Use this parametric representation of the hyperbola to find the equation of its tangent line at $(x_0, y_0) = (a \sec \theta_0, b \tan \theta_0)$.

19. Redraw the figure showing the meaning of Theorem 1 for the case when Δt is negative.

20. Show that the line tangent to the path $\mathbf{x} = \mathbf{g}(t)$ at the point $\mathbf{x}_0 = \mathbf{g}(t_0)$ has the parametric equation

$$\mathbf{x} - \mathbf{x}_0 = \mathbf{g}'(t_0)(t - t_0).$$

21. Show that a path whose tangent vector has a constant *direction* must lie along a straight line. (The hypothesis is that $d\mathbf{x}/dt = f(t)\mathbf{u}$, where $\mathbf{u}$ is a constant unit vector.)

22. Show that a path $\mathbf{x} = \mathbf{g}(t)$ is uniquely determined by its derivative $\mathbf{g}'(t)$ and an initial value $\mathbf{x}_0 = \mathbf{g}(t_0)$. That is, if

$$\mathbf{g}(t) = (g(t), h(t)) \quad \text{and} \quad \mathbf{l}(t) = (l(t), m(t))$$

are two paths such that $\mathbf{g}'(t) = \mathbf{l}'(t)$ for all t and $\mathbf{g}(t_0) = \mathbf{l}(t_0)$, then

$$\mathbf{g}(t) = \mathbf{l}(t) \quad \text{for all } t.$$

23. Let $\mathbf{x} = \mathbf{g}(t) = (g(t), h(t))$ and $\mathbf{x} = \mathbf{p}(s) = (p(s), q(s))$ be two paths that do not intersect. Let d be the shortest distance between them, and suppose that we can find a closest pair of points,

$$\mathbf{x}_1 = \mathbf{g}(t_0) \quad \text{and} \quad \mathbf{x}_2 = \mathbf{l}(s_0),$$

that is, a pair such that

$$|\mathbf{x}_2 - \mathbf{x}_1| = d.$$

Prove that the segment $\mathbf{x}_1\mathbf{x}_2$ is perpendicular to both curves.

24. If one path is a straight line l in the above problem, then the tangent to the other path must be parallel to l at the point closest to l. Use this principle to find the point on $\mathbf{x} = (t, t^2 - 1)$ closest to the line $\mathbf{x} = (-s - 1, 2s)$.

2
VECTOR FORMS OF THE DERIVATIVE RULES

In the beginning sections of Chapter 13 we were sometimes able to prove things about vectors without referring explicitly to their components. For example, we proved in this way that the medians of a triangle are concurrent. Instead of explicitly calculating with components we were able to use the vector laws.

In a similar way, derivative calculations involving vector functions of t are sometimes clearer and frequently more efficient if we use vector forms of the derivative rules, as given below.

1. If $\mathbf{x}$ and $\mathbf{u}$ are differentiable vector functions of t, then so is $\mathbf{x} + \mathbf{u}$ and

$$\frac{d}{dt}(\mathbf{x} + \mathbf{u}) = \frac{d\mathbf{x}}{dt} + \frac{d\mathbf{u}}{dt}.$$

2. If w is a differentiable scalar function of t and $\mathbf{x}$ is a differentiable vector function of t, then $w\mathbf{x}$ is a differentiable vector function of t and

$$\frac{d}{dt}(w\mathbf{x}) = w\frac{d\mathbf{x}}{dt} + \frac{dw}{dt}\mathbf{x}.$$

3. If $\mathbf{x}$ and $\mathbf{u}$ are differentiable vector functions of t, then $\mathbf{x} \cdot \mathbf{u}$ is a differentiable scalar function of t and

$$\frac{d}{dt}(\mathbf{x} \cdot \mathbf{u}) = \mathbf{x} \cdot \frac{d\mathbf{u}}{dt} + \frac{d\mathbf{x}}{dt} \cdot \mathbf{u}.$$

Rules (2) and (3) are simply the product rule for the two products we have for plane vectors.

4. If $\mathbf{x}$ is a differentiable vector function of the parameter r, and if r is a differentiable function of t, then $\mathbf{x}$ is a differentiable vector function of t, and

$$\frac{d\mathbf{x}}{dt} = \frac{d\mathbf{x}}{dr}\frac{dr}{dt}.$$

This is the rule for changing the path parameter; it is another chain rule.

5. If $d\mathbf{x}/dt$ is identically zero, then $\mathbf{x}$ is a constant vector.

These reformulations all follow from the original derivative rules and Theorem 1. For example,

$$\frac{d}{dt}[\mathbf{x} \cdot \mathbf{u}] = \frac{d}{dt}[(x, y) \cdot (u, v)] = \frac{d}{dt}(xu + yv)$$

$$= x\frac{du}{dt} + \frac{dx}{dt}u + y\frac{dv}{dt} + \frac{dy}{dt}v$$

$$= (x, y) \cdot \left(\frac{du}{dt}, \frac{dv}{dt}\right) + \left(\frac{dx}{dt}, \frac{dy}{dt}\right) \cdot (u, v)$$

$$= \mathbf{x} \cdot \frac{d\mathbf{u}}{dt} + \frac{d\mathbf{x}}{dt} \cdot \mathbf{u}.$$

Example   We reconsider Examples 2 and 4 from the last section. In the second example we were given that $d\mathbf{x}/dt$ is the constant vector $\mathbf{a}$. One such vector function of t is $\mathbf{a}t$, since

$$\frac{d}{dt}(\mathbf{a}t) = \mathbf{a},$$

by rule (2) above. Thus

$$\frac{d}{dt}(\mathbf{x} - \mathbf{a}t) = 0$$

(rule (1)), and so

$$\mathbf{x} - \mathbf{a}t = \text{constant vector } \mathbf{c},$$
$$\mathbf{x} = \mathbf{a}t + \mathbf{c}$$

(rule (5)). This was the conclusion of Example 2.

In the fourth example we considered a path $\mathbf{x} = \mathbf{g}(t)$ for which $|\mathbf{x}|$ is constant. We can now compute

$$0 = \frac{d}{dt}|\mathbf{x}|^2 = \frac{d}{dt}(\mathbf{x} \cdot \mathbf{x}) = \mathbf{x} \cdot \frac{d\mathbf{x}}{dt} + \frac{d\mathbf{x}}{dt} \cdot \mathbf{x}$$

$$= 2\mathbf{x} \cdot \frac{d\mathbf{x}}{dt},$$

by rule (3), and so $\mathbf{x} \perp d\mathbf{x}/dt$, as before. $\qquad\square$

PROBLEMS FOR SECTION 2

1. Let $\mathbf{x} = \mathbf{f}(t)$ be a twice-differentiable path such that $d^2\mathbf{x}/dt^2 \equiv 0$. Prove that $\mathbf{x} = \mathbf{a}t + \mathbf{c}$, so the path is simply the parametric representation of a straight line.

2. Prove the rule (4) by a component calculation, the way (3) was proved in the text.

3. Prove the rule (5).

4. Let $\mathbf{x} = \mathbf{f}(t)$ be a path whose tangent vector has a constant direction. Prove that the path lies along a straight line (without resorting to component calculations).

5. The tangent vector of the path $\mathbf{x} = \mathbf{f}(t)$ is everywhere perpendicular to $\mathbf{x}$. Prove that the path runs along a circle about the origin.

6. The tangent vector of the path $\mathbf{x} = \mathbf{f}(t)$ is everywhere perpendicular to the constant vector $\mathbf{a}$ ($\mathbf{a} \neq 0$). Prove that the path runs along a line perpendicular to $\mathbf{a}$.

7. Let $\mathbf{x} = \mathbf{f}(t)$ be a path whose tangent vector $d\mathbf{x}/dt$ is always in the direction of $\mathbf{x}$. Prove that $\mathbf{x}$ runs along a line through the origin. (Write $\mathbf{x} = |\mathbf{x}|\,\mathbf{u}$, where $\mathbf{u}$ is unit vector, presumably varying with t. Apply rule (2) and remember that $d\mathbf{u}/dt$ is necessarily perpendicular to $\mathbf{u}$, as in the second part of the Example.)

8. If $\mathbf{x}$ is a differentiable vector function of t, prove from rule (3) that

$$\frac{d}{dt}|\mathbf{x}| = \frac{\mathbf{x} \cdot \dfrac{d\mathbf{x}}{dt}}{|\mathbf{x}|}.$$

9. If $0 < c < a$, and $\mathbf{f} = (c, 0)$ and $\mathbf{g} = (-c, 0)$, we know (Chapter 10, Section 3, Problem 29) that the set of points $\mathbf{x}$ such that

$$|\mathbf{x} - \mathbf{f}| + |\mathbf{x} - \mathbf{g}| = 2a$$

is the ellipse

$$\frac{x^2}{a^2} + \frac{y^2}{b^2} = 1,$$

where $b^2 = a^2 - c^2$. We can consider the above ellipse to be parameterized in some way, so that $\mathbf{x}$ traces the elliptical path with parameter t. The actual parameterization doesn't matter. Now differentiate the locus identity

$$|\mathbf{x} - \mathbf{f}| + |\mathbf{x} - \mathbf{g}| = 2a$$

with respect to t, using Problem 8, and so prove that the arrows $\overrightarrow{\mathbf{fx}}$ and $\overrightarrow{\mathbf{gx}}$ make equal angles with the tangent to the ellipse at $\mathbf{x}$.

This is the so-called *optical property* of the ellipse, because it shows that the rays from a point source of light placed at $\mathbf{g}$ reflect off the ellipse and all come together (are "focused") at $\mathbf{f}$ (and vice versa). Each of the points $\mathbf{f}$ and $\mathbf{g}$ is called a *focus* of the ellipse.

11. The vector forms of the derivative rules can be proved without resorting to component calculations by going back to the difference quotient definition of the derivative. Show that if $\mathbf{x}$ and $\mathbf{u}$ are functions of the parameter t and if $z = \mathbf{x} \cdot \mathbf{u}$ then

$$\Delta z = \mathbf{x} \cdot \Delta\mathbf{u} + \Delta\mathbf{x} \cdot \mathbf{u} + \Delta\mathbf{x} \cdot \Delta\mathbf{u}$$

just as was the case for the ordinary product of two real-valued functions considered in Chapter 4. Then finish the proof of rule (3) as in Chapter 4.

10. It was shown in Chapter 10, Section 1, that the polar graph of the equation $r(1 - \epsilon \cos \theta) = 1$ is an ellipse if $0 < \epsilon < 1$. An important additional fact is that this ellipse has one focus at the origin. Prove this now from the values of a and b given there. (This is just algebra.)

12. Prove the second rule in this direct component-free manner.

3
ARC LENGTH AS PARAMETER; CURVATURE

We saw in Chapter 10 that the arc length s of a smooth plane path $\mathbf{x} = (x, y) = (g(t), h(t))$ is determined by the equation

$$\frac{ds}{dt} = \sqrt{\left(\frac{dx}{dt}\right)^2 + \left(\frac{dy}{dt}\right)^2},$$

provided that s is measured in the direction of increasing t. The right side is now recognizable as the magnitude of the path tangent vector

$$\frac{d\mathbf{x}}{dt} = \left(\frac{dx}{dt}, \frac{dy}{dt}\right).$$

The equation can thus be rewritten

$$\left|\frac{d\mathbf{x}}{dt}\right| = \frac{ds}{dt},$$

and says:

> *The magnitude of the path tangent vector is the derivative of the path arc length s with respect to the parameter t.*

Note that if the parameter t happens to be equal to the arc length s, or more generally, if $s = t + c$, then

$$\left|\frac{d\mathbf{x}}{dt}\right| = \frac{ds}{dt} = 1$$

and the tangent vector is always a unit vector. The same argument works backward and shows that if the tangent vector is always a unit vector then $ds/dt = 1$ and $s = t + $ constant. Thus

THEOREM 2 *The parameter of a smooth plane path is equal to the path arc length (plus a constant) if and only if the tangent vector is always a unit vector.*

In general the parameter t will not be the arc length s, but we can always reparameterize the path using s. The argument goes as follows. Since the tangent vector to a smooth path is never zero (the definition of smoothness), $ds/dt = |d\mathbf{x}/dt|$ is always positive. Therefore s is an increasing function of t which has an everywhere-differentiable inverse, by the inverse function theorem. That is, t is a differentiable function of s, say

$$t = \phi(s).$$

Then

$$x = g(t) = g(\phi(s))$$
$$y = h(t) = h(\phi(s))$$

is a new smooth parameterization of the path, with the path length s

as the new parameter. The new coordinate functions are $l(s) = g(\phi(s))$ and $m(s) = h(\phi(s))$.

Example 1   Reparameterize the path $\mathbf{x} = (t^3, t^2)$ with the arc length s as parameter.

Solution   We have

$$\frac{ds}{dt} = \sqrt{\left(\frac{dx}{dt}\right)^2 + \left(\frac{dy}{dt}\right)^2} = \sqrt{9t^4 + 4t^2}$$
$$= t\sqrt{9t^2 + 4}.$$

Therefore,

$$s = \int \sqrt{9t^2 + 4} \cdot t \, dt = \frac{1}{27}(9t^2 + 4)^{3/2} + C.$$

We can take $C = 0$. This amounts to choosing the point from which the arc length s is measured. Now we must express the old parameter t as a function of the new parameter s. Here we can explicitly solve for t in terms of s and get

$$t = \sqrt{s^{2/3} - 4/9}.$$

The path is thus

$$\mathbf{x} = (t^3, t^2)$$
$$= \left(\left(s^{2/3} - \frac{4}{9}\right)^{3/2}, s^{2/3} - \frac{4}{9}\right)$$

when its arc length is made the parameter.   □

The unit tangent vector $d\mathbf{x}/ds$ is designated $\mathbf{T}$. The tangent vector $d\mathbf{x}/dt$ can then be written in the form

$$\boxed{\frac{d\mathbf{x}}{dt} = \frac{d\mathbf{x}}{ds}\frac{ds}{dt} = \frac{ds}{dt}\mathbf{T}}$$

by rule (4) from the last section. This is the "polar coordinate" representation for the tangent vector, ds/dt being its magnitude and $d\mathbf{x}/ds = \mathbf{T}$ being the unit vector that we think of as its direction.

Curvature   We turn now to the notion of *curvature*. Some paths are nearly straight while other paths bend more. A very large circle curves only very slowly, while a small circle curves rapidly. This notion of "curviness" of a path is related to how quickly the path is changing its direction, and the following definition should seem reasonable.

DEFINITION   The *curvature* of a path is the rate of change of the direction of the path tangent vector (with respect to arc length s).

In the plane there are two standard ways of specifying direction, so there are two versions of the curvature definition.

First, the direction of the path tangent vector is specified by the angle ϕ measured from the positive x-axis to the tangent vector. Then

$$\text{curvature} = \frac{d\phi}{ds}.$$

We are frequently interested only in the magnitude of the curvature. This is called the *absolute curvature* and is designated κ (Greek "kappa"). Thus, by definition,

$$\kappa = \left| \frac{d\phi}{ds} \right|.$$

Alternatively, the direction of the tangent vector is taken to be its unit vector $\mathbf{T}$. The curvature definition now gives a *vector* curvature, namely,

$$\text{curvature} = \frac{d\mathbf{T}}{ds}$$

However, we shall see that these two versions of curvature agree: the magnitude of the vector $d\mathbf{T}/ds$ is the absolute curvature:

$$\kappa = \left| \frac{d\mathbf{T}}{ds} \right|.$$

For calculating curvature we need the following explicit curvature formula,

$$\frac{d\phi}{ds} = \frac{\dfrac{d^2y}{dt^2}\dfrac{dx}{dt} - \dfrac{d^2x}{dt^2}\dfrac{dy}{dt}}{\left[\left(\dfrac{dx}{dt} \right)^2 + \left(\dfrac{dy}{dt} \right)^2 \right]^{3/2}},$$

which holds under the assumption that s is increasing with t.

We shall look at the proofs of these identities after some further examples.

Example 2 Derive the formula

$$\frac{d\phi}{ds} = \frac{\dfrac{d^2y}{dx^2}}{\left[1 + \left(\dfrac{dy}{dx}\right)^2\right]^{3/2}}$$

for the curvature of a function graph $y = f(x)$.

Solution We assume the graph is traced from left to right, so a possible parameterization is given by $y = f(x)$, $x = x$, that is, by taking $t = x$. It can be seen by inspection that the general formula then reduces to this function graph formula. □

Example 3 Show that the absolute curvature of a parabola is maximum at the vertex.

Solution We consider the parabola $y = x^2$, with vertex at the origin. Then

$$\kappa = \frac{\dfrac{d^2y}{dx^2}}{\left[1 + \left(\dfrac{dy}{dx}\right)^2\right]^{3/2}} = \frac{2}{(1 + 4x^2)^{3/2}}.$$

This is maximum when $x = 0$. □

Example 4 A circle of radius r has constant curvature equal to $\pm 1/r$, the sign depending on which way the circle is traced. We show this here and in Example 5. Here we apply the function formula to the upper semicircle $y = \sqrt{r^2 - x^2}$ and get

$$\frac{dy}{dx} = \frac{-x}{\sqrt{r^2 - x^2}},$$

$$\frac{d^2y}{dx^2} = \frac{-r^2}{(r^2 - x^2)^{3/2}} \quad \text{(after some calculation)}.$$

$$\frac{d\phi}{ds} = \frac{\dfrac{d^2y}{dx^2}}{\left[1 + \left(\dfrac{dy}{dx}\right)^2\right]^{3/2}} = \frac{\dfrac{-r^2}{(r^2 - x^2)^{3/2}}}{\left[\dfrac{r^2}{r^2 - x^2}\right]^{3/2}} = -\frac{1}{r}.$$ □

Newton's Dot Notation

Newton used the following very efficient notation for derivatives with respect to time:

$$\dot{x} = \frac{dx}{dt}, \qquad \ddot{x} = \frac{d^2x}{dt^2}, \qquad \text{etc.}$$

In particular, velocities and accelerations were always expressed this way.

The dot notation is also useful for paths where the parameter isn't necessarily interpreted as time. For example, in this notation, the path curvature formula is

$$\boxed{\text{curvature} = \frac{\ddot{y}\dot{x} - \ddot{x}\dot{y}}{[\dot{x}^2 + \dot{y}^2]^{3/2}}.}$$

Example 5

Show that the circle $\mathbf{x} = (r \cos t, r \sin t)$ has the constant curvature $1/r$.

Solution

$$\frac{d\phi}{ds} = \frac{\ddot{y}\dot{x} - \ddot{x}\dot{y}}{(\dot{x}^2 + \dot{y}^2)^{3/2}} = \frac{(-r \sin t)(-r \sin t) - (-r \cos t)(r \cos t)}{(r^2 \cos^2 t + r^2 \sin^2 t)}$$

$$= \frac{r^2}{r^3} = \frac{1}{r}.$$

Note that this is the same circle as in Example 4. □

Points of Inflection

The path curvature formula can be used to find points of inflection. These are the points across which the curvature changes sign—the curve changes from turning to the right to turning to the left, or vice versa—so they are the points where the *numerator* in the curvature formula changes sign. In particular, they are among the points where the numerator is zero.

Example 6

Find the points of inflection on the path $(x, y) = (t^2 + t, t^3)$.

Solution We have

$$\ddot{y}\dot{x} - \ddot{x}\dot{y} = 6t(2t + 1) - 2(3t^2) = 6(t^2 + t),$$

which is zero at $t = 0$ and $t = -1$. Moreover, $t^2 + t$ changes sign at each of these points. Therefore the corresponding path points $(0, 0)$ and $(0, -1)$ are points of inflection. □

Curvature Proofs

There are two ways of deriving the curvature formula. A vector method will be outlined in the problems for Section 5. A straightforward nonvector derivation starts out with the observation that if $\dot{x} \neq 0$, then $\tan \phi = \dot{y}/\dot{x}$ and

$$\phi = \arctan \frac{\dot{y}}{\dot{x}}.$$

Then

$$\frac{d\phi}{ds} = \frac{\dot{\phi}}{ds/dt},$$

and the curvature formula emerges when the arctan differentiation is carried out and ds/dt is replaced by $\sqrt{(\dot{x})^2 + (\dot{y})^2}$.

If $\dot{x} = 0$, then $\dot{y} \neq 0$, and we start out with

$$\phi = \operatorname{arccot} \frac{\dot{x}}{\dot{y}},$$

with the same final result. The details are left as an exercise.

Finally, we take up the problem of our original double definition of curvature.

We have noted in several contexts that a vector function of constant magnitude is necessarily perpendicular to its derivative. When the vector function is the unit tangent vector $\mathbf{T}$ to a path we can say more.

THEOREM 3

Let $\mathbf{x} = \mathbf{g}(s)$ be a path with arc length as parameter, and suppose that the unit tangent vector $\mathbf{T}$ is a differentiable function of s. Then the vector $d\mathbf{T}/ds$ is perpendicular to the path, and points in the direction in which the path is turning. Moreover, its magnitude is the absolute curvature of the path. Thus $d\mathbf{T}/ds$ has the "polar coordinate" representation

$$\boxed{\frac{d\mathbf{T}}{ds} = \kappa \mathbf{N},}$$

where κ is the absolute path curvature and $\mathbf{N}$ is the unit vector normal to the path in the direction in which the path is turning.

Proof The first step is familiar. We differentiate the identity

$$|\mathbf{T}|^2 = \mathbf{T} \cdot \mathbf{T} = 1$$

by the product law for dot products:

$$\mathbf{T} \cdot \frac{d\mathbf{T}}{ds} + \frac{d\mathbf{T}}{ds} \cdot \mathbf{T} = 0.$$

Therefore,

$$2\left(\mathbf{T} \cdot \frac{d\mathbf{T}}{ds}\right) = 0$$

and the vectors $\mathbf{T}$ and $d\mathbf{T}/ds$ are perpendicular. Since $\mathbf{T}$ is tangent to the path, $d\mathbf{T}/ds$ is perpendicular to the path.

Now consider the approximate equality

$$\frac{\Delta\mathbf{T}}{\Delta s} \approx \frac{d\mathbf{T}}{ds}.$$

Taking Δs positive, this says, in particular, that the direction of $d\mathbf{T}/ds$ is approximately the same as the direction of $\Delta\mathbf{T}$. But this points to the side toward which the curve is turning (Fig. 1). Therefore the direction of $d\mathbf{T}/ds$ is the direction in which the curve is turning.

Figure 1

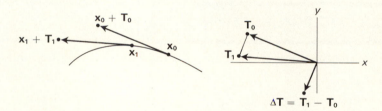

So far nothing has been said about the path being a plane path, and the argument so far will be equally valid in our later discussion of space paths.

For a plane curve we can write the unit vector $\mathbf{T}$ in the form

$$\mathbf{T} = (\cos\phi, \sin\phi),$$

where ϕ is the angle from the positive x-axis to the direction of the path. Then

$$\frac{d\mathbf{T}}{ds} = \frac{d\mathbf{T}}{d\phi}\frac{d\phi}{ds} = (-\sin\phi, \cos\phi)\frac{d\phi}{ds}.$$

But $d\phi/ds$ was our definition of the *curvature* of a path. And

$(-\sin \phi, \cos \phi)$ is another unit vector. Therefore

$$\left| \frac{d\mathbf{T}}{ds} \right| = |(-\sin \phi, \cos \phi)| \left| \frac{d\phi}{ds} \right| = 1 \cdot \kappa = \kappa,$$

where $\kappa = |d\phi/ds|$ is the absolute curvature of the path. The theorem is thus proved.   ∎

Example 7   A path that doesn't curve should lie along a straight line. This conjecture is easy to prove using the formula of the theorem. For if $\kappa = 0$, then $d\mathbf{T}/ds = 0$ and $\mathbf{T}$ is a constant, say $\mathbf{T} = \mathbf{a}$. Since $\mathbf{T} = d\mathbf{x}/ds$ we conclude, as before, that

$$\mathbf{x} = \mathbf{x}_0 + s\mathbf{a}.$$

Here s is arc length. If the path had been given in terms of some other parameter t, then the arc length s is a function of t, say $s = f(t)$, and then the path runs along the above line according to the equation

$$\mathbf{x} = \mathbf{x}_0 + f(t)\mathbf{a}.$$ □

PROBLEMS FOR SECTION 3

Compute the curvature of the graph of each of the following functions.

1. $y = x^2$

2. $y = x^{1/2}$

3. $y = \ln x$

4. $y = \ln(\cos x) \quad (-\pi/2 < x < \pi/2)$

5. $y = mx + b$

6. $y = \sqrt{1 - x^2}$

7. Show that the curvature of $y = \cos x$ oscillates between $+1$ and -1, with period 2π.

8. Find the maximum value of the absolute curvature κ of the graph $y = e^x$.

9. Find the minimum value of the absolute curvature κ of the path

$$x = a \cos^3 t, \qquad y = a \sin^3 t,$$

over the t interval $[0, \pi/2]$.

10. Show that the circle

$$(x, y) = (r \cos \theta, -r \sin \theta)$$

has constant curvature $\dfrac{d\phi}{ds} = 1/r$.

11. Prove that a function graph of zero curvature is a straight line.

12. Consider the function $f(x) = x^a$ on the positive x-axis $(0, \infty)$, where the exponent a can be any real number. Show that the curvature of the graph of f has a finite limit as $x \to 0$ if and only if

$$a \leq \frac{1}{2} \qquad \text{or} \qquad a = 1 \qquad \text{or} \qquad a \geq 2.$$

13. Compute the curvature of the ellipse

$$x = a \cos t, \qquad y = b \sin t.$$

Find its minimum and maximum values.

14. Carry out the proof of the parametric curvature formula.

Find the points of inflection (if any) on the following paths.

15. $(x, y) = (t^3, t^3 - 1)$

16. $(x, y) = (t^3 - 3t, t^2)$

17. $(x, y) = (t^3 + 3t, t^4)$

18. $(x, y) = (t^3 - 3t, t^4)$

19. $(x, y) = (e^{-t}, \ln t)$

20. $(x, y) = (e^t, t^2)$

21. a) Reparameterize the path

$$(x, y) = (e^t \cos t, e^t \sin t)$$

with the arc length s as parameter. (This involves finding s as a function of t, then t as a function of s, then substituting.)

b) Then compute $\mathbf{T} = d\mathbf{x}/ds$ and $d\mathbf{T}/ds$, and verify that they are perpendicular.

c) Compute $\kappa = |d\mathbf{T}/ds|$.

22. Do the same for the path $(x, y) = \frac{2}{3}(\cos^3 t, \sin^3 t)$.

23. The same for the path

$$(x, y) = \left(\frac{4}{3} t^{3/2}, \frac{1}{2} t^2 - t \right).$$

Use the parametric curvature formula to compute the curvatures of the following paths.

24. $x = t^3 + 3t^2, \ y = t^3 - 3t^2$

25. $x = \cos t + t \sin t, \ y = \sin t - t \cos t$

26. The formula for the curvature in polar coordinates is

$$\text{curvature} = \frac{r^2 + 2\left(\dfrac{dr}{d\theta}\right)^2 - r\dfrac{d^2r}{d\theta^2}}{\left[r^2 + \left(\dfrac{dr}{d\theta}\right)^2 \right]^{3/2}}.$$

Prove this formula, starting from the parametric form of the formula given in the text, and using the expressions for $dx/d\theta$ and $dy/d\theta$ from the last section.

Using the above formula, compute the curvature of the following polar graphs.

27. $r = 2a \cos \theta$

28. $r = a(1 + \cos \theta)$

29. $r = e^{a\theta}$

30. $r = a\theta$

4
VELOCITY AND
ACCELERATION

Now let the parameter t be time, so that the path $\mathbf{x} = \mathbf{g}(t)$ is the trajectory of a moving particle and ds/dt is its speed. We could define the velocity of the particle as the time rate of change of its position $\mathbf{x}$, that is, as the tangent vector $d\mathbf{x}/dt$. But it is intuitively more satisfying to define the velocity as the vector $\mathbf{v}$ whose direction is the direction in which the particle is moving, and whose magnitude is the speed ds/dt at which the particle is moving. It then *follows* that

$$\mathbf{v} = \frac{d\mathbf{x}}{dt},$$

because $d\mathbf{x}/dt$ meets both of these requirements.

Acceleration is the *time rate of change of velocity*. It is thus the vector

$$\mathbf{a} = \frac{d\mathbf{v}}{dt} = \frac{d^2\mathbf{x}}{dt^2}.$$

What is interesting about vector acceleration is that it is due *both* to change in speed *and* to change in direction. If *either* the speed is changing *or* the direction is changing, then the particle is accelerating.

Think how one is affected by changes in velocity when riding in a car. If the car speeds up while going down a straight road the back of the seat presses harder against the rider, who attributes the extra pressure to the straight-ahead acceleration of the car. Now consider what happens when the car travels with *constant* speed along a winding road. When riding around a curve, one is thrown sideways. This is a consequence of the *sidewise* acceleration of the car, and the greater or lesser extent that one feels this sidewise pressure is a rough measure of the magnitude of the sidewise acceleration. On this basis we can reach some conclusions about this perpendicular acceleration. One is thrown more to the side when going at the same speed around a sharper curve, or when traveling at a higher speed around the same curve. Thus the acceleration must increase in magnitude if *either* the speed or the curvature of the path is increased.

The exact formula for these different acceleration effects is contained in the following theorem.

THEOREM 4 *If* $\mathbf{x} = \mathbf{g}(t)$ *is the path of a moving particle, with the time t as parameter, then the acceleration* $\mathbf{a}$ *has the formula*

$$\mathbf{a} = \frac{dv}{dt}\mathbf{T} + \kappa v^2\mathbf{N},$$

where v is the speed ds/dt, $\mathbf{T}$ *and* $\mathbf{N}$ *are the unit tangent and normal vectors, and* κ *is the absolute curvature of the path.*

Proof We start with

$$\mathbf{v} = \frac{d\mathbf{x}}{dt} = \frac{d\mathbf{x}}{ds}\frac{ds}{dt} = v\mathbf{T}$$

and differentiate again:

$$\mathbf{a} = \frac{d\mathbf{v}}{dt} = v\frac{d\mathbf{T}}{dt} + \frac{dv}{dt}\mathbf{T}.$$

But

$$\frac{d\mathbf{T}}{dt} = \frac{d\mathbf{T}}{ds}\frac{ds}{dt} = \kappa\mathbf{N}v$$

from Theorem 3, and substituting this value into the line above gives the formula of the theorem. ■

It follows from the theorem that the component of the acceleration vector $\mathbf{a}$ normal to the curve is

$$\mathbf{a} \cdot \mathbf{N} = \kappa v^2,$$

in the direction in which the curve is bending. This shows explicitly how the "sidewise" acceleration varies with curvature κ and speed v.

Example 1 Find the velocity and acceleration vectors, the speed v, and the tangential and normal components of the acceleration, for the motion

$$\mathbf{x} = (x, y) = (2\cos t, \sin t).$$

Solution

$$\mathbf{v} = (\dot{x}, \dot{y}) = (-2\sin t, \cos t);$$

$$\mathbf{a} = (\ddot{x}, \ddot{y}) = (-2\cos t, -\sin t) = -\mathbf{x};$$

$$v = |\mathbf{v}| = \sqrt{(\dot{x})^2 + (\dot{y})^2} = \sqrt{4\sin^2 t + \cos^2 t}$$
$$= \sqrt{1 + 3\sin^2 t};$$

$$\kappa = \frac{|\ddot{y}\dot{x} - \ddot{x}\dot{y}|}{v^3} = \frac{|-\sin t(-2\sin t) - (-2\cos t)\cos t|}{(1 + 3\sin^2 t)^{3/2}}$$

$$= \frac{2}{(1 + 3\sin^2 t)^{3/2}};$$

$$\left.\begin{array}{l} \dfrac{dv}{dt} = \dfrac{3\sin t \cos t}{(1 + 3\sin^2 t)^{1/2}} \\[4mm] \kappa v^2 = \dfrac{2}{(1 + 3\sin^2 t)^{1/2}} \end{array}\right\} \quad \begin{array}{l}\text{the acceleration} \\ \text{components}\end{array}$$

$\square$

Example 2   A circle of radius r has curvature

$$\kappa = \frac{1}{r}.$$

Thus a particle that moves along this circle with *constant* speed v is nevertheless accelerating, according to the formula

$$\mathbf{a} = \kappa v^2 \mathbf{N} = \left(\frac{v^2}{r}\right)\mathbf{N}.$$

That is, its acceleration is directed toward the center of the circle and is of magnitude v^2/r. $\square$

PROBLEMS FOR SECTION 4

Find the velocity and acceleration vectors, the speed v, and the tangential and normal components $(dv/dt$ and $\kappa v^2)$ of the acceleration, for each of the following motions:

1. $(x, y) = (\cos 3t, \sin 3t)$ **2.** $(x, y) = (3\cos t, 3\sin t)$ **3.** $(x, y) = (3\cos t, 2\sin t)$

4. $(x, y) = (t, \sin t)$ **5.** $(x, y) = (t, t^2)$ **6.** $(x, y) = (t^2, t^3)$

7. $(x, y) = (2\cos 3t, 3\sin 2t)$ **8.** $(x, y) = (\cos t \sin 3t, \sin t \sin 3t)$

9. A road lies along the parabola $100y = x^2$. A truck moving along the road is so loaded that the normal component of its acceleration must not exceed 25. What limitation does this impose on its speed as it rounds the vertex of the parabola?

10. A particle moves along the ellipse $(x, y) = (3\cos t, 2\sin t)$ with constant speed 5. What are the maximum and minimum values of the magnitude of its acceleration?

11. Is it possible for a particle to move along a curving path without experiencing any sideways acceleration? Consult the formula of Theorem 4.

13. There are two circular tracks, the large one having twice the radius of the smaller. If you are comfortable driving 50 mph on the smaller track, how much faster can you drive on the larger track without experiencing a greater radial (sidewise) acceleration?

12. If you travel along a road on which the curvature is never zero, how must you adjust your speed so that your sidewise acceleration has a constant magnitude?

14. Suppose that the parameter in the standard parametric equations of the ellipse

$$\frac{x^2}{a^2} + \frac{y^2}{b^2} = 1$$

is the time t, so that the parametric equations

$$x = a \cos t \qquad \text{and} \qquad y = b \sin t$$

describe the motion of a particle moving along the ellipse. Show that the acceleration vector

$$\mathbf{a} = \frac{d^2\mathbf{x}}{dt^2}$$

is always directed to the center of the ellipse (the origin).

**5
PATHS IN
SPACE**

We shall not generally graph space paths, because drawing three-dimensional figures is so complicated. However, three coordinate functions of a parameter t,

$$x = g(t), \qquad y = h(t), \qquad z = k(t),$$

can be viewed as defining a path in space exactly as two functions do in the plane. If t is interpreted as time, the path can be interpreted as the trajectory of a moving particle. The path will be called smooth if the three coordinate functions all have continuous derivatives and if the three derivatives are never simultaneously zero.

A space path can be viewed as a vector-valued function of t,

$$\mathbf{x} = (x, y, z) = (g(t), h(t), k(t)),$$

and, exactly as in the plane, the derivative of this vector-valued function is the vector function

**Tangent
Vectors**

$$\frac{d\mathbf{x}}{dt} = \left(\frac{dx}{dt}, \frac{dy}{dt}, \frac{dz}{dt}\right) = (g'(t), h'(t), k'(t)).$$

As before, $d\mathbf{x}/dt$ can be interpreted as the tangent vector to the path at the point $\mathbf{x} = (g(t), h(t), k(t))$.

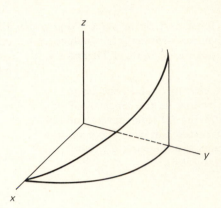

Figure 1

Example 1   The path $\mathbf{x} = (\cos t, \sin t, t/2)$ is a spiral rising over the unit circle in the xy-plane (Fig. 1). Its tangent vector is

$$\frac{d\mathbf{x}}{dt} = (-\sin t, \cos t, 1/2).$$

When $t = \pi/2$, the path point is $(0, 1, \pi/4)$ and the tangent vector is $(-1, 0, 1/2)$.   □

Example 2   Show that a space path $\mathbf{x} = \mathbf{g}(t)$ lies entirely on the surface of a sphere about the origin if and only if

$$\mathbf{x} \perp d\mathbf{x}/dt \qquad \text{for all } t.$$

Solution   To say that the path lies on the surface of a sphere about O is to say that $|\mathbf{x}|^2 = x^2 + y^2 + z^2$ has a constant value, say

$$x^2 + y^2 + z^2 = r^2,$$

along the path. Differentiating with respect to t gives

$$2x\frac{dx}{dt} + 2y\frac{dy}{dt} + 2z\frac{dz}{dt} = 0,$$

or

$$2\mathbf{x} \cdot \frac{d\mathbf{x}}{dt} = 0.$$

Thus $\mathbf{x} \perp d\mathbf{x}/dt$ for such a path. Conversely, if $\mathbf{x}$ and $d\mathbf{x}/dt$ are always perpendicular, then this argument can be reversed to show that $|\mathbf{x}|$ is constant.   □

Just as in the plane, vector derivative calculations like the one above can often be better understood in terms of the vector forms of the derivative rules given in Section 2.

We add the product rule for the cross product.

6. If **a** and **b** are differentiable vector functions of t then so is $\mathbf{x} = \mathbf{a} \times \mathbf{b}$, and

$$\frac{d\mathbf{x}}{dt} = \mathbf{a} \times \frac{d\mathbf{b}}{dt} + \frac{d\mathbf{a}}{dt} \times \mathbf{b}.$$

The proof can be given in terms of components, using the rather complicated definition of the cross product. But it is better to apply the properties of the cross product listed in Problems 16 through 23 in Chapter 13, Section 6. (See Problem 40.)

Path Length and Curvature

Path length is given by the formula

$$\text{length} = \int_a^b \sqrt{\left(\frac{dx}{dt}\right)^2 + \left(\frac{dy}{dt}\right)^2 + \left(\frac{dz}{dt}\right)^2}\, dt.$$

This is the same formula as for the plane, but with three terms under the radical instead of two. To prove it, we assume that a smooth space path has length, and that

$$\frac{\text{chord length}}{\text{arc length}} \to 1$$

as chord length approaches 0. Then the argument that we gave in Section 4 of Chapter 10 applies practically verbatim, and proves that

$$\left(\frac{ds}{dt}\right)^2 = \left(\frac{dx}{dt}\right)^2 + \left(\frac{dy}{dt}\right)^2 + \left(\frac{dz}{dt}\right)^2.$$

Thus the magnitude of the path tangent vector $d\mathbf{x}/dt$ is again

$$\left|\frac{d\mathbf{x}}{dt}\right| = \frac{ds}{dt}.$$

Theorem 2 therefore holds for a space path, with the same proof, as does Theorem 3 down to the final statement

$$\left|\frac{d\mathbf{T}}{ds}\right| = \kappa.$$

There we used a plane argument that does not work in space; the definition of curvature as $d\phi/ds$ is tied to the plane and does not generalize. However, the formula

$$\kappa = \left| \frac{d\mathbf{T}}{ds} \right|$$

was our alternative definition for the curvature of a path. With this definition, Theorem 3 holds for a space path.

The proof of Theorem 4 is also the same. Here are the statements of these three theorems again.

THEOREM 5   *The parameter of a smooth space path is equal to the path arc length (plus a constant) if and only if the tangent vector is always a unit vector.*

THEOREM 6   *Let* $\mathbf{x} = \mathbf{g}(s)$ *be a twice-differentiable space path with arc length as parameter and unit tangent vector* $\mathbf{T}$*. Then*

$$\frac{d\mathbf{T}}{ds} = \kappa \mathbf{N},$$

where $\mathbf{N}$ *is the unit vector normal to the path in the direction in which the path is turning, and* κ *is (by definition) the absolute curvature of the path.*

THEOREM 7   *If* $\mathbf{x} = \mathbf{g}(t)$ *is the path of a moving particle, with time t as parameter, then the tangential and normal components of the acceleration* $\mathbf{a}$ *are given by*

$$\mathbf{a} = \frac{dv}{dt}\mathbf{T} + \kappa v^2 \mathbf{N},$$

where v *is the speed* $ds/dt = |d\mathbf{x}/dt|$*, and* $\mathbf{T}$*,* κ*, and* $\mathbf{N}$ *are as in the above theorem.*

Example 3   Consider again the spiral

$$\mathbf{x} = (\cos t, \sin t, t/2)$$

of Example 1. We see that the tangent vector

$$\frac{d\mathbf{x}}{dt} = (-\sin t, \cos t, 1/2)$$

has constant magnitude

$$\left|\frac{d\mathbf{x}}{dt}\right| = \frac{\sqrt{5}}{2}.$$

Thus $ds/dt = \sqrt{5}/2$, and if we start measuring s at the point when $t = 0$, then

$$s = \frac{\sqrt{5}}{2}\, t.$$

Changing to the parameter s, the path is

$$\mathbf{x} = \left(\cos\frac{2}{\sqrt{5}}\, s, \sin\frac{2}{\sqrt{5}}\, s, \frac{s}{\sqrt{5}}\right).$$

Then the unit tangent vector is

$$\mathbf{T} = \frac{d\mathbf{x}}{ds} = \frac{2}{\sqrt{5}}\left(-\sin\frac{2}{\sqrt{5}}\, s, \cos\frac{2}{\sqrt{5}}\, s, \frac{1}{2}\right),$$

so

$$\frac{d\mathbf{T}}{ds} = \frac{4}{5}\left(-\cos\frac{2}{\sqrt{5}}\, s, -\sin\frac{2}{\sqrt{5}}\, s, 0\right) = \frac{4}{5}\mathbf{N},$$

where $\mathbf{N}$ is a unit vector. Comparing this with

$$\frac{d\mathbf{T}}{ds} = \kappa\mathbf{N},$$

we see that the absolute curvature has the constant value

$$\kappa = \frac{4}{5}. \qquad\qquad \square$$

This constantly rising spiral is called a *circular helix*. We shall see in the problems that it is also *twisting* at a constant rate.

Torsion We conclude with a brief description of this notion of a twisting space curve, leaving the details to the problems. Suppose that $\mathbf{x} = \mathbf{f}(t)$ is a smooth space curve with nonzero curvature κ, so that the unit tangent and normal vectors $\mathbf{T}$ and $\mathbf{N}$ are well defined. Then

$$\boxed{\mathbf{B} = \mathbf{T} \times \mathbf{N}}$$

is a unit vector perpendicular to both $\mathbf{T}$ and $\mathbf{N}$ called the unit *binomial* vector to the curve. The plane containing $\mathbf{T}$ and $\mathbf{N}$ (as arrows drawn from the path point) can be considered to be the plane that "momentarily" contains the curve. It is called the *osculating* plane to the curve at that point. If W is the plane passing through three distinct points $\mathbf{x}_0$, $\mathbf{x}_1$, and $\mathbf{x}_2$ on the curve, then it can be shown that W approaches the osculating plane at $\mathbf{x}_0$ as $\mathbf{x}_1$ and $\mathbf{x}_2$ approach $\mathbf{x}_0$.

The osculating plane is determined by its unit normal vector $\mathbf{B}$, so the *derivative* of $\mathbf{B}$, $d\mathbf{B}/ds$, is the vector rate at which the osculating plane is changing its space orientation, or the (vector) rate at which the curve is *twisting*. It will be seen in the problems that $d\mathbf{B}/ds$ is parallel to $\mathbf{N}$. When the scalar multiple is written with a minus sign,

$$\frac{d\mathbf{B}}{ds} = -\tau\mathbf{N},$$

then τ (Greek tau) is called the *torsion* of the curve. The reason for the minus sign will appear in the problems.

PROBLEMS FOR SECTION 5

Find the tangent vector to each of the following paths at the given point.

1. $\mathbf{x} = (t, t^2, t^3)$; $t = -1$

2. $\mathbf{x} = (2\cos t, 3\sin t, t^2)$; $t = \pi/6$

3. $\mathbf{x} = (3t, t - 2/3, \sin t)$; $t = \pi/3$

4. $\mathbf{x} = (\sec t, \tan t, \sin t)$; $t = \pi/3$

5. $\mathbf{x} = (\ln t, \sqrt{t}, 1/t)$; $t = 4$

Prove the following derivative rules for space paths.

6. If $\mathbf{u}$ and $\mathbf{x}$ are differentiable vector functions of t, then so is $a\mathbf{u} + b\mathbf{x}$ for any constants a and b, and

$$\frac{d}{dt}(a\mathbf{u} + b\mathbf{x}) = a\frac{d\mathbf{x}}{dt} + b\frac{d\mathbf{u}}{dt}.$$

7. If w is a differentiable scalar function of t, and $\mathbf{x}$ is a differentiable vector function of t, then $w\mathbf{x}$ is a differentiable vector function of t, and

$$\frac{d}{dt}w\mathbf{x} = w\frac{d\mathbf{x}}{dt} + \frac{dw}{dt}\mathbf{x}.$$

8. If $\mathbf{x}$ and $\mathbf{u}$ are differentiable vector functions of t, then $\mathbf{x} \cdot \mathbf{u}$ is a differentiable scalar function of t, and

$$\frac{d}{dt}\mathbf{x} \cdot \mathbf{u} = \mathbf{x} \cdot \frac{d\mathbf{u}}{dt} + \frac{d\mathbf{x}}{dt} \cdot \mathbf{u}.$$

9. If $\mathbf{x}$ is a differentiable vector function of t and if $t = \phi(r)$ is differentiable, then $\mathbf{x}$ is a differentiable vector function of r, and

$$\frac{d\mathbf{x}}{dr} = \frac{dt}{dr}\frac{d\mathbf{x}}{dt}.$$

Try to use the above rules in proving the remaining problems.

10. Find the smallest distance between the lines

$$\mathbf{x} = t(1, 2, -1)$$

and

$$\mathbf{u} = (1, 1, 1) + s(3, -2, 2).$$

(If $\mathbf{x}_0$ and $\mathbf{u}_0$ are the closest points on the two lines, show first that $\mathbf{x}_0 - \mathbf{u}_0$ is perpendicular to the directions of both lines. Then solve for t_0 and s_0.)

12. Let $\mathbf{x} = \mathbf{f}(t)$ be a space path that does not pass through the origin, and let $\mathbf{x}_0 = \mathbf{f}(t_0)$ be a point on the path that is closest to the origin. Show that the path tangent vector at $\mathbf{x}_0$ is perpendicular to $\mathbf{x}_0$.

14. Let $\mathbf{x} = \mathbf{f}(t)$ be a path whose tangent vector has a constant direction. Prove that the path lies along a straight line.

16. Let $\mathbf{x} = \mathbf{f}(t)$ be a path whose tangent vector is everywhere perpendicular to the constant vector $\mathbf{a}$ ($\mathbf{a} \neq 0$). Prove that the path lies in a plane perpendicular to $\mathbf{a}$.

18. Let $\mathbf{x} = \mathbf{f}(t)$ be a differentiable vector function of t, and set $\dot{\mathbf{x}} = d\mathbf{x}/dt$ (Newton's dot notation). Prove the rule

$$\frac{d}{dt}|\mathbf{x}| = \frac{\mathbf{x} \cdot \dot{\mathbf{x}}}{|\mathbf{x}|}.$$

[*Hint*: See Problem 8.]

11. If $\mathbf{x}_0$ and $\mathbf{u}_0$ are points on two given space paths at which the paths are closest together, show that the vector $\mathbf{x}_0 - \mathbf{u}_0$ is perpendicular to both paths.

13. Let $\mathbf{x} = \mathbf{f}(t)$ be a twice-differentiable path such that

$$\frac{d^2\mathbf{x}}{dt^2} \equiv 0.$$

Prove that $\mathbf{x} = \mathbf{a}t + \mathbf{k}$, so that the path is simply the standard parametric representation of a straight line.

15. Let $\mathbf{x} = \mathbf{f}(t)$ be a path whose tangent vector is everywhere perpendicular to $\mathbf{x}$. Prove that the path lies on a sphere about the origin. Also prove the converse.

17. Let $\mathbf{x}$ be a path whose tangent vector is always parallel to the vector $\mathbf{x}$. Prove that $\mathbf{x}$ runs along a line through the origin.

19. The absolute curvature $\kappa = |d\mathbf{T}/ds|$ has the following formula in terms of derivatives with respect to the parameter t.

$$\kappa^2 = \frac{|\dot{\mathbf{x}}|^2 |\ddot{\mathbf{x}}|^2 - (\dot{\mathbf{x}} \cdot \ddot{\mathbf{x}})^2}{|\dot{\mathbf{x}}|^6}.$$

Prove this formula, in the following steps:

a) $\dot{\mathbf{x}} = |\dot{\mathbf{x}}|\,\mathbf{T}$

b) $\ddot{\mathbf{x}} = |\dot{\mathbf{x}}|^2 \dfrac{d\mathbf{T}}{ds} + \dfrac{(\dot{\mathbf{x}} \cdot \ddot{\mathbf{x}})}{|\dot{\mathbf{x}}|}\mathbf{T}$

c) Compute $\ddot{\mathbf{x}} \cdot \ddot{\mathbf{x}}$ from (b) and solve for

$$\kappa^2 = \left(\frac{d\mathbf{T}}{ds} \cdot \frac{d\mathbf{T}}{ds}\right).$$

Compute the absolute curvature κ of the following space paths from the formula for κ in Problem 19.

20. $\mathbf{x} = (\cos t, \sin t, t/2)$. (Compare with Example 3.)

21. $\mathbf{x} = (t, t^2/2, t^3/3)$

22. $\mathbf{x} = (\cos t, \sin t, \sin t)$

23. The formula in Problem 19 applies to plane curves as well as to space curves. Show that, in the plane, it reduces to our earlier formula

$$\kappa = \frac{|\ddot{y}\dot{x} - \ddot{x}\dot{y}|}{[(\dot{x})^2 + (\dot{y})^2]^{3/2}}.$$

24. Let $\mathbf{x} = \mathbf{f}(t)$ be a space path of zero curvature. Prove that it necessarily runs along a straight line. (Use the formula in Problem 19, and consider two cases. If $\ddot{\mathbf{x}}$ is identically zero, then the path is the standard parametric representation of a straight line. (This was an earlier problem, but work it out again.) If $\ddot{\mathbf{x}} \neq 0$, then show that $\cos \phi = \pm 1$, where ϕ is the angle between $\dot{\mathbf{x}}$ and $\ddot{\mathbf{x}}$. Finally, apply Problem 17.)

The following problems lean heavily on the vector cross product and its properties given at the end of Chapter 13.

It will be shown below that the absolute curvature κ and the torsion τ of a space curve $\mathbf{x} = \mathbf{f}(t)$ have the following cross-product formulas.

Let $\mathbf{p}$ be the cross product of the first and second derivatives of $\mathbf{x} = \mathbf{f}(t)$:

$$\mathbf{p} = \mathbf{f}'(t) \times \mathbf{f}''(t) = \dot{\mathbf{x}} \times \ddot{\mathbf{x}}.$$

Then

$$\kappa = \frac{|\mathbf{p}|}{|\dot{\mathbf{x}}|^3} \quad \text{and} \quad \tau = \frac{\mathbf{p} \cdot \dddot{\mathbf{x}}}{|\mathbf{p}|^2}.$$

Use these formulas to compute the curvature and torsion of the following space curves.

25. $\mathbf{x} = (t, t^2/2, t^3/3)$

26. $\mathbf{x} = (t^2 + 1, t^2 - 2t, 2t - 1)$

27. $\mathbf{x} = (\sin t, \cos t, bt)$

28. $\mathbf{x} = (\sin at, \cos at, bt)$

29. $\mathbf{x} = (c \sin at, c \cos at, bt)$

30. $\mathbf{x} = (\cos t, \sin t, \sin t)$

31. $\mathbf{x} = (e^t, e^{2t}, e^{-t})$

32. $\mathbf{x} = (\log t, t^3, t^4)$

33. $\mathbf{x} = (\sin t, \tan t, \sec t)$

We turn now to the proofs of the curvature and torsion formulas.

34. Show that if $\mathbf{a}$ and $\mathbf{b}$ are perpendicular unit vectors, then $\mathbf{c} = \mathbf{a} \times \mathbf{b}$ is a unit vector. Thus the binomial to a space curve, $\mathbf{B} = \mathbf{T} \times \mathbf{N}$, is a unit vector.

35. The acceleration formula of Theorem 7 is valid, whether t is considered to be time or not, in the form

$$\ddot{\mathbf{x}} = \left(\frac{d}{dt}\,|\dot{\mathbf{x}}|\right)\mathbf{T} + \kappa\,|\dot{\mathbf{x}}|^2\,\mathbf{N}.$$

Show therefore that

$$\mathbf{p} = \dot{\mathbf{x}} \times \ddot{\mathbf{x}} = \kappa\,|\dot{\mathbf{x}}|^3\,\mathbf{B},$$

and therefore, since $\mathbf{B}$ is a unit vector, that

$$\kappa = \frac{|\mathbf{p}|}{|\dot{\mathbf{x}}|^3}.$$

37. Use the results in Problems 35 and 36 to show that

$$\mathbf{p} \cdot \dddot{\mathbf{x}} = \kappa^2\,|\dot{\mathbf{x}}|^6\,\tau$$

and thus derive the formula for τ.

39. Let $\mathbf{x} = \mathbf{f}(t)$ be a space path having curvature everywhere nonzero (so that the torsion τ is everywhere defined). Show that if $\tau \equiv 0$, then the path lies in a plane.

36. Show, by differentiating the identities

$$\mathbf{B} \cdot \mathbf{T} = 0, \qquad \mathbf{B} \cdot \mathbf{B} = 1, \qquad \mathbf{B} \cdot \ddot{\mathbf{x}} = 0$$

a) That $d\mathbf{B}/ds$ is perpendicular to Both $\mathbf{T}$ and $\mathbf{B}$ and hence can be written in the form

$$\frac{d\mathbf{B}}{ds} = -\tau\mathbf{N};$$

b) that

$$\mathbf{B} \cdot \dddot{\mathbf{x}} = -\dot{\mathbf{B}} \cdot \ddot{\mathbf{x}}.$$

38. Show that the present formula for κ equals the formula given in Problem 19.

40. Prove the derivative rule for the cross product of two vector functions, using properties of the cross product from the list given in Section 6 of Chapter 13. [*Hints*: Note first that $(\mathbf{a}_1 + \mathbf{a}_2) \times \mathbf{b} = \mathbf{a}_1 \times \mathbf{b} + \mathbf{a}_2 \times \mathbf{b}$ (from Problems 20 and 17) and that $|\mathbf{a} \times \mathbf{b}| \leq |\mathbf{a}| \cdot |\mathbf{b}|$ (from Problem 21). Now show that if $\mathbf{x} = \mathbf{a} \times \mathbf{b}$ then

$$\Delta\mathbf{x} = \Delta\mathbf{a} \times \mathbf{b} + \mathbf{a} \times \Delta\mathbf{b} + \Delta\mathbf{a} \times \Delta\mathbf{b},$$

and finish as usual.]

6
KEPLER'S LAWS AND NEWTON'S UNIVERSAL LAW OF GRAVITATION

In Section 4, the acceleration vector of a moving plane particle was resolved into its components parallel and perpendicular to the velocity vector $\mathbf{x}'(t)$. It is also possible to resolve the acceleration vector into components parallel and perpendicular to the path position vector $\mathbf{x}(t)$, and this resolution of the acceleration vector has a spectacular application. Using it, we shall be able to derive Newton's universal law of gravitation from Kepler's laws of planetary motion as an almost routine exercise in calculus. Newton did this in the seventeenth century, and it was one of his greatest achievements.

Kepler's Laws The first two laws concern the orbit of a single planet, while the third law relates the various planetary orbits to each other.

I. *A planet moves in a plane through the sun in such a way that the vector from the sun to the planet sweeps out equal areas in equal times. That is, the vector sweeps out area at a constant rate.*

II. *The path of a planet is an ellipse with the sun at one focus.*

III. *The* **square** *of the period T of a planetary orbit is proportional to the* **cube** *of its semimajor axis a. That is, the ratio T^2/a^3 is the same for all the planets.*

The new resolution of the acceleration vector $\ddot{\mathbf{x}}(t)$ that we need for Kepler's laws can be obtained most easily by using polar coordinates in the plane of the orbit. We write $\mathbf{x}$ in its polar coordinate form,

$$\mathbf{x} = r(\cos\theta, \sin\theta) = r\mathbf{u},$$

where $\mathbf{u}$ is the unit vector of $\mathbf{x}$, $r = |\mathbf{x}|$, and r and θ are functions of t. Then

$$\dot{\mathbf{u}} = \frac{d}{dt}(\cos\theta, \sin\theta) = (-\sin\theta, \cos\theta)\dot{\theta} = \dot{\theta}\mathbf{p},$$

where $\mathbf{p} = (-\sin\theta, \cos\theta)$ is the unit vector perpendicular to $\mathbf{u}$ and leading $\mathbf{u}$ by $\pi/2$ (since $(-\sin\theta, \cos\theta) = (\cos(\theta + \pi/2), \sin(\theta + \pi/2))$). Similarly,

$$\dot{\mathbf{p}} = \frac{d}{dt}(-\sin\theta, \cos\theta) = -\dot{\theta}\mathbf{u}.$$

Therefore, the velocity vector $\mathbf{v} = \dot{\mathbf{x}}$ and acceleration vector $\mathbf{a} = \dot{\mathbf{v}}$ can be written

$$\mathbf{v} = \dot{\mathbf{x}} = \frac{d}{dt}(r\mathbf{u}) = \dot{r}\mathbf{u} + r\dot{\mathbf{u}} = \dot{r}\mathbf{u} + r\dot{\theta}\mathbf{p},$$

$$\mathbf{a} = \dot{\mathbf{v}} = (\ddot{r}\mathbf{u} + \dot{r}\dot{\theta}\mathbf{p}) + (\dot{r}\dot{\theta}\mathbf{p} + r\ddot{\theta}\mathbf{p} - r(\dot{\theta})^2\mathbf{u}).$$

Thus,

$$\mathbf{a} = (\ddot{r} - r\dot{\theta}^2)\mathbf{u} + (r\ddot{\theta} + 2\dot{r}\dot{\theta})\mathbf{p}, \tag{*}$$

and this is the formula we have been looking for.

We can now analyze the consequences of Kepler's first law. If polar coordinates are introduced in the plane containing the planetary orbit, with the sun as the origin, then the area formula in Chapter 10, Section 2, says that the derivative of the area swept out with respect to the polar angle θ is $r^2/2$. Therefore,

$$\frac{d}{dt}(\text{area swept out}) = \frac{r^2}{2}\frac{d\theta}{dt} = \frac{\dot{\theta}r^2}{2}.$$

Kepler's law says that area is being swept out at a constant rate, so

$$\frac{\dot{\theta}r^2}{2} = \text{constant},$$

and

$$\frac{d}{dt}\left(\frac{\dot{\theta}r^2}{2}\right) = \frac{\ddot{\theta}r^2 + 2\dot{\theta}r\dot{r}}{2} = 0.$$

In particular, $\ddot{\theta}r + 2\dot{\theta}\dot{r} = 0$, so (*) reduces to

$$\mathbf{a} = (\ddot{r} - r\dot{\theta}^2)\mathbf{u},$$

which is a vector directed along the line through the origin. We have proved the following theorem.

THEOREM 8   *If a plane path $\mathbf{x}(t)$ has the property that the position vector $\mathbf{x}(t)$ sweeps out area at a constant rate, then the acceleration vector $\ddot{\mathbf{x}}(t)$ is always directed along the line through $\mathbf{x}(t)$ and O (i.e., $\ddot{\mathbf{x}}(t)$ is always a scalar multiple of $\mathbf{x}(t)$).*

Note that this result is *independent of the shape of the path*, and that it includes, as a special case, the result in Section 4 about uniform motion around a circle.

Note also in this situation that the magnitude of the acceleration vector is just the magnitude of its radial component, which by the component formula is

$$|\mathbf{a}| = |\ddot{r} - r\dot{\theta}^2|.$$

Kepler's second law says that a planet moves in an elliptical path with one focus at the sun. If we take the origin at the sun, and the major axis of the ellipse along the x-axis, then the elliptical path has the polar equation

$$r = \frac{k}{1 - \epsilon\cos\theta}.$$

where $0 < \epsilon < 1$ and k is positive. (See Example 6 in Section 1 of

Chapter 10 and Problem 10 in Section 2 of this chapter.) We can now take the major step toward Newton's inverse-square law of gravitational attraction.

THEOREM 9   *Kepler's first and second laws together imply that the acceleration of a planet is directed toward the sun, and is inversely proportional to the square of the distance from the sun in magnitude.*

Proof   We have already seen, in Theorem 8, that the acceleration of the planet is directed along the line to the sun, and that its magnitude is therefore the absolute value of its radial component, which is

$$|\mathbf{a}| = |\ddot{r} - r\dot{\theta}^2|.$$

Here

$$r = \frac{k}{1 - \epsilon \cos \theta},$$

by Kepler's second law (and our discussion above). We differentiate this equation, and use the fact that $r^2\dot{\theta}$ is a constant C, by Kepler's first law, obtaining

$$\dot{r} = \frac{-k\epsilon \sin \theta}{(1 - \epsilon \cos \theta)^2} \dot{\theta} = \frac{-\epsilon \sin \theta}{k} r^2 \dot{\theta}$$

$$= -\frac{C}{k} \epsilon \sin \theta.$$

Differentiating again yields the equation

$$\ddot{r} = -\frac{C}{k} \epsilon(\cos \theta)\dot{\theta}.$$

The radial component of the acceleration can now be computed:

$$\ddot{r} - r\dot{\theta}^2 = -\frac{C}{k} \epsilon(\cos \theta)\dot{\theta} - \frac{C\dot{\theta}}{r} = -C\dot{\theta}\left[\frac{1}{r} + \frac{\epsilon \cos \theta}{k}\right]$$

$$= -C\dot{\theta}\left[\frac{1 - \epsilon \cos \theta}{k} + \frac{\epsilon \cos \theta}{k}\right] = -\frac{C}{k}\dot{\theta}$$

$$= -\left(\frac{C^2}{k}\right)\frac{1}{r^2}.$$

This shows the acceleration to be inversely proportional to r^2 in magnitude, and to be directed toward the origin, completing the proof of the theorem. ∎

THEOREM 10   *Kepler's third law implies that the constant of proportionality C^2/k is the same for all the planets.*

Proof   This calculation will be left as an exercise. It depends on the following facts.

1. The semimajor and semiminor axes of the ellipse $r = k/(1 - \epsilon \cos \theta)$ are

$$a = \frac{k}{1 - \epsilon^2} \quad \text{and} \quad b = \sqrt{1 - \epsilon^2}\, a.$$

   (See Chapter 10, Section 1.)

2. The area of an ellipse is πab.

3. If T is the period of the planet, that is, the time it takes the planet to make one complete revolution, then the area enclosed by its elliptical orbit is $CT/2$. This is because $C/2$ is the constant rate at which the position vector **x** sweeps out area, by the definition of C. ∎

According to Newton's second law, the (vector) acceleration **a** of a moving body is related to the (vector) force **f** (which causes the acceleration) by the equation

$$\mathbf{f} = m\mathbf{a},$$

where m is the mass of the body. Therefore, our three theorems above can be summarized as follows:

The motion of each planet is governed by a force directed toward the sun, of magnitude

$$k\frac{m}{r^2},$$

where r is the distance from the planet to the sun, m is the mass of the planet, and the constant k is the same for all planets.

It is easy to perform a mental experiment showing that the force is also directly proportional to the mass of the attracting body (the sun); for if the sun were divided into two equal parts, then the total force exerted by the sun on the planet would surely be the sum of the two equal forces exerted by its two equal parts, so each part would exert half of the total force. Reasoning in this way, we are led at once to Newton's universal law of gravitation:

> *If two bodies of mass m and M are at a distance r apart, then they exert on each other an attractive (gravitational) force of magnitude*

$$\gamma \frac{mM}{r^2},$$

> *where γ is a universal constant (independent of m, M, and r).*

PROBLEMS FOR SECTION 6

1. Prove Theorem 10.

2. Prove the converse of Theorem 8 (by reversing the chain of reasoning used there). That is, prove that if the acceleration vector $\mathbf{a} = \ddot{\mathbf{x}}(t)$ is always parallel to $\mathbf{x}(t)$, then the position vector $\mathbf{x}$ sweeps out area at a constant rate.

The next three problems are aimed at proving the following theorem.

> **Theorem 11** *If a particle moves in the plane in such a way that its acceleration is directed toward the origin and is proportional to $1/r^2$ in magnitude, then the particle moves along a conic section (ellipse, parabola, or hyperbola) with the origin as a focus.*

We already know from the second problem that then $r^2\dot{\theta}$ is a nonzero constant C, which we shall assume to be positive. In particular, $\dot{\theta}$ is positive, so θ is an increasing function of t, and therefore t is a function of θ. That is, there is a uniquely determined time t_0 at which θ reaches any given value θ_0; this means, in particular, that r is a function of θ along the path. Our problem is to determine the nature of this function.

3. Use the chain rule $\dot{u} = (du/d\theta)\dot{\theta}$ to show that

$$\frac{d\left(\dfrac{1}{r}\right)}{d\theta} = -\frac{\dot{r}}{C}$$

(where C is the constant value of $r^2\dot{\theta}$), and hence that

$$\frac{d^2\left(\dfrac{1}{r}\right)}{d\theta^2} = \frac{-\ddot{r}}{C\dot{\theta}}.$$

4. The theorem hypothesis amounts to the equation

$$\ddot{r} - r(\dot{\theta})^2 = -\frac{k}{r^2},$$

where k is a positive constant. Why is this? Use this equation and the results of the problem above to show that

$$\frac{d^2\left(\dfrac{1}{r}\right)}{d\theta^2} + \frac{1}{r} = \frac{k}{C^2}.$$

5. In Chapter 4 it was shown (essentially) that the differential equation $\ddot{y} + y = C$ has the general solution $y = A \cos t + B \sin t + C$. Assuming this, show that the general solution of the above differential equation can be written in the form

$$kr = \frac{C^2}{1 - \epsilon \cos(\theta - \theta_0)},$$

where ϵ is a positive constant. According to Chapter 10, Section 1, the graph of this equation is an ellipse if $\epsilon < 1$, a parabola if $\epsilon = 1$, and a hyperbola if $\epsilon > 1$.

6. Suppose that $0 < \epsilon < 1$, and that the ellipse

$$kr = \frac{C^2}{1 - \epsilon \cos \theta}$$

is the path of a particle moving in such a way that $r^2\dot{\theta} = C$. Show that

$$\frac{T^2}{a^3} = \frac{4\pi^2}{k},$$

where T is the period of the motion (the time required for one complete revolution) and a is the semimajor axis of the ellipse. (This is Kepler's third law.)

CHAPTER 15
FUNCTIONS OF SEVERAL VARIABLES

More often than not the functional relationships that arise in mathematics and its applications involve several variables. The volume of a pyramid with a square base depends on its base edge b and altitude h, according to the formula

$$V = \frac{1}{3}b^2 h.$$

Thus, V is a function of the two independent variables b and h.

The temperature T indicated on a weather map varies with position and is thus a function of latitude and longitude. If we add the variation of T with time also, then T is a function of three variables, say, x, y, and t.

The state of a small homogeneous body of gas is determined by its pressure p, volume V, and temperature T. According to Boyle's law, these state variables are related by

$$pV = k(T + C),$$

where k and C are constants. This equation determines any one of the state variables as a function of the other two.

We investigate the variation of such a function by a "divide-and-conquer" strategy. We first study the variation that occurs when only one of the independent variables is permitted to change. The tool here is the *partial* derivative, which is just the ordinary derivative with respect to the one changing variable. Then we learn how to combine the information provided by the several partial derivatives in order to understand the way in which the function varies. This "multidimensional" calculus is an enormous subject that we will barely touch upon in this book.

In this chapter we shall consider first the separate partial derivatives, and then their combined action in the chain rule. In Chapter 16 we investigate their vector interpretation and the simplest applications.

1
GRAPHS IN COORDINATE SPACE

The graph of an equation in x, y, and z is the space configuration made up of all the points (x, y, z) satisfying the equation. It will generally be some kind of two-dimensional surface.

Example 1   The graph of

$$x^2 + y^2 + z^2 = 25$$

is the sphere about the origin of radius 5. For, taking square roots and replacing x by $x - 0$, etc., we get the equivalent equation

$$\sqrt{(x - 0)^2 + (y - 0)^2 + (z - 0)^2} = 5,$$

which says that the distance from $(0, 0, 0)$ to (x, y, z) is 5, by the distance formula. The graph consists of all such points (x, y, z) and is thus the sphere. Figure 1 is a drawing of the part of the sphere in the first octant.

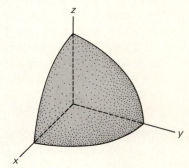

Figure 1

It was shown by a vector argument in Chapter 13, Section 6 that the graph of the equation

$$ax + by + cz + d = 0$$

is always a plane, provided that the coefficients a, b, and c are not all zero. In the simplest cases this can be seen directly, without the use of vectors, and these direct considerations have their own value.

Example 2 A horizontal plane is specified by the constant z-coordinate shared by all of its points. For example, the graph of the equation

$$z = 2$$

is the horizontal plane two units above the xy-coordinate plane. More generally, $z = c$ is the equation of the horizontal plane lying at a distance c above (a distance $|c|$ below if c is negative) the xy-coordinate plane. In particular, $z = 0$ is the equation of the xy-coordinate plane itself. Every horizontal plane is an xy-plane in its own right. For example, the equations

$$x^2 + y^2 = 1, \qquad z = 2$$

specify the unit circle in the plane $z = 2$. See Fig. 2.

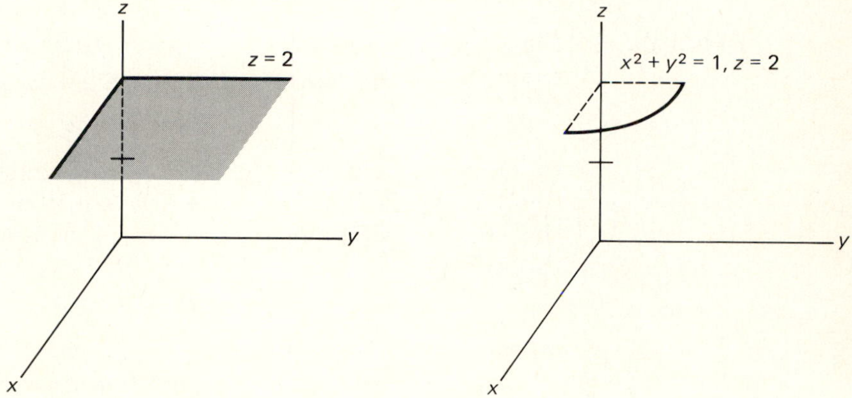

Figure 2

Similarly, $x = a$ is the equation of a vertical plane parallel to the yz-coordinate plane, and $y = b$ is a vertical plane parallel to the xz-coordinate plane (Fig. 3).

The intersection of the latter two planes is the *vertical line* consisting of all points of the form (a, b, z), that is, the vertical line passing through the point $(a, b, 0)$ in the xy-coordinate plane.

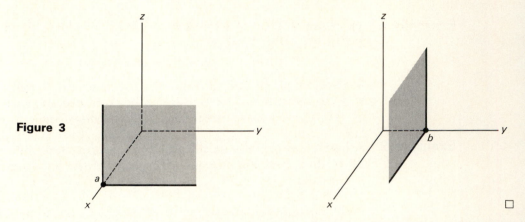

Figure 3

These simple planes can be useful in the study of a more complicated space figure F. For example, if we can determine how F intersects each horizontal plane $z = c$, then by visualizing several of these horizontal plane sections of F, we can build up a picture of F.

The word *section* can be used whenever two configurations intersect: the A section of B is simply the intersection of A and B, but with the implication that the intersection is a piece of B that may reveal information about B.

Example 3   Plane sections can even be used to study other planes. Consider, for example, the space graph of the equation

$$2x + y = 1.$$

Its horizontal plane sections are all the same, being in each case the line having the equation $2x + y = 1$ in the plane $z = c$. This means that the *space* graph of $2x + y = 1$ is the *vertical plane* passing through the *line*

$$2x + y = 1$$

in the xy-coordinate plane.

Alternatively, we can argue that since the equation $2x + y = 1$ does not contain z, a point (a, b, z) lies on its graph if and only if $(a, b, 0)$ also does, for *any* value of z. The graph thus contains the whole vertical line through each of its points. Since its intersection with the xy-coordinate plane is a line l in that plane (having the equation $2x + y = 1$), the space graph consists of the collection of all vertical lines through points of l and hence is the vertical plane through l.   □

Example 4   The same reasoning applies to any other equation involving only x and y. Thus the equation

$$x^2 + y^2 = 1$$

has a circle for its graph in the xy-coordinate plane, but its space graph is the doubly infinite circular *cylinder* made up of all vertical lines through this circle. These lines are called the *rulings* on the cylinder, and their common direction is the direction of the cylinder. Since this is the direction of the z-axis, we say the cylinder is parallel to the z-axis. A piece of such a cylinder is shown at the left in Fig. 4.

The graph of $z = y^2$ is a parabolic cylinder parallel to the x-axis (Fig. 4), and the graph of $x^2 + 2z^2 = 1$ is an elliptical cylinder parallel to the y-axis.

Figure 4

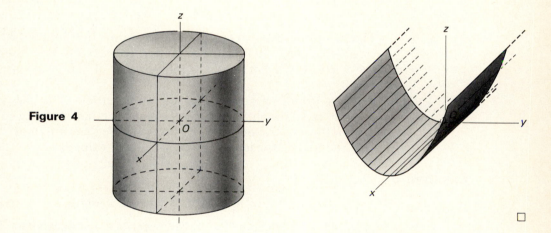

Example 5   The two parabolic cylinders $2z = 4 - y^2$ and $2z = 4 - x^2$, together with the coordinate planes, bound a solid figure in the first octant, as sketched at the right in Fig. 5. It has square cross sections parallel to the xy-plane.   □

So far we have examined only the simplest examples of first- and second-degree equations. The general first-degree equation always has a plane for its graph. This will be a vertical plane if the equation does not contain z, and is thus of the form

$$ax + by + c = 0,$$

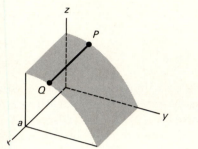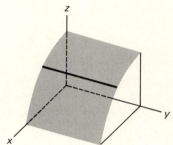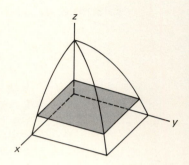

Figure 5

and a nonvertical plane otherwise. In the second case the equation can always be solved for z and thus written in the form

$$z = ax + by + c.$$

That its graph is always a plane was shown in Chapter 13, by vector considerations, but it can also be verified directly by noting that all the vertical and horizontal plane sections of the graph are straight lines.

The graph of any second-degree equation is called a *quadric* surface. It can be shown that *every* plane section of such a surface is a conic (ellipse, parabola, or hyperbola) or a degenerate conic, no matter how the plane is oriented. Here are a few examples.

Example 6   The graph of

$$z = \text{(quadratic in both } x \text{ and } y)$$

is called a *paraboloid*. Its vertical sections are parabolas. There are two types, with standard forms:

$$z = \frac{x^2}{a^2} + \frac{y^2}{b^2} \qquad \text{(elliptic paraboloid)},$$

$$z = \frac{x^2}{a^2} - \frac{y^2}{b^2} \qquad \text{(hyperbolic paraboloid)}.$$

Their horizontal sections are ellipses and hyperbolas, respectively (Fig. 6). If $a = b$ in the first case, so that the horizontal sections are circles, the surface is a *paraboloid* of *revolution*. It can be generated by rotating about the z-axis its section in any plane through the z-axis.

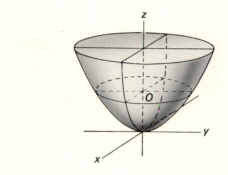

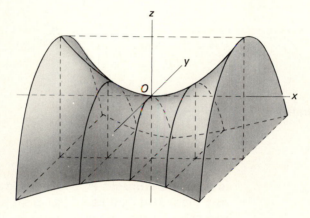

Figure 6

□

Example 7 The graph of

$$z^2 = x^2 + y^2$$

is a *cone*. It is a *circular cone* about the z-axis, meaning that its sections perpendicular to the z-axis are circles centered on the z-axis. In the same way

$$z^2 = \frac{x^2}{a^2} + \frac{y^2}{b^2}$$

is an *elliptical cone*. In each case, the origin is the *vertex* of the cone, and the cone consists of all straight lines (its *rulings*, or *generators*) through the vertex and the points of a suitably chosen plane section. A circular cone is a cone of revolution. See Fig. 7.

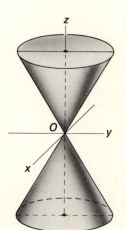

Figure 7

Example 8   The graph of

$$\frac{x^2}{a^2} + \frac{y^2}{b^2} + \frac{z^2}{c^2} = 1$$

is called an *ellipsoid*. It is a bounded surface, and its intersection with *any* plane (that intersects it) is an ellipse in that plane (Fig. 8). This is pretty easy to see for a plane parallel to a coordinate plane, but it is by no means obvious in general.

   If two of the three constants a, b, and c are equal, then we have an *ellipsoid* of *revolution*.

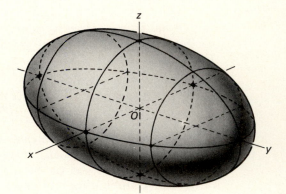

Figure 8

Example 9   Finally, the graph of

$$\frac{x^2}{a^2} + \frac{y^2}{b^2} - \frac{z^2}{c^2} = k$$

is a *hyperboloid*. It has one sheet if $k > 0$ and two sheets if $k < 0$. Its horizontal sections are ellipses and its vertical sections are hyperbolas. If $a = b$ the surface in each case is a hyperboloid of revolution (about the z-axis). See Fig. 9.

Figure 9

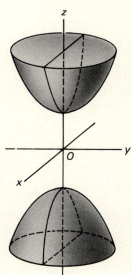

Example 10   The graph of

$$y^2 + z^2 = [f(x)]^2$$

is a surface of revolution about the x-axis. Its section in the plane $x = a$ is the circle

$$y^2 + z^2 = r^2$$

where $r = f(a)$ (assuming $f \geq 0$). The profile is given by its section in any plane through the x-axis. In the plane $z = 0$ it is the function graph

$$y = f(x),$$

and in the plane $y = 0$ it is

$$z = f(x).$$

PROBLEMS FOR SECTION 1

Identify each of the following surfaces, including the nature of the section perpendicular to each axis. (For example: a hyperboloid of two sheets, with elliptical sections perpendicular to the x-axis and hyperbolic sections perpendicular to the y- and z-axes.) If it is a surface of revolution, give the axis.

1. $x^2 + 2y^2 + z^2 = 1$ **2.** $x^2 + 2y^2 - z^2 = 1$ **3.** $x^2 + 2y^2 - z^2 = -1$

4. $x^2 - 2y^2 - z^2 = 1$ **5.** $x^2 - 2y^2 + z^2 = 1$ **6.** $x^2 + 2y^2 + z^2 = 0$

7. $x^2 + 2y^2 - z^2 = 0$ **8.** $x^2 - 2y^2 + z^2 = 0$ **9.** $x^2 + 2y^2 - z = 0$

10. $x^2 - 2y^2 - z = 0$ **11.** $x^2 + 2y + z^2 = 0$ **12.** $x^2 + z^2 = 0$

13. $x^2 - z^2 = 0$

Draw the figure representing the solid in the first octant bounded by the coordinate planes and the following surfaces.

14. $x + y + z = 2$ **15.** $y + z = 1, x = 2$ **16.** $x + z = 1, y = 2$ **17.** $x + y = 1, z = 2$

18. $x^2 + y^2 = 1, z = 2$ **19.** $x^2 + y^2 = 1, x + z = 1$ **20.** $x^2 + y^2 + z^2 = 25, y = 4$

21. $y^2 + z^2 = 16, x^2 + y^2 = 16$ **22.** $x^2 + y^2 = 16, x + y + z = 8$ **23.** $4x^2 + 4y^2 + z^2 = 64, y + z = 4$

24. $x^2 + y^2 = z^2 + 9, z = 4$

If a second-degree equation has no mixed-product terms (xy, yz, xz), then its graph can be identified by completing squares just as in the plane, and it always turns out to be a standard quadric surface of one of the types listed in Section 1, except for being centered at a point other than the origin.

25. By completing squares show that

$$x^2 + y^2 + z^2 + 2x - 4z = 0$$

is a sphere centered at $(-1, 0, 2)$.

26. By completing squares show that

$$x^2 + y^2 - z^2 + 2x + 2z = 0$$

is a cone with vertex at $(-1, 0, 1)$ and with axis parallel to the z-axis.

Identify the graphs of the following equations.

27. $x^2 + y^2 + z + 4 = 0$ **28.** $x^2 + y^2 + 2x + z = 0$ **29.** $x^2 + y^2 + 2y - z = 0$

30. $x^2 - y^2 + 2x + 4y + z = 0$ **31.** $x^2 + 2y^2 + z^2 + 4y + 2z = 0$

32. $x^2 + 2y^2 - z^2 + 4x - 4y = 0$ **33.** $x^2 + 2y^2 - z^2 + 4x - 4y = -7$

34. $x^2 - 2y^2 - z^2 + 4x - 4y = 0$ **35.** $x^2 - 2y^2 - z^2 + 4y - 4z = 0$

36. Prove that the graph of $z^2 = x^2 + y^2$ is a cone with origin as vertex. (Use the fact that the straight line through the origin and the point (a, b, c) has parametric equations

$$x = at, \qquad y = bt, \qquad z = ct.)$$

37. Find the straight lines passing through the origin that lie on the paraboloid

$$z = \frac{x^2}{a^2} - \frac{y^2}{b^2}.$$

2 FUNCTIONS OF TWO VARIABLES; PARTIAL DERIVATIVES

Functions of several variables involve the same language and the same conventions that are used for functions of one variable. Thus, each of the expressions $x^2 + y^2$ and $x \sin y$ defines a function $f(x, y)$ of the two independent variables x and y, since each of them has a uniquely determined value for each pair of numbers x and y. Similarly,

$$g(x, y, z, w) = (x^2 + y^2)zw$$

is a function of four variables. Except for an occasional remark, we shall consider only functions of two variables in this chapter.

In general, a function $f(x, y)$ will not be defined for all pairs of numbers (x, y). For example,

$$f(x, y) = y \ln x$$

is defined only when $x > 0$. Thus,

DEFINITION

A *function of two variables* is an operation that determines a unique output number z when applied to any given input pair of numbers (x, y) chosen from a certain domain.

Each of the equations

$$z = x^2 + y,$$
$$xyz = 1,$$
$$z^3 = x/y,$$

determines z as a function of x and y. Thus a function of *two* variables can be defined by an equation in *three* variables, just as a function of one variable can be defined by an equation in two variables. Similarly,

$$e^w = xyz$$

determines w as a function of the *three* variables x, y, and z.

The domain of a function of two variables $f(x, y)$ will generally be some sort of region in the plane bounded by one or more curves. For example, $f(x, y) = \sqrt{1 - x^2 - y^2}$ is defined when $1 - x^2 - y^2 \geq 0$, that is, when

$$x^2 + y^2 \leq 1.$$

The domain of f is the "unit circular disk," and it is bounded by the unit circle $x^2 + y^2 = 1$.

Figure 1

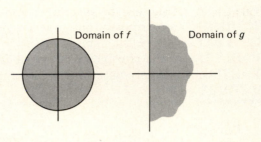

The domain of

$$g(x, y) = y \ln x$$

is the right half-plane, and its boundary is the y-axis (see Fig. 1.)

A point (a, b) is *in the interior* of a domain D if some circular disk about (a, b) lies entirely in D. Any other point of D is a *boundary point*. Thus a point is a boundary point of D if every circle about it contains both points in D and points not in D. As stated above, the boundary points of a domain normally lie along one or more boundary curves (Fig. 2).

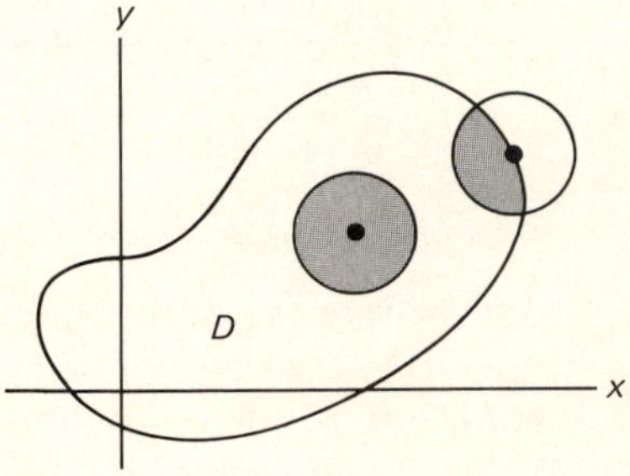

Figure 2

A domain is *closed* if it includes its boundary, *open* if it excludes the boundary. There are also in-between possibilities, but in every case a domain will consist of an open domain plus (possibly) part or all of its boundary. A domain D is *connected* if any two of its points can be joined by a path lying entirely in D.

DEFINITION   A *region* is a connected domain.

A domain that is not connected will consist of two or more separate regions.

Example 1 The domain of

$$f(x, y) = \sqrt{x^2 - 1}$$

consists of two closed half-planes G_1 and G_2, where

G_1 consists of all points (x, y) such that $x \le -1$,

G_2 consists of all points (x, y) such that $x \ge 1$. □

Partial Derivatives If $z = x^2y^3 + xy + y^2$, and if we hold y fixed, then z becomes a function of x alone and we can calculate its derivative. This derivative is called the *partial derivative of z with respect to x*. In Leibniz notation it is written

$$\frac{\partial z}{\partial x},$$

where ∂x is read "del x." Thus, treating y as a constant and computing the derivative of z with respect to x, we have

$$\frac{\partial z}{\partial x} = 2xy^3 + y.$$

Similarly, we obtain $\partial z/\partial y$ by regarding x as a constant and computing the derivative of z with respect to y:

$$\frac{\partial z}{\partial y} = 3x^2y^2 + x + 2y.$$

This procedure is followed for any number of independent variables. For example, if

$$w = x^2 - xy + y^2 + 2yz + 2z^2 + z,$$

then

$$\frac{\partial w}{\partial x} = 2x - y,$$

$$\frac{\partial w}{\partial y} = -x + 2y + 2z,$$

$$\frac{\partial w}{\partial z} = 2y + 4z + 1.$$

We shall also need function notation, analogous to the f' notation for functions of one variable. We shall use

$$D_1 f \qquad \text{and} \qquad D_2 f$$

for the partial derivatives of f with respect to its first and second variables, respectively. Thus, if $z = f(x, y)$, then

$$\frac{\partial z}{\partial x} = D_1 f(x, y),$$

$$\frac{\partial z}{\partial y} = D_2 f(x, y).$$

The value of $\partial z / \partial x$ at $(x, y) = (a, b)$ is given in the two notations by

$$D_1 f(a, b) = \frac{\partial z}{\partial x}\bigg|_{(a,b)}.$$

Example 2 If $f(x, y) = xy + x^2 - y$, then

$$D_1 f(x, y) = y + 2x, \qquad D_2 f(x, y) = x - 1,$$
$$D_1 f(2, 3) = 3 + 2 \cdot 2 = 7, \qquad D_2 f(2, 3) = 2 - 1 = 1. \qquad \square$$

Continuity At a couple of points we shall have to mention continuity, in order to state things correctly. This notion extends naturally to a function of two variables $f(x, y)$:

The function f is continuous at the point (a, b) if its value $f(x, y)$ can be made to be as close to $f(a, b)$ as we wish by taking the point (x, y) suitably close to (a, b).

In the plane, the closeness of two points is measured by the distance between them, so being suitably close to (a, b) means lying within a suitably small circle about (a, b). This is a simultaneous condition on x and y, and the notion of continuity thus involves simultaneous variation in x and y.

Every function that will come up will be continuous wherever it is defined. The principal reason is the theorem below. Its proof is given in Appendix 4.

THEOREM *Suppose that both partial derivatives of $f(x, y)$ exist everywhere inside a circle C, and are everywhere less in magnitude than a fixed constant k. Then f is continuous inside of C.*

The composition of continuous functions is continuous. For example, if $f(x, y)$ is continuous on a region D, and if $(x, y) = (g(t), h(t))$ is a path lying in D, then $f(g(t), h(t))$ is a continuous function of t. This example is important because it implies

THEOREM 1   *If f is continuous on a region D, then the intermediate-value principle holds for f on D. That is, if z_1 and z_2 are values of f on D, then every number between z_1 and z_2 is also a value.*

The details are left to the reader.

Function Graphs   The graph of a function of two variables $f(x, y)$ is the graph of the equation

$$z = f(x, y).$$

It is a surface S in coordinate space. If we interpret the domain D of f as a subset of the xy-coordinate plane, then S is *spread out* over D, in the sense that each vertical line through a point $(x, y, 0)$ of D intersects S exactly once, in the point $(x, y, z) = (x, y, f(x, y))$. See Fig. 3.

Figure 3

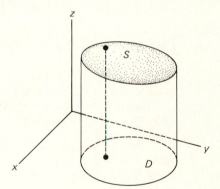

It is very helpful in visualizing such a surface S to consider the *curves* in which it intersects the vertical planes

$$x = a \qquad \text{and} \qquad y = b,$$

that is, the sections of S in these planes. Its section in a plane $x = a$ is the graph in this yz-plane of the function of one variable $f(a, y)$, that is, the graph of the equation

$$z = f(a, y).$$

It is the *slice* in which the plane $x = a$ cuts S (Fig. 4).

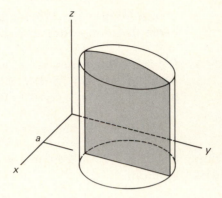

Figure 4

The slope of this curve is the derivative of z with respect to y, while x is held fixed at the value $x = a$, and this is exactly the partial derivative

$$\frac{\partial z}{\partial y}\bigg|_{x=a} = D_2 f(a, y).$$

Thus we can interpret $\partial z/\partial y$ as *the slope of the graph of $z = f(x, y)$ in the y-direction*.

Similarly, the section of S in the plane $y = b$ is the graph in that xz-plane of the equation $z = f(x, b)$, as shown in Fig. 5.

Its slope is $\partial z/\partial x$, evaluated along $y = b$, that is, $D_1 f(x, b)$. Thus, $\partial z/\partial x$ can be interpreted as *the slope of the graph $z = f(x, y)$ in the x-direction*.

Figure 5

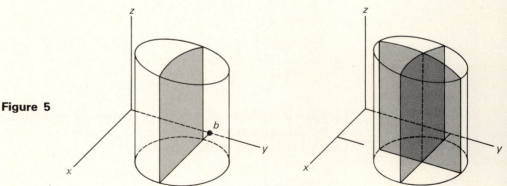

If (a, b) is in the domain of f, then the two sections above necessarily intersect over $(a, b, 0)$, the point of intersection being

$$(a, b, z) = (a, b, f(a, b)).$$

If we sketch two or three curves of each type, even very roughly, we get an impression of how the surface is curving.

Finally, the section of S in the horizontal plane $z = c$ is the graph in that xy-plane of the equation $c = f(x, y)$.
When we draw the graphs

$$c = f(x, y)$$

in the xy-coordinate plane, we call them the *level* curves of f. An xy-figure showing several such level curves is an alternate way of picturing the behavior of $f(x, y)$. Figure 6 shows an example of level curves.

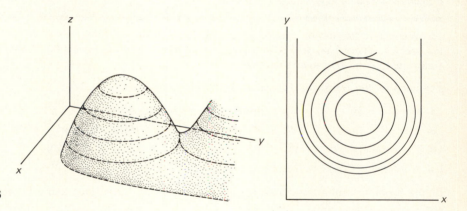

Figure 6

Example 3 The level curves of the parabolic cap

$$5z = 25 - (x^2 + y^2)$$

are just circles. Nevertheless, their spacing can indicate the shape of the space graph of the equation (Fig. 7). This is the principle of a contour map. With a little experience interpreting contour lines, one gets a direct three-dimensional image from looking at the two-dimensional map.

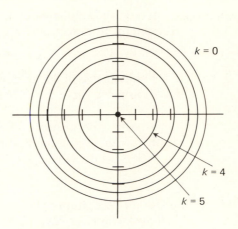

Figure 7

$k = 0$

$k = 4$

$k = 5$

PROBLEMS FOR SECTION 2

1. Write out the definitions of $\partial z/\partial x$ and $\partial z/\partial y$ (where $z = f(x, y)$) as limits of difference quotients, using the increment notation.

For each of the following functions, find $\partial z/\partial x$ and $\partial z/\partial y$.

2. $z = 2x - 3y$ **3.** $z = 3x^2 - 2xy + y$ **4.** $z = \sqrt{x^2 + y^2}$ **5.** $z = e^{ax+by}$

6. $z = e^{xy}$ **7.** $z = \sin(2x + 3y)$ **8.** $z = \cos((x^2 - y^2)/2)$ **9.** $z = xye^{-x^2/2}$

10. $z = y\ln(x + y)$ **11.** $z = y\ln(xy)$ **12.** $z = e^{-x}\cos y$ **13.** $z = \sin ax \cos by$

14. $z = (x/y) - (y/x)$ **15.** $z = \arctan(x/y)$ **16.** $z = x^y$ **17.** $z = (x + y)/(x - y)$

In the problems below, compute the indicated partial derivatives using function notation.

18. $f(u, v) = 3u^2 + 2uv + v^3$, $D_1f(2, -3)$, $D_2f(1, 1)$

19. $f(s, t) = se^{st}$, $D_1f(2, 3)$, $D_2f(2, -1)$

20. $f(x, y) = \sin(3x + y^2)$, $D_1f(\pi, 0)$, $D_2f(0, \sqrt{\pi})$

21. $f(x, y, z) = xy^2z^3$, $D_1f(0, 2, -2)$, $D_2f(0, 2, -2)$, $D_3f(1, 2, -2)$

22. $f(x, u, t) = x^2 + ut^3$, all partials at $(2, 0, 1)$

23. $f(s, t) = s^2 \sin t$, $D_1f(a, b)$, $D_2f(a, 0)$

A function can have a property called "homogeneity of degree n" that we won't define here, but which can be shown to be equivalent to the condition

(*) $$xD_1f(x, y) + yD_2f(x, y) = nf(x, y).$$

24. Show that $z = \arctan(y/x)$ is homogeneous of degree 0.

Each of the following functions is homogeneous. In each case verify (*) and determine the degree of homogeneity.

25. $z = x^4 + y^4 - x^2y^2$ **26.** $z = (x + y)/(x - y)$ **27.** $z = x/(x^2 + y^2)$

28. Polar coordinates and rectangular coordinates are related by the equations

$$x = r \cos \theta, \qquad r = \sqrt{x^2 + y^2},$$

$$y = r \sin \theta, \qquad \theta = \arctan \frac{y}{x}.$$

Find all the partial derivatives of these functions.

29. The boundary of a plane region G need not consist of one or more curves. What is the domain of the function $f(x, y) = 1/(x^2 + y^2)$? What is its boundary?

30. Find a function whose boundary consists of just the two points $(1, 0)$ and $(-1, 0)$.

31. Find a function whose domain boundary consists of the unit circle $x^2 + y^2 = 1$ plus the origin $(0, 0)$.

32. Prove that the intermediate-value principle holds for a continuous function on a plane region D.

Sketch the level curves for each of the following functions at the given values of z.

33. $z = y/x$; $z = 0, 1, 2, 3, 4$ **34.** $z = y/x^2$; $z = 0, 1, 2, 3$

35. $z = x + y$; $z = -2, -1, 0, 1, 2$ **36.** $z = x - y$; $z = -2, -1, 0, 1, 2$

37. $z = xy$; $z = -2, -1, 0, 1, 2$ **38.** $z = x^2 + y^2$; $z = 1, 2, 3, 4$

39. $z = x^2 + 4y^2$; $z = 1, 2, 3, 4$ **40.** $z = x^2 - y^2$; $z = -2, -1, 0, 1, 2$

41. $z^2 = x^2 + y^2 + 1$; $z = 1, 2, 3, 4$ **42.** $z^2 = x^2 + y^2 - 1$; $z = 0, 1, 2, 3$

Assuming that each of the following equations defines z implicitly as a function of x and y, find the partial derivative $\partial z/\partial x$ and $\partial z/\partial y$ by the implicit differentiation.

43. $\dfrac{x^2}{a^2} + \dfrac{y^2}{b^2} + \dfrac{z^2}{c^2} = 1$ **44.** $Ax^2 + Bxy + Cy^2 + Dz^2 + E = 0$

45. $x^3 + y^3 + z^3 = 1$ **46.** $z^3 + xyz = C$

47. $ze^{xyz} = 1$ **48.** $\sin xy + \sin yz + \sin xz = 0$

49. $e^{xz} \sin yz = 1$ **50.** $xy \arctan(z/x) = 1$

Find both partial derivatives of the following functions of x and y.

51. $x^3 + x^2y + xy^2 + y^3$ **52.** $x \sin(y + x^2)$ **53.** $\dfrac{x}{y} \cos \dfrac{y}{x}$

54. $y^2 \ln(x + e^y)$ **55.** $\sin(x^2 + y^2)\cos xy$ **56.** $\dfrac{x^2 + \sec xy}{xy + \tan y^2}$

3
THE CHAIN RULE

Since a partial derivative of a function of several variables is nothing but the ordinary derivative of a suitable restriction of that function, it follows that, by and large, the rules for computing partial derivatives are the same as for ordinary derivatives. Thus the sum, product, and quotient rules are all the same.

The chain rule acquires a new form, however, since now it concerns the composition of an outside function of *two or more variables* with *two or more inside functions.* For example, if z is a function of x and y, and if x and y are each functions of t, then z is a function of t.

Example 1
If $z = x^2 + y^2$ and if $x = 2t$ and $y = t^3$, then
$$z = (2t)^2 + (t^3)^2 = t^2(4 + t^4). \qquad \Box$$

Here is the Leibniz form of the chain rule for a function of two variables. We call $f(x, y)$ *smooth* if it has continuous partial derivatives.

CHAIN RULE
If z is a smooth function of x and y, and if x and y are differentiable functions of t, then z is a differentiable function of t, and

$$\boxed{\frac{dz}{dt} = \frac{\partial z}{\partial x}\frac{dx}{dt} + \frac{\partial z}{\partial y}\frac{dy}{dt}.}$$

Note that the formula is "dimensionally" correct in the same way the old formula was. The positions of x cancel symbolically, one being up and one down. So do the positions of y. The right-hand side thus reduces symbolically to z up and t down, which agrees with the left.

Of course, z will be a function of t only if the domains and ranges overlap. We can picture this overlapping in the xy-plane by considering the two functions

$$x = g(t) \qquad \text{and} \qquad y = h(t)$$

as together defining a path

$$(x, y) = (g(t), h(t)).$$

Then this path must run, for a while at least, in the region D that is the domain of $z = f(x, y)$. See Fig. 1. The path could very well meander in and out of D, and in this case the domain of the composite

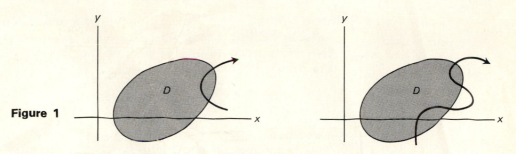

Figure 1

function

$$z = f(x, y) = f(g(t), h(t))$$

is made up of two or more separate t-intervals, corresponding to the separate pieces of the path lying in D. Such diagrams lead us to interpret the composite-function derivative dz/dt as the derivative of $f(x, y)$ *along the path.*

The chain rule will be proved at the end of the section, and its meaning will become clearer in terms of vectors in Chapter 16. Here we shall merely perform some computations with it.

Example 2   We verify the correctness of the chain rule in the situation of Example 1. We had

$$z = x^2 + y^2, \qquad x = 2t, \qquad \text{and} \qquad y = t^3.$$

By the chain rule,

$$\frac{\partial z}{\partial t} = \frac{\partial z}{\partial x}\frac{dx}{dt} + \frac{\partial z}{\partial y}\frac{dy}{dt}$$

$$= 2x \cdot 2 + 2y \cdot 3t^2$$

$$= 2(2t) \cdot 2 + 2t^3 \cdot 3t^2 = 8t + 6t^5.$$

On the other hand,

$$z = (x^2 + y^2) = (2t)^2 + (t^3)^2 = 4t^2 + t^6,$$

from which the same answer,

$$dz/dt = 8t + 6t^5,$$

follows by direct computation.   □

Example 3   What is the formula for $\partial z/\partial u$ if

$$z = f(x, y), \qquad x = g(u, v), \qquad y = h(u, v)?$$

Solution   Here z has become a function of u and v, and we find $\partial z/\partial u$ by holding v constant and taking the ordinary derivative with respect to u. This reduces the present situation to the chain rule, and we have

$$\frac{\partial z}{\partial u} = \frac{\partial z}{\partial x}\frac{\partial x}{\partial u} + \frac{\partial z}{\partial y}\frac{\partial y}{\partial u}.$$

This formula is the chain rule for partial derivatives.   □

Example 4   Calculate $\partial z/\partial u$ in two ways when

$$z = x^2 + y^2, \qquad x = uv^2, \qquad y = u^2v.$$

Solution 1   Substitute the expressions for x and y:

$$z = x^2 + y^2 = (uv^2)^2 + (u^2v)^2.$$
$$= u^2v^4 + u^4v^2.$$

Then compute $\partial z/\partial u$:

$$\frac{\partial z}{\partial u} = 2uv^4 + 4u^3v^2.$$

Solution 2   Use the chain rule for partial derivatives:

$$\frac{\partial z}{\partial u} = \frac{\partial z}{\partial x}\frac{\partial x}{\partial u} + \frac{\partial z}{\partial y}\frac{\partial y}{\partial u}$$
$$= 2xv^2 + 2y2uv$$
$$= 2(uv^2)v^2 + 2(u^2v)2uv$$
$$= 2uv^4 + 4u^3v^2.$$ □

These Leibniz expressions treat z ambiguously. Two quite different functions were being differentiated when we wrote $\partial z/\partial x$ and $\partial z/\partial u$. This ambiguity caused no problems when we were looking at functions of one variable, but here there can be trouble.

Example 5   The following very common situation is a good example of how bad things can be. If z is a function of x and y, and if y is a function of x, then z is a function of x and

$$\frac{dz}{dx} = \frac{\partial z}{\partial x} + \frac{\partial z}{\partial y}\frac{dy}{dx}.$$

Correctly interpreted, this is a valid and important special case of the chain rule. But the correct interpretation requires us to keep in mind that dz/dx and $\partial z/\partial x$ are derivatives of two entirely different functions! In simple situations we can put up with such loose talk, but generally we need more precise formulations of the chain rule involving function notation. □

Good enough for most purposes is the intermediate form, part Leibniz notation and part function notation, as follows:

CHAIN RULE *If $z = f(x, y)$ with f smooth, and if x and y are differentiable functions of t, then z is a differentiable function of t, and*

$$\frac{dz}{dt} = D_1f(x, y)\frac{dx}{dt} + D_2f(x, y)\frac{dy}{dt}.$$

In this form Leibniz notation is permitted for the derivatives with respect to the ultimate independent variable or variables, but function notation is used for the partials with respect to the intermediate variables.

Consider again the important situation of Example 5.

Example 6 If $z = f(x, y)$ and y is a function of x, then we have the special case $t = x$ in the above formula: z becomes a function of x alone, and the formula becomes

$$\frac{dz}{dx} = D_1f(x, y) + D_2f(x, y)\frac{dy}{dx}.$$

Note how this formulation avoids the ambiguity of the pure Leibniz formula. □

Most precise, but also most ungainly, is the pure function form. Leaving out the differentiability hypotheses, it is

If $\phi(t) = f(g(t), h(t))$, then

$$\phi'(t) = D_1f(g(t), h(t))g'(t) + D_2f(g(t), h(t))h'(t).$$

Proof of Chain Rule The proof of the chain rule depends on the fact that if we hold one variable fixed in $f(x, y)$, then the mean-value theorem can be applied to the remaining function of the other variable. For example, if we set $g(x) = f(x, b)$, then the mean-value property

$$g(x + \Delta x) - g(x) = g'(X)\,\Delta x$$

becomes

$$f(x + \Delta x, b) - f(x, b) = D_1 f(X, b)\,\Delta x.$$

With this observation we can prove the theorem.

THEOREM 2   *If the functions $x = g(t)$ and $y = h(t)$ are differentiable at t_0, and if $z = f(x, y)$ is smooth inside some circle about the point*

$$(x_0, y_0) = (g(t_0), h(t_0)),$$

then the composite function

$$k(t) = f(g(t), h(t))$$

is differentiable at t_0, and

$$k'(t_0) = D_1 f(x_0, y_0)g'(t_0) + D_2 f(x_0, y_0)h'(t_0).$$

Proof   If we give t an increment Δt, then x and y acquire increments Δx and Δy, and $z = f(x, y)$ acquires the increment

$$\Delta z = f(x + \Delta x, y + \Delta y) - f(x, y).$$

(We have dropped the zero subscripts to simplify the notation.) What is new in this situation is that the intermediate variables x and y together form a *variable point* (x, y) with new value $(x + \Delta x, y + \Delta y)$. In order to bring in the partial derivatives of f, we make this point change in two steps, first changing x by Δx and *then* y by Δy. That is, we take one step horizontally off the path and then a step vertically back onto the path. Then Δz is the sum of its two partial changes:

$$\Delta z = [f(x + \Delta x, y + \Delta y) - f(x + \Delta x, y)]$$
$$+ [f(x + \Delta x, y) - f(x, y)].$$

Figure 2

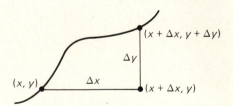

Each bracket involves a change in only one variable and can be rewritten by the mean-value theorem for functions of that variable. Thus,

$$\Delta z = D_2 f(x + \Delta x, Y)\, \Delta y + D_1 f(X, y)\, \Delta x,$$

where Y lies between y and $y + \Delta y$, and X lies between x and $x + \Delta x$. Now divide by Δt,

$$\frac{\Delta z}{\Delta t} = D_2 f(x + \Delta x, Y)\frac{\Delta y}{\Delta t} + D_1 f(X, y)\frac{\Delta x}{\Delta t},$$

and let Δt approach 0. The limit on the right exists and has the value on the right below. Therefore, dz/dt exists and

$$\frac{dz}{dt} = D_2 f(x, y)\frac{dy}{dt} + D_1 f(x, y)\frac{dx}{dt},$$

which is the chain rule. Note that we have needed the continuity of the partial derivative $D_2 f$ in order to conclude that

$$D_2 f(x + \Delta x, Y) \to D_2 f(x, y)$$

as $\Delta t \to 0$. The continuity of D_1 was used, too. ■

REMARK The one-variable chain rule is true even if f' is discontinuous. Here, however, continuous differentiability is essential; there exist pathological functions, with discontinuous partial derivatives, for which the chain rule is false.

PROBLEMS FOR SECTION 3

In Problems 1 through 5 find dz/dt by the chain rule.

1. $z = x^2 - 2xy + y^2$, $x = (t + 1)^2$, $y = (t - 1)^2$

2. $z = x \sin y$, $x = 1/t$, $y = \arctan t$

3. $z = \ln \sqrt{(x - y)/(x + y)}$, $x = \sec t$, $y = \tan t$

4. $z = x^2 y^3$, $x = 2t^3$, $y = 3t^2$

5. $z = \ln(x^2 + y^2)$, $x = e^{-t}$, $y = e^t$

6. The altitude of a right circular cone is 8 inches and is increasing at 0.5 in/min. The radius of the base is 10 inches and is decreasing at the rate of 0.2 in/min. How fast is the volume changing?

Calculate $\partial z/\partial v$ and $\partial z/\partial u$ in two ways in the following problems.

7. $z = x^2 + y^2$, $x = u + v$, $y = u - v$

8. $z = x^2 + 3xy + y^2$, $\quad x = \sin u + \cos v$, $y = \sin u - \cos v$

9. $z = e^{xy}$, $x = u^2 + 2uv$, $y = 2uv + v^2$

10. $z = \sin(4x + 5y)$, $x = u + v$, $y = u - v$

11. Show that $z = f(y/x)$ is homogeneous of degree 0; that is, show that

$$x\frac{\partial z}{\partial x} + y\frac{\partial z}{\partial y} = 0.$$

12. Recall from the last chapter that we call $z = f(x, y)$ *homogeneous of degree n* if

$$(*) \qquad x\frac{\partial z}{\partial x} + y\frac{\partial z}{\partial y} = nz.$$

Suppose that

$$z = f(u, v), \quad u = g(x, y), \quad \text{and} \quad v = h(x, y),$$

where f is homogeneous of degree m and g and h are each homogeneous of degree n. Show that z is homogeneous of degree mn as a function of x and y. That is,

$$z = f(g(x, y), h(x, y))$$

is homogeneous of degree mn.

13. The condition (*) above is not the usual definition of homogeneity. Show that a function $z = f(x, y)$ will satisfy (*) if it has the property

$$(**) \qquad f(tx, ty) = t^n f(x, y).$$

(Differentiate with respect to t.)

14. Show, conversely, that if a function satisfies (*), then it satisfies (**). You will need the following fact:

If $du/dt = n(u/t)$, then $u = ct^n$.

(See Problem 46 in Section 1, Chapter 4.)

15. If x and y are independent variables, then their differentials dx and dy are new independent variables, wholly independent of x and y in their values, but used along with x and y. The definition of the differential of a *dependent* variable is analogous to that given earlier for a function of one independent variable (at the end of Chapter 4).

> **Definition** If $z = f(x, y)$, then
>
> $$dz = D_1 f(x, y)\, dx + D_2 f(x, y)\, dy.$$

Show that this relationship between dx, dy, and dz remains true even if x and y are functions of other variables, say $x = g(s, t)$ and $y = h(s, t)$, so that the underlying independent variables are s and t throughout.

If u and v are both functions of x and y, show that

16. $d(u + v) = du + dv$ 　　**17.** $d(uv) = u\,dv + u\,du$ 　　**18.** $d\!\left(\dfrac{u}{v}\right) = \dfrac{v\,du - u\,dv}{v^2}$

19. If $w = f(u)$, then $dw = f'(u)\,du$ 　　　　**20.** If $w = f(xz, yz)$, show that

$$x\frac{\partial w}{\partial x} + y\frac{\partial w}{\partial y} = z\frac{\partial w}{\partial z}.$$

The Leibniz form for the chain rule for a function of three variables is

If w is a smooth function of x, y, and z, and if x, y, and z are differentiable functions of t, then w is a differentiable function of t, and

$$\frac{dw}{dt} = \frac{\partial w}{\partial x}\frac{dx}{dt} + \frac{\partial w}{\partial y}\frac{dy}{dt} + \frac{\partial w}{\partial z}\frac{dz}{dt}.$$

21. State the pure function form of this rule.

22. Prove this three-variable chain rule, following the plan of the two-variable proof in the text. You will have to express Δw as the sum of *three* "partial" increments.

23. State the chain rule for a function of four variables.

24. State the chain rule for a function of n variables $x_1, x_2, \ldots, x_n$.

25. A smooth path of length L runs from the point $\mathbf{x}_0$ to the point $\mathbf{x}_1$. A function f, defined on a region containing the path, has both partial derivatives bounded by a constant K. Show that

$$|f(\mathbf{x}_1) - f(\mathbf{x}_0)| \le 2KL.$$

26. State and prove the corresponding result for the variation of a function along a space path.

4
IMPLICIT
FUNCTIONS

The chain rule for functions of two variables lets us discuss implicit functions in a general context that was unavailable in Chapter 4.
　　The graph of an equation of the form

$$f(x, y) = k$$

is called a *level set* of the function f, as we saw in Section 2. For example, the level sets of the function

$$f(x, y) = x^2 + y^2$$

are the circles about the origin:

$$x^2 + y^2 = k.$$

Generally, as in this example, a level set of a smooth function of two variables will be some sort of a smooth curve, although probably

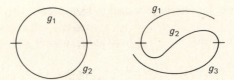

Figure 1

not a function graph (Fig. 1). It may be made up of two or more separate function graphs $y = g(x)$, just as the above level set $x^2 + y^2 = k$ consists of the graphs of the two functions

$$y = g_1(x) = \sqrt{k - x^2} \quad \text{and} \quad y = g_2(x) = -\sqrt{k - x^2}.$$

Implicit Differentiation

By hypothesis, $z = f(x, y)$ has the constant value k along the graph of any such function $y = g(x)$,

$$z = f(x, g(x)) = k,$$

so if f and g are smooth functions then

$$(*) \qquad \frac{dz}{dx} = D_1 f + D_2 f \frac{dy}{dx} = 0.$$

The middle term is just the chain-rule evaluation of dz/dx when $z = f(x, y)$ and y is a function of x. The result is zero because z is constant as a function of x. If $D_2 f$ is not zero, the above equation can then be solved for dy/dx:

$$\boxed{\frac{dy}{dx} = \frac{-D_1 f(x, y)}{D_2 f(x, y)},}$$

or

$$\boxed{g'(x) = \frac{-D_1 f(x, g(x))}{D_2 f(x, g(x))}.}$$

In the language of Chapter 4, a function $y = g(x)$, such that

$$f(x, g(x)) = k$$

is a *function solution* of the equation

$$f(x, y) = k,$$

and is defined *implicitly* by this equation. So we have obtained above a general formula for the derivative of an implicitly defined function.

Note that if $D_2f = 0$ in (*), then also $D_1f = 0$. Such a point—where both partial derivatives of f are zero—is called a *critical point* of f. So if the graph of g contains no critical point of f, then $D_2f \neq 0$ along the graph and we can solve for $g'(x)$ as above.

Altogether we have shown that

THEOREM 3 *If $f(x, y)$ is a smooth function and $y = g(x)$ is a differential function defined implicitly by the equation $f(x, y) = k$, and if there are no critical points of f on the graph of g, then*

$$g'(x) = - \frac{D_1f(x, g(x))}{D_2f(x, g(x))}.$$

This whole analysis works in the same way when we consider a piece of the level set $f(x, y) = k$, along which x is a smooth function of y,

$$x = h(y).$$

Such a piece is not a standard function graph because its independent variable runs vertically (Fig. 2), but we would still call h an implicitly defined function. This time we have

$$z = f(h(y), y) = k,$$

$$\frac{dz}{dy} = D_1f\frac{dx}{dy} + D_2f = 0,$$

$$h'(y) = \frac{dx}{dy} = - \frac{D_2f(x, y)}{D_1f(x, y)} = - \frac{D_2f(h(y), y)}{D_1f(h(y), y)},$$

again provided that $(x, y) = (h(y), y)$ is not a critical point of f.

Implicit Function Theorem

So far, the condition that no critical points of f lie along the piece of the level set we are looking at has been a requirement tacked on at the last minute to permit dividing by D_2f or D_1f in the derivative formula. If (x_0, y_0) is a critical point on the level set $f = k$, then we don't get a derivative formula for an implicitly defined function through that point. However, the no-critical-point condition plays a deeper role than this, as stated below in Theorem 4.

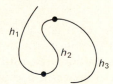

Figure 2

We note, first, that near a critical point the level set may not even *be* a smooth function graph.

Example 1 The function

$$f(x, y) = y^2 - x^2$$

has a critical point at the origin. Its level set through the origin, the graph of $y^2 - x^2 = 0$, is the pair of straight lines $y = x$ and $y = -x$, so in the neighborhood of the origin the level set is not a function graph at all (Fig. 3).

Figure 3

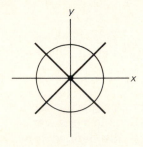

$\square$

Example 2 The function

$$f(x, y) = y^3 - x^2$$

has a critical point at the origin. Its level set through the origin is the graph of the function $y = g(x) = x^{2/3}$, but $g'(0)$ does not exist. The level set has a cusp at the origin (Fig. 4).

Figure 4

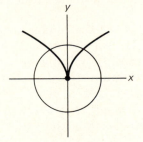

$\square$

Example 3 The level set $y^2 - x^3 = 0$ is the graph of $x = h(y) = y^{2/3}$. It is a nonstandard function graph, and is the image of the graph in Example 2 under the rotation of the plane over the 45° line $y = x$. Again, the origin is a bad point: the graph has a cusp there, h' fails to exist there, and

$$f(x, y) = y^2 - x^3$$

has a critical point there. □

On the other hand, near a noncritical point (x_0, y_0), the equation $f(x, y) = k$ does define y as a smooth function of x, or x as a smooth function of y, depending on whether

$$D_2 f(x_0, y_0) \neq 0 \qquad \text{or} \qquad D_1 f(x_0, y_0) \neq 0.$$

This fact is called the *implicit-function theorem*. Here is the exact statement.

THEOREM 4 *Suppose that $f(x, y)$ is smooth, and that $D_2 f(x_0, y_0) \neq 0$, where (x_0, y_0) is a point on the level set $f(x, y) = k$. Then the level set runs along a smooth function graph as it goes through (x_0, y_0). That is, there is a rectangle about (x_0, y_0) within which the equation*

$$f(x, y) = k$$

determines y as a smooth function of x.

Similarly, if $D_1(x_0, y_0) \neq 0$, then $f(x, y) = k$ defines x as a smooth function of y in a suitably small rectangle about (x_0, y_0).

The theorem is illustrated in Fig. 5. Its proof will be omitted.

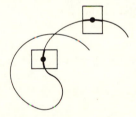

Figure 5

Example 4 Use the implicit-function theorem to determine whether or not the equation

$$x^3 + y - y^3 = 0$$

determines y as a function of x locally about each point (x_0, y_0) on the graph.

Solution The above equation is of the form

$$f(x, y) = 0$$

and we have to determine whether $D_2 f(x, y)$ is ever zero along the level set. We have

$$D_2 f(x, y) = 1 - 3y^2,$$

which is zero at $y = \pm 1/\sqrt{3}$. Substituting these values in the level set equation gives

$$x^3 \pm \left(\frac{1}{\sqrt{3}} - \frac{1}{3\sqrt{3}} \right) = 0$$

or

$$x = \mp \left(\frac{2}{3\sqrt{3}} \right)^{1/3} = \mp \frac{2^{1/3}}{\sqrt{3}}.$$

So there are two points on the level set at which $D_2 f = 0$,

$$\pm \left(\frac{2^{1/3}}{\sqrt{3}}, -\frac{1}{\sqrt{3}} \right).$$

At every other point (x_0, y_0) on the level set the second partial $D_2 f$ is different from zero, so the level set is a function graph near (x_0, y_0).   □

A complete level set may consist of two or more separate paths.

Example 5   The level set

$$5(x^2 + y^2) - (x^2 + y^2)^2 = 4$$

is the pair of concentric circles

$$x^2 + y^2 = 1, \qquad x^2 + y^2 = 4.$$

Figure 6

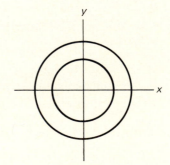

□

PROBLEMS FOR SECTION 4

Use Theorem 4 to determine for each of the following cubic equations $f(x, y) = 0$, whether or not the equation determines y as a function of x locally about each point (x_0, y_0) on the graph. Determine any exceptional points (x_0, y_0) near which the equation may *not* determine y as a function of x. (According to the theorem these are the simultaneous solutions of the two equations $f(x, y) = 0$ and $D_2 f(x, y) = 0$.)

1. $x^3 + y + y^3 = 0$ **2.** $1 - xy + y^3 = 0$ **3.** $1 + xy^2 + y^3 = 0$

4. $x^3 + xy + y^3 = 0$ **5.** $x^3 + xy^2 + y^3 = 0$

6. Study the local solvability of the equation

$$1 - xy + x^2 \ln y = 0$$

for y in terms of x.

CHAPTER 16

VECTORS IN CALCULUS OF SEVERAL VARIABLES; HIGHER PARTIAL DERIVATIVES

1
THE GRADIENT OF A FUNCTION; THE VECTOR FORM OF THE CHAIN RULE

We shall always assume f smooth in this chapter. The two partial derivatives of $z = f(x, y)$ are generally viewed as the components of a vector, called the *gradient* of f, and designated grad f. Thus,

$$\text{grad } f = \left(\frac{\partial z}{\partial x}, \frac{\partial z}{\partial y}\right) = (D_1 f, D_2 f),$$

and

$$\text{grad } f(a, b) = \left(\frac{\partial z}{\partial x}, \frac{\partial z}{\partial y}\right)\bigg|_{(x,y)=(a,b)} = (D_1 f(a, b), D_2 f(a, b)).$$

A *critical point* of f is just a point where the vector grad f is zero. Points where grad $f \neq 0$ are called *regular points* of f.

Example 1   If $z = f(x, y) = (x^2 + 2y)/4$, then

$$\text{grad } f(x, y) = (\partial z/\partial x, \partial z/\partial y) = (x/2, 1/2),$$
$$\text{grad } f(-3, 0) = (x/2, 1/2)_{(-3,0)} = (-3/2, 1/2).$$

There are no critical points of f. (All points are regular.)   □

When we view the point (x, y) in the coordinate plane as a vector

$$\mathbf{x} = (x, y),$$

then a function of two variables $z = f(x, y)$ can be viewed as a function of one vector variable $\mathbf{x}$, and we shall often write

$$\boxed{f(x, y) = f(\mathbf{x}).}$$

For example, grad $f(a, b)$ = grad $f(\mathbf{a})$, where $\mathbf{a} = (a, b)$.

Also, we view grad $f(\mathbf{a})$ as a vector located at the point $\mathbf{a}$, that is, as an arrow drawn from $\mathbf{a}$. Then, for each point $\mathbf{x}$, we have associated the located vector grad $f(\mathbf{x})$, and grad f is a *field of vectors*, or a *vector field*. We can sketch the vector field grad f by drawing some of its arrows.

Example 2 If $(x, y) = (x^2 + y^2)/4$, then

$$\text{grad } f(x, y) = \left(\frac{\partial}{\partial x}\left(\frac{x^2 + y^2}{4}\right), \frac{\partial}{\partial y}\left(\frac{x^2 + y^2}{4}\right)\right) = \left(\frac{x}{2}, \frac{y}{2}\right) = \frac{1}{2}(x, y).$$

Thus

$$\text{grad } f(1, 2) = \frac{1}{2}(1, 2) = \left(\frac{1}{2}, 1\right),$$

$$\text{grad } f(0, -1) = \frac{1}{2}(0, -1) = \left(0, -\frac{1}{2}\right), \qquad \text{etc.}$$

Figure 1 shows some arrows of the vector field.

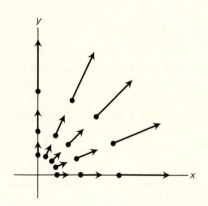

Figure 1

Example 3 In Example 1 we had

$$\operatorname{grad} f(x, y) = \left(\frac{x}{2}, \frac{1}{2}\right),$$

and this vector field is sketched in Fig. 2.

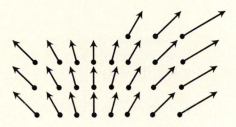

Figure 2

The chain rule

$$\frac{dz}{dt} = \frac{\partial z}{\partial x}\frac{dx}{dt} + \frac{\partial z}{\partial y}\frac{dy}{dt}$$

$$= (D_1 f)\frac{dx}{dt} + (D_2 f)\frac{dy}{dt}$$

can now be given the following very important reformulation:

CHAIN RULE *If $z = f(x, y)$ is smooth, and if x and y are differentiable functions of t, then z is a differentiable function of t, and*

$$\boxed{\begin{aligned} \frac{dz}{dt} &= \operatorname{grad} f \cdot \frac{d\mathbf{x}}{dt} \\ &= \operatorname{grad} f \cdot \text{path tangent vector.} \end{aligned}}$$

This dot-product formula for the derivative of a function along a path has many consequences. To begin with, consider the level curve

$$f(x, y) = k,$$

where k is an arbitrary constant. We saw in the last section that such a curve is locally a smooth path around a regular point of f. More explicitly, if (x_0, y_0) is a point of the level set $f(x, y) = k$ at which grad f is not zero, then there is a small rectangle R about (x_0, y_0) whose intersection with the level set is the locus of a smooth path (Fig. 3).

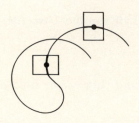

Figure 3

Suppose, then, that $\mathbf{x} = \mathbf{g}(t)$ is a smooth path running along the level curve $f(x, y) = k$. Then $z = f(x, y)$ has the constant value k along the path

$$z = f(g(t), h(t)) = k,$$

and $dz/dt = 0$. Therefore

$$0 = \frac{dz}{dt} = \text{grad } f \cdot \frac{d\mathbf{x}}{dt},$$

so the vector grad f is perpendicular to the path tangent vector. Since the path traces the level curve, its tangent vector is tangent to the curve. We have thus proved the following theorem.

THEOREM 1 *If the function $f(x, y)$ is smooth and $\mathbf{x}_0 = (x_0, y_0)$ is a regular point of f, then grad $f(\mathbf{x}_0)$ is perpendicular to the level curve of f through $\mathbf{x}_0$.*

A vector perpendicular to a curve is said to be *normal* to the curve.

Example 4 According to the theorem, a vector normal to the ellipse

$$\frac{x^2}{6} + \frac{4y^2}{6} = \frac{25}{6}$$

at the point $(x_0, y_0) = (3, 2)$ is given by

$$\text{grad}\left(\frac{x^2}{6} + \frac{4y^2}{6}\right)_{(3,2)} = \left(\frac{2}{6}x, \frac{8}{6}y\right)\Big|_{(3,2)} = \left(1, \frac{8}{3}\right).$$

In Fig. 4, we draw this normal vector located at $\mathbf{x}_0$.

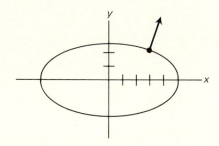

Figure 4

□

Tangent Lines to Level Curves The theorem gives us an easy way to write down the equation of the tangent line to a level curve at any regular point $\mathbf{x}_0$. Since the tangent line is perpendicular to grad $f(\mathbf{x}_0)$, its equation is

$$\boxed{\text{grad } f(\mathbf{x}_0) \cdot (\mathbf{x} - \mathbf{x}_0) = 0,}$$

by the dot product in Section 3 of Chapter 13.

If the gradient of f is zero at $\mathbf{x}_0$, this equation will not give a line because both coefficients will be zero. In fact, we know from examples in the last section that the level curve $f(x, y) = k$ may not *have* a tangent line at a critical point.

Example 5 Find the line tangent to the curve

$$x^3 + xy + y^3 = 5$$

at the point $\mathbf{x}_0 = (2, -1)$.

Solution First verify that $(2, -1)$ is on the graph:

$$2^3 + 2(-1) + (-1)^3 = 8 - 2 - 1 = 5.$$

Then compute grad $f(2, -1)$, where $f(x, y) = x^3 + xy + y^3$:

$$\text{grad } f(2, -1) = (3x^2 + y, x + 3y^2)_{(2, -1)} = (11, 5).$$

Substituting these values in the above formula for the tangent line gives

$$\text{grad } f(\mathbf{x}_0) \cdot (\mathbf{x} - \mathbf{x}_0) = (11, 5) \cdot (x - 2, y + 1)$$
$$= 11(x - 2) + 5(y + 1) = 0$$

or

$$11x + 5y - 17 = 0.$$

This is the equation of the line tangent to the curve $x^3 + xy + y^5 = 5$ at the point $(2, -1)$. $\qquad\square$

Example 6 Show that the general tangent-line formula above agrees with our earlier formula for the line tangent to the graph of a function $y = g(x)$.

Solution This graph is the level curve

$$g(x) - y = 0,$$

and the gradient of this special function $f(x, y) = g(x) - y$ is

$$\text{grad } f(x, y) = (g'(x), -1).$$

The dot-product formula

$$\text{grad } f(\mathbf{x}_0) \cdot (\mathbf{x} - \mathbf{x}_0) = 0$$

reduces in this case to

$$g'(x_0)(x - x_0) - 1(y - y_0) = 0,$$

or

$$y - y_0 = g'(x_0)(x - x_0).$$

This is the original tangent-line formula   □

Two graphs intersect at right angles if and only if their normals at the point of intersection are perpendicular. Therefore a level curve of $f(x, y)$ intersects a level curve of $g(x, y)$ orthogonally (at right angles) if and only if

$$\text{grad } f(x_0, y_0) \perp \text{grad } g(x_0, y_0),$$

where (x_0, y_0) is the point of intersection.

Example 7 Show that every ellipse of the form

$$x^2 + 2y^2 = C$$

is orthogonal to every parabola of the form

$$y = kx^2.$$

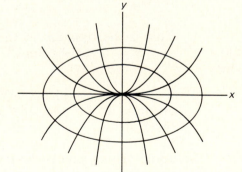

Figure 5

Solution   Set $f(x, y) = x^2 + 2y^2$ and $g(x, y) = y/x^2$. Then

$$\text{grad } f \cdot \text{grad } g = (2x, 4y) \cdot (-2y/x^3, 1/x^2)$$
$$= -4y/x^2 + 4y/x^2 = 0$$

at every point (x, y) (for which $x \neq 0$). So the two level curves through any given point (x_0, y_0) are orthogonal (Fig. 5). $\square$

There is a converse to Theorem 1, as follows.

THEOREM 2   *Suppose that the path* $\mathbf{x} = (g(t), h(t))$ *is perpendicular to the gradient of the function* $z = f(x, y)$ *at every point of the path f. Then the path runs along a level curve of f.*

Proof   The derivative of $z = f(x, y)$ along the path,

$$\frac{dz}{dt} = \text{grad } f \cdot \frac{d\mathbf{x}}{dt},$$

is everywhere zero, by the perpendicularity hypothesis, so z, as a function of the path parameter t, must therefore be a constant k. That is,

$$z = f(g(t), h(t)) = k,$$

and the path $(x, y) = (g(t), h(t))$ runs along the level curve $f(x, y) = k$.   ∎

PROBLEMS FOR SECTION 1

For each of the following functions determine grad f at the given point.

1. $f(x, y) = x^2 + y^2; (3, 4)$

2. $f(x, y) = \arctan \dfrac{x}{y}; (1, 2)$

3. $f(x, y) = \cos(xy); (1, \pi/6)$

4. $f(x, y) = e^x \tan y; (0, \pi/4)$

5. $f(x, y) = (x^3 + 2x^2y - 3xy + 4xy^2 - y^3; (1, 1)$

6. $f(x, y) = x^{2y}; (2, 1)$

7. $f(x, y) = 2x - y; (a, b)$

In the following problems the functions are divided by a constant, just to make the vectors short enough so they won't clutter up the page. You can use a different constant if you wish. Sketch the gradient field of $f(x, y)$ when

8. $f(x, y) = \dfrac{x + y}{4}$

9. $f(x, y) = \dfrac{(y - x^2)}{4}$

10. $f(x, y) = -\dfrac{x}{4y}$

Find a nonzero vector that is normal to each curve at the given point.

11. $\sqrt{x^2 - y^2} = 1; (\sqrt{2}, 1)$

12. $xe^{xy} = 2e; (2, 1/2)$

13. $\arcsin \sqrt{x^2 + y^2} = \pi/6; (1/4, \sqrt{3}/4)$

14. $\arctan xe^y = \pi/4; (1/e, 1)$

Using gradients, find the tangent line to each curve below at the given point.

15. $xy = 2; (1, 2)$

16. $xy^2 = 1; (1/4, 2)$

17. $x^2 + y^2 = 25; (3, 4)$

18. $x^2/9 + y^2/4 = 1; (2, (2/3)\sqrt{5})$

19. $x^2 - 2xy - 3y^2 = 5; (2, -1)$

25. $x^2 + xy + y^2 - 2x + y = 0; (2, -3)$

21. $x^2 - 3xy^2 + 2y^3 = 16; (0, 2)$

22. $xe^{xy} = 2; (2, 0)$

23. $(x + y)\arctan xy = (5/8)\pi; (2, 1/2)$

24. $xy \ln x = 2e; (e, 2)$

25. Show that the level curves of $f(x, y) = x^2 + y^2$ are everywhere perpendicular to the level curves of $g(x, y) = y/x$. Sketch a few curves from each family.

26. Show that the level curves of $f(x, y) = x^2 - y^2$ are everywhere perpendicular to the level curves of $g(x, y) = xy$. Sketch a few curves from each family.

27. Let $f(x, y) = k$ be a level curve not containing the origin, and let (x_0, y_0) be a point on this level set that is closest to the origin. Prove that

$$\mathbf{x}_0 = (x_0, y_0) \quad \text{and} \quad \text{grad } f(\mathbf{x}_0)$$

parallel vectors.

28. Show that the tangent line to the ellipse

$$\frac{x^2}{a^2} + \frac{y^2}{b^2} = 1$$

at the point (x_0, y_0) has the equation

$$\frac{xx_0}{a^2} + \frac{yy_0}{b^2} = 1.$$

29. Prove the corresponding result for the hyperbola

$$\frac{x^2}{a^2} - \frac{y^2}{b^2} = 1.$$

30. Show that the equation of the tangent line to the parabola

$$y = kx^2$$

at the point (x_0, y_0) can be written

$$\frac{1}{2}(y + y_0) = kxx_0.$$

31. Show that the tangent line to the conic

$$ax^2 + 2bxy + cy^2 + d = 0$$

at the point (x_0, y_0) can be written

$$ax_0x + b(x_0y + y_0x) + cy_0y + d = 0.$$

32. Show that the equation of the tangent line to the conic

$$ax^2 + by^2 + cx + dy + e = 0$$

at the point (x_0, y_0) can be written

$$ax_0x + by_0y + \frac{c}{2}(x + x_0) + \frac{d}{2}(y + y_0) + e = 0.$$

33. Let f be a smooth function, and let the domain D of $f(x, y)$ be a region. (That is, a smooth path can be drawn in D between any two points of D.) Show that f is a constant function if and only if grad f is identically 0. (This depends on, and generalizes, Theorem 1 in Chapter 5.)

34. Show that

$$\text{grad}(f + g)(\mathbf{x}) = \text{grad } f(\mathbf{x}) + \text{grad } g(\mathbf{x})$$

$$\text{grad } cf(\mathbf{x}) = c \text{ grad } f(\mathbf{x}).$$

35. Let D be a region and suppose f and g are two functions with domain D such that

$$\operatorname{grad} f = \operatorname{grad} g$$

on D. Use Problems 33 and 34 to show that there is a constant C such that

$$f(x, x) = g(x, y) + C$$

for all (x, y) in D.

37. Show that

$$\operatorname{grad} fg(\mathbf{x}) = f(\mathbf{x})\operatorname{grad} g(\mathbf{x}) + g(\mathbf{x})\operatorname{grad} f(\mathbf{x}).$$

36. Find a function f such that

$$\operatorname{grad} f(\mathbf{x}) \equiv \mathbf{x}.$$

What is the most general such function?

2

DIRECTIONAL DERIVATIVES

Let $\mathbf{u}$ be a given unit vector, $\mathbf{u} = (u, v)$. We want to compute the rate at which $f(x, y)$ changes as the point (x, y) moves through (x_0, y_0) in the direction $\mathbf{u}$. This rate of change will be called *the derivative of f in the direction* $\mathbf{u}$ *at the point* (x_0, y_0), and will be designated

$$D_\mathbf{u}f(x_0, y_0).$$

If we go a distance t from the point $\mathbf{x}_0$ in the direction $\mathbf{u}$, we end up at the point $\mathbf{x} = \mathbf{x}_0 + t\mathbf{u}$. This is true because $\mathbf{u}$ is a unit vector: the distance from $\mathbf{x}_0$ to $\mathbf{x}$ is

$$|\mathbf{x} - \mathbf{x}_0| = |(\mathbf{x}_0 + t\mathbf{u}) - \mathbf{x}_0| = |t\mathbf{u}| = |t|\,|\mathbf{u}| = t.$$

The average rate of change of f from $\mathbf{x}_0$ to $\mathbf{x}$ is thus

$$\frac{f(\mathbf{x}_0 + t\mathbf{u}) - f(\mathbf{x}_0)}{t}.$$

As usual, the true, or "instantaneous," rate of change is defined as the limit of the average rate of change, so;

DEFINITION

$$D_\mathbf{u}f(\mathbf{x}_0) = \lim_{t \to 0} \frac{f(\mathbf{x}_0 + t\mathbf{u}) - f(\mathbf{x}_0)}{t}.$$

In order to evaluate this *directional derivative*, note that it is the ordinary derivative of a certain function of t: if

$$\phi(t) = f(\mathbf{x}_0 + t\mathbf{u}) = f(x_0 + tu, y_0 + tv),$$

then

$$D_\mathbf{u}f(\mathbf{x}_0) = \lim_{t \to 0} \frac{\phi(t) - \phi(0)}{t} = \phi'(0).$$

But ϕ is a composite function. It is the restriction of $f(x, y) = f(\mathbf{x})$ to the straight line path $\mathbf{x} = \mathbf{x}_0 + t\mathbf{u}$. So the derivative $\phi'(t)$ is calculated by the chain rule, and has the value

$$\phi'(t) = \text{grad } f(\mathbf{x}_0 + t\mathbf{u}) \cdot \frac{d\mathbf{x}}{dt}$$

$$= \text{grad } f(\mathbf{x}_0 + t\mathbf{u}) \cdot \mathbf{u}.$$

We only want $\phi'(0)$. Thus,

THEOREM 3   *The directional derivative $D_\mathbf{u}f(\mathbf{x}_0)$ has the value*

$$\boxed{D_\mathbf{u}f(\mathbf{x}_0) = \text{grad } f(\mathbf{x}_0) \cdot \mathbf{u}.}$$

Example 1   Find the derivative of $z = x^2y$ at the point $\mathbf{x}_0 = (2, -3)$ in the direction of the vector $\mathbf{a} = (3, 4)$.

Solution   Since $\mathbf{a} = (3, 4)$ is not a unit vector, we first have to find its direction (the unit vector having the direction of $\mathbf{a}$). This is

$$\mathbf{u} = \frac{\mathbf{a}}{|\mathbf{a}|} = \frac{(3, 4)}{\sqrt{3^2 + 4^2}} = \left(\frac{3}{5}, \frac{4}{5}\right).$$

Next, we compute

$$(\text{grad } f)(2, -3) = \left(\frac{\partial z}{\partial x}, \frac{\partial z}{\partial y}\right)_{(2,-3)}$$

$$= (2xy, x^2)_{(2,-3)} = (-12, 4).$$

Then,

$$D_\mathbf{u}f(\mathbf{x}_0) = \text{grad } f(\mathbf{x}_0) \cdot \mathbf{u}$$

$$= (-12, 4) \cdot \left(\frac{3}{5}, \frac{4}{5}\right)$$

$$= \frac{-36 + 16}{5} = -4. \qquad \square$$

If $\mathbf{u} = (1, 0)$, then $D_\mathbf{u}f$ is simply the first partial derivative of f:

$$D_{(1,0)}f = D_1f = \frac{\partial z}{\partial x}$$

(where $z = f(x, y)$). Similarly $D_{(0,1)}f$ is the second partial derivative $D_2 f = \partial z/\partial y$. These facts can be obtained from the chain-rule formula for $D_{\mathbf{u}}f$ or from the original limit definition of $D_{\mathbf{u}}f$.

Now recall from Section 3 of Chapter 13 that if $\mathbf{u}$ is a unit vector, then the dot product $\mathbf{a} \cdot \mathbf{u}$ can be interpreted as the component of the vector $\mathbf{a}$ in the direction $\mathbf{u}$. Also, $\mathbf{a} \cdot \mathbf{u} = |\mathbf{a}| \cos \phi$, where ϕ is the angle between the direction of $\mathbf{a}$ and $\mathbf{u}$. The formula

$$D_{\mathbf{u}}f(\mathbf{x}) = \operatorname{grad} f(\mathbf{x}) \cdot \mathbf{u}$$

thus shows that

COROLLARY *The directional derivative $D_{\mathbf{u}}f(\mathbf{x})$ is the component of the vector $\operatorname{grad} f(\mathbf{x})$ in the direction $\mathbf{u}$, and can be expressed*

$$\boxed{D_{\mathbf{u}}f(\mathbf{x}) = |\operatorname{grad} f(\mathbf{x})| \cos \phi,}$$

where ϕ is the angle between $\mathbf{u}$ and the vector $\operatorname{grad} f(\mathbf{x})$.

If $\mathbf{x}$ is held fixed, but the direction $\mathbf{u}$ of the derivative is allowed to vary, then $\cos \phi$ varies between -1 and $+1$ in this formula. In particular, $D_{\mathbf{u}}f(\mathbf{x})$ has its maximum value $|\operatorname{grad} f(\mathbf{x})|$, when $\mathbf{u}$ is in the direction of $\operatorname{grad} f(\mathbf{x})$, for then $\cos \phi$ has its maximum value 1. Also

$$D_{\mathbf{u}}f(\mathbf{x}) = 0$$

whenever the direction $\mathbf{u}$ is perpendicular to $\operatorname{grad} f(\mathbf{x})$. This can be thought of as the "local" expression of the fact that f is constant along any path that runs perpendicular to the vector field $\operatorname{grad} f$. (See Theorem 2.)

Example 2 If $f(x, y) = xy^2$, find the maximum value of the directional derivatives of f at $\mathbf{x}_0 = (1, 1)$ and the direction $\mathbf{u}$ in which f has this maximum rate of change.

Solution First,

$$\operatorname{grad} f(1, 1) = (y^2, 2xy)_{(1,1)} = (1, 2).$$

Then, as we saw above, $D_{\mathbf{u}}f(1, 1)$ will have the maximum value

$$|\text{grad } f(1,1)| = |(1,2)| = \sqrt{5}$$

when **u** is the direction of grad $f(1,1)$,

$$\mathbf{u} = \frac{\text{grad } f(1,1)}{|\text{grad } f(1,1)|} = \frac{(1,2)}{\sqrt{5}} = \left(\frac{1}{\sqrt{5}}, \frac{2}{\sqrt{5}}\right). \qquad \square$$

We can interpret the relation between grad $f(\mathbf{x})$ and the directions in which the directional derivative $D_{\mathbf{u}}f(\mathbf{x})$ has its maximum, zero, and minimum values in terms of our experiences on a hilly countryside.

Two straight highways intersecting at right angles can be used as xy-axis for the neighboring countryside, and then the terrain itself is the graph of a certain function $z = f(x,y)$. If we start at a point on a hillside and walk levelly along the side of the hill, we are walking along a level curve of f. (Strictly speaking we are walking *over* a level curve. The level curve itself is normally drawn in the xy-base plane.) If we now turn and start walking upward at right angles to the level curve we are necessarily moving in the direction of *greatest steepness* of the hill, for our new direction, specified as a direction in the base plane, is the direction of the *gradient of f*, and we have just seen that this is the direction in which $z = f(x,y)$ has the maximum rate of change. A stream of water flows down the hillside along the shortest path to the bottom, and this means that it is always running in the direction of greatest steepness downward, which is the direction of $-\text{grad } f$.

The countryside $z = f(x,y)$ can be pictured in the xy-base plane by a series of level curves. These are the contour lines of a geodetic map. We get from one countour line to the next most quickly along the perpendicular direction. This is pretty clear geometrically and is a geometric interpretation of the fact that the maximum rate of change of $z = f(x,y)$ is in the direction of the gradient of f. As we move perpendicularly to the contour lines, the number of lines we cross per unit distance is a rough measure of the magnitude of grad f. When the contour lines bunch together they indicate a steep hillside and a large rate of change of $z = f(x,y)$. With a little experience interpreting geodetic contour lines, one gets a direct three-dimensional image from looking at the two-dimensional map.

PROBLEMS FOR SECTION 2

1. Establish the formula

$$\frac{\text{grad } f(\mathbf{x}_0) \cdot \mathbf{a}}{|\mathbf{a}|}$$

for the derivative of f in the direction of the vector $\mathbf{a}$ at the point $\mathbf{x}_0$.

For each of the functions below, find the derivative at the given point and in the direction of the given vector $\mathbf{a}$. (Either use the formula of Problem 1, or work directly from Theorem 3 as in Example 1 in the text.)

2. $f(x, y) = x^3 y$; $(1, 2)$; $\mathbf{a} = (3/5, 4/5)$

3. $f(x, y) = x^2 - y^2$; $(1/2, -1)$; $\mathbf{a} = (\sqrt{3}/2, 1/2)$

4. $f(x, y) = xe^{xy}$; $(2, 1)$; $\mathbf{a} = (12/13, 5/13)$

5. $f(x, y) = \ln[(x + y)/(x - y)]$; $(1/2, 2)$; $\mathbf{a} = (1/4, 3/4)$

6. $f(x, y) = \cos x \sin y$; $(\pi/3, \pi/3)$; $\mathbf{a} = (1/2, 1/3)$

7. $f(x, y) = e^{x \tan y}$; $(1, \pi/4)$; $\mathbf{a} = (1/e, 2/e)$

8. $f(x, y) = \sqrt{x^2 + y^2}$; $(3, 4)$; $\mathbf{a} = (1/4, 1/3)$

9. $f(x, y) = \arctan x/y$; $(1/2, 1/3)$; $\mathbf{a} = (1/5, 3/4)$

10. $f(x, y) = x^y$; $(e, 0)$; $\mathbf{a} = (0, 1)$

11. $f(x, y) = x^{xy}$; $(1, 2)$; $\mathbf{a} = (2, 0)$

Find the maximum value of the directional derivative of the given function at the point indicated, and the direction of the maximum rate of change.

12. $g(x, y) = xy^2 + yx^2$; $(1/2, 1)$

13. $g(x, y) = x^2 y^2 - 2xy^3 + x$; $(1/2, 2)$

14. $g(x, y) = \tan x \sec y$; $(\pi/4, \pi/3)$

15. $g(x, y) = \ln \dfrac{x^2 - y^2}{x^2 + y^2}$; $(1/2, 1/4)$

16. $g(x, y) = ye^{x/y}$; $(0, 2)$

17. $g(x, y) = \dfrac{x - y}{x + y}$; $(1/4, 3/4)$

18. $g(x, y) = \sqrt{xy}$; $(2, 8)$

19. $g(x, y) = \arctan x/y$; $(1, 1/2)$

20. Write out the definition of the first partial derivative $D_1 f$ as the limit of a difference quotient, and show that it agrees with the definition of the directional derivative $D_{\mathbf{u}} f$ when $\mathbf{u} = \mathbf{e}_1 = (1, 0)$.

21. Suppose that $z = f(x, y)$ and $(x, y) = (g(t), h(t))$. In Section 1 we interpreted the chain-rule derivative dz/dt as the derivative of f along the path $\mathbf{x} = \mathbf{g}(t)$. Show now that

The derivative of f along the path $\mathbf{x} = \mathbf{g}(t)$ is equal to the directional derivative of f in the direction of the path, multiplied by the magnitude $|d\mathbf{x}/dt|$ of the path tangent vector.

22. Given $f(x, y) = x^3 y$, find the directions **u** for which

$$D_u f(2, 1) = 10.$$

23. If we know the derivative of f at $\mathbf{x}_0$ in two independent directions, then we can solve for grad $f(\mathbf{x}_0)$. Suppose that $D_u f(\mathbf{x}_0)$ has the value 6 in the direction of the vector $(3, 4)$ and the value $-\sqrt{5}$ in the direction of the vector $(2, -4)$. Find grad $f(\mathbf{x}_0)$.

24. Find grad $f(\mathbf{x}_0)$ if $D_u f(\mathbf{x}_0) = 1$ in the direction of $(1, -\sqrt{3})$ and $D_u f(\mathbf{x}_0) = -1$ in the direction of $(1, \sqrt{3})$.

Write down the derivative of f at $\mathbf{a} = (a, b)$ in the direction of the unit vector $\mathbf{u} = (u, v)$, for each of the following functions.

25. $f(x, y) = x + y$

26. $f(x, y) = (x^2 + y^2)/2$

27. $f(x, y) = xy$

23. $f(x, y) = xe^{xy}$

When a one-parameter family is given by an equation of the form

$$f(x, y) = C,$$

so that its curves are level curves of the function f, the orthogonal trajectories have the direction of the gradient field of f. Their differential equation is thus

$$\frac{dy}{dx} = \frac{\partial f/\partial y}{\partial f/\partial x}.$$

For example, the orthogonal trajectories of the ellipses

$$x^2 + 2y^2 = C$$

have the direction $(2x, 4y)$, so their differential equation is

$$\frac{dy}{dx} = \frac{2y}{x}.$$

Such orthogonal trajectories are the curves of *steepest descent* for the function f. If $x = f(x, y)$ is a hillside on which rain falls, then the water will run off along paths on the hillside that lie over the one-parameter family of curves of steepest descent in the xy-plane.

Find the curves of steepest descent when

29. $f(x, y) = x^2 - y^2$

35. $f(x, y) = x^2 y^3$

31. $f(x, y) = 1 - (x^2 + y^2)$

32. $f(x, y) = xy$

33. Rain falls on the ellipsoid

$$\frac{x^2}{a^2} + \frac{y^2}{b^2} + \frac{z^2}{c^2} = 1.$$

Along what paths will the water run off?

3
GRADIENTS
IN SPACE

The chain rule generalizes to functions of three or more variables. Thus,

THEOREM 4

Chain Rule *If* $w = f(\mathbf{x}) = f(x, y, z)$ *is smooth, and if*

$$\mathbf{x} = (x, y, z) = (g(t), h(t), k(t))$$

is a differentiable path running through the domain of f, then along the path w is a differentiable function of t and

$$\frac{dw}{dt} = \frac{\partial w}{\partial x}\frac{dx}{dt} + \frac{\partial w}{\partial y}\frac{dy}{dt} + \frac{\partial w}{\partial z}\frac{dz}{dt}$$

$$= D_1 f\frac{dx}{dt} + D_2 f\frac{dy}{dt} + D_3 f\frac{dz}{dt}.$$

The proof is just like the two-dimensional proof, the only change being that Δw is now broken down into the sum of three partial increments corresponding to separate changes in x, y, and z.

The gradient of a function of three variables $f(x, y, z)$ is the three-dimensional vector

$$\operatorname{grad} f = (D_1 f, D_2 f, D_3 f),$$

and the three-dimensional chain rule has the same dot-product formula as before,

Chain Rule

$$\frac{dw}{dt} = \operatorname{grad} f \cdot \frac{d\mathbf{x}}{dt},$$

where, of course, the three-dimensional dot product is involved.

The directional derivatives of a function of three variables $f(x, y, z)$ are defined the same way as in the plane, with the same final formulation:

For any space direction $\mathbf{u}$, *that is, for any three-dimensional unit vector* $\mathbf{u}$, *the derivative of f in the direction* $\mathbf{u}$ *at the point* $\mathbf{x}$ *is*

$$D_{\mathbf{u}}f(\mathbf{x}) = \operatorname{grad} f(\mathbf{x}) \cdot \mathbf{u}.$$

Thus $D_{\mathbf{u}}f(\mathbf{x}_0)$ is the value at $\mathbf{x}_0$ of the derivative of f along the straight line $\mathbf{x} = \mathbf{u}t + \mathbf{x}_0$ through $\mathbf{x}_0$ in the direction $\mathbf{u}$.

Example 1   Find the derivative of $f(x, y, z) = xy^2z^3$ at the point $(1, 1, 1)$ in the direction of the vector $(2, 1, -2)$.

Solution   The gradient of f at $(1, 1, 1)$ is

$$\text{grad } f(1, 1, 1) = (y^2z^3, 2xyz^3, 3xy^2z^2)\big|_{(1,1,1)} = (1, 2, 3).$$

The unit vector $\mathbf{u}$ in the direction of $\mathbf{a} = (2, 1, -2)$ is

$$\mathbf{u} = \frac{\mathbf{a}}{|\mathbf{a}|} = \frac{(2, 1 - 2)}{\sqrt{4 + 1 + 4}} = \left(\frac{2}{3}, \frac{1}{3}, -\frac{2}{3}\right).$$

Then

$$D_\mathbf{u}f(1, 1, 1) = \text{grad } f(1, 1, 1) \cdot \mathbf{u}$$

$$= (1, 2, 3) \cdot \left(\frac{2}{3}, \frac{1}{3}, -\frac{2}{3}\right) = -\frac{2}{3}. \qquad \square$$

The equation

$$\boxed{f(x, y, z) = c}$$

defines (in general) a two-dimensional surface in three-dimensional space, called a *level surface* of f. For example, the level surfaces of

$$f(x, y, z) = x^2 + y^2 + z^2$$

are simply spheres about the origin. Level surfaces of other quadratic functions were discussed and pictured in the first section of Chapter 15. Although it is possible to draw pictures of such surfaces, it is hard to do, and we shall continue to argue without relying on pictures.

It can be shown that a level surface $S: f(\mathbf{x}) = c$ is smooth in the neighborhood of any regular point $\mathbf{x}_0$. This means that S contains smooth curves passing through $\mathbf{x}_0$ in more than one direction.

Now suppose that

$$\mathbf{x} = (x, y, z) = (g(t), h(t), k(t))$$

is any differentiable path lying in the level surface

$$f(x, y, z) = c.$$

Then $w = f(x, y, z)$ has the constant value c along this path, and its path derivative dw/dt is zero. That is,

$$\text{grad } f \cdot \frac{d\mathbf{x}}{dt} = 0,$$

by the chain rule. At a regular point $\mathbf{x}_0$ we can choose two different path directions $d\mathbf{x}/dt$, as noted above, and grad $f(\mathbf{x}_0)$ has the unique direction orthogonal to them both. We therefore have the space analogue of Theorem 1. First a definition:

DEFINITION A vector $\mathbf{n}$ is *perpendicular to a surface S at a point* $\mathbf{x}_0$, or *normal to S at* $\mathbf{x}_0$, if $\mathbf{n}$ is perpendicular to every differentiable curve lying in S and passing through $\mathbf{x}_0$.

In these terms, the discussion above shows that

THEOREM 5 *At every regular point* $\mathbf{x}_0 = (x_0, y_0, z_0)$ *on the level surface* $f(x, y, z) = k$, *the vector* grad $f(\mathbf{x}_0)$ *is perpendicular to the surface and is the only such vector (up to a scalar multiple).*

Example 2 Find a vector normal to the surface $xyz = 1$ at the point $(1, 2, 1/2)$.

Solution This is a level surface of $f(x, y, z) = xyz$ and grad f is a normal vector, by the theorem. Therefore,

$$\text{grad } f(1, 2, 1/2) = (yz, xz, xy)_{(1,2,1/2)} = (1, 1/2, 2)$$

is a normal vector. $\qquad\qquad\qquad\square$

Example 3 Find a parametric equation for the line normal to the surface

$$x^2 + y^2 + z^2 = 14$$

at the point $\mathbf{x}_0 = (1, 2, 3)$.

Solution This is the level surface

$$f(x, y, z) = 14,$$

where $f(x, y, z) = x^2 + y^2 + z^2$. Since grad $f(\mathbf{x}_0)$ is normal to the surface, it is parallel to the line we want. A vector parametric equation of the line is thus

$$\mathbf{x} - \mathbf{x}_0 = t \text{ grad } f(\mathbf{x}_0),$$

or

$$(x, y, z) - (1, 2, 3) = t(2x_0, 2y_0, 2z_0) = t(2, 4, 6).$$

This can be written in the form

$$(x, y, z) = s(1, 2, 3),$$

where $s = 2t + 1$. $\qquad\qquad\qquad\square$

Example 4   Show that the line normal to the surface

$$x^2 + y^2 + z^2 = k$$

at any point $\mathbf{x}_0 = (x_0, y_0, z_0)$ passes through the origin. (This is the general form of the special result in Example 3. We expect it to be true because this level surface is a sphere about the origin, and any line perpendicular to a sphere should be a diameter.)

Solution   If $f(x, y, z) = x^2 + y^2 + z^2$, then

$$\operatorname{grad} f(\mathbf{x}_0) = (2x, 2y, 2z)_{\mathbf{x}_0} = (2x_0, 2y_0, 2z_0) = 2\mathbf{x}_0.$$

The parametric equation for the line is thus

$$\mathbf{x} - \mathbf{x}_0 = t \operatorname{grad} f(\mathbf{x}_0) = 2t\mathbf{x}_0,$$

or

$$\mathbf{x} = (2t + 1)\mathbf{x}_0 = s\mathbf{x}_0,$$

where $s = 2t + 1$. The points $\mathbf{x}$ on the line are the scalar multiples of $\mathbf{x}_0$, and $\mathbf{x} = 0$ when $s = 0$.   □

Example 5   Suppose that the space path

$$\mathbf{x} = (x, y, z) = (g(t), h(t), k(t))$$

is perpendicular to the gradient of the function $F(x, y, z)$ at each path point. Show that the path lies entirely on a level surface of F.

Solution   We simply have to show that F is constant along the curve, that is, that the composite function

$$w = F(g(t), h(t), k(t))$$

is a constant function of t. By the chain rule,

$$\frac{dw}{dt} = \operatorname{grad} F \cdot \frac{d\mathbf{x}}{dt},$$

and this is everywhere 0, by the perpendicularity hypothesis. Thus $dw/dt = 0$, w is constant, say

$$w = F(g(t), h(t), k(t)) = k,$$

and the path lies on the level surface

$$F(x, y, z) = k.$$ □

Tangent Planes The following definition formalizes our intuitive notion of a tangent plane.

DEFINITION Let $f(x, y, z)$ be smooth, and let $\mathbf{x}_0$ be a point on the level surface $f(x, y, z) = k$ at which grad f is not zero. Then the *tangent plane* to the surface at $\mathbf{x}_0$ is the plane through $\mathbf{x}_0$ that is tangent at $\mathbf{x}_0$ to all the smooth paths lying in the surface and passing through $\mathbf{x}_0$.

According to Theorem 5, this is the plane containing $\mathbf{x}_0$ that is perpendicular to grad $f(\mathbf{x}_0)$. Its equation is therefore

$$\boxed{\operatorname{grad} f(\mathbf{x}_0) \cdot (\mathbf{x} - \mathbf{x}_0) = 0,}$$

or

$$\boxed{[D_1 f(\mathbf{x}_0)](x - x_0) + [D_2 f(\mathbf{x}_0)](y - y_0) + [D_3 f(\mathbf{x}_0)](z - z_0) = 0.}$$

Example 6 Find the plane tangent to the surface $x e^y \sin z = 1$ at the point $\mathbf{x}_0 = (2, 0, \pi/6)$.

Solution First verify that $(2, 0, \pi/6)$ is on the surface:

$$2 e^0 \sin\left(\frac{\pi}{6}\right) = 2 \cdot 1 \cdot \frac{1}{2} = 1.$$

Then compute grad $f(2, 0, \pi/6)$, where $f(x, y, z) = x e^y \sin z$:

$$\operatorname{grad} f(2, 0, \pi/6) = (e^y \sin z, x e^y \sin z, x e^y \cos z)_{(2.0, \pi/6)}$$
$$= (1/2, 1, \sqrt{3}).$$

Substituting these values in the above formula for the tangent plane gives

$$\operatorname{grad} f(\mathbf{x}_0) \cdot (\mathbf{x} - \mathbf{x}_0) = (1/2, 1, \sqrt{3}) \cdot (x - 2, y, z - \pi/6)$$
$$= 1/2(x - 2) + y + \sqrt{3}(z - \pi/2) = 0,$$

or

$$x + 2y + 2\sqrt{3}\, z = (2 + \sqrt{3}\pi).$$

This is the equation of the plane tangent to the surface $x e^y \sin z = 1$ at the point $(2, 0, \pi/6)$. □

Example 7 Find the plane tangent to the graph of the function $z = g(x, y)$ at (x_0, y_0).

Solution This graph is the level surface $g(x, y) - z = 0$, and the gradient of this special function

$$f(x, y, z) = g(x, y) - z$$

is

$$\text{grad } f(x, y, z) = (D_1 g(x, y), D_2 g(x, y), -1).$$

The dot-product formula

$$\text{grad } f(\mathbf{x}_0) \cdot (\mathbf{x} - \mathbf{x}_0) = 0$$

thus reduces in this case to

$$[D_1 g(x_0, y_0)](x - x_0) + [D_2 g(x_0, y_0)](y - y_0) - (z - z_0) = 0,$$

or

$$\boxed{z - z_0 = [D_1 g(x_0, y_0)](x - x_0) + [D_2 g(x_0, y_0)](y - y_0).}$$

$\square$

Example 8 Find the equation of the plane tangent to the surface $z = xy^2$ at the point over $(x_0, y_0) = (-3, 2)$.

Solution We compute the remaining coefficients for the above form of the tangent plane equation:

$$z_0 = f(-3, 2) = (-3)(2)^2 = -12,$$

$$D_1 f(-3, 2) = y^2 \big|_{(-3, 2)} = 2^2 = 4,$$

$$D_2 f(-3, 2) = 2xy \big|_{(-3, 2)} = 2(-3)(2) = -12.$$

Substituting these values in the general form above gives

$$z + 12 = 4(x + 3) - 12(y - 2),$$

or

$$-4x + 12y + z = 24,$$

as the equation of the tangent plane. $\square$

PROBLEMS FOR SECTION 3

Determine the gradient vector of each of the following functions at the given point.

1. $f(x, y, z) = \cos xy + \sin xz; P = (0, 2, -1)$

2. $f(x, y, z) = ax + by + cz; P = (x_0, y_0, z_0)$

3. $f(x, y, z) = x^2 + 3xy + y^2 + z^2; P = (1, 0, 2)$

4. $f(x, y, z) = x \cos y + y \cos z + z \cos x;$ $P = (\pi/2, \pi/2, 0)$

5. $f(x, y, z) = \ln(x^2 + y^2) + e^{yz}; P = (3, 4, 0)$

Find the derivative of each of the following functions at the given point in the direction of the given vector.

6. $g(x, y, z) = xyz; P = (1, 1/2, 2); \mathbf{a} = (3, -1, 2)$

7. $g(x, y, z) = x^2 + 2xy - y^2 + xz + z^2;$ $P = (2, 1, 1); \mathbf{a} = (1, 2, 2)$

8. $g(x, y, z) = x^2y + xye^z - 2xze^y; P = (1, 2, 0);$ $\mathbf{a} = (2, -3, 6)$

9. $g(x, y, z) = \ln\sqrt{x^2 + y^2 + z^2};$ $P = (1, 2, -1);$ $\mathbf{a} = (1, -2, 2)$

10. $g(x, y, z) = xe^{yz} + yze^x; P = (-4, 2, -2);$ $\mathbf{a} = (2, 2, -1)$

11. The directional derivative $D_u f(\mathbf{x}_0)$ has the value 3, 1, and -1 in the directions of the positive coordinate axes. Find grad $f(\mathbf{x}_0)$.

12. The directional derivative $D_u f(\mathbf{x}_0)$ has value 3, 1, and -1 in the directions of the vectors

$$(0, 1, 1), \quad (1, 0, 1), \quad \text{and} \quad (1, 1, 0),$$

respectively. Find grad $f(\mathbf{x}_0)$.

13. Find the direction of *maximum rate of change* of

$$f(x, y, z) = x^2 + xy - 2y^2 + 4xz + z^2$$

at the point $(2, 3, -2)$. What is this maximum rate of change? In what direction is the rate of change at f at this point *minimum*?

14. Find the maximum rate of change and the direction in which it occurs for

$$f(x, y, z) = \ln[(x + y)/(x + z)]$$

at the point $(1, 2, 1)$.

In Problems 15 through 19, find a vector normal to the level surface at the given point.

15. $-2x^2 - 6y^2 + 3z^2 = 4; (1, -1, -2)$

16. $x^2y - y^2z + z^2x = 33; (2, 1/2, 4)$

12. $xe^{xy} - ze^{yz} = 0; (-1, 2, -1)$

18. $\sin(x^2yz) = 1/2; (-1, 1/2, \pi/2)$

19. a) $\ln xyz = 0; (1/2, 1/3, 6)$

b) $xyz = 1; (1/2, 1/3, 6)$

Find a parametric equation for the line normal to the surface at the given point.

20. $z = xy; (2, -1, -2)$

21. $z = e^x \sin y; (1, \pi/2, e)$

22. $z = x^2y^2; (-2, 2, 16)$

23. $xy + yz + xz = 1; (2, 3, -1)$

24. $\ln\sqrt{x^2 + y^2 - z^2} = 0; (1, -1, 1)$

Find the plane tangent to the given surface at the given point.

25. $-2x^2 - 6y^2 + 3z^2 = 4; (1, -1, 2)$

26. $x^2y - y^2z + z^2x = 33; (2, 1/2, 4)$

27. $xe^{xy} - ze^{yz} = 0; (-1, 2, -1)$

28. $z = e^x \sin y; (1, \pi/2, e)$

29. $xy + yz + zx = 11, (1, 2, 3)$

30. $xyz = x + y + z, (1/2, 1/3, -1)$

31. $z = \ln\sqrt{x^2 + y^2}, (-3, 4, \ln 5)$

32. $z = e^x \sin y, (1, \pi/2, e)$

33. $z = xy, (2, -1, -2)$

34. $z = 3x^2 + 2y^2 - 11, (2, 1, 3)$

35. $x^2 + y^2 + z^2 = 6; (-1, 2, 1)$

36. $x^2 + 2y^2 + 5z^2 = 6; \left(2, \dfrac{1}{2}, -\dfrac{1}{2}\right)$

37. $x^2 + 2y^2 - z^2 = 2; (1, 1, 1), (1, 1, -1)$

39. $\ln \sqrt{x^2 + y^2 - z^2} = 0; (1, -1, 1)$

38. $x^2 + 2y^2 - z^2 = -1; (1, 1, 2)$

40. Find the point(s) on the level surface

$$\frac{x^2}{9} + \frac{y^2}{4} + z^2 = 1$$

at which the tangent plane is perpendicular to $(1, 1, \sqrt{3})$.

41. The surface

$$z = -(x^{2/3} + y^{2/3})$$

has a "spike" at the origin and has no tangent plane there. Given any nonvertical plane

$$z = ax + by$$

through the origin; show that there is a circle $x^2 + y^2 = r^2$ within which the plane lies above the surface. (That is, any nonvertical plane can be supported on the point of a vertical pin.)

43. Show that the ellipsoid

$$x^2 + y^2 + 2z^2 = C$$

and the paraboloid

$$z = k(x^2 + y^2)$$

intersect orthogonally.

42. Show that the sphere

$$x^2 + y^2 + z^2 = r^2$$

and the cone

$$z^2 = x^2 + y^2$$

intersect orthogonally.

44. Show that the equation of the tangent plane to the ellipsoid

$$\frac{x^2}{a^2} + \frac{y^2}{b^2} + \frac{z^2}{c^2} = 1$$

at the point (x_0, y_0, z_0) can be written

$$\frac{xx_0}{a^2} + \frac{yy_0}{b^2} + \frac{zz_0}{c^2} = 1.$$

45. Prove the analogous result for the hyperboloid

$$\frac{x^2}{a^2} + \frac{y^2}{b^2} - \frac{z^2}{c^2} = k,$$

$(k \neq 0)$.

47. A tangent plane to the surface

$$x^{1/2} + y^{1/2} + z^{1/2} = 1$$

has intercepts a, b, and c on the axes. Show that $a + b + c$ is independent of the point of tangency.

49. Write out the proof of Theorem 4.

46. Prove a similar result for the paraboloid

$$z = \frac{x^2}{a^2} + \frac{y^2}{b^2}.$$

48. A tangent plane to the surface

$$x^{2/3} + y^{2/3} + z^{2/3} = 1$$

has intercepts a, b, and c on the axes. Show that $a + b + c$ is independent of the point of tangency.

4
MAXIMUM–
MINIMUM
PROBLEMS

We shall take up two methods of solving maximum–minimum problems, one that uses vectors and one that does not. We start off without vectors.

Suppose that the function $z = f(x, y)$ has a maximum (or minimum) value at a point (x_0, y_0) in the interior of its domain. If we hold y constant at the value y_0, then $f(x, y_0)$ is a function of the one variable x having its maximum value at $x = x_0$, and so its derivative must be zero there, as in Chapter 5. That is, $\partial z/\partial x = 0$ at the point $(x, y) = (x_0, y_0)$. Similarly, $\partial z/\partial y = 0$ at this point. We have called such a point (x_0, y_0) where both partial derivatives are zero a *critical point* of f. Thus,

THEOREM 6 *If the function f assumes a maximum or minimum value at a point (x_0, y_0) in the interior of its domain, then (x_0, y_0) is necessarily a critical point of f.*

The equations

$$\frac{\partial z}{\partial x} = 0, \qquad \frac{\partial z}{\partial y} = 0$$

are two equations in two unknowns that are satisfied by the maximum point (x_0, y_0), and we may be able to solve them simultaneously and so determine (x_0, y_0).

Example 1 Find the dimensions of the rectangular box with open top and one-cubic-foot capacity that has the smallest surface area.

Solution The box is shown in Fig. 1. Its volume is xyz and this is required to be 1:

$$xyz = 1.$$

The total surface area is

$$A = xy + 2xz + 2yz.$$

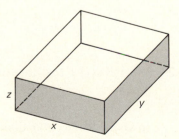

Figure 1

We can eliminate one variable, say z, by solving for z in the first equation and substituting in the second, giving $z = 1/xy$ and

$$A = xy + \frac{2}{y} + \frac{2}{x}.$$

Now A is a function of x and y. Its domain is the first quadrant, because x and y have to be positive, but no other limitation is placed on their values. We assume that A has a minimum value. It has to be at a critical point, that is, a point where

$$\frac{\partial A}{\partial x} = y - \frac{2}{x^2} = 0,$$

$$\frac{\partial A}{\partial y} = x - \frac{2}{y^2} = 0,$$

by Theorem 6. So we solve these equations simultaneously. First,

$$yx^2 = 2,$$

$$xy^2 = 2.$$

Dividing gives $x/y = 1$ and $x = y$. Therefore, $x^3 = 2$ and $x = y = 2^{1/3}$. Then

$$z = \frac{1}{xy} = 2^{-2/3}.$$

The *proportions* of the box are given by

$$x = y = 2z.$$

The box has a square bottom, with edge length twice the height of the box.   □

Example 2   If we restrict the domain of

$$z = f(x, y) = y^3 + xy - x^3$$

to the unit circular disk $x^2 + y^2 \leq 1$, does f have a maximum value at an interior point?

Solution An interior maximum point (x, y) has to be a critical point of f. That is,

$$\frac{\partial z}{\partial x} = y - 3x^2 = 0,$$

$$\frac{\partial z}{\partial y} = 3y^2 + x = 0.$$

at such a point. We can solve these equations simultaneously by substituting from the first into the second, to get

$$27x^4 + x = 0.$$

Factoring this equation,

$$x(27x^3 + 1) = 0,$$

shows its solutions to be $x = 0$, $x = -1/3$. The corresponding y values, obtained from the critical point equations, are $y = 0$ and $y = 1/3$, respectively. The critical points are thus $(0, 0)$ and $(-1/3, 1/3)$, with f values

$$f(0, 0) = 0,$$

$$f\left(-\frac{1}{3}, \frac{1}{3}\right) = -\frac{1}{27}.$$

However,

$$f(0, 1) = 1,$$

so neither critical point yields a maximum value for f. Therefore f does not assume a maximum value at an interior point of its restricted domain. □

This example can be pursued further. We call a domain *closed* if it includes its boundary and *bounded* if it lies entirely inside some circle. The restricted domain in the above example is both closed and bounded. We shall now assume the following generalization of the extreme-value principle:

If a function f is continuous on a bounded, closed domain D, then f assumes a maximum value and a minimum value on D.

Example 3 It follows from this maximum principle that the function f of Example 2 has a maximum value somewhere on its domain $x^2 + y^2 \leq 1$. We saw there that the maximum value could not be at an interior point, and it therefore must be somewhere on the boundary circle $x^2 + y^2 = 1$. We thus want the maximum of $z = y^3 + xy - x^3$ subject to the auxiliary condition $x^2 + y^2 = 1$. But this is now a one-variable problem. We can either solve for y in $x^2 + y^2 = 1$ and substitute in the equation for z, or we can use the method of auxiliary variables. This remaining one-variable problem is rather complicated, and we shall not go any further with it. Instead, we shall work out a simpler problem of the same type.   □

Example 4   Find the maximum value of

$$z = x^2 + 4x^2y^2 - y^2$$

on the closed circular disk $x^2 + y^2 \le 1$.

Solution   The maximum value occurs either at an interior point or at a boundary point. The only possible interior location is a critical point, determined by the critical point equations

$$\frac{\partial z}{\partial x} = 2x + 8xy^2 = 0,$$

$$\frac{\partial z}{\partial y} = 8x^2y - 2y = 0.$$

These equations reduce to

$$x(1 + 4y^2) = 0,$$
$$y(4x^2 - 1) = 0.$$

The first shows that $x = 0$, and then the second shows that $y = 0$. Thus the only critical point is the origin $(x, y) = (0, 0)$. The value of z there is zero, and this is not the maximum value of z since we see by inspection that $z = 1$ at $(1, 0)$.

The maximum therefore occurs on the boundary curve $x^2 + y^2 = 1$, where the problem reduces to a one-variable problem. In this example, the reduction is very simple. On the boundary $y^2 = 1 - x^2$, so

$$z = x^2 + 4x^2y^2 - y^2 = x^2 + (1 - x^2)(4x^2 - 1)$$
$$= 6x^2 - 4x^4 - 1.$$

This is over the x-interval $[-1, 1]$. We saw in Chapter 5 that the maximum value of z on this interval must occur at an endpoint or at a point where $dz/dx = 0$. At the two endpoints

$z = 1$. On the other hand, the equation

$$\frac{dz}{dx} = 12x - 16x^3 = 0$$

has the roots $x = 0$, $\pm\sqrt{3}/2$, and the corresponding values of z are $z = -1$ and $z = 6(3/4) - 4(3/4)^2 - 1 = 5/4$. This second value is larger than the common endpoint value, so z has its maximum value $5/4$ at the boundary points determined by $x = \pm\sqrt{3}/2$. At these points $y^2 = 1 - x^2 = 1/4$ and $y = \pm 1/2$. There are thus four maximum points, symmetric with respect to the two axes. The one in the first quadrant is $(\sqrt{3}/2, 1/2)$. $\square$

We turn now to a vector method of solving maximum–minimum problems.

The Lagrange-Multiplier Method

We often want to find the maximum value of a function $f(\mathbf{x}) = f(x, y, z)$ on a level surface S defined by $g(x, y, z) = c$. That is, we want to maximize $f(\mathbf{x})$ subject to the "constraint" $g(\mathbf{x}) = c$. The following fact can help us make the calculation.

THEOREM 7   *Suppose that $f(\mathbf{x}) = f(x, y, z)$ has a maximum value on the level surface S: $g(\mathbf{x}) = c$ at a point $\mathbf{x}_0$, and suppose that $\mathbf{x}_0$ is not a critical point of g. Then $\operatorname{grad} f(\mathbf{x}_0)$ and $\operatorname{grad} g(\mathbf{x}_0)$ are necessarily parallel. That is,*

$$\boxed{\operatorname{grad} f(\mathbf{x}_0) = k \operatorname{grad} g(\mathbf{x}_0)}$$

for some constant k.

Proof   Let $\mathbf{x} = \mathbf{p}(t)$ be any path lying in S and passing through $\mathbf{x}_0$: $\mathbf{x}_0 = \mathbf{p}(t_0)$. Then $f(\mathbf{p}(t))$ has its maximum value at t_0 and its derivative must be zero there. So, for every such path,

$$0 = \frac{d}{dt} f(\mathbf{p}(t)) = \operatorname{grad} f(\mathbf{x}_0) \cdot \mathbf{p}'(t_0),$$

by the chain rule. Thus $\operatorname{grad} f(\mathbf{x}_0)$ is normal to S at $\mathbf{x}_0$, and hence is of the form $k \operatorname{grad} g(\mathbf{x}_0)$. (See page 679.) ∎

According to the above analysis, the point $\mathbf{x}_0$ where $f(\mathbf{x})$ has its maximum value on the level surface $g(\mathbf{x}) = c$ satisfies the equations

$$g(\mathbf{x}_0) = c,$$
$$\operatorname{grad}(f - kg)(\mathbf{x}_0) = 0,$$

for some constant k. The second equation is a vector equation that is equivalent to three scalar equations, so altogether we have four equations in the four unkowns

$$x_0, \qquad y_0, \qquad z_0, \qquad k.$$

We may be able to solve them to determine the maximum point $\mathbf{x}_0$. This is the method of *Lagrange multipliers*. (The constant k is called the Lagrange multiplier.)

Example 5   Find the maximum value of $f(x) = x + 2y - 2z$ on the sphere $x^2 + y^2 + z^2 = 4$, using the Lagrange multiplier k.

Solution   We suppose that f *has* a maximum value on the sphere, at a point $\mathbf{x}_0$. Then the Lagrange equations for $\mathbf{x}_0$ and the multiplier k are

$$x_0^2 + y_0^2 + z_0^2 = 4,$$
$$(1, 2, -2) - k(2x_0, 2y_0, 2z_0) = 0.$$

The vector equation is equivalent to the three scalar equations

$$2kx_0 = 1, \qquad 2ky_0 = 2, \qquad 2kz_0 = -2,$$

and substituting from them into the first equation gives

$$\frac{1}{4k^2} + \frac{1}{k^2} + \frac{1}{k^2} = 4,$$

Solving this for k gives $k = \pm\frac{3}{4}$. The vector equation then can be solved for $\mathbf{x}_0$:

$$\mathbf{x}_0 = (x_0, y_0, z_0) = \frac{1}{2k}(1, 2, -2) = \pm\frac{2}{3}(1, 2, -2).$$

Finally,

$$f(\mathbf{x}_0) = x_0 + 2y_0 - 2z_0$$
$$= \pm\frac{2}{3}(1 + 4 + 4) = \pm 6.$$

So the maximum value of f on the sphere is 6.   □

Example 6 Redo Example 1 by this method.

Solution We want the maximum value of

$$A = xy + 2xz + 2yz$$

subject to the constraint

$$V = xyz = 1.$$

Here the multiplier equation is grad $A = k$ grad V, or (omitting the subscripts)

$$(y + 2z, x + 2z, 2x + 2y) = k(yz, xz, xy).$$

This vector equation, together with the constraint equation $xyz = 1$, gives four scalar equations in the four unknowns x, y, z, k.

If we equate first components and multiply by x, we get

$$xy + 2xz = kxyz = k \quad \text{(since } xyz = 1\text{)}.$$

In this way, the three component equations become

$$xy + 2xz = k, \qquad xy + 2yz = k, \quad 2xz + 2yz = k.$$

Subtracting these equations in pairs gives

$$x = y = 2z.$$

Then

$$1 = xyz = 4z^3.$$

Thus $z = 2^{-2/3}$, $x = y = 2^{1/3}$, as before. □

The same method works for functions of any number of variables, and it generalizes to situations involving two or more constraints. Some linear algebra is then required, so the more general treatment has to be put off to a later course. For a function of two variables the Lagrange-multiplier method is equivalent to our earlier method of auxiliary variables.

Example 7 Consider again the problem of finding the most efficient proportions for a container in the shape of a circular cylinder (Example 6, Section 5, Chapter 5). We want to minimize the total surface area

$$A = 2\pi r^2 + 2\pi rh,$$

subject to the constraint of constant volume, say

$$V = \pi r^2 h = 1.$$

The Lagrange-multiplier equation grad $A = k$ grad V is then

$$2\pi(2r + h, r) = k\pi(2rh, r^2),$$

or, dividing by π,

$$2(2r + h, r) = kr(2h, r).$$

This vector equation by itself determines the proportions of the container, for it requires that $2 = kr$ (from equating the second components), and then that $2r + h = 2h$, or $h = 2r$ (from equating the first components). The constraint equation $V = 1$ can then be used to obtain the particular values of r and h.   □

PROBLEMS FOR SECTION 4

Each of the following functions is defined on the whole plane and has a minimum or maximum value. Find this extreme value and the point (or points) where it occurs.

1. $z = x^2 + 2xy + 2y^2 - 6y$

2. $z = 4x + 6y - x^2 - y^2$

3. $z = x^4 - x^2y^2 + y^4 + 4x^2 - 6y^2$

4. $z = x^4 + 2xy^2 + y^4$

5. $z = 3x^4 - 4x^3y + y^6$

Let D be the first quadrant. Each of the following functions has a maximum or a minimum on D. Find this extreme value and the point where it occurs.

6. $z = xy + \dfrac{1}{x} + \dfrac{1}{y}$

7. $z = 8y^3 + x^3 - 3xy$

8. $z = y + 2x - \ln xy$

9. Find the minimum distance between the parabola $y = x^2$ and the line $y = x - 1$.

10. A rectangular box with capacity one cubic foot is to be made with an open top. The material for the bottom costs half again as much as the material for the sides. Find the dimensions of the least expensive box.

11. Find the maximum and minimum values of

$$f(x, y) = x^2 + y^2 - xy - x$$

on the unit square $-1 \le x \le 1, -1 \le y \le 1$.

12. Find the maximum and minimum values of

$$f(x, y) = 2x^4 - 3x^2y^2 + 2y^4 - x^2$$

on the closed unit disk $x^2 + y^2 \le 1$.

13. Find the minimum distance between the path $(x, y) = (t, t^2 + 1)$ and the path $(x, y) = (-s - 1, 2s)$.

14. The temperature in degrees Fahrenheit at each point (x, y) in the region $0 \le x \le 1, 0 \le y \le 1$ is given by $T = 48xy - 32x^3 - 24y^2$. Find the points of maximum and minimum temperature and the temperature at each of these points.

15. Find positive numbers x, y, and z such that $x + 3y + 2z = 18$ and xyz is a maximum.

16. Find the point on the plane $x + 2y + 3z = 4$ closest to the origin.

17. Find the point on the surface $xyz = 4$ closest to $(1, 0, 0)$.

18. Find the maximum value of xyz when the point (x, y, z) lies on the plane $2x + y + 3z = 10$.

19. Find the maximum volume of a rectangular box that is inscribed in the ellipsoid

$$x^2 + \frac{y^2}{4} + \frac{z^2}{9} = 1.$$

20. Find the maximum volume of a rectangular box inscribed in the ellipsoid

$$\frac{x^2}{a^2} + \frac{y^2}{b^2} + \frac{z^2}{c^2} = 1.$$

21. Find the minimum value of $f(x, y, z) = x^2 + 3y^2 + 2z^2$ on the plane $x + y - z = 1$.

22. Find the maximum value of $f(x, y, z) = x + y + z$ on the ellipsoid

$$\frac{x^2}{4} + y^2 + \frac{z^2}{9} = 1.$$

23. Find the maximum and minimum values of $f(x, y) = x^2 + x + 2y^2$ on the unit circle $x^2 + y^2 = 1$.

24. Find the maximum and minimum values of $f(x, y) = x^2 + xy - y^2$ on the unit circle $x^2 + y^2 = 1$.

25. Find the point on the ellipse

$$\frac{x^2}{4} + y^2 = 1$$

that is closest to the line $y - x = 4$.

26. Find the point on the hyperbola $x^2 - y^2 = 1$ that is closest to the line $2x + y = 0$.

27. Find the ellipse

$$\frac{x^2}{a^2} + \frac{y^2}{b^2} = 1$$

that passes through the point $(1, 4)$ and has minimum area. (The area of the above ellipse is πab.)

29. Find the rectangle of maximum perimeter (and with sides parallel to the coordinate axes) that can be inscribed in the ellipse $x^2 + 2y^2 = 1$.

23. Find the ellipsoid

$$\frac{x^2}{a^2} + \frac{y^2}{b^2} + \frac{z^2}{c^2} = 1$$

that passes through the point $(2, 1, 3)$ and has minimum volume. (The volume of the above ellipsoid is $\frac{4}{3}\pi abc$.)

30. Find the rectangle of maximum area (and having sides parallel to the coordinate axes) that can be inscribed in the ellipse $x^2 + 2y^2 = 1$.

31. Show that the rectangular box of maximum total surface area that can be inscribed in a sphere is a cube.

32. Find the maximum and minimum values of
$$f(x, y, z) = x - 2y + 5z$$
on the sphere $x^2 + y^2 + z^2 = 30$.

33. Find the minimum value of $x^2 + y^2 + z^2$ on the plane $x - 2y + 5z = 1$.

34. Find the point of the plane $2x - 3y - z = 10$ closest to the origin.

35. Find the maximum and minimum values of $f(x, y, z) = (8x - y + 27z)/2$ on the surface
$$x^4 + y^4 + z^4 = 1.$$

36. Given positive numbers $a_1, a_2, \ldots, a_n$, find the numbers $x_1, \ldots, x_n$ such that
$$\sum_1^n a_i x_i = 1$$
and $\sum_1^n x_i^2$ is as small as possible.

37. Given positive numbers $a_1, a_2, \ldots, a_n$, find the numbers $x_1, x_2, \ldots, x_n$ such that
$$\sum_1^n x_i^2 = 1$$
and $\sum_1^n a_i x_i$ is as large as possible.

38. Find the maximum volume of a rectangular box having three sides on the coordinate planes and one vertex on the plane
$$\frac{x}{a} + \frac{y}{b} + \frac{z}{c} = 1,$$
where a, b, and c are positive.

**5
HIGHER-ORDER
PARTIAL
DERIVATIVES;
RELATIVE
EXTREMA**

Another new situation that arises in the study of partial derivatives is the occurrence of *mixed* higher-order derivatives, like $\partial^2 z/(\partial x \, \partial y)$. The notation means what one would expect

$$\frac{\partial^2 z}{\partial y \, \partial x} = \frac{\partial}{\partial y}\left(\frac{\partial z}{\partial x}\right),$$

$$\frac{\partial^2 z}{\partial x \, \partial y} = \frac{\partial}{\partial x}\left(\frac{\partial z}{\partial y}\right).$$

For example, if

$$z = x^2 y^3 + x^4 y,$$

then

$$\frac{\partial z}{\partial x} = 2xy^3 + 4x^3 y,$$

$$\frac{\partial^2 z}{\partial y \, \partial x} = \frac{\partial}{\partial y}\left(\frac{\partial z}{\partial x}\right) = 6xy^2 + 4x^3,$$

$$\frac{\partial z}{\partial y} = 3x^2 y^2 + x^4,$$

$$\frac{\partial^2 z}{\partial x \, \partial y} = \frac{\partial}{\partial x}\left(\frac{\partial z}{\partial y}\right) = 6xy^2 + 4x^3.$$

Note in the above example that

$$\frac{\partial^2 z}{\partial x\, \partial y} = \frac{\partial^2 z}{\partial y\, \partial x}.$$

Such equality is not accidental, for this "commutative" law is a theorem.

THEOREM 8   *If $f(x, y)$ has continuous partial derivatives of the first and second order, then*

$$\boxed{D_1 D_2 f(x, y) = D_2 D_1 f(x, y).}$$

Proof   We shall show that both mixed partials are approximately equal to the "second-order difference quotient"

(*) $$\frac{f(x + h, y + k) - f(x + h, y) - f(x, y + k) + f(x, y)}{hk}.$$

Consider first the function

$$\phi(x) = f(x, y + k) - f(x, y),$$

where y and $(y + k)$ are both held fixed. The numerator in the above expression (*) is then $\phi(x + h) - \phi(x)$, and hence can be written

$$\phi(x + h) - \phi(x) = h\phi'(X) = h[D_1 f(X, y + k) - D_1 f(X, y)]$$

by the mean-value theorem. Here, X is some number lying between x and $(x + h)$. Since $D_1 f(X, y)$ is differentiable as a function of y, we can apply the mean-value theorem once more:

$$h[D_1 f(X, y + k) - D_1 f(X, y)] = hk D_2 D_1 f(X, Y),$$

where Y lies between y and $y + k$. Thus the double difference quotient (*) has the values $D_2 D_1 f(X, Y)$.

We now start over again with the function

$$\psi(y) = f(x + h, y) - f(x, y),$$

where now x and $x + h$ are held fixed. The numerator in (*) is then $\psi(y + k) - \psi(y)$, and proceeding exactly as before, through two applications of the mean-value theorem, we find this time that the double difference quotient (*) has the value $D_1D_2f(X', Y')$, where Y' lies between y and $y + k$ and X' lies between x and $x + h$.

Altogether we have evaluated (*) in two different ways. In particular, we have proved that there are numbers X and X' both lying between x and $x + h$, and numbers Y and Y' both lying between y and $y + k$, such that

$$D_2D_1f(X, Y) = D_1D_2f(X', Y').$$

Now, let h and k both approach zero. Then X and X' both approach x, and Y and Y' both approach y, and since the partial derivatives D_2D_1f and D_1D_2f are continuous, the above equation becomes

$$D_2D_1f(x, y) = D_1D_2f(x, y)$$

in the limit. This proves the theorem. ∎

REMARK Theorem 8 holds equally well for functions of more than two variables. For example if $z = f(u, v, w, x, y)$, and f has continuous partial derivatives of the first and second order, then

$$\frac{\partial^2 z}{\partial u\, \partial w} = \frac{\partial^2 z}{\partial w\, \partial u}, \qquad \frac{\partial^2 z}{\partial v\, \partial y} = \frac{\partial^2 z}{\partial y\, \partial v}, \qquad \text{etc.}$$

The reason is that all variables but two are held fixed when any one of these mixed partials is investigated, so Theorem 11 applies to the remaining function of two variables.

Relative Maxima and Minima; Saddle Points

In Chapter 5 we saw that if x_0 is a critical point of f at which the second derivative $f''(x_0)$ exists and is nonzero, then f has a relative maximum or minimum at x_0, depending on the sign of $f''(x_0)$. On the other hand, if $f''(x_0) = 0$, then nothing can be concluded.

A similar result holds for a function $z = f(x, y)$ of two variables, the requirement that $f''(x_0) \neq 0$ being replaced by $D(x_0, y_0) \neq 0$, where the "discriminant" function $D(x, y)$ is defined by

$$D(x, y) = \frac{\partial^2 z}{\partial x^2} \frac{\partial^2 z}{\partial y^2} - \left(\frac{\partial^2 z}{\partial x\, \partial y} \right)^2.$$

However, a new possibility occurs now, namely, that f may have a maximum at (x_0, y_0) along one line through this point, and a minimum at (x_0, y_0) along a different line. The resulting configuration

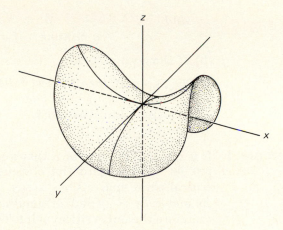

Figure 1

of the graph of f around (x_0, y_0) is called a *saddle point*, a term that Fig. 1 shows to be appropriate.

In this situation, the level set through $\mathbf{x}_0 = (x_0, y_0)$ is a pair of intersecting curves, dividing the neighborhood of $\mathbf{x}_0$ into four quadrant-like pieces on which $f(x, y) - f(x_0, y_0)$ is alternately positive and negative (Fig. 2).

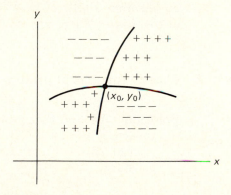

Figure 2

It is understood in the theorem below that f has continuous second partial derivatives.

THEOREM 9   *Let (x_0, y_0) be a critical point of $z = f(x, y)$ at which the discriminant $D(x, y)$ is not zero. Then:*

    a) *If $D(x_0, y_0) > 0$, then f has a relative maximum or a relative minimum at (x_0, y_0),*

$$\text{a maximum if } \partial^2 z/\partial x^2 < 0 \text{ there,}$$

and

$$a\ minimum\ if\ \partial^2 z/dx^2 > 0;$$

b) *If* $D(x_0, y_0) < 0$, *then f has a saddle point at* (x_0, y_0).

If $D(x_0, y_0) = 0$, *then the test fails.*

We shall omit the proof.

Example 1   It is clear on the face of it that $z = x^2 + y^2$ is minimum at the origin, but let us disregard the obvious and see what the machinery of critical-point analysis has to say. The first partial derivatives $\partial z/\partial x = 2x$ and $\partial z/\partial y = 2y$ vanish simultaneously only at the origin, which is therefore the only critical point. The second partial derivatives are all constants, with the values

$$\frac{\partial^2 z}{\partial x^2} = 2, \qquad \frac{\partial^2 z}{\partial y^2} = 2, \qquad \frac{\partial^2 z}{\partial x\,\partial y} = 0.$$

The discriminant thus has the constant value

$$2 \cdot 2 - 0^2 = 4.$$

So Theorem 9 tells us that $f(x, y) = x^2 + y^2$ has a relative minimum at the origin. The graph of f is a paraboloid of revolution. (See page 635.)   □

Example 2   Classify the critical point or points of $z = x^2 - y^2$.

Solution   We find that $\partial z/\partial x = 2x$ and $\partial z/\partial y = -2y$ are simultaneously zero only at the origin, which is thus the only critical point. The second partial derivatives are again all constant,

$$\frac{\partial^2 z}{\partial x^2} = 2, \qquad \frac{\partial^2 z}{\partial y^2} = -2, \qquad \frac{\partial^2 z}{\partial x\,\partial y} = 0,$$

and the discriminant $D(x, y)$ has the constant value

$$\frac{\partial^2 z}{\partial x^2}\frac{\partial^2 z}{\partial y^2} - \left(\frac{\partial^2 z}{\partial x\,\partial y}\right)^2 = 2(-2) - 0^2 = -4.$$

This time the origin is a saddle point by Theorem 9. The graph is a hyperbolic paraboloid. (See page 635.)   □

Example 3 Study the critical points of the function $z = x^3y + 3x + y$.

Solution The equations

$$\frac{\partial z}{\partial x} = 3x^2y + 3 = 0,$$

$$\frac{\partial z}{\partial y} = x^3 + 1 = 0$$

determine the solutions: $x = -1$, from the second equation, and then $y = -1$ from the first. Therefore $(-1, -1)$ is the only critical point. The discriminant

$$\frac{\partial^2 z}{\partial x^2} \frac{\partial^2 z}{\partial y^2} - \left(\frac{\partial^2 z}{\partial x \, \partial y} \right)^2 = (6xy) \cdot 0 - (3x^2)^2 = -9x^4$$

is negative at $(-1, -1)$, so this point is a saddle point, by Theorem 9. $\square$

PROBLEMS FOR SECTION 5

Find *all* of the second partial derivatives of the following functions.

1. $z = 3x^2 - 2xy + y$

2. $z = x \cos y - y \cos x$

3. $z = e^{x^2y}$

4. $z = \sin(2x^2 + 3y)$

5. $z = x^2 + 3xy + y^2$

6. $z = y \ln x$

7. $z = xy$

8. $z = \sin 3x \cos 4y$

9. $z = \dfrac{x}{y^2} - \dfrac{y}{x^2}$

10. $z = \arctan y/x$

11. Let f be a function of one variable and set

$$z = f(x - ct).$$

Show that

$$\frac{\partial^2 z}{\partial t^2} = c^2 \frac{\partial^2 z}{\partial x^2}.$$

12. Show, more generally, that

$$z = f(x - ct) + g(x + ct)$$

satisfies the above partial differential equation.

13. If $z = \ln(x^2 + y^2)$, show that

$$\frac{\partial^2 z}{\partial x^2} + \frac{\partial^2 z}{\partial y^2} = 0.$$

14. If $z = \arctan \dfrac{y}{x}$, show that

$$\frac{\partial^2 z}{\partial x^2} + \frac{\partial^2 z}{\partial y^2} = 0.$$

15. If $z = e^x \sin y$, show that

$$\frac{\partial^2 z}{\partial x^2} + \frac{\partial^2 z}{\partial y^2} = 0.$$

16. If $w = (x^2 + y^2 + z^2)^{-1/2}$, prove that

$$\frac{\partial^2 w}{\partial x^2} + \frac{\partial^2 w}{\partial y^2} + \frac{\partial^2 w}{\partial z^2} = 0.$$

17. Show that if $z = f(x, y)$, then

$$\frac{\partial^2 z}{\partial x \, \partial y \, \partial x} = \frac{\partial^3 z}{\partial y \, \partial x^2}.$$

18. Show that if $z = e^{x^2 - y^2} \sin(2xy)$, then

$$\frac{\partial^2 z}{\partial x^2} + \frac{\partial^2 z}{\partial y^2} = 0.$$

19. If $z = \dfrac{e^{-x^2/4y}}{\sqrt{y}}$, show that

$$\frac{\partial z}{\partial y} = \frac{\partial^2 z}{\partial x^2}.$$

20. If $z = x^4 - 6x^2 y^2 + y^4$, show that

$$\frac{\partial^2 z}{\partial x^2} + \frac{\partial^2 z}{\partial y^2} = 0.$$

Classify the critical points of each of the following functions of x and y.

21. $x^2 + 2y^2 + 2x - y$

22. $x - y + xy$

23. $x^3 - xy + y^2$

24. $x^2 + 3xy + y^2$

25. xy (Draw a few level curves.)

26. $x^2 y^2 + 2x - 2y$

27. $x^2 y + y^2 + x$

28. $x^3 - 3xy + y^3$

29. $x^2 - xy + y^3 - x$

30. $x^4 + y^4$

31. $x^4 + y^4 + 4x - 4y$

32. $x^4 - y^4 + 4x - 4y$

33. $x^2 + y^2 + \dfrac{2}{xy}$

34. $\dfrac{1}{x^2} + \dfrac{4}{y^2} + xy$

35. $e^x + e^y - e^{x+y}$

36. $x \sin y$

37. Prove the following corollary of (part of) Theorem 9.

> **Theorem** *Suppose that the discriminant of $f(x, y)$ is positive at the point (x_0, y_0), and that the second partial derivatives $\partial^2 z/\partial x^2$ and $\partial^2 z/\partial y^2$ are both positive there. Prove that, for points (x, y) sufficiently close to (x_0, y_0), the graph of $z = f(x, y)$ lies entirely above its tangent plane at (x_0, y_0).*

6 WHAT ABOUT HIGHER DIMENSIONS?

We have seen that visualizing and representing equation graphs in three dimensions is much more complicated than in two dimensions, and the corresponding problems for equations in four or more variables are hopeless. It is natural to ask how one can go about trying to understand the behavior of functions of three or more variables.

What we do is to proceed more or less by analogy with the two- and three-dimensional cases, keeping in mind both their similarities and their differences. For example, we consider the set of all quadruples of numbers such as $(1, 3, -4, 2)$ as forming a four-dimensional coordinate space, and we consider the graph of an equation in four variables as forming a three-dimensional "surface" in this 4-space. The graph of a function of three variables $f(x, y, z)$ is the graph of the equation

$$w = f(x, y, z).$$

We view the three-dimensional domain D of f as lying in the three-dimensional coordinate "plane" consisting of all points of the form $(x, y, z, 0)$. Then the graph of f is a three-dimensional "surface" spread out "over" the domain D. If (a, b, c) is in the domain of f, we picture it as the point

$$(a, b, c, 0) \qquad \text{in 4 space}$$

and the graph point over it is

$$(a, b, c, f(a, b, c)).$$

In the same way we view the graph of a function of n variables as forming an n-dimensional "surface" in $(n + 1)$-dimensional coordinate space. We thus discuss higher-dimensional situations in geometric language suggested by our experience in two and three dimensions, and we still reason geometrically, but in a looser, less concrete way.

As our geometric grip loosens, and we reason more by geometric analogy than by direct geometric visualization, we have to bolster our weaker intuition by paying more attention to its algebraic and analytic counterparts. It is here that the theory of vector spaces comes to our aid. The algebraic study of vector spaces gives us very sharp and clear notions of planes and lines and perpendicularity in higher dimensions, and we can then use these flat objects to approximate curving configurations near points of tangency in almost exactly the same way that a tangent line approximates a curve and a tangent plane approximates a surface.

We took a first look at vectors in Chapter 13, but a whole course in linear algebra is needed before one can feel confident about the geometry of $(n + 1)$-dimensional space and the behavior of functions of n variables.

CHAPTER 17
DOUBLE AND
TRIPLE INTEGRALS

A continuous function of two variables $f(x, y)$ can be integrated over a plane region R in much the same way that a continuous function of one variable is integrated over an interval. The number we get is called the *double integral* of f over R, and is designated

$$\iint_R f(x, y) \, dx \, dy.$$

Similarly, a continuous function of three variables $f(x, y, z)$, defined over a finite region G in space, has a *triple integral*

$$\iiint_G f(x, y, z) \, dx \, dy \, dz,$$

and a continuous function of n variables $f(x_1, \ldots, x_n)$ has an n-tuple integral over an analogous domain in "n-dimensional space." Such multiple integrals are as important as partial derivatives for the more advanced theory and applications of calculus. The present chapter is only an introduction to this large subject, and is restricted to some properties and applications of double and triple integrals.

Like the one-variable integral $\int_a^b f$, the double integral $\iint_R f$ has an extensive array of interpretations. Here are a few.

1. If f is positive, then $\iint_R f(x, y)\, dx\, dy$ is the volume of the region that lies between the surface $z = f(x, y)$ and the xy-plane, with the plane region R as its base.

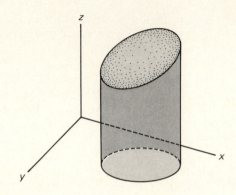

2. For any continuous f,

$$\frac{\displaystyle\iint_R f(xy)\, dx\, dy}{\text{area of } R} \quad \text{is the average value of } f \text{ over } R.$$

3. If R is a plane sheet of material with variable mass density $\rho(x, y)$, then

$$\iint_R \rho(x, y)\, dx\, dy$$

is the mass of R.

4. The surface area of the piece of the graph of $z = f(x, y)$ that lies over R is given by

$$\iint_R \sqrt{1 + \left(\frac{\partial z}{\partial x}\right)^2 + \left(\frac{\partial z}{\partial y}\right)^2}\; dx\, dy.$$

This is only a partial list. In general, if Q is any quantity that can be interpreted as being continuously spread out over the plane, then the amount of Q lying over the plane region R is

$$\iint\limits_{R} \rho(x, y)\, dx\, dy,$$

where ρ is the density function for Q.

Like the definite integral $\int_a^b f$, the double integral $\iint_R f$ will be defined as the limit of certain Riemann sums. But first we take up a related method of calculating volumes.

1
VOLUMES BY
ITERATED
INTEGRATION

If we erect the vertical cylinder based on a horizontal plane region R and then cut it off at its intersection with a surface $z = f(x, y)$, we get *the solid S based on R* and *capped by the surface $z = f(x, y)$*. We propose to calculate the volume of such a solid S, by a method that goes back to Chapters 6 and 9.

Example 1
Find the volume V of the solid S that is based on the unit square in the xy-plane and capped by the surface

$$z = 2 - (x^2 + y^2),$$

as shown in Fig. 1.

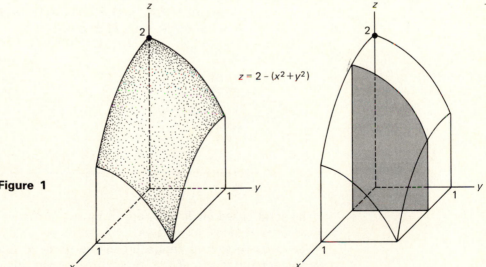

Figure 1

Solution   The cross section of this solid in the yz-plane $x = x_0$ is the plane region under the graph of $z = (2 - x_0^2) - y^2$ from $y = 0$ to $y = 1$, as shown at the right in Fig. 1. Its area is

$$A(x_0) = \int_0^1 [(2 - x_0^2) - y^2]\, dy$$

$$= \left[(2 - x_0^2)y - \frac{y^3}{3} \right]_0^1 = 2 - x_0^2 - \frac{1}{3}.$$

Now we apply the volumes-by-slicing formula (Chapter 9, Section 2) and get

$$V = \int_0^1 A(x)\, dx = \int_0^1 \left(\frac{5}{3} - x^2 \right) dx = \frac{4}{3}. \qquad \square$$

The scheme for the above calculation is given by the formula

$$V = \int_0^1 \left[\int_0^1 f(x, y)\, dy \right] dx,$$

which is called an *iterated* integral. We start with a function of two variables $f(x, y)$ and first "integrate y out," leaving a function of the one variable x, and then "integrate x out," ending up with a number.

Now consider the more general situation shown in Figs. 2 and 3.
The region R in the xy-plane is bounded below and above by the graphs of continuous functions g and h and lies between the x values $x = a$ and $x = b$, as in either part of Fig. 2.

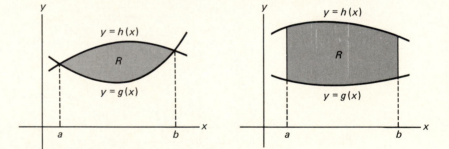

Figure 2

The solid based on R and capped by the surface $z = f(x, y)$ thus has two curved sides, as well as a curving top.

The cross section of this solid in the plane $x = x_0$ is the plane region under the graph of $z = f(x_0, y)$, from $y_1 = g(x_0)$ to $y_2 = h(x_0)$ as shown in Fig. 4.

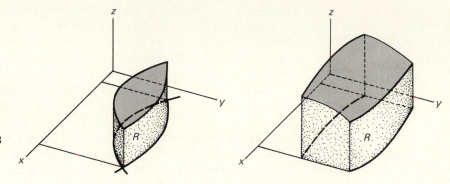

Figure 3

Its area is

$$A(x_0) = \int_{y_1}^{y_2} f(x_0, y)\,dy = \int_{g(x_0)}^{h(x_0)} f(x_0, y)\,dy,$$

by the area considerations in Chapter 6.

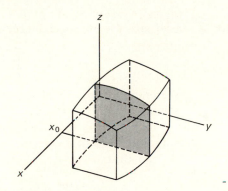

Figure 4

Then

$$V = \int_a^b A(x)\,dx,$$

by the volumes-by-slicing formula. Substituting the above value for $A(x)$, we end up with the volume formula:

$$V = \int_a^b \left[\int_{g(x)}^{h(x)} f(x, y)\,dy \right] dx.$$

Again we have an iterated integral. We first integrate $f(x, y)$ with respect to y, holding x fixed. The limits of integration now depend on the fixed value of x, as does the integrand. The resulting value of the integral was called $A(x)$ above. Then we integrate this function of x between fixed limits a and b, and obtain the number that is called the iterated integral. All in all, we start with a function of two variables $f(x, y)$, and first "integrate y out," leaving a function of the one variable x, and then "integrate x out," ending up with a number.

Here is the theorem we have proved (assuming the volumes-by-slicing formula):

THEOREM 1 *Let S be the solid based on the region R in the xy-plane and capped by the surface $z = f(x, y)$, where f is continuous and nonnegative on R. Suppose, furthermore, that in the xy-plane R is bounded below and above by the graphs of the continuous functions g and h, and left and right by the straight lines $x = a$ and $x = b$. Then the volume V of S is given by the iterated integral*

$$V = \int_a^b \left[\int_{g(x)}^{h(x)} f(x, y) \, dy \right] dx.$$

Example 2 Evaluate the iterated integral

$$\int_0^1 \left(\int_{x^2}^x xy^2 \, dy \right) dx.$$

Sketch the "domain of integration," and interpret the integral as a volume.

Solution We first compute the inside y integral, treating x as a constant:

$$\int_{x^2}^x xy^2 \, dy = \frac{xy^3}{3} \Big]_{x^2}^x = \frac{1}{3}(x^4 - x^7).$$

Then we compute the outside x integral:

$$\frac{1}{3} \int_0^1 (x^4 - x^7) \, dx = \frac{1}{3} \left[\frac{x^5}{5} - \frac{x^8}{8} \right]_0^1 = \frac{1}{3} \left(\frac{1}{5} - \frac{1}{8} \right) = \frac{1}{40}.$$

Thus,

$$\int_0^1 \left(\int_{x^2}^x xy^2 \, dy \right) dx = \frac{1}{40}.$$

For each fixed x_0 between $x = 0$ and $x = 1$ the inside function of y is integrated from $y = x_0^2$ to $y = x_0$. The domain

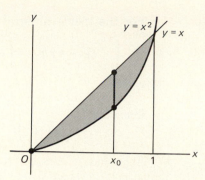

Figure 5

of integration is thus the plane region between the graphs $y = x^2$ and $y = x$, as shown in Fig. 5. The integral gives the volume of the solid based on this region and capped by the surface $z = xy^2$. □

Example 3 Find the volume of the space region capped by the plane $z = x + y$ and based on the triangle in the xy-plane that is cut off from the first quadrant by the line $x + y = 1$.

Solution The figure is shown below (Fig. 6). Such a figure should usually be drawn. The triangular domain is bounded above and below by the graphs $y = 1 - x$ and $y = 0$, and runs left to right from $x = 0$ to $x = 1$. The volume integral is therefore

$$V = \int_0^1 \left[\int_0^{1-x} (x + y) \, dy \right] dx.$$

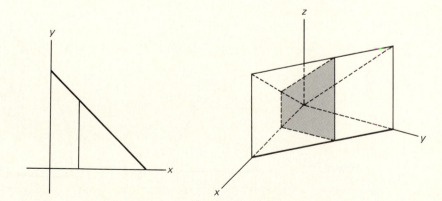

Figure 6

For the inside integral we have

$$\int_0^{1-x} (x + y)\, dy = \left[xy + \frac{y^2}{2} \right]_0^{(1-x)} = x(1 - x) + \frac{(1 - x)^2}{2}$$

$$= \frac{1}{2}(1 - x^2).$$

Therefore,

$$V = \int_0^1 \frac{1 - x^2}{2}\, dx = \frac{1}{2}\left[x - \frac{x^3}{3} \right]_0^1 = \frac{1}{3}. \qquad \square$$

REMARK   The brackets or parentheses setting off the inside integral of an iterated integral are generally omitted, it being understood that the inside integral is computed first. We would thus write the above iterated integral as follows:

$$V = \int_0^1 \int_0^{1-x} (x + y)\, dy\, dx.$$

Example 4   Find the volume of the region in the first octant under the graph of $z = f(x, y) = 1 - y - x^2$.

Solution   The domain is the region in the xy-plane bounded by the coordinate axes and the level curve $f(x, y) = 0$, that is, the parabola $y = 1 - x^2$, as shown in Fig. 7. So the inside y integral runs from $y = 0$ to $y = 1 - x^2$, and the volume formula is

$$V = \int_0^1 \int_0^{1-x^2} (1 - y - x^2)\, dy\, dx.$$

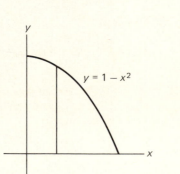

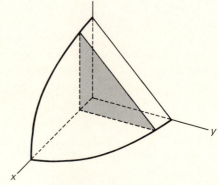

$y = 1 - x^2$

Figure 7

The figure for the solid at the right in Fig. 7 also shows the edge curves in the other two coordinate planes. The edge curve in the xz-coordinate plane is obtained by setting $y = 0$ in the equation

$$z = 1 - y - x^2,$$

and is the parabola $z = 1 - x^2$. In the zy-coordinate plane it is the straight line $z = 1 - y$. These edge curves, together with the cross section at x, are enough to give a good visual image of the solid.

Here is the volume computation:

$$V = \int_0^1 \int_0^{1-x^2} (1 - y - x^2) \, dy \, dx$$

$$= \int_0^1 \left[(1 - x^2)y - \frac{y^2}{2} \right]_0^{1-x^2} dx$$

$$= \int_0^1 \frac{(1 - x^2)^2}{2} \, dx = \frac{1}{2} \left[x - \frac{2x^3}{3} + \frac{x^5}{5} \right]_0^1 = \frac{4}{15}. \qquad \square$$

PROBLEMS FOR SECTION 1

Compute the following iterated integrals. In each case sketch the region that is the domain of integration in the xy-plane.

1. $\int_0^1 \int_0^1 xy \, dy \, dx$

2. $\int_0^1 \int_0^1 dy \, dx$

3. $\int_0^{\pi/2} \int_0^x \cos y \, dy \, dx$

4. $\int_0^{\pi/2} \int_0^{\sin x} dy \, dx$

5. $\int_{-1}^1 \int_x^1 xy \, dy \, dx$

6. $\int_0^1 \int_0^{\sqrt{1-x^2}} x \, dy \, dx$

7. $\int_{-1}^1 \int_{-\sqrt{1-x^2}}^{\sqrt{1-x^2}} y \, dy \, dx$

8. $\int_1^2 \int_x^{2x} y \, dy \, dx$

9. $\int_0^1 \int_{x^2}^x (x + y) \, dy \, dx$

10. $\int_{1/4}^{3/2} \int_{x^2}^x (x + y) \, dy \, dx$

Use iterated integration to find the volume of each of the space regions described in Problems 11 through 21 below. Sketch each configuration.

11. The unit cube $0 \le x \le 1$, $0 \le y \le 1$, $0 \le z \le 1$.

12. The solid based on the rectangle $0 \le x \le a$, $0 \le y \le b$, in the xy-plane, and capped by the plane $z = c$.

13. The solid based on the unit square $0 \le x \le 1$, $0 \le y \le 1$ and capped by the plane $z = x + y$.

14. The solid based on the first quadrant of the unit circle $x^2 + y^2 \le 1$ and capped by the plane $z = y$.

15. The solid based on the rectangle $0 \leq x \leq a$, $0 \leq y \leq b$ in the xy-plane and capped by the surface $z = b^2 - y^2$.

17. The solid bounded by the parabolic cylinders $y = 1 - x^2$, $y = x^2 - 1$, $z = 1 - y^2$, and the plane $z = 0$.

19. The solid bounded by the sphere $x^2 + y^2 + z^2 = 1$ and the square cylinder $|x| + |y| = 1$.

21. The piece of the first octant cut out by the plane $x + y = 1$ and the parabolic cylinder $z = 1 - y^2$.

23. The piece of the first octant cut out by the plane $z = x$ and the cylinder $x^2 + y^2 = 1$.

25. The interior of the ellipsoid

$$\frac{x^2}{a^2} + \frac{y^2}{b^2} + \frac{z^2}{c^2} = 1.$$

27. The solid bounded by the parabolic cylinder $x = 2 - y^2$ and the planes $y = x$, $z = 0$, and $z = 1 - y$. (Here also the base region will have to be divided into two pieces.)

29. Find the volume of the solid lying over the unit square $0 \leq x \leq 1$, $0 \leq y \leq 1$, and bounded above and below by the surface $z = y$ and $z = y^3$. (Apply the formula in Problem 28.)

31. Find the volume of the piece of the first octant lying between the surfaces $z = y$ and $z = 1 - x^2$.

16. The solid cut from the first octant by the plane $x + z = 1$ and the cylinder $x^2 + y^2 = 1$.

18. The solid cut from the first octant by the plane $y + z = 1$ and the ellipse $y^2 + 2x^2 = 1$.

20. The solid bounded by the paraboloid $x^2 + y^2 + z = 1$ and the xy-plane.

22. The piece of the first octant cut out by the plane $x + y = 1$ and the parabolic cylinder $z = 1 - x^2$.

24. The piece of the first octant cut out by the plane

$$\frac{x}{a} + \frac{y}{b} + \frac{z}{c} = 1,$$

where a, b, and c are all positive.

26. The solid bounded by the parabolic cylinders $z = 1 - x^2$, $x = 1 - y^2$, $x = y^2 - 1$, and the plane $z = 0$. (Divide the base region into two pieces and compute the volume as the sum of two integrals.)

28. If $f_1(x, y) \leq f_2(x, y)$ over the region R, then the solid cut out from the cylinder through R by the graphs of f_2 and f_1 has the volume

$$V = \iint_R (f_2(x, y) - f_1(x, y)) \, dy \, dx.$$

Justify this.

30. Find the volume of the solid lying over the quarter of the unit circle in the first quadrant and bounded above and below by the surface $z = y$ and $z = y^3$.

32. For some base regions R, it may be convenient and even necessary to interchange the role of x and y and apply the volumes-by-slicing formula to varying y-slices. Draw the base region R for which the following iterated integral is a correct volume formula, and verify its correctness by giving an argument based on volumes-by-slicing, as we did in the text:

$$V = \int_a^b \left(\int_{g(y)}^{h(y)} f(x, y) \, dx \right) dy.$$

33. Redo Problem 23 by this method.

34. Redo Problem 26 by this method. Now the base region can be left in one piece.

35. Redo Problem 27 by this method. This time the base region can be left intact.

2
RIEMANN SUMS AND THE DOUBLE INTEGRAL

Let $f(x, y)$ be a continuous function defined on a bounded closed plane region R. Then we can define Riemann sums for f over R and the definite integral of f over R much as we did for a function of one variable.

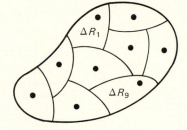 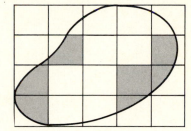

Figure 1

In order to define a Riemann sum for $f(x, y)$ over R, we first subdivide R in any manner into a finite number of small subregions R_i where $i = 1, \ldots, k$, as indicated in Fig. 1. The simplest way to do this is to impose a lattice of rectangles over R, as at the right, but it is desirable for flexibility to allow more general subdivisions. Next, we choose an arbitrary "evaluation point" (x_i, y_i) in the subregion R_i, for each i. Then the Riemann sum for f associated with this subdivision and evaluation set is

$$\sum_{i=1}^{k} f(x_i, y_i)\, \Delta A_i,$$

where ΔA_i is the area of the subregion R_i.

Example 1   We consider $f(x, y) = y/x$ over the unit square R having its lower left corner at $(1, 0)$. We subdivide the square into four congruent subsquares, and use the lower left corner of each as its evaluation point (Fig. 2). Then the Riemann sum for y/x over R associated with this subdivision and this evaluation set has the value

$$f(1, 0)\,\Delta A_1 + f\left(\frac{3}{2}, 0\right)\Delta A_2 + f\left(1, \frac{1}{2}\right)\Delta A_3 + f\left(\frac{3}{2}, \frac{1}{2}\right)\Delta A_4$$

$$= \frac{0}{1}\cdot\frac{1}{4} + \frac{0}{3/2}\cdot\frac{1}{4} + \frac{1/2}{1}\cdot\frac{1}{4} + \frac{1/2}{3/2}\cdot\frac{1}{4}$$

$$= \frac{1}{4}\left[0 + 0 + \frac{1}{2} + \frac{1}{3}\right]$$

$$= \frac{5}{24}.$$

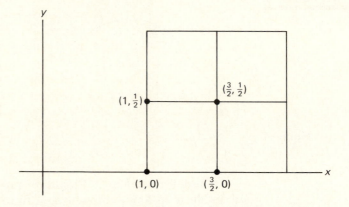

Figure 2

□

Example 2 If we use the *upper right* corners as evaluation points in the above situation, then the Riemann sum has the value

$$f\left(\frac{3}{2}, \frac{1}{2}\right)\Delta A_1 + f\left(2, \frac{1}{2}\right)\Delta A_2 + f\left(\frac{3}{2}, 1\right)\Delta A_3 + f(2, 1)\,\Delta A_4$$

$$= \frac{1/2}{3/2}\cdot\frac{1}{4} + \frac{1/2}{2}\cdot\frac{1}{4} + \frac{1}{3/2}\cdot\frac{1}{4} + \frac{1}{2}\cdot\frac{1}{4}$$

$$= \frac{1}{4}\left[\frac{1}{3} + \frac{1}{4} + \frac{2}{3} + \frac{1}{2}\right]$$

$$= \frac{7}{16}.$$

□

Example 3 Find the Riemann sum for the above function f and region R if R is divided into nine equal squares and the lower left corners are taken as evaluation points.

Solution   The points on the x-axis again contribute zero terms, and the remaining six terms are

$$f\left(1,\frac{1}{3}\right)\Delta A_4 + f\left(\frac{4}{3},\frac{1}{3}\right)\Delta A_5 + f\left(\frac{5}{3},\frac{1}{3}\right)\Delta A_6 + f\left(1,\frac{2}{3}\right)\Delta A_7$$

$$+ f\left(\frac{4}{3},\frac{2}{3}\right)\Delta A_8 + f\left(\frac{5}{3},\frac{2}{3}\right)\Delta A_9$$

$$= \frac{1/3}{1}\cdot\frac{1}{9} + \frac{1/3}{4/3}\cdot\frac{1}{9} + \frac{1/3}{5/3}\cdot\frac{1}{9} + \frac{2/3}{1}\cdot\frac{1}{9} + \frac{2/3}{4/3}\cdot\frac{1}{9} + \frac{2/3}{5/3}\cdot\frac{1}{9}$$

$$= \frac{1}{9}\left[\frac{1}{3} + \frac{1}{4} + \frac{1}{5} + \frac{2}{3} + \frac{2}{4} + \frac{2}{5}\right] = \frac{47}{180}.\qquad\square$$

It is convenient to be able to talk about convergent sequences of Riemann sums in the way we did earlier for functions of one variable. To do so, we first define the *diameter* of a set A to be the diameter of the smallest circle that can be circumscribed about A. Now fix an integer n, and suppose we have subdivided R finely enough so that each subregion R_i has diameter at most $1/n$. Then any associated Riemann sum S will be called an nth Riemann sum and designated S_n.

We also need the following idea. The boundary of R is said to be *piecewise smooth* if it is made up of a finite number of pieces of smooth curves. Then,

THEOREM 2   *Let R be a bounded closed region with a piecewise-smooth boundary, and let f be continuous on R. For each n let S_n be an nth Riemann sum for f over R. Then the limit*

$$\lim_{n\to\infty} S_n$$

exists. Furthermore, every other sequence of Riemann sums formed in this way converges to the same limit.

This theorem is proved in more advanced courses; we shall simply assume it here. This is all in line with the development of integration given in Chapter 6, and we continue in that pattern.

DEFINITION   Let f be *any* function defined on R, continuous or not. Suppose that the above conclusion holds for f. That is, suppose that every sequence of Riemann sums $\{S_n\}$ for f over R converges, and all such sequences converge to the same limit. Then f is said to be *integrable over R*.

Then Theorem 2 becomes, loosely,

THEOREM 2′   *Continuous functions are integrable.*

DEFINITION   If f is integrable over R, then the common limit of all the Riemann sequences $\{S_n\}$ is called the *definite integral of f over R*, and is designated

$$\iint_R f(x, y)\, dA,$$

or simply

$$\iint_R f.$$

Such integrals are called *double* integrals.

Properties of the Double Integral   The basic properties of double integrals follow easily from the definition. Here is a short list. All functions are assumed to be integrable.

1.
$$\iint_R (f + g) = \iint_R f + \iint_R g.$$

2.
$$\iint_R cf = c\iint_R f, \text{ where } c \text{ is a constant.}$$

3. If R is divided into two subregions R_1 and R_2 by a crossing curve, then

$$\iint_R f = \iint_{R_1} f + \iint_{R_2} f.$$

4. If $f \leq g$ on R, then

$$\iint_R f \leq \iint_R g.$$

5.
$$\iint_R 1 = \text{area of } R.$$

As a corollary of these properties, we have

6. If $m \le f(x, y) \le M$ on R, then

$$mA \le \iint_R f \le MA,$$

where A is the area of R.

7. If f is continuous on the region R, then there is a point (x_0, y_0) in R such that

$$\iint_R f = f(x_0, y_0)A,$$

where A is the area of R.

This is the mean-value principle for double integrals. It follows from the intermediate-value principle: (6) says that the double integral lies between the minimum and maximum values of the function $Af(x, y)$ and it must therefore be equal to a value $Af(x_0, y_0)$.

Evaluation of the Double Integral There is no Fundamental Theorem for directly evaluating a double integral. Instead, we show that a double integral can be evaluated by iterated integration.

THEOREM 3 *Let $f(x, y)$ be a continuous function defined over the closed plane region R, and suppose that R is bounded above and below in the xy-plane by the graphs of $y = h(x)$ and $y = g(x)$, and runs from $x = a$ to $x = b$. Then*

$$\iint_R f(x, y)\, dA = \int_a^b \int_{g(x)}^{h(x)} f(x, y)\, dy\, dx.$$

If, instead, the left and right parts of the boundary of R lie along the graphs $x = p(y)$ and $x = q(y)$, respectively, and R extends from

$y = c$ to $y = d$, *then*

$$\iint\limits_{R} f(x, y)\, dA = \int_{c}^{d} \int_{p(y)}^{q(y)} f(x, y)\, dx\, dy.$$

Example 4   Calculate the double integral $\iint_{R} (x \cos y/y)\, dA$, where R is the region in the first quadrant above the parabola $y = x^2$ and below the line $y = 4$.

Solution   We apply Theorem 3, integrating first with respect to x from x to 0 to $x = \sqrt{y}$. See Fig. 3. We get

$$\int_{0}^{4} \int_{0}^{\sqrt{y}} \frac{x \cos y}{y}\, dx\, dy = \int_{0}^{4} \left[\frac{x^2 \cos y}{2y} \right]_{0}^{\sqrt{y}} dy$$
$$= \frac{1}{2} \int_{0}^{4} \cos y\, dy = \frac{\sin 4}{2}.$$

If we set the integral integral up in the other order, we get

$$\int_{0}^{2} \int_{x^2}^{4} \frac{x \cos y}{y}\, dy\, dx,$$

but now we can't even begin, because we can't integrate $\int (\cos y/y)\, dy$.

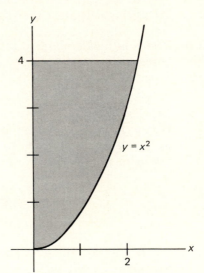

Figure 3

The rigorous proof of Theorem 3 must be postponed to more advanced courses, but we can give a geometric proof, based on the following theorem.

THEOREM 4   *If f is continuous and positive on R, then the volume V of the solid S based on R and capped by the graph of f is given by*

$$V = \iint_R f.$$

Proof   The proof is almost identical to the proof of Theorem 3 in Chapter 6. We start with any subdivision of R into k subregions $R_1, \ldots, R_k$. For each index i, let M_i and m_i be the maximum and minimum values of $f(x, y)$ on the ith subregion R_i and let ΔA_i be the area of R_i.

Let S_i be the incremental solid based on the subregion R_i and capped by the graph of f. Then the cylinders of altitudes m_i and M_i based on R_i are inscribed in, and circumscribed about, S_i. See Fig. 4. Thus the volume ΔV_i of S_i is squeezed between the cylinder volumes,

$$m_i \, \Delta A_i \le V_i \le M_i \, \Delta A_i.$$

We now add these inequalities for $i = 1, \ldots, k$. Then the left and right sums are Riemann sums $\underline{S}$ and $\bar{S}$ for f over R, the

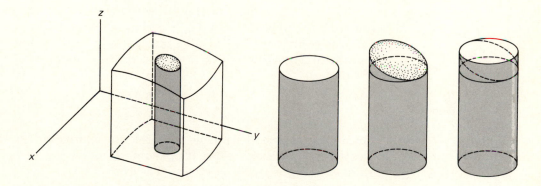

Figure 4

middle sum is the volume V of S, and we have

$$\underline{S} \le V \le \bar{S}.$$

Finally, we vary the subdivision. For a sequence of subdivisions we will have

$$\underline{S}_n \le V \le \bar{S}_n$$

for each n, and also

$$\lim_{n \to \infty} \underline{S}_n = \lim_{n \to \infty} \bar{S}_n = \iint f$$

(assuming these are nth Riemann sums in the sense described earlier). Therefore

$$V = \iint_R f,$$

by the squeeze limit law.   ∎

Combining this formula with Theorem 1 proves Theorem 3 for a positive function.

For any other continuous f we can choose a positive constant C large enough so that $f + C$ is positive on R and write

$$f = (f + C) - C.$$

Since Theorem 3 is true for each term in this difference, it is true for f, by the properties (1) and (2) for integrals.

Theorem 3 is used in the following way. We first show that a quantity Q can be estimated by a certain finite sum. We then observe that this finite sum can also be interpreted as a Riemann sum for a certain function $\rho(x, y)$ over a region R. But this means that

$$Q = \iint_R \rho(x, y) \, dA,$$

since both Q and the double integral are approximated arbitrarily closely by the same estimating sum. Finally, we *compute* Q by iterated integration, by virtue of Theorem 3.

There are several applications of this procedure in the remaining sections of this chapter.

PROBLEMS FOR SECTION 2

1. Evaluate the Riemann sum for $f(x, y) = x/(1 + y)^2$ over the unit square $0 \le x \le 1$, $0 \le y \le 1$, using the subdivision obtained by dividing the square into nine congruent subsquares and choosing the *upper right* vertex of each square as evaluation point.

2. Evaluate the Riemann sum for the same function and subdivision but with the *lower left* vertices as evaluation points.

A corollary of Theorem 3 is that if the base region R can be described in *both* ways specified there, then the two iterated integrals are necessarily equal:

(*)
$$\int_a^b \int_{g(x)}^{h(x)} f(x, y)\, dy\, dx = \int_c^d \int_{p(y)}^{q(y)} f(x, y)\, dx\, dy.$$

In Problems 3 through 8, an iterated integral is given that specifies a base region R in one of the above ways, and you are to work out the correct limits of integration for the other integral. (This may require that the region R be divided into two or more parts, with the given integral expressed as the sum of two or more integrals.) In each case *draw* the region R from the given limits of integration and then describe R the other way.

3. $\int_0^1 \int_0^x f(x, y)\, dy\, dx$

4. $\int_0^1 \int_0^{x^2} \cdots dy\, dx$

5. $\int_0^1 \int_{x^2}^x \cdots dy\, dx$

6. $\int_0^1 \int_{1-x}^{1+x} \cdots dy\, dx$

7. $\int_{-1}^1 \int_0^{y^2} \cdots dx\, dy$

8. $\int_0^1 \int_{y^2}^1 \cdots dx\, dy$

The iterated integral identity (*) can be used to evaluate either integral in terms of the other. This is called *reversing the order of integration*. Sometimes a difficult integration can be made easier by reversing the order of integration. Compute the following iterated integrals by reversing the order of integration.

9. $\int_0^1 \int_y^1 e^{-x^2}\, dx\, dy$

10. $\int_0^1 \int_0^{y^2} y \cos(1 - x)^2\, dx\, dy$

11. $\int_0^1 \int_x^1 \sqrt{1 - y^2}\, dy\, dx$

12. Suppose we choose N points (x_1, y_1), $\ldots, (x_N, y_N)$ that are scattered in a fairly uniform way throughout a plane region R. That is, there must be roughly the same number of chosen points "per unit area" wherever we look in R. If the area of R is A, give some kind of an argument to support the claim that

$$\frac{A}{N} \sum_{i=1}^N f(x_i, y_i)$$

is approximately equal to the double integral of f over R.

13. Let R be the unit square $0 \le x \le 1$, $0 \le y \le 1$.

a) Compute $\iint_R xy\, dA$ by iterated integration.

b) Compute the above estimate of this double integral for the nine vertices obtained by dividing the unit square into four equal subsquares.

▦ **14.** Work out the above problem for $f(x, y) = x + y$.

▦ **16.** The same for $f(x, y) = x^2y^2$.

▦ **18.** Do the above problem for x^2y^2.

20. Prove that if the double integrals of f and g exist over a region R, then so does the double integral of $af(x, y) + bg(x, y)$ and

$$\iint_R [af(x, y) + bg(x, y)] \, dA$$

$$= a \iint_R f(x, y) \, dA + b \iint_R g(x, y) \, dA.$$

Note that this combines the integral properties (1) and (2). See the proof sketched in Chapter 6, Section 5.

▦ **15.** The same for $f(x, y) = x^2 + y^2$.

▦ **17.** The estimates in Problems 15 and 16 were not very good. They would improve with a finer subdivision, of course, but there is something else wrong. In a sense a vertex is surrounded by only 1/4 as much area as an interior point and an edge point that is not a vertex is surrounded by 1/2 as much area as an interior point. We are therefore not "weighting" the effect of these boundary points correctly. In order to adjust for this edge effect, suppose we count each of the four corner values $f(x_i, y_i)$ once, each of the four edge midpoint values twice, and the middle value $f(1/2, 1/2)$ four times. Then we have effectively increased the number of terms in the sum from 9 to 16, so we now divide by 16 rather than by 9. Recompute the estimate for $f(x, y) = x^2 + y^2$, using this weighted sum.

▦ **19.** a) If the unit square is divided into 9 equal subsquares and if their 16 vertices are weighted as in Problem 17, show that we must now divide by $N = 36$.

    b) Recompute the estimate for the double integral of x^2y^2 over the unit square using this new weighted sum.

    c) Compute the double integral again and compare answers.

21. Prove that if $f(x, y)$ has the constant value k over the region R, then its double integral over R exists and

$$\iint_R f = kA,$$

where A is the area of R.

22. Prove that if the base region R is divided into two nonoverlapping subregions R_1 and R_2, then

$$\iint\limits_{R} f \, dA = \iint\limits_{R_1} f \, dA + \iint\limits_{R_2} f \, dA.$$

3
THE MASS OF
A PLANE
LAMINA

We consider first the notion of density in space. Consider a quantity Q, such as a mass distribution or a distribution of charge, that is "spread out" in space. We say that the distribution has the constant density ρ if the amount of Q contained in any region G is exactly ρ times the volume V of G. Thus,

$$\boxed{\rho = \frac{Q(G)}{V(G)},}$$

no matter what space region G we consider. If Q is a mass distribution, and if we measure mass in ounces and volume in cubic inches, then the density ρ is measured in ounces per cubic inch. In the metric system density is measured in grams per cubic centimeter.

Example 1   We find from a handbook that the density of aluminum is 2.7 g/cm^3. This means that the mass of any aluminum object, measured in grams, is 2.7 times the volume it occupies, measured in cubic centimeters. The density of iron is 7.9; the density of gold is 19.3.   □

However, a distribution Q will not normally have a constant density, and we have to consider density as something that varies from point to point. In order to define the density of Q at the point p_0, we first choose some small region ΔG around p_0, such as a small ball or a small cube centered at p_0. Let ΔQ be the amount of Q contained in ΔG and let ΔV be the volume of ΔG. The quotient

$$\frac{\Delta Q}{\Delta V}$$

is then called the *average density* of Q throughout ΔG. For a mass distribution in the above units, $\Delta Q/\Delta V$ is the average number of ounces per cubic inch throughout ΔG. *The space density of Q at p_0 is then defined to be the limit*

$$\rho(p_0) = \lim \frac{\Delta Q}{\Delta V}$$

as the diameter of ΔG approaches 0. This has the look of a derivative, but it is not a derivative in the ordinary sense, and we are in no position to discuss how we might evaluate such a limit. Nevertheless, we can conceive of the limit existing, and if it happened to have the value 5, we would say that the mass density at p_0 is 5 ounces per cubic inch.

Similar remarks apply to two-dimensional distributions. We can consider a distribution of charge spread out over a two-dimensional sheet, or we can think of a flat sheet of material with varying constitution as giving us a two-dimensional mass distribution (called a plane lamina). Now the average mass density in a small plane region ΔR is the quotient

$$\frac{\Delta m}{\Delta A},$$

where ΔA is the *area* of ΔR and Δm is its mass. For a mass distribution with the same units as before, the average density is measured in ounces per *square* inch (or in grams per *square* centimeter). The density of the lamina at the point (x_0, y_0) is the limit

$$\rho(x_0, y_0) = \lim \frac{\Delta m}{\Delta A},$$

where the limit is taken as ΔR shrinks down on this point, that is, as the diameter of ΔR approaches 0.

Density is thus a sort of derivative of mass with respect to area. We shall now see that the mass m can be recovered from its "two-dimensional derivative" $\rho(x, y)$ by two-dimensional integration. That is,

THEOREM 5 *If R is a plane lamina with variable density $\rho(x, y)$, then the total mass m of R is*

$$m = \iint_R \rho(x, y)\, dA.$$

The proof will be given after an example.

Example 2 The density of a square lamina of side length $2a$ is equal to the square of the distance from the center. Find the mass of the lamina.

Solution If we center the square at the origin then the density is given by

$$\rho(x, y) = x^2 + y^2.$$

The mass of the lamina is therefore

$$m = \int_{-a}^{a} \int_{-a}^{a} (x^2 + y^2)\, dx\, dy = \int_{-a}^{a} \left[\frac{x^3}{3} + xy^2 \right]_{-a}^{a} dy$$

$$= 2\int_{-a}^{a} \left(\frac{a^3}{3} + ay^2 \right) dy = 2\left[\frac{a^3 y}{3} + \frac{ay^3}{3} \right]_{-a}^{a} = \frac{8}{3}\, a^4. \qquad \square$$

Example 3 Find the mass of a semicircular lamina occupying the upper half of the circle $x^2 + y^2 \le a^2$ if its density is proportional to y.

Solution By hypothesis, $\rho(x, y) = ky$, where k is the constant of proportionality. Therefore,

$$m = \iint_R \rho\, dA = k\int_{-a}^{a} \int_{0}^{\sqrt{a^2 - x^2}} y\, dy\, dx$$

$$= k\int_{-a}^{a} \left[\frac{y^2}{2} \right]_{0}^{\sqrt{a^2 - x^2}} dx$$

$$= \frac{k}{2} \int_{-a}^{a} (a^2 - x^2)\, dx$$

$$= \frac{k}{2} \left[a^2 x - \frac{x^3}{3} \right]_{-a}^{a} = \frac{2}{3}\, ka^3. \qquad \square$$

The following lemma is the key step in the proof of the theorem.

LEMMA   *If k is a constant such that*

$$\rho(x, y) \le k$$

throughout R, then

$$m \le kA,$$

where m and A are the mass and area of R, respectively.

Sketch of Proof   The proof is by contradiction. Suppose, on the contrary, that $m > kA$, so that

$$m = (k + d)A,$$

where d is positive. If we subdivide R, then at least one of the subregions must have mass Δm and area ΔA satisfying

(*) $\Delta m \ge (k + d)\,\Delta A.$

(Otherwise we would have the inequality $<$ for each subregion, and, adding, we would have $m < (k + d)A$.) Call such a subregion R_1. Now subdivide R_1 and find in a similar way a subregion R_2 of R_1 for which (*) holds. Continuing in this way, we can find a sequence of subregions $\{R_i\}$, each satisfying (*), that squeeze down on some point (x_0, y_0). The inequality

$$\frac{\Delta m_i}{\Delta A_i} \ge k + d$$

holds for each i and therefore holds in the limit. Thus

$$\rho(x_0, y_0) = \lim_{i \to \infty} \frac{\Delta m_i}{\Delta A_i} \ge k + d.$$

This contradicts our beginning hypothesis. So we cannot have $m > kA$. That is, $m \le kA$, as claimed.   ∎

It follows in exactly the same way that if $\rho(x, y) \ge k$ throughout R, then $m \ge kA$. Therefore if $\bar{\rho}$ and $\underline{\rho}$ are the maximum and minimum values of $\rho(x, y)$ on R, then

$$\underline{\rho}A \le m \le \bar{\rho}A.$$

The intermediate-value principle now gives:

COROLLARY *There is a point (x_0, y_0) in the region R such that*

$$m = \rho(x_0, y_0)A.$$

The theorem now follows easily. Given any subdivision $\{R_i\}$ of R, we can choose evaluation points $\{(x_i, y_i)\}$ such that

$$\Delta m_i = \rho(x_i, y_i)\, \Delta A_i$$

for each i, by the corollary. Thus

$$m = \sum \Delta m_i = \sum \rho(x_i, y_i)\, \Delta A_i.$$

The sum on the right is a Riemann sum for ρ over R, and it is an nth Riemann sum if the diameters of the subregions R_i are all at most $1/n$. Thus for each n there is an nth Riemann sum S_n such that $m = S_n$, and, letting $n \to \infty$, we have $m = \iint_R \rho$ in the limit.

PROBLEMS FOR SECTION 3

In each of the following problems find the mass of the plane lamina R having the given shape and density.

1. R is the unit square with lower left vertex at the origin, and $\rho(x, y) = xy$.

2. R is the piece of the unit circle $x^2 + y^2 \le 1$ in the first quadrant and $\rho(x, y) = xy$.

3. R is the lamina between the graphs of $y = x$ and $y = x^2$ and $\rho(x, y) = \sqrt{x}$.

4. R is the parabolic cap bounded by the parabola $y = 4 - x^2$ and the x-axis. The density is proportional to the distance from the x-axis.

5. R is the same as in Problem 4, but the density is proportional to the distance from the y-axis.

6. R is the unit circle $x^2 + y^2 \le 1$ and the density is proportional to the distance from the y-axis.

7. R is the unit circle and the density is proportional to the distance from the x-axis.

8. R is the unit circle and the density is proportional to the sum of the distances from the two coordinate axes.

9. R is the square symmetric about the origin with one vertex at (a, a). Its density is proportional to the distance from the line $x + y = 0$ (that is, $\rho(x, y) = k\,|x + y|$).

10. R is the piece of the first quadrant under the line $x + y = 1$ and the density is e^{x+y}.

11. R is the lamina between the graph of $y = \sin x$ and the x-axis, from $x = 0$ to $x = \pi$. The density is proportional to the distance from the y-axis.

13. R is the lamina between the graph $y = e^{-x}$ and the x-axis, from $x = 0$ to ∞; $\rho(x, y) = x$.

12. R is the above lamina and the density is proportional to the distance from the x-axis.

4
MOMENTS AND CENTER OF MASS OF A PLANE LAMINA

We first define the center of mass of a finite system of point masses on a line. Let us consider the x-axis as a rigid, weightless, horizontal rod pivoting at the point r, and suppose that we place masses m_i at the positions x_i. According to the principle of the lever, or the "seesaw" principle, the system will exactly balance if

$$\sum m_i(x_i - r) = 0.$$

Figure 1

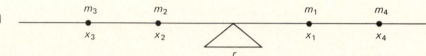

In any case, the sum $\sum m_i(x_i - r)$ measures the tendency of the system to turn about r. It is called the *moment* of the system about r, and if the moment is zero, then the system is in equilibrium.

If we are free to move the pivot point r, then we can find a unique position $r = \bar{x}$ such that the moment about $\bar{x}$ is zero. This holds if and only if

$$\sum m_i(x_i - \bar{x}) = 0,$$

or

$$\sum m_i x_i - \bar{x} \sum m_i - 0,$$

or

$$\boxed{\bar{x} = \frac{\sum m_i x_i}{\sum m_i}.}$$

This balancing point is called the *center of mass* of the system. The formula above says that the center of mass is obtained as the quotient

$$\boxed{\bar{x} = \frac{\text{moment of system about the origin}}{\text{total mass of the system}}.}$$

Now consider the plane. We view the xy-plane as a rigid, weightless sheet lying horizontally. We choose a line $x = r$ parallel to the y-axis, and we consider whether or not a given mass distribution in the plane will balance on this line.

For a finite system of point masses m_i placed at the points (x_i, y_i), the seesaw principle is the same as before: the system will exactly balance if and only if

$$\sum m_i(x_i - r) = 0.$$

Whether the system balances or not, the sum on the left is the moment of the system about the axis $x = r$. Note that only the x-coordinates are involved. The moment depends on the perpendicular distances from the point masses to the moment axis.

Next, suppose that a lamina of density $\rho(x, y)$ occupies the plane region R. The lamina can be approximated by a system of point masses as follows. We subdivide R into very small subregions R_i and then replace the mass Δm_i occupying R_i by an equal point mass at an evaluation point (x_i, y_i) in R_i. By the corollary in the last section, this point can be chosen so that

$$\Delta m_i = \rho(x_i, y_i)\,\Delta A_i.$$

Then the total moment M of the lamina about the axis $x = r$ is given approximately by the moment of its approximating system of point masses,

$$M \approx \sum \Delta m_i(x_i - r) = \sum (x_i - r)\rho(x_i, y_i)\,\Delta A_i.$$

On the other hand, the sum on the right is a Riemann sum for the function $(x - r)\rho(x, y)$ over the region R and hence approximates the double integral of this function over R. Both of these approximations improve as the subdivision is taken finer, so, in the limit, we have:

THEOREM 6 *If a plane lamina with variable density $\rho(x, y)$ occupies the region R, then the moment M of the lamina about the axis $x = r$ is given by*

$$\boxed{M = \iint\limits_{R} (x - r)\rho(x, y)\,dA.}$$

If we now set $M = 0$ and solve for r, we find the balancing axis $x = r = \bar{x}$. A plane lamina R will also have a balancing axis $y = \bar{y}$ parallel to the x-axis, and then $(\bar{x}, \bar{y})$ is the balancing *point* for R. This is not obvious, but it can be proved. The balancing point $(\bar{x}, \bar{y})$ of a plane lamina R is called its *center of mass*. In the case of a *homogeneous* lamina (i.e., one with constant density) the point $(\bar{x}, \bar{y})$ is also called the *centroid* of the region R.

As noted above, we find $r = \bar{x}$ from the equation $M = 0$, which is

$$0 = \iint_R (x - \bar{x})\rho(x, y)\, dA = \iint_R x\rho(x, y)\, dA - \bar{x}\iint_R \rho(x, y)\, dA.$$

Solving for $\bar{x}$ in this equation, and for $\bar{y}$ in the corresponding moment equation about the axis $y = \bar{y}$, we have the corollary:

COROLLARY   *The center of mass $(\bar{x}, \bar{y})$ of the above lamina has the coordinates*

$$\bar{x} = \frac{\displaystyle\iint_R x\rho(x, y)\, dA}{\displaystyle\iint_R \rho(x, y)\, dA} \quad \frac{\text{moment about the } y\text{-axis}}{\text{total mass } m},$$

$$\bar{y} = \frac{\displaystyle\iint_R y\rho(x, y)\, dA}{\displaystyle\iint_R \rho(x, y)\, dA} = \frac{\text{moment about the } x\text{-axis}}{\text{total mass } m}.$$

It is convenient to write M_x and M_y for the moments about the y and x axes, respectively. Then the center-of-mass formulas can be written

$$\bar{x} = \frac{M_x}{m} \quad \text{and} \quad \bar{y} = \frac{M_y}{m},$$

or

$$(\bar{x}, \bar{y}) = \frac{1}{m}(M_x, M_y).$$

Example 1   Find the center of mass of the square R having lower left vertex at the origin and upper right vertex at (a, a), if its density is proportional to the square of the distance from the origin.

Solution   If k is the constant of proportionality for the density,

$$\rho(x, y) = k(x^2 + y^2),$$

then the calculation in Example 2 in the last section shows that the mass of the square lamina in $2ka^4/3$.

For the moment M_x about the y-axis, we have

$$M_x = \int_0^a \int_0^a x \cdot k(x^2 + y^2) \, dx \, dy = k \int_0^a \left[\frac{x^4}{4} + \frac{x^2 y^2}{2} \right]_0^a dy$$

$$= k \int_0^a \left(\frac{a^4}{4} + \frac{a^2 y^2}{2} \right) dy$$

$$= k \left[\frac{a^4 y}{4} + \frac{a^2 y^3}{6} \right]_0^a$$

$$= k \left(\frac{a^5}{4} + \frac{a^5}{6} \right) = \frac{5}{12} ka^5.$$

Thus

$$\bar{x} = \frac{M_x}{m} = \frac{\frac{5}{12} ka^5}{\frac{2}{3} ka^4} = \frac{5}{8} a.$$

Finally, since the whole configuration is symmetric in x and y, we also have $\bar{y} = 5a/8$. ☐

Centroids The center-of-mass formulas simplify if the lamina is homogeneous. First, the constant density ρ can be canceled out from the numerator and denominator, which means we can assume that $\rho = 1$. Then the denominator (the total mass) is just the area A of the region R. Thus,

$$\bar{x} = \frac{M_x}{A} = \frac{\iint\limits_R x \, dA}{A}.$$

Suppose, furthermore, that R is bounded above and below by the graphs of $y = f(x)$ and $y = g(x)$, and left and right by the lines $x = a$ and $x = b$. Then the double integrals for the moments (M_x, M_y) about the axes $x = 0$ and $y = 0$ reduce to

$$M_x = \int_a^b x[f(x) - g(x)] \, dx,$$

$$M_y = \frac{1}{2} \int_a^b [f^2(x) - g^2(x)] \, dx.$$

It will be left to the reader to check this.

If the boundary of R is more conveniently described by making x a function of y, then there are corresponding y integrals for M_x and M_y.

We now use these formulas to calculate the centroid of a region R (the center of mass of a homogeneous lamina occupying R).

Example 2   Find the centroid of a semicircular disk.

Solution   Let R be the right half-disk bounded by the circle $x^2 + y^2 = r^2$ and the y-axis (Fig. 2). Then, by the single-integral formulas above,

$$\bar{x} = \frac{M_x}{A} = \frac{\displaystyle\int_0^r x \cdot 2\sqrt{r^2 - x^2}\,dx}{\tfrac{1}{2}\pi r^2}$$

$$= \frac{-\tfrac{2}{3}(r^2 - x^2)^{3/2}\big]_0^r}{\tfrac{1}{2}\pi r^2} = \frac{\tfrac{2}{3}r^3}{\tfrac{1}{2}\pi r^2} = \frac{4r}{3\pi}.$$

By symmetry $\bar{y}$ must be 0, so $(4r/3\pi, 0)$ is the centroid.

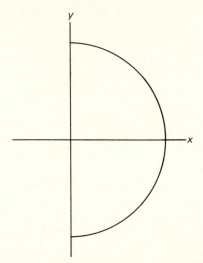

Figure 2

$\square$

Example 3   Find the y coordinate of the centroid of the region bounded by the graphs of $y = e^x$ and $y = 2e^x$, the y-axis and the line $x = 1$.

Solution   The area of R is

$$A = \int_0^1 (2e^x - e^x)\,dx = e^x\bigg]_0^1 = e - 1.$$

According to the new formula above, the moment of R about the x-axis is

$$M_y = \frac{1}{2} \int_0^1 (4e^{2x} - e^{2x})\, dx = \frac{3}{4} e^{2x}\Big]_0^1 = \frac{3}{4}(e^2 - 1).$$

Then

$$\bar{y} = \frac{M_y}{A}$$

$$= \frac{\dfrac{3}{4}(e^2 - 1)}{e - 1} = \frac{3}{4}(e + 1). \qquad \square$$

Example 4   Let R be the region in the first quadrant between the parabola $y = x^2$ and the line $y = 4$. Find the y-coordinate of the centroid of R in two different ways. The area of R is 16/3.

Solution   By the second of the two single-integral formulas above,

$$M_y = \frac{1}{2} \int_0^2 (f^2(x) - g^2(x))\, dx = \frac{1}{2} \int_0^2 (16 - x^4)\, dx$$

$$= \frac{1}{2}\left[16x - \frac{x^5}{5} \right]_0^2 = \frac{1}{2}\left(32 - \frac{32}{5} \right) = \frac{64}{5}.$$

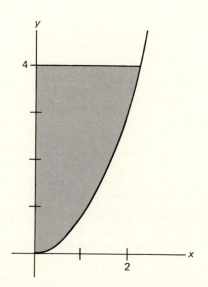

Figure 3

Therefore

$$\bar{y} = \frac{M_y}{A} = \frac{64/5}{16/3} = \frac{12}{5}.$$

But R can also be described as lying between the y-axis and the graph of $x = \sqrt{y}$, from $y = 0$ to $y = 4$ (see Fig. 3). So we can compute $\bar{y}$ by the y integral corresponding to the *first* formula above

$$\bar{y} = \frac{M_y}{A} = \frac{3}{16} \int_c^d y h(y)\, dy = \frac{3}{16} \int_0^4 y^{3/2}\, dy = \frac{3}{16} \cdot \frac{2}{5} y^{5/2} \Big]_0^4 = \frac{12}{5}. \quad \square$$

PROBLEMS FOR SECTION 4

In each of the following problems find the center of mass of the lamina R having the given shape and density. Assume the given value of the mass.

1. R is the unit square with lower left vertex at the origin, and $\rho(x, y) = xy$. Its mass is 1/4.

2. R is the piece of the unit circle $x^2 + y^2 \le 1$ in the first quadrant and $\rho(x, y) = xy$. Its mass is 1/8.

3. R is the lamina between the graphs of $y = x$ and $y = x^2$ and $\rho(x, y) = \sqrt{x}$. The mass is 4/35.

4. R is the parabolic cap bounded by the parabola $y = 4 - x^2$ and the x-axis. The density is proportional to the distance from the x-axis. The mass is $256k/15$, where k is the constant of proportionality for the density.

5. R is the same as in Problem 4, but the density is proportional to the distance from the y-axis. The mass is $8k$.

6. R is the right half of the unit circle $x^2 + y^2 \le 1$ and the density is proportional to the distance from the y-axis. The mass is $2k/3$.

7. R is the right half of the unit circle and the density is proportional to the distance from the x-axis. The mass is $2k/3$.

8. R is the first quadrant of the unit circle and the density is proportional to the sum of the distances from the two coordinate axes. The mass is $2k/3$.

9. R is the square symmetric about the origin with one vertex at (a, a). Its density is proportional to the distance from the line $x + y = 0$ (that is, $\rho(x, y) = k\, |x + y|$). The mass is $8ka^3/3$.

10. R is the piece of the first quadrant under the line $x + y = 1$ and the density is e^{x+y}. the mass is 1.

11. R is the lamina between the graph of $y = \sin x$ and the x-axis, from $x = 0$ to $x = \pi$. The density is proportional to the distance from the y-axis. The mass is πk.

12. R is the above lamina and the density is proportional to the distance from the x-axis. The mass is $\pi k/4$.

13. R is the lamina between the graph $y = e^{-x}$ and the x-axis, from $x = 0$ to infinity; $\rho(x, y) = x$. The mass is 1.

14. Verify the special formulas for M_x and M_y that were used in computing centroids.

In the following problems use either the double integral formulas of Theorem 6 and its corollary, or the reduced single integral formulas.

Find the center of mass of the homogeneous plane lamina R (centroid of the plane of region R) where

15. R is the rectangle with two sides along the axes and a vertex at (a, b).

16. R is an arbitrary rectangle with sides parallel to the axes.

17. R is bounded by the graph $y = |x|$ and the line $y = 2$.

18. R is bounded by the lines $y = x$, $y = 2$, and the y-axis.

19. R is bounded by the y-axis and the parabola $y^2 = 4 - x$.

20. R is the triangle formed by the coordinate axes and the line $x + 2y = 2$.

21. R is bounded by the parabola $y = x^2$ and the line $y = x$.

22. R is bounded by the two parabolas $y = x^2$, $x = y^2$.

23. R is the region in the first quadrant bounded by the cubic $y = x^3$ and the line $y = 2x$.

24. R is bounded by the curve $y^2 = x^3$ and the line $x = 2$.

25. R is bounded by the curves $y = e^x$, $y = -e^x$, $x = 0$, and $x = a$.

26. R is the part of the circular disk of radius r about the origin lying in the first quadrant.

27. R is the right half of the interior of the ellipse

$$\frac{x^2}{a^2} + \frac{y^2}{b^2} = 1.$$

28. R is the region $0 \leq y \leq \ln x$, $x \leq 2$.

29. R is the region $x \geq 0$, $-e^{-x} \leq y \leq e^{-x}$. (Here R extends to infinity and the integrals are improper.)

30. R is the triangle with vertices at $(a, 0)$, $(b, 0)$, and $(0, c)$, where $a < b$ and $c > 0$. [*Hint:* The numerator in the formula for $\bar{x}$ (the moment about the y-axis) can conveniently be evaluated as the sum of two moments.]

31. R is bounded by the parabola $y = x^2$ and the line $y = x + 2$.

32. R is bounded by the x-axis and the first arch of the graph of $\cos x$.

33. R is bounded by the y-axis and the lines $y = 2x$, $y = x + 2$.

34. R is bounded by the parabolas $y = x^2$, $y = 2 - x^2$.

35. R is bounded by the parabolas $y = 2x^2$, $y = x^2 + 1$.

36. Prove the following Theorem of Pappus.

> ***Theorem*** *If a plane region R is revolved about a line in the plane that does not cut into R, then the volume generated is equal to the area of R times the length of the circle traced by the centroid of R.*

Suggestion. Taking the axis of revolution as the y-axis, the center of mass $(\bar{x}, \bar{y})$ traces a circle of

36. (*Continued*)

length $2\pi\bar{x}$ and the formula to be proved is
$$V = 2\pi\bar{x}A,$$
where A is the area of R and V is the volume of the solid of revolution generated by rotating R. In order to prove this formula, write down the formula for $\bar{x}$ and note that its numerator is very close to being the formula for computing V by shells.

38. Compute the formula for the volume of a torus by the Theorem of Pappus. Take the torus to be the solid generated by rotating the circle
$$(x - h)^2 + y^2 = r^2$$
about the y-axis, where $0 < r < h$.

40. Use the Theorem of Pappus to compute the volume of a cylindrical shell with rectangular cross section.

37. We know the formulas for the area of a circle and the volume of a sphere. Use the Theorem of Pappus to compute the centroid of a semicircular disk.

39. The area of an ellipse is πab, where a and b are its semiaxes. Compute the volume of the "elliptical torus" generated by revolving the ellipse
$$\frac{(x - h)^2}{a^2} + \frac{y^2}{b^2} = 1$$
about the y-axis, where $a < h$.

41. The sides of a square S lie along lines having slopes ± 1. Its left vertex is the point $(a, 0)$ and its area is A. Show that the volume generated by rotating S about the y-axis is

$$V = 2\pi A\left(a + \sqrt{\frac{A}{2}}\right).$$

5
THE DOUBLE INTEGRAL IN POLAR COORDINATES

Let R be a region in the xy-plane lying between the polar graphs

$$r = g(\theta) \qquad \text{and} \qquad r = h(\theta),$$

from $\theta = \alpha$ to $\theta = \beta$, as in Fig. 1.

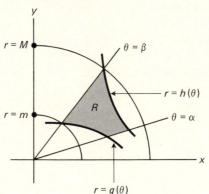

Figure 1

We shall prove that a double integral over R can be evaluated as an iterated integral in *polar* coordinates, as follows:

THEOREM 7 *Let $f(x, y)$ be continuous on a bounded closed region R, and suppose that the boundary of R lies along the polar graphs $r = g(\theta)$, and $r = h(\theta)$, and the rays $\theta = \alpha$ and $\theta = \beta$, where $\alpha < \beta$, and $g(\theta) < h(\theta)$ for all θ between α and β. Then*

$$\iint_R f(x, y)\, dA = \int_\alpha^\beta \int_{g(\theta)}^{h(\theta)} F(r, \theta)\, r\, dr\, d\theta,$$

where $F(r, \theta) = f(r \cos \theta, r \sin \theta)$.

Note the extra factor r in the integral on the right. The proof will be given at the end of the section.

Example 1 Compute the double integral

$$\iint_R e^{-(x^2+y^2)}\, dA,$$

where R is the circular disk $x^2 + y^2 \le a^2$.

Solution In polar coordinates R can be described by the conditions $0 \le r \le a$, $0 \le \theta \le 2\pi$. Moreover,

$$e^{-(x^2+y^2)} = e^{-r^2}.$$

Thus

$$\int_R e^{-(x^2+y^2)}\, dA = \int_0^{2\pi} \int_0^a e^{-r^2} r\, dr\, d\theta$$

$$= \int_0^{2\pi} [-e^{-r^2}/2]_0^a\, d\theta$$

$$= \int_0^{2\pi} \frac{1 - e^{-a^2}}{2}\, d\theta = \pi(1 - e^{-a^2}). \qquad \square$$

This is an example of a double integral that is easier to compute than its one-dimensional analogue. In fact, the integral

$$\int_0^a e^{-x^2}\, dx$$

cannot be computed at all in terms of elementary functions.

Example 2   Compute the volume of a ball of radius a.

Solution   We take the center at the origin, so the ball is bounded by the spherical surface

$$x^2 + y^2 + z^2 = a^2.$$

Moreover, the total volume of the ball is eight times the volume of the piece of the ball in the first octant. Thus

$$V = 8 \iint_R z \, dA = 8 \iint_R \sqrt{a^2 - (x^2 + y^2)} \, dA,$$

where R is the part of the circular disk $x^2 + y^2 \leq a^2$ lying in the first quadrant. By Theorem 7, we can compute this double integral as the following iterated integral in polar coordinates:

$$V = 8 \int_0^{\pi/2} \int_0^a \sqrt{a^2 - r^2} \, r \, dr \, d\theta$$

$$= 8 \int_0^{\pi/2} \left[-\frac{1}{3} (a^2 - r^2)^{3/2} \right]_0^a d\theta$$

$$= \frac{8a^3}{3} \int_0^{\pi/2} d\theta = \frac{8a^3}{3} \cdot \frac{\pi}{2} = \frac{4}{3} \pi a^3.$$

Here again we are helped by the extra r in the integrand of the polar coordinate integral.   □

Example 3   Find the centroid of a semicircular disk R of radius a.

Solution   We can take R to be the right half of the circle $x^2 + y^2 \leq a^2$. Its moment M_x about the y-axis is $\iint_R x \, dA$. Here R can be described in terms of polar coordinates by the conditions that r runs from 0 to a and θ runs from $-\pi/2$ to $\pi/2$. Also $x = r \cos \theta$, so by Theorem 8,

$$M_x = \iint_R x \, dA = \int_{-\pi/2}^{\pi/2} \int_0^a (r \cos \theta) r \, dr \, d\theta$$

$$= \int_{-\pi/2}^{\pi/2} \left[\frac{r^3 \cos \theta}{3} \right]_{r=0}^{r=a} d\theta = \frac{a^3}{3} \int_{-\pi/2}^{\pi/2} \cos \theta \, d\theta$$

$$= \frac{a^3}{3} [\sin \theta]_{-\pi/2}^{\pi/2} = \frac{2}{3} a^3.$$

The x-coordinate of the center of mass is the moment about the y-axis divided by the area:

$$\bar{x} = \frac{M_x}{A} = \frac{\frac{2}{3}a^3}{\frac{\pi a^2}{2}} = \frac{4}{3}\frac{a}{\pi}.$$

The disk will balance on the x-axis, by symmetry, so $\bar{y} = 0$.   □

Example 4 Prove the formula for volumes by shells: If the region A in the xz-plane is bounded above and below by the graphs $z = f(x)$ and $z = g(x)$, and left and right by the line $x = a$ and $x = b$, where $0 \le a < b$, then the volume V generated by revolving A about the z-axis is

$$V = 2\pi \int_a^b x[f(x) - g(x)]\,dx.$$

Solution It is sufficient to consider the case where the lower graph is the x-axis, since the more general volume can be expressed as a difference of two such volumes. (The procedure would be the same as used for areas in Chapter 6, Section 5.) When the graph of f is rotated about the z-axis, the surface it sweeps out is the graph of the function

$$z = f(\sqrt{x^2 + y^2}) = f(r).$$

(The value of f at $r = \sqrt{x^2 + y^2}$ swings around to be the z-coordinate of the surface point over (x, y).) The rotated function has the ring $a \le r \le b$ as its base region R, and its double integral over R is the volume V, by Theorem 4. Therefore,

$$V = \iint_R f(\sqrt{x^2 + y^2})\,dA = \int_0^{2\pi}\int_a^b f(r)r\,dr\,d\theta$$

$$= 2\pi\int_a^b r f(r)\,dr,$$

by Theorem 7. This the desired formula (with r instead of x, and $g = 0$).   □

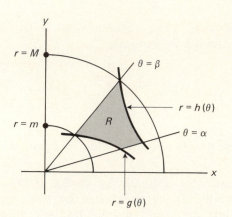
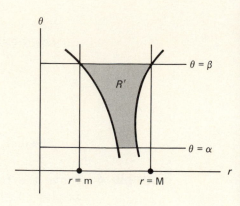

Figure 2

To prove Theorem 7 we must transform the region R, as follows. Each point (x, y) in R determines unique polar coordinates r and θ, where $r = (x^2 + y^2)^{1/2}$ and θ is the unique angle between α and β such that $(\sin \theta, \cos \theta) = (x/r, y/r)$. Thus each point (x, y) in R determines a unique point (r, θ) in the Cartesian plane (with axes labeled r and θ). These (r, θ)-points fill out the region R' bounded by the graphs of $r = g(\theta)$ and $r = h(\theta)$, and the horizontal lines $\theta = \alpha$ and $\theta = \beta$, as shown in the right-hand part of Fig. 2. We say that R' *is the image of R under the mapping* $(x, y) \to (r, \theta)$.

The areas A and A' of these two regions are related by the following lemma.

LEMMA 1   *There is a point (r_0, θ_0) in the region R' such that*

$$A = r_0 A'.$$

Proof   Starting from the polar area formula in Chapter 10, we have

$$A = \int_\alpha^\beta \frac{[h^2(\theta) - g^2(\theta)]}{2} \, d\theta$$

$$= \int_\alpha^\beta \int_{g(\theta)}^{h(\theta)} r \, dr \, d\theta = \iint_{R'} r \, dA$$

by Theorem 3. By the mean-value property for integrals, this double integral is equal to $r_0 A'$, where r_0 is the r coordinate of some point of R'. We thus have proved the lemma.   ∎

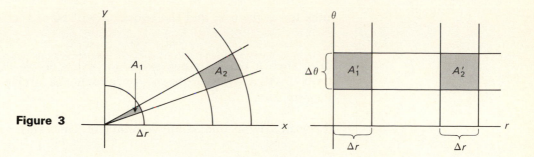

Figure 3

The role played by the conversion factor r_0 in passing from A' to A becomes clear when a small region near the origin is compared with one further away (see Fig. 3).

Lemma 1 leads to the following theorem.

THEOREM 8   *Let $f(x, y)$ be continuous on the bounded closed region R described above, and define $F(r, \theta)$ by*

$$F(r, \theta) = f(r \cos \theta, r \sin \theta).$$

Then

$$\iint\limits_{R} f(x, y)\, dA = \iint\limits_{R'} F(r, \theta) r\, dA',$$

where R' is the image of R under the mapping $(x, y) \to (r, \theta)$.

Proof   If we subdivide R into small subregions R_i, then their images R_i' under the mapping $(x, y) \to (r, \theta)$ form a corresponding subdivision of R'. For each i, we choose an evaluation point (r_i, θ_i) in R_i' such that $\Delta A_i = r_i\, \Delta A_i'$ (by the lemma), and we take (x_i, y_i) as the corresponding point in R_i,

$$(x_i, y_i) = (r_i \cos \theta_i, r_i \sin \theta_i).$$

These choices give us the equation

$$\sum f(x_i, y_i)\, \Delta A_i = \sum F(r_i, \theta_i) r_i\, \Delta A_i'.$$

The left sum is a Riemann sum for $f(x, y)$ over R. The right sum is a Riemann sum for $rF(r, \theta)$ over R'. If the subdivision of R is taken fine enough, both sums will be nth Riemann sums for a given n, and the equation will be of the form

$$S_n = S'_n.$$

Since such a pair of sums exists for every n, we can take limits in this equation as $n \to \infty$ and hence arrive at the theorem.   ∎

Theorem 7 is an immediate corollary of Theorems 3 and 8.

PROBLEMS FOR SECTION 5

1. Find the volume of the solid in the half-space $z \geq 0$ based on the circular disk $x^2 + y^2 \leq 9$ and capped by the sphere $x^2 + y^2 + z^2 = 25$.

2. Find the volume of the solid cut out of the spherical ball $x^2 + y^2 + z^2 \leq b^2$ by the cylinder $x^2 + y^2 = a^2$, where $0 < a \leq b$.

3. Find the volume of the cone $0 \leq z \leq 1 - r$.

4. Find the volume of the cone $0 \leq z \leq (a - r)h/a$.

5. Find the volume of the solid based on the interior of the cardioid $r = 1 + \cos \theta$ and capped by the cone $z = 2 - r$.

6. Find the volume of the solid based on the interior of the circle $r = \cos \theta$ and capped by the cone $z = 1 - r$.

7. Find the volume of the solid based on the interior of the circle $r = \cos \theta$ and capped by the sphere $r^2 + z^2 = 1$.

8. Find the volume of the solid based on the interior of the circle $r = \cos \theta$ and capped by the plane $z = x$.

9. Show that the mass of a circular lamina of radius a whose density is proportional to the distance from the center is $2\pi k a^3/3$, where k is the constant of proportionality.

10. Compute the center of mass of the right half of the above lamina.

11. Compute the mass and center of mass of the semicircular lamina $0 \leq r \leq 1$, $0 \leq \theta \leq \pi$, if the density is proportional to $r(1 - r)$.

12. Compute the mass and center of mass of the semicircular lamina $0 \leq r \leq 1$, $0 \leq \theta \leq \pi$, if its density is proportional to $\sqrt{1 - r^2}$.

13. Find the volume of the solid obtained by intersecting the cylinders

$$x^2 + y^2 = a^2 \quad \text{and} \quad x^2 + z^2 = a^2.$$

14. Find the moments of the plane region bounded by the polar graph of $r = (\cos \theta)^{1/3}$.

15. Find the center of mass of the plane lamina bounded by the cardioid $r = 1 + \cos \theta$ if its density is proportional to $|y|$.

16. Find the centroid of the right-hand leaf of the region bounded by polar graph of $r = \cos 2\theta$.

17. Find the centroid of the cardioid $r = a(1 + \cos \theta)$.

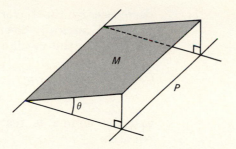

Figure 1

6
SURFACE AREA

Let M and P be two planes that intersect at an angle θ. Their line of intersection will be called their axis.

When we project from M vertically down onto P, lengths parallel to the axis remain unchanged, but lengths perpendicular to the axis are multiplied by $\cos\theta$, and hence are decreased by this constant factor.

It follows that if a region R' on M is projected vertically downward to P, then it is compressed by the factor $\cos\theta$ in the direction perpendicular to the axis, and its area is therefore decreased by the factor $\cos\theta$:

$$\text{area of image region } R = \cos\theta(\text{area of } R').$$

We assume this. However, a proof could be constructed by starting with rectangles having one side parallel to the axis, and then approximating more general regions by finite collections of such rectangles, somewhat in the manner of our Riemann-sum approximations. We shall use this relationship in reverse.

LEMMA *If R is a region in the xy-plane, and if M is the plane $z = ax + by + c$, then the area of the region on M lying over R is*

$$\sqrt{1 + a^2 + b^2}(\text{area of } R).$$

Proof We know that the vector $(-a, -b, 1)$ is normal to M, and that if θ is the angle between this vector and the z-axis, then

$$\cos\theta = \frac{1}{\sqrt{1 + a^2 + b^2}}.$$

(See Section 5 in Chapter 13.) Since θ is also the angle between M and the xy-plane, a region R' on M has its area decreased by the factor $\cos \theta = 1/\sqrt{1 + a^2 + b^2}$ when it is projected vertically downward to the xy-plane, becoming the region R on the xy-plane. That is,

$$\text{area of } R = \frac{1}{\sqrt{1 + a^2 + b^2}} (\text{area of } R'),$$

and the lemma is just the cross-multiplied form of this equation. ∎

Now consider the piece of the surface $z = f(x, y)$ lying over the region R in the xy-plane (Fig. 2). Its tangent plane at $z_0 = f(x_0, y_0)$,

$$z - z_0 = \frac{\partial z}{\partial x}\bigg|_{(x_0, y_0)} (x - x_0) + \frac{\partial z}{\partial y}\bigg|_{(x_0, y_0)} (y - y_0),$$

remains very nearly parallel to the surface near the point of tangency. So if ΔR is an incremental subregion containing the evalu-

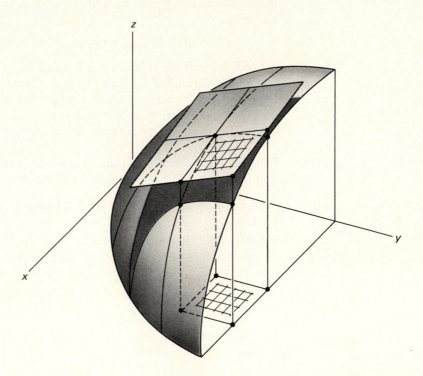

Figure 2

ation point (x_0, y_0), and if ΔS is the area of the little piece of surface lying over ΔR, then ΔS is approximately equal to the corresponding area on the tangent plane. By the lemma, this is

$$\sqrt{1 + \left(\frac{\partial z}{\partial x}\right)_0^2 + \left(\frac{\partial z}{\partial y}\right)_0^2}\, \Delta A,$$

where ΔA is the area of ΔR. We have used the abbreviation $(\partial z/\partial x)_0$ for the value of the derivative at (x_0, y_0). It seems plausible that this approximation improves as ΔA approaches 0, in the sense that the ratio of the two areas approaches 1. This is the same thing as saying that

$$\Delta S \approx \sqrt{1 + \left(\frac{\partial z}{\partial x}\right)_0^2 + \left(\frac{\partial z}{\partial y}\right)_0}\, \Delta A,$$

with an error that is small compared to ΔA. Therefore if we subdivide R into suitably small subregions ΔR_i and add up the above estimates, we see that the area S of the surface over R is given approximately by

$$S \approx \sum \sqrt{1 + \left(\frac{\partial z}{\partial x}\right)_i + \left(\frac{\partial z}{\partial y}\right)_i}\, \Delta A_i,$$

with an error that is small compared to the area A of R. The partial derivatives are evaluated at arbitrarily chosen points (x_i, y_i) in ΔR_i.

Since the sum on the right above is a Riemann sum for the function $\sqrt{1 + (\partial z/\partial x)^2 + (\partial z/\partial y)^2}$, we conclude as usual that S is given exactly by the double integral of this function over R.

We have thus shown that the following theorem is a consequence of plausible geometric assumptions.

THEOREM 9 *If $f(x, y)$ is a continuously differentiable function whose domain includes the region R, then the area of the piece of the surface $z = f(x, y)$ lying over R is*

$$\iint\limits_R \sqrt{1 + \left(\frac{\partial z}{\partial x}\right)^2 + \left(\frac{\partial z}{\partial y}\right)^2}\, dA.$$

Example 1 Find the area of the cap of the paraboloid $2z = x^2 + y^2$ lying under the plane $z = 2$.

Solution   The surface intersects the plane $z = 2$ over the circle $x^2 + y^2 = 4$. The surface area S is therefore given by

$$S = \iint\limits_{x^2+y^2\leq 4} \sqrt{1 + x^2 + y^2}\, dA.$$

We compute the double integral as an iterated integral in polar coordinates and have

$$S = \int_0^{2\pi} \int_0^2 \sqrt{1 + r^2}\, r\, dr\, d\theta$$

$$= 2\pi \cdot \frac{1}{3}(1 + r^2)^{3/2}\Big]_0^2 = \frac{2\pi}{3}(5^{3/2} - 1). \qquad \square$$

Example 2   Find the formula for the area of the surface of a sphere.

Solution   The sphere of radius a about the origin has the equation

$$x^2 + y^2 + z^2 = a^2.$$

In order to compute the partial derivatives $\partial z/\partial x$ and $\partial z/\partial y$, we can solve for z and differentiate, or we can simply differentiate implicitly in the equation as it stands. The latter method gives

$$2x + 2z\frac{\partial z}{\partial x} = 0,$$

so $\partial z/\partial x = -x/z$. Similarly, $\partial z/\partial y = -y/z$. Since the total surface area S is eight times the area in the first octant, we have

$$S = 8\iint\limits_R \sqrt{1 + \left(\frac{\partial z}{\partial x}\right)^2 + \left(\frac{\partial z}{\partial y}\right)^2}\, dA$$

$$= 8\iint\limits_R \sqrt{1 + \left(\frac{x}{z}\right)^2 + \left(\frac{y}{z}\right)^2}\, dA,$$

where R is the quarter circle in the first quadrant. If we write 1 as z/z in the integrand, we see that

$$S = 8 \iint_R \frac{a}{z} \, dA.$$

In polar coordinates, $z = \sqrt{a^2 - (x^2 + y^2)} = \sqrt{a^2 - r^2}$, so

$$\iint_R \frac{a}{z} \, dA = a \int_0^{\pi/2} \int_0^a \frac{r \, dr}{\sqrt{a^2 - r^2}} \, d\theta$$

$$= a \int_0^{\pi/2} [-\sqrt{a^2 - r^2}]_0^a \, d\theta = a^2 \int_0^{\pi/2} d\theta$$

$$= \frac{\pi}{2} a^2.$$

The total area S is eight times this, so the formula for the area of the surface of a sphere of radius a is

$$S = 4\pi a^2. \qquad \square$$

PROBLEMS FOR SECTION 6

Find the area of each of the following surfaces.

1. The piece of the paraboloid $z = 4 - x^2 - y^2$ that lies above the xy-plane. (Polar coordinates are simplest.)

2. The surface cut out of the cone $z = 1 - r$ by the cylinder $r = \cos \theta$.

3. The piece of the sphere $x^2 + y^2 + z^2 = 25$ lying over the circular disk $x^2 + y^2 \le 16$.

4. The piece of the sphere $x^2 + y^2 + z^2 = a^2$ cut out by the cylinder $x^2 + y^2 = b^2$ and lying above the xy-plane, where $0 < b < a$.

5. The piece of the sphere $x^2 + y^2 + z^2 = 1$ lying over the interior of the circle $r = \cos \theta$.

6. The piece of the plane $z = 6 - 2x - 3y$ lying over the unit square with lower left corner at the origin.

7. The piece of the above plane that is cut off by the first octant.

8. The piece of the plane $z = ax + by + c$ lying over the region R. (That is, prove that the Lemma is a special case of Theorem 9.)

9. The cone of altitude h and base radius a.

10. The piece of the parabolic cylinder $z = 4 - x^2$ lying over the triangle bounded by $y = x$, $y = 0$, $x = 2$.

11. When $r \leq a$, the equation

$$az = h(a - r)$$

describes the cone whose altitude is h and whose base is the circular disk $r \leq a$. Let R be any region lying in this disk. Show that the area of the part of the cone lying over R is

$$\sqrt{1 + \left(\frac{h}{a}\right)^2} \text{ (area of } R\text{).}$$

12. Show that the graphs of the functions $f(x, y) = x^2 + y^2$, $g(x, y) = 2xy$ have the same area over any region R in the xy-plane.

13. The same for $f(x, y) = h(x) + k(y)$, $g(x, y) = h(x) - k(y)$.

14. A function $v(x, y)$ is said to be conjugate to a function $u(x, y)$ if

$$\frac{\partial u}{\partial x} = \frac{\partial v}{\partial y}, \qquad \frac{\partial u}{\partial y} = -\frac{\partial v}{\partial x}.$$

Show that if $v(x, y)$ is conjugate to $u(x, y)$ over a region R in the xy-plane, then the graph of u over R has the same surface area as the graph of v over R.

Referring to the above result, show that each of the following pairs of functions have graphs of equal surface area over any region R over which they both are defined.

15. $x^2 - y^2$, $2xy$

16. $x^3 - 3xy^2$, $3x^2y - y^3$

17. $e^x \cos y$, $e^x \sin y$

18. $\frac{1}{2}\ln(x^2 + y^2)$, $\arctan\dfrac{y}{x}$

19. If the graph of a positive function $y = f(x)$ is revolved about the x-axis, the area of the surface of revolution it sweeps out from $x = a$ to $x = b$ given by

$$A = 2\pi\int_a^b f(x)\sqrt{1 + [f'(x)]^2}\, dx.$$

Prove this formula from the general area formula for the graph of a function $z = g(x, y)$.

20. Suppose that $z = f(r)$, where $r = \sqrt{x^2 + y^2}$. Show that the area of the graph of $z = f(r)$ over the circle $x^2 + y^2 \leq a^2$ equals

$$2\pi\int_0^a r\sqrt{1 + [f'(r)]^2}\, dr.$$

21. When the circle $(x - b)^2 + z^2 = a^2, 0 < a < b$ is revolved about the z-axis, the resulting surface is a torus. Its equation can be written

$$(r - b)^2 + z^2 = a, \qquad r = \sqrt{x^2 + y^2}.$$

Find its total surface area.

7
THE TRIPLE
INTEGRAL

Let $f(x, y, z)$ be a continuous function of three variables defined on a bounded space region G. Then f has a *triple* integral over G, designated

$$\iiint_G f(x, y, z)\, dV \qquad \text{or} \qquad \iiint_G f,$$

that is just like the two-dimensional double integral. It is defined in the same way as the limit of Riemann sums, and it has the same properties (1)–(7) that were listed in Section 2. The is just one change: volumes replace areas.

Again, it turns out that the multiple integral can be evaluated by iterated integration, that is by "integrating out" the variables one at a time. We shall assume this and turn to the practical problem of determining the limits of integration. They occur at the boundary of the integration domain G, and a sketch of G is almost essential in order to get them right.

Example 1 Use iterated integration to evaluate the triple integral

$$\iiint_G x\, dV,$$

where G is the region cut off from the first octant by the plane $x + y + z = 2$.

Solution The region G is shown in Fig. 1. Suppose that we decide to integrate first with respect to z. This means holding

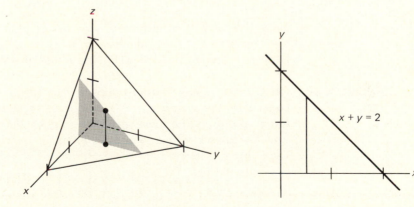

Figure 1

x and y fixed and integrating the resulting function of z. Geometrically, the domain of this integration is a vertical line segment crossing G from the xy-plane to the plane $x + y + z = 2$. Along this segment, z varies from $z = 0$ to $z = 2 - x - y$.

So the first integral is

$$\int_0^{2-x-y} x \, dz = xz \Big]_{z=0}^{z=2-x-y}$$

$$= x(2 - x - y).$$

With z integrated out, we have left a double integral of a function of x and y, over the region R in the xy-plane that is the base of the original space region G. Here R is the triangle cut off in the first quadrant by the line $x + y = 2$. If we choose to integrate next with respect to y, the calculation finishes off as follows:

$$\int_0^2 \int_0^{2-x} x(2 - x - y) \, dy \, dx = \int_0^2 \left[-x \frac{(2 - x - y)^2}{2} \right]_0^{2-x} dx$$

$$= \frac{1}{2} \int_0^2 x(2 - x)^2 \, dx$$

$$= \frac{1}{2} \int_0^2 (4x - 4x^2 + x^3) \, dx$$

$$= \frac{1}{2} \left[2x^2 - \frac{4x^3}{3} + \frac{x^4}{4} \right]_0^2$$

$$= 4 - \frac{16}{3} + 2 = \frac{2}{3}. \qquad \square$$

Note the way the first step was taken above. The fact that $(x + c)^{n+1}/(n + 1)$ is an antiderivative of $(x + c)^n$ often comes in handy. Here is a formal statement of the above procedure.

THEOREM 10   *Let G be the space region based on the region R in the xy-plane and capped by the graph $z = \phi(x, y)$. Suppose also that R is bounded in the xy-plane by the lower graph $y = g(x)$, the upper graph $y = h(x)$, and the lines $x = a$ and $x = b$. Then*

$$\iiint_G f(x, y, z) \, dV = \int_a^b \int_{g(x)}^{h(x)} \int_0^{\phi(x,y)} f(x, y, z) \, dz \, dy \, dx.$$

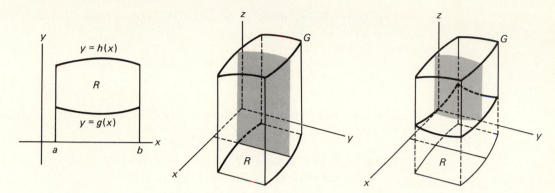

Figure 2

A slightly more general domain is shown at the right in Fig. 2. Here, G lies over the same base region R in the xy-plane, but runs from the lower surface $z = \psi(x, y)$ to the upper surface $z = \phi(x, y)$. Now,

$$\iiint\limits_{G} f \, dV = \int_{a}^{b} \int_{g(x)}^{h(x)} \int_{\psi(x,y)}^{\phi(x,y)} f \, dz \, dy \, dx.$$

Example 2 Suppose G is the region between the parabolic surfaces $z = x^2 + y^2$ and $z = 2 - (x^2 + y^2)$. The intersection curve of these surfaces is the circle $x^2 + y^2 = 1$ in the plane $z = 1$. The part of G in the first octant is shown in Fig. 3. The base region R in the xy-plane is the interior of

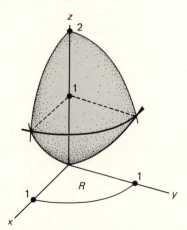

Figure 3

the unit circle, with y ranging therefore from $y = -\sqrt{1-x^2}$ to $y = \sqrt{1-x^2}$. Thus

$$\iiint_G f\, dV = \int_{-1}^{1} \int_{-\sqrt{1-x^2}}^{\sqrt{1-x^2}} \int_{x^2+y^2}^{2-(x^2+y^2)} f\, dz\, dy\, dx. \qquad \square$$

Example 3   Use iterated triple integration to find the volume of the tetrahedral region of Example 1.

Solution   The volume of a space region G is simply the triple integral over G of the constant 1,

$$V = \iiint_G 1\, dV$$

thus

$$
\begin{aligned}
V &= \int_0^2 \int_0^{2-x} \int_0^{2-x-y} dz\, dy\, dx \\
&= \int_0^2 \int_0^{2-x} (2 - x - y)\, dy\, dx \\
&= \int_0^2 \left[-\frac{(2-x-y)^2}{2} \right]_0^{2-x} dx = \frac{1}{2} \int_0^2 (2-x)^2\, dx \\
&= -\frac{1}{6}(2-x)^3 \bigg]_0^2 = \frac{4}{3}. \qquad \square
\end{aligned}
$$

There is nothing sacred about the order in which the variables are integrated when we evaluate a multiple integral by iterated integration, and sometime a different order will make for an easier calculation, as we have already seen in the double-integral sections. Here is a triple-integral example.

Example 4   Compute the volume of the wedge cut from the cylinder $y^2 + z^2 = 1$ in the first octant by the planes $y = x$ and $x = 0$ (Fig. 4).

Solution   In the order we have been using so far, this sets up as the iterated integral

$$V = \int_0^1 \int_x^1 \int_0^{\sqrt{1-y^2}} dz\, dy\, dx = \int_0^1 \int_x^1 \sqrt{1-y^2}\, dy\, dx,$$

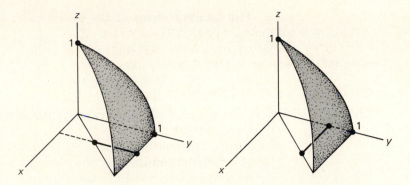

Figure 4

and now we have to compute the rather nasty integral $\int \sqrt{1 - y^2} \, dy$. But if we interchange the order of the x and y integrations, we have

$$V = \int_0^1 \int_0^y \int_0^{\sqrt{1-y^2}} dz \, dx \, dy = \int_0^1 \int_0^y \sqrt{1 - y^2} \, dx \, dy$$

$$= \int_0^1 y\sqrt{1 - y^2} \, dy = -\frac{1}{2} \int_1^0 u^{1/2} \, du \quad (u = 1 - y^2)$$

$$= \frac{1}{3} u^{3/2} \Big]_0^1 = \frac{1}{3}. \qquad \qquad \square$$

A Riemann sum for the triple integral has the form

$$\sum f(x_i, y_i, z_i) \, \Delta V_i,$$

where ΔV_i is the *volume* of the ith incremental piece of G. In terms of such sums, the proof in Section 3 carries over almost word for word and shows: if $\rho(x, y, z)$ is the density of a varying distribution of mass, then

$$m = \iiint\limits_{G} \rho(x, y, z) \, dV$$

is the total mass of G. We skip all the details.

Riemann-sum estimates of turning moments would proceed exactly as in the two-dimensional case, with analogous conclusions:

The moment M about the origin of a mass distribution with density $\rho(x, y, z)$ is the vector

$$M = (M_x, M_y, M_z) = \left(\iiint x\rho \, dV, \iiint y\rho \, dV, \iiint z\rho \, dV \right).$$

The center of mass of the distribution is given by

$$(\bar{x}, \bar{y}, \bar{z}) = \frac{M}{m} = \left(\frac{M_x}{m}, \frac{M_y}{m}, \frac{M_z}{m}\right),$$

where $m = \iiint \rho \, dV$ is the total mass of the distribution.

Example 5 Find the centroid of the tetrahedron in Example 1.

Solution The centroid of a space region G is the center of mass of a distribution in G having a constant density k, and k may as well be taken to have the value $k = 1$, since it factors out and cancels in the quotients defining the center of mass.

   The moment of the above tetrahedron about the origin thus has the x-component $M_x = \iiint_G x \, dV$, which we computed in Example 1 to be 2/3. The mass of the tetrahedron is its volume, which we found in Example 3 to be 4/3. Thus,

$$\bar{x} = \frac{M_x}{m} = \frac{2/3}{4/3} = \frac{1}{2}.$$

Moreover, the configuration is symmetric with respect to the variables x, y, and z, so $\bar{y}$ and $\bar{z}$ must also have the value 1/2. Thus

$$(\bar{x}, \bar{y}, \bar{z}) = \left(\frac{1}{2}, \frac{1}{2}, \frac{1}{2}\right) = \frac{1}{2}(1, 1, 1). \qquad \square$$

Example 6 Find the centroid of the piece of the unit ball lying in the first octant.

Solution The unit sphere has the equation $x^2 + y^2 + z^2 = 1$, so

$$M_z = \iiint_G z \, dV = \int_0^1 \int_0^{\sqrt{1-x^2}} \int_0^{\sqrt{1-x^2-y^2}} z \, dz \, dy \, dx$$

$$= \int_0^1 \int_0^{\sqrt{1-x^2}} \frac{z^2}{2} \Big]_0^{\sqrt{1-x^2-y^2}} dy \, dz$$

$$= \frac{1}{2} \int_0^1 \int_0^{\sqrt{1-x^2}} (1 - x^2 - y^2) \, dy \, dz$$

$$= \frac{1}{2} \int_0^1 \left[(1 - x^2)y - \frac{y^3}{3} \right]_0^{\sqrt{1-x^2}} dx = \frac{1}{3} \int_0^1 (1 - x^2)^{3/2} \, dx$$

$$= \frac{1}{3} \int_0^{\pi/2} \cos^4 u \, du \qquad \text{(from } x = \sin u \text{)}$$

$$= \frac{1}{3} \int_0^{\pi/2} \left(\frac{1 + \cos 2u}{2} \right)^2 du$$

$$= \frac{1}{12} \int_0^{\pi/2} \left(1 + 2 \cos 2u + \left(\frac{1 + \cos 4u}{2} \right) \right) du$$

$$= \frac{1}{12} \int_0^{\pi/2} \frac{3}{2} \, du \qquad \left(\text{since } \int_0^{\pi/2} \cos 2u \, du = 0, \text{ etc.} \right)$$

$$= \frac{\pi}{16}.$$

Since the volume of one eighth of the unit ball is

$$\frac{1}{8} \left(\frac{4}{3} \pi \right) = \frac{\pi}{6},$$

we have

$$\bar{z} = \frac{M_z}{V} = \frac{\pi/16}{\pi/6} = \frac{3}{8}.$$

By symmetry, $\bar{x}$ and $\bar{y}$ have the same value, so the centroid is

$$(\bar{x}, \bar{y}, \bar{z}) = \frac{3}{8} (1, 1, 1). \qquad \square$$

PROBLEMS FOR SECTION 7

Find the volume of each of the following solids by triple (iterated) integration.

1. The tetrahedron cut from the first octant by the plane $3x + 2y + z = 6$.

2. The rectangular solid cut off from the first octant by the planes $x = a$, $y = b$, $z = c$ where a, b, c are positive.

3. The tetrahedron cut off from the first octant by the plane

$$\frac{x}{a} + \frac{y}{b} + \frac{z}{c} = 1.$$

4. The solid cut off from the first octant by the parabolic surfaces $z = 1 - x^2$ and $y = 1 - x^2$.

5. The solid cut off from the first octant by the parabolic surfaces $z = 1 - x^2$ and $x = 1 - y^2$.

6. The solid cut off from the first octant by the plane $z = x$ and the cylinder $x^2 + y^2 = 1$.

7. The solid cut off from the first octant by the plane $z = y$ and the cylinder $x^2 + y^2 = 1$.

8. The solid cut off from the first octant by the cylinders $x^2 + y^2 = 1$ and $x^2 + z^2 = 1$.

9. The solid bounded by the cylinders $x^2 + y^2 = 1$ and $y^2 + z^2 = 1$.

10. The solid bounded above by $z = 1 - x^2$ and below by $z = y^2$.

11. Compute the triple integral of $f(x, y, z) = y$ over the region G lying in the first octant between the xy-plane and the surface $z = 4 - (x^2 + y^2)$.

12. Compute the triple integral of $f(x, y, z) = xy$ over the region G cut off from the first octant by the ellipsoid

$$\frac{x^2}{a^2} + \frac{y^2}{b^2} + \frac{z^2}{c^2} = 1.$$

13. Compute the triple integral of $f(x, y, z) = xyz$ over the solid cut off from the first octant by the cylinders $x^2 + z^2 = 4$ and $x^2 + y^2 = 4$.

14. Calculate the centroid of the wedge cut out of the cylinder $x^2 + y^2 = 1$ by the plane $z = y$ above and the plane $z = 0$ below.

15. Calculate the centroid of the solid cut from the first octant by the ellipsoid

$$\frac{x^2}{a^2} + \frac{y^2}{b^2} + \frac{z^2}{c^2} = 1.$$

16. Calculate the centroid of the tetrahedron in Problem 3.

(Assume that the volume of the ellipsoid is $(4/3)\pi abc$.)

8
CYLINDRICAL
AND
SPHERICAL
COORDINATES

After first integrating out z in a triply-iterated integral, there is left an iterated integral in x and y that is equal to a double integral over the base region R in the xy-plane. We can rewrite the integral this way if we wish.

For example, suppose we want to integrate the function $f(x, y, z)$ over the space region G that is based on the region R in the xy-plane and capped by the graph of $z = \phi(x, y)$. Just as in the last section, we first integrate with respect to z from 0 to $\phi(x, y)$, and have left the double integral over R

$$\int_R \int \left[\int_0^{\phi(x,y)} f(x, y, z)\, dz \right] dA.$$

Now suppose that R is most easily described in polar coordinates, as in Section 5. Then this remaining double integral can be computed in polar coordinates, as in that section.

Example 1 Find the mass of the solid S described by the inequalities $x^2 + y^2 \leq 1$, $0 \leq z \leq x$, if the density of S is given by

$$\rho(x, y, z) = z\sqrt{x^2 + y^2}.$$

Solution

$$m = \iiint_S \rho \, dV = \iiint_S z\sqrt{x^2 + y^2} \, dV$$

$$= \iint_R \int_0^x \sqrt{x^2 + y^2} \, z \, dz \, dx \, dy$$

$$= \iint_R \frac{x^2}{2} \sqrt{x^2 + y^2} \, dx \, dy.$$

Here the integration limits for R are most easily described in polar coordinates, as $-\pi/2 \le \theta \le \pi/2$, $0 \le r \le 1$. So we evaluate this double integral as an iterated integral in polar coordinates, by Theorem 7. This means replacing $dx \, dy$ by $r \, dr \, d\theta$, and setting $x = r \cos \theta$, $x^2 + y^2 = r^2$. We now have

$$\iint_{R'} \frac{(r \cos \theta)^2}{2} \cdot r \cdot r \, dr \, d\theta = \int_0^1 \frac{r^4}{2} \, dr \int_{-\pi/2}^{\pi/2} \cos^2\theta \, d\theta$$

$$= \frac{1}{10}\left[\frac{1 + \cos 2\theta}{2}\right]_{-\pi/2}^{\pi/2} = \frac{1}{20}\,\pi. \qquad \square$$

Example 2 In practice, we would notice on reading the above problem that we want to replace x and y by polar coordinates and would do so from the beginning:

$$m = \iiint_S \rho \, dV = \int_{-\pi/2}^{\pi/2} \int_0^1 \left[\int_0^{r\cos\theta} rz \, dz\right] r \, dr \, d\theta \text{ etc.} \qquad \square$$

Here is the general statement of this procedure.

THEOREM 11 *Suppose that the space region G is based on the region R in the xy-plane and is capped by the surface $z = \phi(x, y)$. Suppose that R is bounded by the polar graphs $r = g(\theta)$ and $r = h(\theta)$ and the rays $\theta = \alpha$ and $\theta = \beta$, where $\alpha < \beta$, and $g(\theta) < h(\theta)$ on $[\alpha, \beta]$. Then*

$$\boxed{\iiint_G f(x, y, z) \, dV = \int_\alpha^\beta \int_{g(\theta)}^{h(\theta)} \left[\int_0^{\psi(r,\theta)} F(r, \theta, z) \, dz\right] r \, dr \, d\theta}$$

where $F(r, \theta, z) = f(r \cos \theta, r \sin \theta, z)$, and $\psi(r, \theta) = \phi(r \cos \theta, r \sin \theta)$.

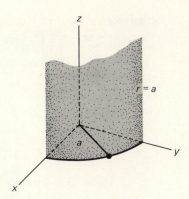

Figure 1

When we set $x = r \cos \theta$, $y = r \sin \theta$, and interpret r and θ as polar coordinates of the point (x, y), then we call the triple (r, θ, z) the "cylindrical" coordinates of the point $(x, y, z) = (r \cos \theta, r \sin \theta, z)$. There is a simple reason for the terminology; in this interpretation of the graph of $r = a$ is the cylinder of radius a about the z-axis (Fig. 1.) In general, the *cylindrical graph* of an equation such as $r = f(\theta, z)$, or $z = g(r, \theta)$, or $h(r, \theta, z) = 0$ consists of the points in *xyz*-space whose cylindrical coordinates satisfy the equation.

Spherical Coordinates The *spherical* coordinates of a point are shown in Fig. 2. Essentially, they combine the polar angle θ in the xy-plane with polar coordinates, ρ, ϕ in the *half-plane* determined by θ and the z-axis.

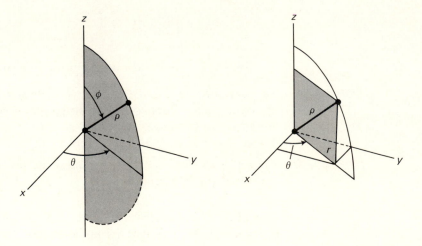

Figure 2

Note that ϕ ranges from 0 to π only. Referring to Fig. 2, we see first that $z = \rho \cos \phi$ and $r = \rho \sin \phi$. Then

$$x = r \cos \theta = \rho \sin \phi \cos \theta,$$
$$y = r \sin \theta = \rho \sin \phi \sin \theta.$$

The change-of-coordinate formulas are thus

$$x = \rho \sin \phi \cos \theta,$$
$$y = \rho \sin \phi \sin \theta,$$
$$z = \rho \cos \phi.$$

REMARK Note that we are using ρ here for the distance to the origin; earlier ρ was our standard symbol for density.

Triple integrals in which the integrand f or the domain of integration G exhibit some elements of spherical symmetry about the origin may be easier to compute by iterated integration in spherical coordinates. Here is the basic theorem.

THEOREM 12 *Let $f(x, y, z)$ be a continuous function on the space region G, and define the function K by*

$$K(\rho, \phi, \theta) = f(\rho \sin \phi \cos \theta, \rho \sin \phi \sin \theta, \rho \cos \phi).$$

Then

$$\iiint\limits_{G} f(x, y, z) \, dV = \iiint K(\rho, \phi, \theta) \rho^2 \sin \phi \, d\rho \, d\phi \, d\theta,$$

where the integral on the right is an iterated integral in which the limits of integration describe the boundary of G in terms of ρ, ϕ, and θ.

Example 3 Compute the volume of the sphere of radius a by using spherical coordinates.

Solution The sphere of radius a is described in terms of the spherical-coordinate variables by the inequalities $0 \le \rho \le a$,

$0 \le \phi \le \pi$, $0 \le \theta \le 2\pi$. Therefore,

$$V = \iiint\limits_{G} 1 \, dV$$

$$= \int_0^{2\pi} \int_0^{\pi} \int_0^{a} \rho^2 \sin \phi \, d\rho \, d\phi \, d\theta$$

$$= \int_0^{2\pi} \int_0^{\pi} \frac{\rho^3}{3} \sin \phi \Big]_{\rho=0}^{a} d\phi \, d\theta$$

$$= \frac{a^3}{3} \int_0^{2\pi} \int_0^{\pi} \sin \phi \, d\phi \, d\theta = \frac{a^3}{3} \int_0^{2\pi} [-\cos \phi]_0^{\pi} \, d\theta$$

$$= \frac{2a^3}{3} \int_0^{2\pi} d\theta = \frac{2a^3}{3} \cdot 2\pi = \frac{4}{3} \pi a^3. \qquad \Box$$

Example 4   Find the mass of the "donut without a hole" that is bounded by the space graph of

$$\rho = \sin \phi,$$

if its density is $1/\rho$. Figure 3 shows the part of the solid in the first octant.

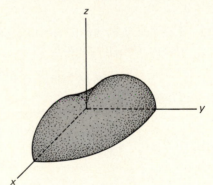

Figure 3

Solution

$$V = \int_0^{2\pi} \int_0^{\pi} \int_0^{\sin\phi} \frac{1}{\rho} \rho^2 \sin \phi \, d\rho \, d\phi \, d\theta$$

$$= \int_0^{2\pi} \int_0^{\pi} \frac{\sin^3 \phi}{2} \, d\phi \, d\theta$$

$$= \frac{4\pi}{3}. \qquad \Box$$

> **REMARK**   This problem can also be solved by the theorem of Pappus. See Problem 36 in Section 4.

PROBLEMS FOR SECTION 8

In Problems 1 through 7 find the volume of the given solid by iterated integration in cylindrical or spherical coordinates.

1. The region between the paraboloid $z = x^2 + y^2$ and the plane $z = 4$.

2. The spherical cap $a < z < \sqrt{b^2 - r^2}$ (where $a \leq b$).

3. The "ice cream cone" cut from the sphere $x^2 + y^2 + z^2 = 4$ by the vertical cone with vertex at the origin and semi-vertex angle equal to $\pi/6$.

4. The region cut from the sphere $x^2 + y^2 + z^2 = a^2$ by the upper half of the cone generated by rotating the line $z = mx$ about the z-axis. Use cylindrical coordinates.

5. The region cut from the sphere $\rho = a$ by the cone $\phi = \alpha$ (spherical coordinates). Interpret your answer for the case $\alpha > \pi/2$, and, in particular for $\alpha = \pi$.

6. The region between the paraboloid $z = x^2 + y^2$ and the plane $z = 2x$.

7. The region bounded by the sphere $x^2 + y^2 + z^2 = 6$ and the paraboloid $z = x^2 + y^2$.

8. Find the mass of a ball of radius a if its density is proportional to the distance from the center.

9. Find the mass of a ball of radius a if its density is inversely proportional to the distance from the center.

10. The density of a ball is $k\rho^\alpha$, where ρ is the distance to the origin and α is a constant. Find what limitation is placed on the exponent α by the requirement that the mass of the ball be finite.

11. The density of a ball centered at the origin is proportional to the distance from the xy-plane. Find the mass of the ball.

12. The density of a ball centered at the origin is proportional to the distance from the z-axis. Find the mass of the ball.

13. A cylinder is bounded by the cylindrical surface $x^2 + y^2 = a^2$ and the planes $z = 0$, $z = b$. Its density is zr. Find its mass.

14. Find the centroid of the region between the xy-plane and the paraboloid $z = 4 - (x^2 + y^2)$.

15. Find the centroid of the "ice cream cone" region of Problem 3.

16. Find the centroid of the region of Problem 7.

17. Find the centroid of the region cut off from the first octant by the sphere $x^2 + y^2 + z^2 = a^2$.

18. Write out the proof of Theorem 11, using the piercing formula.

19. Find the volume of the "cored apple" obtained by removing from the ball $x^2 + y^2 + z^2 \leq b^2$ its intersection with the cylinder $x^2 + y^2 \leq a^2$, where $a < b$.

20. Find the moments about the coordinate axes of the piece of the above solid cut off by the first octant.

21. Given $0 < a < b$, let S be the spherical shell with inner radius a and outer radius b. Find the mass of the piece of S lying in the first octant.

Show that the limit of the centroid as $a \to b$ is the point $(b/2, b/2, b/2)$.

23. Find the volume of the solid bounded by the paraboloid $z = 2 - (x^2 + y^2)$ and the upper sheet of the hyperboloid $z^2 - x^2 - y^2 = 2$.

Same problem, but with the lower sheet of the hyperboloid.

25. A ball centered at the origin is made of a magical material of density

$$\rho(x, y, z) = \frac{1}{x^2 + y^2 + z^2}.$$

Show that its mass is nevertheless finite. Calculate the mass of the ball if its radius is r. (Another improper integral. First find the mass of the shell $a^2 \leq x^2 + y^2 + z^2 \leq r^2$.)

22. Find the volume of the solid bounded by the hyperboloid $z^2 - x^2 - y^2 = 1$ and the plane $z = 2$.

24. Find the volume of the region based on the circular disk $x^2 + y^2 \leq r^2$ and capped by the surface $z = 1/\sqrt{x^2 + y^2}$. (This will involve an improper integral.

CHAPTER 18
GREEN'S THEOREM

This chapter brings together several topics that earlier were developed independently—mixed partial derivatives, paths, double integrals—and shows them to be related in a surprising and powerful way. The culminating theorem, Green's theorem, is important in applications and in other branches of mathematics, such as the theory of functions of a complex variable. Also, it forms a bridge to one of the principal subjects of advanced calculus: the combination of multidimensional integral and differential calculus in multidimensional versions of the Fundamental Theorem of Calculus.

1
WORK AND ENERGY IN ONE-DIMENSIONAL MOTION

Newton's second law, for a body moving along the x-axis, is

$$F = m\frac{d^2x}{dt^2}\left(= m\frac{dv}{dt} = ma\right).$$

Here F is the total force acting on the body (directed along the x-axis), m is the mass of the body, $v = dx/dt$ is the velocity and $a = d^2x/dt^2$ is its acceleration.

The complete history of a particle moving under a given force F is obtained by solving this differential equation. But for many purposes all we need to know is the accumulated effect of the force, in terms either of the length of time over which F acts or of the distance over which F acts, i.e.,

$$\text{either} \quad \int_{t_1}^{t_2} F\,dt \quad \text{or} \quad \int_{x_1}^{x_2} F\,dx.$$

Here x_1 is the initial position of the particle at the initial time t_1, and x_2 and t_2 are the terminal position and time.

We shall confine our attention entirely to the second integral. Also, we shall suppose that the force F is a function of x alone. Thus, as the particle moves along its trajectory, the force acting on it depends only on where it is, and not on its velocity v or the time t.

We can transform the integral $\int_{x_1}^{x_2} F\,dx$ by making the change of variable

$$dx = \left(\frac{dx}{dt}\right) dt = v\,dt$$

and using Newton's law. We have

$$\int_{x_1}^{x_2} F\,dx = \int_{t_1}^{t_2} Fv\,dt = \int_{t_1}^{t_2} m\frac{dv}{dt} v\,dt = \int_{t_1}^{t_2} mv\frac{dv}{dt}\,dt$$

$$= \int_{v_1}^{v_2} mv\,dv = \frac{1}{2}mv^2 \Big]_{v_1}^{v_2} = \frac{1}{2}m(v_2)^2 - \frac{1}{2}m(v_1)^2.$$

The integral $\int_{x_1}^{x_2} F\,dx$ is called the *work* done by the force F during the motion, and the quantity $mv^2/2$ is called the *kinetic energy* of the particle. In these terms, the calculation above shows that

The work done by the total applied force F during a motion is equal to the change in kinetic energy from the beginning to the end of the motion.

This work-energy law lets us answer some questions about a solution to Newton's equation without having to find the solution.

Example 1   If $F = -kx$ in Newton's law, then the body oscillates forever back and forth about the origin in what is called *harmonic motion*. (This differential equation will be studied in the next chapter.) This situation occurs when a mass is attached to the free end of a spring having one end fixed and the free end at the origin. In this case, k is called the *stiffness* of the spring.

Use the work-energy law to find the amplitude of the motion when a spring of stiffness 5 is attached to a particle of mass 3 and given an initial velocity $v_0 = 2$ at $x_0 = 0$.

Solution   The amplitude A is the maximum value of x during the oscillatory motion, and hence is the value of x for which v first becomes zero. When we substitute these values and all the given data in the work-energy equation

$$\int_{x_0}^{x} F\, dx = \frac{1}{2}\, mv^2 - \frac{1}{2}\, mv_0^2,$$

we get

$$\int_{0}^{A} (-5x)\, dx = 0 - \frac{1}{2}\, 3(2)^2.$$

So

$$-\frac{5}{2} A^2 = -6,$$

and

$$A = 2\sqrt{\frac{3}{5}}. \qquad \Box$$

There is a hidden ramification in the work-energy calculation. Suppose that F is everywhere defined and positive, and that the initial velocity v_1 is negative. Since

$$\frac{dv}{dt} = \frac{F}{m} > 0,$$

v is an increasing function of t, and we suppose that it eventually becomes positive, so $v_2 > 0$. The motion thus starts out with x decreasing, then passes through a critical point (t_0, x_0) where $v = 0$ and the direction of motion changes, and finishes with x increasing (Fig. 1). In other words, the particle does not just run straight across the interval $[x_1, x_2]$, as the work integral $\int_{x_1}^{x_2} F(x)\, dx$ might suggest, but first backs up to x_0 and then moves forward from x_0 to x_2. The t integral encompasses this whole motion, but the x integral does not, unless we think of $\int_{x_1}^{x_2}$ as standing for $\int_{x_1}^{x_0} + \int_{x_0}^{x_2}$.

Figure 1

$x_0 \qquad\qquad x_1 \qquad\qquad\qquad\qquad\qquad x_2$

Example 1 is similar. The oscillating mass repeatedly retraces the same x interval.

So our viewpoint should be that the t integral is the basic integral definition of the work done by F along the motion:

DEFINITION   *The work done by the force F along a motion x = g(t) from t = t₁ to*
t = t₂ is

$$\int_{t_1}^{t_2} Fv \, dt = \int_{t_1}^{t_2} F(g(t))g'(t) \, dt.$$

Then the change-of-variable calculation above shows that

If the force F is a function of x alone, then the work done by F over
any motion starting at x = x₁ and ending at x = x₂ is given by

$$\int_{x_1}^{x_2} F(x) \, dx.$$

REMARK 1   If the total force F is the sum of two or more forces, say $F = F_1 + F_2$,
then the work done by F is the sum

$$\int_{x_1}^{x_2} F \, dx = \int_{x_1}^{x_2} F_1 \, dx + \int_{x_1}^{x_2} F_2 \, dx,$$

and we naturally consider the separate F_1 and F_2 integrals to be the
parts of the total work attributable to these constituent forces. Thus
$\int_{x_1}^{x_2} F_1 \, dx$ is the work done by F_1, etc.

REMARK 2   If the force F is constant then

$$\int_{x_1}^{x_2} F \, dx = F(x_2 - x_1).$$

So for a constant force F, the change in kinetic energy is given by F
times how *far* the force acts.

Conservation   Now let $\phi(x)$ be any antiderivative of $F(x)$. Then $V = -\phi$ is called
of Energy   the *potential energy* of the force field $F(x)$; it is uniquely determined
once we have decided what point x_0 is to be assigned zero potential
energy. (Note that the equations $V'(x) = -F(x)$, $V(x_0) = 0$ form an
initial value problem for V.) Since

$$\int_{x_1}^{x_2} F(x) \, dx = V(x_1) - V(x_2),$$

the work-energy identity can now be rewritten

$$V(x_1) + \frac{1}{2} mv_1^2 = V(x_2) + \frac{1}{2} mv_2^2.$$

The sum $E = V(x) + \frac{1}{2}mv^2$ of the potential and kinetic energy is
called the total energy of the particle, and the above equation says

that the total energy remains constant as the particle moves under the force F. This is the law of *conservation of energy*. The work done by F from x_1 to x_2 can be viewed as so much potential energy converted into kinetic energy.

Example 2 The restoring force exerted by a compressed or extended spring is

$$F = -kx.$$

The potential energy of the spring is thus

$$V(x) = \frac{kx^2}{2} + C.$$

When a particle of mass m is attached and the system is set into oscillatory motion, the total energy remains constant:

$$\frac{1}{2}mv^2 + \frac{1}{2}kx^2 \equiv \text{constant}.$$

In Example 1 we had $m = 3$ and $k = 5$, so

$$3v^2 + 5x^2 = C$$

in that motion. The initial condition $v_0 = 2$, $x_0 = 0$ shows that $C = 12$, and since the amplitude A is the value of $|x|$ for which $v = 0$, we end up with the same equation as before,

$$5A^2 = 12. \qquad \square$$

In view of the first remark above, about the work attributable to one of several constituent forces, we can compute the work done by a force F in situations where F is obviously not the total force and no connection with initial and terminal velocity is implied.

Example 3 When a quantity of gas is held at a constant temperature, its pressure p and total volume V are related by the equation

$$pV = \text{constant}.$$

A circular cylinder 1 ft in diameter has a variable length determined by a piston at one end. It is filled with gas at a pressure of 50 lb/in² when its length is 10 ft, and the gas is then further compressed by moving the piston until the length is reduced to 5 ft. Supposing that the gas is held at a constant temperature, how much work is done on the gas by the moving piston?

Solution The volume of the enclosed gas is proportional to the length x of the cylinder, and the force F with which the piston pushes against the gas is proportional to the gas pressure p, so the equation $pV = \text{constant}$ can be replaced by

$$Fx = C \qquad \text{or} \qquad F = C/x.$$

Initially $x_0 = 10$ ft. The force is negative, being in the direction of decreasing x, and

$$F_0 = -pA \ (= -\text{pressure times piston area, in square inches})$$

$$= -50 \times \pi 6^2 \approx -5652 \,\text{lb},$$

so $C = F_0 x_0 = -56{,}520$. The work done is thus

$$W = \int_{10}^{5} F\,dx = -\int_{10}^{5} \frac{C}{x}\,dx = -\int_{10}^{5} \frac{56{,}520}{x}\,dx$$

$$= 56{,}520(\ln 10 - \ln 5)$$

$$= 56{,}520 \ln 2 \approx 56{,}520 \times 0.693$$

$$\approx 39{,}168 \text{ foot-pounds.} \qquad \square$$

Example 4 A tank is in the shape of an inverted cone 10 feet high and 10 feet in diameter at the top (Fig. 2). It is filled with water from an inlet at the bottom. How much work is done in pumping the water in?

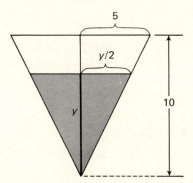

Figure 2

Solution Water weighs 62.4 pounds per cubic foot, so y feet of water presses on the bottom with a pressure of $62.4y$ pounds per square foot. Imagine the water being pushed into the conical tank by a pump having a piston of area A square feet. Neglecting the recovery strokes of the piston, the total distance d that the piston travels in filling the tank is given by

$$dA = \text{volume of tank,}$$

$$d = \frac{1}{A}\left(\frac{1}{3}\pi r^2 h\right) = \frac{250\pi}{3A}.$$

At the point when the piston has traveled x feet, the tank contains Ax cubic feet of water, and its depth y is related to x by

$$\frac{1}{3}\pi(y/2)^2 y = Ax.$$

So

$$y = \left(\frac{12Ax}{\pi}\right)^{1/3}.$$

Since the force exerted by the piston is $pA = (62.4y)A$

$$W = \int_0^d 62.4Ay\,dx$$

$$= \int_0^d 62.4A\left(\frac{12Ax}{\pi}\right)^{1/3} dx$$

$$= 62.4A^{4/3}\left(\frac{12}{\pi}\right)^{1/3}\left(\frac{3}{4}x^{4/3}\right)\Big|_0^d.$$

Using the value of d above, we get

$$W = 62.4\pi \cdot 625 \approx 122{,}460 \text{ ft-lbs.} \qquad \square$$

PROBLEMS FOR SECTION 1

1. How much work is done by a force of magnitude $F(x) = 5x$ on a particle that moves from $x = 2$ to $x = 4$?

2. Same question if the particle moves from $x = -2$ to $x = 2$.

3. How much work is done by a force of magnitude $2/x$ in a motion from $x = 1$ to $x = 2$?

In Problems 4 through 10, first write down the law of conservation of energy for the given system, and then use it to solve the problem.

4. A particle of mass 2 moves along the positive x-axis under a (total) force $F(x) = 1/x^2$. Its velocity at $x = 1$ is -1. (a) At what point will it turn around and start moving in the positive x-direction? (b) What will its subsequent velocity be at $x = 2$? (c) What is its limiting speed as $x \to \infty$?

6. Solve the above problem for a particle of mass m attached to a spring of stiffness k.

8. A particle is repelled from the origin by a force of magnitude $1/x^{3/2}$. A particle of mass 1 starts at $x = 4$ with an initial velocity of $+1$. What is its limiting velocity as $x \to \infty$?

10. A particle of mass m is repelled from the origin with a force of magnitude $1/x^{3/2}$. If it starts from rest at $x = 4$, what is its limiting velocity as $x \to \infty$?

12. a) An irregularly shaped tank is being filled with water from the bottom, in the manner of Example 4. Suppose that the water depth y (in feet) is a known function
$$y = f(v)$$
of the volume v (in cubic feet) of water in the tank. Follow the scheme of Example 4 to show that the work done in filling the tank to a volume V_0 is

$$W = \int_0^{V_0/A} 62.4 A f(Ax)\, dx$$

where A is the area of the pump piston in square feet.

b) Show that the answer above is independent of A.

14. A cylinder with a movable piston at one end contains 300 cubic feet of gas at a pressure of 100 lb/in^2. The gas is compressed to 1/3 of its original volume (at a constant temperature) by moving the piston. How much work does the piston do during this motion?

5. A particle of mass 3 is attached to a spring of stiffness 5. Find the relationship between the amplitude of its motion and its maximum speed.

7. A particle of mass 3 is attached to a spring of stiffness 2. Find its amplitude A if its speed at $x = A/2$ is 5.

9. In the same situation, the initial velocity is -1. How close will the particle come to the origin?

11. A tank has a square cross section, 10 ft on a side, and is 20 ft tall. Find the work done in filling the tank with water from a bottom inlet, as in Example 4.

13. A tank h feet high has a capacity of V cubic feet. It requires W foot-pounds of work to fill the tank with water from the bottom. Show that it requires $62.4Vh - W$ foot-pounds of work to pump out the tank from the top. (*Hint:* the combined act of filling and pumping out the tank amounts to raising V cubic feet of water through h feet.)

15. It can be proved that the work done by gas in an expanding container is always given by

$$W = \int_{V_1}^{V_2} p\, dV \quad \text{foot-pounds,}$$

where p is the pressure in *pounds per square foot*, and V is the volume in cubic feet. This proof is independent of the formula relating p and V. Verify the above formula when the gas is confined to a cylinder with variable length determined by a piston at one end and is held at a constant temperature.

16. Verify the formula in the above problem for gas expanding a spherical balloon (at a constant temperature).

17. The pressure and volume of a certain body of gas are related by

$$pV^\alpha = k,$$

where α and k are positive constants. Find the work done if the gas expands from $V = V_0$ to $V = V_1$.

18. The formula in Problem 15 also holds for the tank-filling problems, with p the pressure at the pump. Show that this is so, by starting with the general formula of Problem 12.

2
WORK ALONG PLANE PATHS: LINE INTEGRALS

In the plane, force and acceleration are vector quantities, and Newton's law is the vector law

$$\mathbf{F} = m\mathbf{a}.$$

Here $\mathbf{F}$ is the total (vector) force acting on a particle, m is the mass of the particle, and $\mathbf{a}$ is the (vector) acceleration of the particle. This law governs all plane motion.

We keep the t-integral definition of work that we decided on in the last section, except that the product of the vector $\mathbf{F}$ with the velocity vector $\mathbf{v}$ has to be the dot product to permit the analogous calculations to go through. The particle trajectory is now a plane path $\mathbf{x} = \boldsymbol{\gamma}(t)$.

DEFINITION   The work W done by the force $\mathbf{F}$ along the motion $\mathbf{x} = \boldsymbol{\gamma}(t)$ over the time interval $[t_1, t_2]$ is

$$W = \int_{t_1}^{t_2} (\mathbf{F} \cdot \mathbf{v})\, dt = \int_{t_1}^{t_2} [\mathbf{F}(\boldsymbol{\gamma}(t)) \cdot \boldsymbol{\gamma}'(t)]\, dt.$$

The work-energy calculation is now the same as before. If $\mathbf{F}$ is the total force acting on the particle, so that $\mathbf{F} = m\, d\mathbf{v}/dt$ by Newton's law, and if $v = |\mathbf{v}| = (\mathbf{v} \cdot \mathbf{v})^{1/2}$ is the speed of the particle, then

$$W = \int_{t_1}^{t_2} (\mathbf{F} \cdot \mathbf{v})\, dt = \int_{t_1}^{t_2} \left(m \frac{d\mathbf{v}}{dt} \cdot \mathbf{v}\right) dt$$

$$= \int_{t_1}^{t_2} \frac{1}{2} m \frac{d}{dt}(\mathbf{v} \cdot \mathbf{v})\, dt = \frac{1}{2} m \int_{t_1}^{t_2} \frac{d}{dt}(v^2)\, dt$$

$$= \frac{1}{2} mv^2 \bigg]_{t_1}^{t_2} = \frac{1}{2} m(v_2)^2 - \frac{1}{2} m(v_1)^2.$$

So again,

> *The work done by the total force* **F** *acting on a particle along its motion is equal to the change in kinetic energy of the particle from the beginning to the end of the motion.*

Line Integrals   As in Section 1, we suppose that the force **F** governing the motion arises from a *field of force*, i.e., from a varying force $\mathbf{F} = \mathbf{F}(\mathbf{x})$ that depends only on the position **x**. In the one-dimensional case we noted a very simple integral formula for the work done by this kind of force, and as a consequence we were led to the notion of potential energy and the law of conservation of energy. It is natural to ask whether similar things happen in the plane. To begin with,

Does the relationship $d\mathbf{x} = \mathbf{v}\, dt$ convert the work integral to a pure **x**-integral,

$$\int_{t_t}^{t_1} (\mathbf{F} \cdot \mathbf{v})\, dt = \int_{\mathbf{x}_1}^{\mathbf{x}_2} \mathbf{F}(\mathbf{x}) \cdot d\mathbf{x}$$

as it did on the line? This turns out to be a more complicated question in the plane, and it leads to some new aspects of integration. First, the notation for the **x**-integral must be modified so as to indicate the *path* of integration rather than just its endpoints. This is necessary because we can go from $\mathbf{x}_0$ to $\mathbf{x}_1$ along many different paths, and the value of the integral may very well change from path to path (Fig. 1). Also, since the same situation arises in other applications, the work-energy context will be discarded at this point. We shall state the definition analytically (i.e., independent of interpretation) as the definition of a *line integral*, and proceed in these neutral terms from now on.

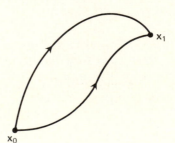

Figure 1

DEFINITION   If $\mathbf{x} = \boldsymbol{\gamma}(t)$ is any smooth path in the plane, defined over the parameter interval $[t_1, t_2]$, and if $\mathbf{p}(\mathbf{x}) = (p(x, y),\ q(x, y))$ is any vector field defined on a region containing the path γ, then the *line integral* of the vector field $\mathbf{p}(\mathbf{x})$ along the path γ, designated

$$\int_\gamma \mathbf{p} \cdot d\mathbf{x} \qquad \text{or} \qquad \int_\gamma p\,dx + q\,dy$$

is defined to be

$$\int_{t_1}^{t_2} \left[\mathbf{p} \cdot \frac{d\mathbf{x}}{dt} \right] dt = \int_{t_1}^{t_2} \left(p\frac{dx}{dt} + q\frac{dy}{dt} \right) dt,$$

where for each t the vector field $\mathbf{p}(\mathbf{x})$ is evaluated at the path point $\mathbf{x} = \gamma(t)$.

The precise definition is thus

$$\boxed{\int_\gamma \mathbf{p} \cdot d\mathbf{x} = \int_{t_1}^{t_2} [\mathbf{p}(\gamma(t)) \cdot \gamma'(t)]\,dt.}$$

Example 1 The line integral $\int_\gamma y\,dx + xy\,dy$ over the path γ given by $(x, y) = (t + t^3, t^2)$ on the parameter interval $[0, 1]$ has the value

$$\int_\gamma y\,dx + xy\,dy = \int_0^1 \left(y\frac{dx}{dt} + xy\frac{dy}{dt} \right) dt$$

$$= \int_0^1 [t^2(1 + 3t^2) + (t + t^3)t^2 2t]\,dt$$

$$= \int_0^1 (2t^6 + 5t^4 + t^2)\,dt$$

$$= \frac{2}{7} + 1 + \frac{1}{3}$$

$$= 1\frac{13}{21}. \qquad \square$$

Example 2 Evaluate the line integral of the vector field $\mathbf{p}(\mathbf{x}) = (x, y^2)$ over the quarter of the unit circle in the first quadrant, parameterized by $(x, y) = (\cos\theta, \sin\theta)$.

Solution

$$\int_\gamma \mathbf{p} \cdot d\mathbf{x} = \int_\gamma x\,dx + y^2\,dy$$

$$= \int_0^{\pi/2} \left(x\frac{dx}{d\theta} + y^2 \frac{dy}{d\theta} \right) d\theta$$

$$= \int_0^{\pi/2} [\cos\theta(-\sin\theta) + \sin^2\theta(\cos\theta)] \, d\theta$$

$$= \left[\frac{\cos^2\theta}{2} + \frac{\sin^3\theta}{3} \right]_0^{\pi/2}$$

$$= \frac{1}{3} - \frac{1}{2}$$

$$= -\frac{1}{6}. \qquad \square$$

Calculations involving line integrals depend on properties that a line integral inherits from its definition as a definite integral. Some properties carry over directly. For example, it follows directly from the definition of a line integral that if $\mathbf{p}(\mathbf{x})$ and $\mathbf{r}(\mathbf{x})$ are any two vector fields, then

$$\int_\gamma (\mathbf{p} + \mathbf{r}) \cdot d\mathbf{x} = \int_\gamma \mathbf{p} \cdot d\mathbf{x} + \int_\gamma \mathbf{r} \cdot d\mathbf{x}.$$

Also, if a path γ is broken up into a succession of "partial paths" γ_i by subdividing the parameter interval $[a, b]$ of γ,

$$a = t_0 < t_1 < \cdots < t_n = b,$$

and if γ_i is the restriction of γ to the parameter interval $[t_{i-1}, t_i]$, then

$$\int_\gamma \mathbf{p} \cdot d\mathbf{x} = \int_{\gamma_1} \mathbf{p} \cdot d\mathbf{x} + \int_{\gamma_2} \mathbf{p} \cdot d\mathbf{x} + \cdots + \int_{\gamma_n} \mathbf{p} \cdot d\mathbf{x}.$$

In this situation γ is sometimes called the sum of the partial paths γ_i, and is written

$$\gamma = \gamma_1 + \gamma_2 + \cdots + \gamma_n.$$

See Fig. 2.

REMARK If a path γ has a finite number of corners, but is smooth between each adjacent pair of corners, then we form the line integral over γ by simply adding up the line integrals over its smooth pieces by the above formula.

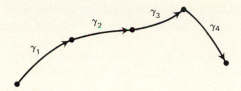

Figure 2

Other properties show up in new forms in the line-integral setting. In the remainder of this section we shall be concerned with ramifications of the substitution rule. We start by showing that a line integral over a path γ is independent of the particular way in which γ happens to be parameterized. This is the first indication that the line-integral concept is fruitful.

THEOREM 1 *A line integral is invariant under any change of path parameter that preserves path endpoints. That is, if*

1. *$\mathbf{x} = \boldsymbol{\lambda}(s)$ is a path with parameter interval $[s_1, s_2]$,*
2. *$s = h(t)$ is a differentiable change of path parameter defined on $[t_1, t_2]$, with $s_1 = h(t_1)$ and $s_2 = h(t_2)$, and*
3. *$\mathbf{x} = \boldsymbol{\gamma}(t)$ is the reparameterized path $\boldsymbol{\gamma}(t) = \boldsymbol{\lambda}(h(t))$ on the parameter interval $[t_1, t_2]$, then*

$$\int_{\gamma} \mathbf{p} \cdot d\mathbf{x} = \int_{\lambda} \mathbf{p} \cdot d\mathbf{x}.$$

Proof This calculation combines the chain rule for functions of one variable with the direct substitution rule for the definite integral:

$$\int_{\gamma} \mathbf{p} \cdot d\mathbf{x} = \int_{t_1}^{t_2} [\mathbf{p}(\boldsymbol{\gamma}(t)) \cdot \boldsymbol{\gamma}'(t)]\, dt$$

$$= \int_{t_1}^{t_2} [\mathbf{p}(\boldsymbol{\lambda}(h(t))) \cdot \boldsymbol{\lambda}'(h(t))h'(t)]\, dt$$

$$= \int_{t_1}^{t_2} [\mathbf{p}(\boldsymbol{\lambda}(h(t))) \cdot \boldsymbol{\lambda}'(h(t))]h'(t)\, dt$$

$$= \int_{s_1}^{s_2} [\mathbf{p}(\boldsymbol{\lambda}(s)) \cdot \boldsymbol{\lambda}'(s)]\, ds$$

$$= \int_{\lambda} \mathbf{p} \cdot d\mathbf{x}. \qquad\blacksquare$$

REMARK If a change of parameter $s = h(t)$ *reverses* the path endpoints, then the line integral changes sign.

Proof The calculation is the same down to the next to the last line, which is now

$$\int_{s_2}^{s_1} [\quad] \, ds = -\int_{s_1}^{s_2} = -\int_{\lambda} \mathbf{p} \cdot d\mathbf{x}. \qquad \blacksquare$$

Now suppose that a path $\mathbf{x} = \boldsymbol{\gamma}(s)$ runs along a curve C from the initial point x_0 to the terminal point x_1. We can draw the curve C, mark the initial and terminal points, and indicate the direction of the path by an arrow. For example, the path might run along the parabola $y = x^2$ from the point $(-1, 1)$ to the point $(2, 4)$ as shown in Fig. 3.

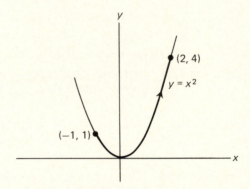

Figure 3

It is reasonable to call such a marked portion of a curve a *directed curve segment*. Theorem 1 says that all paths tracing the same directed curve segment give the same answer in line-integral calculations. For example, the parabolic arc marked as in Fig. 3 contains all the essential information about the path of a line integral: any two paths that run along the parabola $y = x^2$, starting at $(-1, 1)$ and ending at $(2, 4)$, will determine the same line integral (of a given vector field). We can therefore relax the line-integral notation, replacing the reference to a particular parameterization $\boldsymbol{\gamma}$ by the directed curve segment C that it traces, as in

$$\int_C \mathbf{p} \cdot d\mathbf{x}.$$

Example 3 Evaluate the line integral

$$\int_C y^2\,dx - x^3\,dy$$

where C is the above directed curve segment.

> **Solution** By the theorem, we can parameterize the curve segment in any manner consistent with the given initial and terminal points, and a simple choice is to take x as the parameter. The path is thus
>
> $$(x, y) = (x, x^2)$$
>
> over the x-interval $[-1, 2]$, and the integral is
>
> $$\int_C y^2\,dx - x^3\,dy = \int_{-1}^{2} [(x^2)^2 - x^3(2x)]\,dx$$
>
> $$= \int_{-1}^{2} (-x^4)\,dx = -\frac{x^5}{5}\Big]_{-1}^{2} = -\frac{33}{5}.\qquad\square$$

Example 4 Evaluate the line integral of Example 3, where C this time is the unit square traced counterclockwise, as shown in Fig. 4.

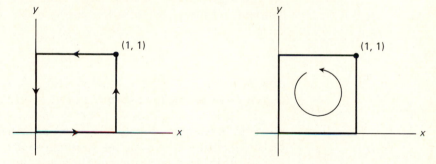

Figure 4

> **Solution** Here it is natural to use x as parameter along the horizontal segments and y along the vertical segments. Note that the integral along the top segment is then $-\int_0^1 y^2\,dx$, because it is traced in the opposite direction. The complete line integral is then
>
> $$\int_C y^2\,dx - x^3\,dy = \int_0^1 0^2\,dx - \int_0^1 1^3\,dy - \int_0^1 1^2\,dx - \int_0^1 0^3\,dy$$
>
> $$= -2.\qquad\square$$

Example 5   A path along a directed curve segment could even reverse directions for a while and then change back. For example, let $\boldsymbol{\lambda}$ be the path along the circular arc $y = 1 - \sqrt{1 - x^2}$, from $x = -a$ to $x = a$, with x as parameter on this interval. Thus $\boldsymbol{\lambda}$ traces the arc once, from left to right. (See Fig. 5.)

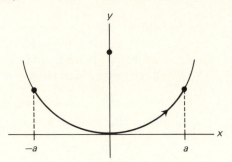

Figure 5

Now let $\boldsymbol{\gamma}$ be a path swept out by three beats of a pendulum swinging along this arc. Thus $\boldsymbol{\gamma}$ might be obtained from $\boldsymbol{\lambda}$ by the change of parameter

$$x = h(t) = a \cos t$$

on the t-interval $[-\pi, 2\pi]$. Note that this parameter change preserves the initial and terminal points of the path, since

$$-a = h(-\pi)$$
$$a = h(2\pi),$$

and Theorem 1 therefore applies. But $\boldsymbol{\gamma}$ traces the arc three times: across, back, and across again.   ☐

It is convenient to use the word *path* ambiguously for either a directed curve segment or (as formerly) for a parametric representation of a directed curve segment. Since the essence of a path is the directed curve segment it traces (according to Theorem 1), this ambiguity causes no trouble.

A path is called *closed* if it ends where it begins (i.e., if its terminal point is its initial point). The unit square path in Fig. 4 was closed. If C' and C'' are two paths running from $\mathbf{x}_0$ to $\mathbf{x}_1$, then $C' - C''$ is a closed path starting and ending at $\mathbf{x}_0$ (Fig. 6).

Conversely, a closed path C starting and ending at $\mathbf{x}_0$ can always be decomposed into two pieces and then written $C = C' - C''$ by cutting it at any intermediate point x_1 and letting C'' be the second

Figure 6

piece traced backwards. A path may cross itself one or more times (Fig. 7). A path that does not cross itself is called *simple*. Thus a *simple closed* path γ on the parameter interval $[a, b]$ has the property that $\gamma(t) = \gamma(t')$ for $t < t'$ if and only if $t = a$ and $t' = b$.

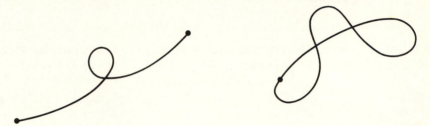

Figure 7

A path along a curve that crosses itself has to be marked more carefully, because there are two possibilities for where the path goes at a crossing point. An example is shown in Fig. 8. Since the loop is traced in opposite directions in the two paths, the line integral along that piece of the path will change signs. In more detail, suppose that we are given a vector field $\mathbf{p}(\mathbf{x})$, and suppose that the line integral of $\mathbf{p}$ along the first path, call it I, is broken up into a sum of three terms

$$I = I_0 + I_1 + I_2,$$

where I_0 is the integral along the piece of the path from $\mathbf{x}_0$ to $\mathbf{x}_1$, I_2

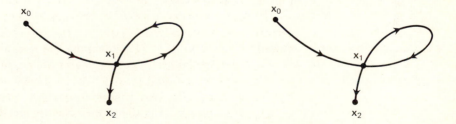

Figure 8

from $\mathbf{x}_1$ to $\mathbf{x}_1$ (i.e., around the loop as first marked), and I_2 from $\mathbf{x}_1$ to $\mathbf{x}_2$. Then the line integral of $\mathbf{p}$ along the *second* path is

$$I_0 - I_1 + I_2.$$

These two line integrals differ by $2I_1$.

PROBLEMS FOR SECTION 2

1. Compute the integral of the vector field $(-y, x)$ around the unit circle $x^2 + y^2 = 1$ traced counterclockwise.

2. Compute the integral of the vector field in the above problem around the unit square (with opposite corners $(0, 0)$ and $(1, 1)$), again traced "counterclockwise."

In Problems 3 through 7, compute $\int_C \mathbf{p} \cdot d\mathbf{x}$ for the given vector field $\mathbf{p}$ and directed curve segment C.

3. $\mathbf{p}(x, y) = (xy, x^2)$; C is the line segment from $(0, 1)$ to $(2, -3)$.

4. $\mathbf{p}(x, y) = (xy, x^2)$; C is the broken line going from $(0, 1)$ to $(2, -3)$ in a vertical step followed by a horizontal step.

5. $\mathbf{p}(x, y) = xy\mathbf{i} + x^2\mathbf{j}$; C is the graph of $y = 1 - x^2$ from $(0, 1)$ to $(2, -3)$.

6. $\mathbf{p}(x, y) = xy\mathbf{i} + x^2\mathbf{j}$; C is given parametrically by $\mathbf{x} = (2t^2, 1 - 4t)$ on $[0, 1]$.

7. a) $\mathbf{p}(\mathbf{x}) = (y^2, 2xy)$; C is the curve of Problem 1.

  b) C is the curve of Problem 3.

8. a) Compute $\dfrac{d}{dt}(xy^2)$ supposing that both x and y are functions of t.

  b) With the above answer in mind, show that

  $$\int_C y^2 \, dx + 2xy \, dy = x_1 y_1^2 - x_0 y_0^2$$

  for *any* path C from (x_0, y_0) to (x_1, y_1).

9. Show that

$$\int_C \frac{x \, dx + y \, dy}{\sqrt{x^2 + y^2}} = \sqrt{x_1^2 + y_1^2} - \sqrt{x_0^2 + y_0^2}$$

where C is any path from (x_0, y_0) to (x_1, y_1) and avoiding the origin. [*Hint*: Compute $d\sqrt{x^2 + y^2}/dt$.]

10. Show that

$$\int_C \frac{dx}{y} - \frac{x \, dy}{y^2} = \frac{x_1}{y_1} - \frac{x_0}{y_0}$$

along any path from $\mathbf{x}_0$ to $\mathbf{x}_1$ that does not touch the x-axis.

11. Compute the work done by the force $\mathbf{F}(\mathbf{x}) = (x + y, x - y)$ along the path $xy^2 = 4$ from $(1, 2)$ to $(4, 1)$.

12. Compute the work done by the force

$$\mathbf{F}(\mathbf{x}) = \frac{\mathbf{i}}{y} - \frac{x\mathbf{j}}{y^2}$$

along the line segment from $(1, 2)$ to $(2, 4)$.

13. Show that if

$$\mathbf{F}(\mathbf{x}) = \left(\frac{x}{r^3}, \frac{y}{r^3}\right), \qquad \text{where} \qquad r = \sqrt{x^2 + y^2},$$

then the work done by $\mathbf{F}$ is the same for all paths from $\mathbf{x}_0$ to $\mathbf{x}_1$. [*Hint*: Compute $d(1/r)/dt$.]

The line integral definition

$$\int_C \mathbf{p} \cdot d\mathbf{x} = \int_{t_1}^{t_2} [\mathbf{p}(\boldsymbol{\gamma}(t)) \cdot \boldsymbol{\gamma}'(t)]\, dt$$

works just as well in three dimensions as in two, with the same proof of independence of parameter. The space integral is conventionally expressed as

$$\int_C p\, dx + q\, dy + r\, dz \qquad \text{or} \qquad \int_C p_1\, dx + p_2\, dx + p_2\, dz.$$

Compute $\int_C \mathbf{p} \cdot d\mathbf{x}$ for the given vector field and path, where

14. $\mathbf{p}(\mathbf{x}) = (x + y, y + z, z + x)$, and C is the line segment from $(1, 2, 1)$ to $(2, 0, 3)$.

15. $\mathbf{p}(\mathbf{x}) = x\mathbf{i} + y\mathbf{j} + z\mathbf{k}$, and C is the curve $(x, y, z) = (t, t^2, t^3)$, from $(0, 0, 0)$ to $(1, 1, 1)$.

16. Show that

$$\int_C yz\, dx + zx\, dy + xy\, dz = x_1 y_1 z_1 - x_0 y_0 z_0$$

along any path from $\mathbf{x}_0$ to $\mathbf{x}_1$.

17. Find the work done by the force

$$\mathbf{F}(\mathbf{x}) = (x, y, x)$$

along the path $y = 2x$, $z = x^2$ from $x = 0$ to $x = 2$.

3
GRADIENT VECTOR FIELDS

In our earlier one-dimensional discussion we defined the *potential energy* $V(x)$ of a varying force $F(x)$ by

$$V'(x) = -F(x).$$

That is, V is the negative of an antiderivative of F, and is uniquely determined up to an additive constant of integration. Since every continuous function has an antiderivative, potential functions always exist.

The analogue for a vector force field $\mathbf{F}(\mathbf{x})$ in the plane would be a function $V(x, y)$ such that

$$\text{grad } V(x, y) = -\mathbf{F}(x, y).$$

In the purely mathematical setting, the minus sign associated with the potential energy is generally discarded:

DEFINITION   A potential function for the vector field $\mathbf{p}(\mathbf{x})$ is any function ϕ such that

$$\boxed{\operatorname{grad} \phi(\mathbf{x}) = \mathbf{p}(\mathbf{x}).}$$

Then the potential energy of the field would be $-\phi$ plus a suitable constant.

But here is where the big difference between one-dimensional and higher dimensional spaces shows itself: in general, a plane vector field $\mathbf{p}(\mathbf{x})$ cannot be expressed as the gradient of a function. The plane situation is thus richer in possibilities than the line, and requires deeper investigation. For example, are there conditions on a vector field $\mathbf{p}(\mathbf{x})$ that will guarantee it to be a gradient field? If it is not, can we find a way of measuring by how much it fails to be a gradient field, and interpret this discrepancy in some meaningful way?

We first note that if $\mathbf{p}(\mathbf{x})$ *is* a gradient field, then the results of Section 1 generalize completely.

THEOREM 2   *Let* $\mathbf{p}(\mathbf{x}) = \mathbf{p}(x, y)$ *be a gradient field on a region G in the plane, with potential function* $\phi(x, y)$ *on G, and let* $\mathbf{x}_0$ *and* $\mathbf{x}_1$ *be any two points of G. Then*

$$\boxed{\int_C \mathbf{p}\, d\mathbf{x} = \phi(\mathbf{x}_1) - \phi(\mathbf{x}_0)}$$

for any path C in G that runs from $\mathbf{x}_0$ *to* $\mathbf{x}_1$.

Proof   If C is given parametrically by $\mathbf{x} = \boldsymbol{\gamma}(t)$ on the interval $[a, b]$, then

$$\int_C \mathbf{p} \cdot d\mathbf{x} = \int_a^b [\mathbf{p}(\boldsymbol{\gamma}(t)) \cdot \boldsymbol{\gamma}'(t)]\, dt$$

$$= \int_a^b [\operatorname{grad} \phi(\boldsymbol{\gamma}(t)) \cdot \boldsymbol{\gamma}'(t)]\, dt$$

$$= \int_a^b \frac{d}{dt} (\phi(\boldsymbol{\gamma}(t))\, dt$$

$$= \phi(\boldsymbol{\gamma}(b)) - \phi(\boldsymbol{\gamma}(a))$$

$$= \phi(\mathbf{x}_1) - \phi(\mathbf{x}_0). \qquad \blacksquare$$

Example 1   Compute the line integral of the field $\mathbf{p}(\mathbf{x}) = (2xy, x^2)$ along the path $(x, y) = (t^3 - t, t^2)$ over the t-interval $[-1, 2]$.

Solution The given field is the gradient of $\phi(x, y) = x^2 y$, so the line integral equals

$$\phi(x_1, y_1) - \phi(x_0, y_0) = \phi(6, 4) - \phi(0, 1)$$
$$= 6^2 \cdot 4 - 0^2 \cdot 1 = 144. \qquad \square$$

COROLLARY *Let $\mathbf{F} = \mathbf{F}(\mathbf{x})$ be a gradient force field, with potential energy $V(\mathbf{x}) = -\phi(\mathbf{x})$, and suppose that $\mathbf{F}$ is the only force acting on a moving particle of mass m. Then the total energy E of the particle remains constant during the motion:*

$$E = \frac{1}{2} mv^2 + V(\mathbf{x}) \equiv constant.$$

Proof The work-energy law from Section 2 combines with Theorem 2 above to give

$$\frac{1}{2} mv^2 - \frac{1}{2} mv_0^2 = \int_C \mathbf{F} \cdot d\mathbf{x} = \phi(\mathbf{x}) - \phi(\mathbf{x}_0) = V(\mathbf{x}_0) - V(\mathbf{x}),$$

where C is the path the particle has followed from $\mathbf{x}_0$ to $\mathbf{x}$. Thus $\frac{1}{2}mv^2 + V(\mathbf{x})$ has the constant value $\frac{1}{2}mv_0^2 + V(\mathbf{x}_0)$. ■

A particle that is subject only to a gradient force field thus obeys the law of conservation of energy, and such a field is said to be *conservative*.

Example 2 A particle of mass 3 moves from the point $(2, 0)$ to the point $(0, \frac{1}{2})$ under the action of the force field $\mathbf{F}(x, y) = (-x/r^3, -y/r^3)$, where $r = \sqrt{x^2 + y^2}$. Its initial speed is 1. What is its speed at the point $(0, \frac{1}{2})$?

Solution The given field is the gradient field of $1/r$, so $V(\mathbf{x}) = -1/r = -1/\sqrt{x^2 + y^2}$. At $(2, 0)$ the total energy of the particle is

$$\frac{1}{2} mv^2 + V(\mathbf{x}) = \frac{3}{2} (1)^2 - \frac{1}{2} = 1.$$

Since the total energy remains constant during its motion, at the point $(0, \frac{1}{2})$ we have

$$1 = \frac{3}{2} v^2 - \frac{1}{1/2}, \qquad v = \sqrt{2}. \qquad \square$$

Theorem 2 shows that the line integral of a gradient field is *independent of path:* it depends on the path endpoints $\mathbf{x}_0$ and $\mathbf{x}_1$, but has the same value for all paths from $\mathbf{x}_0$ to $\mathbf{x}_1$.

The converse is also true.

THEOREM 3   *Let $\mathbf{p}(\mathbf{x}) = (p_1(x), p_2(x))$ be a continuous vector field on a plane region G having line integrals in G that are independent of path in the above sense. Then $\mathbf{p}(\mathbf{x})$ is a gradient field.*

Proof   Choose any fixed point $\mathbf{x}_0 = (x_0, y_0)$ in G and define $\phi(\mathbf{x})$ by

$$\phi(\mathbf{x}_1) = \int_{\mathbf{x}_0}^{\mathbf{x}_1} \mathbf{p}(\mathbf{x}) \cdot d\mathbf{x}$$

where the integral refers to the common value of all line integrals $\int_C \mathbf{p} \cdot d\mathbf{x}$ along paths C running from $\mathbf{x}_0$ to $\mathbf{x}_1$. We claim that then ϕ is a smooth function, and that

$$\operatorname{grad} \phi = \mathbf{p}.$$

It will be sufficient to show that $D_1\phi = p_1$. To do this, note that

$$\phi(x + \Delta x, y) - \phi(x, y) = \int_x^{x + \Delta x} p_1(s, y) \, ds,$$

because the ϕ difference is given by the line integral along *any* path from (x, y) to $(x + \Delta x, y)$, and the horizontal line segment with length s as parameter is the simplest. By the integral form of the mean-value theorem, the integral on the right above equals

$$p_1(X, y) \, \Delta x$$

for some number X between x and $x + \Delta x$. Thus

$$\lim_{\Delta x \to 0} \frac{\phi(x + x, y) - \phi(x, y)}{\Delta x} = \lim_{\Delta x \to 0} p_1(X, y) = p(x, y),$$

so

$$D_1\phi(x, y) = p_1(x, y)$$

as claimed.   ∎

Here is another way of saying "independence of path."

THEOREM 4 *The vector field* $\mathbf{p}(\mathbf{x})$ *is a gradient field if and only if*

$$\int_C p(\mathbf{x}) \cdot d\mathbf{x} = 0$$

for all closed paths C *in* R.

Proof If C is a closed path, then its initial and terminal points are the same, so if $\phi(\mathbf{x})$ is a potential function for $\mathbf{p}(\mathbf{x})$ then

$$\int_C \mathbf{p} \cdot d\mathbf{x} = \phi(\mathbf{x}_1) - \phi(\mathbf{x}_0)$$

$$= \phi(\mathbf{x}_0) - \phi(\mathbf{x}_0) = 0.$$

Conversely, if the integral of $\mathbf{p}$ around every closed path is 0, then its integral is independent of path, because if C' and C'' are any two paths from x_0 to x_1 then $C = C' - C''$ is a closed path, so

$$0 = \int_C \mathbf{p} \cdot d\mathbf{x} = \int_{C'} \mathbf{p} \cdot d\mathbf{x} - \int_{C''} \mathbf{p} \cdot d\mathbf{x}$$

and

$$\int_{C'} = \int_{C''}. \qquad\blacksquare$$

There are now two questions: How do we recognize when a field is a gradient field? And the practical one: How then do we find a potential function ϕ?

The first question has a deceptively easy answer. If $p_1 = \partial\phi/\partial x$ and $p_2 = \partial\phi/\partial y$, then

$$\frac{\partial p_1}{\partial y} = \frac{\partial p_2}{\partial x},$$

because this is just the equality of the mixed partials $\partial^2\phi/\partial y \, \partial x = \partial^2\phi/\partial x \, \partial y$. In other words, supposing that the component functions p_1 and p_2 have continuous partial derivatives:

LEMMA 1 *A necessary condition for the vector field* $\mathbf{p}(\mathbf{x})$ *to be a gradient field is the identity*

$$\frac{\partial p_1}{\partial y} = \frac{\partial p_2}{\partial x}.$$

A field having this property is said to be *closed*. So now the question is whether a closed field is necessarily a gradient field. The answer is tricky: *it depends on whether or not the domain region G has holes in it.* If it does not (and is therefore what is called a *simply connected* domain) then everything works out. We shall consider only the case of a region that is *rectangular,* in the sense that its boundary is a rectangle with sides parallel to the coordinate axes. The crux of the matter is Green's theorem, which will be taken up in the next section. It has the following theorem as a corollary.

THEOREM 5 *Let* $\mathbf{p}(\mathbf{x})$ *be a closed vector field on a region G and let C be the boundary of a rectangular subregion of G. Then*

$$\int_C p(\mathbf{x}) \cdot d\mathbf{x} = 0.$$

From this we can prove

THEOREM 6 *If* $\mathbf{p}(\mathbf{x})$ *is a closed vector field defined on a rectangular region G then* $\mathbf{p}(\mathbf{x})$ *is a gradient field.*

Proof Fix any point (x_0, y_0) in G. Then we can go to any other point (x, y) along a path C_1 consisting of a vertical segment followed by a horizontal segment, and also along a path C_2 consisting of a horizontal segment followed by a vertical segment (Fig. 1). (If $\mathbf{x}$ is directly above or below $\mathbf{x}_0$, then the horizontal leg is missing from each path—or is considered to be present as a trivial segment of length zero. And the vertical leg is missing when $\mathbf{x}$ is directly to the right or left of $\mathbf{x}_0$.)

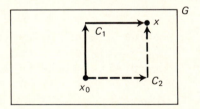

Figure 1

By Green's theorem

$$\int_{C_1} \mathbf{p} \cdot d\mathbf{x} = \int_{C_2} \mathbf{p} \cdot d\mathbf{x}$$

since $\int_{C_2 - C_1} = 0$. We define $\phi(x, y)$ to be the common value of these two line integrals, and can then prove $D_1\phi = p_1$ and $D_2\phi = p_2$ just as in Theorem 3. ∎

Example 3 We can follow the scheme of Theorem 6 to compute a potential function ϕ. For example, we see by inspection that

$$\mathbf{p}(x, y) = (2x + y, x + 2y)$$

is closed, since $\partial p_1/\partial y = 1 = \partial p_2/\partial x$. The theorem then guarantees that we can obtain ϕ by integrating $\mathbf{p}$ "up and over" from (x_0, y_0) to (x, y):

$$\phi(x, y) = \int_{y_0}^{y} (x_0 + 2y)\, dy + \int_{x_0}^{x} (2x + y)\, dx$$

$$= (x_0 y + y^2)_{y_0}^{y} + (x^2 + xy)_{x_0}^{x}$$

$$= (x_0 y + y^2) - (x_0 y_0 + y_0^2) + (x^2 + xy) - (x_0^2 + x_0 y)$$

$$= x^2 + xy + y^2 - (x_0^2 + x_0 y_0 + y_0^2). \qquad \square$$

Finding a Potential Function In a sense, half of the computation in the above example is wasted, because all we need is the fact that

$$\phi(x, y) = x^2 + xy + y^2 + C.$$

The line-integral computation of Theorem 6 and Example 3 is needed theoretically, in order to guarantee the existence of a potential function ϕ, but in practice there is a simpler procedure that eliminates the wasted computation.

Example 4 We recompute the potential function ϕ of the above example. Since $\partial\phi/\partial x = 2x + y$, we can try to find ϕ by "partially integrating" $2x + y$ with respect to x. That is, we treat y as a constant and find an antiderivative with respect to x. Then

$$\phi(x, y) = x^2 + xy + C(y),$$

where the constant of this x integration is a function of y that still must be found. By hypothesis, $\partial\phi/\partial y = x + 2y$, so

$$x + 2y = \frac{\partial}{\partial y}\, \phi(x, y) = x + C'(y).$$

Thus $C'(y) = 2y$, $C(y) = y^2 + C$, and

$$\phi(x, y) = x^2 + xy + y^2 + C.$$

This new procedure is guaranteed to work. Here it is again, in general terms. We first find a function $\alpha(x, y)$ by partially integrating $p_1(x, y)$ with respect to x. In the above example $\alpha(x, y) = x^2 + xy$. Since

$$\frac{\partial}{\partial x}(\phi(x, y) - \alpha(x, y)) = p_1(x, y) - p_1(x, y) = 0,$$

the difference $\phi - \alpha$ is a function of y only, call it $C(y)$. We are thus *guaranteed* that the computation

$$C'(y) = \frac{\partial}{\partial y}(\phi - \alpha) = p_2(x, y) - \frac{\partial}{\partial y}\alpha(x, y)$$

will turn out to involve only y. In the above example, it was

$$C'(y) = (x + 2y) - x = 2y.$$

We then integrate to find $C(y)$ and are done.   □

Example 5   The field $(2xy^3, 3x^2y^2)$ is closed, since $\partial p_1/\partial y = 6xy^2 = \partial p_2/\partial x$. We can therefore find its potential function as above:

$$\phi(x, y) = \int(2xy^3)\,dx = x^2y^3 + C(y),$$

$$C'(y) = \frac{\partial}{\partial y}[\phi(x, y) - x^2y^3] = p_2(x, y) - \frac{\partial}{\partial y}x^2y^3$$

$$= 3x^2y^2 - 3x^2y^2 = 0,$$

so

$$C(y) = C \quad \text{and} \quad \phi(x, y) = x^2y^3 + C.$$ □

Gradient Fields in Space   Gradient fields in space have essentially the same theory as in the plane, but the calculations are more complicated because of the added dimension. We first note

LEMMA 2   *A necessary condition for*

$$\mathbf{p(x)} = (p_1(x, y, z), p_2(x, y, z), p_3(x, y, z))$$

to be a gradient field is the set of three equations:

$$\frac{\partial p_1}{\partial y} = \frac{\partial p_2}{\partial x},$$

$$\frac{\partial p_2}{\partial z} = \frac{\partial p_3}{\partial y},$$

$$\frac{\partial p_3}{\partial x} = \frac{\partial p_1}{\partial z}.$$

Proof If grad $\phi = \mathbf{p}$, then the above equations are the identities for mixed partials:

$$\frac{\partial^2 \phi}{\partial y\,\partial x} = \frac{\partial^2 \phi}{\partial x\,\partial y},$$

$$\frac{\partial^2 \phi}{\partial z\,\partial y} = \frac{\partial^2 \phi}{\partial y\,\partial z},$$

$$\frac{\partial^2 \phi}{\partial x\,\partial z} = \frac{\partial^2 \phi}{\partial z\,\partial x}.$$

∎

A field satisfying the above conditions is called *closed*.

THEOREM 7 *If G is a rectangular box region in space and if $\mathbf{p}(\mathbf{x})$ is a closed vector field on G, then $\mathbf{p}$ is a gradient field.*

Sketch of Proof Fix any point (x_0, y_0, z_0) in G. Then we can go to any other point (x, y, z) in G along a path B_1 consisting of three line segments that are parallel to the x, y, and z axes in that order, or along a path B_2 consisting of line segments parallel to the y, z, and x axes in order, or, finally, along a path B_3 consisting of line segments parallel to the z, x, and y axes in order. We claim that

$$\int_{B_1} \mathbf{p} \cdot d\mathbf{x} = \int_{B_2} \mathbf{p} \cdot d\mathbf{x} = \int_{B_3} \mathbf{p} \cdot d\mathbf{x}.$$

Figure 2 shows that $B_1 - B_2$ can be rewritten as $C_1 + C_2$, where C_1 is a horizontal square traced counterclockwise when viewed from above and C_2 is a vertical square parallel to the xz-plane traced counterclockwise when viewed from the right. But $\int_{C_2} = 0$ by the earlier two-dimensional theory

Figure 2

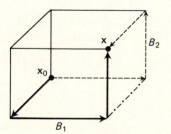
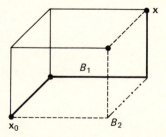

for the variables y and z, and similarly $\int_{C_1} = 0$. Thus

$$\int_{B_1-B_2} \mathbf{p} \cdot d\mathbf{x} = \int_{C_1+C_2} \mathbf{p} \cdot d\mathbf{x} = 0,$$

so

$$\int_{B_1} \mathbf{p} \cdot d\mathbf{x} = \int_{B_2} \mathbf{p} \cdot d\mathbf{x}.$$

In the same manner it can be proved that the line integral along B_2 and B_3 are equal. So all three are equal.

We can now define $\phi(x, y, z)$ as the common value of the above three line integrals, and the proof that $D_i\phi = p_i$, $i = 1, 2, 3$, is the same as before. ∎

In practice we find the potential function ϕ by the second of the two earlier methods.

Example 6   Show that

$$\mathbf{p}(\mathbf{x}) = (2x + y, x + 2y + z, y + 2z)$$

is a gradient field and find its potential function ϕ.

Solution   Since

$$\frac{\partial p_1}{\partial y} = 1 = \frac{\partial p_2}{\partial x},$$

$$\frac{\partial p_2}{\partial z} = 1 = \frac{\partial p_3}{\partial y},$$

$$\frac{\partial p_3}{\partial x} = 0 = \frac{\partial p_1}{\partial z},$$

the field is closed. In order to find ϕ we can partially integrate p_1 with respect to x, and get

$$\phi(x, y, z) = x^2 + xy + C(y, z),$$

where the constant of integration must now be taken to be a function of the two remaining variables. Then

$$C(y, z) = \phi - (x^2 + xy),$$

so

$$\frac{\partial C}{\partial y} = \frac{\partial \phi}{\partial y} - x = p_2(x, y, z) - x$$

$$= (2y + x + z) - x = 2y + z.$$

Therefore

$$C(y, z) = \int (2y + z)\, dy = y^2 + yz + D(z).$$

Finally,

$$D(z) = C(y, z) - (y^2 + yz)$$
$$= \phi(x, y, z) - (x^2 + xy) - (y^2 + yz),$$

so

$$D'(z) = \frac{\partial \phi}{\partial z} - 0 - y = p_3 - y$$
$$= (2z + y) - y = 2z,$$

and

$$D(z) = z^2 + C.$$

Altogether

$$\phi(x, y, z) = x^2 + xy + C(y, z)$$
$$= x^2 + xy + y^2 + yz + D(z)$$
$$= x^2 + xy + y^2 + yz + z^2 + C. \qquad \square$$

Note again how the variables dropped out when they were supposed to, first x in the calculation of $\partial C(y, z)/\partial y$, and then y in the calculation of $C'(z)$. It can be proved, by following the above steps in general language, that this has to happen.

If $r = \sqrt{x^2 + y^2}$, then a "central" potential $\phi(\mathbf{x}) = \psi(r)$ has the gradient

$$\operatorname{grad} \phi(\mathbf{x}) = \left(\frac{x\psi'(r)}{r}, \frac{y\psi'(r)}{r} \right) = \frac{\psi'(r)}{r}(x, y).$$

Vector fields in this form are easily recognizable, and then ψ can be found by a single integration.

Example 7 The field

$$\left(\frac{x}{x^2 + y^2}, \frac{y}{x^2 + y^2} \right) = \left(\frac{x}{r^2}, \frac{y}{r^2} \right) = \frac{1}{r^2}(x, y)$$

is of the above form with $\psi'(r) = 1/r$. Therefore

$$\psi(r) = \ln r, \quad \text{and} \quad \phi(\mathbf{x}) = \tfrac{1}{2}\ln(x^2 + y^2).$$

The same thing works in space. $\qquad \square$

Example 8   The field

$$\left(\frac{x}{r}, \frac{y}{r}, \frac{z}{r}\right) = \frac{1}{r}(x, y, z) = \frac{1}{r}\mathbf{x}$$

is of the form $\dfrac{\psi'(r)}{r}(x, y, z) = \dfrac{\psi'(r)}{r}\mathbf{x}$ with $\psi'(r) = 1$. So $\psi(r) = r$, and $\phi(\mathbf{x}) = \sqrt{x^2 + y^2 + z^2}$. $\square$

PROBLEMS FOR SECTION 3

In each of the following problems verify that the given vector field is closed and find a potential function by the line-integral method of Theorem 6 and Example 3.

1. $\mathbf{p}(x, y) = (x, y)$

2. $\mathbf{p}(x, y) = (y, x)$

3. $\mathbf{p}(x, y) = (2xy, x^2)$

4. $\mathbf{p}(x, y) = (x^2 + 2xy, x^2 + y^2)$

5. $\mathbf{p}(x, y) = (2xy + y^2, 2xy + x^2)$

6. $\mathbf{p}(x, y) = (y/x^2, -1/x)$

In each of the following problems verify that the given vector field is closed, and find a potential function by the method of Examples 4 through 6.

7. $\mathbf{p}(x, y) = (xy^2, x^2y)$

8. $\mathbf{p}(x, y) = (x^2 + 2xy, x^2 + y^2)$

9. $\mathbf{p}(x, y) = (2xy + y^2, 2xy + x^2)$

10. $\mathbf{p}(x, y) = (1/y, -x/y^2)$

11. $\mathbf{p}(x, y, z) = (z, y, x)$

12. $\mathbf{p}(x, y, z) = (y + z, x + y, x + z)$

13. $\mathbf{p}(x, y, z) = \left(\dfrac{-x}{(x^2 + y^2 + z^2)^3}, \dfrac{-y}{(x^2 + y^2 + z^2)^3}, \dfrac{-z}{(x^2 + y^2 + z^2)^3}\right)$

Each of the following integrals is of the form

$$\int_{\mathbf{x}_0}^{\mathbf{x}_2} \mathbf{p} \cdot d\mathbf{x},$$

implying independence of path. Check this in each case, and then compute the integral. (In some cases it may be easier to compute the integral along a simple path than to find a potential function.)

14. $\displaystyle\int_{(1,1)}^{(2,-2)} (x + y)\, dx + (x - y)\, dy$

15. $\displaystyle\int_{(0,0)}^{(1,3)} (2x - y)\, dx + (2y - x)\, dy$

16. $\displaystyle\int_{(1,0)}^{(0,1)} (x^2 + y^2)\, dx + 2xy\, dy$

17. $\displaystyle\int_{(-1,-1)}^{(1,1)} x^2 y^3\, dx + x^3 y^2\, dy$

18. $\displaystyle\int_{(1,1)}^{(e,e)} x \ln y^2\, dx + \frac{x^2}{y}\, dy$

19. $\displaystyle\int_{(1,2)}^{(2,1)} \frac{x\, dx + y\, dy}{(x^2 + y^2)} = \int_{(1,2)}^{(2,1)} \frac{x\, dx + y\, dy}{r^2}$

20. $\displaystyle\int_{(1,2)}^{(2,3)} \frac{x\, dx + y\, dy}{r^3}$

21. $\displaystyle\int_{(0,0)}^{(-1,2)} x(x^2 + y^2)^{3/2}\, dx + y(x^2 + y^2)^{3/2}\, dy$

22. $\int_{(1,0)}^{(0,e)} \frac{\ln r}{r} (x\,dx + y\,dy), \quad r = \sqrt{x^2 + y^2}$

23. Show that the field

$$\mathbf{p}(\mathbf{x}) = \left(\frac{-y}{x^2 + y^2}, \frac{x}{x^2 + y^2} \right)$$

is closed where defined. Show however that it is not a gradient field by computing its integral around the unit circle. (Use $(x, y) = (\cos\theta, \sin\theta)$.)

24. Show that any plane vector field of the form

$$\mathbf{p}(\mathbf{x}) = (xf(r), yf(r)) = f(r)(x, y)$$

is closed.

4
GREEN'S
THEOREM

In this section we consider an *arbitrary* vector field $\mathbf{p}(\mathbf{x})$, presumably not a gradient field. The line integral of $\mathbf{p}(\mathbf{x})$ about a closed path C,

$$\int_C \mathbf{p}(\mathbf{x}) \cdot d\mathbf{x},$$

is then different from zero in general, and we shall see that it can be viewed as the *circulation* of the field about C. To this end we introduce a second interpretation of a vector field.

The flow of a fluid can be described by its *velocity* field, which assigns to each point $\mathbf{x}$ in space and each time t the vector velocity

$$\mathbf{v} = \mathbf{f}(\mathbf{x}, t)$$

of the fluid at that point in space and time. For example, the flow of air in the atmosphere is roughly described by a succession of charts of wind velocities. A better example, for the simplified situation we wish to consider, is provided by a shallow, quiet river. We suppose, first, that the water is in steady-state motion, i.e., that its velocity field is independent of time. Second, we confine our attention to the flow near the surface, and suppose that the motion is (essentially) two-dimensional. This means that the water moves in horizontal planes, and the motion is the same in all planes to the given depth. So the flow is described by a two-dimensional velocity vector field $\mathbf{v}(\mathbf{x})$ defined on the plane region determined by the river bed.

We now raise the following question: Can we tell, by working with the velocity field, whether there is any *circulation* in the flow of the stream? For example, can we recognize a whirlpool?

The first step is to compute the circulation around a closed path. Consider any closed path C and let $\mathbf{x} = \gamma(s)$ be the parametric representation of C by its arc length measured from some field point. Then $\mathbf{v} \cdot \gamma' = \mathbf{v}(\gamma(s)) \cdot \gamma'(s)$ is the scalar component of the velocity vector $\mathbf{v}$ in the direction of the path tangent vector at the point $\gamma(s)$. See page 571. The integral

$$\int_C \mathbf{v} \cdot d\mathbf{x} = \int_{s_0}^{s_1} [\mathbf{v}(\gamma(s)) \cdot \gamma'(s)] \, ds$$

is therefore the accumulation over C of the tendency of the velocity field to be in the direction of C. If the integral is positive then there is more fluid velocity along the positive direction of C than along its negative direction, and the fluid shows net positive rotation about C. Thus the above line integral can be interpreted as the rate at which the fluid is circulating about C. Again:

If a vector field $\mathbf{p}(\mathbf{x})$ is interpreted as the velocity field of a fluid flow, then its line integral around a closed path C can be interpreted as the rate at which the fluid circulates about C. We shall therefore call $\int_C \mathbf{p} \cdot d\mathbf{x}$ the *circulation* of the vector field $\mathbf{p}$ about the closed path C.

REMARK　Although we used arc length as parameter in order to obtain this interpretation, the line integral can be computed with any convenient parameterization, since it is independent of the choice of parameters.

Example 1　Suppose we consider a circular disk of radius a rotating at a constant angular velocity of ω radians per second. (A rigid body doesn't seem much like a fluid, but this might seem to be a reasonable model of pure whirlpool motion.)

We consider the circulation of the "fluid" around the concentric circle of radius r. In the motion around this circle the velocity vector has the direction of the tangent vector and the constant magnitude ωr. So its circulation about the circle is $\omega r (2\pi r) = 2\omega \pi r^2$. In more detail, if we parameterize the circle by following a single point on the rotating disk, we get

$$\mathbf{x} = (x, y) = (r \cos \omega t, r \sin \omega t),$$

with t running from 0 to $2\pi/\omega$. The tangent vector at any point is the field velocity vector at that point and is

$$\mathbf{v} = \frac{d\mathbf{x}}{dt} = r\omega(-\sin \omega t, \cos \omega t) = \omega(-y, x).$$

The circulation integral, with t as parameter, is thus

$$\int_C \mathbf{v} \cdot d\mathbf{x} = \int_0^{2\pi/\omega} \left(\frac{d\mathbf{x}}{dt} \cdot \frac{d\mathbf{x}}{dt} \right) dt = (r\omega)^2 \int_0^{2\pi/\omega} dt = 2\omega\pi r^2.$$

With s as parameter, we get

$$\int_C \mathbf{v} \cdot d\mathbf{x} = \int_0^{2\pi r} \left(\frac{d\mathbf{x}}{dt} \cdot \frac{d\mathbf{x}}{ds} \right) ds = \omega r \int_0^{2\pi r} ds = 2\omega\pi r^2. \qquad \square$$

Example 2 Continuing the above example, we compute the circulation about a sector of radius r and angular opening θ. The path C now consists of a circular arc C_0 of length $r\theta$ and two radii as shown in Fig. 1. But the flow is perpendicular to the radii, so the integrand $\mathbf{v}(\mathbf{x}) \cdot d\mathbf{x}/ds$ is zero along these two pieces of the path. The circulation rate is thus just

$$\int_{C_0} \mathbf{v} \cdot d\mathbf{x} = \omega r \int_0^{\theta r} ds = \omega r \cdot \theta r$$

$$= 2\omega(\tfrac{1}{2}\theta r^2) = 2\omega A,$$

where A is the area of the sector.

Figure 1

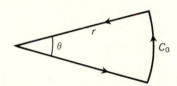

$\qquad \qquad \qquad \qquad \qquad \qquad \qquad \qquad \qquad \qquad \qquad \qquad \qquad \qquad \qquad \qquad \qquad \qquad \qquad \square$

Example 3 Finally, we consider the circulation of the same flow about the "polar rectangle" bounded by circles of radii r_1 and r_2 $(r_1 < r_2)$ and with an angular opening θ (Fig. 2).

Figure 2

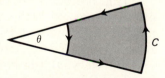

This path C is the difference between two sector paths, $C = C_2 - C_1$, as shown in Fig. 3, so

$$\int_C \mathbf{v} \cdot d\mathbf{x} = \int_{C_2} \mathbf{v} \cdot d\mathbf{x} - \int_{C_1} \mathbf{v} \cdot d\mathbf{x} = 2\omega(A_2 - A_1) = 2\omega A$$

where A is again the area of the region bounded by the path.

Figure 3

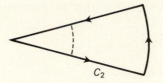   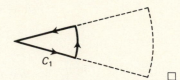

□

In these simple examples the circulation around the boundary of a region turned out to be proportional to the area of the region. Such examples raise the possibility of a two-dimensional aspect to circulation. Figure 4 shows that the circulation of a vector field around the boundary of a rectangle R can always be expressed as a finite sum of incremental circulations around the subregions of a subdivision of R. In the middle figure we have $\int_C = \sum_1^4 \int_{C_i}$ because the parts of the paths C_i interior to R consist of straight line segments that are traced once in each direction and hence cancel in the sum of the four line integrals. The circulation around the whole region is thus the sum of the circulations around its four pieces. The same remains true for any subdivision.

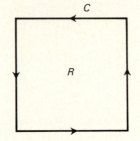   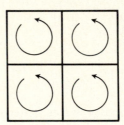   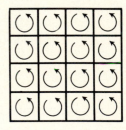

Figure 4

This procedure looks like a two-dimensional Riemann sum approximation and suggests that the circulation of a vector field $\mathbf{p}(\mathbf{x})$ around the boundary of a plane region R can be viewed as the double integral *over R* of a "circulation density" for $\mathbf{p}(\mathbf{x})$.

To see what the density is, consider a single incremental rectangle ΔR, as shown in Fig. 5. The circulation around ΔR can be written

$$\int_y^{y+\Delta y} [q(x + \Delta x, t) - q(x, t)]\, dt + \int_x^{x+\Delta x} [p(t, y) - p(t, y + \Delta y)]\, dt.$$

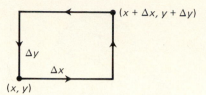

Figure 5

By applying the mean-value theorem twice we can rewrite the first integral, first as

$$\Delta y[q(x + \Delta x, Y) - q(x, Y)],$$

where Y is some number between y and $y + \Delta y$, and then as

$$\Delta y\, \Delta x D_1 q(X, Y),$$

where X is some number between x and $x + \Delta x$. The second integral can be evaluated in the same way, and comes out to be

$$-\Delta x\, \Delta y D_2 p(X', Y'),$$

where X' lies between x and $x + \Delta x$, and Y' between y and $y + \Delta y$.

When the incremental evaluations are added up we have the total circulation of the vector field $\mathbf{p}(\mathbf{x})$ around R equal to a Riemann sum for the *double integral over R* of the *function* $\partial q/\partial x - \partial p/\partial y$. (Actually it is the sum of two Riemann sums, for the functions $\partial q/\partial x$ and $-\partial p/\partial y$ separately.) In any case this means that

$$\int_C \mathbf{p} \cdot d\mathbf{x} = \iint_R \left(\frac{\partial q}{\partial x} - \frac{\partial p}{\partial y}\right) dx\, dy,$$

and this is Greens's theorem for the rectangle R. It says that the circulation *around R* can be expressed as the double integral *over R* of the *circulation density* $\partial q/\partial x - \partial p/\partial y$. In particular, Green's theorem answers our original question: there is circulation in the vector field $\mathbf{p}(\mathbf{x})$ wherever the function $\partial q/\partial x - \partial p/\partial y$ is not zero.

The general form of Green's theorem requires a different approach, and one that is not so directly related to the velocity-circulation interpretation.

We shall call a region R *regular* if its boundary consists of a finite number of smooth arcs, each having only a finite number of critical points (vertical and horizontal tangents) and also, possibly, a finite number of vertical and horizontal line segments.

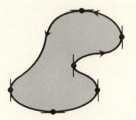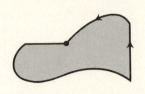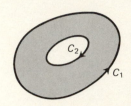

Figure 6

Let C be the boundary of R traced always in the direction that puts R to the left (i.e. traced *counterclockwise* with respect to nearby points of R). For the region shown at the right in Fig. 6, C is the union of two separate closed curves $C = C_1 + C_2$ that are traced in opposite directions, as shown. Then,

THEOREM 8   **Green's Theorem**   *If* $\mathbf{p}(\mathbf{x}) = (p(\mathbf{x}), q(\mathbf{x}))$ *is any smooth vector field defined on a regular region* R *and its boundary* C, *then*

$$\int_C \mathbf{p} \cdot d\mathbf{x} = \iint_R \left(\frac{\partial q}{\partial x} - \frac{\partial p}{\partial y}\right) dx\, dy.$$

Proof   We write $\mathbf{p}(\mathbf{x})$ as the sum of the two vector fields $(p(\mathbf{x}), 0)$ and $(0, q(\mathbf{x}))$ and prove the theorem for them separately. Consider the first of these two "component" fields. By drawing vertical lines through points of vertical tangency (and, possibly, through corners), we can subdivide R into a finite number of simpler regions $R_1, \ldots, R_n$ as shown in Fig. 7. Note that the boundary line integral for R is then just the sum of the boundary line integrals for the subregions R_i, because the extra vertical boundary segments introduced in the subdividing process contribute line integrals that cancel in pairs, each being traced once upward and once downward.

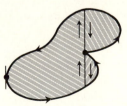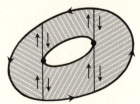

Figure 7

So if we can prove Green's theorem for each of these subregions R_i, $i = 1, \ldots, n$, then the sum of these n equations is Green's theorem for R.

We are thus reduced to proving Green's theorem for the field $(p(\mathbf{x}), 0)$ over a region R of the following simpler type: R is bounded above and below by function graphs, and left and right (possibly) by one or two vertical sides.

Having reduced the problem this far it is suddenly an easy consequence of the Fundamental Theorem of Calculus. We just evaluate the double integral as an iterated integral starting with y. In the calculation below, C_1 is the bottom curve, C_2 the vertical right-hand edge (if any), C_3 the top curve, *traced right to left*, and C_4 the left vertical edge (if any). See Fig. 8.

$$\iint\limits_{R} \left(-\frac{\partial p}{\partial y} \right) dx\, dy = \int_{x_1}^{x_2} \left[\int_{g(x)}^{f(x)} -\frac{\partial p}{\partial y}\, dy \right] dx$$

$$= \int_{x_1}^{x_2} [p(x, g(x)) - p(x, f(x))]\, dx$$

$$= \int_{x_1}^{x_2} p(x, g(x))\, dx + 0 - \int_{x_1}^{x_2} p(x, f(x))\, dx + 0$$

$$= \int_{C_1} p\, dx + \int_{C_2} p\, dx + \int_{C_3} p\, dx + \int_{C_4} p\, dx$$

$$= \int_{C} p\, dx.$$

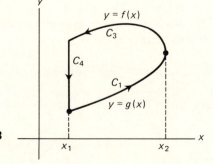
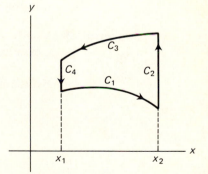

Figure 8

The C_2 and C_4 integrals are zero because x is constant along these edges. In detail, if t is any parameter for C_2, then

$$\int_{C_2} p\, dx = \int_{t_1}^{t_2} p\frac{dx}{dt}\, dt = \int_{t_1}^{t_2} p \cdot 0 \cdot dt = 0.$$

this completes the proof of Green's theorem for the first "component field" $(p(\mathbf{x}), 0)$.

By essentially the same proof we find that

$$\int_R \frac{\partial q}{\partial x}\, dx\, dy = \int_C q\, dy.$$

The principal difference is that this time we use *horizontal* subdividing lines for the preliminary decomposition of R into simpler regions, thus reducing the proof to the special case of a region that is bounded to the *left* and *right* by (nonstandard) function graphs, and, possibly, by horizontal line segments top and/or bottom (Fig. 9).

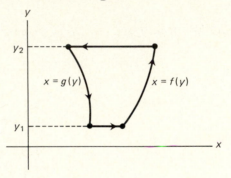

Figure 9

The sum of these two "component" Green's theorems is the general Green's theorem, as stated. ∎

Example 4   Verify Green's theorem for the vector field $\mathbf{p}(\mathbf{x}) = (xy, y^2)$ over the unit circle and its interior.

Solution   a) We use $(x, y) = (\cos\theta, \sin\theta)$ on the unit circle. Then

$$\int_C \mathbf{p} \cdot d\mathbf{x} = \int_0^{2\pi} (\cos\theta\sin\theta(-\sin\theta) + \sin^2\theta\cos\theta)\, d\theta$$

$$= \int_0^{2\pi} (\sin^2\cos\theta - \sin^2\cos\theta)\, d\theta = 0.$$

b)
$$\iint_R \left(\frac{\partial q}{\partial x} - \frac{\partial p}{\partial y}\right) dx\, dy = \iint_R (0 - x)\, dx\, dy.$$

This integral can be calculated by iterated integration, but it must come out to be zero, because x is an odd function (of x) and the region R is symmetric about the y-axis, so the integrals of x over the parts of R to the left and right of the y-axis cancel. □

Example 5 Show that $\int_C - y\, dx$ is the area A of the region R whose boundary curve is C.

Solution The line integral is of the form $\int p\, dx + q\, dy$ with $p(x, y) = -y$ and $q(x, y) = 0$. Then $\partial p/\partial y = -1$ and $\partial q/\partial x = 0$, so, by Green's theorem,

$$\int_C - y\, dx = \iint_R 1\, dx\, dy = A.$$ □

Example 6 Use the above fact to compute the area inside the ellipse

$$\frac{x^2}{a^2} + \frac{y^2}{b^2} = 1.$$

Solution The ellipse is given parametrically by $(x, y) = (a\cos\theta, b\sin\theta)$ on the parameter interval $[0, 2\pi]$. So

$$A = \int_C - y\, dx = \int_0^{2\pi} -(b\sin\theta)(-a\sin\theta)\, d\theta$$

$$= ab\int_0^{2\pi} \sin^2\theta\, d\theta = ab\int_0^{2\pi} \left(\frac{1 - \cos 2\theta}{2}\right) d\theta$$

$$= \frac{ab}{2}\, 2\pi = \pi ab.$$ □

Example 7 Use Green's theorem to show that the circulation of the rotating flow of Example 1 around the boundary of *any* regular region is equal to $2\omega A$, where A is the area of the region.

Solution The velocity field of Example 1 is the vector field

$$\mathbf{v}(\mathbf{x}) = r\omega(-\sin\omega t, \cos\omega t)$$

$$= (-\omega y, \omega x).$$

So

$$\int_C \mathbf{v} \cdot d\mathbf{x} = \iint_G \left(\frac{\partial v_2}{\partial x} - \frac{\partial v_1}{\partial y} \right) dx\, dy = \iint_R (\omega - (-\omega))\, dx\, dy$$

$$= 2\omega \iint_R 1\, dx\, dy = 2\omega A. \qquad \square$$

PROBLEMS FOR SECTION 4

1. Fluid flowing near a boundary is slowed down by viscosity. A simple model of such a flow is given by the field

$$\mathbf{v}(\mathbf{x}) = (0, x)$$

in the right half-plane.

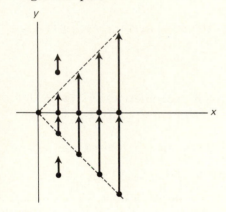

Compute the circulation of this flow about the unit square (with opposite corners at $(0, 0)$ and $(1, 1)$).

2. Compute the circulation of the above flow about the semicircular circumference bounding the right half of the unit circle.

In Problems 3 through 5 verify Green's theorem by directly computing the integrals involved when

3. R is the unit square $0 \le x \le 1, 0 \le y \le 1$ and the vector field is (xy, y).

4. R is the unit disk $x^2 + y^2 \le 1$ and the vector field is $(-y, x)$.

5. $\int_C xy(dx + dy)$, when C is the unit square of Problem 3.

6. Show that the area of a region R can be computed as the line integral $\int_C x\, dy$ where C is the boundary of R.

7. Use the above formula to compute the area of the ellipse $(x^2/a^2) + (y^2/b^2) = 1$, in the manner of Example 6.

8. Problem 6 and Example 5 combine to give the formula

$$A = \frac{1}{2} \int_C x\,dy - y\,dx$$

for the area of a region R in terms of a line integral around the boundary of R. This combined formula seems more complicated than its two ingredient formulas, but in some situations it actually results in an easier computation. Use the combined formula to compute the area inside the ellipse

$$\frac{x^2}{a^2} + \frac{y^2}{b^2} = 1.$$

9. Show that the area formula in Problem 8 can be rewritten

$$A = \frac{1}{2} \int_C x^2\, d\!\left(\frac{y}{x}\right) = -\frac{1}{2} \int_C y^2\, d\!\left(\frac{x}{y}\right).$$

Then use this reformulation to recompute the area inside the ellipse

$$\frac{x^2}{a^2} + \frac{y^2}{b^2} = 1.$$

10. Use the formula in Problem 9 to compute the area of the region between the upper branch of the hyperbola

$$\frac{y^2}{b^2} - \frac{x^2}{a^2} = 2$$

and the line $y = 2b$ $(b > 0)$. (Use $y = \sqrt{2}\, b \sec\theta$, $x = \sqrt{2}\, a \tan\theta$.)

11. As t runs from 0 to $+\infty$, the curve

$$(x, y) = \left(\frac{t}{1 + t^3}, \frac{t^2}{1 + t^3}\right)$$

describes a loop beginning and ending at the origin. Use the formula in Problem 9 to compute the area of this loop.

In Problems 12 through 21 compute the given line integral by Green's theorem. Here C_1 is the unit circle, C_2 is the unit square (with opposite vertices at $(0,0)$ and $(1,1)$), and C_3 is the triangle with vertices $(0,0)$, $(1,1)$, $(2,0)$. In each case state the answer also for a more general contour C if possible.

12. $\displaystyle\int_{C_1} y\,dx + x\,dy$

13. $\displaystyle\int_{C_2} 2y\,dx + x\,dy$

14. $\displaystyle\int_{C_3} 2xy\,dx + x^2\,dy$

15. $\displaystyle\int_{C_1} (y^2 + 1)\,dx + 2x(y + 1)\,dy$

16. $\displaystyle\int_{C_2} e^x \sin y\,dx + e^x \cos y\,dy$

17. $\displaystyle\int_{C_1} y \tan^2 x\,dx + \tan x\,dy$

18. $\displaystyle\int_{C_1} \arctan y\,dx - \left(\frac{xy^2}{1 + y^2}\right) dy$

19. $\displaystyle\int_{C_2} \cos x \sin y\,dx + \sin x \cos y\,dy$

20. $\displaystyle\int_{C_3} \frac{xy\,dx}{1+x} - \ln(1+x)\,dy$

21. $\displaystyle\int_{C_1} 2(2+x)y\,dx + (2x+x^2)\,dy$

22. Let C be any regular simple closed curve that contains the origin its interior, traced counterclockwise, and let $\mathbf{p}(\mathbf{x})$ be the field

$$\left(\frac{-y}{x^2+y^2}, \frac{x}{x^2+y^2}\right).$$

Green's theorem cannot be applied because $\mathbf{p}(\mathbf{x})$ has a singularity at the origin. However, let C' be a small circle about the origin inside of C, also traced counterclockwise. Show that

$$\int_C \mathbf{p}\cdot d\mathbf{x} = \int_{C'} \mathbf{p}\cdot d\mathbf{x}.$$

(Apply Green's theorem to the region R bounded on the outside by C and on the inside by C'). Show therefore that

$$\int_C \mathbf{p}\cdot d\mathbf{x} = 2\pi.$$

CHAPTER 19
FIRST AND SECOND ORDER LINEAR DIFFERENTIAL EQUATIONS

1
THE FIRST-ORDER LINEAR EQUATION

Before starting this chapter it would be well to review the separation-of-variables method for solving first-order differential equations (Chapter 8, Section 8).

A first-order *linear* differential equation is an equation of the form

$$\frac{dy}{dx} + a(x)y + b(x) = 0,$$

where the coefficients $a(x)$ and $b(x)$ are any functions of x. Superficially, the reason for the word *linear* is that we call $ay + b$ a linear expression in y. The deeper meaning of linearity will be discussed a little later.

A linear equation cannot generally be solved by separating the variables. However, if $b(x)$ is the zero function we have the special case

$$\frac{dy}{dx} + a(x)y = 0,$$

called the linear *homogeneous* equation, in which the variables can be separated. Proceeding as usual, we obtain the separated equation

$$\frac{dy}{y} + a(x)\,dx = 0,$$

which has the general solution

$$\int \frac{dy}{y} + \int a(x)\,dx = c,$$

or

$$\ln|y| + A(x) = c,$$

where $A(x)$ is any particular antiderivative of $a(x)$. Then

$$|y| = e^c e^{-A(x)}.$$

And since the right side is never zero, y can never change sign and the general solution can be written

$$y = Ce^{-A(x)}.$$

Note that $y = 0$ is a solution of the homogeneous equation $dy/dx + a(x)y = 0$ that is lost in the separated equation $dy/y + a(x)\,dx = 0$, but is recovered in the general solution by giving the parameter C the value 0.

Note also that the growth equation $dy/dx - ky = 0$ is a special case of the linear homogeneous equation.

Variation of Parameters The solution of the inhomogeneous linear equation requires a new method. We look for a solution y in the form $y = uv$, where v is a known function that we already have in hand for some reason or other, and u is an unknown function to be determined. When uv is substituted for y in the given differential equation, the result is a new differential equation for u, and we hope that the new equation for u will be easier to solve than the old equation for y. It often will be if the multiplier function v is somehow related to the beginning equation.

As a first application of this procedure, we take v to be a nonzero solution of the linear homogeneous equation

$$\frac{dy}{dx} + a(x)y = 0.$$

(We know a formula for v, but there is no need to write it down. We do need the fact that v is never zero, so that we can write down $1/v$.)

Then we try to find a solution y of the *inhomogeneous* equation

$$\frac{dy}{dx} + a(x)y = r(x) = -b(x)$$

in the form $y = vu$, where u is some function of x to be determined. Setting $y = vu$, we see that the inhomogeneous equation becomes

$$v\frac{du}{dx} + u\frac{dv}{dx} + a(x)uv = r(x)$$

or

$$u\left[\frac{dv}{dx} + a(x)v\right] + v\frac{du}{dx} = r(x)$$

or

$$v\frac{du}{dx} = r(x),$$

since, by assumption, the bracket is zero. Thus

$$\frac{du}{dx} = \frac{r(x)}{v}$$

and

$$u = \int \left(\frac{r(x)}{v}\right) dx + C.$$

Therefore,

$$y = uv = v\int \left(\frac{r(x)}{v}\right) dx + Cv.$$

Moreover, if y is *any* solution of the inhomogeneous equation, then we can set $u = y/v$, proceed as above, and conclude that y is given by the above formula. So there are no extraneous solutions. We have thus found a formula for all the solutions of the inhomogeneous linear equation, without having to know explicitly what the homogeneous solution v is. Of course, $v = e^{-A(x)}$, from the separation-of-variables calculation given earlier, so all in all we have proved the following theorem.

THEOREM 1 *The general solution of the linear differential equation*

$$\frac{dy}{dx} + a(x)y = r(x)$$

is given by

$$y = Ce^{-A(x)} + e^{-A(x)} \int e^{A(x)} r(x)\, dx,$$

where $A(x)$ is any particular antiderivative of the coefficient function $a(x)$, and C is the constant of integration. Moreover, every solution is of this form; there are no extraneous solutions.

The solution formula in Theorem 1 seems complicated and awkward, but its general form can be understood once we have looked into the nature of linearity. This will be the topic of the next section. Meanwhile, the formula should not be memorized. It is better to solve any particular inhomogeneous equation by retracing the steps in the proof of the theorem. This is called the *variation-of-parameter* method, for the following reason. The general solution of the homogeneous equation is a one-parameter family of solutions of the form $y = Cv(x)$. Here we replace the parameter C by an unknown function $u(x)$—hence the phrase "variation of parameter"—and try for a solution of the inhomogeneous equation in the form $y = u(x)v(x)$.

Example 1   Solve the linear inhomogeneous equation

$$\frac{dy}{dx} - y = \cos x,$$

by the variation-of-parameter method.

Solution   We first find one solution of the homogeneous equation

$$\frac{dv}{dx} - v = 0$$

by separating variables (or by inspection). It is $v = e^x$. We then set $y = e^x u$ in the given equation, and get

$$ue^x + e^x \frac{du}{dx} - ue^x = \cos x,$$

$$e^x \frac{du}{dx} = \cos x,$$

$$\frac{du}{dx} = e^{-x} \cos x.$$

The integral of $e^{-x}\cos x$ is worked out by integrating by parts twice, and is

$$u = \int e^{-x}\cos x = \frac{1}{2}[e^{-x}\sin x - e^{-x}\cos x] + C.$$

Therefore,

$$y = e^x u = \frac{1}{2}[\sin x - \cos x] + Ce^x$$

is the general solution of the equation

$$\frac{dy}{dx} - y = \cos x. \qquad \square$$

PROBLEMS FOR SECTION 1

Solve the following equations by the variation-of-parameter method.

1. $\dfrac{dy}{dx} - 2y = e^x$ 　　2. $\dfrac{dy}{dx} - 2y = e^{2x}$ 　　3. $\dfrac{dy}{dx} + ay = e^x$ 　　4. $\dfrac{dy}{dx} + ay = e^{bx}$

5. $\dfrac{dy}{dx} + y = x$ 　　6. $\dfrac{dy}{dx} - y = x^2$ 　　7. $\dfrac{dy}{dx} + \dfrac{y}{x} = x$ 　　8. $\dfrac{dy}{dx} - \dfrac{y}{x} = x$

9. $\dfrac{dy}{dx} + \dfrac{y}{x} = \dfrac{1}{x}$ 　　10. $\dfrac{dy}{dx} - \dfrac{y}{x} = \dfrac{1}{x}$ 　　11. $\dfrac{dy}{dx} + ay = \sin x$

12. $\dfrac{dy}{dx} + y\sin x = \sin x$ 　　13. $\dfrac{dy}{dx} + 2xy = x^3$ 　　14. $\dfrac{dy}{dx} - \dfrac{y}{x} = \ln x$

15. $\dfrac{dy}{dx} - y\tan x = 1$ 　　16. $\dfrac{dy}{dx} + y\tan x = 1$

**2
LINEARITY** Some important properties of a linear differential equation were implicit in the results of the last section. These properties explain the general form of the solution formula obtained there, and they will provide a shortcut to the solutions of an important type of inhomogeneous equation. Finally, they will help us to understand the more complicated second-order equations that come up next.

A new notation is helpful here: We set

$$L(y) = \frac{dy}{dx} + a(x)y,$$

so that the linear equations

$$\frac{dy}{dx} + a(x)y = 0, \qquad \frac{dy}{dx} + a(x)y = r(x)$$

abbreviate to

$$\boxed{L(y) = 0, \qquad L(y) = r(x).}$$

Example 1   If the equation is

$$\frac{dy}{dx} + xy = 0,$$

then $L(y) = \dfrac{dy}{dx} + xy$. In particular,

$$L(x^3) = \frac{d}{dx}(x^3) + x(x^3) = 3x^2 + x^4.$$

Similarly,

$$L(-x) = -1 - x^2,$$
$$L(x^3 - x) = -1 + 2x^2 + x^4$$
$$= L(x^3) - L(x),$$
$$L(e^x) = e^x(x + 1),$$
$$L(e^{-x^2}) = -xe^{-x^2},$$
$$L(e^{-x^2/2}) = 0. \qquad\qquad \square$$

THEOREM 2   *For any two differentiable functions $y_1 = f_1(x)$ and $y_2 = f_2(x)$, and any two constants c_1 and c_2,*

$$\boxed{L(c_1 y_1 + c_2 y_2) = c_1 L(y_1) + c_2 L(y_2).}$$

Proof   We just compute.

$$L(c_1 y_1 + c_2 y_2) = \frac{d}{dx}(c_1 y_1 + c_2 y_2) + a(x)(c_1 y_1 + c_2 y_2)$$

$$= c_1 \left[\frac{dy_1}{dx} + a(x)y_1\right] + c_2 \left[\frac{dy_2}{dx} + a(x)y_2\right]$$

$$= c_1 L(y_1) + c_2 L(y_2). \qquad\qquad \blacksquare$$

The theorem says that if
$$z = c_1 y_1 + c_2 y_2,$$
then
$$L(z) = c_1 L(y_1) + c_2 L(y_2).$$

That is, when we apply L to a linear combination of two functions y_1 and y_2, we get *that same* linear combination of the functions $L(y_1)$ and $L(y_2)$: applying L *preserves linear combination relationships*. For this reason we call L a *linear* differential operator, and we call the homogeneous equation
$$L(y) = 0,$$
or
$$\frac{dy}{dx} + a(x)y = 0,$$

a homogeneous *linear* differential equation. Similarly,
$$\frac{dy}{dx} + a(x)y = r(x)$$
or
$$L(y) = r(x),$$

is called an *inhomogeneous linear* differential equation.

THEOREM 3 *Let $y = Cv = Cg(x)$ be the general solution of the linear homogeneous equation*
$$L(y) = 0$$

and let $u = h(x)$ be any particular solution of the linear inhomogeneous equation
$$L(y) = r(x).$$

Then its general solution is

$$\boxed{y = Cv + u = Cg(x) + h(x).}$$

That is, the general solution of the inhomogeneous equation is the sum of the general solution of the homogeneous equation and any particular solution of the inhomogeneous equation.

Proof   We assume that we have found, somehow, one solution $u = h(x)$ of the inhomogeneous equation
$$L(u) = r(x).$$

Let y be any other solution. Then

$$L(y - u) = L(y) - L(u)$$
$$= r(x) - r(x) = 0,$$

by Theorem 2, so $y - u$ is a solution of the homogeneous equation, and hence of the form

$$y - u = Cv = Cg(x).$$

Thus, every solution y of the inhomogeneous equation is of the form

$$y = u + Cv$$
$$= h(x) + Cg(x).$$

Finally, these functions are all solutions, for

$$L(y) = L(u + Cv)$$
$$= L(u) + CL(v)$$
$$= r(x) + C \cdot 0$$
$$= r(x).$$

This completes the proof of the theorem.   ■

Note that the above proof depends almost entirely on the linearity of L as stated in Theorem 2.

If we now reexamine the explicit general solution that we obtained earlier,

$$y = Ce^{-A(x)} + e^{-A(x)} \int e^{A(x)} r(x) \, dx,$$

where $A(x)$ is an integral of $a(x)$, we can see the structure promised by the above theorem. It is the sum

$$y = Cv + u,$$

where

$$Cv = Ce^{-A(x)}$$

is the general solution of the homogeneous equation $L(y) = 0$, and

$$u = e^{-A(x)} \int e^{A(x)} r(x) \, dx$$

is a particular solution of the inhomogeneous equation $L(u) = r(x)$.

Method of Undetermined Coefficients

If a is a constant, we see by inspection that $u = e^{-ax}$ is a solution of the homogeneous equation

$$\frac{dy}{dx} + ay = 0.$$

The general solution of the inhomogeneous equation

$$\frac{dy}{dx} + ay = r(x)$$

then reduces, by virtue of Theorem 3, to finding a particular solution. The variation-of-parameter method is always available, but it can lead to a difficult integration, as we saw in the examples and problems in the last section. There is another, very simple, procedure, called the method of *undetermined coefficients*, that often works here. We illustrate it by some examples.

Example 2

Solve the differential equation

$$\frac{dy}{dx} - 2y = 3 \sin x + \cos x.$$

Solution Here the function $r(x) = 3 \sin x + \cos x$, and each of its derivatives, is a linear combination,

$$c_1 \sin x + c_2 \cos x,$$

of the two functions $\sin x$ and $\cos x$. In this type of situation, with one exception to be mentioned later, the differential equation has a uniquely determined particular solution u of the same form

$$u = c_1 \sin x + c_2 \cos x.$$

In order to find it, we note that

$$L(\sin x) = \cos x - 2 \sin x,$$
$$L(\cos x) = -\sin x - 2 \cos x,$$

so, if $u = c_1 \sin x + c_2 \cos x$, then

$$L(u) = c_1 L(\sin x) + c_2 L(\cos x)$$
$$= (c_1 - 2c_2)\cos x + (-2c_1 - c_2)\sin x.$$

We want $L(u)$ to be equal to $3 \sin x + \cos x$. That is, we want

$$-2c_1 - c_2 = 3 \quad \text{and} \quad c_1 - 2c_2 = 1.$$

Solving these equations simultaneously, we find that $c_1 = c_2 = -1$. Therefore, $-\sin x - \cos x$ is a particular solution. Theorem 4 now tells us that the general solution of the inhomogeneous equation is the sum of this particular solution and the general solution of the homogeneous equation $dy/dx - 2y = 0$, which is Ce^{2x}. Thus

$$y = Ce^{2x} - \sin x - \cos x$$

is the general solution of the inhomogeneous equation

$$\frac{dy}{dx} - 2y = 3\sin x + \cos x. \qquad \qquad \square$$

Example 3   Find a function $y = y(x)$ such that

$$\frac{dy}{dx} + 2y = x^2.$$

Solution   We are asked to find a particular solution of the differential equation, and our scheme is to try a linear combination, with undetermined coefficients, of x^2 and its derivatives $2x$ and 2. So we try for a solution in the form

$$u = ax^2 + b \cdot 2x + c \cdot 2.$$

But $2b$ and $2c$ are just new constants, so we try

$$u = ax^2 + bx + c.$$

Then

$$u' = 2ax + b,$$
$$L(u) = 2u + u'$$
$$= 2ax^2 + (2b + 2a)x + (2c + b).$$

We want $L(u) = x^2$, so we must have

$$\left. \begin{array}{r} 2a = 1 \\ 2b + 2a = 0 \\ 2c + b = 0 \end{array} \right\} a = \frac{1}{2}, b = -\frac{1}{2}, c = \frac{1}{4}.$$

Therefore

$$u = \frac{x^2}{2} - \frac{x}{2} + \frac{1}{4}$$

is a particular solution. $\qquad \qquad \square$

Example 4   Find the general solution of the differential equation

$$\frac{dy}{dx} - 2y = e^{-x}.$$

Solution   Here $r(x)$ and each of its derivatives is of the form ce^{-x}, so we look for a particular solution u of the same form: $u = ce^{-x}$. We see by inspection that $L(e^{-x}) = -3e^{-x}$, so

$$L(u) = L(ce^{-x}) = cL(e^{-x}) = -3ce^{-x},$$

and since this is required to be the function e^{-x}, the coefficient c must be $-1/3$. Thus, $-e^{-x}/3$ is a particular solution of the differential equation

$$\frac{dy}{dx} - 2y = e^{-x},$$

and its general solution is then

$$y = Ce^{2x} - \frac{1}{3} e^{-x},$$

by Theorem 3.   □

Example 5   Find a general solution of the differential equation

$$\frac{dy}{dx} - 2y = e^{2x}.$$

Solution   Here we run into the difficulty alluded to earlier. We note that $L(e^{2x}) = 0$, so if $u = ce^{2x}$, then

$$L(u) = L(ce^{2x}) = cL(e^{2x}) = 0,$$

and it is impossible to find c so that $L(u) = e^{2x}$.

In general, if the function $r(x)$ on the right in the inhomogeneous equation

$$L(y) = r(x)$$

happens to be a solution of the homogeneous equation $L(y) = 0$, then the undetermined-coefficient method cannot be directly applied.

But, just in this case, it will be found that the variation-of-parameter method reduces to the completely trivial integration $u = \int dx = x + C$, and the general solution is

$$uv = (x + C)r(x) = Cr(x) + xr(x).$$

So $xr(x)$ is the particular solution we were looking for.
    Returning to Example 4, the particular solution is xe^{2x} and the general solution is

$$Ce^{2x} + xe^{2x} = (C + x)e^{2x}. \qquad \square$$

Problems 11 and 13 will show a couple of other ways of looking at this special case.

PROBLEMS FOR SECTION 2

In each of the following problems, find the general solution by using the method of undetermined coefficients.

1. $\dfrac{dy}{dx} - 2y = e^x$
    2. $\dfrac{dy}{dx} - ay = e^{bx}$ $(a \neq b)$ 3. $\dfrac{dy}{dx} + y = x$

4. $\dfrac{dy}{dx} + ay = x^2$
    5. $\dfrac{dy}{dx} + ay = 1$
    6. $\dfrac{dy}{dx} + 3y = \sin x$ 7. $\dfrac{dy}{dx} + 2y = xe^x$

8. $\dfrac{dy}{dx} - y = xe^x$
    9. $\dfrac{dy}{dx} + y = x \sin x$
    10. $\dfrac{dy}{dx} + ay = e^x \sin x$

11. Show that the differential equation

$$\frac{dy}{dx} + ay = r(x)$$

is equivalent to

$$\frac{d}{dx}(e^{ax}y) = e^{ax}r(x).$$

Then find a solution by inspection in the case $r(x) = e^{-ax}$.

13. Show that if $L(y) = dy/dx + a(x)y$, then

$$L(xy) = xL(y) + y,$$

so

$$L(xy) = y \quad \text{if} \quad L(y) = 0.$$

12. Show that the solution formula in Theorem 1 is an immediate consequence of the reformulated differential equation in Problem 11.

14. Find the general solution of

$$\frac{dy}{dx} - y = e^x.$$

(Use Problem 11 or Problem 13 for the particular solution.)

15. Find the general solution of

$$\frac{dy}{dx} + 4y = e^{-4x}.$$

16. Show that the general solution of the differential equation

$$L(y) = b_1(x) + b_2(x)$$

is the sum of a particular solution of $L(y) = b_1(x)$, a particular solution of $L(y) = b_2(x)$, and the general solution of $L(y) = 0$.

17. Find the general solution of

$$\frac{dy}{dx} + y = 1 + e^x$$

by following the above prescription.

18. Solve

$$\frac{dy}{dx} - y = x + \sin x.$$

19. Find the general solution of

$$\frac{dy}{dx} - y = x + e^x.$$

20. Find the general solution of

$$\frac{dy}{dx} + y = e^x + e^{-x}.$$

21. If $L(y) = dy/dx - y/x$, compute

a) $L(x^4)$ b) $L(x^2)$

c) $L(x)$ d) $L(xe^x)$.

22. If $L(y) = (dy/dx) + a(x)y$, show that

$$uL(v) - vL(u) = u^2 \frac{d}{dx}\left(\frac{v}{u}\right).$$

23. If $L(y) = (dy/dx) + a(x)y$, show that

a) $L(uv) = uL(v) + v\dfrac{du}{dx}$,

b) $2L(uv) = uL(v) + vL(u) + \dfrac{d}{dx}(uv)$.

24. Let L and M be any two first-order linear differential operators with constant coefficients, say

$$L(y) = y' + ay, \qquad M(y) = y' + by.$$

Show that L and M commute:

$$L[M(f(x))] = M[L(f(x))]$$

for any twice-differentiable function f.

3
THE SECOND-ORDER LINEAR EQUATION

We now turn our attention to certain *second-order* differential equations, i.e., equations involving the derivatives of f up through the order two. In the prime notation,

$$y' = \frac{dy}{dx}, \qquad y'' = \frac{d^2y}{dx^2}, \qquad \text{etc.,}$$

the equations we shall study are

$$\boxed{\begin{aligned} y'' + a(x)y' + b(x)y &= 0, \\ y'' + a(x)y' + b(x)y &= r(x). \end{aligned}}$$

These equations are *linear*, as we shall see. They have important applications to natural phenomena associated with oscillation and vibration, resonance and damping, growth and decay.

Given the coefficient functions $a(x)$ and $b(x)$, we set

$$L(y) = y'' + a(x)y' + b(x)y$$
$$= f''(x) + a(x)f'(x) + b(x)f(x),$$

for any function $y = f(x)$ having two derivatives. Thus, if $a(x) = 2$ and $b(x) = 3$, then

$$L(y) = y'' + 2y' + 3y,$$
$$L(f(x)) = f''(x) + 2f'(x) + 3f(x),$$
$$L(\sin x) = -\sin x + 2\cos x + 3\sin x$$
$$= 2(\sin x + \cos x),$$
$$L(e^x) = e^x + 2e^x + 3e^x = 6e^x,$$
$$L(e^{-x}) = e^{-x} - 2e^{-x} + 3e^{-x} = 2e^{-x}, \quad \text{etc.}$$

THEOREM 4 *L is a linear differential operator. That is, if f_1 and f_2 are any functions having at least two derivatives, and if c_1 and c_2 are any two constants, then*

$$L(c_1f_1 + c_2f_2) = c_1L(f_1) + c_2L(f_2).$$

Proof The proof is just a computational verification, like the proof of Theorem 2. We leave it as an exercise. ∎

The two differential equations we want to study are the linear homogeneous equation

$$L(f(x)) = 0$$

and the linear inhomogeneous equation

$$L(f(x)) = r(x).$$

The general properties of their solution families are stated in the following theorem.

THEOREM 5 *If y_1 and y_2 are solutions of the homogeneous equation $L(y) = 0$, then so is any linear combination $y = c_1y_1 + c_2y_2$. Moreover, if y_1 and y_2*

are independent, in the sense that neither is a constant multiple of the other, then

$$y = c_1 y_1 + c_2 y_2$$

is the general solution of $L(y) = 0$.

If u is any particular solution of the inhomogeneous equation $L(u) = r(x)$, then

$$\boxed{y = (c_1 y_1 + c_2 y_2) + u}$$

is the general solution of the inhomogeneous equation (supposing y_1 and y_2 to be independent). That is, the general solution of the inhomogeneous equation is the sum of a particular solution and the general solution of the homogeneous equation.

Partial Proof The proof of this theorem has an easy part and a hard part, and we are ready now only for the easy part.

If $L(y_1) = 0$ and $L(y_2) = 0$, then

$$L(c_1 y_1 + c_2 y_2) = c_1 L(y_1) + c_2 L(y_2) = 0,$$

by Theorem 4. Thus the family of solutions of the homogeneous equation $L(y) = 0$ is closed under forming linear combinations. What we cannot prove here is that if y_1 and y_2 are independent solutions, then the functions $c_1 y_1 + c_2 y_2$ form the whole solution family. (However, see Problem 18, page 847.)

Now suppose that u is some particular solution of the inhomogeneous equation $L(u) = r(x)$. If y is any other solution, then

$$L(y - u) = L(y) - L(u) = r - r = 0,$$

so $y - u$ is a solution of the homogeneous equation. Because $y = (y - u) + u$, we see that if we have one solution u of the inhomogeneous equation, then any other solution y can be written as u plus a solution of the homogeneous equation. ∎

According to Theorem 5, the general solution of the inhomogeneous equation depends on finding

1. the general solution of the homogeneous equation, and

2. a particular solution of the inhomogeneous equation.

The variation-of-parameter method is the basic theoretical tool in reducing (2) to (1). It now starts off like this: First, we find the general solution of the homogeneous equation in the form

$$y = c_1 v_1 + c_2 v_2.$$

Then we look for a particular solution of the inhomogeneous equation $L(y) = r(x)$, in the form

$$y = u_1 v_1 + u_2 v_2,$$

where u_1 and u_2 are unknown functions that have to be determined. Since we have replaced the parameters c_1 and c_2 by unknown functions u_1 and u_2, as in the first-order case, this method is called *variation of parameters*. But it is now more complicated than in our first-order examples, and it is best understood in the context of linear algebra. Moreover, we may not even be able to get started, because solving the homogeneous equation can be much harder in the second-order case. We shall therefore break off at this point, and leave variation of parameters to a future course.

The situation is much simpler if the coefficient functions $a(x)$ and $b(x)$ are both constants. In this case, we can solve the homogeneous equation completely and will do so in the next section. Moreover, independently of the homogeneous solution, we can find a particular solution of the inhomogeneous equation if the function $r(x)$ on the right lets us use the method of undetermined coefficients. This will be taken up now for the simplest context of the problem: to find a particular solution of the equation

$$L(y) = r(x),$$

when $r(x)$ is *not* itself a solution of the homogeneous equation.

Example 1   Find a particular solution of the equation

$$L(y) = \sin x,$$

where

$$L(y) = y'' + 3y' + 2y.$$

Solution   We try for a solution in the form

$$u = a \sin x + b \cos x.$$

That is, we try for a solution u in the form of a linear combination of $r(x)$ and its derivatives. Remember that, for this to succeed, it is essential that $r(x)$ and its derivatives involve only a finite number of functions. So the device cannot work for a function like $r(x) = 1/x$. Anyway, we see by inspection that

$$L(\sin x) = \sin x + 3 \cos x,$$

$$L(\cos x) = \cos x - 3 \sin x,$$

and

$$L(u) = aL(\sin x) + bL(\cos x)$$

$$= (a - 3b)\sin x + (3a + b)\cos x.$$

Thus, $L(u) = \sin x$ if and only if

$$a - 3b = 1, \qquad 3a + b = 0.$$

The solutions of these equations are $a = 1/10$, $b = -3/10$; so

$$u = \frac{1}{10} \sin x - \frac{3}{10} \cos x$$

is a particular solution of the inhomogeneous equation. $\square$

Example 2 In order to obtain a particular solution of

$$L(y) = y'' + 3y' + 2y = e^t,$$

we try a linear combination of e^t and its derivatives. But they are all the same function, so we just try

$$u = ce^t.$$

Since $L(e^t) = 6e^t$, we then have

$$L(u) = L(ce^t) = 6ce^t,$$

and the equation $L(u) = e^t$ reduces to $6ce^t = e^t$. Thus, $6c = 1$, and a particular solution is $u = e^t/6$. $\square$

Example 3 It is always possible to find a solution of

$$L(y) = r_1(x) + r_2(x),$$

by finding a solution u_1 of $L(y) = r_1(x)$, a solution u_2 of $L(y) = r_2(x)$, and then adding:

$$L(u_1 + u_2) = L(u_1) + L(u_2) = r_1(x) + r_2(x).$$

For an equation like

$$y'' - 2y' + y = \sin x - \cos x,$$

this would be inefficient, since each separate solution would have to be of the form $a \sin x + b \cos x$, and there would be duplication of effort.

In the case of

$$y'' + 2y' - y = e^x + x,$$

however, x and its derivatives remain completely distinct from e^x and its derivatives, and in such a situation the divide-and-conquer strategy may pay off. We first calculate (mentally) that

$$L(e^x) = 2e^x,$$

so we will have $L(ce^x) = e^x$ if $c = 1/2$. That is, $u_1 = e^x/2$ is a particular solution of $L(y) = e^x$.

Next, we calculate what L does to x and its derivatives,

$$L(x) = 2 - x, \qquad L(1) = -1,$$

so $L(ax + b) = a(2 - x) + b(-1) = -ax + (2a - b)$. This will be the function x if $a = -1$ and $b = 2a = -2$. That is,

$$u_2 = -x - 2$$

is a particular solution of $L(y) = x$. Then

$$u = \frac{e^x}{2} - x - 2$$

is a particular solution of $L(y) = e^x + x$. □

PROBLEMS FOR SECTION 3

In each of the following problems find a particular solution by the method of undetermined coefficients.

1. $y'' + y = x$ **2.** $y'' + 4y = e^x$ **3.** $y'' - 4y = e^x$ **4.** $y'' - 4y' = e^{3x}$

5. $y'' + 5y = \cos x$ **6.** $y'' - y = x^2$ **7.** $y'' + y' = x^2$

8. $y'' - y = Ax^2 + Bx + C$ **9.** $y'' + 2y = e^x + 2e^{-x}$ **10.** $y'' - y' = \cos x + 2x$

11. $y'' + y = xe^x$ **12.** $y'' + y' - 2y = x$ **13.** $y'' + y' - 2y = e^{-x}$

14. $y'' + y' - 2y = e^{2x}$ **15.** $y'' + y' - 2y = \sin x$ **16.** $y'' + y' = x^3$

17. $y'' + y' = x + \sin x$

18. $y'' + y' = x^2 e^{-x}$

19. $y'' - 2y' + y = x^2 e^x$

20. $y'' - 2y = xe^x + 1$

21. $y'' - 2y = e^x \sin x$

22. $y'' + 4y = x \sin x$

23. $y'' - 2y' + 3y = e^x \sin x$

24. Supposing that the differential operator L has constant coefficients, show that it commutes with differentiation. That is, show that

$$\frac{d}{dx} L(f(x)) = L(f'(x)).$$

4
THE HOMOGENEOUS EQUATION WITH CONSTANT COEFFICIENTS

In view of our first-order experience, it seems reasonable to check whether $y = e^{kx}$ can ever be a solution of the homogeneous equation

$$L(y) = y'' + ay' + by = 0.$$

Example 1 Can $y = e^{kx}$ be a solution of

$$L(y) = y'' + 2y' + 3y = 0?$$

Solution We have

$$L(e^{kx}) = k^2 e^{kx} + 2ke^{kx} - 3e^{kx}$$
$$= (k^2 + 2k - 3)e^{kx}.$$

This will be the zero function if and only if

$$k^2 + 2k - 3 = 0.$$

Since

$$k^2 + 2k - 3 = (k + 3)(k - 1),$$

the roots of the equation are

$$k = -3 \quad \text{and} \quad k = 1.$$

Thus, $y = e^{-3x}$ and $y = e^x$ are both solutions of $L(y) = 0$. □

Example 2 Continuing the above example,

$$y = c_1 e^{-3x} + c_2 e^x$$

is a solution for any constants c_1 and c_2, by Theorem 5. Moreover, since we cannot express e^{-3x} as a constant times e^x, Theorem 5 says, furthermore, that every solution of $L(y) = 0$ is of the above form.

However, we haven't proved this part of the theorem, so let us see if we can find a direct proof for this example, using the product device again.

Let u be any solution of the equation and write $u = vy$, where v is to be determined and y is a solution we already know. Here we can take y as e^{-3x} or e^x, and the latter looks simpler. So we set $u = ve^x$. Then

$$u' = ve^x + v'e^x,$$
$$u'' = ve^x + 2v'e^x + v''e^x,$$

and

$$0 = L(u) = u'' + 2u' - 3u$$
$$= v(e^x + 2e^x - 3e^x) + 4v'e^x + v''e^x$$
$$= (v'' + 4v')e^x.$$

Therefore,

$$v'' + 4v' = 0,$$

which is a first-order equation in v'. Its general solution is

$$v' = c_1e^{-4x},$$

so

$$v = c_1e^{-4x} + c_2,$$

(with a different constant c_1) and finally

$$u = ve^x = c_1e^{-3x} + c_2e^x.$$

This proves that there are no other solutions than the ones we already had.   □

Now let us consider the general constant coefficient equation, which we shall write in the form

$$\boxed{L(y) = y'' + 2py' + qy = 0.}$$

Note that we are writing $2p$ for the coefficient of y'. For example, if $L(y) = y'' + 2y' - 3y$, then

$$p = 1 \quad \text{and} \quad q = -3.$$

This artificial notation simplifies the algebra. Then

$$L(e^{kx}) = k^2e^{kx} + 2pke^{kx} + qe^{kx}$$
$$= (k^2 + 2pk + q)e^{kx}.$$

The Characteristic Equation This will be the zero function if and only if

$$k^2 + 2pk + q = 0,$$

which is called the *characteristic equation* of L. The solutions of the characteristic equation are given by the quadratic formula

$$k = \frac{-2p \pm \sqrt{4p^2 - 4q}}{2} = -p \pm \sqrt{p^2 - q}.$$

There are three cases.

Case I If $p^2 > q$ and if we set $\Delta = \sqrt{p^2 - q}$, then there are two distinct real roots

$$k_1 = -p + \Delta \qquad \text{and} \qquad k_2 = -p - \Delta,$$

and the situation is exactly like Example 1. In fact, if we replace -3 and 1 by k_1 and k_2, the argument in Example 2 turns into a proof of the following theorem.

THEOREM 6 *If the characteristic equation*

$$k^2 + 2pk + q = 0$$

has distinct real solutions $k = k_1, k_2$, *then the solutions of the linear homogeneous equation*

$$y'' + 2py' + qy = 0$$

are exactly the functions of the form

$$y = c_1 e^{k_1 x} + c_2 e^{k_2 x}.$$

Case II If $p^2 - q = 0$, then there is only one exponential solution e^{kx} given by $k = -p$. We illustrate this situation with an example.

Example 3 For

$$y'' + 2y' + y = 0,$$

the characteristic equation

$$k^2 + 2k + 1 = (k + 1)^2 = 0$$

has only one root, $k = -1$. Thus, e^{-x} is the only pure exponential solution. But this leads to the general solution by the same device as in Example 2. Any other solution u can be written $u = ve^{-x}$, with v to be determined. Then,

$$u' = -ve^{-x} + v'e^{-x},$$
$$u'' = ve^{-x} - 2v'e^{-x} + v''e^{-x},$$

and

$$0 = L(u) = u'' + 2u' + u$$
$$= v(e^{-x} - 2e^{-x} + e^{-x}) + v''e^{-x}$$
$$= v''e^{-x}.$$

Therefore,

$$v'' = 0,$$
$$v' = c_1$$
$$v = c_1 x + c_2,$$

and

$$u = ve^{-x} = c_1 xe^{-x} + c_2 e^{-x}.$$

With general coefficients, the above argument proves the following theorem.   □

THEOREM 7   *If the characteristic equation*

$$k^2 + 2pk + q = 0$$

has exactly one real solution, i.e., if $q = p^2$ and the equation is $(k + p)^2 = 0$, then the solutions of

$$y'' + 2py' + qy = 0$$

are exactly the functions

$$\boxed{y = c_1 xe^{-px} + c_2 e^{-px}.}$$

Case III   If $p^2 < q$, then neither of the roots

$$k = -p \pm \sqrt{p^2 - q}$$

is real, and there are no pure exponential solutions. Let us, nevertheless, proceed as in Case II, and write an arbitrary solution u in the form

$$u = ve^{-px},$$

with v to be determined. Then,

$$u' = -pve^{-px} + v'e^{-px},$$
$$u'' = p^2 ve^{-px} - 2pv'e^{-px} + v''e^{-px},$$

and

$$0 = L(u) = u'' + 2pu' + qu$$
$$= v(p^2 - 2p^2 + q)e^{-px} + v''e^{-px}.$$

The equation for u thus reduces to a simpler equation for v,

$$v'' = -\beta^2 v,$$

where $\beta = \sqrt{q - p^2}$. We know two solutions of this equation, namely

$$v = \sin \beta x \qquad \text{and} \qquad v = \cos \beta x.$$

Therefore, every linear combination

$$v = c_1 \sin \beta x + c_2 \cos \beta x$$

is a solution. We must show that there are no other solutions. The consistent thing to do would be to try the same device once more: set $v = w \cos \beta x$, and then find w by solving a first-order equation. This would work, but not as well as before, because we have to divide by $\cos \beta x$ in the process and then worry about the points where $\cos \beta x = 0$. Instead, we shall resort to the end of Section 3 in Chapter 4, where the proof for the case $\beta = 1$ was sketched. That proof works just as well for arbitrary β, and takes care of our present dilemma.

THEOREM 8   *If the characteristic equation*

$$k^2 + 2pk + q = 0$$

has no real solutions, i.e., if $p^2 - q < 0$, *then the solutions of the equation*

$$y'' + 2py' + qy = 0$$

are exactly the functions

$$\boxed{y = e^{-p(x)}[c_1 \cos \beta x + c_2 \sin \beta x]}$$

where $\beta = \sqrt{q - p^2}$.

Theorems 6 through 8 show that the general solution of the constant coefficient homogeneous equation

$$y'' + 2py' + qy = 0$$

is one of three types, depending on whether q is less than, equal to, or greater than p^2. If $q < p^2$, then we can write $q = p^2 - m^2$, where

$m = \sqrt{p^2 - q}$. Similarly, if $q > p^2$, then q is of the form $q = p^2 + m^2$. We can then summarize the three theorems in the following form.

THEOREM 9   *The general solution of the constant coefficient homogeneous equation*

$$y'' + 2py' + qy = 0$$

is

$$
\begin{array}{ll}
e^{-px}[c_1 e^{mx} + c_2 e^{-mx}] & \text{if } q = p^2 - m^2, \\
e^{-px}[c_1 x + c_2] & \text{if } q = p^2, \\
e^{-px}[c_1 \cos mx + c_2 \sin mx] & \text{if } q = p^2 + m^2.
\end{array}
$$

Finally, Theorem 5 tells us that the general solution of the inhomogeneous equation

$$y'' + ay' + by = r(x)$$

is the sum of the general solution of the homogeneous equation

$$y'' + ay' + by = 0,$$

as determined in the above theorems, and a particular solution of the inhomogeneous equation, determined possibly by the method of undetermined coefficients.

Example 4   Find the general solution of

$$y'' + 3y' + 2y = \sin t.$$

Solution   The characteristic equation

$$k^2 + 3k + 2 = 0$$

factors into

$$(k + 2)(k + 1) = 0,$$

and hence has the two roots $k = -2, -1$. The general solution of the homogeneous equation $y'' + 3y' + 2y = 0$ is therefore

$$c_1 e^{-x} - c_2 e^{-2x}.$$

On the other hand, we found a particular solution of the inhomogeneous equation to be

$$\frac{1}{10}\sin x - \frac{3}{10}\cos x$$

in the first example of the last section. The general solution is therefore

$$c_1 e^{-x} + c_2 e^{-2x} + \frac{1}{10}\sin x - \frac{3}{10}\cos x,$$

by Theorem 5. $\square$

It may happen, when we write down the general solution for the homogeneous equation associated with

$$y'' + ay' + by = r(x),$$

that we discover the right-hand function $r(x)$ to be already one of the homogeneous solutions. In that case, the undetermined-coefficient procedure has to be modified. In view of our experience with the first-order equations, it is natural to try a linear combination of $xr(x)$ and its derivatives. We then have to calculate $L(xr(x))$, and it will shorten our work if we know ahead of time what the result of this calculation is.

LEMMA 1 *If $L(y) = y'' + ay' + by$, then*

$$L(xy) = xL(y) + (2y' + ay).$$

Therefore, if $y = f(x)$ is a solution of the homogeneous equation $L(f) = 0$, then

$$L(xf) = 2f' + af.$$

The proof will be left as an exercise.

The peculiar form of the above result will make more sense after some of the problems are worked out. (See Problem 29.)

Example 5 Find the general solution of

$$y'' - 4y' + 4y = e^{2x}.$$

Solution The characteristic equation is

$$k^2 - 4k + 4 = (k - 2)^2 = 0.$$

which has the single (repeated) root $k = 2$. The general solution of the homogeneous equation is therefore

$$e^{2x}(c_1x + c_2),$$

by Theorem 7 or Theorem 9.

In this case, since $r(x) = e^{2x}$ and $xr(x) = xe^{2x}$ are both solutions of the homogeneous equation, we apply Lemma 1 to $x(xe^{2x}) = x^2e^{2x}$.

$$L(x^2e^{2x}) = xL(xe^{2x}) + 2(xe^{2x})' - 4(xe^{2x}) = 2e^{2x}.$$

So if $u = cx^2e^{2x}$, then

$$L(u) = L(cx^2e^{2x}) = cL(x^2e^{2x}) = 2ce^{2x}.$$

This is required to be equal to $r(x) = e^{2x}$, so $c = 1/2$ and the particular solution is $x^2e^{2x}/2$. The general solution is therefore

$$y = e^{2x}\left[\frac{x^2}{2} + c_1x + c_2\right],$$

by Theorem 5.   □

Example 6   Find a solution of

$$y'' + y = \sin x.$$

Solution   We would normally try a linear combination of $\sin x$ and its derivatives, $u = a \sin x + b \cos x$. But we find here that

$$L(\sin x) = 0 = L(\cos x).$$

So we try

$$u = ax \sin x + bx \cos x,$$

instead. By Lemma 1,

$$L(x \sin x) = xL(\sin x) + 2 \cos x = 0 + 2 \cos x$$
$$= 2 \cos x,$$
$$L(x \cos x) = -2 \sin x;$$

so

$$L(u) = L(ax \sin x + bx \cos x) = aL(x \sin x) + bL(x \cos x)$$
$$= 2a \cos x - 2b \sin x.$$

We want $L(u) = \sin x$, so $a = 0$ and $b = -1/2$. Thus the particular solution is

$$u = -\frac{1}{2}x \cos x,$$

and the general solution is

$$c_1 \sin x + \left(c_2 - \frac{x}{2}\right)\cos x. \qquad \square$$

PROBLEMS FOR SECTION 4

Find the general solution of each of the following equations.

1. $y'' + 2y = 0$ **2.** $y'' + 2y = 1$ **3.** $y'' + 2y = x^2$ **4.** $y'' + 2y = \sin x$

5. $y'' + 4y = \sin 2x$ **6.** $y'' - 2y = x + 1$ **7.** $y'' - y = e^x$

8. $y'' - y' = 1$ **9.** $y'' - y = e^x \sin x$ **10.** $y'' - y = xe^x$

11. $y'' + y' - 2y = 0$ **12.** $y'' + y' - 2y = x$ **13.** $y'' + y' - 2y = \sin x$

14. $y'' + y' - 2y = e^x$ **15.** $y'' - 2y' + y = 0$ **16.** $y'' - 2y' + y = x$

17. $y'' - 2y' + y = \sin x$ **18.** $y'' - 2y' + y = e^x$ **19.** $y'' + 2y' + 2y = 0$

20. $y'' + 2y' + 2y = x$ **21.** $y'' + 2y' + 2y = e^x$ **22.** $y'' + 2y' + 2y = \sin x$

23. Let D be the operation of differentiation, so that

$$Df = f', \qquad Df(x) = f'(x).$$

Then $L(f) = f'' + af' + bf$ can be thought of as a quadratic polynomial in D applied to f,

$$L(f) = D^2 f + aDf + bf$$
$$= (D^2 + aD + b)f = p(D)f,$$

where p is the quadratic polynomial

$$p(x) = x^2 + ax + b.$$

Show that if $p(x)$ can be factored,

$$p(x) = (x - k_1)(x - k_2)$$

then

$$p(D)f = (D - k_1)(D - k_2)f,$$

where the right side is interpreted as

$$(D - k_1)[(D - k_2)f].$$

The above formula gives a systematic way of obtaining the solutions of the homogeneous equation. However, the third case, where the roots k_1 and k_2, obtained from the quadratic formula, involve the imaginary number $i = \sqrt{-1}$, carries us beyond the scope of this course.

24. Supposing that

$$D^2 + aD + b = (D - k_1)(D - k_2)$$

in the above sense, with $k_1 \neq k_2$, show that the general solution of the homogeneous equation

$$L(y) = (D^2 + aD + b)y = 0$$

is

$$y = C_1 e^{k_1 x} + C_2 e^{k_2 x}.$$

[*Hint*: Set $u = (D - k_2)y$. Then the homogeneous second-order equation $L(y) = 0$ reduces to the *first*-order linear equation

$$(D - k_1)u = 0.$$

Solving this gives u, and then $(D - k_2)y = u$ is a first-order inhomogeneous equation for y, that can be solved by variation of parameters, or undetermined coefficients.]

25. Now solve the homogeneous equation by the above scheme in the case where

$$D^2 + aD + b = (D - k)^2$$

(i.e., where $k_1 = k_2 = k$). In this case, use variation of parameters for the last step.

Use the method developed in Problem 24 to find the general solution of each of the following problems, by two applications of the first-order variation-of-parameter calculation.

26. $y'' - y' - 2y = e^x$ **27.** $y'' - y' - 2y = e^{2x}$ **28.** $y'' - y = 1$

29. Show that

$$p(D)(xf) = xp(D)f + p'(D)f,$$

so

if $p(D)f = 0$, then $p(D)(xf) = p'(D)f$.

(Here $p(x) = x^2 + ax + b, p'(x) = 2x + a$.)

30. Show that, if $p(D) = D^2 + aD + b$, then

$$p(D)(fg) = [p(D)f]g + [p'(D)f]g'$$
$$+ \frac{1}{2}[p''(D)f]g''.$$

5
ELASTIC
OSCILLATION

Although all sorts of elastic objects vibrate when they are struck or otherwise disturbed, the vibration normally dies away, as when a note is struck on a piano. We ascribe this attenuation to *resistance* to the motion, such as air resistance in the case of the piano string.

If the object is continuously "excited," then the vibration can continue indefinitely, despite resistance. Violin tones and organ tones are such sustained vibrations. A continuously excited vibration can even *increase* in intensity. This is the phenomenon of *resonance,* and it may be desirable or catastrophic. A suspension bridge is an elastic object, and a group of people marching in step over a suspension bridge at the natural frequency of the bridge can cause the bridge to vibrate with an increasing amplitude that may eventually destroy it. This is why soldiers were trained to break cadence when crossing a bridge. On the other hand, the current in electric circuits oscillates according to the same principles of vibration, and there resonance can be a desirable phenomenon.

We consider first the special case of *simple harmonic motion.* When a weight suspended at the end of a coiled spring is disturbed, it bobs up and down with a definite frequency. The motion dies out as time passes, due to air resistance and internal friction in the spring. Presumably, if the motion were to occur in a perfect vacuum, and if the spring were perfectly elastic, then the oscillatory motion would continue undiminished forever.

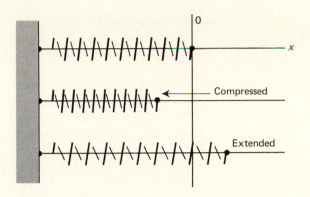

Figure 1

This ideal frictionless model is very instructive and useful. It executes simple harmonic motion.

Consider, then, a coiled spring lying along a frictionless x-axis, with one end held fixed and the free end at the origin (Fig. 1). When we compress the spring it resists and pushes back. When we extend the spring it resists and pulls back. That is, the spring always exerts a "restoring force" when it is deformed. Moreover, we find after careful measurement that the restoring force F is exactly proportional to the amount of deformation x (within the so-called elastic limits). Since F is directed against the deformation, it is given by

$$F = -sx,$$

where s is the constant of proportionality. This is Hooke's Law. If we use a strong spring, s will be larger; we call s the "*stiffness* of the spring."

If we now attach a mass m to the end of the spring and disturb the system somehow, then our ideal oscillatory motion will occur (Fig. 2). In order to study this motion mathematically, we need to know how a mass m moves when it is acted on by a force F. We find that m

Figure 2

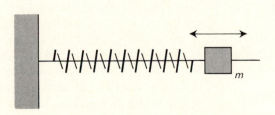

is *accelerated* by F according to Newton's second law of motion

$$F = ma = m\frac{d^2x}{dt^2}$$

(when suitable units are used for all these quantities).

The laws of Newton and Hooke combine to show that the motion of the model mass–spring system satisfies the differential equation

$$m\frac{d^2x}{dt^2} = F = -sx, \qquad \text{or} \qquad \frac{d^2x}{dt^2} + \frac{s}{m}x = 0.$$

We know the general solution of this equation to be

$$x = A\sin \omega t + B\cos \omega t,$$

where $\omega = \sqrt{s/m}$, and A and B are constants of integration. Without knowing A and B we don't know the motion exactly, but we do know its frequency: since sine and cosine are periodic with period 2π, it follows that x is a periodic function of t with period

$$T = \frac{2\pi}{\omega} = 2\pi\sqrt{\frac{m}{s}}.$$

The frequency of the motion is the number of cycles (periods) per unit time:

$$f = \frac{1}{T} = \frac{1}{2\pi}\sqrt{\frac{s}{m}} = \frac{\omega}{2\pi}.$$

These equations show exactly how the period and frequency depend on the physical constants of the elastic system. The frequency varies as the square root of the stiffness of the spring, and is *inversely proportional* to the square root of the mass.

A particular motion is generally determined by known "initial conditions." We shall neglect mentioning units in the examples below.

Example 1   A particle of mass $m = 4$ is attached to a spring of stiffness $s = 2$ in the configuration discussed above. Starting at the equilibrium position, the particle is struck and given an initial velocity $v_0 = -3$. What is the motion?

Solution Here $\omega = \sqrt{s/m} = 1/\sqrt{2}$, and the initial conditions are $x = 0$ and $dx/dt = -3$ when $t = 0$. Substituting these values in the equations

$$x = A \sin \omega t + B \cos \omega t,$$

$$\frac{dx}{dt} = \omega A \cos \omega t - \omega B \sin \omega t,$$

we see that

$$0 = A \cdot 0 + B \cdot 1,$$

$$-3 = \frac{1}{\sqrt{2}} A \cdot 1 - \frac{1}{\sqrt{2}} B \cdot 0.$$

so that $B = 0$ and $A = -3\sqrt{2}$. Thus

$$x = -3\sqrt{2} \sin(t/\sqrt{2})$$

is the equation of motion. The particle oscillates with frequency

$$f = \frac{\omega}{2\pi} = \frac{1}{2\sqrt{2}\,\pi} \approx \frac{1}{9} \quad \text{cycles per unit time,}$$

back and forth across the interval $[-3\sqrt{2}, 3\sqrt{2}]$. Its maximum displacement $3\sqrt{2}$ is called the *amplitude* of its motion. ☐

Example 2 Show that any particular solution

$$x = A \sin \omega t + B \cos \omega t$$

can be written uniquely in the form

$$x = a \sin(\omega t + \alpha,)$$

where a is positive and $0 \le \alpha < 2\pi$.

Solution Expanding the desired form by the sine addition law,

$$x = a[\cos \alpha \sin \omega t + \sin \alpha \cos \omega t],$$

and comparing with the given equation

$$x = A \sin \omega t + B \cos \omega t,$$

we see that the requirements on α and a reduce to

$$a \cos \alpha = A,$$

$$a \sin \alpha = B.$$

Then $A^2 + B^2 = a^2$, so a and α are determined by

$$a = \sqrt{A^2 + B^2},$$
$$(\cos \alpha, \sin \alpha) = (A/a, B/a). \qquad \qquad \square$$

If we consider $x = \sin \omega t$ to be the normal mode of oscillation, then the motion $x = \sin(\omega t + \alpha)$ *leads* the normal oscillation by the fixed angle α. We call α the *leading phase angle*. Recapitulating:

The motion of the ideal frictionless elastic system is always sinusoidal. If it is given by

$$x = A \sin \omega t + B \cos \omega t,$$

then its *frequency* is

$$\boxed{f = \frac{\omega}{2\pi},}$$

it *amplitude* is

$$\boxed{a = \sqrt{A^2 + B^2},}$$

and its *leading phase angle* α is determined by

$$\boxed{(\cos \alpha, \sin \alpha) = \left(\frac{A}{a}, \frac{B}{a}\right).}$$

The motion

$$\boxed{x = a \sin(\omega t + \alpha)}$$

is called *simple harmonic motion*. The development above shows it to be the basic type of periodic motion occurring when elastic objects vibrate.

The ideal mass–spring system becomes realistic when we add a resistance force proportional to the velocity, $-r\,dx/dt$. Then its motion is governed by Newton's law $ma - F = 0$ in the form

$$\boxed{m\frac{d^2x}{dt^2} + \left[r\frac{dx}{dt} + sx\right] = 0,}$$

or

$$\ddot{x} + \frac{r}{m}\dot{x} + \frac{s}{m}x = 0.$$

If, in addition, the system is continuously excited by a periodic external force of the form $c \sin \beta t$, then its motion is governed by Newton's law in the form

$$m\frac{d^2x}{dt^2} + r\frac{dx}{dt} + kx = c \sin \beta t.$$

(This is also the equation for the motion of electricity in a circuit to which a generator has been attached. The earlier equation governs the flow of electricity in a "free" circuit. We shall not analyze those electrical applications, however.)

Here again we have differential equations that state our understanding of a natural phenomenon, and their solutions will show the possible patterns of elastic behavior and permit quantitative predictions.

The analysis of this situation is broken down into Problems 15 through 28. Problems 1 through 14 concern simple harmonic motion.

PROBLEMS FOR SECTION 5

In each of the following motions calculate the period, amplitude, and phase angle. (Rewrite each in the form $x = a \sin(\omega t + \alpha)$, computing a exactly, but possibly only estimating α.)

1. $x = \sin(3t) - \sqrt{3}\cos(3t)$

2. $x = \cos t + \sin t$

3. $x = 5\cos t - 5\sin t$

4. $x = \sin 2t + 2\cos 2t$

5. A particle is moving in simple harmonic motion with amplitude a and frequency f (in cycles per second). Show that its velocity at $x = 0$ is $\pm 2\pi af$.

6. A particle of mass 10 is attached to a spring of stiffness 4 and oscillates freely in simple harmonic motion. Suppose that at time $t = 0$, the position and velocity of the particle are

$$x_0 = 2, \qquad v_0 = -3.$$

Find the motion.

7. The same system is given a new motion with initial conditions $x_0 = 2$, $v_0 = 3$. Find the new motion.

8. Show that a mass m hanging at the bottom of a vertical spring of stiffness k executes simple harmonic motion about its rest position. (Gravity acts with the constant downward force $mg = 32.2m$. Calculate the total force, and then put a new origin at the point where the total force is zero.)

A pendulum consists of a pendulum bob (a weight) suspended at the end of a light rigid rod, and allowed to swing back and forth under the action of gravity. We consider it *ideally* to be a point mass m at the end of a weightless rod of length l. The acceleration caused by gravity in a freely falling body is $g \simeq 32.2 \, \text{ft/sec}^2$, so by Newton's law $F = ma$, the force exerted by gravity on a body of mass m is $mg \simeq 32.2m$. In the case of the pendulum, most of the force of gravity is counteracted by the support of the rod, but a small amount of sidewise force is left over to make the pendulum oscillate. Assuming that forces combine according to the parallelogram law, this residual force is $mg \sin \theta$. See Fig. 3. Newton's second law $F = ma$ thus becomes

$$-mg \sin \theta = m\frac{d^2s}{dt^2} = m\frac{d^2(l\theta)}{dt^2} \quad \text{or} \quad \frac{d^2\theta}{dt^2} = -\left(\frac{g}{l}\right)\sin \theta.$$

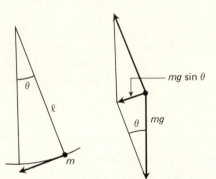

Figure 3

Because of the appearance of $\sin \theta$ rather than θ on the right, we cannot solve this equation. However, when θ is small,

$$\sin \theta \approx \theta$$

(since $\sin \theta/\theta \approx 1$), and small-amplitude motions are therefore governed by the "linearized" equation

$$\frac{d^2\theta}{dt^2} = -\left(\frac{g}{l}\right)\theta.$$

9. The above discussion is important to clockmakers. A one-second pendulum takes one second for each swing, so its period is 2 sec. Show that the length l of such a *second-pendulum* is 39.2 inches.

10. Find the length of a one-half-second pendulum.

11. Find the length of a two-second pendulum.

12. Show that the period of a pendulum is proportional to the square root of its length, and calculate the constant of proportionality.

13. On Planet X the force of gravity is one-third what it is on earth. Find the length of a one-second pendulum on X.

14. Work the same problem for Planet Y, where the force of gravity is twice that on earth.

We turn now to the study of the mass–spring system with resistance, having the equation

$$\ddot{x} + \frac{r}{m}\dot{x} + \frac{s}{m}x = 0.$$

15. Consider first the case where the resistance r dominates the combined effect of the mass m and spring stiffness s, in the sense that

$$r^2 > 4ms.$$

Supposing that the initial velocity (at time $t = 0$) is zero, and the initial position is positive, show that the motion is of the form

$$x = c[ke^{-lt} - le^{-kt}],$$

where $0 < l < k$, and c is positive.

16. Assuming the above formula, show that the graph of x as a function of t has the form shown in Fig. 4 with a single point of inflection at $t = [\ln(k/l)]/(k - l)$.

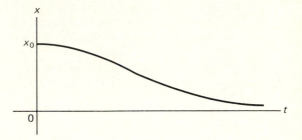

Figure 4

17. In the same situation of a dominant resistance, suppose that the initial position is at $x = 0$ but that the initial velocity is positive. Show that then the motion is of the form

$$x = c[e^{-lt} - e^{-kt}],$$

where $c > 0$ and $k > l > 0$.

18. Show that the graph of the above motion has the general features shown in Fig. 5.

In particular, show that there is one critical point $t = t_0$ at which x has its maximum value, and one point of inflection, located at $2t_0$.

Figure 5

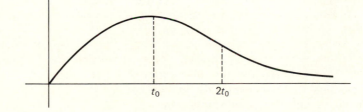

The next two problems analyze in more detail the same situation of a dominant resistance, when the initial position x_0 is positive, say $x_0 = 1$, and the initial velocity v_0, has various values.

19. Show first that if $x_0 = 1$, the motion has the form

$$x \doteq ae^{-lt} + (1 - a)e^{-kt},$$

where $0 < l < k$. Then show that the initial velocity v_0 is less than $-k$ if and only if $a < 0$, and that the graph of the motion in this case has the general form shown in Fig. 6. (One change in sign, one critical point, one point of inflection.)

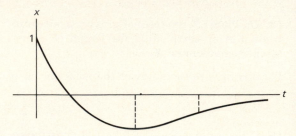

Figure 6

20. Show from the equation for x in the problem above:

a) If $-k \leq v_0 \leq -k/(k + l)$, then the graph of the motion has the general form shown in Fig. 7 (no change of sign, no critical point, no point of inflection).

b) If $v_0 > -k/(k + l)$, then the graph is always positive and has one point of inflection, as in Fig. 8:

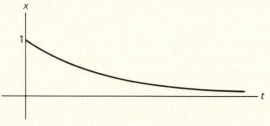

Figure 7

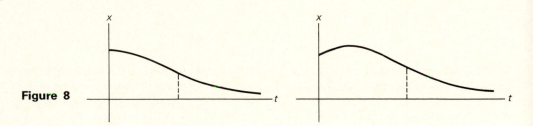

Figure 8

In the next two problems we assume that the resistance r balances the combined effect of the mass m and stiffness s, in the sense that $r^2 = 4ms$.

21. Show that if $r^2 = 4ms$ and if the initial conditions are $(x_0, v_0) = (1, 0)$, then the motion is given by

$$x = (1 + kt)e^{-kt},$$

where $k = r/2m$. Show also that the graph of this motion has the same features as in Problem 16 (but with a different point of inflection).

22. If the initial conditions are $(x_0, v_0) = (0, 1)$, show that the motion is given by

$$x = te^{-kt},$$

and that its graph has the same features as in Problem 18.

23. Consider finally the case where the resistance r is small in comparison with the combined effect of the mass m and spring stiffness s, in the sense that

$$r^2 < 4ms.$$

Show that then the motion is given by

$$x = ae^{-kt}\sin(\omega t + \theta),$$

where the exponential damping rate k has the value $k = r/2m$, and the frequency $\omega/2\pi$ is determined by

$$\omega = \frac{1}{2m}\sqrt{4ms - r^2}.$$

24. Show that the graph of x, as a function of t in the above problem, has the general features shown in Fig. 9.

How is this graph affected by a decrease in the resistance r (m and s remaining unchanged)? What is the limiting configuration as r approaches zero?

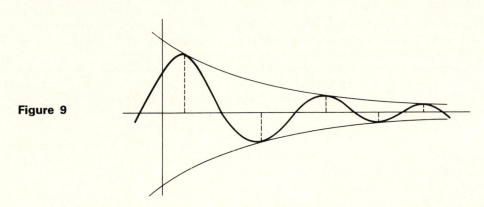

Figure 9

We consider one more situation, in which an extra periodic "driving force" is impressed on the system, say a "sinusoidal" force, proportional to $\sin \alpha t$. The modified differential equation can be taken to be

(*)
$$\ddot{x} + \frac{r}{m}\dot{x} + \frac{s}{m}x = \sin \alpha t,$$

where $r^2 < 4ms$, and where we have written $\sin \alpha t$ instead of the more general $c \sin \alpha t$, for simplicity. The motion is now a superposition (sum) of a transient vibration that dies out exponentially as discussed above (a solution of the homogeneous equation) and a "steady-state" sinusoidal vibration

$$A \sin(\alpha t + \beta)$$

(the particular solution of the inhomogeneous equation obtained by undetermined coefficients). Note that the total motion settles down to this steady-state simple harmonic motion as the transient component wears off. If in the beginning the transient and steady-state components tend to cancel one another, then the total effect is a vibration with an amplitude that builds up as time passes, and if the transient and steady-state frequencies are nearly the same, the limiting amplitude can be very large. This is the phenomenon of resonance.

25. Show that the particular solution of the above equation obtained by the method of undetermined coefficients can be written in the form

$$A \sin(\alpha t + \beta),$$

where the amplitude A is given by

$$A = \frac{1}{\sqrt{\left(\dfrac{s}{m} - \alpha^2\right)^2 + \left(\dfrac{\alpha r}{m}\right)^2}}.$$

Conclude that A is proportional to $1/r$ if the driving frequency $\alpha/2\pi$ is equal to the natural frequency $\sqrt{s/m}/2\pi$ of the undamped and undriven system.

27. If a simple electric circuit contains a total capacitance of amount $1/s$, a total resistance r, and a total inductance m, and if x is the amount of electricity that has moved past a fixed point in a displacement from the equilibrium position, then x fluctuates with time according to the same homogeneous equation

$$m\ddot{x} + r\dot{x} + sx = 0.$$

This is for a "free" circuit, unaffected by any outside influence. If a sinusoidal generator is attached, then the fluctuation of x is determined by the inhomogeneous equation (*), with $\sin \alpha t$ possibly replaced by $c \sin \alpha t$ for some contant c.

But here we are more interested in the behavior of the *current* $i = \dot{x}$. Show that if x satisfies (*), then $i = \dot{x}$ satisfies

(**) $\qquad m\dfrac{d^2i}{dt^2} + r\dfrac{di}{dt} + si = \alpha \cos(\alpha t).$

26. With m, s, and r fixed, but α variable, prove that the maximum value of the amplitude A in the above problem is

$$A_{\max} = \frac{2m^2}{r\sqrt{4ms - r^2}},$$

when $\alpha = (\sqrt{ms - r^2/2})/m$.

28. Show without calculation, by applying a general result from Section 3 (see Problem 24), that if $A \sin(\alpha t + \beta)$ is a particular solution of (*), then $\alpha A \cos(\alpha t + \beta)$ is a particular solution of (**). Then show, from the result in Problem 25, that the amplitude $A' = \alpha A$ of this oscillation is given by

$$A' = \frac{1}{\sqrt{\left(\dfrac{s}{\alpha m} - \alpha\right)^2 + \left(\dfrac{r}{m}\right)^2}},$$

and hence that A' is maximum (for fixed s, m, and r) when

$$\alpha = \sqrt{\frac{s}{m}}.$$

Conclude that the resonant frequency is equal to the frequency of the undamped ($r = 0$) freely oscillating circuit.

**6
SERIES
SOLUTIONS
OF LINEAR
DIFFERENTIAL
EQUATIONS**

Infinite series apply naturally to the solution of linear differential equations, because such a differential equation automatically determines the Maclaurin series of a solution function.

Example 1 Consider the differential equation

$$y' - y = 0.$$

If we assume that a solution $y = f(x)$ is the sum of a power series

$$y = a_0 + a_1x + a_2x^2 + \cdots + a_nx^2 + \cdots,$$

then

$$y' = a_1 + 2a_2x + 3a_3x^2 + \cdots + (n + 1)a_{n+1}x^n + \cdots$$

and

$$y' - y = (a_1 - a_0) + (2a_2 - a_1)x + (3a_3 - a_2)x^2 + \cdots$$
$$+ [(n + 1)a_{n+1} - a_n]x^n + \cdots.$$

Since $y' - y$ is the zero function, its Maclaurin series coefficients must all be zero, so

$$a_1 = a_0,$$

$$a_2 = \frac{a_1}{2} = \frac{a_0}{2},$$

$$a_3 = \frac{a_2}{3} = \frac{a_0}{3!},$$

$$\cdot \quad \cdot$$
$$\cdot \quad \cdot$$
$$\cdot \quad \cdot$$

$$a_{n+1} = \frac{a_n}{n + 1} = \frac{a_0}{(n + 1)!},$$

and

$$y = a_0\left[1 + x + \frac{x^2}{2!} + \cdots + \frac{x^n}{n!} + \cdots\right],$$

where a_0 is an arbitrary constant, the "constant of integration." Of course, we recognize this solution to be a_0e^x. $\square$

What we are doing here is essentially the reverse of our procedure in Section 2 of Chapter 12. There we noted that the function

$$y = 1 + x + \frac{x^2}{2} + \cdots + \frac{x^n}{n!} + \cdots$$

satisfies the differential equation $y' - y = 0$ (and we used this fact to conclude that $y = e^x$). Here we directly compute that, if the sum of a power series satisfies the differential equation $y' - y = 0$, then the series must be $a \sum_0^\infty x^n/n!$ Note that this calculation demands no prior knowledge of the exponential function, and if we have never seen e^x, we could still solve the differential equation $y' - y = 0$ as the sum of a power series in the above way and then use the series to compute the solution function. In this situation it would be necessary to check that the solution series does in fact converge.

Example 2 If $y = f(x)$ is the sum of a power series

$$y = a_0 + a_1 x + \cdots + a_n x^n + \cdots,$$

and satisfies the differential equation

$$y'' + y = 0,$$

then the coefficients are determined by the same kind of recursive procedure as in the first example, except that this time the even coefficients go back to a_0 and the odd coefficients go back to a_1. Thus,

$$y = a_0 + a_1 x + a_2 x^2 + a_3 x^3 + a_4 x^4 + a_5 x^5 + \cdots,$$

$$y'' = 2a_2 + 3 \cdot 2a_3 x + 4 \cdot 3a_4 x^2 + 5 \cdot 4a_5 x^3$$
$$+ 6 \cdot 5a_6 x^4 + 7 \cdot 6a_7 x^5 + \cdots,$$

and, since $y'' + y = 0$, we see that

$$a_2 = -\frac{a_0}{2}, \qquad\qquad a_3 = -\frac{a_1}{3!},$$

$$a_4 = -\frac{a_2}{4 \cdot 3} = \frac{a_0}{4!}, \qquad a_5 = -\frac{a_3}{5 \cdot 4} = \frac{a_1}{5!},$$

$$a_6 = -\frac{a_4}{6 \cdot 5} = -\frac{a_0}{6!}, \qquad a_7 = -\frac{a_5}{7 \cdot 6} = -\frac{a_1}{7!},$$

and so on. Thus

$$y = a_0 \left[1 - \frac{x^2}{2} + \frac{x^4}{4!} - \frac{x^6}{6!} + \cdots \right] + a_1 \left[x - \frac{x^3}{3!} + \frac{x^5}{5!} - \frac{x^7}{7!} + \cdots \right],$$

which we recognize to be $a_0 \cos x + a_1 \sin x$. $\square$

Again, if we knew nothing about the trigonometric functions, we could check that both series above converge everywhere to some functions $C(x)$ and $S(x)$ respectively, and we could compute these functions by the partial sums of the series and make up tables for them. Then the general solution of the differential equation $y'' + y = 0$ is expressed in terms of these two new functions by

$$y = a_0 C(x) + a_1 S(x),$$

as we saw above. This is only an academic exercise, but it has the following point: We can proceed in exactly this way with more complicated differential equations, computing the new functions that are their solutions as the sums of power series determined by the differential equation, and then studying these new solution functions on the basis of these power-series definitions.

The solutions of a second-order linear homogeneous equation with *variable* coefficients,

$$y'' + a(x)y' + b(x)y,$$

are generally not elementary functions, which means that there cannot be any general procedure for turning out solutions as finite combinations of familiar functions. Yet in some circumstances it is easy to compute the solutions as sums of power series in the manner of the above examples.

Example 3   Solve the equation

$$y'' - xy = 0$$

by power series.

Solution   If

$$y = c_0 + c_1 x + c_2 x^3 + \cdots + c_n x^n + \cdots,$$

then

$$y'' = 2c_2 + 3 \cdot 2c_3 x + 4 \cdot 3c_4 x^2 + \cdots$$
$$+ (n + 2)(n + 1)c_{n+2} x^n + \cdots,$$
$$xy = c_0 x + c_1 x^2 + \cdots + c_{n-1} x^n \cdots,$$

and

$$y'' - xy = 2c_2 + (3 \cdot 2c_3 - c_0)x + (4 \cdot 3c_4 - c_1)x^2 + \cdots$$
$$+ ((n + 2)(n + 1)c_{n+2} - c_{n-1})x^n + \cdots.$$

Therefore $y'' - xy = 0$ if and only if

$$2c_2 = 0,$$
$$3 \cdot 2c_3 = c_0,$$
$$4 \cdot 3c_4 = c_1,$$
$$\cdot \qquad \cdot$$
$$\cdot \qquad \cdot$$
$$\cdot \qquad \cdot$$
$$(n + 2)(n + 1)c_{n+2} = c_{n-1}.$$

These equations define the coefficients c_n recursively in terms of c_0 and c_1 (and c_2):

$$c_2 = 0, \quad c_3 = \frac{c_0}{3 \cdot 2}, \quad c_4 = \frac{c_1}{4 \cdot 3}, \quad c_5 = \frac{c_2}{5 \cdot 4} = 0,$$

$$c_6 = \frac{c_3}{6 \cdot 5} = \frac{c_0}{6 \cdot 5 \cdot 3 \cdot 2}, \quad c_7 = \frac{c_4}{7 \cdot 6} = \frac{c_1}{7 \cdot 6 \cdot 4 \cdot 3},$$

and so on. Thus

$$y = c_0 f(x) + c_1 g(x)$$

where

$$f(x) = 1 + \frac{x^3}{3 \cdot 2} + \frac{x^6}{6 \cdot 5 \cdot 3 \cdot 2} + \frac{x^9}{9 \cdot 8 \cdot 6 \cdot 5 \cdot 3 \cdot 2} + \cdots,$$

and

$$g(x) = x + \frac{x^4}{4 \cdot 3} + \frac{x^7}{7 \cdot 6 \cdot 4 \cdot 3} + \frac{x^{10}}{10 \cdot 9 \cdot 7 \cdot 6 \cdot 4 \cdot 3} + \cdots.$$

There is no easy way to write down a formula for the nth coefficient in either of these series, but the pattern is clear. Moreover, it is easy to see, from the ratio test, that both series converge for all x, because the recursive relations give us simple explicit formulas for the ratios of successive terms. For example, $f(x)$ is of the form

$$f(x) = \sum_0^\infty a_n x^{3n},$$

where $a_n / a_{n-1} = 1/(3n(3n - 1))$. The ratio of the corresponding series terms is thus $x^3/(3n(3n - 1))$, which approaches 0 as $n \to \infty$, no matter what x is. Therefore, f and g are everywhere defined functions, and each satisfies the differential equation $y'' - xy = 0$, because the coefficients in its Maclaurin expansion satisfy the recursive relations that were equivalent to this differential equation. That $y = c_0 f(x) + c_1 g(x)$ is the general solution of this differential equation follows from Theorem 5. $\qquad\square$

PROBLEMS FOR SECTION 6

Show that the general solution of each of the following differential equations can be expressed as the sum of convergent power series. In each case, write down the recursion formula for the coefficients, and determine the convergence interval by the ratio test. Write down the first few terms of each series, and, if possible, the general term. (It is also instructive, where possible, to find the solutions by our earlier methods and compare answers.)

1. $y' - xy = 0$

2. $y'' + xy' = 0$

3. $y' - x^2y = x$

4. $(1 - x)y' + y = 0$

5. $(1 - x^2)y' - xy = 0$

6. $y'' - 2y = 1$

7. $y'' - x^2y = 0$

8. $y'' + 2x^2y - xy = 0$

9. $(1 - x^2)y'' - y = 0$

10. Solve the equation

$$xy'' - y = 0$$

by the power series method. What conclusion can you draw about the possibility of expression the general solution

$$y = af(x) + bg(x)$$

as a power series?

12. Show that no solution of the equation

$$x^2y'' - y = 0$$

can be represented by a convergent Maclaurin series.

11. Supposing that $y = f(x)$ is one solution of

$$xy'' - y = 0$$

(the one found above), show that $z = u(x)y$ is another solution if and only if

$$u(x) = C_1 \int \frac{dx}{y^2} + C_2.$$

Solve the following differential equations by power series. (There may be only one series solution.)

13. $y'' + xy + y = 0$

14. $y'' + xy + 2y = 0$

15. $(1 + x^2)y' - 2xy = 0$

16. $(1 + x^2)y'' + 2xy' - 2y = 0$

17. $xy'' + y' + xy = 0$

18. Suppose that y is one solution of the linear homogeneous equation

$$y'' + a(x)y' + b(x)y = 0,$$

and suppose that $y \neq 0$ on an interval I. Use the variation-of-parameter method (as in Examples 2 and 3 of Section 4) to show that the general solution over I is given by

$$u = c_1y \int \frac{e^{-A(x)}}{y^2} \, dx + c_2y,$$

where $A(x)$ is any antiderivative of $a(x)$. This result can be used to complete the proof of Theorem 5 (for the interval I).

αppendix 1
BASIC INEQUALITY LAWS

All properties of inequalities stem from the following two facts about positive numbers:

P1. If x and y are positive, then so are $x + y$ and xy.

We call a number x *negative* if $-x$ is positive.

P2. A number x is either positive, or zero, or negative, and these possibilities are mutually exclusive.

Here is a list of frequently used inequality properties.

1. Law of signs. *If x and y have the same sign (i.e., if both are positive or both are negative), then xy is positive. If x and y have opposite signs, then xy is negative.*

2. *If $x \neq 0$, then x^2 is positive. In particular, $1 = 1 \cdot 1$ is positive.*

3. *If $x \neq 0$, then x and $1/x$ have the same sign.*

DEFINITION   We say that x is less than y, and write $x < y$, if $y - x$ is positive. Similarly $x > y$ (x is greater than y) if $y < x$, that is, if $x - y$ is positive. In particular, $x > 0$ if and only if x is positive.

4. *If $x < y$ and $y < z$, then $x < z$.*

5. *If $x < y$, then $x + c < y + c$, no matter what number c is.*

6. *If $x < y$ and if c is positive, then $cx < cy$.*

7. *If $x < y$ and $a < b$, then $x + a < y + b$.*

8. *If $0 < x < y$ and $0 < a < b$, then $ax < by$.*

9. *If x and y are positive and if $x < y$, then $1/x > 1/y$.*

That is, reciprocals of positive numbers are in opposite order.

10. *If $x < y$, then $-y < -x$.* Changing signs reverses an inequality.

11. *If $x < y$ and c is negative then $cx > cy$.* Multiplying by a negative number reverses an inequality.

DEFINITION We say that *x is less than or equal to y*, and write $x \leq y$ or $x \leqq y$, if either $x < y$ or $x = y$. This so-called *weak* inequality satisfies a corresponding list of laws that we mostly won't give.

12. *If $x \leq y$, then $x + x \leq c + y$, for any number c.*

13. *$|x| \leq a$ if and only if $-a \leq x \leq a$.*

14. *$|x + y| \leq |x| + |y|$.*

SOME PROOFS

1. By hypothesis, x and y are either both positive or both negative. If x and y are both positive, then xy is positive, by P1. If x and y are both negative, then $-x$ and $-y$ are both positive, by P2, so $xy = (-x)(-y)$ is positive, by P1. Thus xy is positive in either case.

The other assertion will be left as an exercise.

2. This is a corollary of (1), since if $x \neq 0$, then x^2 is a product in which both factors have the same sign.

3. Suppose x and $1/x$ have opposite signs. Then $1 = x(1/x)$ is negative, by (1). But 1 is positive, by (2). We therefore have a contradiction, so x and $1/x$ must have the same sign.

4. By hypothesis, $x < y$ and $y < z$. Therefore $y - x$ and $z - y$ are both positive, by the definition of inequality. Therefore $z - x = (z - y) + (y - z)$ is positive, by P1. Therefore $x < z$, by the definition of inequality.

5. If $x < y$, then $y - x$ is positive, by the definition of inequality. Then $(y + c) - (x + c) = y - x$ is positive. Then $x + c < y + c$, by the definition of inequality.

6. By hypothesis and the definition of inequality, $y - x$ and c are both positive. Therefore $cy - cx = c(y - x)$ is positive, by P1, and hence $cx < cy$, by the definition of inequality.

9. We note that

$$\frac{1}{x} - \frac{1}{y} = \frac{y - x}{xy} = (y - x) \cdot \frac{1}{x} \cdot \frac{1}{y}$$

and that all three factors on the right are positive, for various reasons. It follows that $1/x > 1/y$.

12. If $x \leq y$, then either $x < y$ or $x = y$. If $x < y$, then $x + c < y + c$, by (5). If $x = y$, then $x + c = y + c$ (the sum of a pair of numbers is uniquely defined). Therefore, in either case, $x + c \leq y + c$.

13. It follows from the definition of absolute value that $|x|$ is the larger of x and $-x$. The inequality $|x| \leq a$ is thus equivalent to the pair of inequalities $-x \leq a$, $x \leq a$, and hence to the pair $-a \leq x$, $x \leq a$. But the latter pair is what we mean by the "continued" inequality $-a \leq x \leq a$.

14. Since

$$-|x| \leq x \leq |x|,$$

$$-|y| \leq y \leq |y|,$$

it follows that

$$-(|x| + |y|) \leq x + y \leq |x| + |y|,$$

by the weak inequality analogue of (8). But this is equivalent to $|x + y| \leq |x| + |y|$ by (13).

APPENDIX 2
TRIGONOMETRY SUMMARY

**BASIC
TRIGONOMETRIC
IDENTITIES**

$$\sin 0 = \cos \frac{\pi}{2} = 0$$

$$\cos 0 = \sin \frac{\pi}{2} = 1$$

$$\sin(-x) = -\sin x$$

$$\sin\left(x + \frac{\pi}{2}\right) = \cos x$$

$$\cos(-x) = \cos x$$

$$\cos\left(x + \frac{\pi}{2}\right) = -\sin x$$

$$\sin^2 x + \cos^2 x = 1$$

$$\tan^2 x + 1 = \sec^2 x$$

$$1 + \cot^2 x = \csc^2 x$$

Addition Laws

$$\sin(x + y) = \sin x \cos y + \cos x \sin y$$

$$\sin(x - y) = \sin x \cos y - \cos x \sin y$$

$$\cos(x + y) = \cos x \cos y - \sin x \sin y$$

$$\cos(x - y) = \cos x \cos y + \sin x \sin y$$

$$\tan(x + y) = \frac{\tan x + \tan y}{1 - \tan x \tan y}, \qquad \tan(x - y) = \frac{\tan x - \tan y}{1 + \tan x \tan y}$$

Double-angle Formulas

$$\sin 2x = 2 \sin x \cos x$$

$$\cos 2x = \cos^2 x - \sin^2 x = 2 \cos^2 x - 1 = 1 - 2 \sin^2 x$$

$$\tan 2x = \frac{2 \tan x}{1 - \tan^2 x}$$

Half-angle Formulas

$$\cos^2\left(\frac{x}{2}\right) = \frac{1 + \cos x}{2}$$

$$\sin^2\left(\frac{x}{2}\right) = \frac{1 - \cos x}{2}$$

$$\tan\left(\frac{x}{2}\right) = \frac{\sin x}{1 + \cos x} = \frac{1 - \cos x}{\sin x}$$

Sum and Difference Formulas

$$\sin A + \sin B = 2 \sin\frac{A + B}{2} \cos\frac{A - B}{2}$$

$$\sin A - \sin B = 2 \cos\frac{A + B}{2} \sin\frac{A - B}{2}$$

$$\cos A + \cos B = 2 \cos\frac{A + B}{2} \cos\frac{A - B}{2}$$

$$\cos A - \cos B = -2 \sin\frac{A + B}{2} \sin\frac{A - B}{2}$$

Law of Sines

$$\frac{\sin A}{a} = \frac{\sin B}{b} = \frac{\sin C}{c}$$

Law of Cosines

for any triangle

$$c^2 = a^2 + b^2 - 2ab \cos C$$

The proofs of these identities will be discussed later. First we review some other material.

The standard right triangle configuration is shown again in Fig. 1, this time using the unit circle.

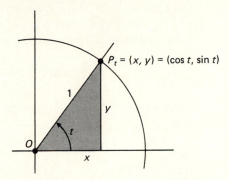

Figure 1

In order to convert from degrees to radians, note that a right angle is one quarter of the way around the circle, that is $(1/4)(2\pi) = \pi/2$ radians. This gives the first line in the table below, and the rest of the table follows from it.

degrees	radians		
90°	$\pi/2$		
1°	$\pi/180$		
$d°$	$(\pi/180)d$		
		$\sin t$	$\cos t$
30°	$\pi/6$	$1/2$	$\sqrt{3}/2$
45°	$\pi/4$	$\sqrt{2}/2$	$\sqrt{2}/2$
60°	$\pi/3$	$\sqrt{3}/2$	$1/2$

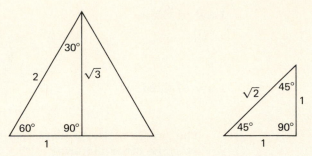

Figure 2

The last three lines record the standard facts for a 30°–60°–90° triangle and for an isosceles right triangle (45°–45°–90°). These values come from the Pythagorean theorem. See Fig. 2.

If an angle vertex is placed at $O' = (a, b)$, then an angle t from the positive x direction around to the ray $O'P'$ satisfies

$$\cos t = \frac{x - a}{r}, \qquad \sin t = \frac{y - b}{r},$$

where $O' = (a, b)$, $P = (x, y)$, and $r = \sqrt{(x - a)^2 + (y - b)^2}$.

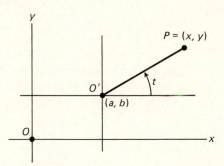

Figure 3

The functions $\sin x$ and $\cos x$ are the basic trigonometric functions, but there are four others, the tangent, cotangent, secant and cosecant functions, defined as follows:

$$\tan x = \frac{\sin x}{\cos x}, \qquad \cot x = \frac{\cos x}{\sin x} \left(= \frac{1}{\tan x} \right),$$

$$\sec x = \frac{1}{\cos x}, \qquad \csc x = \frac{1}{\sin x}.$$

It follows from these definitions that the graphs of $\tan x$ and $\sec x$

Figure 4

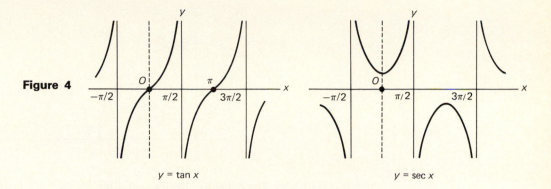

$y = \tan x$ $\qquad\qquad\qquad$ $y = \sec x$

have the general appearance shown in Fig. 4. Some other properties will be derived in the problems.

Now let $P_t = (\cos t, \sin t)$ be the point on the unit circumference obtained by measuring off the signed angle t from the positive x-axis. Underlying all the properties of the trigonometric functions is the following basic fact:

> *If we measure off the signed angle u, starting from the ray OP_t, then we end up at the ray OP_{t+u}.*

This can be checked by looking at a few special cases, and we shall simply assume it. See Fig. 5

Figure 5

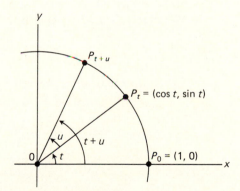

According to this basic principle, if we measure off the same signed arc length $(s - t)$ from each of the two initial points P_t and P_0, we obtain the terminal points P_s and P_{s-t}, respectively (Fig. 6). And

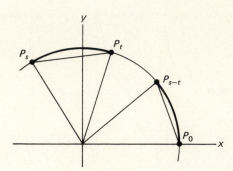

Figure 6

since equal arcs have equal chords, we then have the identity

$$\overline{P_t P_s} = \overline{P_0 P_{s-t}},$$

where we have used the notation $\overline{PQ}$ for the length of the chord from P to Q.

The above identity is equivalent to the trigonometric addition laws! When the chord lengths are expressed by the distance formula, and both sides are squared, the identity becomes

$$(\cos s - \cos t)^2 + (\sin s - \sin t)^2$$
$$= (\cos(s - t) - 1)^2 + (\sin(s - t) - 0)^2.$$

Now just simplify algebraically: expand the squares, use the identity $\sin^2 x + \cos^2 x = 1$ three times, and cancel common terms and factors. What is left is the basic law

$$\cos s \cos t + \sin s \sin t = \cos(s - t).$$

In conjunction with the table

	0	$\pi/2$	π
sin	0	1	0
cos	1	0	-1

this law yields *all* the remaining trigonometric identities. A scheme for working these out is suggested in the problems.

PROBLEMS FOR APPENDIX 2

1. What trigonometric identities are suggested by the following figure?

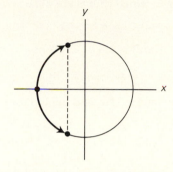

2. What trigonometric identities are suggested by the following figure?

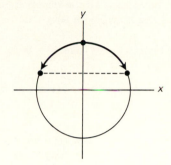

Give the geometric proof (from right triangles) for the following identities:

3. $\sin \pi/6 = 1/2$, $\cos \pi/6 = \sqrt{3}/2$

4. $\sin \pi/4 = \sqrt{2}/2$

5. Let θ be the angle that the segment from $(1, 2)$ to $(2, 4)$ makes with the direction of the positive x-axis. Compute $\sin \theta$ and $\cos \theta$.

6. Same question for the angle made by the segment from $(-1, -2)$ to $(1, 1)$.

7. Same question for the angle θ that the segment from $(1, -1)$ to $(0, 4)$ makes with the direction of the positive x-axis.

Use the $\sin(x + y)$ and $\cos(x + y)$ laws stated in the text, together with the special values for $\sin x$ and $\cos x$ at $x = \pm \pi/2$, π (these can be read off from a diagram), to prove the following laws:

8. $\sin(x + \pi/2) = \cos x$

9. $\sin(x - \pi/2) = -\cos x$

10. $\cos(x + \pi/2) = -\sin x$

11. $\cos(x - \pi/2) = \sin x$

12. $\sin(x + \pi) = -\sin x$

13. $\cos(x + \pi) = -\cos x$

14. Prove the identities:

$$\sin(x + \pi/4) = \frac{\sqrt{2}}{2}(\sin x + \cos x),$$

$$\cos(x + \pi/4) = \frac{\sqrt{2}}{2}(\cos x - \sin x)$$

15. Prove the identities:

$$\sin(x + \pi/3) = \frac{1}{2}(\sin x + \sqrt{3}\cos x),$$

$$\cos(x + \pi/3) = \frac{1}{2}(\cos x - \sqrt{3}\sin x)$$

16. Show that tan x is periodic with period π. That is, prove that

$$\tan(x + \pi) = \tan x$$

for all x.

It was claimed in the text that the $\cos(s - t)$ law, together with the values of $\cos t$ and $\sin t$ at $t = 0$, $\pi/2$, and π, determine all the trigonometric identities. The order in which things are done is important. Here is one sequence, with some hints.

17. $\cos(-t) = \cos t$ (Take $s = 0$ in the formula for $\cos(s - t)$.)

18. $\cos(s - \pi/2) = \sin s$ (Take $t = \pi/2$.)

19. $\cos(s - \pi) = -\cos s$ (Take $t = \pi$.)

20. $\sin(s - \pi/2) = -\cos s$ (Write $s - \pi = (s - \pi/2) - \pi/2$ in (19) and apply (18).)

21. $\sin(x - y) = \sin x \cos y - \cos x \sin y$ (Use (18), the $\cos(s - t)$ law, (18) and (20), in that order.)

22. $\sin(-y) = -\sin y$ (Take $x = 0$.)

23. $\sin(x + y) = \sin x \cos y + \cos x \sin y$

24. $\cos(x + y) = \cos x \cos y - \sin x \sin y$ (From the $\cos(s - t)$ law, using (17) and (22).)

25. $\tan(x + y) = \dfrac{\tan x + \tan y}{1 - \tan x \tan y}$ (Divide (23) by (24).)

26. $\tan(x - y) = \dfrac{\tan x - \tan y}{1 + \tan x \tan y}$

Double-angle Formulas

27. $\sin 2x = 2 \sin x \cos x$

28. $\cos 2x = \cos^2 x - \sin^2 x$
$$= 1 - 2\sin^2 x$$
$$= 2\cos^2 x - 1$$

29. $\tan 2x = \dfrac{2 \tan x}{1 - \tan^2 x}$

Half-angle Formulas

30. $\cos^2\left(\dfrac{y}{2}\right) = \dfrac{1 + \cos y}{2}$

31. $\sin^2\left(\dfrac{y}{2}\right) = \dfrac{1 - \cos y}{2}$

32. $\tan\left(\dfrac{y}{2}\right) = \dfrac{\sin y}{1 + \cos y} = \dfrac{1 - \cos y}{\sin y}$

(Verify by setting $y = 2x$ and using double-angle formulas.)

Finally, if we set $x + y = A$ and $x - y = B$, so that

$$A + B = 2x \quad \text{and} \quad A - B = 2y,$$

the addition formulas give us:

33. $\sin A - \sin B = 2 \cos\left(\dfrac{A + B}{2}\right) \sin\left(\dfrac{A - B}{2}\right)$

34. $\cos A - \cos B = -2 \sin\left(\dfrac{A + B}{2}\right) \sin\left(\dfrac{A - B}{2}\right)$

35. Show that $a \sin x + b \cos x$ can be written in the form

$$A \sin(x + \theta).$$

Here a and b are arbitrary constants, and A and θ are to be determined from them.

36. What trigonometric identities are suggested by the following figures?

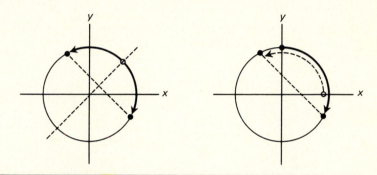

APPENDIX 3
THE CONIC SECTIONS

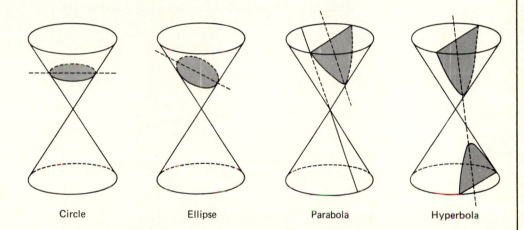

Circle   Ellipse   Parabola   Hyperbola

Properties of the parabola, ellipse, and hyperbola are treated in the problem sets at various places in the text, mostly on the basis of parametric and polar-coordinate representations involving the trigonometric functions. Here we shall give a straightforward algebraic treatment of some of these questions, starting with traditional locus definitions of the curves.

1
THE PARABOLA

DEFINITION   A *parabola* is the locus traced by a point whose distance from a fixed point (called the *focus*) is equal to its distance from a fixed line (called the *directrix*).

In order to obtain the equation of the parabola in standard form, we take the focus at $(0, p)$ on the y-axis and the directrix parallel to the x-axis with intercept $-p$, as shown in Fig. 1. Then a point satisfies the locus condition if and only if

$$\sqrt{x^2 + (y - p)^2} = |y + p|.$$

Squaring and simplifying, we obtain first

$$x^2 + y^2 - 2py + p^2 = y^2 + 2py + p^2,$$

and then

$$x^2 = 4py.$$

This is the standard equation of the parabola.

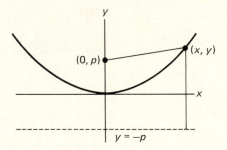

Figure   1

Example 1   Identify the graph of the equation $y - 2x^2 = 0$.

Solution   We solve for x^2,

$$x^2 = \left(\frac{1}{2}\right)y,$$

and interpret the result as the standard form. Thus, $1/2 = 4p$, so $p = 1/8$. The graph is the parabola with focus at $(0, 1/8)$ and directrix $y = -1/8$.

Note that we never did write the given equation explicitly in the standard form $x^2 = 4py$. This would be

$$x^2 = 4\left(\frac{1}{8}\right)y.$$

Instead, we found the right value of p by setting $1/2 = 4p$, and that was all we needed to identify the graph.   □

The parabola $x^2 = 4py$ is symmetric about the y-axis. In terms of the locus definition, the axis of symmetry is the line through the focus perpendicular to the directrix. The *vertex* of a parabola is the point where it intersects its axis of symmetry; i.e., the point halfway from the focus to the directrix. The vertex of the parabola $y^2 = 4px$ is at the origin.

In the derivation above we assumed that p is positive. If p is negative, the algebra is the same, but the picture is different. The focus at $(0, p)$ and directrix $y = -p$ now give us a parabola opening *downward*. Thus, $x^2 = -4y$ is the equation $x^2 = 4py$ with $p = -1$; so its graph is the parabola with focus at $(0, -1)$ and directrix $y = 1$, a downward opening parabola with vertex at the origin, as shown in Fig. 2.

Figure   2

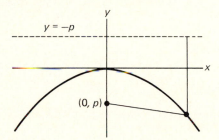

In the same way, the parabola with focus $(p, 0)$ on the x-axis and directrix $x = -p$ parallel to the y-axis has the equation

$$y^2 = 4px.$$

It is symmetric about the x-axis, with vertex at the origin, and opens to the right or to the left depending on whether p is positive or negative (Fig. 3).

The proof does not have to be repeated. If we interchange the roles of x and y throughout an argument, then we automatically

Figure   3

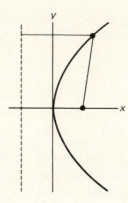 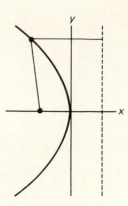

prove a new result, the statement of which is obtained from the old statement by simply interchanging the roles of x and y.

In summary;

THEOREM 1   *The equation of a parabola with vertex at the origin is*

$$x^2 = 4py$$

if its axis is the y-axis, and

$$y^2 = 4px$$

if its axis is the x-axis. In each case, p is the coordinate of the focus on the axis.

Example 2   Identify the graph of $x + y^2 = 0$.

Solution   Solving for y^2,

$$y^2 = -x,$$

we have the form $y^2 = 4px$ with $p = -1/4$. So the graph is the parabola with focus at $(-1/4, 0)$ and directrix at $x = 1/4$. It opens to the left.   □

Vertex at (h, k)   Now consider a parabola with vertex at (h, k). If we choose a new XY-axis system, with new origin at the point having old coordinates (h, k), then the vertex of the parabola is at the new origin; so the equation of the parabola in this new system will be

$$X^2 = 4pY$$

if its axis is vertical, and

$$Y^2 = 4pX$$

if its axis is horizontal, by Theorem 1. Since $X = x - h$ and $Y = y - k$ (Chapter 1, Section 7), we have the following generalization of Theorem 1.

THEOREM 2   *The equation of a parabola with vertex at the point (h, k) is*

$$(x - h)^2 = 4p(y - k)$$

if its axis is vertical, and

$$(y - k)^2 = 4p(x - h)$$

if its axis is horizontal. In each case p is the signed distance from the vertex to the focus along the axis of the parabola. Thus, the focus is at $(h, k + p)$ in the first case and $(h + p, k)$ in the second case.

Example 3   Find the equation of the parabola with focus at $(4, 1)$ and directrix $x = 2$ (see Fig. 4).

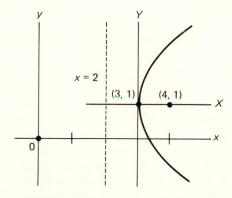

Figure 4

Solution   The vertex is halfway from the focus to the directrix and hence is at $(h, k) = (3, 1)$. The focus at $(4, 1)$ is one unit to the right of the vertex at $(3, 1)$, so $p = 1$. And the parabola opens in the x-direction, so it has the y-squared equation. Its equation is thus:

$$(y - k)^2 = 4p(x - h)$$

with $(h, k) = (3, 1)$ and $p = 1$; that is,

$$(y - 1)^2 = 4(x - 3),$$

or

$$y^2 - 2y - 4x + 13 = 0. \qquad \square$$

Example 4   Find the equation of the parabola with focus at $(-2, 0)$ and with the line $y = 2$ as directrix.

Solution   The vertex is $(-2, 1)$, the point halfway between focus and directrix. The focus is one unit *below* the vertex, so $p = -1$, and the axis is vertical. So the equation is

$$(x - h)^2 = 4p(y - k)$$

with $(h, k) = (-2, 1)$ and $p = -1$, which works out to

$$x^2 + 4x + 4y = 0. \qquad \square$$

The equation $(x - h)^2 = 4p(y - k)$ can be solved for y, leading to an equivalent equation of the form

$$y = ax^2 + bx + c,$$

with $a \neq 0$. Thus,

Each vertical parabola is the graph of a quadratic function

$$\boxed{y = f(x) = ax^2 + bx + c.}$$

Conversely, every quadratic function $f(x) = ax^2 + bx + c$, with $a \neq 0$, is the graph of a vertical parabola. To identify the parabola we transform the equation $y = ax^2 + bx + c$ into the standard form $(x - h)^2 = 4p(y - k)$ by completing the square on the x-terms, just as in Section 7 of Chapter 1.

Example 5   Identify the graph of the equation $y = 2x^2 - 4x$.

Solution   We start by completing the square:

$$y + 2 = 2(x^2 - 2x + 1) = 2(x - 1)^2.$$

This equation can be written

$$(x - 1)^2 = (1/2)(y + 2),$$

which is in the standard form

$$(x - h)^2 = 4p(y - k),$$

with $(h, k) = (1, -2)$ and $p = 1/8$. The graph is thus the parabola with vertex at $(1, -2)$ and focus 1/8 unit above the vertex, at $(9/8, -2)$. □

More generally, any equation in x and y that is quadratic in one variable and linear in the other has a parabolic graph that can be identified by completing a square.

Example 6 Identify the graph of $y^2 - 6y - 2x + 7 = 0$.

Solution We convert the equation to standard form, starting by completing the square on the y-terms, as follows:

$$y^2 - 6y + 9 - 2x + 7 = 9,$$
$$(y - 3)^2 = 2x + 2,$$
$$(y - 3)^2 = 2(x + 1).$$

This is of the form

$$(y - k)^2 = 4p(x - h)$$

with $(h, k) = (-1, 3)$ and $p = 1/2$. The graph is thus the parabola with vertex at $(-1, 3)$, opening to the right, with focus 1/2 unit to the right of the vertex at $(-1/2, 3)$ and with directrix $x = -3/2$. □

A unique "vertical" parabola, with equation of the form

$$y = ax^2 + bx + c,$$

can be passed through any given set of three points having distinct x-coordinates, provided they are not collinear, for the three sets of numerical coordinates give us three equations in the three unknowns a, b, and c, and can be solved for these coefficients in the standard manner.

Example 7 Find the vertical parabola that contains the points $(-1, 2)$, $(1, 1)$, $(2, 3)$.

Solution The equation of the parabola, of the form

$$y = ax^2 + bx + c,$$

must be satisfied by each of the above three pairs of values. That is,

$$2 = a - b + c,$$
$$1 = a + b + c,$$
$$3 = 4a + 2b + c.$$

Subtracting the first equation from each of the other two gives

$$-1 = 2b,$$
$$1 = 3a + 3b.$$

Therefore, $b = -1/2$, $a = 5/6$, $c = 2/3$, so the parabola has the equation

$$y = \frac{5}{6}x^2 - \frac{1}{2}x + \frac{2}{3}.$$ □

The Tangent Line The geometry of the parabola is concerned largely with properties of the tangent line. We have to know how to differentiate in order to compute the slope of the tangent line, but from then on it is pure analytic geometry.

THEOREM 3 *The tangent line to the parabola*

$$x^2 = 4py$$

at the point (x_0, y_0) on the parabola has the equation

$$\boxed{x_0 x = 2p(y + y_0).}$$

Proof The parabola $x^2 = 4py$ has the slope

$$\frac{dy}{dx} = \frac{2x}{4p} = \frac{x}{2p}.$$

The tangent line at the point (x_0, y_0) on the parabola thus has the slope $x_0/2p$, and its point–slope equation is

$$y - y_0 = \frac{x_0}{2p}(x - x_0).$$

We multiply out the right side and use the fact that $x_0^2 = 4py_0$ to get the form given in the theorem.   ■

The tangent-line equation and the slope formula $m_0 = x_0/2p$ will be used in the problems. Here we shall check only the optical property of the parabola.

THEOREM 4   *The tangent line to the parabola $x^2 = 4py$ at (x_0, y_0) makes equal angles with the focal radius to (x_0, y_0) and the vertical line at that point.*

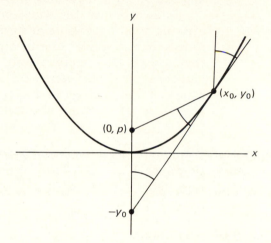

Figure 5

Proof   The tangent line $xx_0 = 2p(y + y_0)$ has y-intercept $y = -y_0$. The distance from the y-intercept to the focus $(0, p)$ is thus $p + y_0$. But this is also the distance from the point (x_0, y_0) to the directrix $y = -p$, and therefore is equal to the distance from (x_0, y_0) to the focus, by the locus definition of the parabola. The triangle in Fig. 5 is thus isosceles and has equal base angles. Since the tangent meets all vertical lines at the same angle, this proves the theorem.   ■

The theorem shows that the light rays emitted by a point source of light placed at the focus will all reflect off the parabola along vertical lines, forming a beam of light shining vertically without any weakening caused by spreading out. This is why a searchlight (or automobile headlight) has a small intense light source at the focus

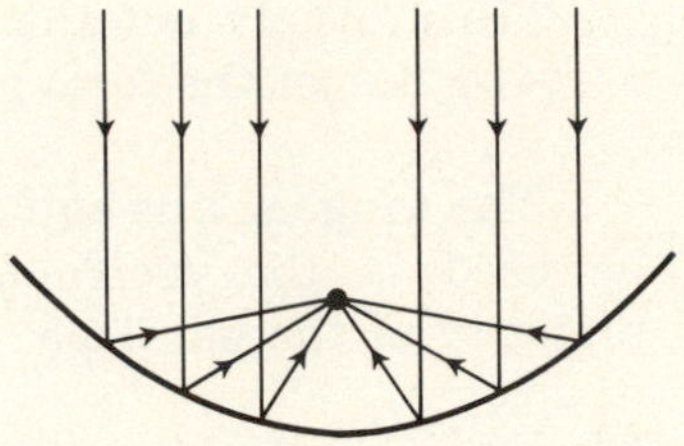

Figure 6

and a reflector that is parabolic in cross section. The ideal parallel beam will not quite be realized because the light source can only approximate a point source of light.

Telescopes use the same principle in reverse. The light rays arriving from a distant star form a parallel beam and are therefore focused onto a single point by a parabolic mirror (Fig. 6). (Hence the word *focus*.) A camera placed (approximately) at the focus thus gathers all the light intercepted by the mirror, and this huge magnification of the light-gathering power of the camera lens makes it possible to take pictures of objects too faint to be recorded otherwise.

PROBLEMS FOR SECTION 1

Find the equations for each of the following parabolas.

1. Vertex at the origin; focus at $(2, 0)$

2. Vertex at the origin; focus at $(0, -1)$

3. Directix $x = 1$; focus $(-1, 0)$

4. Vertex at the origin; directrix $y = 1/4$

5. Directrix the y-axis; focus $(1, 0)$

6. Directrix the y-axis; focus $(-2, 0)$

7. Directrix the x-axis; vertex $(0, 2)$

8. Directrix $x = 1$; focus at $(3, -1)$

9. Directrix $x = 1$; focus at the origin

10. Directrix $y = -2$; focus at the origin

11. Directrix $y = -2$; vertex at $(-2, 0)$

Identify each of the following parabolas by finding its vertex and focus.

12. $y = 2x^2$

13. $y^2 = 2x$

14. $y^2 + 2x = 0$

15. $y = 2x^2 + 1$

16. $y^2 + 2x + 1 = 0$

17. $3y + x^2 - 6 = 0$

18. $y = x^2 + 2x + 1$

19. $y = x^2 + 2x$

20. $y^2 + x + y = 0$

21. $2y^2 - 4y - x + 2 = 0$

22. $2y^2 - 4y - x = 0$

23. A chord of the parabola $x^2 = 4py$ has endpoints with x-coordinates x_1 and x_2. Show that the slope of the chord is $(x_1 + x_2)/4p$.

24. A focal chord (chord through the focus) of the parabola $x^2 = 4py$ runs from (x_1, y_1) to (x_2, y_2). Show that its length is $y_1 + y_2 + 2p$.

In each of the following problems find the equation of and identify the vertical parabola passing through the three given points.

25. $(1, 0), (2, -1), (3, 0)$ **26.** $(1, 0), (2, 1), (4, 0)$ **22.** $(-1, 1), (0, 0), (2, -1)$

28. There can be no parabola having a horizontal axis and passing through the three points in Problem 25. Why?

29. Find the parabola having a horizontal axis and passing through the three points in Problem 27.

30. Find the equation of the parabola with vertical axis, vertex at the origin, and tangent to the line $y = x - 2$.

31. Find the lines tangent to the parabola $y = x^2$ and passing through the point $(2, 1)$.

32. Show that the tangent line to the parabola $y^2 = 4px$ at the point (x_0, y_0) has slope $2p/y_0$.

33. The normal to a parabola at a point P on the parabola intersects the axis of the parabola at the point N. Show that P and N are at the same distance from the focus.

34. The tangent to a parabola at a point P intersects the directrix at the point T. Show that the segment PT subtends a right angle at the focus.

35. The tangent to a parabola at a point P intersects the tangent at the vertex at the point Q. If F is the focus, show that the segments FQ and PQ are perpendicular.

36. Prove that the midpoints of all chords of a parabola parallel to a given chord form a straight line that is parallel to the axis of the parabola.

37. Show that the lines tangent to a parabola at the ends of a focal chord (a chord through the focus) meet at right angles.

38. Show that the lines tangent to a parabola at the ends of a focal chord meet on the directrix.

39. Prove that the vertical line $x = x_0$ bisects every chord of the parabola $x^2 = 4py$ that is drawn parallel to the tangent line at (x_0, y_0).

40. Show that $x^2 = 4p(y + p)$ is the standard equation for a parabola with vertical axis and focus at the origin.

41. Show that every upward-opening parabola with focus at the origin intersects at right angles every downward-opening parabola with focus at the origin. (Use the equation in the above problem with two different values of p, one positive and one negative.)

2
THE ELLIPSE

Ellipses and hyperbolas also have focus–directrix locus definitions, but we shall start instead from another locus problem.

DEFINITION   An *ellipse* is the locus traced by a point P whose distances from two fixed points F and F'' have a constant sum, k,

$$\boxed{\overline{PF} + \overline{PF''} = k,}$$

where k is greater than the distance $\overline{FF''}$. Each of the fixed points F

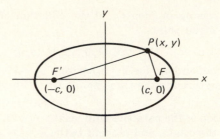

Figure   1

and F' is called a *focus* of the ellipse (for reasons having to do with the optical property, which will be discussed later).

In order to obtain the simplest equation for this locus we take the x-axis along the segment FF' and the y-axis as its perpendicular bisector (Fig. 1).

Then $F = (c, 0)$ and $F' = (-c, 0)$, where c is one half the segment length. Also, if a is the intercept of the locus on the positive x-axis, then the locus condition says that

$$(a - c) + (a + c) = k,$$

so $k = 2a$. The locus condition for $P = (x, y)$ can thus be rewritten

$$\sqrt{(x - c)^2 + y^2} + \sqrt{(x + c)^2 + y^2} = 2a,$$

where a and c are positive constants such that $a > c$. This, then, is the basic equation of the locus. In order to simplify it, we follow the usual procedure for eliminating radicals: solve for a radical, and square. If we solve for the second radical, square everything, and then simplify, we end up with

$$\sqrt{(x - c)^2 + y^2} = a - \left(\frac{c}{a}\right)x.$$

The ratio c/a occurring here is called the *eccentricity* of the ellipse and is designated e; note that $e = c/a$ is less than 1. The first simplifying step thus results in a formula for the length of the right-hand "focal radius" PF. This is an important formula, and we record it as a theorem.

THEOREM 5   *Let E be the ellipse having foci $F = (c, 0)$ and $F' = (-c, 0)$, and x-intercepts at $\pm a$. Then for any point $P = (x, y)$ on E, the focal radii*

PF and PF′ have the lengths

$$\overline{PF} = a - ex, \qquad \overline{PF'} = a + ex,$$

where $e = c/a$.

Note that the second formula follows from the first, because $\overline{PF} + \overline{PF'} = 2a$. Continuing, we square the equation

$$\sqrt{(x - c)^2 + y^2} = a - \left(\frac{c}{a}\right)x$$

and simplify, this time ending up with

$$\left(\frac{a^2 - c^2}{a^2}\right)x^2 + y^2 = a^2 - c^2.$$

Setting $x = 0$ determines the y-intercepts as the points $\pm\sqrt{a^2 - c^2}$. We call the y-intercepts $\pm b$, as usual. That is, we set

$$b^2 = a^2 - c^2.$$

Then, after dividing by b^2, the above equation turns into the standard equation for the ellipse:

THEOREM 6 *An ellipse with foci located on the x-axis and centered at the origin has the equation*

$$\frac{x^2}{a^2} + \frac{y^2}{b^2} = 1.$$

Actually, the reasoning above does only half the job. We have shown that if a point (x, y) is on the locus, then its coordinates satisfy the standard equation. It must also be shown that the argument can be traced backward, so that if a point (x, y) satisfies the standard equation, then it is on the locus. This backwards process involves taking square roots twice, and it must be checked that no sign ambiguity arises. For example, we must check that if (x, y) is on the graph of

$$\frac{x^2}{a^2} + \frac{y^2}{b^2} = 1,$$

then $(a - ex)$ is positive. This will be left as an exercise.

Remember that $a > b$ in the standard equation above, the foci being located by $c^2 = a^2 - b^2$. If we start instead with foci at the points $(0, c)$ and $(0, -c)$ on the y-axis, then we end up with the same standard equation in terms of the intercepts a and b, but this time $b > a$ and $c^2 = b^2 - a^2$. The graph of the standard equation is thus

1. An ellipse with foci on the x-axis if $a > b$;
2. A circle of radius $r = a = b$ if $a = b$;
3. An ellipse with foci on the y-axis if $a < b$.

We consider that we have identified an ellipse if we have found its standard equation, and hence know a and b. The larger of a and b is (the length of) the *semimajor axis* of the ellipse, and the smaller is (the length of) its *semiminor axis*. We then know the location of the foci, as we noted above, and everything else follows.

Example 1 The graph of

$$\frac{x^2}{25} + \frac{y^2}{9} = 1$$

is the ellipse with foci at $\pm\sqrt{25 - 9} = \pm 4$ on the x-axis, and intercepts $\pm a = \pm 5$ on the x-axis and $\pm b = \pm 3$ on the y-axis. However, in view of the symmetry about the axes, we normally record only the positive intercepts $a = 5$, $b = 3$.   □

Example 2 Identify the graph of

$$x^2 + 4y^2 = 9.$$

Solution 1 We throw the equation into the standard form

$$\frac{x^2}{9} + \frac{y^2}{9/4} = 1,$$

and then read off $a^2 = 9$, $b^2 = 9/4$. The graph is an ellipse with foci at

$$\pm c = \pm\sqrt{9 - \frac{9}{4}} = \pm\frac{3}{2}\sqrt{3}$$

on the x-axis, and positive intercepts $a = 3$, $b = 3/2$.

Solution 2 We recognize that the graph is an ellipse symmetric in the coordinate axes, so we compute the intercepts $a = 3$,

$b = 3/2$ directly from the equation. Since $a > b$, we conclude that the foci are on the x-axis with $c = \sqrt{a^2 - b^2} = 3\sqrt{3}/2$. $\square$

Example 3 The graph of $4x^2 + y^2 = 1$ is clearly going to be a central ellipse. Its positive intercepts are $a = 1/2$, $b = 1$; so its foci are on the y-axis and $c = \sqrt{1 - (1/2)^2} = \sqrt{3}/2$. $\square$

Example 4 Compute the lengths of the focal radii to the point(s) on the ellipse

$$\frac{x^2}{4} + \frac{y^2}{2} = 1$$

having x-coordinate 1.

Solution We could solve for the y-coordinates and then use the distance formula, but the formulas for the focal radii are much simpler. We see that $a = 2$, $b = \sqrt{2}$, $c = \sqrt{4 - 2} = \sqrt{2}$, $e = c/a = \sqrt{2}/2$, and

$$\overline{PF} = a - ex = 2 - \frac{\sqrt{2}}{2} \cdot 1 = \frac{4 - \sqrt{2}}{2},$$

$$\overline{PF'} = a + ex = 2 + \frac{\sqrt{2}}{2} \cdot 1 = \frac{4 + \sqrt{2}}{2}. \qquad \square$$

The standard equation for an ellipse centered at (h, k) and having axes parallel to the coordinate axes is

$$\boxed{\frac{(x - h)^2}{a^2} + \frac{(y - k)^2}{b^2} = 1.}$$

The reasoning goes exactly as it did for the parabola.

Example 5 Find the equation of the ellipse whose foci are at $(1, -2)$ and $(5, -2)$, and with $2a = 6$.

Solution The center is midway between the foci, so $(h, k) = (3, -2)$ and $c = 2$. Since we are given that $a = 3$, it follows that $b = \sqrt{a^2 - c^2} = \sqrt{5}$. The standard equation of the ellipse is therefore

$$\frac{(x - 3)^2}{9} + \frac{(y + 2)^2}{5} = 1.$$

If we expand the squares and collect coefficients, the equation becomes

$$5x^2 + 9y^2 - 30x + 36y + 36 = 0.$$ □

Example 6 Identify the graph of the equation

$$x^2 + 2y^2 + 4x - 4y = 0.$$

Solution After completing the squares the equation is

$$(x + 2)^2 + 2(y - 1)^2 = 6,$$

from which we get standard equation

$$\frac{(x + 2)^2}{6} + \frac{(y - 1)^2}{3} = 1.$$

Thus, the graph is an ellipse centered at $(-2, 1)$ with $a = \sqrt{6}$ and $b = \sqrt{3}$. Then $c = \sqrt{6 - 3} = \sqrt{3}$, and since the major axis is parallel to the x-axis, the foci are at $(-2 - \sqrt{3}, 1)$ and $(-2 + \sqrt{3}, 1)$. □

Consider again the formula for the right focal radius

$$PF = a - ex.$$

If we factor out e, the resulting expression $(a/e) - x$ can be interpreted as the distance from the point (x, y) to the vertical line $x = a/e$. See Fig. 2. That is, the distance from P to F is e times its distance from the line $x = a/e$. So,

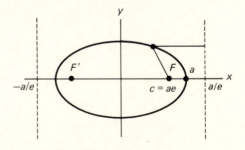

Figure 2

THEOREM 7 *An ellipse can be characterized as the locus traced by a point whose distance from a fixed point F (the focus) is a constant e times its distance from a fixed line (the directrix), the constant e being less than 1.*

The ellipse $(x^2/a^2) + (y^2/b^2) = 1$ is also the focus–directrix locus defined by the left focus F' and a left directrix at $x = -a/e$. Many geometric properties of an ellipse involve its directrices and foci.

We shall now prove:

THEOREM 8   *The tangent line to the ellipse*

$$\frac{x^2}{a^2} + \frac{y^2}{b^2} = 1$$

at the point (x_0, y_0) on the ellipse has the standard equation

$$\boxed{\frac{x_0 x}{a^2} + \frac{y_0 y}{b^2} = 1.}$$

Proof   Calculus is needed to compute the slope of the tangent line. Differentiating the equation of the ellipse implicitly (Chapter 4, Section 2), we get

$$\frac{2x}{a^2} + \frac{2y}{b^2}\frac{dy}{dx} = 0$$

and

$$\frac{dy}{dx} = -\frac{b^2 x}{a^2 y}.$$

The tangent line at (x_0, y_0) thus has the slope $-b^2 x_0/a^2 y_0$, and its point–slope equation is

$$y - y_0 = -\frac{b^2 x_0}{a^2 y_0}(x - x_0).$$

This simplifies to the standard equation when we make use of the fact that

$$\frac{x_0^2}{a^2} + \frac{y_0^2}{b^2} = 1. \qquad\blacksquare$$

The optical property of the ellipse involves the tangent line, as shown in Fig. 3.

THEOREM 9   *The focal radii to the point P on the ellipse make equal angles with the tangent line at P.*

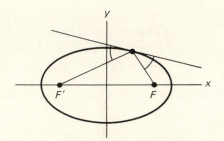

Figure 3

Thus, if a point source of light is placed at one focus, say F', then all of its reflected rays will pass through the other focus F. The ellipse "focuses" the light on the point F, which is thus one of the two focal points of the optical system.

This focusing property of the ellipse explains a startling acoustical phenomenon that has been observed in some buildings. If a hall with walls made of stone, or some other acoustically reflecting material, is in the shape of an ellipse, then a person standing at one focus and speaking quietly can be heard with great clarity at the other focus, but not at points in between.

Proof of Theorem   In order to prove the optical property, we make use of the focal radius formulas and the formula

$$d = \frac{|Ax_1 + By_1 + C|}{\sqrt{A^2 + B^2}}$$

for the distance from the point (x_1, y_1) to the line $Ax + By + C = 0$ (Problem 42, in Chapter 1, Section 3).

According to this formula, the distances from the foci to the tangent line $xx_0/a^2 + yy_0/b^2 = 1$ have the ratio

$$\frac{F'A'}{FA} = \frac{\left|\dfrac{-cx_0}{a^2} - 1\right|}{\left|\dfrac{cx_0}{a^2} - 1\right|}$$

$$= \frac{a^2 + cx_0}{a^2 - cx_0}$$

$$= \frac{a + ex_0}{a - ex_0},$$

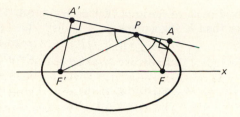

Figure 4

which is the ratio $F'P/FP$ of the focal radii. The two right triangles are thus similar, and hence have equal angles at P. See Fig. 4. ∎

PROBLEMS FOR SECTION 2

In Problems 1 through 9, find the equation of the ellipse determined by the given conditions.

1. The foci are $(-1, 0)$ and $(1, 0)$, and $2a = 3$.

2. The foci are $(0, -1)$ and $(0, 1)$, and $2b = 3$.

3. The foci are at $(-1, 0)$ and $(1, 0)$, and $b = 1$.

4. The foci are at $(-1, 0)$ and $(1, 0)$, and the eccentricity ($e = c/a$) is 1/2.

5. The foci are at $(0, 1)$ and $(0, 3)$, and $e = 1/2$.

6. The foci are at $(0, 1)$ and $(4, 1)$, and $e = 3/4$.

7. The intercepts on the major and minor axes are $(-1, 1)$, $(3, 1)$, $(1, 2)$, and $(1, 0)$.

8. The foci are $(-1, 0)$, and $(1, 0)$, and the line $x = 2$ is a directrix.

9. The center is the origin, one focus is at $(0, 2)$, and the line $y = 3$ is a directrix.

10. Find the foci of the ellipse $x^2 + 4x + 2y^2 = 0$.

11. Same question for $2x^2 + 4x + y^2 = 0$.

12. Find the equation of the ellipse having the y-axis as a directrix, the point $(1, 0)$ as the corresponding focus, and eccentricity 1/2.

13. Same question for the point $(1, 0)$, line $x = 3$, and $e = 2/3$.

14. Let P be a point on the ellipse $(x^2/a^2) + (y^2/b^2) = 1$ that is not an intercept. Supposing that $b < a$, show that the distance from P to the origin is greater than b and less than a. (This does not require calculus.)

15. Show that if $a^2 = b^2 + c^2$ and if the point (x, y) satisfies the equation $(x^2/a^2) + (y^2/b^2) = 1$, then the distance from (x, y) to $(c, 0)$ is $(a - ex)$, and the distance from (x, y) to $(-c, 0)$ is $(a + ex)$, where $e = c/a$. That is, work out the algebra for Theorem 5.

16. Complete the proof of Theorem 8 by supplying the missing algebraic calculation at the end of the proof.

17. Let k be a fixed positive constant. For each value of a greater than k, the equation

$$\frac{x^2}{a^2} + \frac{y^2}{a^2 - k^2} = 1$$

represents an ellipse; so, regarding a as a parameter, we thus have a one-parameter family of ellipses. Show that this is a *confocal* one-parameter family; that is, show that all ellipses of this form have the same foci.

18. The tangent to an ellipse at a point P intersects a directrix at the point D. Prove that the segment PD subtends a right angle at the corresponding focus.

19. Show that the product of the distances from the foci of an ellipse to a tangent is a constant, independent of the tangent.

20. The *vertices* of an ellipse are its intersections with its major axis. The lines from the vertices A' and A of an ellipse, through a point P on the ellipse, meet a directrix at D' and D. Show that the segment $D'D$ subtends a right angle at the corresponding focus.

21. A normal to an ellipse at a point P has intercepts Q_1 and Q_2 on the axes. Show that the product of the lengths of the segments PQ_1 and PQ_2 equals the product of the focal radii to P.

22. The tangent to an ellipse at a point P intersects the tangent at a vertex at the point Q. Show that the line through Q and the other vertex is parallel to the line through P and the center.

23. Show that every ellipse of the form
$$x^2 + 2y^2 = C$$
is orthogonal to every parabola of the form
$$y = kx^2.$$

(Let (x_0, y_0) be a point of intersection. Compute the two slopes at this point, and show that their product is -1.)

3
THE HYPERBOLA

DEFINITION   A *hyperbola* is the locus traced by a point P whose distances from two fixed points F and F' have a constant positive difference k, where k is less than the distance between F' and F.

The locus condition is thus

$$PF' - PF = k \qquad \text{or} \qquad PF - PF' = k,$$

that is,

$$PF' = PF \pm k.$$

The requirement $k < \overline{F'F}$ is necessary if there are to be locus points P that are not on the line through F and F', because any such point P satisfies triangle inequalities like $\overline{PF'} < \overline{PF} + \overline{FF'}$, so

$$k = PF' - PF < \overline{FF'}.$$

A little thought (or experimentation) will show that the locus has the general features shown in Fig. 1.

Figure 1

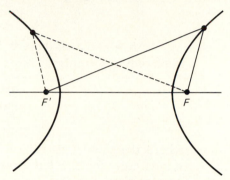

It has two branches, each of which cuts between F' and F. The right branch, bending around F, is the locus of the equation $PF' - PF = k$, while the left branch, bending around F', is the locus of the equation $PF - PF' = k$.

If we take the x-axis along the segment $F'F$ and the origin at its midpoint, then the coordinates of F and F' are $F = (c, 0)$ and $F' = (-c, 0)$, where c is one half the segment length (Fig. 2). If a is the positive x-intercept of the locus, then $a < c$, and for this point the locus condition becomes

$$(a + c) - (c - a) = k.$$

So $k = 2a$, and the locus equation is

$$\sqrt{(x + c)^2 + y^2} = \sqrt{(x - c)^2 + y^2} \pm 2a.$$

The procedure now is exactly as it was in the case of the ellipse. The radicals are eliminated by twice squaring the equation and simplifying.

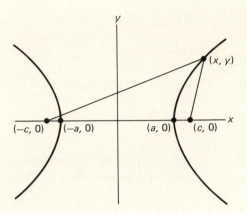

Figure 2

With $e = c/a$ as before, the first of these two steps yields the focal radius formulas

$$PF = \pm(ex - a),$$
$$PF' = \pm(ex + a),$$

where the plus signs are used for the right branch $(x > 0)$ and the minus sign for the left branch $(x < 0)$. Here the eccentricity e is greater than 1, since $e = c/a$ and $c > a$.

The second step reduces the equation

$$\sqrt{(x - c)^2 + y^2} = \pm\left(\frac{c}{a}x - a\right)$$

to

$$\frac{x^2}{a^2} - \frac{y^2}{b^2} = 1,$$

where $b^2 = c^2 - a^2$. This is the standard equation of the hyperbola.

It can be checked, conversely, that the graph of any equation of this form is a hyperbola with foci on the x-axis. Setting $c = \sqrt{a^2 + b^2}$, the graph is the hyperbola with foci at $F = (c, 0)$ and $F' = (-c, 0)$, and with $k = 2a$.

Certain aspects of the graph can be read directly from the standard equation $(x^2/a^2) - (y^2/b^2) = 1$. It is symmetric with respect to both coordinate axes. It has x-intercepts at $\pm a$, but no y-intercepts. Moreover, $x^2 \geq a^2$ on the graph. This is because the y^2 term is negative, so x^2/a^2 is at least 1. The graph thus consists of two separate pieces (branches), a part to the right of $x = a$ and a part to the left of $x = -a$.

Example 1 The hyperbola with foci at $(2, 0)$ and $(-2, 0)$ and with $k = 2a = 2$ has the equation

$$x^2 - \frac{y^2}{3} = 1,$$

because the given data are $a = 1$, $c = 2$, and hence $b = \sqrt{c^2 - a^2} = \sqrt{4 - 1} = \sqrt{3}$. $\square$

Example 2 For the hyperbola

$$\frac{x^2}{16} - \frac{y^2}{9} = 1$$

we have $a^2 = 16$ and $b^2 = 9$, so $c^2 = a + b^2 = 25$, and the foci are at $(c, 0) = (5, 0)$ and $(-c, 0) = (-5, 0)$. Its eccentricity e is 5/4 (since $e = c/a$). $\square$

Example 3 Find the equation of the hyperbola with foci on the x-axis, x-intercepts at ± 2, and eccentricity 3/2.

Solution We are given that $a = 2$ and $e = c/a = 3/2$. Therefore, $c = 3$ and $b^2 = c^2 - a^2 = 9 - 4 = 5$. The equation is, therefore,

$$\frac{x^2}{4} - \frac{y^2}{5} = 1. \qquad \square$$

The Asymptotes The branching of the hyperbola

$$\frac{x^2}{a^2} - \frac{y^2}{b^2} = 1$$

is related to another distinctive feature of the graph; it is asymptotic to each of the straight lines

$$y = \left(\frac{b}{a}\right)x, \qquad y = -\left(\frac{b}{a}\right)x,$$

in the manner shown in Fig. 3.

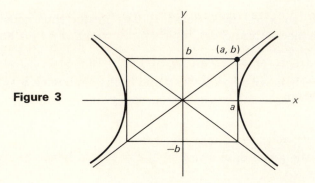
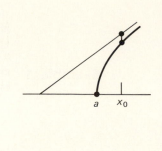

Figure 3

For example, the top of the right branch, whose equation is

$$y = \frac{b}{a}\sqrt{x^2 - a^2},$$

is asymptotic to the line

$$y = \left(\frac{b}{a}\right)x;$$

the two graphs come arbitrarily close together as they run out "to infinity." In order to see this, consider the vertical distance from the hyperbola to the line at a given value of x. It is

$$\frac{b}{a}x - \frac{b}{a}\sqrt{x^2 - a^2} = \frac{b}{a}(x - \sqrt{x^2 - a^2})$$

$$= \frac{b}{a}\frac{(x - \sqrt{x^2 - a^2})(x + \sqrt{x^2 - a^2})}{x + \sqrt{x^2 - a^2}}$$

$$= \frac{ab}{x + \sqrt{x^2 - a^2}},$$

and this is less than ab/x for all x larger than a. The vertical distance between the curves thus approaches zero as x tends to infinity. That is, the curves are asymptotic to each other as $x \to \infty$.

If we interchange the roles of x and y in the preceding discussion, we see that a hyperbola with foci at $(0, c)$ and $(0, -c)$ on the y-axis has the standard equation

$$\boxed{\frac{y^2}{b^2} - \frac{x^2}{a^2} = 1}$$

with $c^2 = a^2 + b^2$.

Note that *the axis containing the foci is not determined by the relative size of a and b*, as it was in the case of the ellipse, *but rather by where the minus sign occurs.* So a and b can be any relative size, and in particular they can be equal. In this case, the hyperbola is called *rectangular* or *equilateral.* Thus, an equilateral hyperbola has the equation

$$x^2 - y^2 = a^2 \quad \text{or} \quad y^2 - x^2 = a^2.$$

For given a and b, the equation

$$\frac{x^2}{a^2} - \frac{y^2}{b^2} = \pm 1$$

represents a pair of hyperbolas, one with foci on the x-axis and one with foci on the y-axis. Each of these hyperbolas is said to be the *conjugate* of the other. A pair of conjugate hyperbolas shares the same asymptotes; and they fit together to enclose the origin almost as though they formed a single closed curve (Fig. 4).

Figure 4

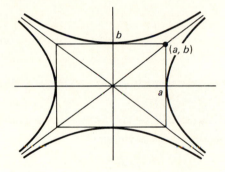

Example 4   The hyperbola

$$x^2 - \frac{y^2}{4} = 1$$

has the asymptotes $y = 2x$ and $y = -2x$, because from the standard equation above we see that $a^2 = 1$ and $b^2 = 4$, so the asymptotes $y = \pm(b/a)x$ are the lines $y = \pm 2x$. The conjugate of the above hyperbola is the hyperbola

$$-x^2 + \frac{y^2}{4} = 1. \qquad \qquad \square$$

Example 5 A hyperbola has asymptotes $y = \pm x$ and passes through the point $(2, 1)$. Find its equation.

Solution Since the asymptote $y = (b/a)x$ is the line $y = x$, we have $b/a = 1$ and $a = b$, so the hyperbola is equilateral. Since the hyperbola contains a point (namely $(2, 1)$) "to the right of" the lines $y = \pm x$, its branches are right and left rather than up and down, and its equation has the form

$$\frac{x^2}{a^2} - \frac{y^2}{a^2} = 1.$$

And, since it contains $(2, 1)$,

$$\frac{4}{a^2} - \frac{1}{a^2} = 1.$$

So $a = \sqrt{3}$, and the hyperbola has the equation

$$\frac{x^2}{3} - \frac{y^2}{3} = 1,$$

or

$$x^2 - y^2 = 3. \qquad \square$$

PROBLEMS FOR SECTION 3

Find the foci, eccentricity, and asymptotes for each of the following hyperbolas.

1. $x^2 - y^2 = 1$

2. $y^2 - x^2 = 1$

3. $x^2 - y^2 = 2$

4. $x^2 - \dfrac{y^2}{3} = 3$

5. $3y^2 - x^2 = 3$

6. $y^2 - 4x^2 = 1$

7. $2x^2 - 3y^2 = 6$

Find the equation of the hyperbola where

8. The foci are at $(\pm 3, 0)$ and the eccentricity is $3/2$.

9. It is symmetric about the coordinate axes; its x-intercepts are $(\pm 1, 0)$ and its eccentricity is 2.

10. Its foci are at $(\pm 3, 0)$ and its x-intercepts are $(\pm 2, 0)$.

11. Its foci are at $(0, \pm 3)$ and $a = 2$.

12. Its x-intercepts are $(\pm 1, 0)$ and it contains the two points $(3, \pm 1)$.

Find the equation of the hyperbola with asymptotes $y = \pm 2x$ if

13. It passes through $(2, 0)$.

14. It has y-intercepts $(0, \pm 2)$.

15. Its foci are $(\pm 5, 0)$.

16. It passes through the point $(2, 2)$.

17. It passes through the point $(1, 1)$.

18. It passes through the point $(1, 3)$.

19. Two hyperbolas have the same asymptotes, one with foci on the x-axis and the other with foci on the y-axis. If their eccentrities are e and f, show that

$$\frac{1}{e^2} + \frac{1}{f^2} = 1.$$

20. Carry out the algebra (completing the square twice and simplifying each time) leading from the locus definition of the hyperbola to its standard equation.

21. Show by algebraic computation that if

$$\frac{x^2}{a^2} - \frac{y^2}{b^2} = 1$$

and if $c^2 = a^2 + b^2$, then the distance from (x, y) to $(c, 0)$ is $|ex - a|$, where $e = c/a$.

22. Show that the hyperbola

$$\frac{x^2}{2} - \frac{y^2}{2} = 1$$

is the locus of a point whose distance from $(2, 0)$ is $\sqrt{2}$ times its distance from line $x = 1$.

23. Show that the hyperbola

$$\frac{x^2}{a^2} - \frac{y^2}{b^2} = 1$$

is the locus of a point whose distance from the point $(c, 0)$ is e times its distance from the line $x = a/e$, where $c = \sqrt{a^2 + b^2}$ and $e = c/a$. [*Hint*: Use the focal radius formulas from the text discussion.]

24. Show that the hyperbola

$$\frac{x^2}{a^2} - \frac{y^2}{b^2} = 1$$

is the locus of a point whose distance from the point $(-c, 0)$ is e times its distance from the line $x = -a/e$.

25. Show that a line through a focus perpendicular to an asymptote intersects the asymptote at a point on a directrix.

26. A line through a point P on a hyperbola parallel to an asymptote intersects a directrix at Q. If F is the corresponding focus, show that the triangle FPQ is isosceles.

27. Prove that the tangent line to the hyperbola

$$\frac{x^2}{a^2} - \frac{y^2}{b^2} = 1$$

at the point (x_0, y_0) has the (standard) equation

$$\frac{xx_0}{a^2} - \frac{yy_0}{b^2} = 1.$$

28. *Optical property of the hyperbola.* Prove that the focal radii to a point P on a hyperbola make equal angles with the tangent line at P. (Follow the proof of the optical property of the ellipse, using the above equation of the tangent line.)

29. Prove that each hyperbola having foci F' and F intersects every ellipse having the same foci at right angles. [*Hint*: Combine the optical properties of the ellipse and hyperbola.]

30. Illustrate the above problem by taking the foci at $(\pm 3, 0)$ and sketching the hyperbolas whose eccentricities are 3 and 3/2 and the ellipse whose eccentricities are 3/4 and 1/2.

31. Show that the product of the distances from the foci of a hyperbola to the tangent at a point P on the hyperbola is constant; i.e., independent of P.

32. Show that the product of the focal radii to a point P on an equilateral hyperbola is equal to the square of the distance from P to the center of the hyperbola.

33. Show that the product of the distances from a point P on a hyperbola to the asymptotes is constant; i.e., independent of P.

34. The tangent to a hyperbola at a point P intersects the asymptotes at Q_1 and Q_2. Show that P is the midpoint of the segment Q_1Q_2.

APPENDIX 4
INTRODUCTION TO THE THEORY OF LIMITS

This appendix contains a brief introduction to the ϵ, δ characterizations of continuity and limits that are needed for advanced technical work.

Suppose we want to calculate π^3 accurately to six decimal places. Our problem is to determine how many places we have to take in the decimal expansion of π. That is, if a_n is the n-place truncation (or round off) of the decimal expansion of π, then how large must n be to ensure that

$$|\pi^3 - a_n^3| \leq 10^{-6}/2?$$

In somewhat more general terms, the question is: given the tolerable error ϵ, how close must a number x be to π in order to ensure that

$$|\pi^3 - x^3| \leq \epsilon?$$

We can ask this question in neutral terms, and for a general function f, as follows:

Given a positive number ϵ, how close do we have to take x to a in order to ensure that

$$\boxed{|f(a) - f(x)| < \epsilon?}$$

Or

How small should we take Δx in order to ensure that

$$\boxed{|\Delta y| < \epsilon,}$$

where $\Delta y = f(x + \Delta x) - f(x)$?

This question concerns the quantitative aspect of continuity. We said, in Chapter 2, that a function f is continuous at $x = a$ if $f(x)$ can be made to be as close to $f(a)$ as we wish by taking x suitably close to a. This was a purely qualitative description. But in making computations we have to find out *how close* is "suitably close."

The smooth functions of calculus give us the simplest answer to this question, because of the following criterion. (It is the general statement of the procedure we were following in the examples in Section 7 of Chapter 5.)

The Derivative Criterion for Continuity *If $|f'|$ is bounded by the constant k on an interval I, that is, if $|f'(x)| \le k$ for every x in I, then*

$$\boxed{|\Delta y| \le k\,|\Delta x|}$$

for any two points x and $(x + \Delta x)$ in I.

Proof By the mean-value principle,

$$\Delta y = f(x + \Delta x) - f(x) = f'(X) \cdot \Delta x,$$

where X is some number between x and $x + \Delta x$. But then $|f'(X)| \le k$ by hypothesis, so

$$|\Delta y| = |f'X| \cdot |\Delta x| \le k\,|\Delta x|,$$

as asserted. ∎

By virtue of this inequality, if now we want to ensure that

$$|\Delta y| < \epsilon,$$

it is sufficient to see to it that

$$k\,|\Delta x| < \epsilon,$$

that is, to take

$$|\Delta x| < \frac{\epsilon}{k}.$$

This is the simplest conceivable solution to the continuity problem.

Example 1 If we consider $y = f(x) = x^3$ on the interval $I = [-4, 4]$, then

$$f'(x) = 3x^2 \le 3 \cdot 4^2 = 48,$$

and hence

$$|\Delta y| \le 48 \,|\Delta x|,$$

by the criterion above. Therefore, in order to ensure that $|\Delta y| < \epsilon$, it is sufficient to take

$$48 \,|\Delta x| < \epsilon, \qquad \text{or} \qquad |\Delta x| < \frac{\epsilon}{48}.$$

In other words, if x and a are both on the interval $I = [-4, 4]$, then

$$|a^3 - x^3| < \epsilon \qquad \text{whenever } |a - x| < \frac{\epsilon}{48}. \qquad \square$$

Although the above criterion furnishes a very simple way to make the continuity calculation, it is important to find ways that are independent of the derivative, for two reasons.

First, we have to know that some functions are continuous before we can prove that they have derivatives.

Example 2 Look back at the calculation of $d\sqrt{x}/dx = 1/2\sqrt{x}$. The difference quotient was

$$\frac{\Delta y}{\Delta x} = \frac{\sqrt{(x + \Delta x)} - \sqrt{x}}{\Delta x} = \frac{1}{\sqrt{x + \Delta x} + \sqrt{x}}.$$

In order to get the answer $1/2\sqrt{x}$, we have to know that $\sqrt{x + \Delta x}$ approaches $\sqrt{x}$ as $\Delta x \to 0$, i.e., that $\sqrt{x}$ is continuous. A prior continuity calculation is thus required, and to avoid circularity in our logic, we cannot base this calculation on the derivative of $\sqrt{x}$ and so can't use the derivative criterion.

So what do we do? Generally speaking, we make the same calculation and algebraic simplification of Δy that we did for the difference quotient in Chapter 3, but now we derive an inequality instead of taking a limit. In the above case, where

$$\frac{\Delta y}{\Delta x} = \frac{1}{\sqrt{x + \Delta x} + \sqrt{x}},$$

we note that since $\sqrt{x + \Delta x}$ is nonnegative, we make the right side larger (if anything) by omitting it. That is,

$$\frac{\Delta y}{\Delta x} \le \frac{1}{\sqrt{x}}.$$

Multiplying by Δx, and allowing for the possibility that Δx is negative, we get

$$|\Delta y| \le \frac{1}{\sqrt{x}}|\Delta x|.$$

Now we fix a positive number a and restrict x to the interval $[a, \infty)$. Then $1/\sqrt{x} \le 1/\sqrt{a}$, so

$$|\Delta y| \le k\,|\Delta x|,$$

where $k = 1/\sqrt{a}$. We have thus obtained an inequality like that of the derivative criterion by a direct calculation.   □

Example 3   Let us see how this direct continuity calculation turns out for our original function $y = f(x) = x^3$ on the interval $[-4, 4]$. Here, the factoring procedure works most smoothly because it uses the two numbers a and x that we are restricting to $[-4, 4]$. We have

$$f(a) - f(x) = a^3 - x^3 = (a - x)(a^2 + ax + x^2),$$

so

$$|f(a) - f(x)| \le |a - x|(16 + 16 + 16) = 48\,|a - x|.$$

Thus, $|\Delta y| \le 48\,|\Delta x|$, as before.

If we try the increment approach, we get

$$\Delta y = f(x + \Delta x) - f(x) = (x + \Delta x)^3 - x^3$$
$$= x^3 + 3x^2\,\Delta x + 3x(\Delta x)^2 + (\Delta x)^3 - x^3$$
$$= \Delta x(3x^2 + 3x\,\Delta x + \Delta x^2).$$

In this form, it isn't useful to assume that both x and $(x + \Delta x)$ lie in $[-4, 4]$. Let us assume that x lies there, and that $|\Delta x|$ is at most 1; then $|3x^2 + 3x\,\Delta x + \Delta x^2| \le 48 + 12 + 1 = 61$, so

$$|\Delta y| \le 61\,|\Delta x|,$$

a different inequality stemming from different assumptions as to where the two points x and $(x + \Delta x)$ lie. In this case, in order to ensure that

$$|\Delta y| < \epsilon,$$

it is sufficient to see to it that

$$61\,|\Delta x| < \epsilon,$$

that is, to take

$$|\Delta x| < \frac{\epsilon}{61}. \qquad \square$$

The second reason for making the direct continuity calculation is that we frequently need to establish the continuity of f at points where f' does not exist.

Example 4 Consider, again, $y = x^{1/3}$ around the origin. We know that $f'(x) = x^{-2/3}$ blows up at $x = 0$. However, since

$$\Delta y = f(0 + \Delta x) - f(0) = (0 + \Delta x)^{1/3} - 0^{1/3} = (\Delta x)^{1/3},$$

the inequality

$$|\Delta y| < \epsilon$$

will hold if

$$|\Delta x|^{1/3} < \epsilon, \qquad \text{or} \qquad |\Delta x| < \epsilon^3.$$

Thus Δy does go to 0 with Δx, but much more slowly. For example, in order to make $|\Delta y| < 1/100$, we have to take

$$|\Delta x| < \left(\frac{1}{100}\right)^3 = \frac{1}{1,000,000}. \qquad \square$$

Example 5 The upper semicircle $y = f(x) = \sqrt{1 - x^2}$ is another example. It is clear from geometry that this graph has vertical tangents at $x = 1$ and $x = -1$, and that f' fails to exist at these two points. Nevertheless, we can work out by simple algebra that

$$|\Delta y| \le \sqrt{2}\,|\Delta x|^{1/2}$$

at $x_0 = \pm 1$. Therefore,

$$|\Delta y| < \epsilon$$

if

$$\sqrt{2}\,|\Delta x|^{1/2} < \epsilon,$$

that is, if

$$|\Delta x| < \frac{\epsilon^2}{2}.$$

Here again, $|\Delta y|$ approaches 0 much more slowly than Δx. For example, in order to ensure that $|\Delta y| < .01$, using the above inequalities, we have to take $|\Delta x| < (0.01)^2/2 = 0.00005$.   □

Let us summarize our various continuity calculations. In each case we wanted to know how small to take $|\Delta x|$ in order to ensure that $|\Delta y| < \epsilon$, where ϵ is some arbitrary preassigned positive number which can be thought of as the tolerable error. In each case, we found a number δ (Greek delta) depending on ϵ which served as an adequate control for the size of Δx. Our various control numbers δ were

$$\delta = \frac{\epsilon}{k}, \qquad \delta = \epsilon^3, \qquad \delta = \frac{\epsilon^2}{2}.$$

A general description of the continuity calculation would therefore run as follows: We first obtain an inequality which limits $|\Delta y|$ in terms of $|\Delta x|$. From this inequality, we see that in order to ensure that $|\Delta y| < \epsilon$, it is sufficient to take $|\Delta x|$ less than a certain control number δ depending on ϵ.

This description of "effective" continuity is the modern *definition* of continuity.

DEFINITION   The function $y = f(x)$ is *continuous* at the point $x = x_0$ in its domain if for each positive "tolerable error" ϵ we can find an adequate "control" number δ. This means: we can ensure that $|\Delta y| < \epsilon$ by taking $|\Delta x| < \delta$, or

$$\boxed{|\Delta x| < \delta \Rightarrow |\Delta y| < \epsilon,}$$

($\Rightarrow$ means *implies*), or

$$\boxed{|x - x_0| < \delta \Rightarrow |f(x) - f(x_0)| < \epsilon.}$$

Let us consider next the "effective" continuity of the multiplication operation $y = uv$. In the examples above, the direct proof of continuity always started off as though its objective were a derivative, by setting up a formula for the dependent increment Δy, and manipulating it so that its dependence on Δx becomes apparent. The same scheme works here for the product function $y = uv$. If we start with fixed values $y_0 = u_0 v_0$ and give u and v increments Δu and Δv, then the new value of y is

$$y_0 + \Delta y = (u_0 + \Delta u)(v_0 + \Delta v).$$

Multiplying this out and subtracting $y = uv$, we see that

$$\Delta y = u_0 \, \Delta v + v_0 \, \Delta u + \Delta u \, \Delta v.$$

This increment formula is not new. It was the basis for our proof of the product rule for derivatives in Chapter 4. Here we use it differently, to get an inequality limiting $|\Delta y|$. First,

$$|\Delta y| \le |u_0| \cdot |\Delta v| + |v_0| \cdot |\Delta u| + |\Delta u| \cdot |\Delta v|.$$

Then suppose we choose some positive number δ less than 1, and restrict Δu and Δv to be at most δ in magnitude. Substituting the inequalities $|\Delta u| \le \delta$, $|\Delta v| \le \delta$, in the right side above, and noting that $\delta^2 < \delta$ because $\delta < 1$, we find

$$|\Delta y| \le |u_0| \, \delta + |v_0| \, \delta + \delta^2$$
$$< [|u_0| + |v_0| + 1]\delta.$$

If we want $|\Delta y| < \epsilon$, it is thus sufficient to make

$$[|u_0| + |v_0| + 1]\delta = \epsilon,$$

that is, to take

$$\delta = \frac{\epsilon}{|u_0| + |v_0| + 1}.$$

We have shown that the increment in the product function $y = uv$ will be less than ϵ in magnitude if the increments Δu and Δv in the two factors are each bounded by

$$\delta = \frac{\epsilon}{|u_0| + |v_0| + 1}.$$

We have thus effectively demonstrated the continuity of multiplication.

Now consider how the continuity calculation would go for the product of two continuous functions. Suppose that $u = f(x)$ and $v = g(x)$ are both continuous at $x = x_0$. Given ϵ, we want to find how close to take x to x_0 in order to guarantee that

$$|f(x)g(x) - f(x_0)g(x_0)| < \epsilon,$$

or, more simply,

$$|uv - u_0 v_0| < \epsilon.$$

We saw above that this inequality will hold if

$$\Delta u = u - u_0 \qquad \text{and} \qquad \Delta v = v - v_0$$

are each less than

$$\epsilon' = \frac{\epsilon}{|u_0| + |v_0| + 1}$$

in magnitude. Since $u = f(x)$ is continuous at x_0, we can find a positive control number δ_1 for the inequality

$$|\Delta u| < \epsilon'.$$

That is, the inequality will hold whenever we restrict $\Delta x = x - x_0$ to be less than δ_1 in magnitude. Similarly, we can find a positive control number δ_2 for the continuous function $v = g(x)$, so that

$$|\Delta v| < \epsilon'$$

whenever $|\Delta x| < \delta_2$. We want *both* $|\Delta u| < \epsilon'$ and $|\Delta v| < \epsilon'$, and this double requirement will be met if $|\Delta x|$ meets both control requirements,

$$|\Delta x| < \delta_1 \qquad \text{and} \qquad |\Delta x| < \delta_2.$$

We can state this final condition more simply by setting δ equal to the *smaller* of the two control numbers δ_1 and δ_2. So if $|\Delta x| < \delta$, then Δx meets both control requirements. Backtracking, we see that then both $|\Delta x| < \epsilon'$ and $|\Delta v| < \epsilon'$ hold, and therefore that $|uv - u_0 v_0| < \epsilon$. That is,

$$if\ |x - x_0| < \delta,\ then\ |f(x)g(x) - f(x_0)g(x_0)| < \epsilon.$$

We have thus found a suitable control number δ for the product function $f(x)g(x)$ at $x = x_0$.

In this way, by explicitly tracking down how the continuity calculation goes, we have proved

if f and g are continuous at $x = x_0$, then so is the product function fg.

This is the way all the limit and continuity laws are established. In every case we prove that a function is continuous, or that a limit exists and has a certain value, by exhibiting an explicit scheme that will furnish a control number δ for any given positive number ϵ. This means, of course, that we use the corresponding "effective" definition of a function limit:

DEFINITION The function f has the limit l at $x = x_0$ if for every positive number ϵ, we can find a positive number δ such that

$$0 < |x - x_0| < \delta \Rightarrow |f(x) - l| < \epsilon.$$

Note that we don't allow x to take on the value x_0 when we consider whether or not $f(x)$ has the limit l as x approaches x_0.

The notion of sequential convergence is made precise in the same quantitative way.

DEFINITION The sequence $\{a_n\}$ converges to the number l as its limit as n tends to infinity if, given any positive number ϵ, we can find an integer N such that

$$n > N \Rightarrow |a_n - l| < \epsilon.$$

Sooner or later one should work through all of this material in a systematic way. It is probably not appropriate to take the time to do this in a first calculus course, and we shall just spotlight a few additional results in the exercises.

Finally, we consider the counterpart of the derivative continuity criterion for functions of two variables.

Suppose that on a circular disk R both partial derivatives of $z = f(x, y)$ exist and are bounded by k in magnitude. That is,

$$|D_1 f(x, y)| \le k,$$
$$|D_2 f(x, y)| \le k,$$

for all points (x, y) in R. Then

$$|\Delta z| \le k(|\Delta x| + |\Delta y|)$$

for any two points (x, y) and $(x + \Delta x, y + \Delta y)$ in R.

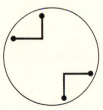

Figure 1

Proof Up to a point we just repeat the proof of the chain rule from Chapter 15. We go from the first point (x, y) to the second point $(x + \Delta x, y + \Delta y)$ in two steps, first a horizontal step to $(x + \Delta x, y)$ and then a vertical step. (It may be necessary to make the first step vertical and the second horizontal, as Fig. 1 shows.) Then Δz can be written as the sum of the two corresponding partial increments,

$$\Delta z = f(x + \Delta x, y + \Delta y) - f(x, y)$$
$$= [f(x + \Delta x, y + \Delta y) - f(x + \Delta x, y)]$$
$$+ [f(x + \Delta x, y) - f(x, y)],$$

and we can apply the mean-value principle for functions of one variable to each term on the right, obtaining

$$\Delta z = D_2 f(x + \Delta x, Y) \cdot \Delta y + D_1 f(X, y) \cdot \Delta x.$$

All this is exactly as before. But now we use the assumed inequalities on the partial derivatives and have at once the inequality we want:

$$|\Delta z| \le k\,|\Delta y| + k\,|\Delta x| = k(|\Delta x| + |\Delta y|).$$

It was because of this derivative criterion that we could be sure that the functions we met in Chapter 15 were all continuous.   ∎

REMARK We can restate this conclusion in terms of the length $|\Delta \mathbf{x}|$ of the vector $\Delta \mathbf{x} = (\Delta x, \Delta y)$, because of the fact that

$$a + b \le \sqrt{2}\,\sqrt{a^2 + b^2}$$

for any two numbers. (Square both sides and see what happens.) The inequality

$$|\Delta z| \le k(|\Delta x| + |\Delta y|)$$

thus implies that

$$|\Delta z| \le \sqrt{2}\,k\,|\Delta \mathbf{x}|.$$

Example 6 The two-variable criterion provides another way to establish the continuity inequality for the product operation $z = f(x, y) = xy$. The partial derivatives

$$\frac{\partial z}{\partial x} = y, \qquad \frac{\partial z}{\partial y} = x$$

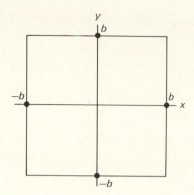

Figure 2

are both less than b in magnitude on the square $|x| \leq b$, $|y| \leq b$. Therefore,

$$|\Delta z| \leq b(|\Delta x| + |\Delta y|)$$

for any two points (x, y) and $(x + \Delta x, y + \Delta y)$ in the square. If we set $(r, s) = (x + \Delta x, y + \Delta y)$, then we can rewrite this new version of the continuity inequality as

$$|rs - xy| \leq b(|r - x| + |s - y|).$$

It is closely related to the earlier version, the differences being due to different assumptions about the magnitudes of x, y, and their increments.   □

PROBLEMS FOR APPENDIX 4

1. How many places in the decimal expansion

$$\ln 4 = 1.3 \cdots$$

will be needed to approximate $(\ln 4)^2$ with an error less than 0.01? (Use the derivative criterion.)

2. How many places in the decimal expansion

$$\log_{10} \pi = 0.4 \cdots$$

are needed to approximate $(\log_{10} \pi)^3$ with an error less than 0.005?

3. How many places in the expansion

$$\ln 100 = 4.6 \cdots$$

are needed to approximate $(\ln 100)^4$ with an error less than 0.005?

4. How many places in the decimal expansion

$$e = 2.7 \cdots$$

are needed to approximate e^5 with an error less than 0.002?

5. Prove the following counterpart of the derivative criterion.

If

$$|f'(x)| \geq b > 0$$

for every x in an interval I, then

$$|\Delta y| \geq b\,|\Delta x|$$

for any two points x and $(x + \Delta x)$ in I.

6. Show that if $e = 2.7\cdots$, then $e^3 < 23$, *without* computing $(2.8)^3$. (Use Problem 5 and the obvious inequalities $e > 8/3$, $3 - e > 0.2$.)

Prove the inequalities in Problems 7 through 11 by direct arguments (that is, without using calculus).

7. If $0 < x < a$, then

$$4x^3(a - x) < a^4 - x^4 < 4a^3(a - x).$$

8. If x and y are both greater than the positive number a, then

$$\left|\frac{1}{x} - \frac{1}{y}\right| < \frac{1}{a^2}|x - y|.$$

9. If $0 < x < a$ then $a^n - x^n < na^{n-1}(a - x)$.

10. Assuming that $|\sin x| \leq |x|$ for all x, show that

$$|\cos x - \cos y| \leq |x - y|.$$

11. If $0 < b < y$ then $y^{1/4} - b^{1/4} < (1/4)b^{-3/4}(y - b)$. (This can be proved directly from Problem 1 by relabeling.)

12. If $y = 1/x$, show by a direct estimation of Δy, as in Examples 2 and 3, that if $|x|$ and $|x + \Delta x|$ are both larger than a positive number b, then

$$|\Delta y| \leq \epsilon \qquad \text{whenever } |\Delta x| \leq b^2\epsilon.$$

13. If $y = x^4$, show by a direct algebraic estimate of Δy that if $|x|$ and $|x + \Delta x|$ are both less than a, then

$$|\Delta y| \leq \epsilon \qquad \text{whenever } |\Delta x| \leq \frac{\epsilon}{4a^3}.$$

14. Show that if $e = 2.7182\cdots$, then

$$e^4 - (2.718)^4 < 0.05.$$

(Use the preceding problem or Problem 7.)

15. Show that if $y = \sqrt{1 - x^2}$, then $y \leq \sqrt{2}(1 - x)^{1/2}$. Similarly, show that $y \leq \sqrt{2}(1 + x)^{1/2}$. This completes Example 5 in the text.

In each of the following four problems, find how small to take $\Delta x = x - a$ in order to ensure that $\Delta y = f(x) - f(a)$ will be less than ϵ in magnitude.

16. $f(x) = (x - 2)^{2/3}$; $a = 2$

17. $f(x) = (4 - x^2)^{4/5}$; $a = 2$

18. $f(x) = \sqrt{x - 1}$; $a = 1$

19. $f(x) = \sqrt{x^2 - 1}$; $a = 1$

20. Prove that if f is continuous and $f(a) \neq 0$, then $1/f$ is continuous at $x = a$.

21. Show by a direct algebraic argument that $f(x, y) = x + y$ is a continuous function of two variables. (This is much easier than the proof for the product in the text.)

22. Prove that if f and g are continuous at $x = a$, then so is $f + g$.

24. Prove the sign-preserving law: If $a_n \to a$ as $n \to \infty$, and if $a_n \geq 0$ for all n, then $a \geq 0$. [*Hint*: Suppose, on the contrary, that $a < 0$, and take $\epsilon = |a| = -a$, in the definition of convergence. You should be able to show that, then,

$$a_n < 0 \qquad \text{for all } n \text{ larger than } N,$$

which is a contradiction.]

23. Give a rigorous ϵ, N proof of the law: If $a_n \to a$ and $b_n \to b$ as $n \to \infty$, then $a_n b_n \to ab$ as $n \to \infty$. (Imitate the proof that fg is continuous, replacing the control numbers δ_1 and δ_2 by control integers N_1 and N_2, etc.)

ΛPPENDIX 5

THE PRINCIPLES UNDERLYING CALCULUS

In this appendix we shall complete the chain of proofs showing that all of calculus rests on the nested-interval property of the real numbers. There are many possible routes through this material. Here we shall repeatedly construct a nested sequence of intervals by the "bisection procedure." This device is not always feasible, but we use it whenever it is efficient. On other occasions we need variants of the completeness property, and two are considered in Section 2.

1
THE INTERMEDIATE-VALUE THEOREM

This is the theorem:

THEOREM   *If f is continuous on the closed interval [a, b] and if l is any number between f(a) and f(b), then there is at least one point X in [a, b] for which f(X) = l.*

If we set $g(x) = f(x) - l$, and possibly change sign, then the intermediate-value theorem reduces to the zero-crossing theorem for a continuous function.

THEOREM 1   *If f is continuous on the closed interval [a, b], and if f(a) < 0 < f(b), then f(X) = 0 for some number X in [a, b].*

Let us see how we might go about computing such a number X. Suppose, to be specific, that $[a, b] = [1, 2]$. As a first step we divide $[1, 2]$ into 10 equal subintervals with endpoints

$$1.0, \qquad 1.1, \qquad 1.2, \qquad \cdots, \qquad 1.9, \qquad 2.0.$$

Then we run through the corresponding values of f:

$$f(1.0), \qquad f(1.1), \qquad \cdots, \qquad f(1.9), \qquad f(2.0).$$

We shall suppose that we are never lucky enough to land exactly on a number where f is zero, so each value is either positive or negative. Since we start with the negative value $f(1.0)$ and end with the positive value $f(2.0)$, there will be a first sign change in this sequence of eleven values, say, between 1.2 and 1.3. Then

$$f(1.2) < 0 < f(1.3).$$

Our second step is to divide $[1.2, 1.3]$ into ten equal subintervals, with endpoints

$$1.20, \qquad 1.21, \qquad \ldots, \qquad 1.29, \qquad 1.30$$

and again select the first pair where f changes sign, say,

$$f(1.26) < 0 < f(1.27).$$

Continuing in this way, we construct an infinite decimal, beginning, say, 1.26745..., such that

$$f(1) < 0 < f(2)$$
$$f(1.2) < 0 < f(1.3)$$
$$f(1.26) < 0 < f(1.27)$$
$$\cdot$$
$$\cdot$$
$$\cdot$$
$$f(1.26745) < 0 < f(1.26746).$$

Now let X be the number having this infinite decimal expansion. Then $f(X)$ has to be 0. To see why, we note that the left-hand numbers 1, 1.2, 1.26, ... form an infinite sequence $\{a_n\}$ converging to X, and since f is continuous at X, we have

$$f(X) = \lim_{n \to \infty} f(a_n).$$

But $f(a_n) < 0$ for all n. Therefore $f(X) \leq 0$ by the sign-preserving property of a convergent sequence. Similarly, since the sequence $\{b_n\}$ of right-hand numbers converges to X from above, and since $f(b_n) > 0$ for all n, it follows that

$$f(X) = \lim_{n \to \infty} f(b_n) \geq 0.$$

That is, $f(X)$ must be simultaneously ≤ 0 and ≥ 0 and hence must be 0.

The above process actually constitutes a proof of the zero-crossing theorem. Of course, to be called a proof it has to be described in more general terms. Moreover, it is somewhat more efficient and also more suitable for computers to use successive bisections rather than successive division into ten parts. This means that the number being computed is not now described by an infinite decimal, but instead is specified by an infinite sequence of closed intervals each of which is one half of its predecessor.

Proof of Theorem Set $I_0 = [a, b]$ and let $I_1 = [a_1, b_1]$ be the left or right half of I_0 depending on whether the value of f at the midpoint $c = (a + b)/2$ is positive or negative. That is, if $f(c) > 0$, we set $a_1 = a$ and $b_1 = c$, and if $f(c) < 0$, we set $a_1 = c$ and $b_1 = b$ In either case,

$$f(a_1) < 0 < f(b_1).$$

(We suppose we are not lucky enough to find that $f(c) = 0$.) We continue bisecting in this way, at the nth step choosing $I_n = [a_n, b_n]$ as that half of I_{n-1} for which

$$f(a_n) < 0 < f(b_n).$$

Since $b_n - a_n = (b - a)/2^n$, the conditions for the nested-interval form of the completeness property are met, and the sequences $\{a_n\}$ and $\{b_n\}$ therefore converge to a common limit x. Then

$$f(x) = \lim_{n \to \infty} f(a_n) \leq 0,$$

$$f(x) = \lim_{n \to \infty} f(b_n) \geq 0,$$

and so $f(x) = 0$. ∎

2
COMPLETENESS

The nested-interval principle is only one of several mutually equivalent versions of the completeness property of the real numbers, and other versions are sometimes easier to use. Two of these alternative forms are proved here.

A set of numbers A is *bounded above* by a number b if $a \le b$ for every a in A. For example, the interval $[0, 1]$ is bounded above by 1 (and also by 2, and by any other number greater than 1). So is the open interval $(0, 1)$. The set of all positive integers doesn't have an upper bound; it is *unbounded from above*. Note that if b is an upper bound to the set A, then so is any number larger than b.

In the examples above, the number 1 is not only an upper bound to both the closed interval $[0, 1]$ and the open interval $(0, 1)$, but in each case it is the *smallest possible upper bound*—any smaller number fails to be an upper bound. It is also the smallest upper bound to the sequence $\{(n - 1)/n\}$, where n runs through all positive integers.

The least upper bound principle says that such a smallest upper bound always exists.

THE LEAST UPPER BOUND PRINCIPLE

If A is a nonempty set of numbers that is bounded above, then A has a least upper bound. That is, the collection B of all of the upper bounds of A contains a smallest number.

Proof   Let b_0 be an upper bound of A and let a_0 be any number that is *not* an upper bound. For example, we could choose some number a from A and set $a_0 = a - 1$. Let I_0 be the closed interval $[a_0, b_0]$. Let $I_1 = [a_1, b_1]$ be the left or right half of I_0, depending on whether the midpoint $c = (a_0 + b_0)/2$ is or is not an upper bound of A. That is, if c is an upper bound, then we set $a_1 = a_0$ and $b_1 = c$, and if c is not an upper bound, then we set $a_1 = c$ and $b_1 = b_0$. In either case we have a new interval $I_1 = [a_1, b_1]$ for which the right endpoint b_1 is an upper bound of A and the left endpoint a_1 is not. We continue bisecting in this way, at the nth step choosing $I_n = [a_n, b_n]$ as that half of I_{n-1} for which b_n is an upper bound of A and a_n is not. Since $b_n - a_n = (b_0 - a_0)/2^n$, the nested-interval principle guarantees that the two endpoint sequences $\{a_n\}$ and $\{b_n\}$ converge to a common limit X. We now show that X is the least upper bound of A.

Consider any number a in A. Then $a \le b_n$ for every n, since every b_n is an upper bound to A, so $a \le \lim b_n = X$.

Thus X *is* an upper bound to A. On the other hand, if c is any number smaller than X, then $c < a_n$ for all sufficiently large n (because $a_n \to X$ as $n \to \infty$), and since a_n is never an upper bound to A, then neither is c. So X is the smallest upper bound to A and we are done. ∎

Notation We write

$$\text{lub } A$$

for the least upper bound to A.

Suppose now that A is a nonempty set that is bounded *below*, and let B be the (nonempty) collection of all its *lower* bounds. Then every a in A is an upper bound to B, so if

$$X = \text{lub } B,$$

then $X \leq a$ for every a in A. That is, X is also a lower bound to A. It is therefore the maximum element of B. Thus,

If A is a nonempty set that is bounded below, then A has a greatest lower bound (glb A).

This complementary principle can be proved in other ways. We can imitate the proof of the lub property, with inequalities reversed. Or we can verify that

$$-\text{lub}(-A) = \text{glb } A.$$

A third useful form of completeness is

MONOTONE-LIMIT PRINCIPLE *If the sequence $\{a_n\}$ is increasing and bounded above, then it necessarily converges.*

If the numbers a_n increase with n and are blocked from above by a number b, then it seems obvious that they must pile up at some point $l \leq b$, in which case l must be the limit of the sequence. So this version of completeness is very intuitive.

Proof We define l as the least upper bound of the sequence terms a_n and then prove that we can cause a_n to be as close to l as we wish simply by taking n large enough. So suppose we want a_n to be closer to l than the distance ϵ, where ϵ is some given positive number. Since l is the smallest upper bound to the sequence, it follows that $l - \epsilon$ is *not* an upper bound, i.e., that

$$l - \epsilon < a_N$$

for at least one integer N. But then

$$l - \epsilon < a_n$$

for all n beyond N, since $\{a_n\}$ is increasing. So we have

$$l - \epsilon < a_n \leq l,$$

or

$$0 \leq l - a_n < \epsilon,$$

for all such n, and this is exactly what we wanted to show. Thus $a_n \to l$ as $n \to \infty$. ∎

3
THE RANGE OF A CONTINUOUS FUNCTION

If f is a continuous function on an interval I, then, by the intermediate-value theorem, the range of f over I is a set A having the following property:

(*) *If x and y are any two points of A, with $x < y$, then A includes the whole closed interval $[x, y]$.*

We can now show that such a set A is necessarily an interval, at which point we will have proved that the range of a continuous function over an interval is always itself an interval (Theorem 4 in Chapter 2).

To start with, choose any fixed point x_0 in A, and let B be the intersection of A with the semi-infinite interval $[x_0, \infty)$. Note that B also satisfies (*).

Suppose first that B is bounded above and let b be its least upper bound. We claim that B is then either the closed interval $[x_0, b]$ or the semiopen interval $[x_0, b)$. To see this, let x be any number in $[x_0, b)$:

$$x_0 \leq x < b.$$

Then x is not an upper bound to B, so B contains a number y greater than x. Then B includes the interval $[x_0, y]$, by (*), and hence contains x. So every point of $[x_0, b)$ is in B. On the other hand, since $b = \text{lub } B$, the only possibility for a point of B not being in $[x_0, b)$ is the point b itself. So

$$B = [x_0, b) \quad \text{or} \quad B = [x_0, b].$$

If B is not bounded above, the same argument shows that every point of $[x_0, \infty)$ is in B, so $B = [x_0, \infty)$ in this case.

We have thus shown that B is either a closed interval $[x_0, b]$ or a semiopen interval $[x_0, b)$, where b may be $+\infty$ in the second case. In exactly the same way, the intersection of A with the semi-infinite interval $(-\infty, x_0]$ is of the form $[a, x_0]$ or $(a, x_0]$, where a is permitted to be $-\infty$ in the second case. So A itself is one of the intervals

$$[a, b], \qquad [a, b), \qquad (a, b], \qquad (a, b),$$

where a may be $-\infty$ and/or b may be $+\infty$, whenever these possibilities make sense.

4
THE EXTREME-VALUE THEOREM

LEMMA   *If f is continuous on a closed interval $[a, b]$, then f is bounded there.*

Proof   We show that f has an upper bound. The crucial observation is that if $a < c < b$ and if f is bounded above (by B_1) on $[a, c]$ and (by B_2) on $[c, b]$, then f is bounded above (by the larger of B_1 and B_2) on $[a, b]$. So if we assume that f is *not* bounded above on $[a, b]$ then f must be unbounded from above on at least one of the two halves of $[a, b]$. We can then repeatedly bisect, in the manner of Sections 1 and 2, and thus determine a nested sequence of closed intervals $I_n = [a_n, b_n]$, each being one half of its predecessor, such that f is unbounded from above on every I_n. Let X be the common limit of the endpoint sequences $\{a_n\}$ and $\{b_n\}$ (by the nested-interval principle), and choose any number B larger than $f(X)$. Since f is continuous at X, there is an interval I about the point X on which $f(x) < B$. But I contains all the left endpoints a_n from some point on (since $a_n \to X$ as $n \to \infty$), and similarly for the right endpoints, so I includes all the intervals I_n from some point on. In particular, f is bounded above by B on any such I_n, contradicting the fact that f is unbounded from above on I_n. So the assumption that f is unbounded from above on $[a, b]$ leads to a contradiction, and we are done.   ■

The proof that f has a lower bound on $[a, b]$ is essentially the same.

We now know that the values of f on $[a, b]$ form a nonempty set that is bounded. Let B be the least upper bound of this set. We claim that the number B is itself a value, i.e., that $B = f(X)$ for some X in $[a, b]$, in which case B is the *maximum* value of f on $[a, b]$. In order to see this, note first that B must be the least upper bound of f on one or the other (or perhaps both) of the two halves of $[a, b]$. It is left to the reader to check this. We can then repeatedly bisect as before, always choosing $I_n = [a_n, b_n]$ as one of the halves of I_{n-1} on which f has B as its least upper bound. If X is the unique number common to all the intervals I_n (by the nested-interval principle), then $f(X) = B$, and we are done. (The details are left to the reader. Show by an argument like that used in the Lemma that if $f(X) < B$, then f has an upper bound smaller than B on some of the intervals I_n, contradicting the fact that $B = \text{lub} f$ on each I_n.)

Alternative Trick Proof   If f never assumes the value B, then $B - f(x)$ is never zero and

$$g(x) = \frac{1}{B - f(x)}$$

is everywhere defined and continuous on $[a, b]$. Then g is bounded above, say by C, so we have

$$\frac{1}{B - f(x)} \leq C,$$

$$B - f(x) \geq \frac{1}{C},$$

and hence

$$B - \frac{1}{C} \geq f(x).$$

Thus $B - (1/C)$ is an upper bound to f, contradicting the fact that B is the *smallest* upper bound. Therefore f must assume the value B.

We have thus proved, incompletely:

THE EXTREME-   *If f is continuous on the closed interval $[a, b]$, then f assumes max-*
VALUE THEOREM   *imum and minimum values there.*

APPENDIX 6
TABLES

Natural Logarithms of Numbers

n	$\log_e n$	n	$\log_e n$	n	$\log_e n$
0.0	*	4.5	1.5041	9.0	2.1972
0.1	7.6974	4.6	1.5261	9.1	2.2083
0.2	8.3906	4.7	1.5476	9.2	2.2192
0.3	8.7960	4.8	1.5686	9.3	2.2300
0.4	9.0837	4.9	1.5892	9.4	2.2407
0.5	9.3069	5.0	1.6094	9.5	2.2513
0.6	9.4892	5.1	1.6292	9.6	2.2618
0.7	9.6433	5.2	1.6487	9.7	2.2721
0.8	9.7769	5.3	1.6677	9.8	2.2824
0.9	9.8946	5.4	1.6864	9.9	2.2925
1.0	0.0000	5.5	1.7047	10	2.3026
1.1	0.0953	5.6	1.7228	11	2.3979
1.2	0.1823	5.7	1.7405	12	2.4849
1.3	0.2624	5.8	1.7579	13	2.5649
1.4	0.3365	5.9	1.7750	14	2.6391
1.5	0.4055	6.0	1.7918	15	2.7081
1.6	0.4700	6.1	1.8083	16	2.7726
1.7	0.5306	6.2	1.8245	17	2.8332
1.8	0.5878	6.3	1.8405	18	2.8904
1.9	0.6419	6.4	1.8563	19	2.9444
2.0	0.6931	6.5	1.8718	20	2.9957
2.1	0.7419	6.6	1.8871	25	3.2189
2.2	0.7885	6.7	1.9021	30	3.4012
2.3	0.8329	6.8	1.9169	35	3.5553
2.4	0.8755	6.9	1.9315	40	3.6889
2.5	0.9163	7.0	1.9459	45	3.8067
2.6	0.9555	7.1	1.9601	50	3.9120
2.7	0.9933	7.2	1.9741	55	4.0073
2.8	1.0296	7.3	1.9879	60	4.0943
2.9	1.0647	7.4	2.0015	65	4.1744
3.0	1.0986	7.5	2.0149	70	4.2485
3.1	1.1314	7.6	2.0281	75	4.3175
3.2	1.1632	7.7	2.0412	80	4.3820
3.3	1.1939	7.8	2.0541	85	4.4427
3.4	1.2238	7.9	2.0669	90	4.4998
3.5	1.2528	8.0	2.0794	95	4.5539
3.6	1.2809	8.1	2.0919	100	4.6052
3.7	1.3083	8.2	2.1041		
3.8	1.3350	8.3	2.1163		
3.9	1.3610	8.4	2.1282		
4.0	1.3863	8.5	2.1401		
4.1	1.4110	8.6	2.1518		
4.2	1.4351	8.7	2.1633		
4.3	1.4586	8.8	2.1748		
4.4	1.4816	8.9	2.1861		

Exponential Functions

x	e^x	e^{-x}	x	e^x	e^{-x}
0.00	1.0000	1.0000	2.5	12.182	0.0821
0.05	1.0513	0.9512	2.6	13.464	0.0743
0.10	1.1052	0.9048	2.7	14.880	0.0672
0.15	1.1618	0.8607	2.8	16.445	0.0608
0.20	1.2214	0.8187	2.9	18.174	0.0550
0.25	1.2840	0.7788	3.0	20.086	0.0498
0.30	1.3499	0.7408	3.1	22.198	0.0450
0.35	1.4191	0.7047	3.2	24.533	0.0408
0.40	1.4918	0.6703	3.3	27.113	0.0369
0.45	1.5683	0.6376	3.4	29.964	0.0334
0.50	1.6487	0.6065	3.5	33.115	0.0302
0.55	1.7333	0.5769	3.6	36.598	0.0273
0.60	1.8221	0.5488	3.7	40.447	0.0247
0.65	1.9155	0.5220	3.8	44.701	0.0224
0.70	2.0138	0.4966	3.9	49.402	0.0202
0.75	2.1170	0.4724	4.0	54.598	0.0183
0.80	2.2255	0.4493	4.1	60.340	0.0166
0.85	2.3396	0.4274	4.2	66.686	0.0150
0.90	2.4596	0.4066	4.3	73.700	0.0136
0.95	2.5857	0.3867	4.4	81.451	0.0123
1.0	2.7183	0.3679	4.5	90.017	0.0111
1.1	3.0042	0.3329	4.6	99.484	0.0101
1.2	3.3201	0.3012	4.7	109.95	0.0091
1.3	3.6693	0.2725	4.8	121.51	0.0082
1.4	4.0552	0.2466	4.9	134.29	0.0074
1.5	4.4817	0.2231	5	148.41	0.0067
1.6	4.9530	0.2019	6	403.43	0.0025
1.7	5.4739	0.1827	7	1096.6	0.0009
1.8	6.0496	0.1653	8	2981.0	0.0003
1.9	6.6859	0.1496	9	8103.1	0.0001
2.0	7.3891	0.1353	10	22026	0.00005
2.1	8.1662	0.1225			
2.2	9.0250	0.1108			
2.3	9.9742	0.1003			
2.4	11.023	0.0907			

Natural Trigonometric Functions

Angle					Angle				
Degree	Radian	Sine	Cosine	Tangent	Degree	Radian	Sine	Cosine	Tangent
0°	0.000	0.000	1.000	0.000					
1°	0.017	0.017	1.000	0.017	46°	0.803	0.719	0.695	1.036
2°	0.035	0.035	0.999	0.035	47°	0.820	0.731	0.682	1.072
3°	0.052	0.052	0.999	0.052	48°	0.838	0.743	0.669	1.111
4°	0.070	0.070	0.998	0.070	49°	0.855	0.755	0.656	1.150
5°	0.087	0.087	0.996	0.087	50°	0.873	0.766	0.643	1.192
6°	0.105	0.105	0.995	0.105	51°	0.890	0.777	0.629	1.235
7°	0.122	0.122	0.993	0.123	52°	0.908	0.788	0.616	1.280
8°	0.140	0.139	0.990	0.141	53°	0.925	0.799	0.602	1.327
9°	0.157	0.156	0.988	0.158	54°	0.942	0.809	0.588	1.376
10°	0.175	0.174	0.985	0.176	55°	0.960	0.819	0.574	1.428
11°	0.192	0.191	0.982	0.194	56°	0.977	0.829	0.559	1.483
12°	0.209	0.208	0.978	0.213	57°	0.995	0.839	0.545	1.540
13°	0.227	0.225	0.974	0.231	58°	1.012	0.848	0.530	1.600
14°	0.244	0.242	0.970	0.249	59°	1.030	0.857	0.515	1.664
15°	0.262	0.259	0.966	0.268	60°	1.047	0.866	0.500	1.732
16°	0.279	0.276	0.961	0.287	61°	1.065	0.875	0.485	1.804
17°	0.297	0.292	0.956	0.306	62°	1.082	0.883	0.469	1.881
18°	0.314	0.309	0.951	0.325	63°	1.100	0.891	0.454	1.963
19°	0.332	0.326	0.946	0.344	64°	1.117	0.899	0.438	2.050
20°	0.349	0.342	0.940	0.364	65°	1.134	0.906	0.423	2.145
21°	0.367	0.358	0.934	0.384	66°	1.152	0.914	0.407	2.246
22°	0.384	0.375	0.927	0.404	67°	1.169	0.921	0.391	2.356
23°	0.401	0.391	0.921	0.424	68°	1.187	0.927	0.375	2.475
24°	0.419	0.407	0.914	0.445	69°	1.204	0.934	0.358	2.605
25°	0.436	0.423	0.906	0.466	70°	1.222	0.940	0.342	2.748
26°	0.454	0.438	0.899	0.488	71°	1.239	0.946	0.326	2.904
27°	0.471	0.454	0.891	0.510	72°	1.257	0.951	0.309	3.078
28°	0.489	0.469	0.883	0.532	73°	1.274	0.956	0.292	3.271
29°	0.506	0.485	0.875	0.554	74°	1.292	0.961	0.276	3.487
30°	0.524	0.500	0.866	0.577	75°	1.309	0.966	0.259	3.732
31°	0.541	0.515	0.857	0.601	76°	1.326	0.970	0.242	4.011
32°	0.559	0.530	0.848	0.625	77°	1.344	0.974	0.225	4.332
33°	0.576	0.545	0.839	0.649	78°	1.361	0.978	0.208	4.705
34°	0.593	0.559	0.829	0.675	79°	1.379	0.982	0.191	5.145
35°	0.611	0.574	0.819	0.700	80°	1.396	0.985	0.174	5.671
36°	0.628	0.588	0.809	0.727	81°	1.414	0.988	0.156	6.314
37°	0.646	0.602	0.799	0.754	82°	1.431	0.990	0.139	7.115
38°	0.663	0.616	0.788	0.781	83°	1.449	0.993	0.122	8.144
39°	0.681	0.629	0.777	0.810	84°	1.466	0.995	0.105	9.514
40°	0.698	0.643	0.766	0.839	85°	1.484	0.996	0.087	11.43
41°	0.716	0.656	0.755	0.869	86°	1.501	0.998	0.070	14.30
42°	0.733	0.669	0.743	0.900	87°	1.518	0.999	0.052	19.08
43°	0.750	0.682	0.731	0.933	88°	1.536	0.999	0.035	28.64
44°	0.768	0.695	0.719	0.966	89°	1.553	1.000	0.017	57.29
45°	0.785	0.707	0.707	1.000	90°	1.571	1.000	0.000	

Powers and Roots

No.	Sq.	Sq. Root	Cube	Cube Root	No.	Sq.	Sq. Root	Cube	Cube Root
1	1	1.000	1	1.000	51	2,601	7.141	132,651	3.708
2	4	1.414	8	1.260	52	2,704	7.211	140,608	3.733
3	9	1.732	27	1.442	53	2,809	7.280	148,877	3.756
4	16	2.000	64	1.587	54	2,916	7.348	157,464	3.780
5	25	2.236	125	1.710	55	3,025	7.416	166,375	3.803
6	36	2.449	216	1.817	56	3,136	7.483	175,616	3.826
7	49	2.646	343	1.913	57	3,249	7.550	185,193	3.849
8	64	2.828	512	2.000	58	3,364	7.616	195,112	3.871
9	81	3.000	729	2.080	59	3,481	7.681	205,379	3.893
10	100	3.162	1,000	2.154	60	3,600	7.746	216,000	3.915
11	121	3.317	1,331	2.224	61	3,721	7.810	226,981	3.936
12	144	3.464	1,728	2.289	62	3,844	7.874	238,328	3.958
13	169	3.606	2,197	2.351	63	3,969	7.937	250,047	3.979
14	196	3.742	2,744	2.410	64	4,096	8.000	262,144	4.000
15	225	3.873	3,375	2.466	65	4,225	8.062	274,625	4.021
16	256	4.000	4,096	2.520	66	4,356	8.124	287,496	4.041
17	289	4.123	4,913	2.571	67	4,489	8.185	300,763	4.062
18	324	4.243	5,832	2.621	68	4,624	8.246	314,432	4.082
19	361	4.359	6,859	2.668	69	4,761	8.307	328,509	4.102
20	400	4.472	8,000	2.714	70	4,900	8.367	343,000	4.121
21	441	4.583	9,261	2.759	71	5,041	8.426	357,911	4.141
22	484	4.690	10,648	2.802	72	5,184	8.485	373,248	4.160
23	529	4.796	12,167	2.844	73	5,329	8.544	389,017	4.179
24	576	4.899	13,824	2.884	74	5,476	8.602	405,224	4.198
25	625	5.000	15,625	2.924	75	5,625	8.660	421,875	4.217
26	676	5.099	17,576	2.962	76	5,776	8.718	438,976	4.236
27	729	5.196	19,683	3.000	77	5,929	8.775	456,533	4.254
28	784	5.292	21,952	3.037	78	6,084	8.832	474,552	4.273
29	841	5.385	24,389	3.072	79	6,241	8.888	493,039	4.291
30	900	5.477	27,000	3.107	80	6,400	8.944	512,000	4.309
31	961	5.568	29,791	3.141	81	6,561	9.000	531,441	4.327
32	1,024	5.657	32,768	3.175	82	6,724	9.055	551,368	4.344
33	1,089	5.745	35,937	3.208	83	6,889	9.110	571,787	4.362
34	1,156	5.831	39,304	3.240	84	7,056	9.165	592,704	4.380
35	1,225	5.916	42,875	3.271	85	7,225	9.220	614,125	4.397
36	1,296	6.000	46,656	3.302	86	7,396	9.274	636,056	4.414
37	1,369	6.083	50,653	3.332	87	7,569	9.327	658,503	4.431
38	1,444	6.164	54,872	3.362	88	7,744	9.381	681,472	4.448
39	1,521	6.245	59,319	3.391	89	7,921	9.434	704,969	4.465
40	1,600	6.325	64,000	3.420	90	8,100	9.487	729,000	4.481
41	1,681	6.403	68,921	3.448	91	8,281	9.539	753,571	4.498
42	1,764	6.481	74,088	3.476	92	8,464	9.592	778,688	4.514
43	1,849	6.557	79,507	3.503	93	8,649	9.644	804,357	4.531
44	1,936	6.633	85,184	3.530	94	8,836	9.695	830,584	4.547
45	2,025	6.708	91,125	3.557	95	9,025	9.747	857,375	4.563
46	2,116	6.782	97,336	3.583	96	9,216	9.798	884,736	4.579
47	2,209	6.856	103,823	3.609	97	9,409	9.849	912,673	4.595
48	2,304	6.928	110,592	3.634	98	9,604	9.899	941,192	4.610
49	2,401	7.000	117,649	3.659	99	9,801	9.950	970,299	4.626
50	2,500	7.071	125,000	3.684	100	10,000	10.000	1,000,000	4.642

Table of Integrals

1. $\int u\,dv = uv - \int v\,du$

2. $\int a^u\,du = \dfrac{a^u}{\ln a} + C, \qquad a \neq 1, \qquad a > 0$

3. $\int \cos u\,du = \sin u + C$

4. $\int \sin u\,du = -\cos u + C$

5. $\int (ax + b)^n\,dx = \dfrac{(ax + b)^{n+1}}{a(n + 1)} + C, \qquad n \neq -1$

6. $\int (ax + b)^{-1}\,dx = \dfrac{1}{a} \ln |ax + b| + C$

7. $\int x(ax + b)^n\,dx = \dfrac{(ax + b)^{n+1}}{a^2}\left[\dfrac{ax + b}{n + 2} - \dfrac{b}{n + 1}\right] + C, \qquad n \neq -1, -2$

8. $\int x(ax + b)^{-1}\,dx = \dfrac{x}{a} - \dfrac{b}{a^2} \ln |ax + b| + C$

9. $\int x(ax + b)^{-2}\,dx = \dfrac{1}{a^2}\left[\ln |ax + b| + \dfrac{b}{ax + b}\right] + C$

10. $\int \dfrac{dx}{x(ax + b)} = \dfrac{1}{b} \ln \left|\dfrac{x}{ax + b}\right| + C$

11. $\int (\sqrt{ax + b})^n\,dx = \dfrac{2}{a}\dfrac{(\sqrt{ax + b})^{n+2}}{n + 2} + C, \qquad n \neq -2$

12. $\int \dfrac{\sqrt{ax + b}}{x}\,dx = 2\sqrt{ax + b} + b\int \dfrac{dx}{x\sqrt{ax + b}}$

13. (a) $\int \dfrac{dx}{x\sqrt{ax + b}} = \dfrac{2}{\sqrt{-b}} \tan^{-1} \sqrt{\dfrac{ax + b}{-b}} + C, \qquad$ if $\qquad b < 0$

    (b) $\int \dfrac{dx}{x\sqrt{ax + b}} = \dfrac{1}{\sqrt{b}} \ln \left|\dfrac{\sqrt{ax + b} - \sqrt{b}}{\sqrt{ax + b} + \sqrt{b}}\right| + C, \qquad$ if $\qquad b > 0$

14. $\int \dfrac{\sqrt{ax + b}}{x^2}\,dx = -\dfrac{\sqrt{ax + b}}{x} + \dfrac{a}{2}\int \dfrac{dx}{x\sqrt{ax + b}} + C$

15. $\int \dfrac{dx}{x^2\sqrt{ax + b}} = -\dfrac{\sqrt{ax + b}}{bx} - \dfrac{a}{2b}\int \dfrac{dx}{x\sqrt{ax + b}} + C$

16. $\int \dfrac{dx}{a^2 + x^2} = \dfrac{1}{a} \tan^{-1} \dfrac{x}{a} + C$

17. $\int \dfrac{dx}{(a^2 + x^2)^2} = \dfrac{x}{2a^2(a^2 + x^2)} + \dfrac{1}{2a^3} \tan^{-1} \dfrac{x}{a} + C$

18. $\displaystyle\int \frac{dx}{a^2 - x^2} = \frac{1}{2a} \ln \left| \frac{x + a}{x - a} \right| + C$

19. $\displaystyle\int \frac{dx}{(a^2 - x^2)^2} = \frac{x}{2a^2(a^2 - x^2)} + \frac{1}{2a^2} \int \frac{dx}{a^2 - x^2}$

20. $\displaystyle\int \frac{dx}{\sqrt{a^2 + x^2}} = \ln \left| x + \sqrt{a^2 + x^2} \right| + C$

21. $\displaystyle\int \sqrt{a^2 + x^2}\, dx = \frac{x}{2} \sqrt{a^2 + x^2} + \frac{a^2}{2} \log(x + \sqrt{a^2 + x^2}) + C$

22. $\displaystyle\int x^2 \sqrt{a^2 + x^2}\, dx = \frac{x(a^2 + 2x^2)\sqrt{a^2 + x^2}}{8} - \frac{a^4}{8} \log(x + \sqrt{a^2 + x^2}) + C$

23. $\displaystyle\int \frac{\sqrt{a^2 + x^2}}{x}\, dx = \sqrt{a^2 + x^2} - a \log(x + \sqrt{a^2 + x^2}) + C$

24. $\displaystyle\int \frac{\sqrt{a^2 + x^2}}{x^2}\, dx = \log(x + \sqrt{a^2 + x^2}) - \frac{\sqrt{a^2 + x^2}}{x} + C$

25. $\displaystyle\int \frac{x^2}{\sqrt{a^2 + x^2}}\, dx = -\frac{a^2}{2} \log(x + \sqrt{a^2 + x^2}) + \frac{x\sqrt{a^2 + x^2}}{2} + C$

26. $\displaystyle\int \frac{dx}{x\sqrt{a^2 + x^2}} = -\frac{1}{a} \ln \left| \frac{a + \sqrt{a^2 + x^2}}{x} \right| + C$

27. $\displaystyle\int \frac{dx}{x^2 \sqrt{a^2 + x^2}} = -\frac{\sqrt{a^2 + x^2}}{a^2 x} + C$

28. $\displaystyle\int \frac{dx}{\sqrt{a^2 - x^2}} = \sin^{-1} \frac{x}{a} + C$

29. $\displaystyle\int \sqrt{a^2 - x^2}\, dx = \frac{x}{2} \sqrt{a^2 - x^2} + \frac{a^2}{2} \sin^{-1} \frac{x}{a} + C$

30. $\displaystyle\int x^2 \sqrt{a^2 - x^2}\, dx = \frac{a^4}{8} \sin^{-1} \frac{x}{a} - \frac{1}{8} x \sqrt{a^2 - x^2}(a^2 - 2x^2) + C$

31. $\displaystyle\int \frac{\sqrt{a^2 - x^2}}{x}\, dx = \sqrt{a^2 - x^2} - a \ln \left| \frac{a + \sqrt{a^2 - x^2}}{x} \right| + C$

32. $\displaystyle\int \frac{\sqrt{a^2 - x^2}}{x^2}\, dx = -\sin^{-1} \frac{x}{a} - \frac{\sqrt{a^2 - x^2}}{x} + C$

33. $\displaystyle\int \frac{x^2}{\sqrt{a^2 - x^2}}\, dx = \frac{a^2}{2} \sin^{-1} \frac{x}{a} - \frac{1}{2} x \sqrt{a^2 - x^2} + C$

34. $\displaystyle\int \frac{dx}{x\sqrt{a^2 - x^2}} = -\frac{1}{a} \ln \left| \frac{a + \sqrt{a^2 - x^2}}{x} \right| + C$

35. $\displaystyle\int \frac{dx}{x^2 \sqrt{a^2 - x^2}} = -\frac{\sqrt{a^2 - x^2}}{a^2 x} + C$

36. $\displaystyle\int \frac{dx}{\sqrt{x^2 - a^2}} = \ln \left|x + \sqrt{x^2 - a^2}\right| + C$

37. $\displaystyle\int \sqrt{x^2 - a^2}\, dx = \frac{x}{2}\sqrt{x^2 - a^2} - \frac{a^2}{2} \log \left|x + \sqrt{x^2 - a^2}\right| + C$

38. $\displaystyle\int (\sqrt{x^2 - a^2})^n\, dx = \frac{x(\sqrt{x^2 - a^2})^n}{n + 1} - \frac{na^2}{n + 1}\int (\sqrt{x^2 - a^2})^{n-2}\, dx, \qquad n \neq -1$

39. $\displaystyle\int \frac{dx}{(\sqrt{x^2 - a^2})^n} = \frac{x(\sqrt{x^2 - a^2})^{2-n}}{(2 - n)a^2} - \frac{n - 3}{(n - 2)a^2}\int \frac{dx}{(\sqrt{x^2 - a^2})^{n-2}}, \qquad n \neq 2$

40. $\displaystyle\int x(\sqrt{x^2 - a^2})^n\, dx = \frac{(\sqrt{x^2 - a^2})^{n+2}}{n + 2} + C, \qquad n \neq -2$

41. $\displaystyle\int x^2\sqrt{x^2 - a^2}\, dx = \frac{x}{8}(2x^2 - a^2)\sqrt{x^2 - a^2} - \frac{a^4}{8}\log \left|x + \sqrt{x^2 - a^2}\right| + C$

42. $\displaystyle\int \frac{\sqrt{x^2 - a^2}}{x}\, dx = \sqrt{x^2 - a^2} - a \sec^{-1}\left|\frac{x}{a}\right| + C$

43. $\displaystyle\int \frac{\sqrt{x^2 - a^2}}{x^2}\, dx = \log \left|x + \sqrt{x^2 - a^2}\right| - \frac{\sqrt{x^2 - a^2}}{x} + C$

44. $\displaystyle\int \frac{x^2}{\sqrt{x^2 - a^2}}\, dx = \frac{a^2}{2}\log \left|x + \sqrt{x^2 - a^2}\right| + \frac{x}{2}\sqrt{x^2 - a^2} + C$

45. $\displaystyle\int \frac{dx}{x\sqrt{x^2 - a^2}} = \frac{1}{a}\sec^{-1}\left|\frac{x}{a}\right| + C = \frac{1}{a}\cos^{-1}\left|\frac{a}{x}\right| + C$

46. $\displaystyle\int \frac{dx}{x^2\sqrt{x^2 - a^2}} = \frac{\sqrt{x^2 - a^2}}{a^2 x} + C$ 　　47. $\displaystyle\int \frac{dx}{\sqrt{2ax - x^2}} = \sin^{-1}\left(\frac{x - a}{a}\right) + C$

48. $\displaystyle\int \sqrt{2ax - x^2}\, dx = \frac{x - a}{2}\sqrt{2ax - x^2} + \frac{a^2}{2}\sin^{-1}\left(\frac{x - a}{a}\right) + C$

49. $\displaystyle\int (\sqrt{2ax - x^2})^n\, dx = \frac{(x - a)(\sqrt{2ax - x^2})^n}{n + 1} + \frac{na^2}{n + 1}\int (\sqrt{2ax - x^2})^{n-2}\, dx,$

50. $\displaystyle\int \frac{dx}{(\sqrt{2ax - x^2})^n} = \frac{(x - a)(\sqrt{2ax - x^2})^{2-n}}{(n - 2)a^2} + \frac{(n - 3)}{(n - 2)a^2}\int \frac{dx}{(\sqrt{2ax - x^2})^{n-2}}$

51. $\displaystyle\int x\sqrt{2ax - x^2}\, dx = \frac{(x + a)(2x - 3a)\sqrt{2ax - x^2}}{6} + \frac{a^3}{2}\sin^{-1}\frac{x - a}{a} + C$

52. $\displaystyle\int \frac{\sqrt{2ax - x^2}}{x}\, dx = \sqrt{2ax - x^2} + a \sin^{-1}\frac{x - a}{a} + C$

53. $\displaystyle\int \frac{\sqrt{2ax - x^2}}{x^2}\, dx = -2\sqrt{\frac{2a - x}{x}} - \sin^{-1}\left(\frac{x - a}{a}\right) + C$

54. $\displaystyle\int \frac{x\, dx}{\sqrt{2ax - x^2}} = a \sin^{-1}\frac{x - a}{a} - \sqrt{2ax - x^2} + C$

55. $\displaystyle\int \frac{dx}{x\sqrt{2ax - x^2}} = -\frac{1}{a}\sqrt{\frac{2a - x}{x}} + C$

56. $\displaystyle\int \sin ax\, dx = -\frac{1}{a}\cos ax + C$

57. $\displaystyle\int \cos ax\, dx = \frac{1}{a}\sin ax + C$

58. $\displaystyle\int \sin^2 ax\, dx = \frac{x}{2} - \frac{\sin 2ax}{4a} + C$

59. $\displaystyle\int \cos^2 ax\, dx = \frac{x}{2} + \frac{\sin 2ax}{4a} + C$

60. $\displaystyle\int \sin^n ax\, dx = \frac{-\sin^{n-1} ax \cos ax}{na} + \frac{n-1}{n}\int \sin^{n-2} ax\, dx$

61. $\displaystyle\int \cos^n ax\, dx = \frac{\cos^{n-1} ax \sin ax}{na} + \frac{n-1}{n}\int \cos^{n-2} ax\, dx$

62. (a) $\displaystyle\int \sin ax \cos bx\, dx = -\frac{\cos(a+b)x}{2(a+b)} - \frac{\cos(a-b)x}{2(a-b)} + C, \qquad a^2 \neq b^2$

    (b) $\displaystyle\int \sin ax \sin bx\, dx = \frac{\sin(a-b)x}{2(a-b)} - \frac{\sin(a+b)x}{2(a+b)}, \qquad a^2 \neq b^2$

    (c) $\displaystyle\int \cos ax \cos bx\, dx = \frac{\sin(a-b)x}{2(a-b)} + \frac{\sin(a+b)x}{2(a+b)}, \qquad a^2 \neq b^2$

63. $\displaystyle\int \sin ax \cos ax\, dx = -\frac{\cos 2ax}{4a} + C$

64. $\displaystyle\int \sin^n ax \cos ax\, dx = \frac{\sin^{n+1} ax}{(n+1)a} + C, \qquad n \neq -1$

65. $\displaystyle\int \frac{\cos ax}{\sin ax}\, dx = \frac{1}{a}\ln|\sin ax| + C$

66. $\displaystyle\int \cos^n ax \sin ax\, dx = -\frac{\cos^{n+1} ax}{(n+1)a} + C, \qquad n \neq -1$

67. $\displaystyle\int \frac{\sin ax}{\cos ax}\, dx = -\frac{1}{a}\ln|\cos ax| + C$

68. $\displaystyle\int \sin^n ax \cos^m ax\, dx = -\frac{\sin^{n-1} ax \cos^{m+1} ax}{a(m+n)} + \frac{n-1}{m+n}\int \sin^{n-2} ax \cos^m ax\, dx,$

                                              $n \neq -m$ (If $n = -m$, use No. 86.)

69. $\displaystyle\int \sin^n ax \cos^m ax\, dx = \frac{\sin^{n+1} ax \cos^{m-1} ax}{a(m+n)} + \frac{m-1}{m+n}\int \sin^n ax \cos^{m-2} ax\, dx,$

                                              $m \neq -n$ (If $n = -m$, use No. 86.)

70. $\displaystyle\int \frac{dx}{b + c\sin ax} = \frac{-2}{a\sqrt{b^2 - c^2}}\tan^{-1}\left[\sqrt{\frac{b-c}{b+c}}\tan\left(\frac{\pi}{4} - \frac{ax}{2}\right)\right] + C, \qquad b^2 > c^2$

71. $\displaystyle\int \frac{dx}{b + c\sin ax} = \frac{-1}{a\sqrt{c^2 - b^2}}\ln\left|\frac{c + b\sin ax + \sqrt{c^2 - b^2}\cos ax}{b + c\sin ax}\right| + C, \qquad b^2 < c^2$

72. $\displaystyle\int \frac{dx}{1 + \sin ax} = -\frac{1}{a}\tan\left(\frac{\pi}{4} - \frac{ax}{2}\right) + C$

73. $\displaystyle\int \frac{dx}{1 - \sin ax} = \frac{1}{a} \tan\left(\frac{\pi}{4} + \frac{ax}{2}\right) + C$

74. $\displaystyle\int \frac{dx}{b + c \cos ax} = \frac{2}{a\sqrt{b^2 - c^2}} \tan^{-1}\left[\sqrt{\frac{b - c}{b + c}} \tan \frac{ax}{2}\right] + C, \qquad b^2 > c^2$

75. $\displaystyle\int \frac{dx}{b + c \cos ax} = \frac{1}{a\sqrt{c^2 - b^2}} \ln\left|\frac{c + b \cos ax + \sqrt{c^2 - b^2}\, \sin ax}{b + c \cos ax}\right| + C, \qquad b^2 < c^2$

76. $\displaystyle\int \frac{dx}{1 + \cos ax} = \frac{1}{a} \tan \frac{ax}{2} + C$

77. $\displaystyle\int \frac{dx}{1 - \cos ax} = -\frac{1}{a} \cot \frac{ax}{2} + C$

78. $\displaystyle\int x \sin ax\, dx = \frac{1}{a^2} \sin ax - \frac{x}{a} \cos ax + C$

79. $\displaystyle\int x \cos ax\, dx = \frac{1}{a^2} \cos ax + \frac{x}{a} \sin ax + C$

80. $\displaystyle\int x^n \sin ax\, dx = -\frac{x^n}{a} \cos ax + \frac{n}{a} \int x^{n-1} \cos ax\, dx$

81. $\displaystyle\int x^n \cos ax\, dx = \frac{x^n}{a} \sin ax - \frac{n}{a} \int x^{n-1} \sin ax\, dx$

82. $\displaystyle\int \tan ax\, dx = -\frac{1}{a} \ln |\cos ax| + C$

83. $\displaystyle\int \cot ax\, dx = \frac{1}{a} \ln |\sin ax| + C$

84. $\displaystyle\int \tan^2 ax\, dx = \frac{1}{a} \tan ax - x + C$

85. $\displaystyle\int \cot^2 ax\, dx = -\frac{1}{a} \cot ax - x + C$

86. $\displaystyle\int \tan^n ax\, dx = \frac{\tan^{n-1} ax}{a(n - 1)} - \int \tan^{n-2} ax\, dx, \qquad n \neq 1$

87. $\displaystyle\int \cot^n ax\, dx = -\frac{\cot^{n-1} ax}{a(n - 1)} - \int \cot^{n-2} ax\, dx, \qquad n \neq 1$

88. $\displaystyle\int \sec ax\, dx = \frac{1}{a} \ln |\sec ax + \tan ax| + C$

89. $\displaystyle\int \csc ax\, dx = -\frac{1}{a} \ln |\csc ax + \cot ax| + C$

90. $\displaystyle\int \sec^2 ax\, dx = \frac{1}{a} \tan ax + C$

91. $\displaystyle\int \csc^2 ax\, dx = -\frac{1}{a} \cot ax + C$

92. $\displaystyle\int \sec^n ax\ dx = \frac{\sec^{n-2}ax\ \tan\ ax}{a(n-1)} + \frac{n-2}{n-1}\int \sec^{n-2}ax\ dx, \qquad n \neq 1$

93. $\displaystyle\int \csc^n ax\ dx = -\frac{\csc^{n-2}ax\ \cot\ ax}{a(n-1)} + \frac{n-2}{n-1}\int \csc^{n-2}ax\ dx, \qquad n \neq 1$

94. $\displaystyle\int \sec^n ax\ \tan\ ax\ dx = \frac{\sec^n ax}{na} + C, \qquad n \neq 0$

95. $\displaystyle\int \csc^n ax\ \cot\ ax\ dx = -\frac{\csc^n ax}{na} + C, \qquad n \neq 0$

96. $\displaystyle\int \sin^{-1}ax\ dx = x\ \sin^{-1}ax + \frac{1}{a}\sqrt{1 - a^2x^2} + C$

97. $\displaystyle\int \cos^{-1}ax\ dx = x\ \cos^{-1}ax - \frac{1}{a}\sqrt{1 - a^2x^2} + C$

98. $\displaystyle\int \tan^{-1}ax\ dx = x\ \tan^{-1}ax - \frac{1}{2a}\ln(1 + a^2x^2) + C$

99. $\displaystyle\int x^n\sin^{-1}ax\ dx = \frac{x^{n+1}}{n+1}\sin^{-1}ax - \frac{a}{n+1}\int \frac{x^{n+1}\ dx}{\sqrt{1 - a^2x^2}}, \qquad n \neq -1$

100. $\displaystyle\int x^n\cos^{-1}ax\ dx = \frac{x^{n+1}}{n+1}\cos^{-1}ax + \frac{a}{n+1}\int \frac{x^{n+1}\ dx}{\sqrt{1 - a^2x^2}}, \qquad n \neq -1$

101. $\displaystyle\int x^n\tan^{-1}ax\ dx = \frac{x^{n+1}}{n+1}\tan^{-1}ax - \frac{a}{n+1}\int \frac{x^{n+1}\ dx}{1 + a^2x^2}, \qquad n \neq -1$

102. $\displaystyle\int e^{ax}\ dx = \frac{1}{a}e^{ax} + C$

103. $\displaystyle\int b^{ax}\ dx = \frac{1}{a}\frac{b^{ax}}{\ln b} + C, \qquad b > 0, \qquad b \neq 1$

104. $\displaystyle\int xe^{ax}\ dx = \frac{e^{ax}}{a^2}(ax - 1) + C$

105. $\displaystyle\int x^n e^{ax}\ dx = \frac{1}{a}x^n e^{ax} - \frac{n}{a}\int x^{n-1}e^{ax}\ dx$

106. $\displaystyle\int x^n b^{ax}\ dx = \frac{x^n b^{ax}}{a\ln b} - \frac{n}{a\ln b}\int x^{n-1}b^{ax}\ dx, \qquad b > 0, \qquad b \neq 1$

107. $\displaystyle\int e^{ax}\sin bx\ dx = \frac{e^{ax}}{a^2 + b^2}(a\sin bx - b\cos bx) + C$

108. $\displaystyle\int e^{ax}\cos bx\ dx = \frac{e^{ax}}{a^2 + b^2}(a\cos bx + b\sin bx) + C$

109. $\displaystyle\int \ln ax\ dx = x\ln ax - x + C$

110. $\displaystyle\int x^n \ln ax\ dx = \frac{x^{n+1}}{n+1}\ln ax - \frac{x^{n+1}}{(n+1)^2} + C, \qquad n \neq -1$

111. $\displaystyle\int x^{-1}\ln ax\ dx = \frac{1}{2}(\ln ax)^2 + C$ \qquad 112. $\displaystyle\int \frac{dx}{x\ln ax} = \ln|\ln ax| + C$

113. $\displaystyle\int_0^\infty x^{n-1} e^{-x}\, dx = \Gamma(n) = (n-1)!, \qquad n > 0.$

114. $\displaystyle\int_0^\infty e^{-ax^2}\, dx = \frac{1}{2}\sqrt{\frac{\pi}{a}}, \qquad a > 0$

115. $\displaystyle\int_0^{\pi/2} \sin^n x\, dx = \int_0^{\pi/2} \cos^n x\, dx = \begin{cases} \dfrac{1 \cdot 3 \cdot 5 \cdots (n-1)}{2 \cdot 4 \cdot 6 \cdots n} \cdot \dfrac{\pi}{2}, & \text{if } n \text{ is an even integer} \geq 2, \\[2ex] \dfrac{2 \cdot 4 \cdot 6 \cdots (n-1)}{3 \cdot 5 \cdot 7 \cdots n}, & \text{if } n \text{ is an odd integer} \geq 3 \end{cases}$

Greek Alphabet

Capital Letters	Name of Letter	Lower-case Letters
A	Alpha	α
B	Beta	β
Γ	Gamma	γ
Δ	Delta	δ
E	Epsilon	ϵ
Z	Zeta	ζ
H	Eta	η
Θ	Theta	θ, ϑ
I	Iota	ι
K	Kappa	κ
Λ	Lambda	λ
M	Mu	μ
N	Nu	ν
Ξ	Xi	ξ
O	Omicron	o
Π	Pi	π
P	Rho	ρ
Σ	Sigma	σ
T	Tau	τ
Υ	Upsilon	υ
Φ	Phi	ϕ, φ
X	Chi	χ
Ψ	Psi	ψ
Ω	Omega	ω

ANSWERS TO SELECTED ODD-NUMBERED PROBLEMS

CHAPTER 1

Section 1

1.

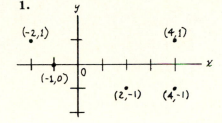

3. This is the straight line parallel to, and three units below, the x-axis.

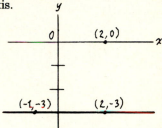

5.

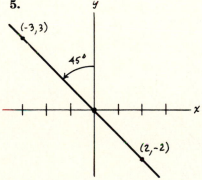

7.

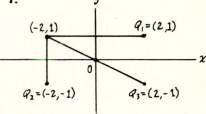

9.

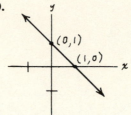

11.

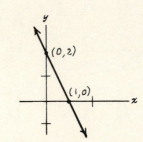

13.

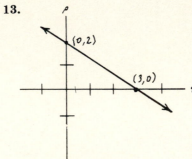

15.

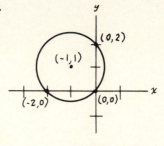

17.

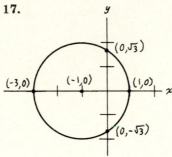

19.

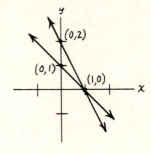

21.

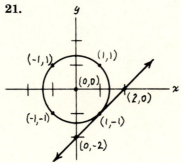

23.

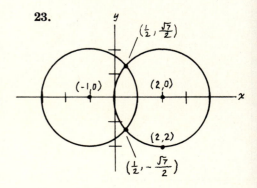

25. If $y = 3 - x$ and $x^2 + y^2 = 4$, then

$$x^2 + (3 - x)^2 = 4$$

$$\Rightarrow 2x^2 - 6x + 5 = 0$$

$$\Rightarrow x^2 - 3x + 5/2 = 0$$

$$\Rightarrow (x - 3/2)^2 + 5/2 - 9/4 = 0$$

$$\Rightarrow (x - 3/2)^2 = -1/4,$$

which is impossible.

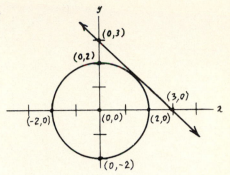

27.

x	y
± 2	2
$\pm 3/2$	1/4
± 1	-1
$\pm 1/2$	$-7/4$
0	-2

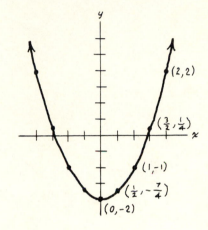

29.

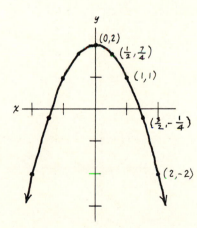

31.

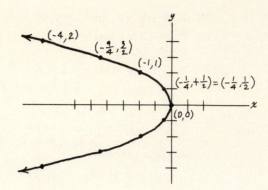

33.

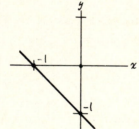

35.

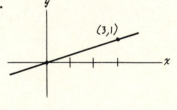

Section 2   **1.** $y = 3x - 6$; $(2, 0)$, $(1, -3)$, $(7/3, 1)$

3. $m = -1/2$, $b = 2$ **5.** $m = -2/3$, $b = 2$ **7.** $m = 1$, $b = 1$

9.

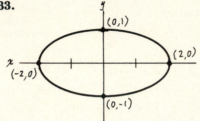

11.

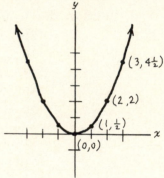

13.

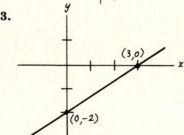

15.

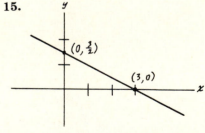

Powers and Roots

No.	Sq.	Sq. Root	Cube	Cube Root	No.	Sq.	Sq. Root	Cube	Cube Root
1	1	1.000	1	1.000	51	2,601	7.141	132,651	3.708
2	4	1.414	8	1.260	52	2,704	7.211	140,608	3.733
3	9	1.732	27	1.442	53	2,809	7.280	148,877	3.756
4	16	2.000	64	1.587	54	2,916	7.348	157,464	3.780
5	25	2.236	125	1.710	55	3,025	7.416	166,375	3.803
6	36	2.449	216	1.817	56	3,136	7.483	175,616	3.826
7	49	2.646	343	1.913	57	3,249	7.550	185,193	3.849
8	64	2.828	512	2.000	58	3,364	7.616	195,112	3.871
9	81	3.000	729	2.080	59	3,481	7.681	205,379	3.893
10	100	3.162	1,000	2.154	60	3,600	7.746	216,000	3.915
11	121	3.317	1,331	2.224	61	3,721	7.810	226,981	3.936
12	144	3.464	1,728	2.289	62	3,844	7.874	238,328	3.958
13	169	3.606	2,197	2.351	63	3,969	7.937	250,047	3.979
14	196	3.742	2,744	2.410	64	4,096	8.000	262,144	4.000
15	225	3.873	3,375	2.466	65	4,225	8.062	274,625	4.021
16	256	4.000	4,096	2.520	66	4,356	8.124	287,496	4.041
17	289	4.123	4,913	2.571	67	4,489	8.185	300,763	4.062
18	324	4.243	5,832	2.621	68	4,624	8.246	314,432	4.082
19	361	4.359	6,859	2.668	69	4,761	8.307	328,509	4.102
20	400	4.472	8,000	2.714	70	4,900	8.367	343,000	4.121
21	441	4.583	9,261	2.759	71	5,041	8.426	357,911	4.141
22	484	4.690	10,648	2.802	72	5,184	8.485	373,248	4.160
23	529	4.796	12,167	2.844	73	5,329	8.544	389,017	4.179
24	576	4.899	13,824	2.884	74	5,476	8.602	405,224	4.198
25	625	5.000	15,625	2.924	75	5,625	8.660	421,875	4.217
26	676	5.099	17,576	2.962	76	5,776	8.718	438,976	4.236
27	729	5.196	19,683	3.000	77	5,929	8.775	456,533	4.254
28	784	5.292	21,952	3.037	78	6,084	8.832	474,552	4.273
29	841	5.385	24,389	3.072	79	6,241	8.888	493,039	4.291
30	900	5.477	27,000	3.107	80	6,400	8.944	512,000	4.309
31	961	5.568	29,791	3.141	81	6,561	9.000	531,441	4.327
32	1,024	5.657	32,768	3.175	82	6,724	9.055	551,368	4.344
33	1,089	5.745	35,937	3.208	83	6,889	9.110	571,787	4.362
34	1,156	5.831	39,304	3.240	84	7,056	9.165	592,704	4.380
35	1,225	5.916	42,875	3.271	85	7,225	9.220	614,125	4.397
36	1,296	6.000	46,656	3.302	86	7,396	9.274	636,056	4.414
37	1,369	6.083	50,653	3.332	87	7,569	9.327	658,503	4.431
38	1,444	6.164	54,872	3.362	88	7,744	9.381	681,472	4.448
39	1,521	6.245	59,319	3.391	89	7,921	9.434	704,969	4.465
40	1,600	6.325	64,000	3.420	90	8,100	9.487	729,000	4.481
41	1,681	6.403	68,921	3.448	91	8,281	9.539	753,571	4.498
42	1,764	6.481	74,088	3.476	92	8,464	9.592	778,688	4.514
43	1,849	6.557	79,507	3.503	93	8,649	9.644	804,357	4.531
44	1,936	6.633	85,184	3.530	94	8,836	9.695	830,584	4.547
45	2,025	6.708	91,125	3.557	95	9,025	9.747	857,375	4.563
46	2,116	6.782	97,336	3.583	96	9,216	9.798	884,736	4.579
47	2,209	6.856	103,823	3.609	97	9,409	9.849	912,673	4.595
48	2,304	6.928	110,592	3.634	98	9,604	9.899	941,192	4.610
49	2,401	7.000	117,649	3.659	99	9,801	9.950	970,299	4.626
50	2,500	7.071	125,000	3.684	100	10,000	10.000	1,000,000	4.642

Table of Integrals

1. $\int u \, dv = uv - \int v \, du$

2. $\int a^u \, du = \dfrac{a^u}{\ln a} + C, \qquad a \neq 1, \qquad a > 0$

3. $\int \cos u \, du = \sin u + C$

4. $\int \sin u \, du = -\cos u + C$

5. $\int (ax + b)^n \, dx = \dfrac{(ax + b)^{n+1}}{a(n + 1)} + C, \qquad n \neq -1$

6. $\int (ax + b)^{-1} \, dx = \dfrac{1}{a} \ln |ax + b| + C$

7. $\int x(ax + b)^n \, dx = \dfrac{(ax + b)^{n+1}}{a^2}\left[\dfrac{ax + b}{n + 2} - \dfrac{b}{n + 1}\right] + C, \qquad n \neq -1, -2$

8. $\int x(ax + b)^{-1} \, dx = \dfrac{x}{a} - \dfrac{b}{a^2} \ln |ax + b| + C$

9. $\int x(ax + b)^{-2} \, dx = \dfrac{1}{a^2}\left[\ln |ax + b| + \dfrac{b}{ax + b}\right] + C$

10. $\int \dfrac{dx}{x(ax + b)} = \dfrac{1}{b} \ln \left|\dfrac{x}{ax + b}\right| + C$

11. $\int (\sqrt{ax + b})^n \, dx = \dfrac{2}{a}\dfrac{(\sqrt{ax + b})^{n+2}}{n + 2} + C, \qquad n \neq -2$

12. $\int \dfrac{\sqrt{ax + b}}{x} \, dx = 2\sqrt{ax + b} + b\int \dfrac{dx}{x\sqrt{ax + b}}$

13. (a) $\int \dfrac{dx}{x\sqrt{ax + b}} = \dfrac{2}{\sqrt{-b}} \tan^{-1}\sqrt{\dfrac{ax + b}{-b}} + C, \qquad \text{if} \quad b < 0$

    (b) $\int \dfrac{dx}{x\sqrt{ax + b}} = \dfrac{1}{\sqrt{b}} \ln \left|\dfrac{\sqrt{ax + b} - \sqrt{b}}{\sqrt{ax + b} + \sqrt{b}}\right| + C, \qquad \text{if} \quad b > 0$

14. $\int \dfrac{\sqrt{ax + b}}{x^2} \, dx = -\dfrac{\sqrt{ax + b}}{x} + \dfrac{a}{2}\int \dfrac{dx}{x\sqrt{ax + b}} + C$

15. $\int \dfrac{dx}{x^2\sqrt{ax + b}} = -\dfrac{\sqrt{ax + b}}{bx} - \dfrac{a}{2b}\int \dfrac{dx}{x\sqrt{ax + b}} + C$

16. $\int \dfrac{dx}{a^2 + x^2} = \dfrac{1}{a} \tan^{-1}\dfrac{x}{a} + C$

17. $\int \dfrac{dx}{(a^2 + x^2)^2} = \dfrac{x}{2a^2(a^2 + x^2)} + \dfrac{1}{2a^3} \tan^{-1}\dfrac{x}{a} + C$

17. $m = 1/2$, $y = x/2$ **19.** No slope, $x = -2$ **21.** $m = 0$, $y = 1$

23. Point–slope equation: $y - 2 = (-1/2)(x - 1)$, which is equivalent to $2y + x - 5 = 0$.

25. $3y - 2x - 1 = 0$ **27.** $y = x + 1$ **29.** $(-2, 3)$

31. $y = -x + 1$ **33.** $y = (-1/6)x + 4/3$

35. Yes; the three slopes are all 4/5.   **37.** No; the slopes are 2/3, 1/2, 3/5.

39. $(2, 1)$

41. Is the given equation the equation of some line? If so, what are its intercepts?

43. Both points are on the side of the line opposite from the origin.

Section 3   **1.** $\sqrt{5}$ **3.** 4   **5.** $2\sqrt{26}$ **7.** $\sqrt{5}$ **9.** $2\sqrt{2}$

11. $AB = 2\sqrt{2}$, $BC = 3\sqrt{5}$, $CA = \sqrt{41}$ **13.** $AB = \sqrt{26}$, $BC = 5$, $CA = \sqrt{13}$

15. $P_1P_2 = P_2P_3 = \sqrt{41}$

17. $(2, 1)$ and $(1, 3)$ have slope -2; $(2, 1)$ and $(8, 4)$ have slope $1/2$.

19. $y = 3x - 6$ **21.** $8y + 14x - 15 = 0$ **23.** $(a + 2b)/3$

25. a) Show that the midpoint makes the slope $(b_2 - b_1)/(a_2 - a_1)$ with each endpoint.

27. $3y - 4x + 1 = 0$ **29.** $12y - 5x + 59 = 0$

31. P_1P_2 has midpoint $(7/2, -1/2)$. P_1P_3 has midpoint $(0, 3/2)$. The midpoints have slope $-4/7$. So does P_2P_3.

33. If the two points are (x_1, y_1) and (x_2, y_2), then the locus condition is

$$\sqrt{(x - x_1)^2 + (y - y_1)^2} = \sqrt{(x - x_2)^2 + (y - y_2)^2}.$$

Square; expand the squares, and simplify.

35. 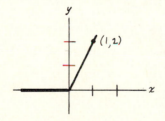   **37.**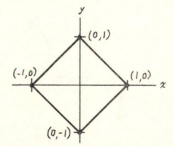

39. Choose the x-axis along one side of the triangle and the y-axis through the opposite vertex. The triangle then has vertices $(a, 0)$, $(b, 0)$, $(0, c)$. The perpendicular bisectors of the sides meet at the point

$$\left(\frac{a + b}{2}, \frac{c}{2} + \frac{ab}{2c}\right).$$

41. $4/\sqrt{5}$

Section 4

1. $(x - 2)^2 + (y - 1)^2 = 9$; $x^2 + y^2 - 4x - 2y - 4 = 0$

3. $(x - 2)^2 + (y + 3)^2 = 1$; $x^2 + y^2 - 4x + 6y + 12 = 0$

5. $(x + 2)^2 + (y - 4)^2 = 25$; $x^2 + y^2 + 4x - 8y - 5 = 0$

7. $x^2 + y^2 = 16$; $x^2 + y^2 - 16 = 0$ 9. Center $(-3, 4)$; radius 5

11. Center $(-3/2, 5/2)$; radius 3   13. $c(-1, 0)$; $r = 1$

15. $C(0, -1/2)$; $r = \sqrt{5}/2$ 17. No locus

19. $(x - 5)^2 + (y + 2)^2 = 85$ 21. $(x - 5/2)^2 + y^2 = 25/4$

23. $(x - 1/2)^2 + (y - 3/2)^2 = 10/4$ 25. The circle $(x - 4)^2 + y^2 = 4$.

27. The line through the origin and the point (a, b) has the equation $y = bx/a$. Solving this equation simultaneously with $x^2 + y^2 = 1$ (say by substituting $y = bx/a$ into the quadratic equation) gives $x = \pm a/\sqrt{a^2 + b^2}$.

29. If a point lies inside a circle, then its distance from the center is less than the radius of the circle. Write this inequality out for the given point and circle; square, and simplify. The result should be the required inequality.

Section 5

1. Symmetric in both axes and the origin. Reflection image in $y = x$: $9y^2 + 16x^2 = 1$.

3. The origin. Image in the x-axis: $2x^3 - 3y - 2y^3 = 0$; in the y-axis: the same; in $y = x$: $2x^3 + 3x + 2y^3 = 0$.

5. The y-axis. Image in the x-axis: $2x^2 - y = 0$; in the origin: the same; in $y = x$: $2y^2 + x = 0$.

7. Both axes, the origin and the line $y = x$.

9. The origin and $y = x$. Image in the x-axis: $x^2 - xy + y^2 = 1$; in the y-axis: the same.

11. The x-axis. Image in the y-axis: $x(1 - y^2) = -1$; in the origin: the same; in $y = x$: $y(1 - x^2) = 1$.

13. The origin. Image in the x-axis: $xy^2 + y = x^3$; in the y-axis: the same; in $y = x$: $yx^2 - x = y^3$.

15. Reflecting over the line $y = -x$ can be accomplished indirectly by reflecting first over the y-axis, then over the line $y = x$, then over the y-axis again. This takes the point (a, b) to the point $(-b, -a)$ in three stages:

$$(a, b) \to (-a, b) \to (b, -a) \to (-b, -a).$$

Section 6   1.

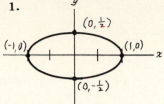

3.

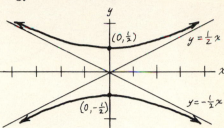

5.

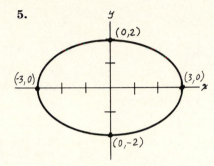

7.

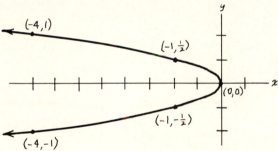

9.

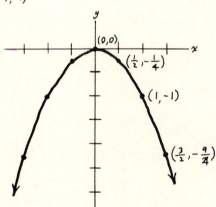

11.

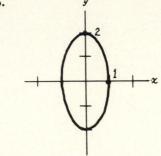

13.

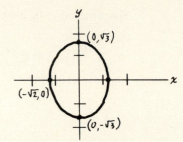

15.

17.

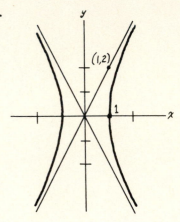

19.

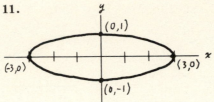

21.

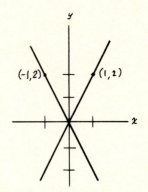

Section 7　**1.** a) $(2, -1)$　　b) $X = x - 1;\ Y = y - 4$

3. a) $(3, 9)$　　b) $X = x - 2;\ Y = y + 4$

5. a) $(-2, -1)$　　b) $X = x - 3;\ Y = y - 2$

7. a) $(2, 1)$　　b) $X = x + 3;\ Y = y + 2$

9. New origin at $(-2, 3)$; $X^2 + Y^2 = 18$.

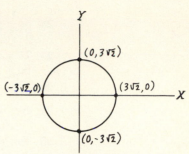

11. New origin at $(0, 0)$
(See answer to Problem 11 of Section 6.)

13. New origin at $(-2, 1)$; $\dfrac{X^2}{3^2} + \dfrac{Y^2}{2^2} = 1$

(See answer to Problem 5 of Section 6.)

15. New origin at $(1, 1)$; $X^2 - \dfrac{Y^2}{2^2} = -1$

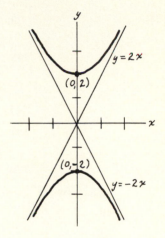

17. New origin at $(1, 2)$; $\dfrac{X^2}{2^2} + Y^2 = 1$

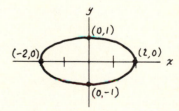

19. $\bar{X}^2 + \bar{Y}^2 + 4\bar{X} - 2\bar{Y} + 1 = 0$ **21.** $\overline{XY} = 0$

Section 8   **1.** $[0, \infty)$ **3.** $(-\infty, 2) \cup (2, \infty)$; $\cup$ is the symbol for union
5. $x \neq \pm 2$; $(-\infty, 2) \cup (-2, 2) \cup (2, \infty)$ **7.** $x \neq 2, 3$
9. $(-\infty, -4] \cup [4, \infty)$ **11.** -1 **13.** $a^3 + 6a^2 + 11a + 6$

15. $u - 2$ **17.** $H(1/t) = \dfrac{t - 1}{t + 1} = -H(t)$
19. $H(H(t)) = t$ **21.** $1/x - 1/y = (x - y)(-1/xy)$

23. $\dfrac{1}{x^2 + 1} - \dfrac{1}{y^2 + 1} = (x - y)(x + y)(-1/(x^2 + 1)(y^2 + 1))$

25. $D = -1 \le x \le 1.$ For $0 \le x \le 1,$ $y(y(x)) = \sqrt{1 - (\sqrt{1 - x^2})^2} = \sqrt{x^2} = |x| = x$

27. $g(y) = y^3$

29. No; when $x = 0, y = \pm 1/2.$

31. No; when $x = 0, y = \pm 1/2.$

33. $f(x) = x - 1$

35. No; when $x = -1, y = \pm 1.$

37. $f(x) = \dfrac{x^2 - x}{x + 1}; x \neq -1, x \neq 0$

39. $f(s) = \dfrac{s^2 \sqrt{3}}{4}$

41. $1 + x + x^2 = \underbrace{1 + x^2}_{\text{even}} + \underbrace{x}_{\text{odd}}$

43. $g_1,$ g_2 odd functions implies $g_1(-x)g_2(-x) = (-g_1(x))(-g_2(x)) = g_1(x)g_2(x).$ So $g_1 g_2$ is an even function; even + even = even; even × odd = odd; even × even = even.

45. $f(x) = 1/(1 - x)^2$ **47.** $h(x) = \sqrt{x}$ **49.** $g(x) = x + 1/x$

51. $g(x) = \sqrt{x - 1}, h(x) = x^3,$ or $g(x) = \sqrt{x}, h(x) = x^3 - 1$

53. $g(x) = x\sqrt{x^2 + 1}, h(x) = x - 1$

Section 9 **1.** Parabola

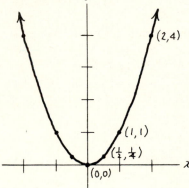

3. The graph is a semicircle, because $(f(x))^2 + x^2 = 1$ and because $f(x) \ge 0.$

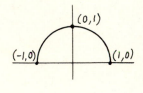

5. 0, 1

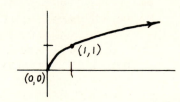

7.

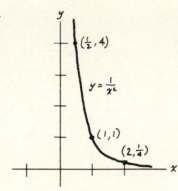

9. Symmetric about the origin.

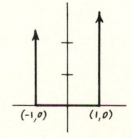

11.

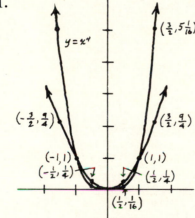

For large n, the graph approaches:

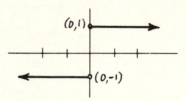

13. $p(x)$ behaves like x^3 for large $|x|$, so for large positive (negative) x, $p(x)$ is large positive (negative). Therefore the graph of $p(x)$ must cross the x-axis at least once. If the graph of $p(x)$ crosses or touches the x-axis at a, then $p(a) = 0$, so $(x - a)$ is a factor of $p(x)$. Since $p(x)$ is third-degree, there can be at most three such a's. In all, the graph crosses the x-axis at least once and crosses or touches at most three times, and $p(x)$ is positive (negative) for large positive (negative) values of x. The only four

possibilities are

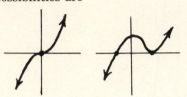
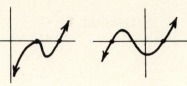

15.

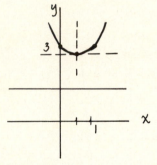

17.

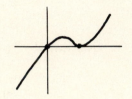

19. $y = \frac{11}{4} + (\frac{1}{2} - x)^2$: graph of $Y = X^2$ when new origin is placed at $(\frac{1}{2}, \frac{11}{4})$.

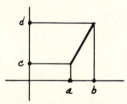

21. $y = (x + 1)^3$ so there is a triple root at $x = -1$. Graph looks just like x^3 moved one unit to the left. That is, graph of $Y = X^3$ when new origin is placed at $(-1, 0)$.

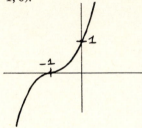

25. $(y - c)/(x - a) = (d - c)/(b - a)$. Solve for y.

29. At $x = -4$

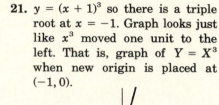

31. If we take a new origin at $x = 1$, then the new coordinate of x is $X = x - 1$ (Section 7). The given condition turns into

$$f(X) = f(-X)$$

$$(f(X) = f(x - 1) = f(2 - (x - 1)) = f(1 - x) = f(-X)).$$

So f is even about the new origin.

CHAPTER 2

Section 1

1. $\sqrt{2}$ **3.** 2   **5.** 1   **7.** 4   **9.** 1/4

11. 0.0481   **13.** 12   **15.** 2/3   **17.** 3/4   **19.** 1/2

21. 1/4   **23.** −3   **25.** $\sqrt{3}/36$ **27.** 1/7   **29.** 1/5

31. No limit   **33.** No limit   **35.** −2   **37.** $m = 2$

39. $m = -2$, $y = -2x - 1$ **41.** $m = -1$, $y = -x + 2$

43. $m = 1$, $y = x$

45. $m = 1 - 2a$, $y = (1 - 2a)x + a^2$. The tangent line is horizontal ($m = 0$) when $a = 1/2$; its equation is then $y = 1/4$.

Section 2

1. 3, 5, and x are continuous by L0, so by the sum and product rules of L1′, $3x^2 + 5x$ is continuous.

In Problems 3, 5, and 7, L1′a refers to the first operation listed in L1′, L1′b refers to the second, etc.

3. By L0 and L1′d, $x^{1/2}$ is continuous ($x \geq 0$); so by L1′c, $1/\sqrt{x}$ is continuous for $x > 0$.

5. $\sqrt{x}$ continuous by L0 and L1′d

   $\sqrt{x} + 1/\sqrt{x}$ continuous by L1′c,a

7. 1, $\sqrt{x}$, x continuous by L0, L1′d

          x^2 continuous by L1′b

     $1 + x^2$ continuous by L1′a

    $\dfrac{1}{1 + x^2}$ continuous by L1′c

    $\dfrac{\sqrt{x}}{1 + x^2}$ continuous by L1′b   (product rule)

9. $f(x) = x^2$ and $g(x) = x^{1/2}$ are continuous as above, so $\sqrt{x^2} = g(f(x))$ is continuous by L3.

11. $g(x) = \sqrt{x^2} = |x|$ is continuous (by Problem 5); $f(x) = x^{1/3}$ is continuous by Example 6, so $g(f(x)) = |x^{1/3}|$ is continuous by L3.

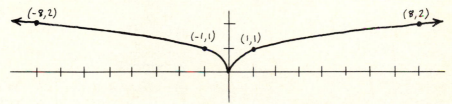

17. a) $\dfrac{f(3 + h) - f(3)}{h} = -4 - h$; -0.41, -0.0401, -0.004001, -0.00040001

    b) $\dfrac{f(2 + h) - f(2)}{h} = -2 - h$; -0.21, -0.0201, -0.002001, -0.00020001

19.
$$|x - 2| \le d < 1 \Rightarrow |x - 2| < 1 \Rightarrow -1 < x - 2 < 1$$
$$\Rightarrow 1 < x < 3 \Rightarrow |x + 2| < 5;$$
$$|x^2 - 4| = |(x - 2)(x + 2)| = |x - 2| \, |x + 2| < 5d$$

21. $f(1) = 1, \ f(x) = x$

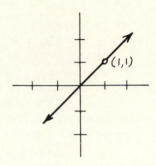

25. The assumptions translate into $0 \le g(x) - f(x) \le h(x) - f(x)$, where $h - f$ has the limit 0 at $x = a$. Therefore,

$$\lim_{x \to a} (g(x) - f(x)) = 0,$$

by L2′, and

$$\lim_{x \to a} g(x) = \lim_{x \to a} (g(x) - f(x)) + \lim_{x \to a} f(x)$$
$$= \lim_{x \to a} f(x)$$

by L1.

27. Since $\sqrt{1 + x^2}$ and $1 + |x|$ are both continuous and equal to 1 at $x = 0$, it follows from the above problem that $f(x) \to 1$ as $x \to 0$.

29. For the given function f,

$$f(f(x)) = f(0) = 1 \qquad \text{for } x \ne 0,$$

so $f(f(x)) \to 1$ as $x \to 0$. However, $f(x) \to 0$ as $x \to 0$, so the conjectured rule would give $f(f(x)) \to 0$ as $x \to 0$, which is false.

Section 3   **1.** $\lim_{x \to 0^-} \dfrac{|x|}{x} = -1; \ \lim_{x \to 0^+} \dfrac{|x|}{x} = +1$ **3.** -1

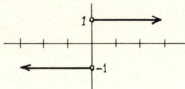

5. 1   **7.** -2

9. $f(x) = [x]$ is continuous from the right everywhere. It is continuous from the left everywhere except at the integer points.

11. $\lim_{x \to 4^-}[x] + p(x) = 7$; $\lim_{x \to 4^+}[x] + p(x) = 9$; but $[4] + p(4) = 8$. The same holds at $x = n$ for any positive integer n.

13. $3t + 5$ is continuous, so $\lim_{t \to -1^-} h(t) = 3(-1) + 5 = 2$.
$t^2 + 1$ is continuous, so $\lim_{t \to -1^+} h(t) = h(-1) = 2$.

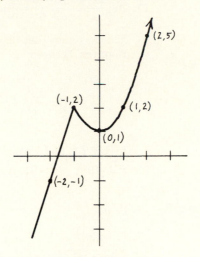

15. Here is one example. Draw a different one.

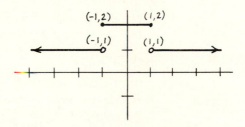

17. Note that $g(x) = f(x)$ for $x \ge 0$, so g and f have the same limit at 0^+ by the one-sided version of L2. Moreover, g has this same limit at 0^-, since $g(x) = g(-x)$. Thus g has a limit at $0 \Leftrightarrow f$ has a limit at 0^+.

19. If $g(x) = f^3(x) + 4f(x) + 1 = y^3 - 4y + 1$, then

$$\lim_{x \to a^-} g(x) = \lim_{x \to a^-} (y^3 - 4y + 1) = (-2)^3 - 4(-2) + 1 = 1,$$

$$\lim_{x \to a^+} g(x) = \lim_{x \to a^+} (y^3 - 4y + 1) = 2^3 - 4(2) + 1 = 1,$$

$$g(a) = (y^3 - 4y + 1)_{y=0} = 1.$$

Thus $\lim_{x \to a^-} g(x) = g(a) = \lim_{x \to a^+} g(x) = 1$, and g is continuous at $x = a$.

Section 4

1. -7 **3.** 1 **5.** $-2/3$ **7.** 1

9. $1/2$ **11.** $2/3$ **13.** $-\infty$ **15.** $+\infty$

17. $\displaystyle \lim_{x \to 3} \frac{2x^2 - 5x - 3}{x^2 - x - 6} = \lim_{x \to 3} \frac{(2x + 1)(x - 3)}{(x + 2)(x - 3)} \underset{L2}{=} \lim_{x \to 3} \frac{2x + 1}{x + 2} \underset{L0}{=} \frac{7}{5}.$

This assumes that $(2x + 1)/(x + 2)$ is continuous at $x = 3$, which in turn requires Theorem 1.

$$\lim_{x \to \infty} \frac{2x^2 - 5x - 3}{x^2 - x - 6} = \lim_{x \to \infty} \frac{2 - \dfrac{5}{x} - \dfrac{3}{x^2}}{1 - \dfrac{1}{x} - \dfrac{6}{x^2}} = \frac{2 - 0 - 0}{1 - 0 - 0} = 2,$$

by L2, L1, and Example 3.

19. $x > 10$ **21.** $x > d^{-1/3}$ **23.** $0 < x < 10^{-6}$ **25.** $M = d^{-1/2}$

Section 5

1. $[0, \infty)$ **3.** $[0, \infty)$ **5.** $(-\infty, 0) \cup \left[\dfrac{1}{9}, \infty\right)$

7. $(-\infty, 36]$ **9.** $(-\infty, 2]$ **11.** $(-\infty, -2] \cup [2, \infty)$

13. $f(1) = -1$, $f(2) = 12$, and $-1 < 0 < 12$

15. $f(0) = 0 < 2 < 2 + 4^{1/3} = f(4)$

17. If $f(x) = x^3 - 3(1 + \sqrt{x})$, then $f(0) = -3$, $f(2) = 5 - 3\sqrt{2} > 0$.

CHAPTER 3

Section 1

1. $\dfrac{f(x + h) - f(x)}{h} = \dfrac{(x + h)^2 - x^2}{h} = 2x + h.$

Now take the limit as $h \to 0$.

3. $\dfrac{f(x + h) - f(x)}{h} = \dfrac{1}{h}\left[\dfrac{1}{(x + h)^2} - \dfrac{1}{x^2}\right] = -\dfrac{2x + h}{(x + h)^2 x^2}.$

As $h \to 0$, this approaches $-2x/x^4 = -2/x^3$.

5. $\dfrac{f(s + h) - f(s)}{h} = \dfrac{-3(s + h) - (-3s)}{h} = -3$

7. Your calculation should show that the difference quotient for the linear function $mx + b$ has the constant value m, regardless of the values of x and r.

9. If $f(x) = x^\alpha$, then $f'(x) = \alpha x^{\alpha - 1}.$ **11.** $2x$

13. $1/\sqrt{2x}$ **15.** $-1/(y-1)^2$ **17.** $1 - 1/x^2$

19. $x/\sqrt{x^2+1}$ **21.** $-1/(2(x+1)^{3/2})$ **23.** $(1-x^2)/(1+x^2)^2$

Section 2   **1.** $7x^6$, $42x^5$ **3.** $3x^2 + 6x$, $6x + 6$ **5.** m, 0

7. $16x^3 - 2x$, $48x^2 - 2$ **9.** $2x - 1$, 2 **11.** 0

13. $\dfrac{d^{m+1}}{dx^{m+1}}\,x^{m+1} = \dfrac{d^m}{dx^m}\left(\dfrac{d}{dx}\,x^{m+1}\right)$

Compute the inner derivative and then use the assumed formula.

The formula holds for $n = 1$. Since it holds for $n = 1$ it holds for $n = 2$. Since it holds for $n = 2$ it holds for $n = 3$. So?

15. $1 - x^{-2}$ **17.** $1 - 2x^{-3}$

23. $dy/dx = 0$ at $x = 1$ **25.** $(3, -8)$

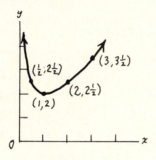

27. $y = -\frac{1}{9}x + 20\frac{1}{3}$ **29.** $(-b/2a, c - b^2/4a)$

31. Your tangent-line equation should be $y - y_0 = 2x_0(x - x_0)$, or an equivalent form using $y_0 = x_0^2$.

33. $y = 6x - 9$, $y = 2x - 1$

37. The problem is to show that

$$\lim_{r \to 0} \frac{f(r) - f(0)}{r - 0} = g(0).$$

This should follow immediately upon setting $f(r) = rg(r)$ in the difference quotient, and simplifying.

39. Show first that the difference quotients for g, f, and h must satisfy a similar inequality, and then apply the squeeze limit law.

Section 3   **1.** $\dfrac{\Delta y}{\Delta x} = 2x + \Delta x$ **3.** $\dfrac{\Delta y}{\Delta x} = \dfrac{1}{\sqrt{x + \Delta x} + \sqrt{x}}$

5. $\dfrac{\Delta y}{\Delta x} = 1 + 2x + \Delta x$ **7.** $\dfrac{\Delta y}{\Delta x} = \dfrac{-2x - \Delta x}{[1 + (x + \Delta x)^2][1 + x^2]}$

9. $\Delta y = 3x^2 \Delta x + 3x(\Delta x)^2 + (\Delta x)^3$, which can be thought of as a sum of seven terms. Your figure should show each of these seven terms as the volume of a rectangular solid.

11. $v = 3t^2 - 4t + 1$; $t = \frac{1}{3}$, 1. Forward $(-\infty, \frac{1}{3})$, $(1, \infty)$; backward $(\frac{1}{3}, 1)$.

13. 144 ft   15. 3/4 sec; 16 ft/sec (up or down)

17. a) -11.73 b) 1.8   c) -10.73 19. -80 ft/sec; $t = 1\frac{1}{4}$ sec

21. Write $\Delta s = s - s_0$, $\Delta t = t - t_0$ in the increment equation, and solve for s.

Section 4   1. a) $4\pi r^2 \, \text{unit}^3/\text{unit}$

3. a) $2\pi r \, \text{unit}^2/\text{unit}$ b) $10\pi \, \text{in.}^2/\text{in.}$ c) $20\pi \, \text{cm}^2/\text{cm}$

5. $(75\pi/4)$ cubic inches per year; $(1875\pi/4)$ cubic inches per year.

7. $\dfrac{1}{l_0} \cdot \dfrac{dl_t}{dt} = \alpha + 2\beta t$.

Here β is small compared to α, so β can be ignored if t varies less than 20°. (See Example 2.)

13. $dA/dV = 4V^{-1/3} \, \text{unit}^2/\text{unit}^3$

CHAPTER 4   1. $(x - 1)^2(x + 2)^3(7x + 2)$ 3. $3t^5(7t + 12)$

5. $5x^4 + 36x^3 + 33x^2 - 70x + 13$ 7. $-4(1 + 2v)^3/v^5$

Section 1   9. 1

11. $y' = y\left[\dfrac{3}{x} + \dfrac{8}{2x - 1} + \dfrac{2}{x + 2}\right]$

$\qquad = x^2(2x - 1)^3(x + 2)[18x^2 + 23x - 6]$

13. $1/(1 + 4x)^2$ 15. $6x/(x^2 + 2)^2$ 17. $2(1 - 3t^2)/(1 + t^2)^3$

19. $(6v + 10)/(2 - 3v)^3$ 21. $\dfrac{(x^2 - 2x - 2)}{(1 + x + x^2)^2}$ 23. $2x/(1 - x^2)^2$

25. $-4(1 - x)/(1 + x)^3$ 27. $\dfrac{-2f'(x)}{(1 + f(x))^2}$ 29. $\dfrac{2(gf' - fg')}{(f + g)^2}$

31. By Problem 30,

$$\frac{y'}{y} = \frac{[u_1 \cdots u_m]'}{u_1 \cdots u_m} - \frac{[v_1 \cdots v_n]'}{v_1 \cdots v_n}.$$

Now expand each term on the right by the general form of the product law.

33. $y\left[\dfrac{-2}{1 - 2x} - \dfrac{1}{x + 3} - \dfrac{3}{3x + 1}\right] = \dfrac{6x^2 - 6x - 10}{(x + 3)^2(3x + 1)^2}$

35. $2xy\left[\dfrac{1}{x^2} + \dfrac{1}{x^2 + 2} - \dfrac{1}{x^2 - 1} - \dfrac{1}{x^2 + 3}\right] = \dfrac{-6(x^2 + 1)}{(x^2 - 1)^2(x^2 + 3)^2}$

37. $u\left[\dfrac{2v}{v^2 + 1} - \dfrac{10}{v^2 + v} + \dfrac{3}{1 - 3v}\right]$

39. $y\left[\dfrac{1}{x} + \dfrac{4}{(x - x^2)} + \dfrac{5}{1 - x} - \dfrac{6}{3x - 4}\right]$

41. $s\left[\dfrac{3t^2}{1 + t^3} - \dfrac{2t}{1 + t^2} - \dfrac{4t^3}{1 + t^4}\right] = \dfrac{-3t^8 - t^6 - 6t^5 + t^4 - 4t^3 + 3t^2 - 2t}{(1 + t^2)^2(1 + t^4)^2}$

43. $y\left[\dfrac{10x}{1 + x^2} + \dfrac{10x^4 + 8x^3 + 6x^2 + 4x}{x^5 + x^4 + x^3 + x^2 + 1} - \dfrac{20}{1 + x}\right]$

47. $2(3x^2 - 1)/(1 + x^2)^3$

49. $\left(\dfrac{x^n + 1}{x^n}\right)' = \left(1 + \dfrac{1}{x^n}\right)' = \dfrac{-n}{x^{n+1}}$ (if $n \neq 0$)

51. $\dfrac{\Delta u}{u} + \dfrac{\Delta v}{v} + \dfrac{\Delta w}{w} + \dfrac{(\Delta u)(\Delta v)}{uv} + \dfrac{(\Delta u)(\Delta w)}{uw} + \dfrac{(\Delta v)(\Delta w)}{vw} + \dfrac{(\Delta u)(\Delta v)(\Delta w)}{uvw}$

53.

$$\lim_{\Delta x \to 0} \frac{\Delta y}{\Delta x} = \lim_{\Delta x \to 0} \frac{v\dfrac{\Delta u}{\Delta x} - u\dfrac{\Delta v}{\Delta x}}{v(v + \Delta v)} \text{ (algebra)}$$

$$= \frac{\displaystyle\lim_{\Delta x \to 0}\left[v\dfrac{\Delta u}{\Delta x} - u\dfrac{\Delta v}{\Delta x}\right]}{\displaystyle\lim_{\Delta x \to 0}(v(v + \Delta v))} \quad \text{L?}$$

$$= \frac{\left[\displaystyle\lim_{\Delta x \to 0} v\dfrac{\Delta u}{\Delta x} - \lim_{\Delta x \to 0} u\dfrac{\Delta v}{\Delta x}\right]}{\displaystyle\lim_{\Delta x \to 0}(v(v + \Delta v))} \quad \text{L?}$$

$$= \frac{\left[\left(\displaystyle\lim_{\Delta x \to 0} v\right)\left(\lim_{\Delta x \to 0}\dfrac{\Delta u}{\Delta x}\right) - \left(\lim_{\Delta x \to 0} u\right)\left(\lim_{\Delta x \to 0}\dfrac{\Delta v}{\Delta x}\right)\right]}{\left(\displaystyle\lim_{\Delta x \to 0} v\right)\left(\lim_{\Delta x \to 0}(v + \Delta v)\right)} \quad \text{L?}$$

$$= \frac{(vu' - uv')}{v\displaystyle\lim_{\Delta x \to 0}(v + \Delta v)} \quad \begin{array}{l}\text{(By Theorem 4, Chapter 3,}\\ \lim_{\Delta x \to 0} v + \Delta v = v.)\end{array}$$

$$= \frac{(vu' - uv')}{v^2}$$

55. At $x = (mb + na)/(m + n)$. Show that this lies between a and b.

Section 2

1. $-y/x = 5/2x^2$ **3.** $-x/y = -x/\sqrt{16 - x^2}$

5. $(-3x - 1)/y = (-3x - 1)/\sqrt{10 - 2x - 3x^2}$

7. $2x/y(x^2 + 1)^2 = 2x/(x^2 - 1)^{1/2}(x^2 + 1)^{3/2}$

9. $-\sqrt{x/y} = -x^{1/2}/(2 - x^{3/2})^{1/3}$

11. $(y^2 - 2xy - 2x)/(x^2 - 2xy + 2y)$

13. $-(3y^2x^2 + 5x^4)/(2yx^3 + 5y^4)$

15. $36x^5y^{7/6}/(2y^{1/2} + 3y^{2/3})$

17. $-\dfrac{1 - uy}{1 - ux}$, where $u = 4(xy)^3 + \dfrac{1}{3}(xy)^{-2/3}$

19. $-25/16y^3$

21. $y = (-1/2)x + 3/2$

23. Slope at $(x_0, y_0) = m = (-x_0b^2)/(y_0a^2)$. Solve for k in $y_0 = mx_0 + k$, and recall that $(x_0^2/a^2) + (y_0^2/b^2) = 1$ to reduce $y = mx + k$ to

$$(xx_0/a^2) + (yy_0/b^2) = 1.$$

25. $y = (28)^{1/3}$; $y = 3$

27. $-1/2\sqrt{1 - x}$

29. $(1 + 2x^2)/\sqrt{1 + x^2}$

31. $-\dfrac{5}{3}(5x + 1)^{-4/3}$

33. $4(1 + x)/3(2x + x^2)^{1/3}$

35. $(5x^2 - 3)/2\sqrt{x - (1/x)}$

37. $y\dfrac{(5 + x)}{6(1 - x^2)}$

39. $(1 - 5u^4)/(1 + 5u^4)^{3/2}$

41. $y\left[\dfrac{1}{x} - \dfrac{1}{1 + 2x} - \dfrac{6x}{2 + 3x^2}\right] = \dfrac{2 + 2x - 3x^2 - 9x^3}{(1 + 2x)^{3/2}(2 + 3x^2)^2}$

43. $y\left[\dfrac{x - 4 - 1/2x^2}{x^2 - 8x + 1/x} + \dfrac{63x^{1/2}/8 - 9x^2}{7x^{3/2} - 4x^3 + 1}\right]$

45. $\dfrac{y}{6}\left[\dfrac{1}{(1 + x^{1/2})^{2/3} + x + x^{1/2}} - \dfrac{1}{(1 + x^{1/3})^{1/2} + x + x^{2/3}}\right]$

Section 3

1. $\dfrac{d}{dx}\sec x = \dfrac{d}{dx}\left(\dfrac{1}{\cos x}\right) = \cdots$

3. $\dfrac{d}{dx}\csc x = \dfrac{d}{dx}\left(\dfrac{1}{\sin x}\right) = \cdots$

5. $(\cos x - 2x \sin x)/2\sqrt{x}$

7. $\sin 2x$

9. 0

11. $1/(1 + \cos x) = \csc x \tan\dfrac{x}{2} = \dfrac{1}{2}\sec^2\dfrac{x}{2}$

13. $2 \tan x \sec^2 x$

15. $-\sin x/2\sqrt{1 + \cos x}$

17. $3 \sin^2 x \cos^2 x(\cos^2 x - \sin^2 x)$

19. 0

21. $\sin^{(a-1)}x \cos^{(b-1)}x[a \cos^2 x - b \sin^2 x]$

23. $\cos^2 x$

25. $\dfrac{-2}{(\sin x - \cos x)^2}$

27. $\dfrac{\sec x}{(1 + x^2)^2}[(1 + x^2)\tan x - 2x]$

29. $\cos 2x$

31. $\cos^5 x$

33. $\sec^6 x$

35. $\dfrac{x \cos x - \sin x}{2x^2\sqrt{1 + (\sin x)/x}}$

37. $\dfrac{\sin x - 3x \cos x + 2x}{3x^{2/3}(\sin x - x)^2}$

39. $(k^2 - 1)\sin^3 x \cos^{k-1} x$ **41.** $\dfrac{x^2}{(x \cos x - \sin x)^2}$

43. $\dfrac{1 - kv \sec v}{(\tan v + \sec v)^k}$

45. For $t > 0$, $0 < \sin t < t$, so $\sin^2 x < x^2$ for $x \neq 0$. Manipulate this into the desired inequality.

51. $\cos h = \cos\left(\dfrac{h}{2} + \dfrac{h}{2}\right) = 1 - 2\sin^2\dfrac{h}{2}$.

$$\frac{1 - \cos h}{h} = \frac{2}{h}\sin^2\frac{h}{2} = \frac{\sin(h/2)}{h/2}\sin(h/2),$$

which $\to 1 \cdot 0 = 0$ as $h \to 0$.

Section 4   **1.** $3\cos 3x$ **3.** $4x\sec^2(1 + 2x^2)$ **5.** $\dfrac{2}{9}\dfrac{(2 + 5x^4)}{(2x + x^5)^{7/9}}$

7. $a\cos(ax + b)$ **9.** $ab(\cos b\theta + \cos a\theta)$ **11.** $\dfrac{\cos\sqrt{x}}{2\sqrt{x}}$

13. $-\dfrac{1}{2x^2}\sec\dfrac{1}{2x}\tan\dfrac{1}{2x}$ **15.** $\dfrac{2 + 3x\sin 4x + 4x^2\cos 4x}{2\sqrt{1 + x\sin 4x}}$

17. $-\dfrac{2}{x^3}\sec^2\dfrac{1}{x^2}$ **19.** $-(\sec^2\theta)[\sin(\tan\theta)]$ **21.** $\dfrac{x\sin\sqrt{1 - x^2}}{\sqrt{1 - x^2}}$

23. $(ax^2 + bx + c)^{k-1}[(2k + 1)2ax(ax + b) + (kb^2 + 2ac)]$

25. $\dfrac{y}{2}\left[\dfrac{mx^{m-1}}{1 + x^m} + \dfrac{nx^{n-1}}{1 - x^n}\right] = \dfrac{mx^{m-1} + nx^{n-1} + (n - m)x^{m+n-1}}{(1 + x^m)^{1/2}(1 - x^n)^{3/2}}$

27. $(\cos x)[\cos(\sin x)]$

29. $y\left[\dfrac{1}{x} + \dfrac{1}{1 + 2x} - \dfrac{2x}{1 + x^2}\right] = \dfrac{1 + 3x - x^2 - x^3}{\sqrt{1 + 2x}(1 + x^2)^2}$

31. $(\sec x^2)[1 + 2x^2\tan x^2]$ **33.** $\sqrt{1 + \left(\dfrac{a}{x}\right)^{2/3}}$

35. $\dfrac{y}{3}\left[\dfrac{\cos x}{\sin x} + \dfrac{\sin x}{\cos^{2/3}x + \cos x}\right] = \dfrac{1}{3}\dfrac{\cos^{5/3}x + 1}{\sin^{2/3}x\cos^{2/3}x(1 + \cos^{1/3}x)^2}$

37. $y\left[\dfrac{1}{2x} + \dfrac{\cot\sqrt{x}}{2\sqrt{x}} + \tan x\right]$ **39.** $u[\cot v - 2v\tan v^2 + 3v^2\cot v^3]$

41. $[x^n g'(x) + nx^{n-1}g(x)]f'(x^n g(x))$ **43.** $\dfrac{-g(x)g'(x)}{\sqrt{1 - g^2(x)}}$

45. $6x\sin^2(x^2)\cos(x^2)$

47. $-6x^2\cos(x^3)\sin(x^3) = -3x^2\sin(2x^3)$

49. $2\sec\left(\dfrac{2}{x}\right)\left[x - \tan\left(\dfrac{2}{x}\right)\right]$

51. $(\sec x)[\cos(\tan x)] - (\sin x)[\sin(\tan x)]$

53. $\dfrac{\sin\sqrt{t}\cos\sqrt{t}}{2\sqrt{t}\sqrt{1 + \sin^2\sqrt{t}}}$

55. $\dfrac{yx}{(1 + x^2)^2}\tan\left(\dfrac{1}{1 + x^2}\right)$

57. $\dfrac{2a\sin ax\cos ax(\sin ax - \cos ax)}{(\sin^3 ax + \cos^3 ax)^{1/3}}$

59. $\dfrac{-y}{2x^2}\left(\cos\dfrac{4}{x}\right)\left[\tan\left(\sin\dfrac{4}{x}\right)\right]$

61. $\dfrac{x}{4\sqrt{1 + x^2}\,\sqrt{1 + \sqrt{1 + x^2}}\sqrt{1 + \sqrt{1 + \sqrt{1 + x^2}}}}$

63. $-\csc y$ **65.** $y/(\cos y - x)$ **67.** $\cos x/(3y^2 + 1)$

69. By the chain rule, $(\cos g(x))g'(x) = 1$. Since $\cos = \sqrt{1 - \sin^2}$ and $\sin g(x) = x$, we have

$$g'(x) = \frac{1}{\cos g(x)} = \frac{1}{\sqrt{1 - \sin^2 g(x)}} = \frac{1}{\sqrt{1 - x^2}}.$$

71. Let f be even, and set $h(x) = -x$. Then

$$\frac{d}{dx}f(x) = \frac{d}{dx}f(-x) = \frac{d}{dx}f(h(x))$$

$$= f'(h(x))h'(x) = f'(-x)(-1).$$

So $f'(x) = -f'(-x)$: f' is odd. Work out what happens when f is odd in a similar way.

Section 5

1. $3x^2\,dx + 2\,dx$

3. $2(2x + 1)^2(5x^2 + x + 6)\,dx$

5. $-\dfrac{x}{\sqrt{1 + x^2}}\sin\sqrt{1 + x^2}\,dx$

7. $[(x^4 + 7x^2 - 2x + 6)/(x^2 + 3)^2]\,dx$

9. $(-1/x^2 + 2)\,dx$

11. $dy = 2\,dx; dy/dx = 2$

13. $dy/dx = x(x + 2)^{-1/3}(x - 1)^{-2/3}$

15. First, $\sec^2 y\,dy = 2x\,dx$. So

$$\frac{dy}{dx} = \frac{2x}{\sec^2 y}, \qquad \frac{dx}{dy} = \frac{\sec^2 y}{2x}.$$

17. $(3y^2 + x)\,dy + (y - 5x^4)\,dx = 0$. Solve for dy/dx and for dx/dy.

CHAPTER 5

Section 1

1. 5/2   **3.** $1/\sqrt{3}$ **5.** $2/3^{1/3}$ **7.** $\sqrt{6} - 1$

9. If $g(x) = x^3$, then $(f - g)'(x) = 0$ for all x. By Theorem 1, $(f - g)(x) = c$. $f(x) = g(x) + c = x^3 + c$.

11. Set $g(x) = mx$ and follow the argument of Problem 9 ($b = c$).

13. a) By 11, $v = ds/dt = -32t + b$ for some constant b. Let $g(t) = -16t^2 + bt$; then

$$\frac{d}{dt}(s(t) - g(t)) = v(t) - (-32t + b) = 0.$$

Then by Theorem 1, $s(t) = g(t) + c$ for some constant c. That is, $s = -16t^2 + bt + c$.

b) $s = -16t^2 + 50t$.

15. $X = a^{1/(1-a)}x$ **17.** Use Problem 10.

19. The motorist was driving at an average velocity of $140/2 = 70$ mph and, hence, must have been going 70 mph at some time, by the mean-value principle.

Section 2   **1.**

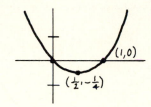

3.

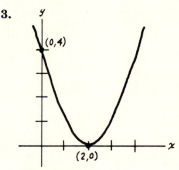

5.

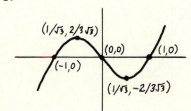

7.

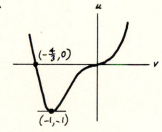

13.

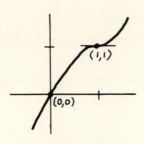

15.

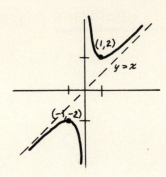

17.

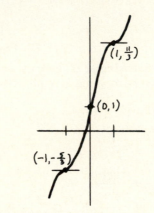

21. $dy/dx = 0$ at $x = -b/2a$. If $a > 0$, then $y \to +\infty$ as $x \to \pm\infty$. Sketch the graph in this case.

Section 3   **3.** Critical points: $x = \pm 1/2$, concave down, $(-\infty, 0)$; concave up, $(0, \infty)$.

5. No critical points; concave up, $(0, +\infty)$; concave down, $(-\infty, 0)$; slope $= 1$ at point of inflection.

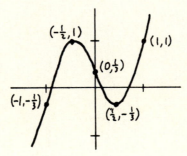

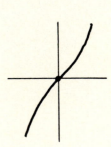

7. Critical point 0; concave up $(-\infty, 4)$, $(4, \infty)$; concave down $(-4, 4)$. Asymptotic to the line $x = -4$, $x = +4$, $y = 2$.

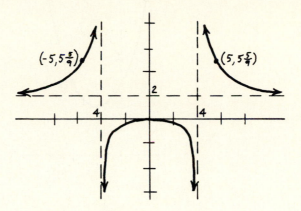

9. Critical points $\pi/6 + n\pi/3$; concave up $((2n + 1)\pi/3, (2n + 2)\pi/3)$; concave down $(2n\pi/3, (2n + 1)\pi/3)$.

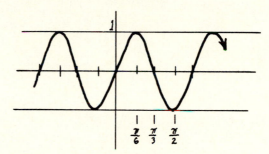

11.

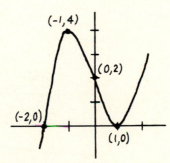

13. Critical point 1; concave up everywhere.

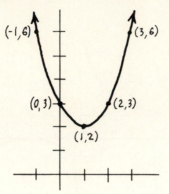

15. Critical points $x = \frac{3}{2}$, 2; concave down $(-\infty, 7/4)$; concave up $(7/4, \infty)$.

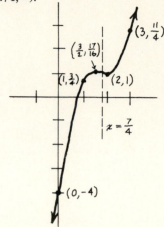

17. Critical points $x = -1, 2$; concave up $(-\infty, 0)$, $(2, \infty)$; concave down $(0, 2)$.

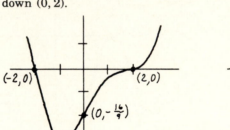

19. Critical points $x = 2, 6$; concave down, $(-\infty, 4)$; concave up, $(4, \infty)$.

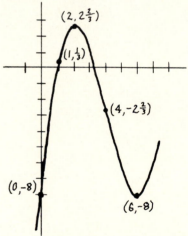

21.

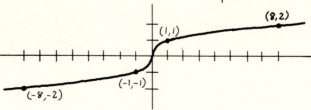

23. Critical points 0, 1; concave up, $(-\infty, -1/2)$; concave down, $(-1/2, 0)$, $(0, \infty)$.

25. Singular points 0, 1; critical point 1/2; concave up $(-\infty, 0)$, $(1, \infty)$; concave down $(0, 1)$.

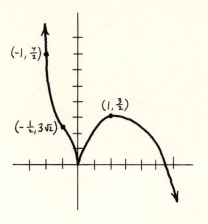

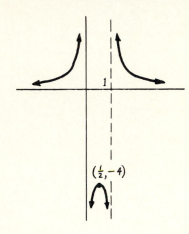

27. Critical points 0, $(2n + 1)\pi/2$; concave up, $((2n + 1)\pi, (2n + 2)\pi)$, $n \geq 0$; $(2n\pi, (2n + 1)\pi)$, $n < 0$; concave down, $(2n\pi, (2n + 1)\pi)$, $n \geq 0$; $((2n + 1)\pi, (2n + 2)\pi)$, $n < 0$.

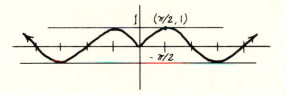

29. Critical points -1, 1; concave up, $(-\infty, -1)$, $(-1, 0)$, concave down, $(0, 1)$, $(1, +\infty)$.

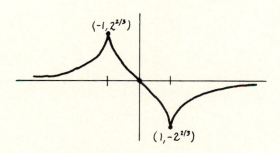

31. Critical points $x = \pi/4,\ \pi,\ 7\pi/4$; concave down $(0, 7\pi/12)$, $(\pi, 17\pi/12)$, approx.; concave up $(7\pi/12, \pi)$, $(17\pi/12, 2\pi)$, approx.

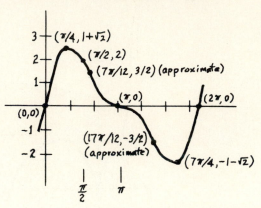

33. Critical point, $x = 0$; concave up, $(-\infty, \sqrt{3})$, $(0, \sqrt{3})$; concave down, $(-\sqrt{3}, 0)$, $(\sqrt{3}, \infty)$.

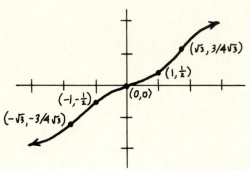

35.

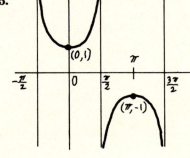

37. Concave up, $(0, \infty)$; concave down, $(-\infty, 0)$.

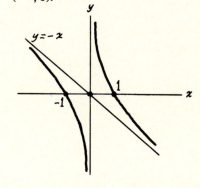

39. Critical point, $2^{2/3}$; concave up, $(0, 4)$; concave down, $(4, +\infty)$.

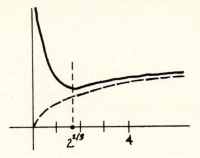

41. $a^2 < 3b$ means no critical points;

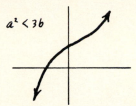

$a^2 = 3b$ means one critical point;

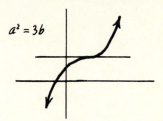

$a^2 > 3b$ means two critical points.

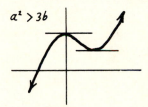

43. Take two nonoverlapping chords C_1 and C_2. By MVP there are tangents T_1 and T_2 with the same slope as C_1 and C_2 (see figure). Since the graph lies above T_1 and T_2 in particular, Y lies above T_1 and X lies above T_2 so $\angle XzY$ is less than π, in other words, slope T_1 = slope C_1 < slope C_2 = slope T_2.

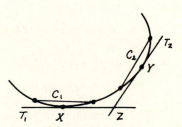

45. Say $x < y$. Let

$$s(u, v) = \frac{f(v) - f(u)}{v - u}.$$

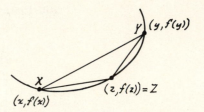

Thus $s(u, v)$ is the slope of the chord from $(u, f(u))$ to $(v, f(v))$. The given is that Z is below XY so that slope $XZ <$ slope $XY <$ slope ZY: $s(x, z) < s(x, y) < s(y, z)$. Now fix z_0 as the midpoint of the interval $[x, y]$, and repeat the above argument for Z between X and z_0. Letting Z approach X, show that $f'(x) \le s(x, z_0)$. In a similar manner, show that $f'(y) \ge s(z_0, y)$. Conclude that $f'(x) < f'(y)$.

Section 4   **1.** $3\frac{1}{4}$, 1   **3.** $22\frac{22}{27}$, -28 **5.** $0, -6$ **7.** $3\sqrt{3}/4, 0$

9. $3, -3/8$ **11.** ± 12 **13.** $5/4, -1$ **15.** max $= 3\frac{1}{4}$

17. max $= 2\frac{5}{27}$ **19.** min $= -3/8$ **21.** max $= 1/2$

27. Relative minimum at $x = 2$; maximum at $x = 0$.

29. $x = -1/3$, relative maximum; $x = 1$, relative minimum.

31. $x = -2$, relative maximum; $x = 2$, relative minimum.

33. $x = -3$, relative maximum; $x = -9/5$, relative minimum; $x = 0$, case of Problem 23.

35. Relative minimum at $x = 1$; decreasing through the critical point $x = 0$.

Section 5   **1.** $\sqrt{2}$ **3.** $1/2\sqrt{2}$ **5.** $(2, -22)$ **7.** -4

9. $r = \frac{2}{3}R$, $h = \frac{1}{3}H$

11. 6×8 yards (8 yard length borders the neighbor's lot)

13. $\sqrt{2A} \times \sqrt{A/2}$

15. Square piece $(96/(4 + \pi))$ in.; circle piece $24\pi/(4 + \pi)$

17. $\sqrt{3}$ **19.** Width:depth $= 1:\sqrt{2}$

21. $(x, y) = (\pm\sqrt{3/2}, 3/2)$

23. D is maximum when D^2 is maximum, and $D^2 = (x - 1)^2 + (y - 4)^2 = (x - 1)^2 + (x^2/3 - 4)^2$. $dD^2/dx = (2/9)x^3 - (5/3)x - 1$, which equals 0 at $x = 3$. The minimum distance is $(2^2 + (-1)^2)^{1/2} = \sqrt{5}$.

25. Let r = radius of cylinder, h = height. Then

$$4\pi r + 4r + 4h = 16, \qquad h = 4 - (\pi + 1)r.$$

Volume = $V = r^2 h = 4\pi r^2 - \pi(\pi + 1)r^3$, $dV/dr = 8\pi r - 3\pi(\pi + 1)r^2$
The maximum value of V occurs at the critical point $r = 8/3(\pi + 1)$.

27. L = length of wire = $a + b$, where $a^2 = 2000 - 80x + x^2$ and $b^2 = 900 + x^2$. Then $da/dx = (x - 40)/a$ and $db/dx = x/b$, and $dL/dx = (x - 40)/a + x/b$. Since this can be written $dL/dx = -\cos\theta + \cos\phi$, the critical point can most simply be determined by $\cos\theta = \cos\phi$, $\theta = \phi$.

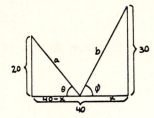

Then $20/(40 - x) = \tan\theta = \tan\phi = 30/x$, and so $x = 24$ ft, which should be the answer. (Then $a + b = 4\sqrt{41} + 6\sqrt{41} = 10\sqrt{41}$. The "endpoint" values for $a + b$ are $20 + 50 = 70$ and $30 + 20\sqrt{5} > 70$, each of which is larger than the critical point value $10\sqrt{41}$, which therefore is the minimum.)

29. Let r = radius of semicircle, h = height of rectangle. $2h + 2r + \pi r = 24$, $h = 12 - ((\pi + 2)/2)r$, and $A = 2rh + \pi r^2/2 = 24r - (\pi/2 + 2)r^2$. $dA/dr = 24 - (\pi + 4)r$. The max area occurs when $r = 24/(\pi + 4)$ ft. $h = 12 - 12(\pi + 2)/(\pi + 4) = 24/(\pi + 4)$ ft.

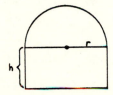

31. $D^2 = (10 - 5m)^2 + (5 - 10/m)^2 = (10 - 5m)^2(1 + 1/m^2)$. D is minimum when $m = -\sqrt[3]{2}$, and the path meets the two roads $10 + 5\sqrt[3]{2}$ and $5 + 10/\sqrt[3]{2}$ yds from the intersection, respectively.

33. L = light = $2rh + \pi r^2/4 = 24r - (3\pi/4 + 2)r^2$. $dL/dr = 24 - (3\pi/2 + 4)r$; L is max when $r = 48/(3\pi + 8)$ ft. $h = 12 - 24(\pi + 2)/(3\pi + 8) = (12\pi + 48)/(3\pi + 8) = ((\pi + 4)/4)r$.

35. Cost = $C = 5\pi r^2 + 2\pi rh$, so $0 = 10\pi r + 2\pi h + 2\pi r\,dh/dr$ and $dh/dr = -(5 + h/r)$. Then (as in above problem) $0 = dV/dr = \pi r[2r + 2h - r(5 + h/r)] = \pi r(h = 3r)$. Thus the maximum volume for a given total cost occurs when $h = 3r$.

37. Let A be the given surface area, r the radius of the cylinder, h its height. Then $A = 4\pi r^2 + 2\pi rh$ and $0 = 8\pi r + 2\pi h + 2\pi r \, dh/dr$; $dh/dr = -(4 + h/r)$. $V = (4/3)\pi r^3 + \pi r^2 h$, so $0 = dV/dr = 4\pi r^2 + 2\pi rh + \pi r^2 \, dh/dr = \pi r[4r + 2h - r(4 + h/r)] = \pi rh$. So $r = 0$ or $h = 0$. Since $r = 0 \Rightarrow V = 0$, max V occurs when $h = 0$ and the tank is spherical.

39. $P = 10x - (x^3 - 3x^2 + 4x + 1)$, $dP/dx = -3x^2 + 6x + 6$; $d^2P/dx^2 = 6x + 6$. Thus the max profit P is made when $x = 1 + \sqrt{3} \approx 2.73$. Thus he should produce 2,730 items/week to realize a maximum profit of about 1,739 dollars per week.

41. $L = k \sin\theta/r^2 = k \sin\theta \cos^2\theta/(25)^2$. $dL/d\theta = (k/625)(-\sin^2\theta \cos\theta + \cos^3\theta)$. This is zero when $(\tan\theta)^2 = 1/2$ ($\cos\theta = 0$ being ruled out), so with proper height is $25 \tan\theta = 25/\sqrt{2}$.

Section 6

1. $-3/4$

3. $25/12$ ft/sec

5. $9/5$ ft/sec

7. 65 mi/hr, 65 mi/hr

9. $20\sqrt{2}$ ft/min

11. 0 ft/sec

13. 10π cu.in./sec

15. Volume of water in trough $= 8$ $bh/2 = 8h^2/\sqrt{3}$, where h is depth of water and $b =$ width of water surface. $16 = dV/dt = (16h/\sqrt{3}) \, dh/dt$. Thus $dh/dt = \sqrt{3}/h$, and when $h = 1$, $dh/dt = \sqrt{3}$ ft/min.

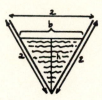

17. $u \, dV/dt + V \, du/dt = 0$, $du/dt = (-u/V) \, dV/dt = -2u/V = -32$ when $u = 4(V = 1/4)$.

19. $dy/dt = 2x \, dx/dt$, $dx/dt = (1/2x) \, dy/dt$. When $x = 2$, $dx/dt = (1/4)3 = 3/4$.

21. $dy/dt = 5 \cos\theta \cdot d\theta/dt = 5 \cos\theta = x$

23. $dV/dt = kA = k^4\pi r^2$ where k is a positive constant. But also $dV/dt = (dV/dr) \, dr/dt = 4\pi r^2 \, dr/dt$. Thus $4\pi k r^2 = 4\pi r^2 \, dr/dt$ and $dr/dt = k$.

Section 7

1. If $f(x) = x^5$, then $(3.1416)^5 - \pi^5 = f'(X)(3.1416 - \pi)$ for some X between π and 3.1416. Show that the right side above is less than $5(10)^{-3}$, using the assumptions stated in the problem.

3. Here $f(x) = x^{1/5}$, and $(100,001)^{1/5} - 10 = f'(X)(100,001 - 100,000) = f'(X)$ where $100,000 < X < 100,001$. Show that $f'(X) < 10^{-4}/5$.

5. Given $|r - \pi| < 10^{-n}/2$ and $f(x) = x^{1/3}$. Then for some X between r and π,

$$|r^{1/3} - \pi^{1/3}| = |f'(X)(r - \pi)| < \frac{1}{3}X^{-2/3}\frac{1}{2}(10^{-n}).$$

Show that this is less than $10^{-(n+1)}$, since $X > 3$. (Note: $\sqrt{3} > 5/3$.)

7. $|\pi - r| < 9e$ (assuming $r \geq 3$)

9. $b^n > b^n - 1 > n(b - 1)$. So b^n will be larger than M if $n(b - 1) > M$, and hence if $n > M/(b - 1)$.

11. $\sqrt{1 + x} - 1 = 1/2\sqrt{1 + X} \cdot x$, etc. Here is a direct proof: $1 + x < 1 + x + x^2/4 = (1 + x/2)^2$. Taking the square root gives $\sqrt{1 + x} < 1 + x/2$.

Section 8   **1.** $8\dfrac{1}{16}$ **3.** $2\dfrac{107}{108}$ **5.** 1.02

         7. $3 - \dfrac{1}{27} \approx 2.96$ **9.** $\dfrac{33}{128} \approx 0.258$ **11.** $0.08\pi r^3$

15. The tangent-line error is less than 0.001. Also, $1/16 \approx 0.06$, with an error of 0.0025. So, $\sqrt{65} \approx 8.06$, with an error less than $0.004 = 4(10)^{-3}$.

17. $|E| \leq 0.002$. Also, $1/108 \approx 0.01$, with an error less than 0.001. So $(80)^{1/4} \approx 2.99$, with an error less than 0.003.

19. $(0.98)^{-1} \approx 1.0200$ with an error less than $4(10)^{-5}$.

21. $26^{1/3} \approx 3 - 1/27 = 2.962962 \cdots$ with a negative error less than $5(10)^{-4}$ in magnitude. Thus

$$26^{1/3} = 2.962 \cdots.$$

23. Let $f(x) = x^{1/4}$, $f'(x) = (1/4)x^{-3/4}$, $f''(x) = (-3/16)x^{-7/4}$. Then $5^{1/4} = f(5) \simeq f(5\ 1/16) + f'(5\ 1/16)(5 - 5\ 1/16) = 3/2 + (1/4)(2/3)^3(-1/16) = 3/2 - (1/64)8/27 = 3/2 - 1/216$. If $x \geq 5$, then $|f''(x)| \leq (3/16)1/5^{7/4} \leq (3/16) \times (1/5)(1/3) = 1/80$, hence $|E| \leq (1/2)(1/80)(1/16)^2 = 1/10 \cdot 2^{12} = 1/40{,}960 < 3 \times 10^{-5}$.

25. Let $f(x) = \cos x$. Then $\cos x = f(x) \simeq f(0) + f'(0)(x - 0) = 1$, and since $|f''(x)| = |-\cos x| \leq 1$, $|E| \leq (1/2)(1)x^2 = x^2/2$.

27. If $f(h) = (1 + h)^a$, then $f'(h) = a(1 + h)^{a-1}$ and $f''(h) = a(a - 1)(1 + h)^{a-2}$. The tangent line approximation is $(1 + h)^a \simeq 1 + ah$. If $h > 0$ and $a < 2$, then $|f''(h)| \leq |a(a - 1)|$, and $|E| \leq (1/2)|a(a - 1)|\,h^2$.

29.

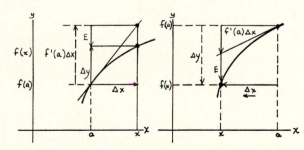

31. Let $f(x) = x^{1/2}$. $\sqrt{3} \simeq f(25/9) + f'(25/9)(3 - 25/9) = 5/3 + (1/2)(3/5)(2/9) =$
$5/3 + 1/15 = 26/15 = 1.733 \cdots$. Since $f''(x)$ is negative for $x > 0$, E is
also negative. Since $|f''(x)| \leq (1/4)(3/5)^3 = 27/500$ for $x \geq 5/3$, $|E| \leq$
$(1/2)(27/500)(2/9)^2 = 2/15{,}000 < 1.3(10)^{-3}$. Thus $1.732 < \sqrt{3} < 1.734$.

33. $S \simeq 4\pi r^2$; $\Delta S \simeq (8\pi r)(\Delta r) = 8\pi$;
$S \simeq 4\pi(3V/4\pi)^{2/3}$;
$S \simeq (8\pi/3)(3V/4\pi)^{-1/3}(3/4\pi)(\Delta V) = (-2\pi/3)(4/3)^{1/3}$

35. 3%　　　　　　　　　　　　　　**37.** 0.053

Section 9　　**1.** $3.1\bar{6}$　　　　　　　　　　**3.** 1, 3/4

7. Let $f(x) = \sin x = 2x/3$, $a = \pi/2$. Then $f'(x) = \cos x - 2/3$, and

$$x_1 = \frac{\pi}{2} - \frac{f(\pi/2)}{f'(\pi/2)} = \frac{3}{2}.$$

Between 3/2 and $\pi/2$, $\cos x = \sin(\pi/2 - x) < \pi/2 - x < (\pi - 3)/2 < 0.1$,
and $|f'(x)| = |\cos x - 2/3| > 1/2 = L$. Also $|f''(x)| = \sin x \leq 1 = B$. So

$$|E| \leq \frac{B}{2L}(x_1 - a)^2 - (x_1 - a)^2 < 0.01.$$

9. $f(x) = x^2 - 10$, $x_0 = 3$, $x_1 = (19/6)$, $x_2 = (721/228) = 3.162280 \cdots$,
$|f''(x)| = 2 = B$, $|f'(x)| = 2x \geq 6 = L$,

$$|E| \leq \frac{B}{2L}(x_2 - x_1)^2 = \frac{2}{2 \cdot 6}\left(\frac{1}{19 \cdot 12}\right)^2 < \frac{1}{6(200)^2} = \frac{(0.005)^2}{6} < 5(10)^{-6}.$$

11. $x_1 = 10.050$, $|E| \leq 1.25(10)^{-4}$

13.
$$x_2 = \frac{17}{12} - \frac{1}{24 \cdot 17} = 1.414215 \cdots;$$

$$|E| < \frac{1}{2}\left(\frac{1}{24 \cdot 17}\right)^2 < 5 \cdot (10)^{-6}.$$

Thus $\sqrt{2} = 1.41421$ to the nearest 5 decimal places.

15.
$$x_2 = \frac{97}{56} - \frac{1}{2 \cdot 56 \cdot 97};$$

$$|E| < \frac{1}{2}\left(\frac{1}{2 \cdot 56 \cdot 97}\right)^2 < 5 \cdot (10)^{-9}.$$

25.
$$x_{n+1} = \frac{1}{k}\left[(k - 1)x_n + \frac{c}{x_n^{k-1}}\right]$$

Section 10　　**3.** Let

$$g(x) = f(x) - \left[\frac{f(b) - f(a)}{b - a}\right](x - a).$$

Assume there is no point X; then g' is not 0 in (a, b), so either $g' > 0$ on all (a, b) or $g' < 0$ on all (a, b). In the former case, g is increasing. In the latter case, g is decreasing. But $g(a) = g(b) = f(a)$, so g can be neither decreasing nor increasing. Contradiction. So the point X exists after all.

7. Say p has roots $x_1, \ldots, x_n$ with $x_1 < x_2 < \cdots < x_n$. By Rolle's theorem, p' is 0 in each interval (x_k, x_{k+1}) $k = 1, \ldots, n - 1$. p' is of degree $n - 1$, so this accounts for all its roots and forces them to be distinct.

CHAPTER 6

1. $\dfrac{1}{7}x^7 + C$ 3. $\dfrac{1}{2}x^2 + C$ 5. $\dfrac{3}{4}x^{4/3} + C$

Section 1   7. $\dfrac{3}{4}x^4 + \dfrac{1}{6}x^{-2} + C$ 9. $\dfrac{2}{3}x^3 - \dfrac{1}{2}x^2 - 6x + C$

11. $2\sqrt{x} + C$ 13. $ay^3 + C$

15. $2x^2 - 4\sqrt{x} + C$ 17. $3t^3 + \dfrac{25}{2}t^2 + 14t + C$

19. $3\sqrt[3]{x} + C$ 23. $2(x + 1)^{1/2} + C$

25. $(3x + 2)^{1/3} + C$ 27. $-\cos t + C$

29. $-\cos x + x^2/2 + Cx + C'$ 31. $\tan 3\theta/3 + \sec \theta + C$

33. $f(x) = x^2/2 - 3x + 13$ 35. $f(x) = (1/4)y^4 - (b^2/2)y^2 + 2b^2 - 4$

37. $f(t) = (2/3)t^{3/2} + 2t^{1/2} - 28/3$ 39. $8p^2/3$

41. 80 ft/sec   43. $126\dfrac{2}{3}$ ft

Section 2   1. 2   3. 0   5. -5 7. 2   9. 1/2

11.

$$\lim n \sin \frac{1}{n} = \lim \frac{\sin \dfrac{1}{n}}{\dfrac{1}{n}} = \lim_{x \to 0} \frac{\sin x}{x} = 1$$

The last limit was fundamental in Chapter 4.

13. 0   15. 1/3   17. 2.72   19. 2.718

21. $\displaystyle\sum_{n=2}^{17} n$ 23. $\displaystyle\sum_{n=1}^{99} \frac{n}{n+1}$ 25. $\displaystyle\sum_{j=0}^{18} a_{2j+1}$

Section 3   9. $1.846 < \displaystyle\int < 1.892$ 11. $0.64 < \displaystyle\int < 0.75$ 13. $0.73 < \displaystyle\int < 0.84$

Section 4 1. 1/6   3. −4/15   5. $y^3 - x^3$ 7. −1/2

9. $\int 2x \, dx = x^2 + c$. When $x = 0$, $A = 0$, so $c = 0$; $A = 4^2 = 16$.

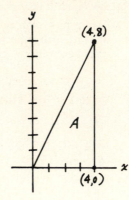

11. 64   13. 16/3   15. $15\dfrac{3}{4}$ 17. 2

19. The top line has the equation

$$y = \frac{b_2 - b_1}{h}x + b_1; \quad A = \frac{b_1 + b_2}{2}h.$$

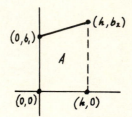

21. 2   23. 0   25. 4/3   27. 3   29. 30 mph

31. Average of instantaneous velocity = $1/(t_1 - t_0)\int_{t_0}^{t_1} f'(t)\, dt = \cdots$.

Section 5 1. 8/3   3. 18   5. 8   7. 32/3

9. 8   11. 3/4   13. 1/12   15. $2\sqrt{2}/3$

17. 8 (The sum of two pieces each having area = 4)

19. 27/4   21. 81/4   23. 32/3   25. 9/2

Section 6 1. The graph being rotated is $y = f(x) = \sqrt{r^2 - x^2}$, so $\pi f^2(x) = \pi(r^2 - x^2)$.

3. $\pi/7$ 5. $\pi(6^4/5)$ 7. 3π

9. $\dfrac{8}{5}\pi$ 11. $\dfrac{2}{9}\pi(2\sqrt{2} - 1)$ 13. π

Section 7 1. $1/(1 + x^2)$ 3. $2/x - 1/x = 1/x$ 5. $-\sqrt{a^2 - x^2}$

7. $\sqrt{1 - \sin^2 x}\cos x = \cos^2 x$ 9. $\dfrac{1}{\sqrt{\sec^2 x - 1}}\sec x \tan x = \sec x$

11. $2x \sin |x|$ **13.** $(\sec \sqrt{x})/2\sqrt{x}$

15. $\left(\dfrac{1}{1 + \tan^2 x}\right) 2 \tan x \sec^2 x = 2 \tan x$

17. $\dfrac{1}{1 + x^2} - \dfrac{1}{1 + (1/x)^2}\left(-\dfrac{1}{x^2}\right) = \dfrac{2}{1 + x^2}$

19. $f'(x) = \dfrac{1}{1 + x^2}$, so $\dfrac{d}{dx} f(\tan x) = \dfrac{1}{1 + \tan^2 x} \sec^2 x = 1$. Thus $f(\tan x) = x + C$, and setting $x = 0$ shows that $C = 0$.

21. This is like Problem 19.

Section 8   **3.** $S_4 = 21/32$; the integral is $5/8 = 20/32$, so the error is $1/32$. This is the same as given by the error estimate

$$\frac{1}{2} \Delta x[f(b) - f(a)] = \frac{1}{2} \cdot \frac{1}{8} \cdot \left[\frac{3}{2} - 1\right] = \frac{1}{32}.$$

7. $n = 40{,}000$

9. The sum listed is the midpoint evaluation S_8 for the integral $\ln 2 = \int_1^2 dx/x$. A more recognizable form is

$$\frac{1}{8}\left[\frac{1}{17/16} + \frac{1}{19/16} + \cdots + \frac{1}{31/16}\right].$$

It follows that $\ln 2 = 0.693$ with an error less than one in the third place.

11. One must show that, in the accompanying figure, the shaded region R_1 with corners CDA is larger than the shaded region R_2 with corners DEB, because if that is the case, then the rectangle $ABZX$ misses more of the area under the curve than the extra amount it contains. To show this, use the fact that the graph lies above its tangent line at D, and that the tangent line cuts off congruent triangles.

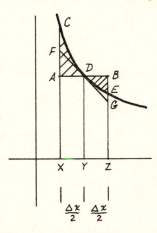

13. The sum is $\bar{S}_5$ in the integral $\pi/4 = \int_1^2 (dx/(1 + x^2))$. So work out the value of the error formula. The result: $0.782 < \pi/4 < 0.79$.

15. $|E| \le \dfrac{2}{24}(0.4)(0.2)^2 + \dfrac{2}{3.24}(0.6)(0.2)^2 = 0.002$.

CHAPTER 7   **1.** $\ln(x^2 - 1)$ **3.** $2\ln(x - 1)$

5. Already shown for $x > 0$. Let $h(x) = -x$. For $x < 0$,

Section 1
$$\frac{d}{dx}\ln|x| = \frac{d}{dx}\ln(-x) = \frac{d}{dx}\ln(h(x)) = \frac{1}{h(x)}h'(x) \qquad \text{(chain rule)}$$

$$= \frac{-1}{(-x)} = \frac{1}{x}.$$

7. $2x/(x^2 + 1)$ **9.** $1 + \ln x$ **11.** $(1 - \ln x)/x^2$

13. 2 **15.** $x^n \ln x$

25. We require a to be rational because at the moment we have the power rule only for rational exponents.

27. $-1/e$ **29.** $(2ax + b)/(ax^2 + bx + c)$

31. $(1 - x^2)/(x + x^3)$ **33.** $1/\sin x \cos x$

35. $-\dfrac{1}{[1 + \ln(ax + b)]^2}\left(\dfrac{a}{ax + b}\right)$ **37.** $\dfrac{-2\ln x \sin(\ln^2 x)}{x}$

39. $\dfrac{\ln(1 - x)}{1 + x} - \dfrac{\ln(1 + x)}{1 - x}$

41. $2\sec^2(x + \ln x)\tan(x + \ln x)\left[1 + \dfrac{1}{x}\right]$

43. $-8x\tan(x^2)\ln(\cos^2(x^2))$ **45.** $\dfrac{-\sin(\sqrt{\ln(\ln x)})}{2x\ln x\sqrt{\ln(\ln x)}}$

47. $f'(x) = x^{-2}(1 - \ln x)$; critical point at $x = e$; $f(e) = 1/e$. $f''(x) = x^{-3}(2\ln x - 3)$; inflection point at $x = e^{3/2}$; $f(e^{3/2}) = 3/2e^{3/2}$. Concave down on $(0, e^{3/2})$, up on $(e^{3/2}, \infty)$. $f(x) \to -\infty$ as $x \to 0$; $f(x) \to 0$ as $x \to \infty$ (from Problem 26).

49. f has only one critical point, at $x = e^{1/k}$, and $f''(e^{1/k}) < 0$. Therefore $f(e^{1/k}) = 1/ke$ is the maximum value of f on $(0, \infty)$, by Theorem 7 in Chapter 5. Then

$$f(x) = \frac{\ln x}{x^k} = \left(\frac{\ln x}{x^{k/2}}\right)\frac{1}{x^{k/2}} \le \frac{2}{kex^{k/2}},$$

which $\to 0$ as $x \to \infty$.

Section 2   **7.** $f'(x) = 3(x - 1)^2$ **9.** $f'(x) = 3\left(x - \dfrac{1}{3}\right)^2 + \dfrac{2}{3}$

11. $f'(x) = 1/(1 + x^2)^{3/2}$ **19.** They are inverses, where defined.

21. $x = (1/3)(y + 5)$ **23.** $x = y^{5/3}$

25. $x = (1 - y)/(1 + y)$ **27.** $x = y/\sqrt{1 - y^2}$

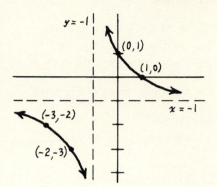

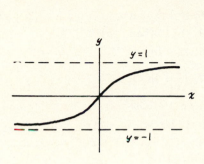

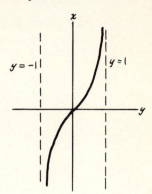

29. Domain $f = [1, \infty)$; domain $g = [0, \infty)$

31. Domain $f = [0, \infty)$

39. Example: $f(x) = \sqrt{1 - x^2}$, on $[0, 1]$

Section 3

1. xe^x **3.** x^2e^x **5.** $-\dfrac{e^{1/x}}{x^2}$

7. $e^x/2\sqrt{e^x - 1}$ **9.** $\left(\dfrac{x}{2}e^{x/2}\right)\Big/\sqrt{x - 1}$ **11.** ae^{ax}

13. $\dfrac{k}{x}e^{k\ln x} = kx^{k-1}$ **15.** $\dfrac{1}{3}(3e^{3x} + 1)(e^{3x} + x)^{-2/3}$

17. $e^x - e^{-x}$ **19.** $e^{-x}(1 - x - x^2 - x^3)/(1 + x^2)^2$

21. $e^x \cos x$ **23.** $-\sin x e^{\cos x}$ **25.** $e^{(e^x + x)}$

27. $-\dfrac{xe^{\sqrt{1-x^2}}}{\sqrt{1 - x^2}}$ **29.** $e^{-x^2}(1 - 2x - 2x^2)$ **31.** $\dfrac{ae^{-ax}\cos\sqrt{1 - e^{-ax}}}{2\sqrt{1 - e^{-ax}}}$

33. $\dfrac{e^{3x}(3 + 3\cos^2 5x - 5\sin 10x)}{\sqrt{1 + \cos^2 5x}}$

35. $y\left[1 + \dfrac{2\sin 4x}{1 + \cos^2 2x} - \dfrac{2x}{(1 + x^2)\ln(1 + x^2)}\right]$

37. $1, -2$ **39.** $1, -3$

45. Show that f has a single critical point, and apply Theorem 7 from Chapter 5.

47. Critical point at $x = 1$, $y = 1$. Inflection point at $x = 2$, $y = 2/e$. Concave down on $(0, 2/e]$, up on $[2/e, \infty)$. $f(x) \to 0$ as $x \to \infty$, by Problem 45. Draw the graph from these data.

49. Min $= -1/e$, at $x = -1$ **51.** $x = 1/e$

Section 4

1. $\left(\dfrac{1}{2}\ln x + 1\right)x^{x^{1/2}-1/2}$

3. $9^x \ln 9$

5. $(a \ln x + a)x^{ax}$

7. $(x \cot x + \ln(\sin x))(\sin x)^x$

9. $(\ln 10)(2ax + b)10^{ax^2+bx+c}$

11. $(\ln x \sec^2 x + x^{-1}\tan x)x^{\tan x}$

13. $(1 - \ln x)x^{(1/x)-2}$

15. $[\ln(\ln x)](\ln x)^x + (\ln x)^{x-1}$

17. $\left(\dfrac{1}{x} + \ln x\right)e^x x^{(e^x)}$

19. See Problem 41 of Section 2.

21. $x/(1 + x^2)^{1/2}$

23. $(m/n)x^{m/n-1}$

25. $(-1/x^3)4^{1/x}(x + \ln 4)$

27. $(\sec^2 x \ln x + x^{-1}\tan x)x^{\tan x}$

29. $-(1 - x)^{-1/2}(1 + x)^{-3/2}$

31. $-\dfrac{y}{4}\left[\dfrac{1}{1-x} + \dfrac{2}{1-2x} + \dfrac{3}{1-3x} + \dfrac{4}{1-4x}\right]$

Section 5

1. $\ln 2/0.01 \approx 70$ hrs   **3.** $\ln 2/24 \simeq 2.9\%$ **5.** 6.6 yrs

7. $y_0 = ce^{kt_0}$, $y_1 = ce^{kt_1}$; $\ln y_0 = \ln c + kt_0$; $\ln y_1 = \ln c + kt_1$. Subtracting: $k(t_1 - t_0) = \ln y_1 - \ln y_0 = \ln(y_1/y_0)$; $k = (\ln(y_1/y_0))/(t_1 - t_0)$.

11. The temperature will reach $170°$ at time $t = 5 \ln 1.3/\ln(13/12) \approx 16.5$ min.

13. $5700 \times 2 = 11{,}400$ yrs

Section 6

1. $1/2\sqrt{x - x^2}$

3. $2x/\sqrt{1 - x^4}$

5. ∓ 1 (-1 from 0 to π, $+1$ from π to 2π, etc.)

7. $\arcsin x$

9. $\dfrac{-1}{x\sqrt{x^2 - 1}\arcsin(1/x)}$

11. $\dfrac{1}{1 + ax}\sqrt{\dfrac{1 - a^2}{1 - x^2}}$

13. $2\sqrt{1 - x^2}$

15. $-1/\sqrt{2\sin x + 2\sin^2 x}$

17. 0

19. The problem is to furnish a value of y such that $\arcsin(\sin y) \neq y$.

21. Show that it must be of the form $2n\pi + \arcsin x$ or $(2n + 1)\pi - \arcsin x$ for some integer n.

23. The identity involved is $\cos 2\theta = 1 - 2\sin^2\theta$.

Section 7

1. $2x/(1 + x^4)$

3. $a/(1 + a^2x^2)$

5. $1/x\sqrt{x^2 - 1}$

7. $(1 + x)/(1 + x^2)$

11. Show that

$$\cot\left(\frac{\pi}{2} - \arctan x\right) = \frac{\cos\left(\dfrac{\pi}{2} - \arctan x\right)}{\sin\left(\dfrac{\pi}{2} - \arctan x\right)}$$

reduces to x (by the $\cos(s - t)$ identity, etc.).

13. Show that $\operatorname{arcsec} x = \arccos(1/x)$, and then differentiate.

15. -1 **17.** $1/(1 + x^2)$ **19.** $-1/(x(\ln x)^2 + x)$

21. $1 + \dfrac{2x \arctan x}{1 + x^2}$ **23.** $\dfrac{1}{\sqrt{1 - x^2}}$ **25.** $-\dfrac{x}{1 + x^2}$

27. Show that the graph looks like this:

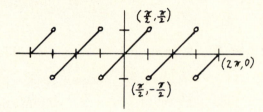

31. $1/\sqrt{2}$ **35.** $x = \sqrt{ab + b^2}$

CHAPTER 8

Section 1

1. $-\dfrac{1}{2}\cos 2x + C$ **3.** $\dfrac{1}{3}\sin^3 x + C$

5. $\dfrac{-1}{3}e^{-t^3} + C$ **7.** $-\sin(2 - x) + C$

9. $-\ln(1 + \cos x) + C$ **11.** $\dfrac{1}{2}(\ln x)^2 + C$

13. $\ln|x^2 + x - 1| + C$ **15.** $\dfrac{3}{7}(a + y)^{7/3} - \dfrac{3a}{4}(a + y)^{4/3} + C$

17. $(4/\sqrt{3})\arctan\sqrt{\dfrac{1}{3}x - 1} + 2\sqrt{x - 3} + C$

19. $-\dfrac{4}{3}(x^3 + 2)^{-2} + C$ **21.** $2e^{\sqrt{x}} + C$

23. $\dfrac{1}{4}(e^x + 1)^4 + C$ **25.** $\dfrac{1}{3}(\ln x)^3 + C$

27. $\dfrac{1}{2}(\arctan t)^2 + C$ **29.** $-2\cos\sqrt{x} + C$

31. $-2\sqrt{1 - \sin t} + C$

33. $\dfrac{1}{3}(x^2 + 1)^{3/2} + C$ (if $x > 0$)

35. $\arctan(e^t) + C$

37. $\dfrac{1}{n + 1}\tan^{n+1}\theta + C$

39. $\dfrac{1}{(n + 1)}\sin^{n+1}\theta + C$

41. $\dfrac{1}{1 - a}(\ln x)^{1-a} + C$

43. $\dfrac{1}{1 - a}[\ln(\ln x)]^{1-a} + C$

Section 2

1. $\sin x - x \cos x + C$

3. $\dfrac{2}{3}x(x + 1)^{3/2} - \dfrac{4}{15}(x + 1)^{5/2} + C$

5. $-y^2 \cos x + 2y \sin x + 2 \cos x + C$

7. $-x^2 e^{-x} - 2xe^{-x} - 2e^{-x} + C$

9. $\dfrac{1}{2}(x - \sin x \cos x) + C$

11. $x \arcsin x + \sqrt{1 - x^2} + C$

13. $\dfrac{1}{2}(\ln x)^2 + C$

15. $2 \sin \sqrt{x} - 2\sqrt{x} \cos \sqrt{x} + C$

17. $x \ln(1 + x^2) - 2x + 2 \arctan x + C$

19. $\dfrac{1}{2}[(x^2 + 1)\arctan x - x] + C$

21. $(x/(x + 1))\ln x - \ln(x + 1) + C$

23. $(x + 1)\arctan \sqrt{x} - \sqrt{x} + C$

25. $e^{ax}(a \cos bx + b \sin bx)/(a^2 + b^2)$

27. $\dfrac{1}{n - 1}\dfrac{\sin x}{\cos^{n-1}x} + \dfrac{n - 2}{n - 1}\displaystyle\int \sec^{n-2}x \, dx$

29. $x(\ln x)^n - n\displaystyle\int (\ln x)^{n-1}\, dx$

Section 3

1. $\dfrac{1}{6}\ln \left| \dfrac{3 - x}{3 + x} \right| + C$

3. $\dfrac{1}{9}\ln \left| 1 - \dfrac{9}{x} \right| + C$

5. $[1/(2\sqrt{3})]\ln \left| \dfrac{x + 1 - \sqrt{3}}{x + 1 + \sqrt{3}} \right| + C$

7. $\ln \left| \dfrac{x + 1}{x + 2} \right| + C$

9. $\dfrac{1}{6}\arctan \dfrac{2}{3}x + C$

11. $\dfrac{1}{6}\ln |x - 1| + \dfrac{5}{6}\ln |x + 5| + C$

13. $x - 2 \ln |x + 1| + (-1)/(x + 1) + C$

15. $\dfrac{2}{3}\ln |x + 5| + \dfrac{1}{3}\ln |x - 1| + C$

17. $(-5)/(x - 1) + 2 \ln |x| + C$

19. $\ln |\sec \theta + \tan \theta| + C$

21. $\ln\left(\dfrac{e^t + 1}{e^t + 2}\right) + C$

23. $\dfrac{1}{8}\ln\left(1 - \dfrac{4}{x^2}\right) + C$

25. $(-1/16)\ln|x| + (17/16)\ln|x - 4| + (1/4x) + C$

27. $6x^{1/6} - 6\arctan(x^{1/6}) + C$ **29.** $x^2 - \dfrac{3}{\sqrt{2}}\arctan\dfrac{x}{\sqrt{2}} + C$

31. $\dfrac{1}{10}\ln\dfrac{\sqrt{x^2 + 4x + 5}}{x + 2} + 3\arctan(x + 2) + C$

33. $\dfrac{1}{3}(x + 1)^3 - \dfrac{5}{2}(x + 1)^2 + 10(x + 1)$

$\qquad - 10\ln(x + 1) + 5(x + 1)^{-1} + \dfrac{1}{2}(x + 1)^{-2} + C$

35. $\dfrac{1}{2}\left[\ln\dfrac{x - 1}{\sqrt{x^2 + 1}} + \arctan x\right] + C$

37. $\ln\dfrac{(x - 2)^{1/2}}{x^{1/6}(x + 3)^{1/3}} + C$

39. $\ln\sqrt{x^2 + x + 1} - \dfrac{1}{\sqrt{3}}\arctan\dfrac{2x + 1}{\sqrt{3}} + C$

Section 4   **1.** $\ln\left|\dfrac{x - 1}{x}\right| + \dfrac{1}{x} + \dfrac{1}{2x^2} + C$ **3.** $x^{-1} + \dfrac{1}{2}\ln\left|\dfrac{x - 1}{x + 1}\right| + C$

      5. $-x^{-1} - \arctan x + C$ **7.** $\dfrac{1}{4}\ln\left|\dfrac{1 + x}{1 - x}\right| + \dfrac{x}{2(1 - x^2)} + C$

      9. $\dfrac{1}{6}\ln\left(\dfrac{x + 1}{x - 1}\right) + \dfrac{1}{12}\ln\left(\dfrac{x - 2}{x + 2}\right) + C$

     11. $\dfrac{1}{3}\arctan x - \dfrac{1}{6}\arctan\dfrac{x}{2} + C$

     13. $\dfrac{1}{4}\ln\left|\dfrac{(x - 1)^3}{(x + 1)(x^2 + 1)}\right| - \arctan x + C$

     15. $x - \dfrac{2}{3}\arctan x + \dfrac{2\sqrt{2}}{3}\ln\left|\dfrac{x - \sqrt{2}}{x + \sqrt{2}}\right| + C$

     17. $\dfrac{1}{4}\ln\left(\dfrac{x + 1}{x - 1}\right) - \dfrac{1}{6(x - 1)} + \dfrac{2}{3\sqrt{3}}\arctan\left(\dfrac{2x + 1}{\sqrt{3}}\right) + C$

     19. $-\dfrac{1}{2x} - \arctan x + \dfrac{1}{2\sqrt{2}}\arctan\dfrac{x}{\sqrt{2}} + C$

     21. a) $a = 1;\ b = 0,\ c = -1,\ d = 0$

          b) $\dfrac{1}{2}\ln(x^2 + 1) + 1/2(x^2 + 1) + C$

23. $\dfrac{1}{2(n-1)a^2}\dfrac{x}{(x^2+a^2)^{n-1}}+\dfrac{2n-3}{2(n-1)a^2}\displaystyle\int\dfrac{dx}{(x^2+a^2)^{n-1}}$

Section 5

1. $x/2-\dfrac{\sin 2x}{4}+C=\dfrac{1}{2}(x-\sin x\cos x)+C$

3. $(-1/4)\sin^3 x\cos x-(3/8)\sin x\cos x+(3/8)x+C$

5. $(1/24)\sin^4 6x+C$ 　　　　　　　　　　**7.** $(-1/3)\csc^3 x+\csc x+C$

9. $(1/6)\cos^5\theta\sin\theta+(5/24)\cos^3\theta\sin\theta+(5/16)\cos\theta\sin\theta+(5/16)x+C$

11. $(-3/4)(\cos 2x)^{2/3}+(3/16)(\cos 2x)^{8/3}+C$

13. $\ln(\tan x)+C$ 　　　　　　　　　　**15.** $(-1/3)\cot^3 x-(1/5)\cot^5 x+C$

17. $-4\cot\dfrac{x}{4}-(4/3)\cot^3\dfrac{x}{4}+C$ 　　**19.** $+\dfrac{2}{9}\sec^{9/2}x-\dfrac{2}{5}\sec^{5/2}x+C$

21. $+\dfrac{1}{2}\sec^2 x+\ln(\cos x)+C$ 　　**23.** $\dfrac{1}{4}\tan^4 x-\dfrac{1}{2}\sec^2 x+\ln(\sec x)+C$

25. $-x+\tan x-\dfrac{1}{3}\tan^3 x+\dfrac{1}{5}\tan^5 x+C$

29. $-\dfrac{15}{8}x+\dfrac{1}{3}\tan 3x+\dfrac{1}{6}\sin 6x-\dfrac{1}{96}\sin 12x+C$

31. $3\tan x+\dfrac{1}{3}\tan^3 x-\dfrac{1}{3}\cot^3 x-3\cot x+C$

33. $\dfrac{1}{2}\left[x^2+\cot(x^2)-\dfrac{1}{3}\cot^3(x^2)\right]+C$

35. $x\tan x+\ln(\cos x)+C$

37. $\left(\tan x+\dfrac{1}{3}\tan^3 x\right)\ln(\sin x)+\dfrac{2}{3}x+\dfrac{1}{3}\tan x+C$

39. $\left(\tan x+\dfrac{1}{3}\tan^3 x\right)\ln(\cos x)-\dfrac{2}{3}x+\dfrac{2}{3}\tan x+\dfrac{1}{9}\tan^3 x+C$

41. $\dfrac{3}{4}x-\dfrac{\sqrt{2}}{4}\sin(2\sqrt{2}x)+\dfrac{\sqrt{2}}{32}\sin(4\sqrt{2}x)+C$

43. $\dfrac{e^x}{1+n^2}(\sin^n x-n\sin^{n-1}x\cos x)+\dfrac{n(n-1)}{1+n^2}\displaystyle\int e^x\sin^{(n-2)}x\,dx$

Section 6

1. $\dfrac{1}{3}(4-x^2)^{3/2}-4(4-x^2)^{1/2}+C$

3. $x/a^2\sqrt{a^2+x^2}+C$ 　　　　　　　**5.** $-\dfrac{1}{3}(16-x^2)^{3/2}+C$

7. $\ln\left(\dfrac{\sqrt{1+x^2}-1}{|x|}\right)+C=-\ln\left(\dfrac{\sqrt{1+x^2}+1}{|x|}\right)+C$

9. $\dfrac{1}{5}\ln\left(\dfrac{5 - \sqrt{25 - x^2}}{|x|}\right) + C = -\dfrac{1}{5}\ln\left(\dfrac{5 + \sqrt{25 - x^2}}{|x|}\right) + C$

11. $\dfrac{1}{3}\ln\left(\dfrac{\sqrt{9 + 4x^2} - 3}{|2x|}\right) + C$ 　　　**13.** $x/\sqrt{a^2 - x^2} - \arcsin(x/a) + C$

15. $-\dfrac{1}{4}(x - 1)(x^2 - 2x - 3)^{-1/2} + C$ 　　**17.** $-(1 - x^2)^{1/2} + C$

19. $(1 + x^2)^{1/2} + C$ 　　　　　　**21.** $(x^2 - a^2)^{1/2} + C$

23. $2\sqrt{x} - 2\ln(1 + \sqrt{x}) + C$

25. $\dfrac{6}{7}x^{7/6} - \dfrac{6}{5}x^{5/6} + \dfrac{3}{2}x^{2/3} + 2x^{1/2} - 3x^{1/3} - 6x^{1/6} + 3\ln(x^{1/3} + 1)$

　　　 $+ 6\arctan(x^{1/6}) + C$

27. $2\sqrt{1 - x} + \sqrt{2}\ln\left|\dfrac{\sqrt{1 - x} - \sqrt{2}}{\sqrt{1 - x} + \sqrt{2}}\right| + C$

29. $-\arctan(\cos t) + C$ 　　　　**31.** $-(x^2)/2 - (a^2/2)\ln|a^2 - x^2| + C$

33. $2\sqrt{x} - 2\arctan(\sqrt{x}) + C$

35. $\dfrac{1}{2}\ln(\sqrt{1 + x^4} + x^2) - \dfrac{\sqrt{1 + x^4}}{2x^2} + C$

Section 7
Extra Integration
Problems

1. $\dfrac{1}{2}e^{2x} + C$ 　　　　　　**3.** $\dfrac{1}{2}\sec^2 x + C$

5. $e^{(e^x)} + C$ 　　　　　　　**7.** $4\sqrt{x}(\ln\sqrt{x}) - 4\sqrt{x} + C$

9. $\dfrac{1}{2}\arcsin x - \dfrac{x}{2}\sqrt{1 - x^2} + C$ 　　**11.** $\dfrac{2}{3bn}(a + bx^n)^{3/2} + C$

13. $-\dfrac{\sqrt{x^2 + 4}}{4x} + C$ 　　　　**15.** $-\dfrac{1}{2(\ln x)^2} + C$

17. $\dfrac{1}{3}(x^2 + 1)^{3/2} - (x^2 + 1)^{1/2} + C$ 　　**19.** $2\ln|\sec\sqrt{x} + \tan\sqrt{x}| + C$

21. $-\cos x[\ln(\cos x)] + \cos x + C$ 　　**23.** $-\dfrac{1}{4}\ln|x| + \dfrac{5}{8}\ln|x^2 - 4| + C$

25. $\ln(1 + \sqrt{x})^2 + C$ 　　　　**27.** $\dfrac{\tan^7 x}{7} + C$

29. $\dfrac{(\arctan x)^3}{3} + C$ 　　　　**31.** $-\ln\cos\sqrt{x} + C$

33. $\dfrac{2}{15}(1 - \sin t)^{3/2}(7 + 3\sin t) + C$

35. $\dfrac{1}{3}(x^2 - 1)^{3/2} + C$ 　　　　**37.** $\ln\left|\dfrac{1 + c^t}{1 - e^{2t}}\right| + C$

39. $\dfrac{\tan^{n+3}\theta}{n+3} + \dfrac{\tan^{n+1}\theta}{n+1} + C$

41. $\dfrac{\sin^{n+1}\theta}{n+1} - \dfrac{\sin^{n+3}\theta}{n+3} + C$

43. $-\dfrac{1}{9}\dfrac{\sqrt{9+4x^2}}{x} + C$

45. $(a^2 - x^2)^{-1/2} + C$

42. $\dfrac{1}{2}\dfrac{x-1}{\sqrt{x^2-2x+3}} + C$

49. $\sin t + \dfrac{1}{2\sqrt{2}}\ln\left(\dfrac{\sqrt{2}-\sin t}{\sqrt{2}+\sin t}\right) + C$

51. $\dfrac{3\sqrt{2}}{2}\arcsin\sqrt{\dfrac{2}{3}(1+x)} + \sqrt{1+x}\,\sqrt{1-2x} + C$

53. $-a^4(a^2-x^2)^{1/2} + \dfrac{2}{3}a^2(a^2-x^2)^{3/2} - \dfrac{1}{5}(a^2-x^2)^{5/2} + C$

55. $x - \dfrac{3}{2}x^{2/3} + 3x^{1/3} - 3\ln(1+x^{1/3}) + C$

57. $\dfrac{1}{3}x^3\arcsin x + \dfrac{1}{3}(1-x^2)^{1/2} - \dfrac{1}{9}(1-x^2)^{3/2} + C$

59. $x^3\sin x + 3x^2\cos x - 6x\sin x - 6\cos x + C$

61. $\dfrac{x^{a+1}}{a+1}\ln x - \dfrac{x^a}{a} + C$

63. $x[(\ln x)^3 - 3(\ln x)^2 + 6\ln x - 6] + C$

65. $\pi^2/2$

67. $\displaystyle\int_0^1 \arctan x\,dx = \left(x\arctan x - \dfrac{1}{2}\ln(x^2+1)\right)\Bigg]_0^1$ (by Problem 10, Section 2)

$$= \dfrac{\pi}{4} - \dfrac{1}{2}\ln 2$$

Section 8

1. $y = Ce^{2x}$ **3.** $y = Ce^{kx}$ **5.** $y = -2/(x^2 + C)$

7. $y = \ln(2/(C - x^2))$ **9.** $y = \sin(x + C)$ **11.** $y^2 - 2x^2 = C$

13. $y = x^2/(Cx^2 + x - 2)$

17. $y = -\arctan(\ln cx)$ or $y = \pi/2 + n\pi$ for fixed n.

19. $y = \arctan(c - \cos x)$ or $y = \pi/2 + n\pi$ for fixed n.

21. $\dfrac{dv}{f(v)-v} = \dfrac{dx}{x}$ **23.** $dy = dx$

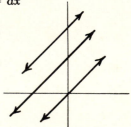

25. $dy/dx = -x/y$

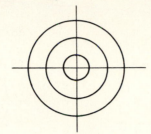

27. $\dfrac{dy}{dx} = \dfrac{-x}{\sqrt{1-x^2}}$

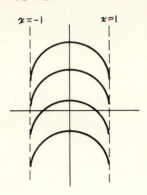

29. $\dfrac{dy}{dx} = \pm 2\sqrt{y};\ \left(\dfrac{dy}{dx}\right)^2 = 4y$

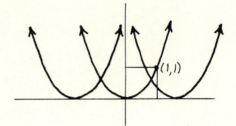

Section 9   **1.** Circles: $x^2 + y^2 = C$

3. $y = Ce^{-2x}$

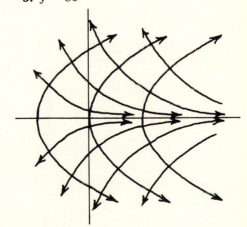

5. $y = Cx$ (see (1))

7. $x^2 + Cy + y^2 = 0$

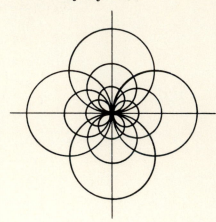

11. $y = e^{Ce^{-kt}}$

13. $P = \dfrac{1}{4}k^2t^2 + kP_0^{1/2}t + P_0$

15. $\ln\left(\dfrac{c}{c-y}\right) = y + kt$

17. $T = 60 + 150e^{-0.6} \approx 142.35$

19. $xy = C$

21. $y^2 = Cx$

23. $x = Cae^{at}/(b + bCe^{at})$

25. $50e^{-1.2} \approx 15.06\,\text{lb}$

27. $z = \dfrac{x_0 Ce^{(2y_0-x_0)kt} - y_0}{\frac{2}{3}Ce^{(2y_0-x_0)kt} - \frac{1}{3}}, \qquad C > 0$

29. $t = 100(\sqrt{15} - \sqrt{10}) + \dfrac{400}{\pi}\ln\dfrac{(\pi\sqrt{15}/4 - 1)}{(\pi\sqrt{10}/4 - 1)} \approx 112\,\text{sec.}$

31. $M = Ce^{k(t-t_0)^2/2}$

CHAPTER 9

Section 1

1. 2/3

3. 9/4

5. $(18 - 10\sqrt{3})/27$

7. $\dfrac{1}{2}\ln 3$

9. $\ln(4/3)$

11. $\dfrac{3}{2}\arctan(1/\sqrt{2}) - 1/\sqrt{2}$

13. $(14\sqrt{3}/5)a^5$

15. $\displaystyle\int_a^{\sqrt{3a}} \dfrac{dx}{x^2\sqrt{a^2 + x^2}} = \dfrac{1}{a^2}\left(\dfrac{\sqrt{6} - 2}{\sqrt{3}}\right)$

17. a

19. $1/\sqrt{2} + \dfrac{1}{2}\ln(1 + \sqrt{2})$

21. $\pi/2$

23. 1

25. Apply Theorem 1.

Section 2   **1.** $2\pi/3$ **3.** $4\sqrt{3}/3$ **5.** $4/3$ **7.** $14/15$ **9.** $a^2h/3$

              13. $2e^{2\pi} - 8e^{\pi} + 4\pi + 6$ **15.** $4/3$

              17. $\dfrac{2\sqrt{3}}{9}\, r^3$ **19.** $\dfrac{a^2h}{3}$ **21.** $\dfrac{16}{3}\, r^3$

Section 3   **1.** $\dfrac{2}{3}\,\pi a^3$ **3.** $1024\pi/7 = 2^{10}\pi/7$

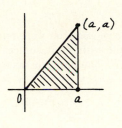

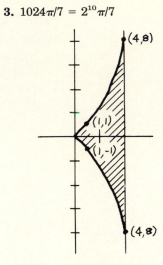

       5. $8\pi/15$ **7.** $2\pi^2$

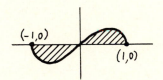

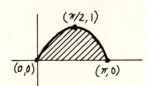

       9. π **11.** $28\pi\sqrt{21}$

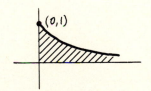

       13. $\dfrac{\pi}{2}(\pi - 1)$ **15.** $\dfrac{\pi}{2}(\sqrt{2} + \ln(\sqrt{2} + 1))$

       17. $\dfrac{3}{2} - 2\ln 2$

19. $\pi\left[\dfrac{1}{3} - 2(\ln 2 - 1)^2\right]$ (or $\pi/3 - 2\pi(1 - \ln 2)^2$)

Section 4
1. $\ln(\sqrt{2} + 1)$
3. $(1/a^2)((4/9 + a^2)^{3/2} - 8/27)$
5. $(8/27)(10\sqrt{10} - 1)$
7. $\sqrt{1 + e^2} - \sqrt{2} - \ln[(\sqrt{1 + e^{-2}} + e^{-1})/(\sqrt{2} + 1)]$
9. $(b - a)\sqrt{1 + m^2}$
11. $33/16$
13. $3/8 + \ln 2$
15. $7/3$
17. $g(b) - g(a)$
19. Any smooth curve may be approximated arbitrarily closely by an inscribed polygonal path; so it suffices to prove the inequalities for such a polygonal path P. The curve is decreasing and $0 = x_0 < x_1 < \cdots < x_n = 1$; therefore $1 - y_0 > y_1 > \cdots > y_n = 0$.

$$\sqrt{(x_k - x_{k-1})^2 + (y_k - y_{k-1})^2} \le (x_k - x_{k-1}) + (y_{k-1} - y_k). \quad \text{(Triangle Law)}$$

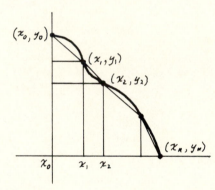

Show, therefore, that

$$\text{the length of } P \le (x_n - x_0) + (y_0 - y_n) = 2.$$

The distance from $(0, 1)$ to $(1, 0)$ is $\sqrt{2}$, and the shortest path between two points is a straight line, so $\sqrt{2} \le \text{length of } P \le 2$.

Section 5
1. $\dfrac{\pi}{9}(2\sqrt{2} - 1)$
3. $\dfrac{515}{64}\pi$
5. $\pi\left[\ln\dfrac{2 + \sqrt{5}}{1 + \sqrt{2}} + 2\sqrt{5} - \sqrt{2}\right]$
7. $9\pi\left[2\sqrt{17} + \dfrac{1}{2}\ln(4 + \sqrt{17})\right]$
9. $3\pi\left[10\sqrt{5} + \dfrac{2}{5}\right]$
11. $\dfrac{506}{1215}\pi$
13. $\dfrac{67}{10}\pi$
15. $\dfrac{4\pi}{15}(11\sqrt{2} - 4)$
17. $\pi\left[2\ln\left(\dfrac{4}{3}\right) - \dfrac{1}{4}\right]$
19. $6\pi/5$

Section 6

1. The total depth is $h = 6(\sqrt{3}/2) = 3\sqrt{3}$. The triangle width at depth x is

$$f(x) = (3\sqrt{3} - x)\frac{2}{\sqrt{3}} = 6 - \frac{2}{\sqrt{3}}x.$$

Therefore,

$$F = 62.4 \int_0^{3\sqrt{3}} x\left(6 - \frac{2}{\sqrt{3}}x\right) dx \approx 1685 \text{ pounds.}$$

3. The side strip between x and $x + \Delta x$ has width

$$\Delta s = \frac{5\sqrt{5}}{10}\Delta x = \frac{\sqrt{5}}{2}\Delta x,$$

circumference $\approx \pi(10 - x)$, and area $\approx$

$$\pi(10 - x)\frac{\sqrt{5}}{2}\Delta x.$$

Therefore

$$F = 62.4 \int_0^{10} x\pi(10 - x)\frac{\sqrt{5}}{2} dx$$

$$= \frac{(62.4)\sqrt{5}\,\pi}{2} \int_0^{10} (10x - x^2) \, dx \approx 73{,}000 \text{ pounds.}$$

5. 400π lb   **7.** $800/3$ lb   **9.** $80{,}640\pi$ lb

Section 7

1. $1/2$ **3.** Diverges   **5.** Diverges   **7.** Diverges

9. Diverges   **11.** Diverges   **13.** $V = 3\pi$ **19.** Converges

21. Converges   **23.** Diverges   **25.** Converges   **27.** Converges

29. ∞ **31.** $2/e$

33. $1/(a - 1)(\ln 2)^{a-1}$ if $a > 1$; ∞ if $0 < a \le 1$.

Section 8

1. $\dfrac{1}{10}\left[\dfrac{1 + 1/2}{2} + \displaystyle\sum_{k=1}^{9} \dfrac{10}{10 + k}\right] = \dfrac{3}{40} + \dfrac{1}{11} + \dfrac{1}{12} + \dfrac{1}{13} + \cdots + \dfrac{1}{19} = 0.6937 \cdots;$

$K\dfrac{(b - a)(\Delta x)^2}{12} = \dfrac{1}{600};$ thus $0.691 < \ln 2 < 0.694$.

3. $S = \dfrac{1}{10}\left[\dfrac{3}{4} + \dfrac{100}{101} + \dfrac{25}{26} + \dfrac{100}{109} + \dfrac{25}{29} + \dfrac{4}{5} + \dfrac{25}{34} + \dfrac{100}{149} + \dfrac{25}{41} + \dfrac{100}{181}\right]$

$= 0.78498 \cdots.$

The main problem in the error estimate is finding a reasonable bound K for the second derivative of $f(x) = 1/(1 + x^2)$. Compute $f''(x)$ and show that $|f''(x)| \le 2$ (because $|1 - 3x^2| \le 1 + 3x^2$).

7. $10^{-5}/3$; $10^{-5}/48$

9. p a polynomial of degree at most 5;

$$p(b) - p(a) = \frac{p'(b) + p'(a)}{2}(b - a)$$

$$- \frac{p''(b) - p''(a)}{12}(b - a)^2 + \frac{p^{(5)}(X)(b - a)^5}{720}$$

Section 9   **1.** A polynomial of degree three or less has fourth derivative 0, so one can set $K = 0$ in the error estimate.

CHAPTER 10   **1.** [r, θ-plane]

Section 1

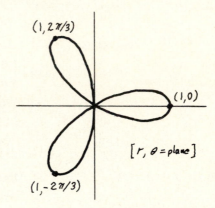

3.

5.

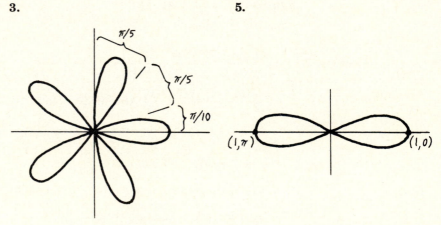

7. Circle of radius 1; center at (0, 1)   **9.**
(rectangular coordinates)

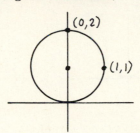

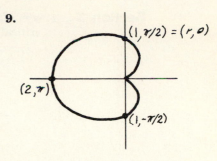

11.

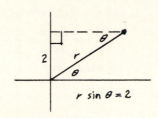

13. The equation in Cartesian coordinates is $y^2 = 1 + 2x$.

17. $(1/2, 2\pi/3)$, $(1/2, 4\pi/3)$ **19.** $(1 + \sqrt{2}/2, \pi/4)$, $(1 - \sqrt{2}/2, 5\pi/4)$

23. $\cot \psi = 1$. So ψ has the constant value $\pi/4$. The curve is turning counterclockwise at a constant rate equal to the turning rate of a ray from the origin, $\phi = \theta + \pi/4$.

25. $\cot \psi = -\tan \theta$, so $\psi = \theta + \pi/2$.
Erect the perpendicular to the tangent line T at the point of tangency P, intersecting the axis $\theta = 0$ at C. The fact that $\psi = \theta + \pi/2$ means that $\triangle OCP$ is isosceles consistent with the fact that the graph is a circle centered at C.

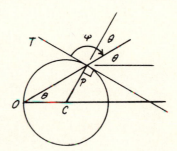

Section 2   **1.** $16\pi^5/5$ **3.** $\pi/8$

5. Area in the quadrant: $n\pi/2 \le \theta \le (n + 1)\pi/2 = \dfrac{e^{n\pi}}{4}(e^\pi - 1)$

7. $3\pi/2$ **9.** 11π **11.** a/n

13. $a^2\pi/4n$ **15.** $\pi - 3\sqrt{3}/2$ **17.** $\pi/2(1 + \pi)$

19. $\pi/2$ **21.** $5\pi/6 + 7\sqrt{3}/8$ **23.** $\pi/3$

Section 3

1. $y = x^2$; parabola $dy/dx = 2t$; critical point $t = 0$

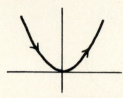

3. $y = (x + 1)^3$; $dy/dt = 3t^2$, critical point at $t = 0$

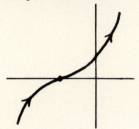

5. $y = 3x/2$; straight line;

$$\frac{dy}{dx} = \frac{dy/dt}{dx/dt} = \frac{3}{2}\frac{t^{-2/3}}{t^{-2/3}} = \frac{3}{2},$$

except at $t = 0$, where undefined

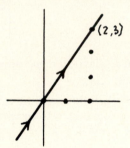

7. $\dfrac{x^2}{4} + \dfrac{y^2}{9} = 1$; ellipse;

$$\frac{dy}{dx} = -\frac{3}{2}\tan t;$$

critical points at $t = n\pi/2$

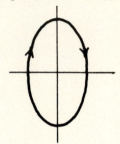

9. Ellipse; $\dfrac{x^2}{4} + \dfrac{y^2}{16} = 1$;

$$\frac{dy}{dx} = -2\cot t;$$

critical points $t = n\pi/2$

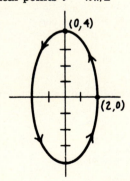

11. Circle; $x^2 + y^2 = 4$;

$$\frac{dy}{dx} = -\cot t;$$

critical points $t = n\pi/2$

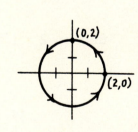

13.

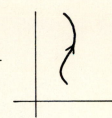

15.

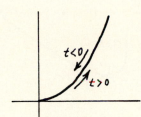

17. 5 sec; 300 ft; 100 ft

19.

t	$-\infty$	0	$+\infty$
x	$+\infty$	0	$+\infty$
y	$+\infty$	0	$+\infty$

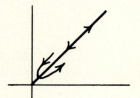

Singular point $t = 0$.

21.

t	$-\infty$	-1	0	1	2	$+\infty$
x	$-\infty$	2	0	-2	2	$+\infty$
y	$-\infty$	-4	0	-2	-4	$+\infty$

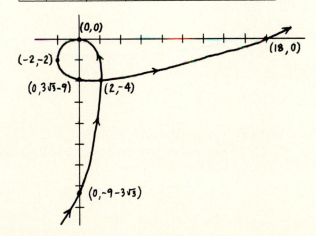

23.

t	$-\infty$	0	1	2	$+\infty$
x	$-\infty$	0	1	2	$+\infty$
y	$-\infty$	0	-2	-4	$+\infty$

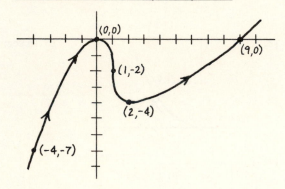

25. Singular point at $t = 0$

t	$-\infty$	-1	0	1/2	1	$+\infty$
x	$-\infty$	-7	0	$-1/4$	1	$+\infty$
y	$+\infty$	-1	0	$-7/16$	-1	$+\infty$

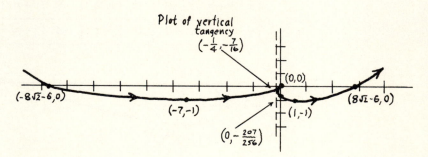

29. $(\overline{PF})^2 = (a\cos\theta - c)^2 + b^2\sin^2\theta$

$\qquad\quad = a^2\cos^2\theta - 2ac\cos\theta + c^2 + (a^2 - c^2)\sin^2\theta$

$\qquad\quad = a^2 - 2cx + c^2\cos^2\theta$

$\qquad\quad = \left(a - \dfrac{c}{a}x\right)^2 = (a - ex)^2$

31. $(\overline{PF}^2) = (a\sec\theta - c)^2 + b^2\tan^2\theta$

$\qquad = a^2\sec^2\theta - 2ac\sec\theta + c^2 + (c^2 - a^2)\tan^2\theta$

$\qquad = a^2 - 2cx + c^2\sec^2\theta$

$\qquad = \left(a - \dfrac{c}{a}x\right)^2 = (a - ex)^2$

Thus $\overline{PF} = \pm(a - ex)$. On the right branch $ex > a$, since $e > 1$ and $x > a$, so then $\overline{PF} = ex - a$.

Section 4

1. $(1 + 4t^2)^{1/2}$

3. $(9\cos^2 t + \sin^2 t)^{1/2} = \sqrt{1 + 8\cos^2 t}$

5. $a^2/2 + a$ **7.** a) $5\sqrt{10} - 8\sqrt{2}$ b) $32 - 8\sqrt{2}$

9. $\sqrt{2}(e - 1)$ **11.** $1/2$

15. $(8/3)((\pi^2 + 1)^{3/2} - 1)$ **17.** 1

19. $\ln(2[\sqrt{1 + \pi^2} + \pi]/[\sqrt{4 + \pi^2} + \pi]) + \dfrac{1}{\pi}[\sqrt{4 + \pi^2} - \sqrt{1 + \pi^2}]$

Section 6

1. 0 **3.** 1 **5.** $1/2$ **7.** $1/2$

9. $-1/2$ **11.** $-1/4$ **13.** 1

15. 0 (but not by l'Hôpital; see page 319)

17. 1 **19.** 0

21. The definition of $f'(0)$ leads to the following direct proof.

$$\lim_{x\to 0}\frac{f(x) - f(-x)}{x} = \lim_{x\to 0}\frac{f(x) - f(0) - [f(-x) - f(0)]}{x}$$

$$= \lim_{x\to 0}\frac{f(x) - f(0)}{x} + \lim_{x\to 0}\frac{f(-x) - f(0)}{-x}$$

$$= f'(0) + f'(0) = 2f'(0)$$

But work the problem by l'Hôpital's rule, assuming that $f'(x)$ is continuous.

23. L'Hôpital gives

$$\lim_{x\to 0}\frac{\ln(1 + x)}{x} = 1.$$

25. 1 **27.** ∞ **29.** 0 **31.** ∞ **33.** $1/4$

35. $7/5$ **37.** 1 **39.** e **41.** 1

43. If we define $H(0) = 0$, then H becomes continuous at the origin, and $H'(y)/G'(y)$ turns out to be $h'(1/y)/g'(1/y)$.

CHAPTER 11

Section 1

1. $\dfrac{1}{7} - \dfrac{10}{71} = \dfrac{1}{497}$

3. $\dfrac{385}{536} - \dfrac{334}{465} = \dfrac{1}{249{,}240}$

5. 3.8 (correct expansion; next place between 2 and 7)

7. 1.4 (closest; within 0.02)

9. 3.1 (closest; within 0.025)

11. 1.0 (correct); 1.1 (closest)

13. Within 0.006

15. Within 0.005; within 0.011

19. Since $I_{n+1} \subset I_n$, $n = 1, 2, 3 \ldots$, the sequence $\{a_n\}$ is increasing and $\{b_n\}$ is decreasing. Also $a_n \le b_k$ for every n and k (for if p is larger than both n and k, then $a_n \le a_p \le b_p \le b_k$). Thus b_k is an upper bound for $\{a_n\}$, so $\{a_n\}$ converges, by the monotone limit property, and its limit a satisfies $a \le b_k$ for each k, by a variant of Limit Law (D), page 252. Thus $\{b_k\}$ converges and its limit b is $\ge a$, by Problem 18 and (D).

21. For a fixed positive integer m, the m-place decimals divide $[0, 1]$ into 10^m closed intervals of length $1/10^m$. Since there is a finite number of intervals, we may determine the last interval (on the right) that contains any of the numbers a_n. Call its right endpoint b_m. By construction, b_m is the smallest m-place decimal that is an upper bound of $\{a_n\}$.

23. Fix N. Since the sequence $\{a_n\}$ is increasing, with upper bound b_N, and since some a_{n_0} is in the interval $J_N = [b_N - 1/10^N, b_N]$, it follows that J_N contains every a_n with $n \ge n_0$. It is also clear from Problem 22 that if $m \ge N$ then b_m is contained in J_N. Thus *both* a_n and b_n are in J_N if $n \ge$ both n_0 and N, and then $b_n - a_n < 1/10^N$ for any such n. Thus, for any integer N, the sequence $\{b_n - a_n\}$ contains terms less than 10^{-N}. Since $\{b_n - a_n\}$ is decreasing, and since its terms eventually become less than 10^{-N}, no matter what N is, it follows that $b_n - a_n \to 0$ as $n \to \infty$. (We have shown that its terms can be made as small as desired simply by taking n large enough.)

25. Since $\{f(n)\}$ is increasing and bounded above by B, it converges, by the monotone limit property, and its limit l is $\le B$. We claim that $f(x) \to l$ as $x \to \infty$. First, note that $f(x) \le l$ for every x (because if we choose an integer $n \ge x$ then we have $f(x) \le f(n) \le l$). We can make $l - f(n)$ as small as we wish simply by choosing a large enough integer n. Since for any $x \ge n$ we have $0 \le l - f(x) \le l - f(n)$, it follows that we can make $l - f(x)$ as small as we wish simply by choosing x larger than a suitably chosen integer n. That is, $f(x) \to l$ as $x \to \infty$.

Section 2

1. 4/3

3. 3/5

5. 3

7. 1/6

9. 3/4

11. 27/16

13. 12

15. 1/6

19. Multiply by 2, and compare to $\sum 1/n$.

21. Multiply by 4 and compare to $\sum 1/n$.

23. $\sum_{n=1}^{\infty} \dfrac{23}{(100)^n} = \dfrac{23}{99}$ **25.** $\sum_{n=1}^{\infty} \dfrac{315}{(1000)^n} = \dfrac{35}{111}$ **27.** $2 + \sum_{n=1}^{\infty} \dfrac{1}{100^n} = 2\dfrac{1}{99}$

31. $2s_n = \dfrac{2}{1 \cdot 3} + \dfrac{2}{3 \cdot 5} + \dfrac{2}{5 \cdot 7} + \cdots + \dfrac{2}{(2n-1)(2n+1)}$

$$= \left[\dfrac{1}{1} - \dfrac{1}{3}\right] + \left[\dfrac{1}{3} - \dfrac{1}{5}\right] + \left[\dfrac{1}{5} - \dfrac{1}{7}\right] + \cdots + \left[\dfrac{1}{2n-1} - \dfrac{1}{2n+1}\right]$$

$$= 1 - \dfrac{1}{2n+1} = \dfrac{2n}{2n+1}.$$

Therefore $s = \frac{1}{2}$.

33. This series can be obtained as the sum of the series in Problem 31 and the series in Problem 32. Its sum is 3/4.

35. 390 ft

Section 3   **1.** Diverges   **3.** Diverges   **5.** Converges   **7.** Diverges

9. Converges (because $\ln n < n$)

11. Converges ($\sin x < x$ for positive x)

13. Converges ($n!/n^n \leq 2/n^2$)   **15.** Diverges

17. Converges   **19.** Diverges   **21.** Converges   **23.** Converges

Section 4   **1.** Converges   **3.** Converges

5. Test fails (but the series converges by comparison with $\sum 1/n^2$)

7. Test fails (but have divergence by comparison with harmonic series)

9. Converges   **11.** Converges   **13.** Diverges

15. $1/x^2$ is decreasing and continuous, so, by Theorem 7,

$$\sum_{n=2}^{\infty} \dfrac{1}{n^2} < \int_1^{\infty} \dfrac{dx}{x^2} = 1.$$

17. $\sum_2^{\infty} 1/n^3 < 1/2$, as in Problem 15

19. Diverges   **21.** Converges   **23.** Diverges   **25.** Diverges

27. Converges   **29.** Converges   **31.** Converges   **33.** Diverges

35. Converges if $p > 1$ **37.** Converges   **39.** Converges

43. The trapezoidal approximation to $\int_a^b f$ is less than the integral if f is concave down. Therefore,

$$\dfrac{\ln 1}{2} + \ln 2 + \cdots + \ln(n-1) + \dfrac{\ln n}{2} < \int_1^n \ln x \, dx$$

or $\ln[(n-1)! \sqrt{n}] < n \ln n - n + 1$.

Section 5

1. Converges absolutely
3. Diverges
5. Diverges
7. Converges conditionally
9. Converges conditionally
11. Diverges
13. Converges conditionally
15. $\sum (-1)^n \dfrac{n+1}{n}$

Extra Problems for Chapter 11

1. Converges absolutely
3. Converges absolutely
5. Converges absolutely
7. Converges absolutely
9. Converges conditionally
11. Diverges
13. Converges absolutely
15. Converges absolutely
17. Converges absolutely
19. Converges conditionally
21. Converges absolutely
23. Converges absolutely
25. Converges absolutely
27. Converges conditionally if x is not a multiple of π; converges absolutely if x is a multiple of π.
29. Converges conditionally
31. Converges absolutely if $|r| < 1$; diverges otherwise
33. Diverges
35. Converges absolutely
37. Diverges
39. Converges absolutely if $|r| < 1$; diverges otherwise
41. Converges absolutely
43. Converges conditionally
45. Converges conditionally
49. 0.40546510

CHAPTER 12

1. $[-1, 1)$
3. $(-\infty, \infty)$
5. $(-1, 1)$
7. $[-1, 1]$
9. $(-1, 1]$
11. $[-1, 1)$
13. $(0, 4]$
15. $[-2, 4]$

Section 1

17. $(0, 4)$
19. $\left(\dfrac{1}{2}, \dfrac{3}{2}\right)$
21. $[0, 6]$
23. 6

25. $1/(1 - x)^2$
27. $(1 + x)/(1 - x)^3$

Section 2

1. $1 + x + x^2 + x^3 + \cdots$
3. $0 + 0 + 0 + x^3 + 0 + \cdots$

5. $5 - 2x + 0 + x^3 + 0 + \cdots$
7. $\displaystyle\sum_{n=0}^{\infty} \dfrac{(-1)^n x^{2n}}{n!}$; $(-\infty, \infty)$

9. $\displaystyle\sum_{n=0}^{\infty} \dfrac{(-1)^n x^{6n+3}}{(2n+1)}$; $[-1, 1]$
11. $\displaystyle\sum_{n=0}^{\infty} \dfrac{(-1)^n x^{2n+1}}{2^n (2n+1) n!}$; $(-\infty, \infty)$

13. $\displaystyle\sum_{n=0}^{\infty} \dfrac{(-1)^n x^{2n+1}}{(2n+1)(2n+2)!}$; $(-\infty, \infty)$

19. $1 + \frac{1}{2}(x-1) - \frac{1}{8}(x-1)^2 + \cdots + \dfrac{\frac{1}{2}\left(-\frac{1}{2}\right)\left(-\frac{3}{2}\right)\cdots((3-2n)/2)}{n!}$
$\times (x-1)^n + \cdots$

21. $\displaystyle\sum_{n=0}^{\infty} \frac{(-1)^n(x-\pi/2)^{2n}}{(2n)!}$ **23.** $\displaystyle\sum_{n=0}^{\infty}(-1)^{n+1}\frac{(x-\pi)^{2n+1}}{(2n+1)!}$

29. $\dfrac{2x^2}{2!} - \dfrac{2^3 x^4}{4!} + \dfrac{2^5 x^6}{6!} - \cdots + \dfrac{(-1)^{n+1}2^{2n-1}x^{2n}}{(2n)!} + \cdots$

31. $x - \dfrac{x^3}{3^2} + \dfrac{x^5}{5^2} - \cdots + (-1)^{n+1}\dfrac{x^{2n+1}}{(2n+1)^2}$

Section 4

1. $p_3(x) = x + x^3/3!$ **3.** $p_3(x) = 1 + x^2/2$

5. $p_4(x) = (\sqrt{2}/2)(1 + x - (x^2/2!) - (x^3/3!) + (x^4/4!))$

7. $p_4(x) = x + (x^3/3)$

9. $(x-1) - ((x-1)^2/2) + ((x-1)^3/3)$

11. $1/2 + (\sqrt{3}/2)(x - \pi/6) - (1/4)(x-\pi/6)^2 - (\sqrt{3}/12)(x-\pi/6)^3$
$+ (1/48)(x-\pi/6)^4$

13. $e^a + e^a(x-a) + (e^a/2)(x-a)^2 + (e^a/6)(x-a)^3$

15. $4 + 7(x-1) + 5(x-1)^2 + (x-1)^3$

17. We proved earlier that the derivative of an odd function is even and vice versa, so if f is odd, then $f^{(2n)}$ is odd for all $n = 0, 1, \ldots$. So f has no even terms in its expansion. The argument for f even is similar.

19. $A = 1/2$, $B = -1/8$, $C = 1/16$ **21.** $x + x^3/3$

23. Let $g(x) = f(x) - p_n(x)$. Then g has derivatives up through order $n - 1$ on an interval about the origin, $g^n(0)$ exists, and $g(0) = g'(0) = \cdots = g^n(0) = 0$.

25. Suppose f is even and suppose p_n is one of its Maclaurin polynomials.

$$\lim_{x\to 0}\frac{f(x) - p_n(x)}{x^n} = 0; \quad \lim_{x\to 0}\frac{f(-x) - p_n(-x)}{(-x)^n} = 0.$$

$$f(x) = f(-x), \text{ so } \lim_{x\to 0}\frac{f(x) - p_n(-x)}{x^n} = 0.$$

Therefore, by Problem 24, $p_n(-x)$ is also the Maclaurin polynomial of f of degree n; that is, $p_n(-x) = p_n(x)$. So p_n is even. The proof for f odd is similar.

Section 5

1. $f(x) = -\dfrac{x^2}{2} - 2(3\sec^4 X - 2\sec^2 X)\dfrac{x^4}{4!}$

We can take $K = 5.1$. (Use the fact that $1/2 < \pi/6$.)

3. $f(x) = 1 + \dfrac{x^2}{2} + \dfrac{(5 \sec^3 X \tan X + \sec X \tan^2 X)}{6} x^3$

We can take $K = 1$. (Use the fact that $1/2 < \pi/6$.)

5. $f(x) = 1 - x + (x^2/2) - (x^3/3!) + (x^4/4!) - (e^{-X}/5!)x^5$ can take $K = 1/60$ (because $e^{1/2} < 2$).

7. $f(x) = 1 + (3/2)x + (3/8)x^2 - (1/16)x^3 + \left(\dfrac{9}{16}(1 + X)^{-5/2}/4!\right)x^4$;

$K = 3\sqrt{2}/32$

9. $\dfrac{\sqrt{2}}{2}\left[1 - (x - \pi/4) - \dfrac{(x - \pi/4)^2}{2} + \dfrac{(x - \pi/4)^3}{3!} + \dfrac{(x - \pi/4)^4}{4!}\right]$

$- \dfrac{\sin X}{5!}(x - \pi/4)^5$

11. p_4 as in Problem 11 of Section 4. $r_4(x) = (\cos X/5!)(x - \pi/6)^5$

13. $1 + (1/3)(x - 1) - (1/9)(x - 1)^2 + (5/81)(x - 1)^3 - (10/243)X^{-11/3}(x - 1)^4$

15. $1 + r(x - 1) + \dfrac{r(r - 1)}{2}(x - 1)^2 + \dfrac{r(r - 1)(r - 2)}{3!}(x - 1)^3$

$+ \dfrac{r(r - 1)(r - 2)(r - 3)}{4!} X^{r-4}(x - 1)^4$

17. $-\dfrac{1}{2}\ln 2 - (x - \pi/4) - (x - \pi/4)^2 - \dfrac{2}{3}(x - \pi/4)^3$

$- \dfrac{4 \sec^2 X \tan^2 X + 2 \sec^4 X}{4!}(x - \pi/4)^4$

19. $R_{2n} = \pm(\cos X)\dfrac{x^{2n+1}}{(2n + 1)!}$.

Since $|\cos X| \leq 1$, and since $x^n/n! \to 0$ as $n \to \infty$, it follows that $R_{2n} \to 0$ as $n \to \infty$.

21. $|r_n(x)| = \dfrac{\left(\dfrac{\sin}{\cos}\right)(X)(x - \pi/2)^{n+1}}{(n + 1)!} \leq \dfrac{(x - \pi/2)^{n+1}}{(n + 1)!}$.

Since x is fixed, you must show that $(K^n/n!) \to 0$ as $n \to \infty$.

23. If $x > a$, then $|r_n(x)| \leq e^x \left|\dfrac{(x - a)^{n+1}}{(n + 1)!}\right| = \dfrac{K_1(K_2)^{n+1}}{(n + 1)!}$.

25. $\sin 0.2 \simeq 0.19866933$ (The next digit is between 0 and 3.)

27. Integrate $e^{-t^2/2} = \sum_{n=0}^{\infty} \frac{(-t^2/2)^n}{n!}$ termwise; 0.86.

29. $\int_0^1 \cos \sqrt{x}\, dx \approx 1 - \frac{1}{4} + \frac{1}{3 \cdot 4!} - \frac{1}{4 \cdot 6!}$

with a positive error less than $1/5 \cdot 8!$ Doing some arithmetic shows that the integral is

$$0.76354 \cdots \approx 0.7635$$

to the nearest four decimal places.

CHAPTER 13

Section 1

1.

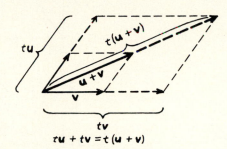

$$t\mathbf{u} + t\mathbf{v} = t(\mathbf{u} + \mathbf{v})$$

3. Force along bow string $\approx$ 57 pounds

5. $\theta = 45°$ **7.** speed $\approx$ 196 mph

9. $\mathbf{a} + \mathbf{x} = (a, b) + (x, y) = (a + x, b + y)$. Thus,

$$\mathbf{a} + \mathbf{x} = \mathbf{a} \leftrightarrow (a + x, b + y) = (a + b)$$
$$\leftrightarrow a + x = a \quad \text{and} \quad b + y = b$$
$$\leftrightarrow x = 0 \quad \text{and} \quad y = 0.$$

11. $(\mathbf{u} + \mathbf{x}) + \mathbf{a} = ((u, v) + (x, y)) + (a + b)$
$$= (u + x, v + y) + (a + b)$$
$$= ((u + x) + a, (v + y) + b)$$
$$= (u + (x + a), v + (y + b)), \text{ etc.}$$

13. $0\mathbf{a} = 0(a, b) = (0a, 0b) = (0, 0) = \mathbf{0}$

15. $\overrightarrow{\mathbf{a_1 x_1}} = \overrightarrow{\mathbf{a_2 x_2}} \Rightarrow \mathbf{x_1} - \mathbf{a_1} = \mathbf{x_2} - \mathbf{a_2}$
$$\Rightarrow \mathbf{a_2} - \mathbf{a_1} = \mathbf{x_2} - \mathbf{x_1}$$
$$\Rightarrow \overrightarrow{\mathbf{a_1 a_2}} = \overrightarrow{\mathbf{x_1 x_2}}$$

Section 2

1. $(x, y) = (1, 2) + t(-3, 0); \; x = 1 - 3t, \; y = 2$

3. $(x, y) = t(3, 4); \; x = 3t, \; y = 4t$

5. $x = x_0 + t, \; y = y_0 + t$ **7.** $(-13/8, -15/4)$

9. $\mathbf{x} = (1, 1) + t[(3, 0) - (1, 1)] = (1, 1) + t(2, -1); \; x = 1 + 2t, \; y = 1 - t$

11. $\mathbf{x} = (2, 0) + t(0, 1); x = 2, y = t$

13. $(-11/3, 1/3)$ **15.** $(5/6, 2/5)$ **17.** $(-1/6, 9/2)$

Section 3   **1.** $|\mathbf{v}| = 5; \dfrac{\mathbf{v}}{|\mathbf{v}|} = (3/5, 4/5)$ **3.** $|\mathbf{a}| = \sqrt{10}; \dfrac{\mathbf{a}}{|\mathbf{a}|} = \left(\dfrac{-3}{\sqrt{10}}, \dfrac{1}{\sqrt{10}}\right)$

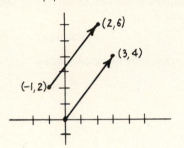   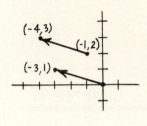

5. $|\mathbf{v}| = \sqrt{5}; \dfrac{\mathbf{v}}{|\mathbf{v}|} = \left(\dfrac{1}{\sqrt{5}}, \dfrac{-2}{\sqrt{5}}\right)$ **7.** $\mathbf{a} \cdot \mathbf{x} = 5/2; \phi = \arccos(5/\sqrt{34})$

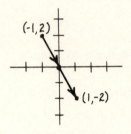   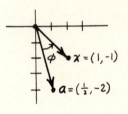

9. $\mathbf{a} \cdot \mathbf{x} = 0; \phi = \pi/2$ **11.** $\pm(2/\sqrt{5}, -1\sqrt{5})$

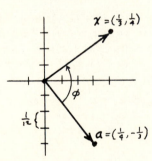

13. $\left(\dfrac{\sqrt{2} - \sqrt{6}}{4}, \dfrac{\sqrt{6} + \sqrt{2}}{4}\right)$ or $\left(\dfrac{\sqrt{6} + \sqrt{2}}{4}, \dfrac{\sqrt{2} - \sqrt{6}}{4}\right)$

15. $t(1, \sqrt{3}), s(1, -\sqrt{3})$ **17.** $-2x + 3y = 0$

19. The equation of l is $\mathbf{a} \cdot (x - x_0) = 0$. The point x_p closest to the origin is on the line through the origin perpendicular to l, so $x_p = t\mathbf{a}$. So

$$\mathbf{a} \cdot (t\mathbf{a} - x_0) = 0 \Rightarrow t\mathbf{a} \cdot \mathbf{a} - \mathbf{a} \cdot x_0 = 0$$
$$\Rightarrow t = \mathbf{a} \cdot x_0 / \mathbf{a} \cdot \mathbf{a}.$$

21. $-3/\sqrt{13}$ **23.** 0 **25.** $10/\sqrt{101}$ **27.** $1/\sqrt{101}$

Section 4   **1.**

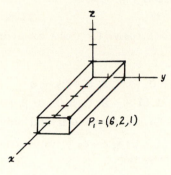

$P_1 = (6, 2, 1)$

3.

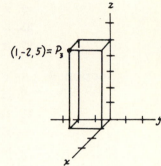

$(1, -2, 5) = P_3$

5.

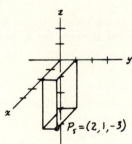

$P_5 = (2, 1, -3)$

7.

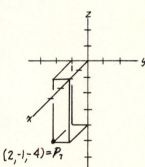

$(2, -1, -4) = P_7$

9.

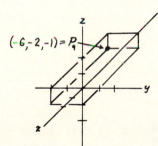

$(-6, -2, -1) = P_9$

11. $\sqrt{26}$ **13.** $\sqrt{42}$ **15.** $3\sqrt{10}$

17. The plane perpendicular to the y-axis (parallel to the xz-coordinate plane) and cutting the y-axis at $y = 3$

Section 5   **1.** $(2/3, 1/3, 2/3)$ **3.** $(7/11, -6/11, -6/11)$

5. $(2/3, 2/3, 1/3)$ **7.** $0, \phi = \pi/2$

9. $0, \phi = \pi/2$ **11.** $\mathbf{x} = (1 + t, -2 + 3t, 1 - 2t)$

13. $\mathbf{x} = (-2 + 3t, -1 + 4t, 2 + 2t)$ **15.** $\mathbf{x} = (-1 + 3t, -2 + 2t, 1 + t)$

17. $x = 1, y = -1 + (3/2)t, z = 4 - 4t$

19. $x = -2 + (5/2)t, y = -1, z = -2t$

21. $\mathbf{x} = (t, -2t, -t)$ **23.** $\mathbf{x} = (1 + 2t, 1, 1 - t)$

25. $\mathbf{a} = (\cos \alpha, \cos \beta, \cos \gamma); \mathbf{l} = (\cos \lambda, \cos \mu, \cos \nu); \mathbf{a} \cdot \mathbf{l} = \cos \phi$

27. $(t, -t, -t)$

Section 6   **1.** $x + 2y + z = 0$ **3.** $x + 2y = 6$ **5.** $4x + 3y - 2z = 48$

7. $(1, 0, 1), (2, -1, 0)$ **9.** $(1, 5, 0), (0, 3, 1)$

11. $t\mathbf{a}_1 + s\mathbf{a}_2$ is in the xy-plane if and only if $t + s = 0$. So $s = -t$, and points
$$t\mathbf{a}_1 + s\mathbf{a}_2 = t(\mathbf{a}_1 - \mathbf{a}_2) = t(1, -4, 0)$$
form the intersection.

13. $\mathbf{x} = t(3, -5, -7)$ **15.** $x + 9y + 7z = 12$

17. Note that reversing the roles of $\mathbf{a}_1$ and $\mathbf{a}_2$ reverses the sign of each expression in the formula for $\mathbf{a}_1 \times \mathbf{a}_2$.

19. $(\mathbf{a}_1 \times \mathbf{a}_2) \cdot \mathbf{a}_2 = \mathbf{a}_1 \cdot (\mathbf{a}_2 \times \mathbf{a}_2)$
$$= \mathbf{a}_1 \cdot \mathbf{0}$$
$$= 0$$

21. Expand both sides in coordinate notation to verify the formula.

23. The parallelogram with base $|\mathbf{a}_2|$ has altitude $|\mathbf{a}_1| \sin \phi$. So its area is $|\mathbf{a}_1| \, |\mathbf{a}_2| \sin \phi = |\mathbf{a}_1 \times \mathbf{a}_2|$.

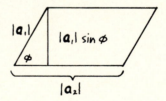

CHAPTER 14   **1.** $\dfrac{d\mathbf{f}}{dt} = (1, -1/t^2); \ t = \pm 1$ $(\mathbf{f}(1) = \pm(1, 1))$

Section 1   **3.** $\dfrac{d\mathbf{f}}{dt} = (2t, 1); \ t = -1$ $(\mathbf{f}(-1) = (1, 2))$

5. $\dfrac{d\mathbf{x}}{dt} = (-\sin t, \cos t);$ everywhere

7. $(d\mathbf{g}/dt) = (-3\sin t, 2\cos t)$; $t = (n\pi/2)$ $(\mathbf{g} = (0, \pm 2), (\pm 3, 0))$

9. $(40, -2)$; $\mathbf{x} = (40, 14) + t(40, -2) = (40 + 40t, 14 - 2t)$,
or $x = 40 + 40t$, $y = 14 - 2t$

11. $(1, 1)$; $\mathbf{x} = (1, 0) + u(1, 1) = (1 + u, u)$, or $x = 1 + u$, $y = u$

13. $(0, 2)$; $\mathbf{x} = (2, 0) + u(0, 2) = (2, 2u)$

15. $(1, -2)$; $x = u$, $y = 2 - 2u$

17. $x = x_0 - at\sin\theta_0$, $y = y_0 + bt\cos\theta_0$

19. The increment arrows are drawn to points on the other side of $\mathbf{x}_0$, but when divided by the *negative* Δt, the adjusted arrows approach the same limiting arrow (representing the tangent vector).

21. Let $h(t) = \int_0^t f(s)\,ds$; then $h'(t) = f(t)$. By hypothesis, $x'(t) = uf(t)$, $y'(t) = vf(t)$, so $x(t) = uh(t) + k$, $y(t) = vh(t) + l$, and $\mathbf{x}(t) = \mathbf{k} + h(t)\mathbf{u}$. So $\mathbf{x}(t)$ is a straight line.

23. If d is the shortest distance between the two paths, then it is certainly the shortest distance from $\mathbf{x}_2$ to path $\mathbf{x} = \mathbf{g}(t)$. Therefore, by a slight generalization of Example 5, $\mathbf{x}_1\mathbf{x}_2$ is perpendicular to $\mathbf{x} = \mathbf{g}(t)$. By a similar argument, $\mathbf{x}_1\mathbf{x}_2$ is perpendicular to the other curve.

Section 2

1. By rule (5), $d\mathbf{x}/dt = \mathbf{a}$, a constant vector. Then by the example in this section $\mathbf{x} = \mathbf{a}t + \mathbf{c}$.

5. $\mathbf{x}\cdot d\mathbf{x}/dt = 0$ implies

$$0 = \mathbf{x}\cdot\frac{d\mathbf{x}}{dt} + \frac{d\mathbf{x}}{dt}\cdot\mathbf{x} = \frac{d}{dt}\mathbf{x}\cdot\mathbf{x} = \frac{d}{dt}|\mathbf{x}|^2.$$

So $|\mathbf{x}|^2 = c$, a constant. So $\mathbf{x} = \mathbf{f}(t)$ runs along the circle of radius $\sqrt{c}$ about O.

7. Assume that the path does not contain the origin (since the zero vector does not specify a direction). Solve for $|\mathbf{x}|\,d\mathbf{u}/dt$ in the identity

$$\frac{d\mathbf{x}}{dt} = \frac{d|\mathbf{x}|}{dt}\mathbf{u} + |\mathbf{x}|\frac{d\mathbf{u}}{dt}$$

and use the hypothesis of the problem, to conclude that $d\mathbf{u}/dt$ must be zero. Therefore, . . .

9. Your differentiated equation should be equivalent to $\cos\theta_1 + \cos\theta_2 = 0$, where θ_1 and θ_2 are the angles that $\mathbf{x} - \mathbf{f}$ and $\mathbf{x} - \mathbf{g}$ make with the tangent vector. Therefore $\theta_2 = \pi - \theta_1$, so the angle of incidence = angle of reflection = θ_1.

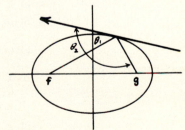

Section 3

1. $2/(1 + 4x^2)^{3/2}$ **3.** $-x/(1 + x^2)^{3/2}$ **5.** 0

7. The curvature takes on the value -1 at $x = 0$ and $+1$ at $x = \pi$. It is periodic, of period 2π, and is bounded by 1 in absolute value.

9. $2/3\,|a|$

11. If a function graph has 0 curvature, then $d^2y/dx^2 = 0$. Integrate this equation twice.

13. $K = ab/(a^2 \sin^2 t + b^2 \cos^2 t)^{3/2}$. Max and min: a/b^2, b/a^2 (supposing $0 < b < a$).

15. At $t = 0$ **17.** None **19.** $t = 1$

21. $\dot{x} = t\cos(t)$, $\ddot{x} = -t\sin(t) + \cos t$; $\dot{y} = t\sin(t)$, $\ddot{y} = t\cos(t) + \sin t$; so $\dfrac{d\phi}{ds} = \dfrac{1}{t}$.

23. $r = 2a\cos\theta$, $\dfrac{dr}{d\theta} = -2a\sin\theta$, $\dfrac{d^2r}{d\theta^2} = -2a\cos\theta$;

$$k = \frac{4a^2\cos^2\theta + 8a^2\sin^2\theta + 4a^2\cos^2\theta}{(4a^2\cos^2\theta + 4a^2\sin^2\theta)^{3/2}} = \frac{1}{a}$$

25. $\dfrac{1}{e^{a\theta}}\dfrac{1}{\sqrt{1 + a^2}}$

27. a) $(x, y) = \left(\dfrac{s}{\sqrt{2}}\cos\left(\ln\dfrac{s}{\sqrt{2}}\right), \dfrac{s}{\sqrt{2}}\sin\left(\ln\dfrac{s}{\sqrt{2}}\right)\right)$

   b) $\mathbf{T} = \left(\dfrac{1}{\sqrt{2}}\left(\cos\left(\ln\dfrac{s}{\sqrt{2}}\right) - \sin\left(\ln\dfrac{s}{\sqrt{2}}\right)\right), \dfrac{1}{\sqrt{2}}\left(\sin\left(\ln\dfrac{s}{\sqrt{2}}\right)\right.\right.$

   $\left.\left. + \cos\left(\ln\dfrac{s}{\sqrt{2}}\right)\right)\right)$

   $\dfrac{d\mathbf{T}}{ds} = \dfrac{1}{s\sqrt{2}}\left(-\sin\left(\ln\dfrac{s}{\sqrt{2}}\right) - \cos\left(\ln\dfrac{s}{\sqrt{2}}\right), \cos\left(\ln\dfrac{s}{\sqrt{2}}\right)\right.$

   $\left. - \sin\left(\ln\dfrac{s}{\sqrt{2}}\right)\right)$

   c) $\kappa = 1/s$ (Note that in all the above, $0 < s < \infty$.)

29. Note: $0 \le t < \infty$, $0 \le s < \infty$.

   a) $(x, y) = \left(\dfrac{4}{3}((1 + 2s)^{1/2} - 1)^{3/2}, 2 - 2(1 + 2s)^{1/2} + s\right)$

   b) $\mathbf{T} = (2(1 + 2s)^{1/2} - 1)^{1/2}(1 + 2s)^{-1/2}, -2(1 + 2s)^{-1/2} + 1)$;
   $(d\mathbf{T}/ds) = (((1 + 2s)^{1/2} - 1)^{-1/2}(1 + 2s)^{-1} - 2((1 + 2s)^{1/2} - 1)^{1/2}$
   $\times (1 + 2s)^{-3/2}, 2(1 + 2s)^{-3/2})$

   c) $\kappa = 1/(1 + 2s)\sqrt{\sqrt{1 + 2s} - 1}$

Section 4

1. $\mathbf{v} = (-3\sin 3t, 3\cos 3t)$; $\mathbf{a} = (-9\cos 3t, -9\sin 3t)$; $v = 3$, $dv/dt = 0$. We know that the unit circle has curvature $\kappa = 1$ (however, see the next answer); so $\kappa v^2 = 9$.

3. $\mathbf{v} = (-3\sin t, 2\cos t)$; $\mathbf{a} = (-3\cos t, -2\sin t)$; $v = (4 + 5\sin^2 t)^{1/2}$, $dv/dt = (5/2)\sin 2t/(4 + 5\sin^2 t)^{1/2}$. The absolute curvature κ can be computed from the formula $\kappa = |\ddot{x}\dot{y} - \ddot{y}\dot{x}|/v^3$ (Section 3); so $\kappa v^2 = |\ddot{x}\dot{y} - \ddot{y}\dot{x}|/v$. Here, $\kappa v^2 = 6/(4 + 5\sin^2 t)^{1/2}$.

5. $\mathbf{v} = (1, 2t)$; $\mathbf{a} = (0, 2)$; $v = (1 + 4t^2)^{1/2}$, $dv/dt = 4t(1 + 4t^2)^{-1/2}$. $\kappa v^2 = |\ddot{x}\dot{y} - \ddot{y}\dot{x}|/v = 2(1 + 4t^2)^{-1/2}$

7. $\mathbf{v} = (-6\sin 3t, 6\cos 2t)$; $\mathbf{a} = (-18\cos 3t, -12\sin 2t)$; $v = 6(\sin^2 3t + \cos^2 2t)^{1/2}$, $dv/dt = (9\sin 6t - 6\sin 4t)(\sin^2 3t + \cos^2 2t)^{-1/2}$; $kv^2 = (18\cos 3t\cos 2t + 12\sin 3t\sin 2t)(\sin^2 3t + \cos^2 2t)^{-1/2}$

9. $v \le 25\sqrt{2}$ ($\kappa = |d^2 y/dx^2|/(1 + (dy/dx)^2)^{3/2} = 1/50$ and κv^2 must be ≤ 25.)

11. If the particle is moving ($v \ne 0$) and the path is curved, $\kappa \ne 0$, then the normal component $\kappa v^2 \ne 0$.

13. $50\sqrt{2}$ mph ≈ 70 mph

Section 5

1. $(1, -2, 3)$ 3. $(3, 1, 1/2)$ 5. $(1/4, 1/4, -1/16)$

7. $\dfrac{d}{dt} w\mathbf{x} = \left(\dfrac{d}{dt} wx_1, \dfrac{d}{dt} wx_2, \dfrac{d}{dt} wx_3 \right)$

$= \left(w\dfrac{dx_1}{dt} + \dfrac{dw}{dt} x_1, w\dfrac{dx_2}{dt} + \dfrac{dw}{dt} x_2, w\dfrac{dx_3}{dt} + \dfrac{dw}{dt} x_3 \right)$

$= w\dfrac{d\mathbf{x}}{dt} + \dfrac{dw}{dt} \mathbf{x}$

9. Change to coordinate notation and use the ordinary chain rule, as in (7).

11. The function $|\mathbf{u} - \mathbf{x}_0|^2$ attains a minimum at $\mathbf{u}_0$, so

$$\frac{d}{dt} |\mathbf{u} - \mathbf{x}_0|^2 = 0$$

at this point. Apply Problem 8, and conclude that

$$2\frac{d\mathbf{u}}{dt} \cdot [\mathbf{u}_0 - \mathbf{x}_0] = 0.$$

$(d\mathbf{u}/dt)$ is the vector tangent to the curve $\mathbf{u}$ at $\mathbf{u}_0$, so $\mathbf{u}_0 - \mathbf{x}_0$ is perpendicular to $\mathbf{u}$. Similarly, $\mathbf{u}_0 - \mathbf{x}_0$ is perpendicular to $\mathbf{x}$ at $\mathbf{x}_0$.

13. Exactly similar to Section 2, Problems 1 and 3.

15. Prove that $|\mathbf{x}|^2 = \mathbf{x} \cdot \mathbf{x}$ is a constant.

$$0 = \frac{d\mathbf{x}}{dt} \cdot \mathbf{x}$$

implies

$$0 = \frac{d\mathbf{x}}{dt} \cdot \mathbf{x} + \mathbf{x} \cdot \frac{d\mathbf{x}}{dt} = \frac{d}{dt}[\mathbf{x} \cdot \mathbf{x}]$$

$$= \frac{d}{dt}(|\mathbf{x}|^2).$$

So $|\mathbf{x}|^2$ is constant.

17. $(d\mathbf{x}/dt) = \mathbf{x}f(t)$, where f is a scalar function. Let $\mathbf{u}\,|\mathbf{x}| = \mathbf{x}$; then follow the argument of Problem 7, Section 2.

21. $\kappa = (1 + 4t^2 + t^4)^{1/2}/(1 + t^2 + t^4)^{3/2}$

25. $\kappa = \dfrac{(t^4 + 4t^2 + 1)^{1/2}}{(t^4 + t^2 + 1)^{3/2}}; \ \tau = \dfrac{2}{t^4 + 4t^2 + 1}$

27. $\kappa = \dfrac{1}{b^2 + 1}; \ \tau = -\dfrac{b}{b^2 + 1}$

29. $\kappa = \dfrac{(a^4b^2c^2 + a^6c^4)^{1/2}}{(a^2c^2 + b^2)^{3/2}}; \ \tau = \dfrac{-ab}{b^2 + a^2c^2}$

31. $\kappa = \dfrac{(4e^{6t} + 36e^{2t} + 4)^{1/2}}{(4e^{4t} + e^{2t} + e^{-2t})^{3/2}}; \ \tau = \dfrac{-12e^{2t}}{4e^{6t} + 36e^{2t} + 4}$

33. $\kappa = \dfrac{\sec t\sqrt{9\sec^4 t - 8\sec^2 t + 1}}{(2\sec^4 t - \sec^2 t + 1)^{3/2}}; \ \tau = \dfrac{-2\sec t(\sec^2 t + 1)}{9\sec^4 t - 8\sec^2 t + 1}$

CHAPTER 15

Section 1

1. Ellipsoid of revolution about the y-axis

3. Hyperboloid of two sheets; elliptical sections perpendicular to z-axis, and hyperbolic sections perpendicular to x- and y-axes

5. Hyperboloid of one sheet, of revolution about the y-axis; elliptical sections perpendicular to y-axis, and hyperbolic sections perpendicular to x- and z-axes

7. Elliptical cone about z-axis with vertex at the origin

9. Elliptic paraboloid; elliptic sections perpendicular to z-axis, and parabolic sections perpendicular to x- and y-axes

11. Paraboloid of revolution about y-axis with vertex at origin; circular sections perpendicular to y-axis, and parabolic sections perpendicular to x- and z-axes

13. Two planes intersecting in the y-axis, and both forming a 45° dihedral angle with the xy-plane

15.

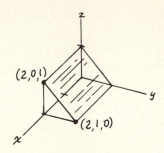

17.

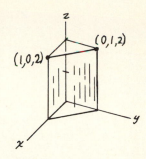

19.

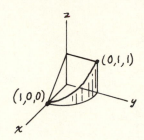

21. The intersection of the cylinders of radius 4 with axes the x- and z-axes. The curve lies in the plane $x = z$.

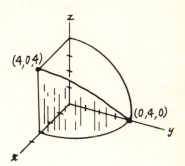

23. The intersection of the ellipsoid with semiaxes 4, 4, and 8, the first octant, and the region under the plane $y + z = 4$

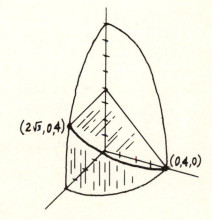

27. $x^2 + y^2 = -(z + 4)$. Paraboloid of revolution about z-axis, opening down, with vertex at $(0, 0, -4)$

29. $x^2 + (y + 1)^2 = z + 1$. Paraboloid of revolution about line $x = 0$, $y = -1$, opening up, with vertex at $(0, -1, -1)$

31. $x^2 + 2(y + 1)^2 + (z + 1)^2 = 3$. Ellipsoid of revolution centered at $(0, -1, -1)$ about the axis parallel to the y-axis through $(0, -1, -1)$

33. $(x + 2)^2 + 2(y - 1)^2 - z^2 = -1$. Hyperboloid of two sheets, centered at $(-2, 1, 0)$

35. $-x^2 + 2(y - 1)^2 + (z + 2)^2 = 6$. Hyperboloid of one sheet, centered at $(0, 1, -2)$

37. $z = 0$, $y = \pm \dfrac{b}{a} x$

Section 2   **1.** $\dfrac{\partial z}{\partial y} = \lim\limits_{\Delta y \to 0} \dfrac{f(x, y + \Delta y) - f(x, y)}{\Delta y}$ **3.** $\dfrac{\partial z}{\partial x} = 6x - 2y$; $\dfrac{\partial z}{\partial y} = -2x + 1$

5. $\dfrac{\partial z}{\partial x} = ae^{ax+by}$; $\dfrac{\partial z}{\partial y} = bc^{ax+by}$

7. $\dfrac{\partial z}{\partial x} = 2\cos(2x + 3y)$; $\dfrac{\partial z}{\partial y} = 3\cos(2x + 3y)$

9. $\dfrac{\partial z}{\partial x} = ye^{-x^2/2} - x^2 y e^{-x^2/2}$; $\dfrac{\partial z}{\partial y} = xe^{-x^2/2} = (1 - x^2)ye^{-x^2/2}$

11. $\dfrac{\partial z}{\partial x} = \dfrac{y}{x}$; $\dfrac{\partial z}{\partial y} = \ln(xy) + 1$

13. $\dfrac{\partial z}{\partial x} = a\cos ax \cos by$; $\dfrac{\partial z}{\partial y} = -b\sin ax \sin by$

15. $\dfrac{\partial z}{\partial x} = \dfrac{y}{x^2 + y_2}$; $\dfrac{\partial z}{\partial y} = \dfrac{-x}{x^2 + y_2}$ **17.** $\dfrac{\partial z}{\partial x} = \dfrac{-2y}{(x - y)^2}$; $\dfrac{\partial z}{\partial y} = \dfrac{2x}{(x - y)^2}$

19. $D_1 f(s, t) = (1 + st)e^{st}$, $D_2 f(s, t) = s^2 e^{st}$; $D_1(2, 3) = 7e^6$, $D_2(-2, 1) = 4e^{-2}$

21. $D_1 f(x, y) = y^2 z^3$, $D_2 f(x, y) = 2xyz^3$, $D_3 f(x, y) = 3xy^2 z^2$; -32; 0; 48

23. $2a \sin b$; a^2 **25.** 4   **27.** -1

29. The boundary consists of just one point, the origin.

31. One example is

$$\ln\left(\frac{1}{x^2 + y^2} - 1\right).$$

Find another (possibly by modifying this one).

33.

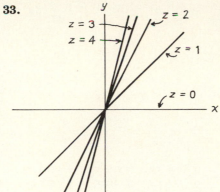

35.

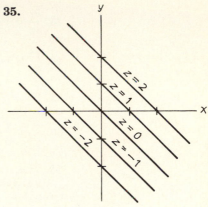

37.

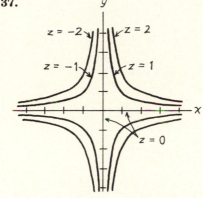

39.

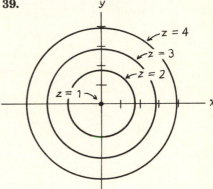

41.

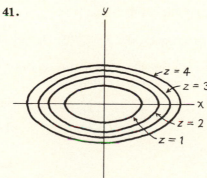

43. $\dfrac{\partial z}{\partial x} = -\dfrac{c^2 x}{a^2 z}; \ \dfrac{\partial z}{\partial y} = \dfrac{-c^2 y}{b^2 z}$

45. $\dfrac{\partial z}{\partial x} = -\dfrac{x^2}{z^2}; \dfrac{\partial z}{\partial y} = -\dfrac{y^2}{z^2}$ **47.** $\dfrac{\partial z}{\partial x} = -\dfrac{yz^2}{1+xyz}; \dfrac{\partial z}{\partial y} = -\dfrac{xz^2}{1+xyz}$

49. $\dfrac{\partial z}{\partial x} = -\dfrac{z}{y}\tan yz - \dfrac{z}{x}; \dfrac{\partial z}{\partial y} = -\dfrac{z}{y} - \dfrac{z}{x}\cot yz$

51. $3x^2 + 2xy + y^2, x^2 + 2xy + 3y^2$

53. $\dfrac{1}{y}\cos\dfrac{y}{x} + \dfrac{1}{x}\sin\dfrac{y}{x}, -\dfrac{x}{y^2}\cos\dfrac{y}{x} - \dfrac{1}{y}\sin\dfrac{y}{x}$

55. $2x\cos(x^2 + y^2)\cos xy - y\sin(x^2 + y^2)\sin xy,$
$2y\cos(x^2 + y^2)\cos xy - x\sin(x^2 + y^2)\sin xy$

Section 3   **1.** $32t$ **3.** $-\sec t$

5. $2(e^{2t} - e^{-2t})/(e^{2t} + e^{-2t})$ **7.** $\dfrac{\partial z}{\partial u} = 4u; \dfrac{\partial z}{\partial v} = 4v$

9. $\dfrac{\partial z}{\partial u} = 2[v^3 + 5v^2u + 3vu^2]e^{(u^2+2uv)(2uv+v^2)};$

   $\dfrac{\partial z}{\partial v} = 2[u^3 + 5u^2v + 3uv^2]e^{(u^2+2uv)(2uv+v^2)}$

11. $\dfrac{\partial z}{\partial x} = f'\left(\dfrac{y}{x}\right)\left(\dfrac{-y}{x^2}\right); \dfrac{\partial z}{\partial y} = f'\left(\dfrac{y}{x}\right)\left(\dfrac{1}{x}\right)$

21. If the functions $x = g(t)$, $y = h(t)$, and $z = k(t)$ are differentiable at $t = t_0$, and if $f(x, y, z)$ is smooth (continuously differentiable) inside a small sphere about $(x_0, y_0, z_0) = (g(t_0), h(t_0), k(t_0))$, then $F(t) = f(g(t), h(t), k(t))$ is a differentiable function of t at $t = t_0$, and

$$F'(t_0) = D_1f(x_0, y_0, z_0)g'(t_0) + D_2f(x_0, y_0, z_0)h'(t_0) + D_3f(x_0, y_0, z_0)k'(t_0).$$

23. If w is a smooth function of x_1, x_2, x_3, x_4 which in turn are differentiable functions of t, then w is a differentiable function of t and

$$\frac{dw}{dt} = \frac{\partial w}{\partial x_1}\frac{dx_1}{dt} + \frac{\partial w}{\partial x_2}\frac{dx_2}{dt} + \frac{\partial w}{\partial x_3}\frac{dx_3}{dt} + \frac{\partial w}{\partial x_4}\frac{dx^4}{dt}.$$

Section 4   **1.** Yes, everywhere   **3.** Everywhere but $(-3/4^{1/3}, 2^{1/3})$

5. Everywhere but $(0, 0)$

CHAPTER 16

Section 1

1. $(6, 8)$ **3.** $(-\pi/12, -1/2)$

5. $(8, 4)$ **7.** $(2, -1)$

9.   **11.** $(\sqrt{2}, -1)$

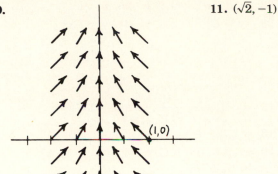

13. $(1, \sqrt{3})$ **15.** $2x + y = 4$

17. $3x + 4y = 25$ **19.** $3x + y = 5$

21. $-x + 2y = 4$

23. $(5/8 + \pi/4)(x - 2) + (5/2 + \pi/4)(y - 1/2) = 0$

25. $\operatorname{grad} f = (2x, 2y)$, $\operatorname{grad} g = (-y/x^2, 1/x)$.
So $\operatorname{grad} f \cdot \operatorname{grad} g = 0$.

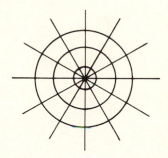

27. The vector $\mathbf{x}_0$ is perpendicular to the level curve at the point $\mathbf{x}_0$, by Example 5 of Section 1 of Chapter 14. So is $\operatorname{grad} f(\mathbf{x}_0)$.

Section 2

1. Set $\mathbf{u} = \mathbf{a}/|\mathbf{a}|$; then apply Theorem 3.

3. $\sqrt{3}/2 + 1$ **5.** $2\sqrt{10}/75$ **7.** $e\sqrt{5}$

9. $54/169$ **11.** 2

13. $(11\sqrt{2}; (1/\sqrt{2})(-1, -1)$ **15.** $32\sqrt{5}/15; (1/\sqrt{5})(1, -2)$

17. $\sqrt{10}/2; (1/\sqrt{10})(3, -1)$ **19.** $2/\sqrt{5}; (1/\sqrt{5})(1, -2)$

21. Let $\mathbf{a} = \dfrac{d\mathbf{x}}{dt}$; $\dfrac{dz}{dt} = \operatorname{grad} f \cdot \dfrac{d\mathbf{x}}{dt} = \dfrac{\operatorname{grad} f \cdot \mathbf{a}}{|\mathbf{a}|}\, |\mathbf{a}| = (D_{a/|a|} f)\, |\mathbf{a}|.$

23. $\operatorname{grad} f(\mathbf{x}_0) = (4, 9/2)$ **25.** $u + v$ **27.** $bu + av$

29. $xy = C$ **31.** $y = Cx$ **33.** $y = C\,|x|^{a^2/b^2}$

Section 3

1. $(-1, 0, 0)$ **3.** $(2, 3, 4)$ **5.** $(6/25, 8/25, 4)$

7. $19/3$ **9.** $-5/18$ **11.** $(3, 1, -1)$

13. $\mathbf{u} = \left(\dfrac{-1}{\sqrt{117}}, \dfrac{-10}{\sqrt{117}}, \dfrac{4}{\sqrt{117}} \right);\ \sqrt{117};\ -\mathbf{u}$

15. $(-1, 3, -3)$ **17.** $(-1, 0, 1)$

19. a) $(2, 3, 1/6)$ if $(1/2, 1/3, 6)$ **21.** $\mathbf{x} = (1 + et, \pi/2, e - t)$

     b) $(2, 3, 1/6)$

23. $\mathbf{x} = (2 + 2t, 3 + t, -1 + 5t)$ **25.** $-x + 3y + 3z = 2$

27. $x = z$ **29.** $5x + 4y + 3z = 22$

31. $3x - 4y + 25z = 25(\ln 5 - 1)$ **33.** $x - 2y + z = 2$

35. $z = x - 2y + 6$ **37.** $z = x + 2y - 2$

39. $x - y - z = 1$

41. Choose any positive r less than min $(1/|a|^3, 1/|b|^3)$. Show that if $x^2 + y^2 < r^2$, then, in particular, $x^2 < 1/a^{2/3}$ and $|x|^{1/3}\,|a| < 1$. Similarly, $|y|^{1/3}\,|b| < 1$. Show that, therefore, $|ax + by| < x^{2/3} + y^{2/3}$. Thus, within the circle of radius r, $ax + by > -(x^{2/3} + y^{2/3})$.

Section 4

1. $(-3, 3);\ -9$ (min)   **3.** $(0, \pm\sqrt{3});\ -9$ (min)

5. $\pm\left(\sqrt{\dfrac{2}{3}}, \sqrt{\dfrac{2}{3}} \right);\ -4/27$ (min)   **7.** $\left(\dfrac{1}{2}, \dfrac{1}{4} \right);\ -1/8$ (min)

9. $\dfrac{3}{8}\sqrt{2}$ **11.** $4,\ -\dfrac{1}{3}$

13. $2/\sqrt{5}$ **15.** $x = 6,\ y = 2,\ z = 3$

17. $(2, \sqrt{2}, \sqrt{2}),\ (2, -\sqrt{2}, -\sqrt{2})$ **19.** $16/\sqrt{3}$

21. $66/25$ **23.** $9/4;\ 0$

25. The perpendicularity condition in Problem 23 of Chapter 14, Section 1, can be written $(x_1 - x_2, y_1 - y_2) = \lambda(2x_1/4, 2y_1) = \mu(-1, 1)$. The last equation requires $x_1 = -4y_1$, and the ellipse equation then gives

$$(x_1, y_1) = \pm\left(\frac{4}{\sqrt{5}}, \frac{-1}{\sqrt{5}} \right).$$

These are the points that maximize and minimize the distance to the line. A quick sketch shows the closer to be

$$\left(-\frac{4}{\sqrt{5}}, \frac{1}{\sqrt{5}}\right).$$

27. $(x^2/2) + (y^2/32) = 1$

29. Corner point $(2/\sqrt{6}, 1/\sqrt{6})$; perimeter $2\sqrt{6}$

33. $1/30$ **35.** $\pm(98)^{3/4}/2$

37. $x_i = a_i/\sqrt{\sum_1^n a_i^2}$, $i = 1, \dots, n$

Section 5   **1.** $\dfrac{\partial^2 z}{\partial x^2} = 6$; $\dfrac{\partial^2 z}{\partial x\,\partial y} = -2 = \dfrac{\partial^2 z}{\partial y\,\partial x}$; $\dfrac{\partial^2 z}{\partial y^2} = 0$

3. $\dfrac{\partial^2 z}{\partial x^2} = (4x^2 y + 2y)e^{x^2 y}$; $\dfrac{\partial^2 z}{\partial x\,\partial y} = (2x^3 y + 2x)e^{x^2 y} = \dfrac{\partial^2 z}{\partial y\,\partial x}$; $\dfrac{\partial^2 z}{\partial y^2} = x^4 e^{x^2 y}$

5. $\dfrac{\partial^2 z}{\partial x^2} = 2$; $\dfrac{\partial^2 z}{\partial x\,\partial y} = 3 = \dfrac{\partial^2 z}{\partial y\,\partial x}$; $\dfrac{\partial^2 z}{\partial y^2} = 2$

7. $\dfrac{\partial^2 z}{\partial x^2} = 0 = \dfrac{\partial^2 z}{\partial y^2}$; $\dfrac{\partial^2 z}{\partial x\,\partial y} = \dfrac{\partial^2 z}{\partial y\,\partial x} = 1$

9. $\dfrac{\partial^2 z}{\partial x^2} = -\dfrac{6y}{x^4}$; $\dfrac{\partial^2 z}{\partial x\,\partial y} = \dfrac{2}{x^3} - \dfrac{2}{y^3} = \dfrac{\partial^2 z}{\partial y\,\partial x}$; $\dfrac{\partial^2 z}{\partial y^2} = \dfrac{6x}{y^4}$

11. Set $g(x) = x - ct$ (t fixed), $h(t) = x - ct$ (x fixed). Then:

$$\frac{\partial z}{\partial x} = f'(g(x))g'(x) = f'(g(x))$$

$$\frac{\partial^2 t}{\partial x^2} = f''(g(x))g'(x) = f''(g(x))$$

$$\frac{\partial z}{\partial t} = f'(h(t))h'(t) = -cf'(h(t))$$

$$\frac{\partial^2 z}{\partial t^2} = -cf''(h(t))h'(t) = c^2 f''(h(t))$$

$$c^2 \frac{\partial^2 z}{\partial x^2} = \frac{\partial^2 z}{\partial t^2}$$

13. $\dfrac{\partial^2 z}{\partial y^2} = \dfrac{2x^2 - 2y^2}{(x^2 + y^2)^2}$; $\dfrac{\partial^2 z}{\partial x^2} = \dfrac{2y^2 - 2x^2}{(x^2 + y^2)^2}$

19. $\dfrac{\partial z}{\partial y} = \dfrac{\partial^2 z}{\partial x^2} = (x^2 - 2y)e^{-x^2/4y}/4y^2\sqrt{y}$

21. Rel. min. at $(-1, -1/4)$

23. Saddle point at $(0, 0)$; rel. min. at $(1/6, 1/12)$

25. Saddle point at $(0, 0)$ **27.** Saddle point at $(1, -1/2)$

29. Saddle point at $(1/3, -1/3)$; rel. min. at $(3/4, 1/2)$

31. Rel. min. at $(-1, 1)$ **33.** Rel. min. at $(1, 1)$ and $(-1, -1)$

35. Saddle point at $(0, 0)$

CHAPTER 17

Section 1

1. $1/4$ **3.** 1 **5.** 0 **7.** 0 **9.** $3/20$

11. 1

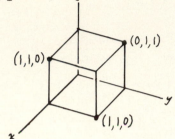

13. 1

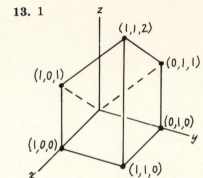

15. $\dfrac{2}{3} ab^3$ **17.** $\dfrac{72}{35}$ **19.** $4\pi\left(\dfrac{5\sqrt{2}}{12} - \dfrac{1}{3}\right)$

21. $5/12$

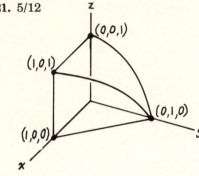

23. $1/3$

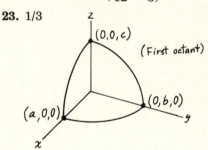

25. $\dfrac{4}{3}\pi abc$

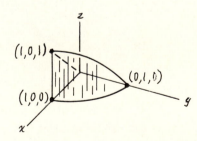

27. 27/4

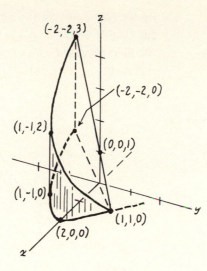

29. 1/4   **31.** 4/15   **33.** 1/3   **35.** 27/4

Section 2   **1.** 469/1800

3. $\displaystyle\int_0^1 \int_y^1 f(x, y)\, dx\, dy$

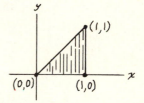

5. $\displaystyle\int_0^1 \int_y^{\sqrt{y}} \cdots dx\, dy$

7. $\displaystyle\int_0^1 \int_{\sqrt{x}}^1 \cdots dy\, dx + \int_0^1 \int_{-1}^{\sqrt{x}} \cdots dy\, dx$

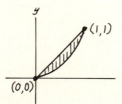

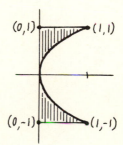

9. $1/2 - 1/2e$ **11.** 1/3

13. a) 1/4   b) 1/4   **15.** a) 2/3   b) 5/6   **17.** 3/4

19. a) $4 \cdot 1 + 8 \cdot 2 + 4 \cdot 4 = 36$

     b) $(1/36)(19/9)^2 = (19/18)^2(1/9)$

     c) 1/9

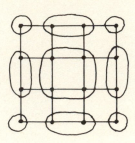

Section 3   **1.** 1/4   **3.** 4/35   **5.** $8k$ **7.** $4k/3$

     9. $8ka^3/3$ **11.** πk **13.** 1

Section 4   **1.** $(2/3, 2/3)$ **3.** $(5/9, 5/11)$ **5.** $(0, 4/3)$ **7.** $(3/8, 0)$

     9. $(0, 0)$ **11.** $((\pi^2 - 4)/\pi, \pi/8)$ **13.** $(2, 1/8)$

     15. $(a/2, b/2)$ **17.** $(0, 4/3)$ **19.** $(8/5, 0)$

     21. $(1/2, 2/5)$ **23.** $(8\sqrt{2}/15, 16\sqrt{2}/21)$ **25.** $([ae^a/(e^a - 1)] - 1, 0)$

     27. $(4a/3\pi, 0)$ **29.** $(1, 0)$ **31.** $(1/2, 8/5)$

33. $\bar{x} = \dfrac{\displaystyle\int_0^2 x(x + 2 - 2x)\, dx}{\displaystyle\int_0^2 (x + 2 - 2x)\, dx} = \dfrac{\left(-\frac{1}{3}x^3 + x^2\right)\Big]_0^2}{\left(-\frac{1}{2}x^2 + 2x\right)\Big]_0^2} = \dfrac{4/3}{2} = \dfrac{2}{3};$

$$\bar{y} = \frac{\dfrac{1}{2}\displaystyle\int_0^2 (x + 2 - 2x)(x + 2 + 2x)\, dx}{\displaystyle\int_0^2 (x + 2 - 2x)\, dx}$$

$$= \frac{1}{2}\int_0^2 \frac{4 + 4x - 3x^2}{2}\, dx$$

$$= \frac{1}{4}\, 4x + 2x^2 - x^3\Big]_0^2 = 2.$$

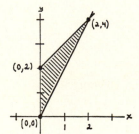

The centroid is at $(2/3, 2)$.

35. By symmetry $\bar{x} = 0$.

$$\bar{y} = \frac{\dfrac{1}{2}\displaystyle\int_{-1}^{1}[(x^2+1)^2 - (2x^2)^2]\,dx}{\displaystyle\int_{-1}^{1}[x^2+1-2x^2]\,dx}$$

$$= \frac{\dfrac{1}{2}\displaystyle\int_{-1}^{1}[-3x^4+2x^2+1]\,dx}{\displaystyle\int_{-1}^{1}[1-x^2]\,dx}$$

$$= \frac{\dfrac{1}{2}\left[-\dfrac{3}{5}x^5+\dfrac{2}{3}x^3+x\right]_{-1}^{1}}{\left(x-\dfrac{1}{3}x^3\right)\Big]_{-1}^{1}} = \frac{16/15}{4/3} = \frac{4}{5}.$$

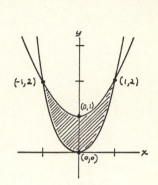

The centroid is at $(0, 4/5)$.

37. $(4r/3\pi, 0)$ 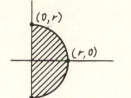 **39.** $2\pi^2 Rab$

41. The centroid is at $(a + \sqrt{A/2}, 0)$. Apply the Theorem of Pappus.

Section 5   **1.** $122/3\pi$ **3.** $\pi/3$ **5.** $4\pi/3$ **7.** $\pi/3 - 4/9$

9. $\displaystyle\int_{0}^{2\pi}\int_{0}^{R}(kr)r\,dr\,d\theta = 2\pi kR^3/3$

11. Mass $k\pi/12$; center of mass $(0, 6/5\pi)$ (rectangular coordinates)

13. $16a^3/3$

Section 6   **1.** $(\pi/6)(17\sqrt{17}-1)$ **3.** 20π **5.** $\pi - 2$

7. $3\sqrt{14}$ **9.** $\pi a\sqrt{a^2+h^2}$ **21.** $4\pi^2 ab$

Section 7   **1.** 6   **3.** $abc/6$ **5.** $18/35$ **7.** $1/3$

9. $16/3$ **11.** $64/15$ **13.** $8/3$ **15.** $(3/8)(a, b, c)$

Section 8   **1.** 8π

3. $\dfrac{16\pi}{3}\left(1 - \dfrac{\sqrt{3}}{2}\right) \approx \dfrac{4\pi}{5}$

5. $\dfrac{2}{3}\pi a^3 (1 - \cos\alpha)$

7. $\dfrac{2}{3}\pi[6^{3/2} - 11]$

9. $2\pi a^2 k$, where k is the constant of proportionality

11. $\dfrac{1}{2}\pi a^4 k$, where a is the radius, and k is the proportionality constant

13. $\dfrac{1}{3}\pi a^3 b^2$

15. $\bar{x} = \bar{y} = 0;\ \bar{z} = 3/(16(1 - \sqrt{3}/2)) \approx 5/4$

17. $(3a/8)(1, 1, 1)$

19. $\dfrac{4\pi}{3}(b^2 - a^2)^{3/2}$

21. Mass $= \dfrac{1}{6}\pi(b^3 - a^3)$

23. $\dfrac{49}{24} - \left[\dfrac{17^{3/2}}{24} - \dfrac{2^{3/2}}{3}\right];\ \dfrac{49}{24} + \left[\dfrac{17^{3/2}}{24} - \dfrac{2^{3/2}}{3}\right]$

25. $2\pi r^2$

CHAPTER 18

Section 1

1. 30

3. $2\ln 2$

5. $\dfrac{5}{2}A^2 = \dfrac{3}{2}V_{\max}^2$

7. $A = 5\sqrt{2}$

9. $16/(2 + m)^2$

11. 1,248,000 ft-lb

13. $62.4Vh - W$

17. $k\ln(V_1/V_0)$ if $\alpha = 1;\ \dfrac{k}{\alpha - 1}(V_1^{\alpha-1} - V_0^{\alpha-1})$ if $\alpha \neq 1$

Section 2

1. 2π

3. $-26/3$

5. -10

7. a) 0   b) 18   **11.** 11

13. $1/r_0 - 1/r_1$, for all paths

15. 3/2

17. 18

Section 3

1. $\phi(x, y) = \dfrac{1}{2}(x^2 + y^2 - x_0^2 - y_0^2)$

3. $\phi(x, y) = x^2 y - x_0^2 y_0$

5. $\phi(x, y) = x^2 y + xy^2 - x_0 y_0^2 - x_0^2 y_0$

7. $\phi(x, y) = \dfrac{1}{2}x^2 y^2 + C$

9. $\phi(x, y) = x^2 y + xy^2 + C$

11. $\phi(x, y, z) = zx + y^2/2 + C$

13. $\phi(x, y, z) = 1/4r^4$

15. 7

17. 0

19. 0

21. $5^{3/2}$

23. 2π

Section 4

1. 1   **3.** $-1/2$ **5.** 0

7. πab **9.** πab **11.** 1/6

13. -1; $-$(area of G), for any regular G **15.** 2π; 2(area of G)

17. 1; area of G for any regular G not intersecting any line $x = \dfrac{\pi}{2} + k\pi$

19. 0 (for any G)   **21.** -2π; -2(area of G)

CHAPTER 19

Section 1

1. $y = -e^x + Ce^{+2x}$

3. $y = e^x/(1 + a) + Ce^{-ax}$, for $a \ne -1$; $y = xe^x + Ce^x$ for $a = -1$

5. $y = x - 1 + Ce^{-x}$ **7.** $y = x^2/3 + C/x$ **9.** $y = 1 + C/x$

11. $y = (1/(1 + a^2))(a \sin x - \cos x) + Ce^{-ax}$

13. $y = (1/2)(x^2 - 1) + Ce^{-x^2}$ **15.** $y = \tan x + C \sec x$

Section 2

1. $y = -e^x + Ce^{2x}$ **3.** $y = x - 1 + Ce^{-x}$

5. $y = (1/a) + Ce^{-ax}$ $(a \ne 0)$; $y = x + C$ $(a = 0)$

7. $y = (1/3)xe^x - (1/9)e^x + Ce^{-2x}$

9. $y = (1/2)(x \sin x = x \cos x + \cos x) + Ce^{-x}$

11. $\dfrac{d}{dx}(e^{ax}y) = e^{ax}r(x) \Leftrightarrow e^{ax}\dfrac{dy}{dx} + ae^{ax}y = e^{ax}r(x)$ (product formula)

13. Use the derivative product rule to expand $L(xy)$.

15. $y = xe^{-4x} + Ce^{-4x}$ **17.** $y = \dfrac{1}{2}e^x + 1 + Ce^{-x}$

19. $y = -x - 1 + xe^x + Ce^x$ **21.** a) $3x^3$ b) x c) 0   d) xe^x

23. Use the derivative product rule to expand $L(u, v)$.

Section 3

1. $y = x$ **3.** $y = -\dfrac{1}{3}e^x$

5. $y = \dfrac{1}{4}\cos x$ **7.** $y = \dfrac{1}{3}x^3 - x^2 + 2x$

9. $y = \dfrac{1}{3}e^x + \dfrac{2}{3}e^{-x}$ **11.** $y = \dfrac{1}{2}(xe^x - e^x)$

13. $y = -\dfrac{1}{2}e^{-x}$ **15.** $y = -\dfrac{3}{10}\sin x - \dfrac{1}{10}\cos x$

17. $y = \dfrac{1}{2}x^2 - x - \dfrac{1}{2}\sin x - \dfrac{1}{2}\cos x$ **19.** $y = \dfrac{1}{12}x^4e^x$

21. $y = -\dfrac{1}{4}e^x(\sin x + \cos x)$ **23.** $y = e^x \sin x$

Section 4

1. $y = c_1 \sin\sqrt{2} + c_2 \cos\sqrt{2}\,x$

3. $y = \dfrac{1}{2}(x^2 - 1) + c_1 \sin\sqrt{2}\,x + c_2 \cos\sqrt{2}\,x$

5. $y = -\dfrac{1}{4}x \cos 2x + c_1 \sin 2x + c_2 \cos 2x = c_1 \sin 2x + (c_2 - (x/4))\cos 2x$

7. $y = \dfrac{1}{2}xe^x + c_1 e^x + c_2 e^{-x} = (c_1 + (x/2))e^x + c_2 e^{-x}$

9. $y = -\dfrac{1}{5}e^x \sin x - \dfrac{2}{5}e^x \cos x + c_1 e^x + c_2 e^{-x}$

11. $y = c_1 e^x + c_2 e^{-2x}$

13. $-\dfrac{3}{10}\sin x - \dfrac{1}{10}\cos x + c_1 e^x + c_2 e^{-2x}$

15. $y = +c_1 e^x + c_2 xe^x$ 17. $y = \dfrac{1}{2}\cos x + c_1 e^x + c_2 xe^x$

19. $y = c_1 e^{-x} \sin x + c_2 e^{-x} \cos x$

21. $y = \dfrac{1}{5}e^x + c_1 e^{-x} \sin x + c_2 e^{-x} \cos x$

23. $(D - k_1)[(D - k_2)f] = (D - k_1)[f' - k_2 f]$, etc.

25. $y = c_1 e^{kx} + c_2 xe^{kx}$ 27. $y = \dfrac{1}{3}xe^{2x} + c_1 e^{-x} + c_2 e^{2x}$

Section 5

1. $x = 2\sin(3t - \pi/3)$ 3. $x = 5\sqrt{2}\sin(t + 3\pi/4)$

5. $x = a\sin(2\pi ft + \alpha)$, so $dx/dt = ?$

7. $x = \sqrt{53/2}\sin(\sqrt{2/5}\,t) + \alpha$, where tangent $\alpha = 2\sqrt{2}/3\sqrt{5} \approx 0.42$. Appendix 6, Table 3 gives $\alpha \approx 0.4$ radians.

9. The motion $(d^2x/dt^2) = -kx$ has period $2\pi/\sqrt{k}$. Here $2 = 2\pi\sqrt{l/g}$.

11. $13\,\text{ft } \tfrac{1}{2}\,\text{in. (approx)}$ 13. $13\,\text{in. (approx)}$

15. For $m^2 > 4ms$, $m\ddot{x} + r\dot{x} + sx = 0$ has solutions $x = c_1 e^{-kt} + c_2 e^{-lt}$, where

$$k = \frac{r + \sqrt{r^2 - 4ms}}{2m} \quad \text{and} \quad l = \frac{r - \sqrt{r^2 - 4ms}}{2m}, \quad (0 < l < k).$$

$0 = \dot{x}(0) = -kc_1 e^{-k\cdot 0} - lc_2 e^{-l\cdot 0}$, so $c_2 = (-k/l)c_1$;
$0 < x(0) = c_1 e^{-k\cdot 0} - (k/l)c_1 e^{-l\cdot 0} = (1 - (k/l))c_1$; so $c_1 < 0$.
Let $c = -c_1/l$; then $c > 0$ and $x = c[ke^{-lt} - le^{-kt}]$.

17. Similarly, as in Problem 1, $x = c_1 e^{-kt} + c_2 e^{-lt}$ with $0 < l < k$.
$0 = x(0) = c_1 e^{-k\cdot 0} + c_2 e^{-l\cdot 0} = c_1 + c_2$, so $c_1 = -c_2$ and $x = c_2[e^{-lt} - e^{-kt}]$.
$0 < \dot{x}(0) = c_2[-le^{-l\cdot 0} + ke^{-k\cdot 0}] = c_2(k - l)$, so $c_2 = c > 0$.

19. As in (1) and (3), $x = c_1 e^{-kt} + c_2 e^{-lt}$, $0 < l < k$. $1 = x(0) = c_1 + c_2$, so $c_1 = 1 - c_2$. Let $a = c_2$; then $x = ae^{-lt} + (1-a)e^{-kt}$.
$\dot{x}(0) = a(-l)e^{-l \cdot 0} + (1-a)(-k)e^{-k \cdot 0} = a(k-l) - k$. So $a < 0$ implies $\dot{x}(0) = v_0 = a(k-l) - k < -k$, and $v_0 = a(k-l) - k < -k$ implies $a < 0$.

21. $x = c_1 e^{-kt} + c_2 t e^{-kt}$; $x_0 = 1 = c_1 + (c_2)(0) = c_1$, so $x = e^{-kt} + c_2 t e^{-kt}$.
$0 = v_0 = \dot{x}(0) = -ke^{-k \cdot 0} + c_2(-k)(0)e^{-k(0)} + c_2 e^{-k \cdot 0} = -k + c_2$, so $c_2 = k$ and $x = (1 + kt)e^{-kt}$

23. This is a result of Theorem 9 (Section 4).

27. Differentiate (*) to get (**).

Section 6   **1.** $c_1 = 0$, $c_n = (n+2)c_{n+2}$; converges for all x;

$$y = c_0\left(1 + \frac{x^2}{2} + \frac{x^4}{2 \cdot 4} + \frac{x^6}{2 \cdot 4 \cdot 6} + \cdots + \frac{x^{2n}}{2^n \cdot n!} + \cdots\right)$$

3. $c_2 = 0$, $2c_2 = 1$, $(n+3)c_{n+3} = c_n$;

$$y = c_0\left(1 + \frac{x^3}{3} + \frac{x^6}{3 \cdot 6} + \cdots + \frac{x^{3n}}{3^n n!} + \cdots\right)$$

$$+ \left(\frac{x^2}{2} + \frac{x^5}{2 \cdot 5} + \frac{x^8}{2 \cdot 5 \cdot 8} + \cdots + \frac{x^{3n-1}}{2 \cdot 5 \cdot 8 \cdots (3n-1)} + \cdots\right)$$

$$= c_0 f(x) = g(x).$$

Each series converges for all x.

5. c_0, $(n+1)c_{n+1} = nc_{n-1}$;

$$y = c_1\left(1 + \frac{x^2}{2} + \frac{3}{2 \cdot 4}x^4 + \frac{3 \cdot 5}{2 \cdot 4 \cdot 6}x^6 + \cdots\right); \text{ converges on } (-1, 1).$$

7. $c_3 = c_2 = 0$, $(n+2)(n+1)c_{n+2} = c_{n-2}$;

$$y = c_0\left[1 + \frac{x^4}{4 \cdot 3} + \frac{x^8}{8 \cdot 7 \cdot 4 \cdot 3} + \cdots\right] + c_1\left[x + \frac{x^5}{5 \cdot 4} + \frac{x^9}{9 \cdot 8 \cdot 5 \cdot 4} + \cdots\right].$$

Each series converges for all x.

9. $(n+2)(n+1)c_{n+2} = [n(n-1) + 1]c_n$;

$$y = c_0\left[1 + \frac{x^2}{2} + \frac{x^4}{8} + \cdots\right] + c_1\left[x + \frac{x^3}{6} + \frac{7x^5}{120} + \cdots\right].$$

Converges on $(-1, 1)$, by the ratio test.

13. The coefficients c_0 and c_1 are arbitrary, and $c_2 = -c_0/2$, $c_3 = -(c_0 + c_1)/6$. After that, the recursion formula

$$c_{n+2} = \frac{c_{n-1} + c_n}{(n+2)(n+1)}$$

determines the coefficients. There is no obvious pattern, but it can be shown that if k is the larger of $|c_0|$ and $|c_1|$, then

$$|c_n| < \frac{k}{\sqrt{n!}},$$

as follows. It is true by inspection for c_0, c_1, c_2, and c_3. Suppose that it is true up through an integer m. Then

$$|c_{m+1}| = \frac{|c_{m-2} + c_{m-1}|}{(m+1)m} \leq \frac{k}{(m+1)m}\left[\frac{1}{\sqrt{(m-2)!}} + \frac{1}{\sqrt{(m-1)!}}\right]$$

$$= \frac{k}{\sqrt{(m+1)!}}\left[\frac{\sqrt{m-1}+1}{\sqrt{(m+1)m}}\right] < \frac{k}{\sqrt{(m+1)!}}.$$

So it is true for $m + 1$, too. Starting from our original list, the inequality has now been proved for $m + 1 = 4$, then (taking $m = 4$) for $m + 1 = 5$, and so on forever. This method of proof is called *mathematical induction*.

The solution series is thus dominated term by term by $k \sum |x|^n/\sqrt{n!}$ and hence converges for all x.

15. Let $y = c_0 + c_1x + c_2x^2 + c_3x^3 + \cdots$. Then

$$y' = c_1 + 2c_2x + 3c_3x^2 + 4c_4x^3 + 5c_5x^4 + \cdots,$$
$$x^2y' = \qquad\qquad c_1x^2 + 2c_2x^3 + 3c_3x^4 + \cdots,$$
$$-2yx = \quad -2c_0x - 2c_1x^2 - 2c_2x^3 - 2c_3x^4 + \cdots.$$

So

$$(1 + x^2)y' - 2xy = 0$$
$$= c_1 + (2c_2 - 2c_0)x + (3c_3 - c_1)x^2$$
$$+ (4c_4)x^3 + (5c_5 + c_3)x^4$$
$$+ (6c_6 + 2c_4)x^5 + (7c_7 + 3c_5)x^6 + \cdots.$$

Solving: $c_1 = 0$, $c_2 = c_0$, $c_3 = c_1/3 = 0$, $c_4 = 0$, $c_5 = -c_3/5 = 0$, $c_6 = -2c_4/6 = 0$, etc. So $y = c + cx^2$. (This may be checked by solving the differential equation by separating variables.)

17. Let $y = c_0 + c_1x + c_2x^2 + c_3x^3 + \cdots$. Then

$$xy'' = \qquad 2c_2x + 6c_3x^2 + 12c_4x^3 + 20c_5x^4 + \cdots,$$
$$y' = c_1 + 2c_2x + 3c_3x^2 + \quad 4c_4x^3 + \quad 5c_5x^4 + \cdots,$$
$$xy = \qquad\quad c_0x + \quad c_1x^2 + \quad c_2x^3 + \quad c_3x^4 + \cdots.$$

So

$$xy'' + y' + xy = 0$$
$$= c_1 + (4c_2 + c_0)x + (9c_3 + c_1)x^2$$
$$+ (16c_4 + c_2)x^3 + (25c_5 + c_3)x^4$$
$$+ (36c_6 + c_4)x^5 + \cdots.$$

Solving: $c_1 = 0$, $c_2 = -c_0/4$, $c_3 = -c_1/9 = 0$, $c_4 = -c_2/16 = c_0/(16 \cdot 4)$, $c_5 = 0$, etc.

Choose $c_0 = 1$:

$$f(x) = 1 - \frac{x^2}{2^2} + \frac{x^4}{2^2 4^2} - \frac{x^6}{2^2 4^2 6^2} + \frac{x^8}{2^2 4^2 6^2 8^2} - \cdots .$$

General solution $= c \cdot f(x)$.

Interval of convergence:

$$\text{ratio of consecutive terms} = \frac{x^{2n+2}}{2^2 \cdots (2n+2)^2} \cdot \frac{2^2 \cdots (2n)^2}{x^{2n}}$$

$$= \frac{x^2}{(2n+2)^2} \cdot \to 0.$$

So $f(x)$ converges for all x.

inDex